0451995 - 7

KU-405-408

CHEN, HOLLIS C
THEORY OF ELECTROMAGNETIC WAVE
000451995

621.371 C51

WITHDRAWN FROM STOCK

IDENTITIES

$\mathbf{b} \cdot (\bar{\mathbf{I}} \times \mathbf{a}) = \mathbf{b} \cdot (\mathbf{a} \times \bar{\mathbf{I}}) = \mathbf{b} \times \mathbf{a}$

$\mathbf{a} \times \bar{\mathbf{I}} = \bar{\mathbf{I}} \times \mathbf{a}$

$(\mathbf{a} \times \bar{\mathbf{I}})^T = -\mathbf{a} \times \bar{\mathbf{I}}$

$(\mathbf{a} \times \mathbf{b}) \times \bar{\mathbf{I}} = \mathbf{ba} - \mathbf{ab}$

$(\mathbf{a} \times \bar{\mathbf{I}}) \cdot (\mathbf{b} \times \bar{\mathbf{I}}) = \mathbf{a} \times (\mathbf{b} \times \bar{\mathbf{I}}) = \mathbf{ba} - (\mathbf{a} \cdot \mathbf{b})\bar{\mathbf{I}}$

$(\mathbf{a} \times \bar{\mathbf{I}}) \cdot (\mathbf{b} \times \bar{\mathbf{I}}) \cdot (\mathbf{c} \times \bar{\mathbf{I}}) = \mathbf{b}(\mathbf{a} \times \mathbf{c}) - (\mathbf{a} \cdot \mathbf{b})(\mathbf{c} \times \bar{\mathbf{I}})$

$(\mathbf{a} \times \bar{\mathbf{I}}) \cdot (\text{adj } \bar{\mathbf{A}}) \cdot (\mathbf{b} \times \bar{\mathbf{I}}) = (\tilde{\bar{\mathbf{A}}} \cdot \mathbf{b})(\mathbf{a} \cdot \tilde{\bar{\mathbf{A}}}) - (\mathbf{a} \cdot \tilde{\bar{\mathbf{A}}} \cdot \mathbf{b})\tilde{\bar{\mathbf{A}}}$

$(\text{adj } \bar{\mathbf{A}}) \cdot (\mathbf{a} \times \mathbf{b}) = (\tilde{\bar{\mathbf{A}}} \cdot \mathbf{a}) \times (\tilde{\bar{\mathbf{A}}} \cdot \mathbf{b}) = (\mathbf{a} \cdot \bar{\mathbf{A}}) \times (\mathbf{b} \cdot \bar{\mathbf{A}})$

If $\bar{\mathbf{A}}$ is symmetric, then

$$\bar{\mathbf{A}} \cdot (\mathbf{a} \times \bar{\mathbf{I}}) + (\mathbf{a} \times \bar{\mathbf{I}}) \cdot \bar{\mathbf{A}} = (\bar{\mathbf{A}}_t \mathbf{a} - \bar{\mathbf{A}} \cdot \mathbf{a}) \times \bar{\mathbf{I}}$$

TRACE

$(\bar{\mathbf{A}} + \bar{\mathbf{B}})_t = \bar{\mathbf{A}}_t + \bar{\mathbf{B}}_t$

$(\lambda\bar{\mathbf{A}})_t = \lambda\bar{\mathbf{A}}_t \qquad (\lambda\bar{\mathbf{I}})_t = 3\lambda$

$(\bar{\mathbf{A}} \cdot \bar{\mathbf{B}})_t = (\bar{\mathbf{B}} \cdot \bar{\mathbf{A}})_t \qquad (\tilde{\bar{\mathbf{A}}})_t = \bar{\mathbf{A}}_t$

$(\bar{\mathbf{A}} \cdot \bar{\mathbf{B}} \cdot \bar{\mathbf{C}})_t = (\bar{\mathbf{B}} \cdot \bar{\mathbf{C}} \cdot \bar{\mathbf{A}})_t = (\bar{\mathbf{C}} \cdot \bar{\mathbf{A}} \cdot \bar{\mathbf{B}})_t$

$$(\text{adj } \bar{\mathbf{A}})_t = \frac{(\bar{\mathbf{A}}_t)^2 - (\bar{\mathbf{A}}^2)_t}{2}$$

$[\text{adj } (\bar{\mathbf{A}} + \bar{\mathbf{B}})]_t = (\text{adj } \bar{\mathbf{A}})_t + (\text{adj } \bar{\mathbf{B}})_t + \bar{\mathbf{A}}_t \bar{\mathbf{B}}_t - (\bar{\mathbf{A}} \cdot \bar{\mathbf{B}})_t$

$(\mathbf{a} \times \bar{\mathbf{I}})_t = 0$

$(\mathbf{ab})_t = \mathbf{a} \cdot \mathbf{b} \qquad (\mathbf{ab} \cdot \bar{\mathbf{A}})_t = \mathbf{b} \cdot \bar{\mathbf{A}} \cdot \mathbf{a}$

Q29-95p

THEORY OF ELECTROMAGNETIC WAVES

A Coordinate-Free Approach

McGraw-Hill Series in Electrical Engineering

Consulting Editor
Stephen W. Director, Carnegie-Mellon University

Networks and Systems
Communications and Information Theory
Control Theory
Electronics and Electronic Circuits
Power and Energy
Electromagnetics
Computer Engineering and Switching Theory
Introductory and Survey
Radio, Television, Radar, and Antennas

Previous Consulting Editors

Ronald M. Bracewell, Colin Cherry, James F. Gibbons, Willis W. Harman, Hubert Heffner, Edward W. Herold, John G. Linvill, Simon Ramo, Ronald A. Rohrer, Anthony E. Siegman, Charles Susskind, Frederick E. Terman, John G. Truxal, Ernst Weber, and John R. Whinnery

Electromagnetics

Consulting Editor
Stephen W. Director, *Carnegie-Mellon University*

Chen: *Theory of Electromagnetic Waves: A Coordinate-Free Approach*
Dearhold and McSpadden: *Electromagnetic Wave Propagation*
Goodman: *Introduction to Fourier Optics*
Hayt: *Engineering Electromagnetics*
Jenkins and White: *Fundamentals of Optics*
Johnson: *Transmission Lines and Networks*
Kraus and Carver: *Electromagnetics*
Paris and Hurd: *Basic Electromagnetic Theory*
Paul and Nasar: *Introduction to Electromagnetic Fields*
Plonsey and Collin: *Principles and Applications of Electromagnetic Fields*
Plonus: *Applied Electromagnetics*
Siegman: *Introduction to Lasers and Masers*
White: *Basic Quantum Mechanics*
Young: *Fundamentals of Waves, Optics, and Modern Physics*

UNIVERSITY LIBRARY LIVERPOOL

THEORY OF ELECTROMAGNETIC WAVES

A Coordinate-Free Approach

Hollis C. Chen

Professor of Electrical Engineering
Ohio University

McGraw-Hill Book Company

New York St. Louis San Francisco Auckland Bogotá Hamburg
Johannesburg London Madrid Mexico Montreal New Delhi
Panama Paris São Paulo Singapore Sydney Tokyo Toronto

This book was set in Times Roman by Santype-Byrd.
The editors were T. Michael Slaughter and Madelaine Eichberg;
the production supervisor was Leroy A. Young.
The drawings were done by Allyn-Mason, Inc.
The cover was designed by Carla Bauer.
R. R. Donnelley & Sons Company was printer and binder.

THEORY OF ELECTROMAGNETIC WAVES
A Coordinate-Free Approach

Copyright © 1983 by McGraw-Hill, Inc. All rights reserved.
Printed in the United States of America. Except as permitted under the United States Copyright Act of 1976, no part of this publication may be reproduced or distributed in any form or by any means, or stored in a data base or retrieval system, without the prior written permission of the publisher.

1234567890 DOCDOC 89876543

ISBN 0-07-010688-6

Library of Congress Cataloging in Publication Data

Chen, Hollis C.
Theory of electromagnetic waves.
(McGraw-Hill series in electrical engineering. Electromagnetics)
Bibliography: p.
Includes index.
1. Electromagnetic waves. I. Title. II. Series.
QC661.C47 1983 530.1'41 82-14054
ISBN 0-07-010688-6

CONTENTS

101363

PREFACE

This book is devoted to the theory of propagation and excitation of electromagnetic waves in isotropic as well as in anisotropic media. In addition, it also treats the reflection and transmission of waves from the surface of such media. It is intended, in particular, to present a coherent and comprehensive account of the subject matter from a coordinate-free point of view.

In applied electromagnetics, the standard approach to solutions of various problems has been the coordinate method; that is, during the processes of obtaining solutions, one or more coordinate systems are used. Indeed, during the past several decades much effort has gone into applying vector calculus to electromagnetism as evidenced by comparing Maxwell's original work and the recent books on electromagnetism. Nevertheless, these attempts have been limited largely to the Maxwell equations and some of their direct consequences, such as the laws of conservation of energy and of charge, potentials, etc. In presenting basic physical concepts or in solving concrete electromagnetic problems, we have invariably resorted to the use of coordinate systems. For example, in defining the polarization of a plane wave in isotropic media, we choose a cartesian coordinate system with x and y axes lying on the plane of constant phase. The polarization is then determined by the ratio $\mathscr{E}_y/\mathscr{E}_x$, where $\mathscr{E}_x$ and $\mathscr{E}_y$ are, respectively, the x and y components of the electric field intensity $\boldsymbol{\mathscr{E}}$. This ratio, which depends upon the choice of the directions of the coordinate axes, is not an invariant, although the shape, the size, the sense of rotation and the orientation of the polarization ellipse should not depend on such a choice. Another example would be the problem of wave propagation in an anisotropic medium. In such a medium, the problem is commonly formulated and solved through the principal coordinate system of the dielectric tensor because the latter takes the simplest diagonal form with respect to the former. However, when a boundary surface exists, the problem becomes more complex. In this case, two generally inconsistent requirements govern the choice. Inside the anisotropic medium the principal coordinate is preferred; but on the boundary surface, a coordinate system with one of its coordinate planes coinciding with the surface is preferred. Using either system leads to a large number of simultaneous equations and ends in greatly cumbersome results.

Thus an objective of this book is to present an alternative approach—a coordinate-free approach—in solving various wave propagation and excitation problems. Based on the direct manipulation of vectors, dyadics, and their invariants, the approach eliminates the use of coordinate systems. It facilitates solutions, condenses exposition, and provides results in a greater generality. It further renders physical concepts more tangible and easier to grasp.

The organization of the book is somewhat different from that normally found in books on electromagnetics. The material is arranged according to the electric and magnetic properties of the media: isotropic media, crystals, plasmas, ferrites, moving media, etc. Such an arrangement reflects another objective of the book, namely, to unify the various topics in applied electromagnetics with a single approach, thus making the contents of the book easy to learn and convenient to teach.

The reader is assumed to have had an intermediate course in electromagnetics, including Maxwell's equations and the wave equations. It is my experience that a student familiar with vector and matrix algebras should find the coordinate-free method no more difficult than the coordinate method. However, to make the book self-contained, a detailed discussion of the related topics in vector and dyadic analysis is given in Chapter 1. The object here is to provide a review for those who are not familiar with the subject and a common basis of terminology and notation for those who already know the subject. Chapter 2 is devoted to a brief review of the Maxwell field equations and those parts of the electromagnetic theory which will be needed to understand the rest of the book. Also included in the chapter is the transformation of field vectors to moving systems. Chapter 3 considers some general concepts and properties of monochromatic plane waves, such as the states of polarization and the laws of reflection and refraction, by the coordinate-free method.

Chapter 4 deals with the standard topics of uniform and nonuniform plane wave propagation in an isotropic medium, and reflection and transmission from an interface of two different isotropic media. Even here we see the advantages of the coordinate-free method; for example, the polarization of the incident wave can be arbitrary. Chapter 5 examines the propagation of electromagnetic waves in nonmagnetic crystals. Step by step, the dispersion equation, the polarizations, and the directions of field vectors are derived. This chapter also shows explicitly that the energy is no longer propagated along the direction of wave normal. Finally, the reflection and transmission coefficients from an interface of an isotropic medium and non-magnetic crystal are determined. All results are obtained in coordinate-free forms. Chapter 6 applies the results of Chapter 5 to the case of simpler uniaxially anisotropic media. Included in the chapter are some interesting and important special orientations of the optic axis with respect to the plane of incidence and the interface. Chapter 7 begins with derivations of a dielectric tensor of a plasma and a permeability tensor of a ferrite, with the biasing constant magnetic field arbitrarily oriented. These are then followed by a treatment of wave propagation in an unbounded plasma, and the reflection and transmission of waves from a plasma half-space. Chapter 8 extends the discussion to wave propagation in a moving medium. Here the Minkowski constitutive relations are first derived. The effects of motion on wave propagation, reflection and transmission are then examined.

The final three chapters are devoted to sources radiating in various media. Rather than the usual approach through the scalar and vector potentials, we use the method of Green's functions. Chapter 9 presents dyadic Green's function of an isotropic medium and demonstrates its use in computing radiation from electromagnetic sources. Included in the discussion are radiations from an electric dipole and thin-wire antenna. Chapter 10 derives the dyadic Green function of a uniaxial medium. Detailed evaluation of some important integrals are given; radiation from an electric dipole is also included. Finally, Chapter 11 explores the effects of the motion of a medium on the radiation from an electric dipole. Here again, the dyadic Green function of a moving medium is explicitly evaluated. Two special orientations of the dipole with respect to the motion of the medium are discussed.

This book is intended primarily as a text for a graduate course in electromagnetics or optics. The book contains more than enough material for a one-semester or two-quarter course. This allows for a certain amount of flexibility in the choice of topics. The first three chapters provide the basic concepts, principles, and mathematical techniques for the remainder of the book. A number of problems are given at the end of each chapter; some are designed to supplement the text and others are important extensions of it. The book provides answers to most of the problems, making it especially suitable for self-study or reference use.

The International System (SI) of units is used throughout the text. For a monochromatic field, we choose the time-dependent factor exp $(-i\omega t)$. Those readers who are more comfortable with the notation exp $(j\omega t)$ need only replace $-i$ by j in any of the final expressions. Boldface letters denote vectors, and boldface letters with an overbar denote matrices or "tensors." The development of this book relies heavily on published work but some of the results in the text appear in book form for the first time. References cited at the end of each chapter and the bibliography at the end of the text are, at best, representative but by no means exhaustive.

I am pleased to acknowledge my appreciation to Ohio University for a sabbatical leave in the academic year 1978–79 and to the Stocker endowment funds for one summer of partial support. I am grateful to Professor David K. Cheng of Syracuse University for the support and encouragement to pursue research in electromagnetics. I am indebted to Professor Markus Zahn of Massachusetts Institute of Technology for his review of the manuscript and suggestions for improvement, and to several anonymous reviewers whose comments and criticisms helped refine the text. I wish to thank Dr. Raymond Luebbers for his reading of portions of the manuscript and also many graduate students for valuable comments. In addition, thanks are due to Mrs. Katie Cline for her expert typing of the manuscript.

Finally, I would like to express gratitude to my family: to my children, Desiree and Hollis, Jr., for their understanding of the many hours of my time required for this project; and especially to my wife, Donna, to whom this book is dedicated.

Hollis C. Chen

CHAPTER

ONE

LINEAR ANALYSIS

In this book we shall be concerned with the generation and propagation of electromagnetic waves in various media. The basic laws which govern these phenomena are the Maxwell equations, and they will be reviewed in the next chapter. Every well-behaved solution of the Maxwell equations represents a possible electromagnetic wave which can be supported by a medium. Hence our main objective is to find such solutions under various conditions and in various media.

Since the electric and magnetic fields are vector quantities and since these fields are related by the vector Maxwell equations, we shall seek vector solutions directly from the vector equations. To this end, matrices of three-dimensional space and linear algebra will be used extensively. The reader is assumed to be familiar with the more elementary aspects of vector and matrix algebras. In this chapter, we shall develop the mathematical techniques needed and establish a common basis of terminology and notation. However, no attempt is made at completeness or mathematical rigor.

1.1 INDEX AND DIRECT NOTATIONS

Two kinds of notations are generally used: index notation of tensor analysis and direct notation of vectors and matrices. We shall adopt the latter throughout the book because it not only expresses physical quantities such as electric field intensity, conductivity tensor, etc., by single letters instead of components, but also reflects their nature of coordinate independence. On the other hand, we shall use the index notation to establish and justify results which are expressed in direct notation. In this section, we shall review both notations and rules for their conversions. Because of applications in the following chapters, we shall limit our discussion of vectors and matrices to three-dimensional space only.

Let v_i ($i = 1, 2, 3$) be the cartesian components of a three-dimensional vector **v**. A linear transformation transforms this vector to a new vector **u** with cartesian components u_i. The components of the two vectors are related by the linear equations

$$\begin{aligned} a_{11}v_1 + a_{12}v_2 + a_{13}v_3 &= u_1 \\ a_{21}v_1 + a_{22}v_2 + a_{23}v_3 &= u_2 \\ a_{31}v_1 + a_{32}v_2 + a_{33}v_3 &= u_3 \end{aligned} \tag{1.1}$$

where a_{ij} ($i, j = 1, 2, 3$) are the coefficients of the transformation. Using the summation sign, we may write this more briefly as

$$\sum_{j=1}^{3} a_{ij}v_j = u_i \qquad i = 1, 2, 3 \tag{1.2}$$

As a further simplification, we adopt the *summation convention* for the repeated indices. That is, whenever an index is repeated once in any term, it is understood that the index is to be summed from 1 to 3 unless otherwise stated. According to this convention, Eq. (1.2) becomes

$$a_{ij}v_j = u_i \qquad \begin{matrix} i = 1, 2, 3 \\ j = 1, 2, 3 \end{matrix} \tag{1.3}$$

where the summation over j is understood. The repeated index j is called the *dummy index* because of the fact that the particular letter used for the index is not important; thus $a_{ij}v_j = a_{ik}v_k$. An index which appears only once in each term of an equation such as the index i in Eq. (1.3) is called a *free index*. A free index takes on the integer 1, 2, or 3 one at a time. Thus Eq. (1.3) is a shorthand of three equations in Eq. (1.1).

The coefficients of the transformation form a square array called a *matrix* and are denoted, in direct notation, by $\bar{\mathbf{A}}$:

$$\bar{\mathbf{A}} = \begin{bmatrix} a_{11} & a_{12} & a_{13} \\ a_{21} & a_{22} & a_{23} \\ a_{31} & a_{32} & a_{33} \end{bmatrix} \tag{1.4}$$

which is often abbreviated as $[a_{ij}]$. In this notation, a_{ij} represents a typical element (or component) of the matrix $\bar{\mathbf{A}}$ and appears in the ith row and the jth column. $\bar{\mathbf{A}}$ may also be regarded as a linear operator transforming the vector **v** into the vector **u**. In direct notation, we may express this fact by writing Eq. (1.1) as

$$\bar{\mathbf{A}} \cdot \mathbf{v} = \mathbf{u} \tag{1.5}$$

This form has several advantages over the index form (1.3). First, the symbols $\mathbf{v}$, $\mathbf{u}$, and $\bar{\mathbf{A}}$ themselves, not their components, correspond to physical quantities. For example, $\mathbf{v}$, $\mathbf{u}$, and $\bar{\mathbf{A}}$ may represent, respectively, the electric field intensity, the electric current density, and the conductivity of a medium. Secondly, direct notation records a physical statement which clearly shows the relationship among physical quantities. If $\mathbf{v}$, $\mathbf{u}$, and $\bar{\mathbf{A}}$ represent physical quantities stated in the previous example, then Eq. (1.5) is a statement of Ohm's law. Thirdly, direct notation provides a natural way of describing a geometrical or physical quantity that is independent of coordinate systems. The components v_i or u_i are the projections of vector $\mathbf{v}$ or $\mathbf{u}$ on a given set of base vectors. For a different choice of base vectors, the components of the given vector $\mathbf{v}$ or $\mathbf{u}$ are different. In other words, a given physical property characterized by a vector may be represented by different sets of numbers because of the different choices of base vectors. Similarly, the components a_{ij} of the matrix $\bar{\mathbf{A}}$ depend on the choice of basis. In this sense, the index form does not reflect the coordinate-independent nature of a physical quantity but the direct form does. Finally, rules of operations on vectors and matrices may apply to the direct form to obtain solutions without resorting to components. Such operations frequently prove to be simpler and less laborious than operations using components.

In converting from direct to index notation or vice versa, we shall adopt the following general rules:

1. The dot in direct notation denotes *contraction* in index notation, i.e., it means summation over the adjacent repeated indices from 1 to 3. For example, the *scalar product* (or *dot product*) of two vectors $\mathbf{u}$ and $\mathbf{v}$ expressed in both notations is

$$\mathbf{u} \cdot \mathbf{v} = u_i v_i \tag{1.6}$$

   Another example is the left-hand side of Eq. (1.5), whose meaning is given by the left-hand side of Eq. (1.3).
2. The dummy indices must always be adjacent in conversion from the index to direct notation or vice versa. When there are several pairs of dummy indices in an expression or an equation, each pair must be denoted by a different letter to avoid confusion.
3. The free index appearing in every term of an equation must be the same. For example, the expression

$$\bar{\mathbf{D}} = \bar{\mathbf{A}} \cdot \bar{\mathbf{B}} \cdot \bar{\mathbf{C}} \tag{1.7}$$

   in direct form, becomes

$$d_{ij} = a_{ik} b_{kl} c_{lj} \tag{1.8}$$

   in index form, where d_{ij}, a_{ik}, b_{kl}, and c_{lj} are typical elements of matrices $\bar{\mathbf{D}}$, $\bar{\mathbf{A}}$, $\bar{\mathbf{B}}$, and $\bar{\mathbf{C}}$ respectively; i, j are free indices and k, l are dummy indices which may be replaced by any other letters.

In applying the above rules, we sometimes need to interchange the subscripts in a_{ij}. From Eq. (1.4), we see that $a_{ij} \neq a_{ji}$ in general, and the interchange of subscripts is equivalent to the interchange of the row and column. The matrix obtained from $\bar{\mathbf{A}}$ by interchanging its rows and columns is called the *transpose* of $\bar{\mathbf{A}}$ and is denoted by $\tilde{\bar{\mathbf{A}}}$ or $\bar{\mathbf{A}}^T$.† In terms of a typical element a_{ij}, we have

$$\tilde{a}_{ij} = a_{ji} \tag{1.9}$$

From this definition, we obviously have $\tilde{\tilde{\bar{\mathbf{A}}}} = \bar{\mathbf{A}}$. A three-dimensional vector $\mathbf{v}$ may be represented by a *column matrix* (or *column vector*)

$$\mathbf{v} = \begin{bmatrix} v_1 \\ v_2 \\ v_3 \end{bmatrix} \tag{1.10}$$

the transpose of which gives a *row matrix* (or *row vector*) $\tilde{\mathbf{v}} = [v_1, v_2, v_3]$. However, in index notation both vectors are represented by v_i, the ith component of vector $\mathbf{v}$. In other words, we will not distinguish a column vector from a row vector, i.e., we consider them the same: $\tilde{\mathbf{v}} = \mathbf{v}$. In converting from index to direct notation involving vectors, special care is required. A typical case is the identity

$$a_{ij} v_j = v_j a_{ij} \tag{1.11}$$

which simply states the fact that multiplication of numbers is commutative. Expressing the right-hand side of Eq. (1.11) in direct form, we first bring the dummy indices j adjacent to each other, that is, $v_j a_{ij} = v_j \tilde{a}_{ji}$ and then convert to direct form. Thus, the identity (1.11) becomes

$$\bar{\mathbf{A}} \cdot \mathbf{v} = \mathbf{v} \cdot \tilde{\bar{\mathbf{A}}} \tag{1.12}$$

Similarly, from the identity

$$v_j a_{ji} = a_{ji} v_j = \tilde{a}_{ij} v_j \tag{1.13}$$

we obtain

$$\mathbf{v} \cdot \bar{\mathbf{A}} = \tilde{\bar{\mathbf{A}}} \cdot \mathbf{v} \tag{1.14}$$

Relations (1.12) and (1.14) are valid for any matrix $\bar{\mathbf{A}}$. That is, in the product of a vector by a matrix, we may change the order of the vector and the matrix provided that the matrix is replaced by its transpose.

1.2 ALGEBRA OF MATRICES

In this section, we shall review some definitions and basic operations of matrices. Both direct and index forms of definitions and operations will be given. Their correspondence will be placed side by side and indicated by $\leftrightarrow$.

† For convenience, we shall denote the transpose of a single matrix by using the tilde ~ and the transpose of the product of a number of matrices by using the superscript ^{T}.

The *sum* of two matrices $\bar{\mathbf{A}} = [a_{ij}]$ and $\bar{\mathbf{B}} = [b_{ij}]$ is defined as the matrix $\bar{\mathbf{C}} = [c_{ij}]$ such that

$$\bar{\mathbf{C}} = \bar{\mathbf{A}} + \bar{\mathbf{B}} \leftrightarrow c_{ij} = a_{ij} + b_{ij} \tag{1.15}$$

Similarly, the *difference* is defined by

$$\bar{\mathbf{C}} = \bar{\mathbf{A}} - \bar{\mathbf{B}} \leftrightarrow c_{ij} = a_{ij} - b_{ij} \tag{1.16}$$

Multiplication of a matrix $\bar{\mathbf{A}}$ by a scalar α is to multiply each element of the matrix by α. Namely,

$$\alpha\bar{\mathbf{A}} \leftrightarrow \alpha a_{ij} \tag{1.17}$$

Operating the vector $\mathbf{u}$ in Eq. (1.5) by $\bar{\mathbf{B}}$, we get

$$\mathbf{w} = \bar{\mathbf{B}} \cdot \mathbf{u} = \bar{\mathbf{B}} \cdot \bar{\mathbf{A}} \cdot \mathbf{v} \tag{1.18}$$

On the other hand, we may obtain vector $\mathbf{w}$ directly from vector $\mathbf{v}$ by a linear transformation characterized by a matrix $\bar{\mathbf{C}}$; that is,

$$\mathbf{w} = \bar{\mathbf{C}} \cdot \mathbf{v} \tag{1.19}$$

Comparing Eqs. (1.18) and (1.19), and noting that $\mathbf{v}$ is an arbitrary vector, we obtain

$$\bar{\mathbf{C}} = \bar{\mathbf{B}} \cdot \bar{\mathbf{A}} \leftrightarrow c_{ij} = b_{ik}\, a_{kj} \tag{1.20}$$

which is called the *product of matrices* $\bar{\mathbf{B}}$ and $\bar{\mathbf{A}}$. In other words, the product of two matrices $\bar{\mathbf{B}} = [b_{ij}]$ and $\bar{\mathbf{A}} = [a_{ij}]$, denoted by a dot between them in direct notation, is the matrix $\bar{\mathbf{C}} = [c_{ij}]$. The typical element c_{ij} of the product matrix is obtained by multiplying the elements in the ith row of $\bar{\mathbf{B}}$ by the corresponding elements in the jth column of $\bar{\mathbf{A}}$ and adding. In general, the product of matrices is not commutative. That is, in general, $\bar{\mathbf{B}} \cdot \bar{\mathbf{A}} \neq \bar{\mathbf{A}} \cdot \bar{\mathbf{B}}$. If, however, $\bar{\mathbf{A}} \cdot \bar{\mathbf{B}}$ is equal to $\bar{\mathbf{B}} \cdot \bar{\mathbf{A}}$, the two matrices $\bar{\mathbf{A}}$ and $\bar{\mathbf{B}}$ are said to *commute* with each other. From Eqs. (1.20) and (1.8), we may conclude that the product of three or more matrices is associative:

$$(\bar{\mathbf{A}} \cdot \bar{\mathbf{B}}) \cdot \bar{\mathbf{C}} = \bar{\mathbf{A}} \cdot (\bar{\mathbf{B}} \cdot \bar{\mathbf{C}}) \tag{1.21}$$

According to Eqs. (1.15) and (1.17), we obviously have

$$(\bar{\mathbf{A}} + \bar{\mathbf{B}})^T = \tilde{\bar{\mathbf{A}}} + \tilde{\bar{\mathbf{B}}} \tag{1.22}$$

That is, the transpose of the sum of two matrices is the sum of the transposed matrices, and

$$(\alpha\bar{\mathbf{A}})^T = \alpha\tilde{\bar{\mathbf{A}}} \tag{1.23}$$

By taking the transpose on both sides of Eq. (1.20), we obtain

$$\tilde{c}_{ij} = c_{ji} = b_{jk}\, a_{ki} = \tilde{a}_{ik}\, \tilde{b}_{kj} \tag{1.24}$$

In direct notation, this means that $\tilde{\bar{\mathbf{C}}} = \tilde{\bar{\mathbf{A}}} \cdot \tilde{\bar{\mathbf{B}}}$; hence

$$(\bar{\mathbf{B}} \cdot \bar{\mathbf{A}})^T = \tilde{\bar{\mathbf{A}}} \cdot \tilde{\bar{\mathbf{B}}} \tag{1.25}$$

The result may easily be extended to the product of more than two matrices. From Eqs. (1.21) and (1.25), we obtain

$$(\bar{\mathbf{A}} \cdot \bar{\mathbf{B}} \cdot \bar{\mathbf{C}})^T = [(\bar{\mathbf{A}} \cdot \bar{\mathbf{B}}) \cdot \bar{\mathbf{C}}]^T = \tilde{\bar{\mathbf{C}}} \cdot (\bar{\mathbf{A}} \cdot \bar{\mathbf{B}})^T = \tilde{\bar{\mathbf{C}}} \cdot \tilde{\bar{\mathbf{B}}} \cdot \tilde{\bar{\mathbf{A}}} \tag{1.26}$$

In other words, the transpose of the product of several matrices is equal to the product of their transposes taken in the reverse order. A matrix $\bar{\mathbf{A}}$ is said to be *symmetric* if

$$\tilde{\bar{\mathbf{A}}} = \bar{\mathbf{A}} \leftrightarrow a_{ji} = a_{ij} \tag{1.27}$$

and *antisymmetric* if

$$\tilde{\bar{\mathbf{A}}} = -\bar{\mathbf{A}} \leftrightarrow a_{ji} = -a_{ij} \tag{1.28}$$

Hence, it follows that the *diagonal elements* (i.e., elements with $i = j$) of an antisymmetric matrix must be zero, for $a_{ii} = -a_{ii}$ implies $a_{ii} = 0$. Any matrix $\bar{\mathbf{C}}$ can be decomposed uniquely into the sum of a symmetric matrix $\bar{\mathbf{A}}$ and an antisymmetric matrix $\bar{\mathbf{B}}$. Indeed

$$\bar{\mathbf{C}} = \bar{\mathbf{A}} + \bar{\mathbf{B}} \tag{1.29}$$

where

$$\bar{\mathbf{A}} = \tfrac{1}{2}(\bar{\mathbf{C}} + \tilde{\bar{\mathbf{C}}}) = \tilde{\bar{\mathbf{A}}}$$

is symmetric, and

$$\bar{\mathbf{B}} = \tfrac{1}{2}(\bar{\mathbf{C}} - \tilde{\bar{\mathbf{C}}}) = -\tilde{\bar{\mathbf{B}}}$$

is antisymmetric. Also, if $\bar{\mathbf{A}}$ and $\bar{\mathbf{B}}$ are symmetric matrices, their product is not necessarily symmetric. For according to Eq. (1.25),

$$(\bar{\mathbf{B}} \cdot \bar{\mathbf{A}})^T = \tilde{\bar{\mathbf{A}}} \cdot \tilde{\bar{\mathbf{B}}} = \bar{\mathbf{A}} \cdot \bar{\mathbf{B}} \neq \bar{\mathbf{B}} \cdot \bar{\mathbf{A}}$$

Thus, the product of two symmetric matrices is symmetric if they commute. Since a matrix always commutes with itself, any integer power of a symmetric matrix is also a symmetric matrix. Similarly, the even power of an antisymmetric matrix is symmetric and the odd power of an antisymmetric matrix is antisymmetric.

The *trace* of a matrix $\bar{\mathbf{A}}$, denoted by $\bar{\mathbf{A}}_t$, is defined to be the sum of the diagonal elements of $\bar{\mathbf{A}}$; that is,

$$\bar{\mathbf{A}}_t = a_{ii} = a_{11} + a_{22} + a_{33} \tag{1.30}$$

Thus, the trace of a matrix is a scalar, and since the operation is linear, we have

$$(\bar{\mathbf{A}} \pm \bar{\mathbf{B}})_t = \bar{\mathbf{A}}_t \pm \bar{\mathbf{B}}_t \tag{1.31}$$

and

$$(\alpha\bar{\mathbf{A}})_t = \alpha\bar{\mathbf{A}}_t \tag{1.32}$$

Furthermore, since the diagonal elements remain unchanged during transposition,

$$(\tilde{\bar{\mathbf{A}}})_t = \bar{\mathbf{A}}_t \tag{1.33}$$

Also

$$(\bar{\mathbf{A}} \cdot \bar{\mathbf{B}})_t = (\bar{\mathbf{A}} \cdot \bar{\mathbf{B}})_{ii} = a_{ij} b_{ji} = b_{ji} a_{ij}$$
$$= (\bar{\mathbf{B}} \cdot \bar{\mathbf{A}})_{jj} = (\bar{\mathbf{B}} \cdot \bar{\mathbf{A}})_t \quad (1.34)$$

which may easily be extended to the product of three matrices:

$$(\bar{\mathbf{A}} \cdot \bar{\mathbf{B}} \cdot \bar{\mathbf{C}})_t = [(\bar{\mathbf{A}} \cdot \bar{\mathbf{B}}) \cdot \bar{\mathbf{C}}]_t = (\bar{\mathbf{C}} \cdot \bar{\mathbf{A}} \cdot \bar{\mathbf{B}})_t$$
$$= (\bar{\mathbf{B}} \cdot \bar{\mathbf{C}} \cdot \bar{\mathbf{A}})_t \quad (1.35)$$

In words, the trace of the product of matrices is unaltered in value by any cyclic rearrangement of the matrices. The trace of an antisymmetric matrix is zero since all the diagonal elements are zero. The trace of the product of a symmetric matrix $\bar{\mathbf{A}}$ (that is, $\tilde{\bar{\mathbf{A}}} = \bar{\mathbf{A}}$) and an antisymmetric matrix $\bar{\mathbf{B}}$ (that is, $\tilde{\bar{\mathbf{B}}} = -\bar{\mathbf{B}}$) is also zero, for

$$(\bar{\mathbf{A}} \cdot \bar{\mathbf{B}})_t = [(\bar{\mathbf{A}} \cdot \bar{\mathbf{B}})^T]_t = (\tilde{\bar{\mathbf{B}}} \cdot \tilde{\bar{\mathbf{A}}})_t$$
$$= -(\bar{\mathbf{B}} \cdot \bar{\mathbf{A}})_t = -(\bar{\mathbf{A}} \cdot \bar{\mathbf{B}})_t$$

Hence

$$(\bar{\mathbf{A}} \cdot \bar{\mathbf{B}})_t = 0$$

which proves the statement.

1.3 DETERMINANT AND ADJOINT OF A MATRIX

Since the Kronecker delta and permutation symbol play a very important role in the manipulation of three-dimensional determinants and matrices, we shall first state their definitions and proceed with applications.

The *Kronecker delta*, δ_{ij}, is defined by

$$\delta_{ij} = \begin{cases} 1 & \text{for } i = j \quad \text{(no sum)} \\ 0 & \text{for } i \neq j \end{cases} \quad (1.36)$$

Thus

$$\delta_{11} = \delta_{22} = \delta_{33} = 1$$
$$\delta_{12} = \delta_{21} = \delta_{13} = \delta_{31} = \delta_{23} = \delta_{32} = 0$$

The matrix with elements δ_{ij}, denoted by $\bar{\mathbf{I}}$ in direct notation, is called the *unit matrix*; that is,

$$\bar{\mathbf{I}} = [\delta_{ij}] = \begin{bmatrix} 1 & 0 & 0 \\ 0 & 1 & 0 \\ 0 & 0 & 1 \end{bmatrix} \quad (1.37)$$

From the definition (1.36) it follows that

$$a_{ik} \delta_{kj} = \delta_{ik} a_{kj} = a_{ij} \quad (1.38)$$

Converting this into direct notation, we have

$$\bar{\mathbf{A}} \cdot \bar{\mathbf{I}} = \bar{\mathbf{I}} \cdot \bar{\mathbf{A}} = \bar{\mathbf{A}} \tag{1.39}$$

In other words, multiplication of any matrix $\bar{\mathbf{A}}$ either from the right or from the left by the unit matrix leaves the matrix unchanged. The same is true for any vector **u**, for

$$u_j \delta_{ji} = \delta_{ij} u_j = u_i \tag{1.40}$$

or

$$\mathbf{u} \cdot \bar{\mathbf{I}} = \bar{\mathbf{I}} \cdot \mathbf{u} = \mathbf{u} \tag{1.41}$$

The matrix

$$\alpha \bar{\mathbf{I}} = [\alpha \delta_{ij}] \tag{1.42}$$

is called the *scalar matrix* since the multiplication of any matrix or vector by the scalar α gives the same result as multiplication by the scalar matrix.

The *permutation symbol*, denoted by ε_{ijk}, is defined by

$$\varepsilon_{ijk} = \begin{cases} +1 & \text{if } i, j, k \text{ is an even permutation of } 1, 2, 3 \\ -1 & \text{if } i, j, k \text{ is an odd permutation of } 1, 2, 3 \\ 0 & \text{if any two of } i, j, k \text{ are equal} \end{cases} \tag{1.43}$$

A *permutation* is even or odd depending on whether an even or odd number of exchanges are required to rearrange the i, j, k into the natural order of 1, 2, 3. For instance 312 has an even number of permutations (312 → 132 → 123), whereas 213 has an odd number of them (213 → 123); thus

$$\varepsilon_{312} = +1 \qquad \varepsilon_{213} = -1$$

From the definition of ε_{ijk}, we may write down all the $3^3 = 27$ components:

ε_{ij1}

i \ j	1	2	3
1	0	0	0
2	0	0	+1
3	0	−1	0

ε_{ij2}

i \ j	1	2	3
1	0	0	−1
2	0	0	0
3	+1	0	0

ε_{ij3}

i \ j	1	2	3
1	0	+1	0
2	−1	0	0
3	0	0	0

(1.44)

Thus, 21 components of ε_{ijk} are zero, three components (ε_{123}, ε_{231}, ε_{312}) are equal to +1, and the other three components (ε_{321}, ε_{132}, ε_{213}) are equal to −1. Note that interchanging any two subscripts changes the sign of the permutation symbol. That is,

$$\varepsilon_{ijk} = \varepsilon_{jki} = \varepsilon_{kij} = -\varepsilon_{jik} = -\varepsilon_{ikj} = -\varepsilon_{kji} \tag{1.45}$$

Hence ε_{ijk} is antisymmetric.

With the aid of the permutation symbol, we may express compactly the determinant of a matrix. Let $\bar{\mathbf{A}} = [a_{ij}]$ be the given matrix. Then the *determinant* of $\bar{\mathbf{A}}$, denoted by $|\bar{\mathbf{A}}|$ (or det $\bar{\mathbf{A}}$), may be written as

$$|\bar{\mathbf{A}}| = \det \bar{\mathbf{A}} = \begin{vmatrix} a_{11} & a_{12} & a_{13} \\ a_{21} & a_{22} & a_{23} \\ a_{31} & a_{32} & a_{33} \end{vmatrix} = \varepsilon_{lmn} a_{l1} a_{m2} a_{n3}$$

$$= \varepsilon_{lmn} a_{1l} a_{2m} a_{3n} \tag{1.46}$$

Expression (1.46) may be justified by carrying out the indicated sums in detail. For instance,

$$\begin{aligned} \varepsilon_{lmn} a_{1l} a_{2m} a_{3n} &= \varepsilon_{123} a_{11} a_{22} a_{33} + \varepsilon_{132} a_{11} a_{23} a_{32} \\ &\quad + \varepsilon_{213} a_{12} a_{21} a_{33} + \varepsilon_{231} a_{12} a_{23} a_{31} \\ &\quad + \varepsilon_{312} a_{13} a_{21} a_{32} + \varepsilon_{321} a_{13} a_{22} a_{31} \\ &= a_{11}(a_{22} a_{33} - a_{23} a_{32}) - a_{12}(a_{21} a_{33} - a_{23} a_{31}) \\ &\quad + a_{13}(a_{21} a_{32} - a_{22} a_{31}) \end{aligned} \tag{1.47}$$

A direct expansion of the determinant gives the same result. Another useful equivalent expression is

$$\begin{vmatrix} a_{i1} & a_{i2} & a_{i3} \\ a_{j1} & a_{j2} & a_{j3} \\ a_{k1} & a_{k2} & a_{k3} \end{vmatrix} = \varepsilon_{lmn} a_{il} a_{jm} a_{kn} = \varepsilon_{ijk} |\bar{\mathbf{A}}| \tag{1.48}$$

Here if the rows have been interchanged an odd or even number of times, then ε_{ijk} automatically provides the correct sign. Similarly, we have

$$\begin{vmatrix} a_{1i} & a_{1j} & a_{1k} \\ a_{2i} & a_{2j} & a_{2k} \\ a_{3i} & a_{3j} & a_{3k} \end{vmatrix} = \varepsilon_{lmn} a_{li} a_{mj} a_{nk} = \varepsilon_{ijk} |\bar{\mathbf{A}}| \tag{1.49}$$

As an application, let us show that the determinant of the product of two matrices $\bar{\mathbf{A}} = [a_{ij}]$ and $\bar{\mathbf{B}} = [b_{ij}]$ is the product of the determinants $|\bar{\mathbf{A}}|$ and $|\bar{\mathbf{B}}|$. Indeed, if $\bar{\mathbf{C}} = [c_{ij}] = \bar{\mathbf{A}} \cdot \bar{\mathbf{B}}$, then according to Eqs. (1.46), (1.20) and (1.48), we have

$$\begin{aligned} |\bar{\mathbf{C}}| = |\bar{\mathbf{A}} \cdot \bar{\mathbf{B}}| &= \varepsilon_{lmn} c_{1l} c_{2m} c_{3n} \\ &= (\varepsilon_{lmn} b_{il} b_{jm} b_{kn})(a_{1i} a_{2j} a_{3k}) \\ &= |\bar{\mathbf{B}}| \, (\varepsilon_{ijk} a_{1i} a_{2j} a_{3k}) \\ &= |\bar{\mathbf{B}}| \, |\bar{\mathbf{A}}| \end{aligned} \tag{1.50}$$

which proves the result.

Setting $\bar{\mathbf{A}} = \bar{\mathbf{I}}$ or $a_{ij} = \delta_{ij}$ in Eqs. (1.48) and (1.49), we obtain

$$\begin{vmatrix} \delta_{i1} & \delta_{i2} & \delta_{i3} \\ \delta_{j1} & \delta_{j2} & \delta_{j3} \\ \delta_{k1} & \delta_{k2} & \delta_{k3} \end{vmatrix} = \begin{vmatrix} \delta_{1i} & \delta_{1j} & \delta_{1k} \\ \delta_{2i} & \delta_{2j} & \delta_{2k} \\ \delta_{3i} & \delta_{3j} & \delta_{3k} \end{vmatrix} = \varepsilon_{ijk} \tag{1.51}$$

Hence

$$\varepsilon_{ijk}\varepsilon_{lmn} = \begin{vmatrix} \delta_{i1} & \delta_{i2} & \delta_{i3} \\ \delta_{j1} & \delta_{j2} & \delta_{j3} \\ \delta_{k1} & \delta_{k2} & \delta_{k3} \end{vmatrix} \begin{vmatrix} \delta_{1l} & \delta_{1m} & \delta_{1n} \\ \delta_{2l} & \delta_{2m} & \delta_{2n} \\ \delta_{3l} & \delta_{3m} & \delta_{3n} \end{vmatrix}$$

$$= \begin{vmatrix} \delta_{ip}\delta_{pl} & \delta_{ip}\delta_{pm} & \delta_{ip}\delta_{pn} \\ \delta_{jp}\delta_{pl} & \delta_{jp}\delta_{pm} & \delta_{jp}\delta_{pn} \\ \delta_{kp}\delta_{pl} & \delta_{kp}\delta_{pm} & \delta_{kp}\delta_{pn} \end{vmatrix}$$

$$= \begin{vmatrix} \delta_{il} & \delta_{im} & \delta_{in} \\ \delta_{jl} & \delta_{jm} & \delta_{jn} \\ \delta_{kl} & \delta_{km} & \delta_{kn} \end{vmatrix} \tag{1.52}$$

This is a useful identity that relates the permutation symbol and Kronecker delta. By contracting the subscripts, we may obtain other identities. Thus, setting $n = k$ in Eq. (1.52) and expanding the resulting determinant by its third row, we find that

$$\varepsilon_{ijk}\varepsilon_{lmk} = \delta_{kl}\begin{vmatrix} \delta_{im} & \delta_{ik} \\ \delta_{jm} & \delta_{jk} \end{vmatrix} - \delta_{km}\begin{vmatrix} \delta_{il} & \delta_{ik} \\ \delta_{jl} & \delta_{jk} \end{vmatrix} + \delta_{kk}\begin{vmatrix} \delta_{il} & \delta_{im} \\ \delta_{jl} & \delta_{jm} \end{vmatrix}$$

$$= \begin{vmatrix} \delta_{im} & \delta_{il} \\ \delta_{jm} & \delta_{jl} \end{vmatrix} - \begin{vmatrix} \delta_{il} & \delta_{im} \\ \delta_{jl} & \delta_{jm} \end{vmatrix} + 3\begin{vmatrix} \delta_{il} & \delta_{im} \\ \delta_{jl} & \delta_{jm} \end{vmatrix}$$

Interchanging the columns of the first 2×2 determinant, we obtain

$$\varepsilon_{ijk}\varepsilon_{lmk} = \begin{vmatrix} \delta_{il} & \delta_{im} \\ \delta_{jl} & \delta_{jm} \end{vmatrix} = \delta_{il}\delta_{jm} - \delta_{im}\delta_{jl} \tag{1.53}$$

In Eq. (1.53), on taking $m = j$, we get

$$\varepsilon_{ijk}\varepsilon_{ljk} = 2\delta_{il} \tag{1.54}$$

Finally, letting $l = i$ gives

$$\varepsilon_{ijk}\varepsilon_{ijk} = 2\delta_{ii} = 6 \tag{1.55}$$

To express the determinant of a matrix $\bar{\mathbf{A}}$ in direct form, let us multiply Eq. (1.48) by ε_{ijk}. Making use of the result in Eq. (1.55), we obtain

$$|\bar{\mathbf{A}}| = \tfrac{1}{6}\varepsilon_{ijk}\varepsilon_{lmn}a_{il}a_{jm}a_{kn} \tag{1.56}$$

Next, substitution of Eq. (1.52) into Eq. (1.56) yields

$$|\bar{\mathbf{A}}| = \frac{1}{6}\begin{vmatrix} \delta_{il} & \delta_{im} & \delta_{in} \\ \delta_{jl} & \delta_{jm} & \delta_{jn} \\ \delta_{kl} & \delta_{km} & \delta_{kn} \end{vmatrix} a_{il}a_{jm}a_{kn}$$

Multiplying the first row by a_{il}, the second row by a_{jm}, the third row by a_{kn}, and then expanding the resulted determinant, we finally obtain the expansion of the determinant of $\bar{\mathbf{A}}$ in terms of traces of $\bar{\mathbf{A}}$:

$$|\bar{\mathbf{A}}| = \tfrac{1}{6}[(\bar{\mathbf{A}}_t)^3 - 3(\bar{\mathbf{A}}_t)(\bar{\mathbf{A}}^2)_t + 2(\bar{\mathbf{A}}^3)_t] \tag{1.57}$$

For the given matrix $\bar{\mathbf{A}} = [a_{ij}]$, we now define the *adjoint matrix* of $\bar{\mathbf{A}}$, denoted by adj $\bar{\mathbf{A}}$, by

$$\text{adj } \bar{\mathbf{A}} = [(\text{adj } \bar{\mathbf{A}})_{ns}] \tag{1.58}$$

where

$$(\text{adj } \bar{\mathbf{A}})_{ns} = \tfrac{1}{2}\varepsilon_{ijs}\,\varepsilon_{lmn}\,a_{il}\,a_{jm} \tag{1.59}$$

It can easily be checked that the transpose of (adj $\bar{\mathbf{A}})_{ns}$ is the cofactor of the element a_{sn}. Thus the adjoint matrix is the transpose of the matrix obtained from $\bar{\mathbf{A}}$ by replacing each element a_{sn} by its cofactor. Using the identity (1.52) in Eq. (1.59), we have

$$(\text{adj } \bar{\mathbf{A}})_{ns} = \frac{1}{2}\begin{vmatrix} \delta_{il} & \delta_{im} & \delta_{in} \\ \delta_{jl} & \delta_{jm} & \delta_{jn} \\ \delta_{sl} & \delta_{sm} & \delta_{sn} \end{vmatrix} a_{il}\,a_{jm} \tag{1.60}$$

Multiplying the first row by a_{il}, the second row by a_{jm}, and then expanding the determinant by the last row, we obtain an alternative form:

$$(\text{adj } \bar{\mathbf{A}})_{ns} = (a_{nl}\,a_{ls} - a_{ii}\,a_{ns}) + \tfrac{1}{2}(a_{ii}\,a_{jj} - a_{lm}\,a_{ml})\delta_{ns} \tag{1.61}$$

or, in direct notation,

$$\text{adj } \bar{\mathbf{A}} = \bar{\mathbf{A}}^2 - \bar{\mathbf{A}}_t\bar{\mathbf{A}} + \tfrac{1}{2}[(\bar{\mathbf{A}}_t)^2 - (\bar{\mathbf{A}}^2)_t]\bar{\mathbf{I}} \tag{1.62}$$

Setting $s = n$ in Eq. (1.61), we obtain the trace of adj $\bar{\mathbf{A}}$:

$$(\text{adj } \bar{\mathbf{A}})_t = (\text{adj } \bar{\mathbf{A}})_{nn} = \tfrac{1}{2}[(\bar{\mathbf{A}}_t)^2 - (\bar{\mathbf{A}}^2)_t] \tag{1.63}$$

Hence, we may also write adj $\bar{\mathbf{A}}$ as

$$\text{adj } \bar{\mathbf{A}} = \bar{\mathbf{A}}^2 - \bar{\mathbf{A}}_t\bar{\mathbf{A}} + (\text{adj } \bar{\mathbf{A}})_t\bar{\mathbf{I}} \tag{1.64}$$

From Eqs. (1.59), (1.48) and (1.54), we find

$$\begin{aligned} a_{kn}(\text{adj } \bar{\mathbf{A}})_{ns} &= a_{kn}(\tfrac{1}{2}\varepsilon_{ijs}\,\varepsilon_{lmn}\,a_{il}\,a_{jm}) \\ &= \tfrac{1}{2}\varepsilon_{ijs}\,\varepsilon_{ijk}\,|\bar{\mathbf{A}}| \\ &= |\bar{\mathbf{A}}|\,\delta_{ks} \end{aligned} \tag{1.65}$$

and

$$(\text{adj } \bar{\mathbf{A}})_{ns}\,a_{sk} = |\bar{\mathbf{A}}|\,\delta_{nk} \tag{1.66}$$

and thus the following useful identity in direct form

$$\bar{\mathbf{A}} \cdot (\text{adj } \bar{\mathbf{A}}) = (\text{adj } \bar{\mathbf{A}}) \cdot \bar{\mathbf{A}} = |\bar{\mathbf{A}}|\,\bar{\mathbf{I}} \tag{1.67}$$

That is, the product of the matrix $\bar{\mathbf{A}}$ and its adjoint matrix is commutative and is equal to the scalar matrix $|\bar{\mathbf{A}}|\,\bar{\mathbf{I}}$. A matrix $\bar{\mathbf{A}}$ is said to be *nonsingular* if $|\bar{\mathbf{A}}| \neq 0$, and *singular* if $|\bar{\mathbf{A}}| = 0$. For a nonsingular matrix $\bar{\mathbf{A}}$, we may divide Eq. (1.67) by $|\bar{\mathbf{A}}|$:

$$\bar{\mathbf{A}} \cdot \frac{\operatorname{adj} \bar{\mathbf{A}}}{|\bar{\mathbf{A}}|} = \frac{\operatorname{adj} \bar{\mathbf{A}}}{|\bar{\mathbf{A}}|} \cdot \bar{\mathbf{A}} = \bar{\mathbf{I}} \tag{1.68}$$

From this equation it seems natural to define the matrix $\operatorname{adj} \bar{\mathbf{A}}/|\bar{\mathbf{A}}|$ as the *inverse* of $\bar{\mathbf{A}}$, denoted by $\bar{\mathbf{A}}^{-1}$, such that

$$\bar{\mathbf{A}} \cdot \bar{\mathbf{A}}^{-1} = \bar{\mathbf{A}}^{-1} \cdot \bar{\mathbf{A}} = \bar{\mathbf{I}} \tag{1.69}$$

where

$$\bar{\mathbf{A}}^{-1} = \frac{\operatorname{adj} \bar{\mathbf{A}}}{|\bar{\mathbf{A}}|} \tag{1.70}$$

In summary, the inverse $\bar{\mathbf{A}}^{-1}$ of a nonsingular matrix $\bar{\mathbf{A}}$ is unique and is commutative with $\bar{\mathbf{A}}$, the product being the unit matrix $\bar{\mathbf{I}}$. Note that the adjoint of $\bar{\mathbf{A}}$ exists whether $\bar{\mathbf{A}}$ is singular or nonsingular. However, only nonsingular matrices have inverses.

From Eq. (1.69), we may easily establish the following properties of the inverses of matrices $\bar{\mathbf{A}}$ and $\bar{\mathbf{B}}$:

$$\begin{aligned}
&(a)\quad (\alpha\bar{\mathbf{A}})^{-1} = \frac{1}{\alpha}\,\bar{\mathbf{A}}^{-1} \\
&(b)\quad (\bar{\mathbf{A}}^{-1})^{-1} = \bar{\mathbf{A}} \\
&(c)\quad (\bar{\mathbf{A}} \cdot \bar{\mathbf{B}})^{-1} = \bar{\mathbf{B}}^{-1} \cdot \bar{\mathbf{A}}^{-1} \\
&(d)\quad (\bar{\mathbf{A}}^{-1})^{T} = (\tilde{\bar{\mathbf{A}}})^{-1}
\end{aligned} \tag{1.71}$$

To prove (*a*) we note that

$$\alpha\bar{\mathbf{A}} \cdot \left(\frac{1}{\alpha}\,\bar{\mathbf{A}}^{-1}\right) = \alpha\left(\frac{1}{\alpha}\right)\bar{\mathbf{A}} \cdot \bar{\mathbf{A}}^{-1} = \bar{\mathbf{A}} \cdot \bar{\mathbf{A}}^{-1} = \bar{\mathbf{I}}$$

and hence the result of (*a*). Property (*b*) follows directly from Eq. (1.69). To show (*c*) we note that

$$(\bar{\mathbf{B}}^{-1} \cdot \bar{\mathbf{A}}^{-1}) \cdot (\bar{\mathbf{A}} \cdot \bar{\mathbf{B}}) = \bar{\mathbf{B}}^{-1} \cdot \bar{\mathbf{I}} \cdot \bar{\mathbf{B}} = \bar{\mathbf{B}}^{-1} \cdot \bar{\mathbf{B}} = \bar{\mathbf{I}}$$

and

$$(\bar{\mathbf{A}} \cdot \bar{\mathbf{B}}) \cdot (\bar{\mathbf{B}}^{-1} \cdot \bar{\mathbf{A}}^{-1}) = \bar{\mathbf{A}} \cdot \bar{\mathbf{I}} \cdot \bar{\mathbf{A}}^{-1} = \bar{\mathbf{A}} \cdot \bar{\mathbf{A}}^{-1} = \bar{\mathbf{I}}$$

Thus the inverse of the product of two matrices is equal to the product of their inverses taken in the reverse order. To demonstrate (*d*) we start by taking the transpose of Eq. (1.69) and obtain

$$(\bar{\mathbf{A}} \cdot \bar{\mathbf{A}}^{-1})^{T} = (\bar{\mathbf{A}}^{-1})^{T} \cdot \tilde{\bar{\mathbf{A}}} = \tilde{\bar{\mathbf{I}}} = \bar{\mathbf{I}}$$

Thus the transpose of the inverse of $\bar{\mathbf{A}}$ is equal to the inverse of its transpose.

Next we list some formulas of the determinant and adjoint of matrices for easy future reference.

$$
\begin{aligned}
&(a)\quad |\tilde{\bar{\mathbf{A}}}| = |\bar{\mathbf{A}}| \\
&(b)\quad |\alpha\bar{\mathbf{A}}| = \alpha^3 |\bar{\mathbf{A}}| \qquad |\alpha\bar{\mathbf{I}}| = \alpha^3 \\
&(c)\quad |\bar{\mathbf{A}} \cdot \bar{\mathbf{B}}| = |\bar{\mathbf{A}}||\bar{\mathbf{B}}| \\
&(d)\quad |\bar{\mathbf{A}}^{-1}| = |\bar{\mathbf{A}}|^{-1} \\
&(e)\quad |\text{adj } \bar{\mathbf{A}}| = |\bar{\mathbf{A}}|^2 \\
&(f)\quad \text{adj}(\alpha\bar{\mathbf{A}}) = \alpha^2(\text{adj } \bar{\mathbf{A}}) \qquad \text{adj}(\alpha\bar{\mathbf{I}}) = \alpha^2\bar{\mathbf{I}} \\
&(g)\quad \text{adj } \bar{\mathbf{A}}^{-1} = (\text{adj } \bar{\mathbf{A}})^{-1} = \frac{\bar{\mathbf{A}}}{|\bar{\mathbf{A}}|} \\
&(h)\quad \text{adj}(\text{adj } \bar{\mathbf{A}}) = |\bar{\mathbf{A}}|\bar{\mathbf{A}} \\
&(i)\quad (\text{adj } \bar{\mathbf{A}})^T = \text{adj } \tilde{\bar{\mathbf{A}}} \\
&(j)\quad \text{adj}(\bar{\mathbf{A}} \cdot \bar{\mathbf{B}}) = \text{adj } \bar{\mathbf{B}} \cdot \text{adj } \bar{\mathbf{A}}
\end{aligned}
\tag{1.72}
$$

The result of (*a*) follows from the definition of the determinant of a matrix. Formula (*b*) can easily be verified by using Eq. (1.56). Formula (*c*) was proved in Eq. (1.50). To show (*d*) let us take the determinant on both sides of the identity (1.69). Making use of the result of (*c*) and noting that $|\bar{\mathbf{I}}| = 1$, we obtain $|\bar{\mathbf{A}} \cdot \bar{\mathbf{A}}^{-1}| = |\bar{\mathbf{A}}||\bar{\mathbf{A}}^{-1}| = 1$ or $|\bar{\mathbf{A}}^{-1}| = |\bar{\mathbf{A}}|^{-1}$. To prove (*e*) we note that adj $\bar{\mathbf{A}} = |\bar{\mathbf{A}}|\bar{\mathbf{A}}^{-1}$; thus

$$|\text{adj } \bar{\mathbf{A}}| = ||\bar{\mathbf{A}}|\bar{\mathbf{A}}^{-1}| = |\bar{\mathbf{A}}|^3|\bar{\mathbf{A}}|^{-1} = |\bar{\mathbf{A}}|^2$$

Formula (*f*) follows from definition (1.59) or Eq. (1.70). In fact,

$$
\begin{aligned}
\text{adj}(\alpha\bar{\mathbf{A}}) &= |\alpha\bar{\mathbf{A}}|\,(\alpha\bar{\mathbf{A}})^{-1} = \alpha^3|\bar{\mathbf{A}}|\left(\frac{1}{\alpha}\right)\bar{\mathbf{A}}^{-1} \\
&= \alpha^2|\bar{\mathbf{A}}|\bar{\mathbf{A}}^{-1} = \alpha^2\,(\text{adj } \bar{\mathbf{A}})
\end{aligned}
$$

Also since $|\bar{\mathbf{I}}| = 1$ and $\bar{\mathbf{I}}^{-1} = \bar{\mathbf{I}}$, adj $(\alpha\bar{\mathbf{I}}) = \alpha^2\bar{\mathbf{I}}$. To prove (*g*), we again use adj $\bar{\mathbf{A}} = |\bar{\mathbf{A}}|\bar{\mathbf{A}}^{-1}$. Thus

$$
\begin{aligned}
\text{adj } \bar{\mathbf{A}}^{-1} &= |\bar{\mathbf{A}}^{-1}|\,(\bar{\mathbf{A}}^{-1})^{-1} = |\bar{\mathbf{A}}|^{-1}\bar{\mathbf{A}} \\
&= |\bar{\mathbf{A}}^{-1}|\left(\frac{\text{adj } \bar{\mathbf{A}}}{|\bar{\mathbf{A}}|}\right)^{-1} = (\text{adj } \bar{\mathbf{A}})^{-1}
\end{aligned}
$$

Formula (*h*) follows from the fact that

$$
\begin{aligned}
\text{adj}(\text{adj } \bar{\mathbf{A}}) &= |\text{adj } \bar{\mathbf{A}}|(\text{adj } \bar{\mathbf{A}}^{-1}) \\
&= |\bar{\mathbf{A}}|^2|\bar{\mathbf{A}}|^{-1}\bar{\mathbf{A}} = |\bar{\mathbf{A}}|\bar{\mathbf{A}}
\end{aligned}
$$

Formula (*i*) follows from Eqs. (1.71*d*) and (1.72*a*). Formula (*j*) follows from Eqs. (1.71*c*) and (1.72*c*).

To conclude this section, let us apply Eqs. (1.57) and (1.62) to find the determinant and adjoint of the sum of two matrices. If $\bar{\mathbf{A}}$ and $\bar{\mathbf{B}}$ are two arbitrary

3×3 matrices, then

$$(\bar{\mathbf{A}} + \bar{\mathbf{B}})^2 = (\bar{\mathbf{A}} + \bar{\mathbf{B}}) \cdot (\bar{\mathbf{A}} + \bar{\mathbf{B}})$$
$$= \bar{\mathbf{A}}^2 + \bar{\mathbf{B}} \cdot \bar{\mathbf{A}} + \bar{\mathbf{A}} \cdot \bar{\mathbf{B}} + \bar{\mathbf{B}}^2 \tag{1.73}$$

and

$$(\bar{\mathbf{A}} + \bar{\mathbf{B}})^3 = \bar{\mathbf{A}}^3 + \bar{\mathbf{B}}^3 + \bar{\mathbf{B}} \cdot \bar{\mathbf{A}}^2 + \bar{\mathbf{A}} \cdot \bar{\mathbf{B}} \cdot \bar{\mathbf{A}} + \bar{\mathbf{A}}^2 \cdot \bar{\mathbf{B}}$$
$$+ \bar{\mathbf{B}}^2 \cdot \bar{\mathbf{A}} + \bar{\mathbf{B}} \cdot \bar{\mathbf{A}} \cdot \bar{\mathbf{B}} + \bar{\mathbf{A}} \cdot \bar{\mathbf{B}}^2 \tag{1.74}$$

Taking the trace on both sides of Eqs. (1.73) and (1.74), we find

$$[(\bar{\mathbf{A}} + \bar{\mathbf{B}})^2]_t = (\bar{\mathbf{A}}^2)_t + 2(\bar{\mathbf{A}} \cdot \bar{\mathbf{B}})_t + (\bar{\mathbf{B}}^2)_t \tag{1.75}$$

and

$$[(\bar{\mathbf{A}} + \bar{\mathbf{B}})^3]_t = (\bar{\mathbf{A}}^3)_t + 3(\bar{\mathbf{B}} \cdot \bar{\mathbf{A}}^2)_t + 3(\bar{\mathbf{A}} \cdot \bar{\mathbf{B}}^2)_t + (\bar{\mathbf{B}}^3)_t \tag{1.76}$$

Substituting the above into Eqs. (1.57) and (1.62) respectively, after some simplification, we find the determinant of the sum of two matrices:

$$|\bar{\mathbf{A}} + \bar{\mathbf{B}}| = |\bar{\mathbf{A}}| + |\bar{\mathbf{B}}| + (\text{adj } \bar{\mathbf{A}} \cdot \bar{\mathbf{B}})_t + (\bar{\mathbf{A}} \cdot \text{adj } \bar{\mathbf{B}})_t \tag{1.77}$$

and the adjoint of $\bar{\mathbf{A}} + \bar{\mathbf{B}}$ as

$$\text{adj}(\bar{\mathbf{A}} + \bar{\mathbf{B}}) = \text{adj } \bar{\mathbf{A}} + \text{adj } \bar{\mathbf{B}} + \bar{\mathbf{A}} \cdot \bar{\mathbf{B}} + \bar{\mathbf{B}} \cdot \bar{\mathbf{A}}$$
$$- \bar{\mathbf{A}}_t \bar{\mathbf{B}} - \bar{\mathbf{B}}_t \bar{\mathbf{A}} + \bar{\mathbf{A}}_t \bar{\mathbf{B}}_t \bar{\mathbf{I}} - (\bar{\mathbf{A}} \cdot \bar{\mathbf{B}})_t \bar{\mathbf{I}} \tag{1.78}$$

Furthermore, by taking the trace on both sides of Eq. (1.78), we obtain

$$[\text{adj}(\bar{\mathbf{A}} + \bar{\mathbf{B}})]_t = (\text{adj } \bar{\mathbf{A}})_t + (\text{adj } \bar{\mathbf{B}})_t - (\bar{\mathbf{A}} \cdot \bar{\mathbf{B}})_t + \bar{\mathbf{A}}_t \bar{\mathbf{B}}_t \tag{1.79}$$

Equations (1.77) and (1.78) prove to be useful in later applications since they express the determinant and adjoint of the sum of two matrices in terms of the matrices themselves, not their components.

1.4 DYAD AND ANTISYMMETRIC MATRIX $\mathbf{u} \times \bar{\mathbf{I}}$

There are three types of products defined between two vectors. We have already encountered one of them, the dot product. In this section we consider the other two, namely, the dyadic and cross products. The *dyadic product* (or simply dyad) of vectors **u** and **v**, denoted by **uv**, is defined by

$$\mathbf{uv} \leftrightarrow u_i v_j \tag{1.80}$$

We note that in the dyadic product, there is no dot between vectors **u** and **v** in direct notation, and the indices for u_i and v_j are different in index notation. The vector **u** is called *antecedent* and the vector **v** *consequent*. The sum of two or more dyads is called a *dyadic*. The nine components of **uv** may be represented by the matrix†

$$\mathbf{uv} = [u_i v_j] = \begin{bmatrix} u_1v_1 & u_1v_2 & u_1v_3 \\ u_2v_1 & u_2v_2 & u_2v_3 \\ u_3v_1 & u_3v_2 & u_3v_3 \end{bmatrix} \tag{1.81}$$

We see that all rows (or columns) in Eq. (1.81) are in proportion. The transpose of a dyad is the interchange of order of vectors in the dyad:

$$(\mathbf{uv})_{ij}^T = (\mathbf{uv})_{ji} = u_j v_i$$

$$= v_i u_j = (\mathbf{vu})_{ij}$$

or

$$(\mathbf{uv})^T = \mathbf{vu} \tag{1.82}$$

The trace of a dyad is equal to the dot product of vectors forming the dyad:

$$(\mathbf{uv})_t = (\mathbf{uv})_{ii} = u_i v_i = \mathbf{u} \cdot \mathbf{v} \tag{1.83}$$

From which it follows that

$$(\mathbf{uv} \cdot \bar{\mathbf{A}})_t = \mathbf{v} \cdot \bar{\mathbf{A}} \cdot \mathbf{u} \tag{1.84}$$

Furthermore, both the determinant and the adjoint of a dyad are zero since, if $\bar{\mathbf{A}} = \mathbf{uv}$, then $\bar{\mathbf{A}}_t = \mathbf{u} \cdot \mathbf{v}$ and

$$\bar{\mathbf{A}}^2 = (\mathbf{u} \cdot \mathbf{v})\mathbf{uv} \qquad (\bar{\mathbf{A}}^2)_t = (\mathbf{u} \cdot \mathbf{v})^2$$

$$\bar{\mathbf{A}}^3 = (\mathbf{u} \cdot \mathbf{v})^2\mathbf{uv} \qquad (\bar{\mathbf{A}}^3)_t = (\mathbf{u} \cdot \mathbf{v})^3$$

Substituting the above into Eqs. (1.57) and (1.62), we prove the stated results

$$|\mathbf{uv}| = 0 \tag{1.85}$$

and

$$\operatorname{adj}(\mathbf{uv}) = \bar{\mathbf{0}} \tag{1.86}$$

Next we consider the *cross product* of vectors **u** and **v**. With the aid of the permutation symbol, the cross product is defined as

$$\mathbf{s} = \mathbf{u} \times \mathbf{v} \leftrightarrow s_i = \varepsilon_{ijk} u_j v_k \tag{1.87}$$

where s_i, u_i, and v_i are the ith component of vectors **s**, **u**, and **v** respectively. In order to verify that the correspondence of Eq. (1.87) furnishes the cross product, we check the components of vector **s**. Thus

$$s_1 = \varepsilon_{1jk} u_j v_k = \varepsilon_{123} u_2 v_3 + \varepsilon_{132} u_3 v_2$$

$$= u_2 v_3 - u_3 v_2$$

Similarly,

$$s_2 = u_3 v_1 - u_1 v_3 \qquad s_3 = u_1 v_2 - u_2 v_1$$

† In terms of matrix multiplication, the dyad may be written as $\mathbf{u}\tilde{\mathbf{v}}$, a product of a column matrix **u** with a row matrix $\tilde{\mathbf{v}}$. On the other hand, the dot product may be expressed as $\tilde{\mathbf{u}}\mathbf{v}$, a product of a row matrix with a column matrix.

The scalar $\mathbf{u} \times \mathbf{v} \cdot \mathbf{w}$ formed by vectors $\mathbf{u}$, $\mathbf{v}$, and $\mathbf{w}$ is called a *scalar triple product.* From Eq. (1.45), we see that $\varepsilon_{ijk} = \varepsilon_{jki}$, hence

$$\begin{aligned} \mathbf{u} \times \mathbf{v} \cdot \mathbf{w} &= (\mathbf{u} \times \mathbf{v})_i w_i = \varepsilon_{ijk} u_j v_k w_i \\ &= \varepsilon_{jki} u_j v_k w_i = \mathbf{u} \cdot \mathbf{v} \times \mathbf{w} \end{aligned} \tag{1.88}$$

which shows that in any scalar triple product the position of the dot and cross can be interchanged without altering the value of the product.

Alternatively, we may write $\varepsilon_{ijk} u_j v_k$ as $(\varepsilon_{ijk} u_j)v_k$. The nine components of $\varepsilon_{ijk} u_j$ form a matrix, denoted in direct notation by either $\mathbf{u} \times \bar{\mathbf{I}}$ or $\bar{\mathbf{I}} \times \mathbf{u}$. That is,

$$(\mathbf{u} \times \bar{\mathbf{I}})_{ik} = (\bar{\mathbf{I}} \times \mathbf{u})_{ik} = \varepsilon_{ijk} u_j \tag{1.89}$$

or

$$\mathbf{u} \times \bar{\mathbf{I}} = \bar{\mathbf{I}} \times \mathbf{u} = \begin{bmatrix} 0 & -u_3 & u_2 \\ u_3 & 0 & -u_1 \\ -u_2 & u_1 & 0 \end{bmatrix} \tag{1.90}$$

In other words, the cross product of vectors $\mathbf{u}$ and $\mathbf{v}$ may be considered as the dot product of the matrix $(\mathbf{u} \times \bar{\mathbf{I}})$ and vector $\mathbf{v}$, namely,

$$\mathbf{u} \times \mathbf{v} = (\mathbf{u} \times \bar{\mathbf{I}}) \cdot \mathbf{v} = (\bar{\mathbf{I}} \times \mathbf{u}) \cdot \mathbf{v} \tag{1.91}$$

The operation defined by Eq. (1.89) is linear, i.e.,

$$(\alpha\mathbf{u} + \beta\mathbf{v}) \times \bar{\mathbf{I}} = \alpha(\mathbf{u} \times \bar{\mathbf{I}}) + \beta(\mathbf{v} \times \bar{\mathbf{I}}) \tag{1.92}$$

for any scalars α, β, and vectors $\mathbf{u}$, $\mathbf{v}$. From definition (1.89) we see that the matrix $(\mathbf{u} \times \bar{\mathbf{I}})$ is antisymmetric. In fact,

$$\begin{aligned} (\mathbf{u} \times \bar{\mathbf{I}})^T_{ik} &= (\mathbf{u} \times \bar{\mathbf{I}})_{ki} = \varepsilon_{kji} u_j \\ &= -\varepsilon_{ijk} u_j = -(\mathbf{u} \times \bar{\mathbf{I}})_{ik} \end{aligned}$$

or

$$(\mathbf{u} \times \bar{\mathbf{I}})^T = -(\mathbf{u} \times \bar{\mathbf{I}}) \tag{1.93}$$

The trace and the determinant of the antisymmetric matrix $\mathbf{u} \times \bar{\mathbf{I}}$ are zero since

$$(\mathbf{u} \times \bar{\mathbf{I}})_t = (\mathbf{u} \times \bar{\mathbf{I}})_{ii} = \varepsilon_{iji} u_j = 0 \tag{1.94}$$

and

$$|\mathbf{u} \times \bar{\mathbf{I}}| = |(\mathbf{u} \times \bar{\mathbf{I}})^T| = |-(\mathbf{u} \times \bar{\mathbf{I}})| = (-1)^3 |\mathbf{u} \times \bar{\mathbf{I}}|$$

Here we have used Eqs. (1.72*a*), (1.93), and (1.72*b*). Thus

$$|\mathbf{u} \times \bar{\mathbf{I}}| = 0 \tag{1.95}$$

From definition (1.89) and Eq. (1.53), the dot product of $\mathbf{u} \times \bar{\mathbf{I}}$ and $\mathbf{v} \times \bar{\mathbf{I}}$ yields

$$\begin{aligned} [(\mathbf{u} \times \bar{\mathbf{I}}) \cdot (\mathbf{v} \times \bar{\mathbf{I}})]_{in} &= (\mathbf{u} \times \bar{\mathbf{I}})_{ij} (\mathbf{v} \times \bar{\mathbf{I}})_{jn} \\ &= (\varepsilon_{ikj} u_k)(\varepsilon_{jln} v_l) \\ &= (\delta_{il}\delta_{kn} - \delta_{in}\delta_{kl}) u_k v_l \\ &= v_i u_n - u_l v_l \delta_{in} \end{aligned}$$

or, in direct form,

$$(\mathbf{u} \times \bar{\mathbf{I}}) \cdot (\mathbf{v} \times \bar{\mathbf{I}}) = \mathbf{u} \times (\mathbf{v} \times \bar{\mathbf{I}}) = \mathbf{vu} - (\mathbf{u} \cdot \mathbf{v})\bar{\mathbf{I}} \tag{1.96}$$

Dot-multiplying both sides from the right by vector $\mathbf{w}$, we obtain an expanded form of the *vector triple product*:

$$\mathbf{u} \times (\mathbf{v} \times \mathbf{w}) = (\mathbf{u} \cdot \mathbf{w})\mathbf{v} - (\mathbf{u} \cdot \mathbf{v})\mathbf{w} \tag{1.97}$$

On the other hand, making use of Eqs. (1.87), (1.89), and (1.53), we obtain another useful identity:

$$\begin{aligned}[(\mathbf{u} \times \mathbf{v}) \times \mathbf{I}]_{lm} &= \varepsilon_{lim}(\varepsilon_{ijk} u_j v_k) \\ &= -\varepsilon_{lmi} \varepsilon_{jki} u_j v_k \\ &= (\delta_{lk} \delta_{mj} - \delta_{lj} \delta_{mk}) u_j v_k \\ &= v_l u_m - u_l v_m\end{aligned}$$

or, in direct notation,

$$(\mathbf{u} \times \mathbf{v}) \times \bar{\mathbf{I}} = \mathbf{vu} - \mathbf{uv} \tag{1.98}$$

The trace of Eq. (1.96) gives

$$[(\mathbf{u} \times \bar{\mathbf{I}}) \cdot (\mathbf{v} \times \bar{\mathbf{I}})]_t = -2\mathbf{u} \cdot \mathbf{v} \tag{1.99}$$

In the special case when $\mathbf{u} = \mathbf{v}$, Eqs. (1.96) and (1.99) reduce to

$$(\mathbf{u} \times \bar{\mathbf{I}})^2 = \mathbf{uu} - \mathbf{u}^2\bar{\mathbf{I}} \tag{1.100}$$

and

$$[(\mathbf{u} \times \bar{\mathbf{I}})^2]_t = -2\mathbf{u}^2 \tag{1.101}$$

respectively. Substituting Eqs. (1.100) and (1.101) into Eq. (1.62), we find

$$\operatorname{adj}(\mathbf{u} \times \bar{\mathbf{I}}) = \mathbf{uu} \tag{1.102}$$

Any antisymmetric matrix $\bar{\mathbf{A}} = [a_{ij}]$ (that is, $\tilde{\bar{\mathbf{A}}} = -\bar{\mathbf{A}}$) can always be represented as $(\mathbf{u} \times \bar{\mathbf{I}})$, where the ith component of vector $\mathbf{u}$ is related to the components of $\bar{\mathbf{A}}$ by

$$u_i = \tfrac{1}{2}\varepsilon_{ikl} a_{lk}$$

To prove this, we multiply both sides of the above by ε_{min}. Noting that $a_{ij} = -a_{ji}$, we obtain

$$\begin{aligned}\varepsilon_{min} u_i &= (\mathbf{u} \times \bar{\mathbf{I}})_{mn} \\ &= \tfrac{1}{2}\varepsilon_{min} \varepsilon_{ikl} a_{lk} \\ &= \tfrac{1}{2}(\delta_{ml} \delta_{nk} - \delta_{mk} \delta_{nl}) a_{lk} \\ &= \tfrac{1}{2}(a_{mn} - a_{nm}) = (\bar{\mathbf{A}})_{mn}\end{aligned}$$

In other words, $(\mathbf{u} \times \bar{\mathbf{I}})$ is the most general antisymmetric matrix. As an example, let us show that

$$\bar{\mathbf{A}} \cdot (\mathbf{u} \times \bar{\mathbf{I}}) + (\mathbf{u} \times \bar{\mathbf{I}}) \cdot \tilde{\bar{\mathbf{A}}} = (\bar{\mathbf{A}}_t \mathbf{u} - \tilde{\bar{\mathbf{A}}} \cdot \mathbf{u}) \times \bar{\mathbf{I}} \tag{1.103}$$

for any matrix $\bar{\mathbf{A}}$. First, we note that the matrix $\bar{\mathbf{B}} = \bar{\mathbf{A}} \cdot (\mathbf{u} \times \bar{\mathbf{I}}) + (\mathbf{u} \times \bar{\mathbf{I}}) \cdot \tilde{\bar{\mathbf{A}}}$ is antisymmetric. Indeed

$$\tilde{\bar{\mathbf{B}}} = -(\mathbf{u} \times \bar{\mathbf{I}}) \cdot \tilde{\bar{\mathbf{A}}} - \bar{\mathbf{A}} \cdot (\mathbf{u} \times \bar{\mathbf{I}}) = -\bar{\mathbf{B}}$$

Thus, we let

$$\bar{\mathbf{B}} = \mathbf{c} \times \bar{\mathbf{I}} \tag{1.104}$$

where $\mathbf{c}$ is a vector to be determined. Dot-multiplying both sides of Eq. (1.104) by $\mathbf{u}$ from the right, we get

$$(\mathbf{u} \times \bar{\mathbf{I}}) \cdot (\tilde{\bar{\mathbf{A}}} \cdot \mathbf{u}) = \mathbf{c} \times \mathbf{u}$$

or

$$(\tilde{\bar{\mathbf{A}}} \cdot \mathbf{u} + \mathbf{c}) \times \mathbf{u} = \mathbf{0}$$

which means that vector $\tilde{\bar{\mathbf{A}}} \cdot \mathbf{u} + \mathbf{c}$ is parallel to $\mathbf{u}$, or

$$\tilde{\bar{\mathbf{A}}} \cdot \mathbf{u} + \mathbf{c} = \alpha \mathbf{u} \tag{1.105}$$

To determine the scalar α, we substitute Eq. (1.105) back in Eq. (1.104) to give

$$\bar{\mathbf{A}} \cdot (\mathbf{u} \times \bar{\mathbf{I}}) + (\mathbf{u} \times \bar{\mathbf{I}}) \cdot \tilde{\bar{\mathbf{A}}} = (\alpha \mathbf{u} - \tilde{\bar{\mathbf{A}}} \cdot \mathbf{u}) \times \bar{\mathbf{I}} \tag{1.106}$$

Dot-multiplying both sides of Eq. (1.106) by $\mathbf{u} \times \bar{\mathbf{I}}$ and then taking the trace of the resulting equation, we obtain

$$\alpha = \bar{\mathbf{A}}_t$$

Thus

$$\mathbf{c} = \bar{\mathbf{A}}_t \mathbf{u} - \tilde{\bar{\mathbf{A}}} \cdot \mathbf{u}$$

which in turn proves Eq. (1.103).

To conclude this section, we define the *cross product* of a vector $\mathbf{u} = [u_i]$ and a matrix $\bar{\mathbf{A}} = [a_{ij}]$, denoted by $\mathbf{u} \times \bar{\mathbf{A}}$, by

$$\begin{aligned}(\mathbf{u} \times \bar{\mathbf{A}})_{ij} &= \varepsilon_{ilm} u_l a_{mj} \\ &= (\mathbf{u} \times \bar{\mathbf{I}})_{im} a_{mj} \\ &= [(\mathbf{u} \times \bar{\mathbf{I}}) \cdot \bar{\mathbf{A}}]_{ij}\end{aligned} \tag{1.107}$$

or, in direct notation,

$$\mathbf{u} \times \bar{\mathbf{A}} = (\mathbf{u} \times \bar{\mathbf{I}}) \cdot \bar{\mathbf{A}} = (\bar{\mathbf{I}} \times \mathbf{u}) \cdot \bar{\mathbf{A}} \tag{1.108}$$

Similarly,

$$\bar{\mathbf{A}} \times \mathbf{u} = \bar{\mathbf{A}} \cdot (\bar{\mathbf{I}} \times \mathbf{u}) = \bar{\mathbf{A}} \cdot (\mathbf{u} \times \bar{\mathbf{I}}) \tag{1.109}$$

In the special case when $\bar{\mathbf{A}} = \mathbf{ab}$, Eqs. (1.108) and (1.109) reduce to

$$\begin{aligned}\mathbf{u} \times (\mathbf{ab}) &= (\mathbf{u} \times \bar{\mathbf{I}}) \cdot \mathbf{ab} = (\mathbf{u} \times \mathbf{a})\mathbf{b} \\ (\mathbf{ab}) \times \mathbf{u} &= \mathbf{ab} \cdot (\bar{\mathbf{I}} \times \mathbf{u}) = \mathbf{a}(\mathbf{b} \times \mathbf{u})\end{aligned} \tag{1.110}$$

We note that $(\mathbf{u} \times \mathbf{a})\mathbf{b}$ is a dyad, the dyadic product of the vectors $\mathbf{u} \times \mathbf{a}$ and $\mathbf{b}$. Similarly, $\mathbf{a}(\mathbf{b} \times \mathbf{u})$ is a dyad.

As in the case of a scalar triple product where the dot and cross are interchangeable, we have the formulas

$$(\bar{\mathbf{A}} \times \mathbf{u}) \cdot \mathbf{v} = \bar{\mathbf{A}} \cdot (\mathbf{u} \times \mathbf{v}) \tag{1.111}$$

and

$$\mathbf{u} \cdot (\mathbf{v} \times \bar{\mathbf{A}}) = (\mathbf{u} \times \mathbf{v}) \cdot \bar{\mathbf{A}} \tag{1.112}$$

Here we interchange brackets and at the same time interchange the position of the dot and cross. To prove Eq. (1.111), we use the definition (1.109) and the property that multiplication of matrices is associative. Thus

$$(\bar{\mathbf{A}} \times \mathbf{u}) \cdot \mathbf{v} = \bar{\mathbf{A}} \cdot (\mathbf{u} \times \bar{\mathbf{I}}) \cdot \mathbf{v} = \bar{\mathbf{A}} \cdot (\mathbf{u} \times \mathbf{v})$$

As a consequence of Eq. (1.111), we see that $(\bar{\mathbf{A}} \times \mathbf{u}) \cdot \mathbf{v}$ vanishes if vectors $\mathbf{u}$ and $\mathbf{v}$ are parallel. Equation (1.112) may be proved in the same way.

1.5 SOME IDENTITIES

In this section we shall establish some identities which will prove to be useful in future applications. First let us consider a matrix $\bar{\mathbf{A}}$ and vectors $\mathbf{u}$ and $\mathbf{v}$ for which we have

$$\begin{aligned}(\operatorname{adj} \bar{\mathbf{A}}) \cdot (\mathbf{u} \times \mathbf{v}) &= (\tilde{\bar{\mathbf{A}}} \cdot \mathbf{u}) \times (\tilde{\bar{\mathbf{A}}} \cdot \mathbf{v}) \\ &= (\mathbf{u} \cdot \bar{\mathbf{A}}) \times (\mathbf{v} \cdot \bar{\mathbf{A}})\end{aligned} \tag{1.113}$$

To show this, we rewrite the vector on the left-hand side in index form and make use of Eq. (1.53):

$$\begin{aligned}[(\operatorname{adj} \bar{\mathbf{A}}) \cdot (\mathbf{u} \times \mathbf{v})]_n &= (\operatorname{adj} \bar{\mathbf{A}})_{ns} (\mathbf{u} \times \mathbf{v})_s \\ &= \left(\tfrac{1}{2}\varepsilon_{ijs}\varepsilon_{lmn} a_{il} a_{jm}\right)(\varepsilon_{spq} u_p v_q) \\ &= \tfrac{1}{2}\varepsilon_{lmn}(\delta_{ip}\delta_{jq} - \delta_{iq}\delta_{jp}) a_{il} a_{jm} u_p v_q \\ &= \tfrac{1}{2}\varepsilon_{lmn}(u_p a_{pl})(v_q a_{qm}) - \tfrac{1}{2}\varepsilon_{lmn}(u_p a_{pm})(v_q a_{ql}) \\ &= \varepsilon_{lmn}(\mathbf{u} \cdot \bar{\mathbf{A}})_l (\mathbf{v} \cdot \bar{\mathbf{A}})_m = [(\mathbf{u} \cdot \bar{\mathbf{A}}) \times (\mathbf{v} \cdot \bar{\mathbf{A}})]_n \\ &= \varepsilon_{lmn}(\tilde{\bar{\mathbf{A}}} \cdot \mathbf{u})_l (\tilde{\bar{\mathbf{A}}} \cdot \mathbf{v})_m = [(\tilde{\bar{\mathbf{A}}} \cdot \mathbf{u}) \times (\tilde{\bar{\mathbf{A}}} \cdot \mathbf{v})]_n\end{aligned}$$

Converting to direct notation, we thus establish the identity (1.113). When $\bar{\mathbf{A}}$ is symmetric or antisymmetric, Eq. (1.113) becomes

$$(\operatorname{adj} \bar{\mathbf{A}}) \cdot (\mathbf{u} \times \mathbf{v}) = (\bar{\mathbf{A}} \cdot \mathbf{u}) \times (\bar{\mathbf{A}} \cdot \mathbf{v}) \tag{1.114}$$

On the other hand, if $\bar{\mathbf{A}}$ is nonsingular, we may write Eq. (1.113) as

$$\bar{\mathbf{A}}^{-1} \cdot (\mathbf{u} \times \mathbf{v}) = \frac{1}{|\bar{\mathbf{A}}|} [(\tilde{\bar{\mathbf{A}}} \cdot \mathbf{u}) \times (\tilde{\bar{\mathbf{A}}} \cdot \mathbf{v})] \tag{1.115}$$

Replacing $\bar{\mathbf{A}}^{-1}$ by $\bar{\mathbf{A}}$ in the above identity, we obtain

$$\bar{\mathbf{A}} \cdot (\mathbf{u} \times \mathbf{v}) = |\bar{\mathbf{A}}| (\tilde{\bar{\mathbf{A}}}^{-1} \cdot \mathbf{u}) \times (\tilde{\bar{\mathbf{A}}}^{-1} \cdot \mathbf{v}) \tag{1.116}$$

which may also be written as

$$\bar{\mathbf{A}} \cdot (\mathbf{u} \times \bar{\mathbf{I}}) \cdot \mathbf{v} = |\bar{\mathbf{A}}|[(\tilde{\bar{\mathbf{A}}}^{-1} \cdot \mathbf{u}) \times \bar{\mathbf{I}}] \cdot \tilde{\bar{\mathbf{A}}}^{-1} \cdot \mathbf{v}$$

But since $\mathbf{v}$ is an arbitrary vector, we have

$$\bar{\mathbf{A}} \cdot (\mathbf{u} \times \bar{\mathbf{I}}) = [(\mathbf{u} \cdot \bar{\mathbf{A}}^{-1}) \times \bar{\mathbf{I}}] \cdot (\text{adj } \tilde{\bar{\mathbf{A}}}) \tag{1.117}$$

Hence

$$\bar{\mathbf{A}} \cdot (\mathbf{u} \times \bar{\mathbf{I}}) \cdot \tilde{\bar{\mathbf{A}}} = (\mathbf{u} \cdot \text{adj } \bar{\mathbf{A}}) \times \bar{\mathbf{I}} \tag{1.118}$$

Now, dot-multiplying Eq. (1.117) from the left by $(\mathbf{v} \times \bar{\mathbf{I}})$ and then expanding the resulted right-hand side according to Eq. (1.96), we obtain

$$(\mathbf{v} \times \bar{\mathbf{I}}) \cdot \bar{\mathbf{A}} \cdot (\mathbf{u} \times \bar{\mathbf{I}}) = (\text{adj } \tilde{\bar{\mathbf{A}}} \cdot \mathbf{u})(\bar{\mathbf{A}}^{-1} \cdot \mathbf{v}) - (\mathbf{u} \cdot \bar{\mathbf{A}}^{-1} \cdot \mathbf{v})\,(\text{adj } \tilde{\bar{\mathbf{A}}}) \tag{1.119}$$

or, after a simple substitution,

$$(\mathbf{v} \times \bar{\mathbf{I}}) \cdot \text{adj } \bar{\mathbf{A}} \cdot (\mathbf{u} \times \bar{\mathbf{I}}) = \tilde{\bar{\mathbf{A}}} \cdot \mathbf{uv} \cdot \tilde{\bar{\mathbf{A}}} - (\mathbf{u} \cdot \bar{\mathbf{A}} \cdot \mathbf{v})\tilde{\bar{\mathbf{A}}} \tag{1.120}$$

Next let us establish the following identities:

$$\begin{aligned}(\mathbf{a} \times \mathbf{b})(\mathbf{c} \times \mathbf{d}) = (\mathbf{a} \times \mathbf{b}) \cdot (\mathbf{c} \times \mathbf{d})\bar{\mathbf{I}} + (\mathbf{a} \cdot \mathbf{d})\mathbf{cb} + (\mathbf{b} \cdot \mathbf{c})\mathbf{da} \\ - (\mathbf{a} \cdot \mathbf{c})\mathbf{db} - (\mathbf{b} \cdot \mathbf{d})\mathbf{ca}\end{aligned} \tag{1.121}$$

and

$$(\mathbf{a} \times \mathbf{b}) \cdot (\mathbf{c} \times \mathbf{d}) = (\mathbf{a} \cdot \mathbf{c})(\mathbf{b} \cdot \mathbf{d}) - (\mathbf{a} \cdot \mathbf{d})(\mathbf{b} \cdot \mathbf{c}) \tag{1.122}$$

To prove Eq. (1.121) we use Eq. (1.96) and find

$$[(\mathbf{c} \times \mathbf{d}) \times \bar{\mathbf{I}}] \cdot [(\mathbf{a} \times \mathbf{b}) \times \bar{\mathbf{I}}] = (\mathbf{a} \times \mathbf{b})(\mathbf{c} \times \mathbf{d}) - (\mathbf{a} \times \mathbf{b}) \cdot (\mathbf{c} \times \mathbf{d})\bar{\mathbf{I}} \tag{1.123}$$

Expanding each square bracket on the left-hand side according to Eq. (1.98) and then rearranging the terms, we obtain the desired result, Eq. (1.121). The trace of Eq. (1.121) yields Eq. (1.122).

We shall now find the traces, the determinants, and the adjoints and their traces of the following matrices. We shall present them in the form of examples for easy future reference.

Example 1.1 If $\bar{\mathbf{C}} = \bar{\mathbf{A}} \pm \lambda\bar{\mathbf{I}}$, then

$$\begin{aligned}&(a)\quad \bar{\mathbf{C}}_t = \bar{\mathbf{A}}_t \pm 3\lambda \\ &(b)\quad |\bar{\mathbf{C}}| = \pm\lambda^3 + \bar{\mathbf{A}}_t\lambda^2 \pm (\text{adj } \bar{\mathbf{A}})_t\lambda + |\bar{\mathbf{A}}| \\ &(c)\quad \text{adj } \bar{\mathbf{C}} = \lambda^2\bar{\mathbf{I}} \pm \lambda(\bar{\mathbf{A}}_t\bar{\mathbf{I}} - \bar{\mathbf{A}}) + \text{adj } \bar{\mathbf{A}} \\ &(d)\quad (\text{adj } \bar{\mathbf{C}})_t = 3\lambda^2 \pm 2\bar{\mathbf{A}}_t\lambda + (\text{adj } \bar{\mathbf{A}})_t\end{aligned} \tag{1.124}$$

SOLUTION Part (*a*) follows from the linearity property of the trace. To find the determinant of matrix $\bar{\mathbf{C}}$, we use Eq. (1.77) and note that $|\pm\lambda\bar{\mathbf{I}}| = \pm\lambda^3$ and adj $(\pm\lambda\bar{\mathbf{I}}) = \lambda^2\bar{\mathbf{I}}$. Thus the result of (*b*) follows. Using Eq. (1.78) and noting that $(\pm\lambda\bar{\mathbf{I}})_t = \pm 3\lambda$, we prove (*c*). Finally, taking the trace of (*c*), we obtain (*d*).

Example 1.2 If $\bar{\mathbf{C}} = \bar{\mathbf{A}} + \mathbf{uv}$, then

$$\begin{aligned}
&(a)\quad \bar{\mathbf{C}}_t = \bar{\mathbf{A}}_t + \mathbf{u}\cdot\mathbf{v}\\
&(b)\quad |\bar{\mathbf{C}}| = |\bar{\mathbf{A}}| + \mathbf{v}\cdot(\text{adj }\bar{\mathbf{A}})\cdot\mathbf{u}\\
&(c)\quad \text{adj }\bar{\mathbf{C}} = \text{adj }\bar{\mathbf{A}} + (\bar{\mathbf{A}} - \bar{\mathbf{A}}_t\bar{\mathbf{I}})\cdot(\mathbf{v}\times\bar{\mathbf{I}})\cdot(\mathbf{u}\times\bar{\mathbf{I}})\\
&\qquad\qquad + [(\mathbf{v}\cdot\bar{\mathbf{A}})\times\bar{\mathbf{I}}]\cdot(\mathbf{u}\times\bar{\mathbf{I}})\\
&(d)\quad (\text{adj }\bar{\mathbf{C}})_t = (\text{adj }\bar{\mathbf{A}})_t + (\mathbf{u}\cdot\mathbf{v})\bar{\mathbf{A}}_t - \mathbf{v}\cdot\bar{\mathbf{A}}\cdot\mathbf{u}
\end{aligned}\tag{1.125}$$

Solution Part (*a*) follows directly from the linearity property of the trace. Part (*b*) can be proved by using Eq. (1.77) and the facts that $|\mathbf{uv}| = 0$ and adj $(\mathbf{uv}) = \bar{\mathbf{0}}$. Part (*c*) follows from Eqs. (1.78) and (1.96). The trace of part (*c*) gives (*d*).

Example 1.3 As a special case of Example 1.2, when $\bar{\mathbf{A}} = \lambda\bar{\mathbf{I}}$, we have $\bar{\mathbf{C}} = \lambda\bar{\mathbf{I}} + \mathbf{uv}$. Thus

$$\begin{aligned}
&(a)\quad \bar{\mathbf{C}}_t = 3\lambda + \mathbf{u}\cdot\mathbf{v}\\
&(b)\quad |\bar{\mathbf{C}}| = \lambda^2(\lambda + \mathbf{u}\cdot\mathbf{v})\\
&(c)\quad \text{adj }\bar{\mathbf{C}} = \lambda[(\lambda + \mathbf{u}\cdot\mathbf{v})\bar{\mathbf{I}} - \mathbf{uv}]\\
&(d)\quad (\text{adj }\bar{\mathbf{C}})_t = \lambda(3\lambda + 2\mathbf{u}\cdot\mathbf{v})
\end{aligned}\tag{1.126}$$

Solution The results follow from Eq. (1.125) and the facts that $\bar{\mathbf{A}}_t = 3\lambda$, $|\bar{\mathbf{A}}| = \lambda^3$, adj $\bar{\mathbf{A}} = \lambda^2\bar{\mathbf{I}}$, and $(\text{adj }\bar{\mathbf{A}})_t = 3\lambda^2$.

In the case when $\bar{\mathbf{C}} = \lambda\bar{\mathbf{I}} + \mathbf{uv}\cdot\bar{\mathbf{B}}$, Eq. (1.126) becomes

$$\begin{aligned}
&(a)\quad \bar{\mathbf{C}}_t = 3\lambda + \mathbf{v}\cdot\bar{\mathbf{B}}\cdot\mathbf{u}\\
&(b)\quad |\bar{\mathbf{C}}| = \lambda^2(\lambda + \mathbf{v}\cdot\bar{\mathbf{B}}\cdot\mathbf{u})\\
&(c)\quad \text{adj }\bar{\mathbf{C}} = \lambda[(\lambda + \mathbf{v}\cdot\bar{\mathbf{B}}\cdot\mathbf{u})\bar{\mathbf{I}} - \mathbf{uv}\cdot\bar{\mathbf{B}}]\\
&(d)\quad (\text{adj }\bar{\mathbf{C}})_t = \lambda(3\lambda + 2\mathbf{v}\cdot\bar{\mathbf{B}}\cdot\mathbf{u})
\end{aligned}\tag{1.127}$$

Example 1.4 As an extension of Example 1.3, we consider $\bar{\mathbf{C}} = \lambda\bar{\mathbf{I}} + \mathbf{uv} + \mathbf{mn}$. In this case

$$\begin{aligned}
&(a)\quad \bar{\mathbf{C}}_t = 3\lambda + \mathbf{u}\cdot\mathbf{v} + \mathbf{m}\cdot\mathbf{n}\\
&(b)\quad |\bar{\mathbf{C}}| = \lambda[\lambda^2 + \lambda(\mathbf{u}\cdot\mathbf{v} + \mathbf{m}\cdot\mathbf{n}) + (\mathbf{v}\times\mathbf{n})\cdot(\mathbf{u}\times\mathbf{m})]\\
&(c)\quad \text{adj }\bar{\mathbf{C}} = \lambda[(\lambda + \mathbf{u}\cdot\mathbf{v} + \mathbf{m}\cdot\mathbf{n})\bar{\mathbf{I}} - \mathbf{uv} - \mathbf{mn}] + (\mathbf{v}\times\mathbf{n})(\mathbf{u}\times\mathbf{m})\\
&(d)\quad (\text{adj }\bar{\mathbf{C}})_t = \lambda[3\lambda + 2(\mathbf{u}\cdot\mathbf{v} + \mathbf{m}\cdot\mathbf{n})] + (\mathbf{v}\times\mathbf{n})\cdot(\mathbf{u}\times\mathbf{m})
\end{aligned}\tag{1.128}$$

SOLUTION Again (*a*) and (*d*) follow from the linearity property of the trace. To prove (*b*), we expand the determinant according to Eq. (1.77). Recalling the results of Eqs. (1.85) and (1.86), we obtain

$$|\bar{\mathbf{C}}| = |(\lambda\bar{\mathbf{I}} + \mathbf{uv}) + \mathbf{mn}|$$
$$= |\lambda\bar{\mathbf{I}} + \mathbf{uv}| + [\text{adj } (\lambda\bar{\mathbf{I}} + \mathbf{uv}) \cdot \mathbf{mn}]_t$$

Next substituting Eqs. (1.126*b*) and (1.126*c*) into the above equation and using the result of Eq. (1.122), we prove the required result. Similarly, part (*c*) can be proved by expanding the adjoint of the sum of the matrices according to Eq. (1.78) followed by the application of Eqs. (1.126*c*) and (1.121).

Example 1.5 If $\bar{\mathbf{C}} = \bar{\mathbf{A}} + \mathbf{c} \times \bar{\mathbf{I}}$, then

$$\begin{aligned}
&(a)\quad \bar{\mathbf{C}}_t = \bar{\mathbf{A}}_t \\
&(b)\quad |\bar{\mathbf{C}}| = |\bar{\mathbf{A}}| + [\text{adj } \bar{\mathbf{A}} \cdot (\mathbf{c} \times \bar{\mathbf{I}})]_t + \mathbf{c} \cdot \bar{\mathbf{A}} \cdot \mathbf{c} \\
&(c)\quad \text{adj } \bar{\mathbf{C}} = \text{adj } \bar{\mathbf{A}} + \mathbf{cc} + (\bar{\mathbf{A}} - \bar{\mathbf{A}}_t\bar{\mathbf{I}}) \cdot (\mathbf{c} \times \bar{\mathbf{I}}) \\
&\qquad\qquad + (\mathbf{c} \times \bar{\mathbf{I}}) \cdot \bar{\mathbf{A}} + [\bar{\mathbf{A}} \cdot (\mathbf{c} \times \bar{\mathbf{I}})]_t\bar{\mathbf{I}} \\
&(d)\quad (\text{adj } \bar{\mathbf{C}})_t = (\text{adj } \bar{\mathbf{A}})_t + \mathbf{c}^2 - [\bar{\mathbf{A}} \cdot (\mathbf{c} \times \bar{\mathbf{I}})]_t
\end{aligned} \tag{1.129}$$

SOLUTION Part (*a*) follows from the fact that $(\mathbf{c} \times \bar{\mathbf{I}})_t = 0$. Using Eqs. (1.77), (1.95), and (1.102), we prove (*b*). Similarly, using Eqs. (1.78), (1.94), and (1.102), we establish (*c*). Finally, the trace of (*c*) gives (*d*).

In the special case when $\bar{\mathbf{A}}$ is a symmetric matrix (that is, $\tilde{\bar{\mathbf{A}}} = \bar{\mathbf{A}}$), Eq. (1.129) reduces to

$$\begin{aligned}
&(a)\quad |\bar{\mathbf{C}}| = |\bar{\mathbf{A}}| + \mathbf{c} \cdot \bar{\mathbf{A}} \cdot \mathbf{c} \\
&(b)\quad \text{adj } \bar{\mathbf{C}} = \text{adj } \bar{\mathbf{A}} + \mathbf{cc} - (\bar{\mathbf{A}} \cdot \mathbf{c}) \times \bar{\mathbf{I}} \\
&(c)\quad (\text{adj } \bar{\mathbf{C}})_t = (\text{adj } \bar{\mathbf{A}})_t + \mathbf{c}^2
\end{aligned} \tag{1.130}$$

Here we note that if $\bar{\mathbf{A}}$ is symmetric, so are $\bar{\mathbf{A}}^{-1}$ and adj $\bar{\mathbf{A}}$, for the transpose of the identity $\bar{\mathbf{A}} \cdot \bar{\mathbf{A}}^{-1} = \bar{\mathbf{I}}$ gives

$$(\bar{\mathbf{A}} \cdot \bar{\mathbf{A}}^{-1})^T = (\bar{\mathbf{A}}^{-1})^T \cdot (\tilde{\bar{\mathbf{A}}}) = \tilde{\bar{\mathbf{I}}} = \bar{\mathbf{I}}$$

or

$$(\bar{\mathbf{A}}^{-1})^T = (\tilde{\bar{\mathbf{A}}})^{-1} = \bar{\mathbf{A}}^{-1}$$

That is, $\bar{\mathbf{A}}^{-1}$ is symmetric if $\bar{\mathbf{A}}$ is symmetric. But $|\tilde{\bar{\mathbf{A}}}| = |\bar{\mathbf{A}}|$: hence

$$(\text{adj } \bar{\mathbf{A}})^T = \text{adj } \bar{\mathbf{A}}$$

That is, adj $\bar{\mathbf{A}}$ is also symmetric. The results of Eq. (1.130) then follow from the identity (1.103) and the fact that the trace of the product of a symmetric and an antisymmetric matrix is zero [cf. Sec. 1.2].

When $\bar{\mathbf{A}} = \lambda\bar{\mathbf{I}}$ or $\bar{\mathbf{C}} = \lambda\bar{\mathbf{I}} + \mathbf{c} \times \bar{\mathbf{I}}$, Eq. (1.130) becomes

$$
\begin{aligned}
&(a) \quad |\bar{\mathbf{C}}| = \lambda(\lambda^2 + \mathbf{c}^2) \\
&(b) \quad \operatorname{adj} \bar{\mathbf{C}} = \lambda(\lambda\bar{\mathbf{I}} - \mathbf{c} \times \bar{\mathbf{I}}) + \mathbf{cc} \qquad (1.131) \\
&(c) \quad (\operatorname{adj} \bar{\mathbf{C}})_t = 3\lambda^2 + \mathbf{c}^2
\end{aligned}
$$

As another special case of Eq. (1.129), if $\bar{\mathbf{C}} = \mathbf{ab} + \mathbf{c} \times \bar{\mathbf{I}}$, then

$$
\begin{aligned}
&(a) \quad \bar{\mathbf{C}}_t = \mathbf{a} \cdot \mathbf{b} \\
&(b) \quad |\bar{\mathbf{C}}| = (\mathbf{a} \cdot \mathbf{c})(\mathbf{b} \cdot \mathbf{c}) \\
&(c) \quad \operatorname{adj} \bar{\mathbf{C}} = \mathbf{cc} - (\mathbf{b} \cdot \mathbf{c})(\mathbf{a} \times \bar{\mathbf{I}}) - \mathbf{c}(\mathbf{a} \times \mathbf{b}) \qquad (1.132) \\
&(d) \quad (\operatorname{adj} \bar{\mathbf{C}})_t = \mathbf{c}^2 - \mathbf{c} \cdot (\mathbf{a} \times \mathbf{b})
\end{aligned}
$$

The truth of Eq. (1.132) follows from Eqs. (1.83), (1.85), and (1.86).

Example 1.6 As an application of Examples 1.1 and 1.5, we consider $\bar{\mathbf{C}} = \lambda\bar{\mathbf{I}} + \mathbf{ab} + \mathbf{c} \times \bar{\mathbf{I}}$. In this case,

$$
\begin{aligned}
&(a) \quad \bar{\mathbf{C}}_t = 3\lambda + \mathbf{a} \cdot \mathbf{b} \\
&(b) \quad |\bar{\mathbf{C}}| = \lambda^3 + (\mathbf{a} \cdot \mathbf{b})\lambda^2 + (\mathbf{c}^2 - \mathbf{a} \times \mathbf{b} \cdot \mathbf{c})\lambda + (\mathbf{a} \cdot \mathbf{c})(\mathbf{b} \cdot \mathbf{c}) \\
&(c) \quad \operatorname{adj} \bar{\mathbf{C}} = \lambda^2\bar{\mathbf{I}} - \lambda[\mathbf{ab} - (\mathbf{a} \cdot \mathbf{b})\bar{\mathbf{I}} + \mathbf{c} \times \bar{\mathbf{I}}] \qquad (1.133) \\
&\qquad\qquad + \mathbf{cc} - (\mathbf{b} \cdot \mathbf{c})(\mathbf{a} \times \bar{\mathbf{I}}) - \mathbf{c}(\mathbf{a} \times \mathbf{b}) \\
&(d) \quad (\operatorname{adj} \bar{\mathbf{C}})_t = 3\lambda^2 + 2(\mathbf{a} \cdot \mathbf{b})\lambda + \mathbf{c}^2 - \mathbf{c} \cdot (\mathbf{a} \times \mathbf{b})
\end{aligned}
$$

Solution Parts (*a*) and (*d*) follow from the definition of trace. To prove (*b*) and (*c*), we expand the determinant and adjoint of the sum of two matrices according to Eqs. (1.77) and (1.78). Making use of Eqs. (1.124) and (1.132), we obtain the desired results.

1.6 DYADIC DECOMPOSITION OF A MATRIX; SOLUTIONS OF HOMOGENEOUS EQUATIONS

Now we shall show that any three-dimensional matrix can be decomposed into a sum of no more than three dyads. This decomposition has important theoretical consequences and leads to solutions of homogeneous equations.

First, we introduce the following definition: three vectors **a**, **b**, and **c** are said to be *linearly independent* if no scalars α, β, and γ exist (except $\alpha = \beta = \gamma = 0$) for which

$$\alpha\mathbf{a} + \beta\mathbf{b} + \gamma\mathbf{c} = \mathbf{0} \qquad (1.134)$$

Consequently, three non-coplanar vectors are linearly independent. To prove this, let us assume that the relation (1.134) exists among three non-coplanar vectors **a**, **b** and **c**, where α, β, and γ are not all zero. To be specific, we assume $\alpha \neq 0$. Dot-multiplying both sides of Eq. (1.134) by $\mathbf{b} \times \mathbf{c}$, we have $\alpha \mathbf{a} \cdot \mathbf{b} \times \mathbf{c} = 0$. But $\alpha \neq 0$; hence $\mathbf{a} \cdot \mathbf{b} \times \mathbf{c} = 0$ which means that vector **a** must be perpendicular to $\mathbf{b} \times \mathbf{c}$ and therefore lie in the plane formed by vectors **b** and **c**. In other words, vectors **a**, **b**, and **c** are coplanar, contradicting our initial assumption. Conversely, if **a**, **b**, and **c** are linearly independent, they are non-coplanar and the scalar triple product $\mathbf{a} \cdot \mathbf{b} \times \mathbf{c} \neq 0$.

Now, let us consider a given set of three linearly independent vectors $\mathbf{a}_1$, $\mathbf{a}_2$, and $\mathbf{a}_3$. We construct a new set of three vectors:

$$\mathbf{b}_1 = \frac{\mathbf{a}_2 \times \mathbf{a}_3}{V_a} \qquad \mathbf{b}_2 = \frac{\mathbf{a}_3 \times \mathbf{a}_1}{V_a} \qquad \mathbf{b}_3 = \frac{\mathbf{a}_1 \times \mathbf{a}_2}{V_a} \tag{1.135}$$

where

$$V_a = \mathbf{a}_1 \cdot \mathbf{a}_2 \times \mathbf{a}_3$$

is the volume of the parallelepiped having the vectors $\mathbf{a}_1$, $\mathbf{a}_2$, and $\mathbf{a}_3$ as concurrent edges. Upon forming all possible dot products of $\mathbf{a}_1$, $\mathbf{a}_2$, and $\mathbf{a}_3$ and $\mathbf{b}_1$, $\mathbf{b}_2$, and $\mathbf{b}_3$, we see that they satisfy the *biorthonormal condition*:

$$\mathbf{b}_i \cdot \mathbf{a}_j = \delta_{ij} \tag{1.136}$$

The set of vectors $\mathbf{b}_1$, $\mathbf{b}_2$, and $\mathbf{b}_3$ is said to be *reciprocal* to the set of vectors $\mathbf{a}_1$, $\mathbf{a}_2$, and $\mathbf{a}_3$ since

$$V_b = \mathbf{b}_1 \cdot \mathbf{b}_2 \times \mathbf{b}_3 = \frac{(\mathbf{a}_2 \times \mathbf{a}_3) \cdot (\mathbf{a}_3 \times \mathbf{a}_1) \times (\mathbf{a}_1 \times \mathbf{a}_2)}{V_a^3} = \frac{1}{V_a} \tag{1.137}$$

Also, from Eqs. (1.135) and (1.137) we find

$$\mathbf{a}_1 = \frac{\mathbf{b}_2 \times \mathbf{b}_3}{V_b} \qquad \mathbf{a}_2 = \frac{\mathbf{b}_3 \times \mathbf{b}_1}{V_b} \qquad \mathbf{a}_3 = \frac{\mathbf{b}_1 \times \mathbf{b}_2}{V_b} \tag{1.138}$$

Hence, the relation between the two sets of vectors is a reciprocal one. Furthermore, Eq. (1.137) implies that the vectors $\mathbf{b}_1$, $\mathbf{b}_2$, and $\mathbf{b}_3$ are also linearly independent.

In three-dimensional space if $\mathbf{a}_1$, $\mathbf{a}_2$, and $\mathbf{a}_3$ are three non-coplanar vectors, any fourth vector **u** can be expressed as a linear combination of them, namely,

$$\mathbf{u} = \alpha_i \mathbf{a}_i = \alpha_1 \mathbf{a}_1 + \alpha_2 \mathbf{a}_2 + \alpha_3 \mathbf{a}_3 \tag{1.139}$$

To determine the constants α_i, we dot-multiply both sides of Eq. (1.139) by $\mathbf{b}_j$, make use of the biorthonormal condition (1.136), and obtain

$$\alpha_j = \mathbf{u} \cdot \mathbf{b}_j = \mathbf{b}_j \cdot \mathbf{u}$$

Substitution of the above into Eq. (1.139) yields

$$\mathbf{u} = \mathbf{u} \cdot \bar{\mathbf{I}} = \mathbf{u} \cdot \mathbf{b}_i \mathbf{a}_i = \mathbf{a}_i \mathbf{b}_i \cdot \mathbf{u} \tag{1.140}$$

But **u** is arbitrary; thus we obtain the *completeness relation*

$$\begin{aligned} \bar{\mathbf{I}} &= \mathbf{b}_i \mathbf{a}_i = \mathbf{b}_1 \mathbf{a}_1 + \mathbf{b}_2 \mathbf{a}_2 + \mathbf{b}_3 \mathbf{a}_3 \\ &= \mathbf{a}_i \mathbf{b}_i = \mathbf{a}_1 \mathbf{b}_1 + \mathbf{a}_2 \mathbf{b}_2 + \mathbf{a}_3 \mathbf{b}_3 \end{aligned} \tag{1.141}$$

That is, the unit matrix may be decomposed as a sum of three dyads in which the antecedents and consequents form reciprocal sets. Using Eq. (1.135), we may express Eq. (1.141) as

$$\bar{\mathbf{I}} = \frac{\mathbf{a}_1(\mathbf{a}_2 \times \mathbf{a}_3) + \mathbf{a}_2(\mathbf{a}_3 \times \mathbf{a}_1) + \mathbf{a}_3(\mathbf{a}_1 \times \mathbf{a}_2)}{\mathbf{a}_1 \cdot \mathbf{a}_2 \times \mathbf{a}_3}$$

$$= \frac{(\mathbf{a}_2 \times \mathbf{a}_3)\mathbf{a}_1 + (\mathbf{a}_3 \times \mathbf{a}_1)\mathbf{a}_2 + (\mathbf{a}_1 \times \mathbf{a}_2)\mathbf{a}_3}{\mathbf{a}_1 \cdot \mathbf{a}_2 \times \mathbf{a}_3} \tag{1.142}$$

Dot-multiplying both sides of Eq. (1.141) from the left by a nonsingular matrix $\bar{\mathbf{A}}$, we obtain

$$\bar{\mathbf{A}} = \mathbf{m}_i \mathbf{a}_i = \mathbf{m}_1 \mathbf{a}_1 + \mathbf{m}_2 \mathbf{a}_2 + \mathbf{m}_3 \mathbf{a}_3 \tag{1.143}$$

where $\mathbf{m}_i = \bar{\mathbf{A}} \cdot \mathbf{b}_i$

Similarly, dot-multiplication of Eq. (1.141) from the right by $\bar{\mathbf{A}}$ yields

$$\bar{\mathbf{A}} = \mathbf{a}_i \mathbf{n}_i = \mathbf{a}_1 \mathbf{n}_1 + \mathbf{a}_2 \mathbf{n}_2 + \mathbf{a}_3 \mathbf{n}_3 \tag{1.144}$$

where $\mathbf{n}_i = \mathbf{b}_i \cdot \bar{\mathbf{A}}$

If $\bar{\mathbf{A}}$ is symmetric, then $\mathbf{m}_i = \mathbf{n}_i$. Equations (1.143) and (1.144) state that any nonsingular matrix may be decomposed as a sum of three dyads in which either antecedents or consequents may be an arbitrary set of three linearly independent vectors.

Example 1.7 Use the three vectors

$$\mathbf{a}_1 = \begin{bmatrix} 3 \\ -4 \\ 0 \end{bmatrix} \qquad \mathbf{a}_2 = \begin{bmatrix} 0 \\ 3 \\ 4 \end{bmatrix} \qquad \mathbf{a}_3 = \begin{bmatrix} -1 \\ 1 \\ 2 \end{bmatrix} \tag{1.145}$$

as consequents, and express the nonsingular matrix

$$\bar{\mathbf{A}} = \begin{bmatrix} 0 & 0 & 1 \\ 8 & 0 & 2 \\ 0 & -1 & 5 \end{bmatrix} \tag{1.146}$$

as a sum of three dyads.

Solution We first evaluate the scalar triple product of vectors $\mathbf{a}_1$, $\mathbf{a}_2$, and $\mathbf{a}_3$:

$$V_a = \mathbf{a}_1 \cdot \mathbf{a}_2 \times \mathbf{a}_3 = \begin{vmatrix} 3 & -4 & 0 \\ 0 & 3 & 4 \\ -1 & 1 & 2 \end{vmatrix} = 22$$

Since V_a does not vanish, the vectors $\mathbf{a}_1$, $\mathbf{a}_2$, and $\mathbf{a}_3$ are linearly independent. Using Eq. (1.135), we find the three reciprocal vectors:

$$\mathbf{b}_1 = \frac{1}{22}\begin{bmatrix} 2 \\ -4 \\ 3 \end{bmatrix} \qquad \mathbf{b}_2 = \frac{1}{22}\begin{bmatrix} 8 \\ 6 \\ 1 \end{bmatrix} \qquad \mathbf{b}_3 = \frac{1}{22}\begin{bmatrix} -16 \\ -12 \\ 9 \end{bmatrix} \tag{1.147}$$

Also

$$V_b = \mathbf{b}_1 \cdot \mathbf{b}_2 \times \mathbf{b}_3 = \frac{1}{22^3}\begin{vmatrix} 2 & -4 & 3 \\ 8 & 6 & 1 \\ -16 & -12 & 9 \end{vmatrix} = \frac{1}{22} = \frac{1}{V_a}$$

The reciprocal sets of vectors (1.145) and (1.147) satisfy the completeness relations

$$\mathbf{b}_1\mathbf{a}_1 + \mathbf{b}_2\mathbf{a}_2 + \mathbf{b}_3\mathbf{a}_3 = \frac{1}{22}\begin{bmatrix} 2 \\ -4 \\ 3 \end{bmatrix}[3, \quad -4, \quad 0] + \frac{1}{22}\begin{bmatrix} 8 \\ 6 \\ 1 \end{bmatrix}[0, \quad 3, \quad 4]$$

$$+\frac{1}{22}\begin{bmatrix} -16 \\ -12 \\ 9 \end{bmatrix}[-1, \quad 1, \quad 2]$$

$$= \begin{bmatrix} 1 & 0 & 0 \\ 0 & 1 & 0 \\ 0 & 0 & 1 \end{bmatrix} = \bar{\mathbf{I}}$$

and

$$\mathbf{a}_1\mathbf{b}_1 + \mathbf{a}_2\mathbf{b}_2 + \mathbf{a}_3\mathbf{b}_3 = \frac{1}{22}\begin{bmatrix} 3 \\ -4 \\ 0 \end{bmatrix}[2, \quad -4, \quad 3] + \frac{1}{22}\begin{bmatrix} 0 \\ 3 \\ 4 \end{bmatrix}[8, \quad 6, \quad 1]$$

$$+\frac{1}{22}\begin{bmatrix} -1 \\ 1 \\ 2 \end{bmatrix}[-16, \quad -12, \quad 9]$$

$$= \begin{bmatrix} 1 & 0 & 0 \\ 0 & 1 & 0 \\ 0 & 0 & 1 \end{bmatrix} = \bar{\mathbf{I}}$$

To express matrix (1.146) as a sum of dyads, we evaluate

$$\mathbf{m}_1 = \bar{\mathbf{A}} \cdot \mathbf{b}_1 = \frac{1}{22}\begin{bmatrix} 3 \\ 22 \\ 19 \end{bmatrix}$$

$$\mathbf{m}_2 = \bar{\mathbf{A}} \cdot \mathbf{b}_2 = \frac{1}{22}\begin{bmatrix} 1 \\ 66 \\ -1 \end{bmatrix}$$

$$\mathbf{m}_3 = \bar{\mathbf{A}} \cdot \mathbf{b}_3 = \frac{1}{22}\begin{bmatrix} 9 \\ -110 \\ 57 \end{bmatrix}$$

which indeed yield

$$\mathbf{m}_1\mathbf{a}_1 + \mathbf{m}_2\mathbf{a}_2 + \mathbf{m}_3\mathbf{a}_3 = \frac{1}{22}\begin{bmatrix} 3 \\ 22 \\ 19 \end{bmatrix}[3, \quad -4, \quad 0] + \frac{1}{22}\begin{bmatrix} 1 \\ 66 \\ -1 \end{bmatrix}[0, \quad 3, \quad 4]$$

$$+\frac{1}{22}\begin{bmatrix} 9 \\ -110 \\ 57 \end{bmatrix}[-1, \quad 1, \quad 2]$$

$$= \begin{bmatrix} 0 & 0 & 1 \\ 8 & 0 & 2 \\ 0 & -1 & 5 \end{bmatrix} = \bar{\mathbf{A}}$$

A matrix which can not be reduced to a sum of less than three dyads is said to be *complete*. For the complete matrix

$$\bar{\mathbf{A}} = \mathbf{a}_i\mathbf{c}_i = \mathbf{a}_1\mathbf{c}_1 + \mathbf{a}_2\mathbf{c}_2 + \mathbf{a}_3\mathbf{c}_3 \tag{1.148}$$

the antecedents $\mathbf{a}_1$, $\mathbf{a}_2$, and $\mathbf{a}_3$ and consequents $\mathbf{c}_1$, $\mathbf{c}_2$, and $\mathbf{c}_3$ are two sets of non-coplanar vectors. Dot-multiplication of Eq. (1.148) from the right by any vector $\mathbf{v}$ yields a new vector

$$\bar{\mathbf{A}} \cdot \mathbf{v} = (\mathbf{c}_1 \cdot \mathbf{v})\mathbf{a}_1 + (\mathbf{c}_2 \cdot \mathbf{v})\mathbf{a}_2 + (\mathbf{c}_3 \cdot \mathbf{v})\mathbf{a}_3$$

Hence, a complete matrix transforms a three-dimensional vector into another three-dimensional vector.

If, however, $\mathbf{a}_1$, $\mathbf{a}_2$, and $\mathbf{a}_3$ (or $\mathbf{c}_1$, $\mathbf{c}_2$, and $\mathbf{c}_3$), are coplanar, but not collinear, we can express each $\mathbf{a}_i$ in terms of two non-parallel vectors, for example, $\mathbf{a}_3 = \alpha_1\mathbf{a}_1 + \alpha_2\mathbf{a}_2$, and reduce $\bar{\mathbf{A}}$ to a sum of two dyads:

$$\begin{aligned} \bar{\mathbf{A}} &= \mathbf{a}_1\mathbf{c}_1 + \mathbf{a}_2\mathbf{c}_2 + (\alpha_1\mathbf{a}_1 + \alpha_2\mathbf{a}_2)\mathbf{c}_3 \\ &= \mathbf{a}_1\mathbf{h}_1 + \mathbf{a}_2\mathbf{h}_2 \end{aligned} \tag{1.149}$$

where $\qquad \mathbf{h}_1 = \mathbf{c}_1 + \alpha_1\mathbf{c}_3 \qquad \mathbf{h}_2 = \mathbf{c}_2 + \alpha_2\mathbf{c}_3$

Dot-multiplication of Eq. (1.149) from the right by any vector $\mathbf{v}$ gives

$$\bar{\mathbf{A}} \cdot \mathbf{v} = (\mathbf{h}_1 \cdot \mathbf{v})\mathbf{a}_1 + (\mathbf{h}_2 \cdot \mathbf{v})\mathbf{a}_2$$

That is, the matrix (1.149) projects any three-dimensional vector into a vector on the plane formed by the antecedents $\mathbf{a}_1$ and $\mathbf{a}_2$. A matrix having this property is said to be *planar*. A planar matrix cannot be reduced to a single dyad $\mathbf{ab}$, for then all vectors $\mathbf{v}$ would transform into vectors $(\mathbf{b} \cdot \mathbf{v})\mathbf{a}$ parallel to $\mathbf{a}$.

If $\mathbf{a}_1$, $\mathbf{a}_2$, and $\mathbf{a}_3$ (or $\mathbf{c}_1$, $\mathbf{c}_2$, and $\mathbf{c}_3$) are collinear, we can express each $\mathbf{a}_i$ as a constant multiple of a single vector $\mathbf{a}$:

$$\mathbf{a}_1 = \alpha_1\mathbf{a} \qquad \mathbf{a}_2 = \alpha_2\mathbf{a} \qquad \mathbf{a}_3 = \alpha_3\mathbf{a}$$

and reduce $\bar{\mathbf{A}}$ to a single dyad:

$$\bar{\mathbf{A}} = \mathbf{ac} \tag{1.150}$$

This matrix projects all three-dimensional vectors $\mathbf{v}$ into vectors $(\mathbf{c} \cdot \mathbf{v})\mathbf{a}$ parallel to $\mathbf{a}$. Such a matrix is called *linear*.

Finally, if $\mathbf{a}_1$, $\mathbf{a}_2$, and $\mathbf{a}_3$ are all zero, $\bar{\mathbf{A}} = \bar{\mathbf{0}}$.

From the above, we conclude that a three-dimensional matrix $\bar{\mathbf{A}}$ can fall into one of the four groups: zero, linear (single dyad), planar (sum of two dyads), or complete (sum of three dyads). We shall now examine some properties of each group. First, for a zero matrix, both the determinant and adjoint of the matrix are zero. Also, according to Eqs. (1.85) and (1.86), the determinant and adjoint of a linear matrix are zero. The converse is also true; that is, if $|\bar{\mathbf{A}}| = 0$, adj $\bar{\mathbf{A}} = \bar{\mathbf{0}}$ but $\bar{\mathbf{A}} \neq \bar{\mathbf{0}}$, then $\bar{\mathbf{A}} = \mathbf{ab}$ is a single dyad.

From Eq. (1.149), we see that a planar matrix may be expressed as $\bar{\mathbf{A}} = \mathbf{al} + \mathbf{bm}$, the determinant of which can be found by expanding the sum according to Eq. (1.77). Using the results of Eqs. (1.85) and (1.86), we have

$$|\mathbf{al} + \mathbf{bm}| = 0 \tag{1.151}$$

To obtain adj $\bar{\mathbf{A}}$, we use the definition (1.59) and the fact that $\varepsilon_{ijs} a_i a_j = \varepsilon_{ijs} b_i b_j = 0$; thus

$$\begin{aligned}(\text{adj } \bar{\mathbf{A}})_{ns} &= \tfrac{1}{2}\, \varepsilon_{ijs}\, \varepsilon_{pqn}(a_i l_p + b_i m_p)(a_j l_q + b_j m_q) \\ &= (\varepsilon_{pqn} l_p m_q)(\varepsilon_{ijs} a_i b_j) \\ &= (\mathbf{l} \times \mathbf{m})_n(\mathbf{a} \times \mathbf{b})_s\end{aligned}$$

or, in direct form,

$$\text{adj } (\mathbf{al} + \mathbf{bm}) = (\mathbf{l} \times \mathbf{m})(\mathbf{a} \times \mathbf{b}) \tag{1.152}$$

That is, the planar matrix is singular and its adjoint matrix is a single dyad. Conversely, if $|\bar{\mathbf{A}}| = 0$ and adj $\bar{\mathbf{A}} = \mathbf{cn} \neq \bar{\mathbf{0}}$, then $\bar{\mathbf{A}}$ is a planar matrix.

According to Eq. (1.148), a complete matrix may be represented as $\bar{\mathbf{A}} = \mathbf{al} + \mathbf{bm} + \mathbf{cn}$. To find the determinant, we expand the sum according to Eq. (1.77). Using the results of Eqs. (1.85), (1.86), (1.151), and (1.152), we obtain

$$\begin{aligned}|\bar{\mathbf{A}}| = |\mathbf{al} + \mathbf{bm} + \mathbf{cn}| &= [\text{adj } (\mathbf{al} + \mathbf{bm}) \cdot \mathbf{cn}]_t \\ &= [(\mathbf{l} \times \mathbf{m})(\mathbf{a} \times \mathbf{b}) \cdot \mathbf{cn}]_t \\ &= (\mathbf{a} \cdot \mathbf{b} \times \mathbf{c})(\mathbf{l} \cdot \mathbf{m} \times \mathbf{n})\end{aligned} \tag{1.153}$$

Since $\mathbf{a}$, $\mathbf{b}$, and $\mathbf{c}$ and $\mathbf{l}$, $\mathbf{m}$, and $\mathbf{n}$ are two sets of non-coplanar vectors ($\mathbf{a} \cdot \mathbf{b} \times \mathbf{c} \neq 0$, $\mathbf{l} \cdot \mathbf{m} \times \mathbf{n} \neq 0$), $|\bar{\mathbf{A}}| \neq 0$. That is, a complete matrix is nonsingular. Conversely, if $|\bar{\mathbf{A}}| \neq 0$, then $\bar{\mathbf{A}}$ is a complete matrix. To find the adjoint of $\bar{\mathbf{A}}$, we use definition (1.59). Thus

$$\begin{aligned}(\text{adj } \bar{\mathbf{A}})_{ks} &= \tfrac{1}{2}\, \varepsilon_{ijs}\, \varepsilon_{pqk}(a_i l_p + b_i m_p + c_i n_p)(a_j l_q + b_j m_q + c_j n_q) \\ &= (\varepsilon_{pqk} m_p n_q)(\varepsilon_{ijs} b_i c_j) + (\varepsilon_{pqk} n_p l_q)(\varepsilon_{ijs} c_i a_j) \\ &\quad + (\varepsilon_{pqk} l_p m_q)(\varepsilon_{ijs} a_i b_j) \\ &= (\mathbf{m} \times \mathbf{n})_k(\mathbf{b} \times \mathbf{c})_s + (\mathbf{n} \times \mathbf{l})_k(\mathbf{c} \times \mathbf{a})_s + (\mathbf{l} \times \mathbf{m})_k(\mathbf{a} \times \mathbf{b})_s\end{aligned}$$

or, in direct form,

$$\operatorname{adj} \bar{\mathbf{A}} = \operatorname{adj} (\mathbf{al} + \mathbf{bm} + \mathbf{cn})$$
$$= (\mathbf{m} \times \mathbf{n})(\mathbf{b} \times \mathbf{c}) + (\mathbf{n} \times \mathbf{l})(\mathbf{c} \times \mathbf{a}) + (\mathbf{l} \times \mathbf{m})(\mathbf{a} \times \mathbf{b}) \tag{1.154}$$

It is interesting to note from the above that the adjoint of a matrix can never be planar; if it is different from zero, it must either be linear or complete. The planar, linear, and zero matrices are singular.

In summary, we may state the following necessary and sufficient conditions:

1. If $|\bar{\mathbf{A}}| \neq 0$, $\bar{\mathbf{A}}$ is complete. That is, both $\bar{\mathbf{A}}$ and adj $\bar{\mathbf{A}}$ may be expressed as sums of three dyads. In this case, the inverse of $\bar{\mathbf{A}}$ exists.
2. If $|\bar{\mathbf{A}}| = 0$ but adj $\bar{\mathbf{A}} \neq \bar{\mathbf{0}}$, then $\bar{\mathbf{A}}$ is planar. In this case, $\bar{\mathbf{A}}$ may be expressed as a sum of two dyads and adj $\bar{\mathbf{A}}$ as a single dyad. Since $\bar{\mathbf{A}}$ is singular, the inverse does not exist.
3. If $|\bar{\mathbf{A}}| = 0$, adj $\bar{\mathbf{A}} = \bar{\mathbf{0}}$ but $\bar{\mathbf{A}} \neq \bar{\mathbf{0}}$, then $\bar{\mathbf{A}}$ is linear. That is, $\bar{\mathbf{A}}$ is a single dyad. Again, since $\bar{\mathbf{A}}$ is singular, the inverse does not exist.

Next, let us consider the problem of finding solutions of the homogeneous equation

$$\bar{\mathbf{A}} \cdot \mathbf{u} = \mathbf{0} \tag{1.155}$$

where $\bar{\mathbf{A}}$ is a given 3×3 coefficient matrix and $\mathbf{u}$ is an unknown vector with components u_1, u_2, and u_3. The matrix equation (1.155) is equivalent to a set of three homogeneous linear equations in three unknowns u_1, u_2, and u_3. Clearly $\mathbf{u} = \mathbf{0}$ is a solution. This solution is called *trivial.* In applications, we are interested in *nontrivial* solutions (that is, $\mathbf{u} \neq 0$). Moreover, if a nontrivial solution does exist, it is not unique. For if $\mathbf{u} \neq 0$ satisfies the equation (1.155), then for any scalar α

$$\bar{\mathbf{A}} \cdot \alpha\mathbf{u} = \alpha\bar{\mathbf{A}} \cdot \mathbf{u} = \mathbf{0}$$

so that $\alpha\mathbf{u}$ is also a solution. In other words, the homogeneous equation (1.155) determines only the direction of vector $\mathbf{u}$ not its magnitude. The existence of nontrivial solutions of Eq. (1.155) and the method of finding such solutions clearly depend on the coefficient matrix $\bar{\mathbf{A}}$. If $\bar{\mathbf{A}}$ is nonsingular (that is, $|\bar{\mathbf{A}}| \neq 0$), then $\bar{\mathbf{A}}^{-1}$ exists. Dot-multiplying both sides of Eq. (1.155) from the left by $\bar{\mathbf{A}}^{-1}$, we obtain $\mathbf{u} = \mathbf{0}$. Consequently, if $|\bar{\mathbf{A}}| \neq 0$, the homogeneous equation (1.155) has only the trivial solution $\mathbf{u} = \mathbf{0}$.

On the other hand, if $\bar{\mathbf{A}}$ is singular (that is, $|\bar{\mathbf{A}}| = 0$), two possibilities occur, namely , $\bar{\mathbf{A}}$ is either planar or linear. In the first case, when the coefficient matrix $\bar{\mathbf{A}}$ is planar (that is, $|\bar{\mathbf{A}}| = 0$, adj $\bar{\mathbf{A}} \neq \bar{\mathbf{0}}$), we dot-multiply both sides of the identity

$$\bar{\mathbf{A}} \cdot (\operatorname{adj} \bar{\mathbf{A}}) = |\bar{\mathbf{A}}|\bar{\mathbf{I}} = \bar{\mathbf{0}}$$

from the right by an arbitrary constant vector $\mathbf{c}$ and obtain

$$\bar{\mathbf{A}} \cdot [(\operatorname{adj} \bar{\mathbf{A}}) \cdot \mathbf{c}] = \mathbf{0}$$

Comparing this with Eq. (1.155), we find the nontrivial solution

$$\mathbf{u} = (\text{adj } \bar{\mathbf{A}}) \cdot \mathbf{c} \tag{1.156}$$

Since $\bar{\mathbf{A}}$ is planar, adj $\bar{\mathbf{A}} = \mathbf{ab}$ is a single dyad, and Eq. (1.156) becomes

$$\mathbf{u} = (\mathbf{b} \cdot \mathbf{c})\mathbf{a}$$

That is, the nontrivial solution of the homogeneous equation (1.155) is proportional to vector $\mathbf{a}$, a nonzero column of adj $\bar{\mathbf{A}}$. In the second case when the coefficient matrix $\bar{\mathbf{A}}$ is linear, then $\bar{\mathbf{A}}$ itself must be a single dyad, that is, $\bar{\mathbf{A}} = \mathbf{ab}$. The homogeneous equation (1.155) becomes

$$\bar{\mathbf{A}} \cdot \mathbf{u} = \mathbf{ab} \cdot \mathbf{u} = \mathbf{0} \tag{1.157}$$

Thus any vector $\mathbf{u}$ perpendicular to vector $\mathbf{b}$ is a solution of the homogeneous equation. In summary, the homogeneous equation $\bar{\mathbf{A}} \cdot \mathbf{u} = \mathbf{0}$ has nontrivial solutions if and only if the determinant of the coefficient matrix vanishes. A method of obtaining such solutions is described in Eq. (1.156) or (1.157).

1.7 EIGENVALUE PROBLEMS

Closely related to the solution of a homogeneous equation is the *eigenvalue problem* defined by

$$\bar{\mathbf{A}} \cdot \mathbf{u} = \lambda \mathbf{u} \leftrightarrow a_{ij} u_j = \lambda u_i \tag{1.158}$$

where $\bar{\mathbf{A}} = [a_{ij}]$ is a 3×3 matrix and $\mathbf{u}$ is a vector. That is, in an eigenvalue problem, we seek those nonzero vectors $\mathbf{u}$ associated with the given matrix $\bar{\mathbf{A}}$ such that only their magnitudes are altered by the multiplication of the matrix while their directions remain unchanged. The scalar λ is called an *eigenvalue* of the matrix $\bar{\mathbf{A}}$ and $\mathbf{u}$ is called an *eigenvector* of $\bar{\mathbf{A}}$ corresponding to the eigenvalue λ.

Equation (1.158) may be written in the form

$$(\bar{\mathbf{A}} - \lambda \bar{\mathbf{I}}) \cdot \mathbf{u} = \mathbf{0} \tag{1.159}$$

The matrix $\bar{\mathbf{A}} - \lambda \bar{\mathbf{I}}$ is called the *characteristic matrix* of $\bar{\mathbf{A}}$. The homogeneous equation (1.159) has nontrivial solutions if and only if the determinant of the characteristic matrix vanishes:

$$|\bar{\mathbf{A}} - \lambda \bar{\mathbf{I}}| = 0 \tag{1.160}$$

This equation is called the *characteristic equation* of $\bar{\mathbf{A}}$. The expansion of the determinant in Eq. (1.160) according to Eq. (1.124*b*) yields a polynomial of degree 3 in λ:

$$f(\lambda) = |\bar{\mathbf{A}} - \lambda \bar{\mathbf{I}}| = -[\lambda^3 - \bar{\mathbf{A}}_t \lambda^2 + (\text{adj } \bar{\mathbf{A}})_t \lambda - |\bar{\mathbf{A}}|] \tag{1.161}$$

This polynomial is called the *characteristic polynomial* of $\bar{\mathbf{A}}$. The eigenvalues are, therefore, solutions of the characteristic equation:

$$\lambda^3 - \bar{\mathbf{A}}_t \lambda^2 + (\text{adj } \bar{\mathbf{A}})_t \lambda - |\bar{\mathbf{A}}| = 0 \tag{1.162}$$

where

$$\bar{\mathbf{A}}_t = a_{11} + a_{22} + a_{33}$$
$$(\text{adj } \bar{\mathbf{A}})_t = \tfrac{1}{2}[(\bar{\mathbf{A}}_t)^2 - (\bar{\mathbf{A}}^2)_t] \tag{1.163}$$
$$|\bar{\mathbf{A}}| = \tfrac{1}{6}[(\bar{\mathbf{A}}_t)^3 - 3(\bar{\mathbf{A}}_t)(\bar{\mathbf{A}}^2)_t + 2(\bar{\mathbf{A}}^3)_t]$$

The cubic equation (1.162) may have three real roots which need not all be distinct or one real root and two conjugate complex roots. On writing Eq. (1.162) in terms of its three roots λ_1, λ_2, and λ_3:

$$(\lambda - \lambda_1)(\lambda - \lambda_2)(\lambda - \lambda_3) = 0$$

and comparing coefficients of the different powers of λ, we find that

$$\bar{\mathbf{A}}_t = \lambda_1 + \lambda_2 + \lambda_3$$
$$(\text{adj } \bar{\mathbf{A}})_t = \lambda_1\lambda_2 + \lambda_2\lambda_3 + \lambda_3\lambda_1 \tag{1.164}$$
$$|\bar{\mathbf{A}}| = \lambda_1\lambda_2\lambda_3$$

That is, the sum of the eigenvalues is the trace of the matrix and the product of the eigenvalues is the determinant of the matrix. Hence if $\bar{\mathbf{A}}$ is nonsingular, none of its eigenvalues is equal to zero, and vice versa.

To find the eigenvector $\mathbf{u}_1$ of Eq. (1.159) corresponding to an eigenvalue λ_1 of Eq. (1.162), we will consider in turn the three cases in which the characteristic matrix $\bar{\mathbf{A}} - \lambda_1\bar{\mathbf{I}}$ is singular.

1. If $\bar{\mathbf{A}} - \lambda_1\bar{\mathbf{I}}$ is planar, the adjoint of $\bar{\mathbf{A}} - \lambda_1\bar{\mathbf{I}}$ must be a single dyad; that is,

   $$\text{adj } (\bar{\mathbf{A}} - \lambda_1\bar{\mathbf{I}}) = \mathbf{ab} \neq \bar{\mathbf{0}}$$

   and the eigenvector $\mathbf{u}_1$ is given by

   $$\mathbf{u}_1 = [\text{adj } (\bar{\mathbf{A}} - \lambda_1\bar{\mathbf{I}})] \cdot \mathbf{c} = (\mathbf{b} \cdot \mathbf{c})\mathbf{a} \tag{1.165}$$

   where $\mathbf{c}$ is a constant vector. That is, the eigenvector $\mathbf{u}_1$ corresponding to the eigenvalue λ_1 is parallel to the vector $\mathbf{a}$, a column of adj $(\bar{\mathbf{A}} - \lambda_1\bar{\mathbf{I}})$. Also, for a real matrix $\bar{\mathbf{A}}$, if λ_1 is real, $\mathbf{u}_1$ is a real vector. But if λ_1 is complex, then $\mathbf{u}_1$ is complex. In this case, Eq. (1.158) shows that the vector $\mathbf{u}_1^*$, the complex conjugate of the complex vector $\mathbf{u}_1$ corresponding to the eigenvalue λ_1^*, is also an eigenvector.
2. If $\bar{\mathbf{A}} - \lambda_1\bar{\mathbf{I}}$ is linear, we may reduce it to a single dyad; that is,

   $$\bar{\mathbf{A}} - \lambda_1\bar{\mathbf{I}} = \mathbf{ab} \tag{1.166}$$

   Then any vector $\mathbf{u}_1$ in the plane perpendicular to $\mathbf{b}$ is an eigenvector corresponding to the eigenvalue λ_1. Since

   $$\bar{\mathbf{A}} \cdot \mathbf{a} = (\lambda_1\bar{\mathbf{I}} + \mathbf{ab}) \cdot \mathbf{a}$$
   $$= (\lambda_1 + \mathbf{a} \cdot \mathbf{b})\mathbf{a} \tag{1.167}$$

   thus we see that $\mathbf{u}_2 = \mathbf{a}$ is also an eigenvector of $\bar{\mathbf{A}}$. The corresponding eigenvalue is $\lambda_2 = \lambda_1 + \mathbf{a} \cdot \mathbf{b}$.

3. If $\bar{\mathbf{A}} - \lambda_1 \bar{\mathbf{I}}$ is zero, then

$$\bar{\mathbf{A}} \cdot \mathbf{u}_1 = \lambda_1 \mathbf{u}_1$$

for any vector $\mathbf{u}_1$. In this case all vectors are eigenvectors with the eigenvalue λ_1.

In summary, we have shown that for any matrix $\bar{\mathbf{A}}$, there exists at least one eigenvector. We have also shown how, in general, the eigenvectors may be determined if the eigenvalues are known.

Example 1.8 Find the characteristic equation, the eigenvalues and eigenvectors of the following matrix:

$$\bar{\mathbf{A}} = \begin{bmatrix} 0 & 1 & 0 \\ 0 & 0 & 1 \\ 1 & 0 & 0 \end{bmatrix}$$

SOLUTION From the given matrix $\bar{\mathbf{A}}$, we have

$$\bar{\mathbf{A}}^2 = \begin{bmatrix} 0 & 0 & 1 \\ 1 & 0 & 0 \\ 0 & 1 & 0 \end{bmatrix} \qquad \bar{\mathbf{A}}^3 = \bar{\mathbf{I}}$$

Hence

$$\bar{\mathbf{A}}_t = 0 \qquad (\bar{\mathbf{A}}^2)_t = 0 \qquad (\bar{\mathbf{A}}^3)_t = \bar{\mathbf{I}}_t = 3$$

Substitution of the above into Eq. (1.163) yields

$$(\text{adj } \bar{\mathbf{A}})_t = 0 \qquad |\bar{\mathbf{A}}| = 1$$

The characteristic equation (1.162) becomes

$$\lambda^3 - 1 = (\lambda - 1)(\lambda^2 + \lambda + 1) = 0$$

and has roots

$$\lambda_1 = 1$$
$$\lambda_2 = \tfrac{1}{2}(-1 + i\sqrt{3}) = \omega$$
$$\lambda_3 = \tfrac{1}{2}(-1 - i\sqrt{3}) = \omega^2 = \omega^*$$

where the asterisk * denotes the complex conjugate. Also

$$\text{adj } \bar{\mathbf{A}} = \bar{\mathbf{A}}^2 - \bar{\mathbf{A}}_t \bar{\mathbf{A}} + (\text{adj } \bar{\mathbf{A}})_t \bar{\mathbf{I}} = \bar{\mathbf{A}}^2$$

and

$$\begin{aligned} \text{adj } (\bar{\mathbf{A}} - \lambda \bar{\mathbf{I}}) &= \text{adj } \bar{\mathbf{A}} + (\bar{\mathbf{A}} - \bar{\mathbf{A}}_t \bar{\mathbf{I}})\lambda + \lambda^2 \bar{\mathbf{I}} \\ &= \lambda^2 \bar{\mathbf{I}} + \lambda \bar{\mathbf{A}} + \bar{\mathbf{A}}^2 \end{aligned}$$

Hence

$$\operatorname{adj}(\bar{\mathbf{A}} - \lambda_1\bar{\mathbf{I}}) = \bar{\mathbf{I}} + \bar{\mathbf{A}} + \bar{\mathbf{A}}^2 = \begin{bmatrix} 1 & 1 & 1 \\ 1 & 1 & 1 \\ 1 & 1 & 1 \end{bmatrix} = \begin{bmatrix} 1 \\ 1 \\ 1 \end{bmatrix} [1, \quad 1, \quad 1]$$

$$\operatorname{adj}(\bar{\mathbf{A}} - \lambda_2\bar{\mathbf{I}}) = \omega^2\bar{\mathbf{I}} + \omega\bar{\mathbf{A}} + \bar{\mathbf{A}}^2 = \begin{bmatrix} \omega^2 & \omega & 1 \\ 1 & \omega^2 & \omega \\ \omega & 1 & \omega^2 \end{bmatrix} = \begin{bmatrix} 1 \\ \omega \\ \omega^2 \end{bmatrix} [\omega^2, \quad \omega, \quad 1]$$

$$\operatorname{adj}(\bar{\mathbf{A}} - \lambda_3\bar{\mathbf{I}}) = \omega\bar{\mathbf{I}} + \omega^2\bar{\mathbf{A}} + \bar{\mathbf{A}}^2 = \begin{bmatrix} \omega & \omega^2 & 1 \\ 1 & \omega & \omega^2 \\ \omega^2 & 1 & \omega \end{bmatrix} = \begin{bmatrix} 1 \\ \omega^2 \\ \omega \end{bmatrix} [\omega, \quad \omega^2, \quad 1]$$

Choosing vector $\mathbf{c}$ with components $c_1 = c_2 = 0$, and $c_3 = 1$, we obtain the eigenvectors corresponding to eigenvalues $\lambda_1 = 1$, $\lambda_2 = \omega$, and $\lambda_3 = \omega^2$:

$$\mathbf{u}_1 = \begin{bmatrix} 1 \\ 1 \\ 1 \end{bmatrix} \qquad \mathbf{u}_2 = \begin{bmatrix} 1 \\ \omega \\ \omega^2 \end{bmatrix} \qquad \mathbf{u}_3 = \begin{bmatrix} 1 \\ \omega^2 \\ \omega \end{bmatrix}$$

We note that when λ_1 is real, so is $\mathbf{u}_1$. The eigenvalue λ_3 is the complex conjugate of λ_2; thus the eigenvector $\mathbf{u}_3$ is the complex conjugate of $\mathbf{u}_2$.

1.8 SYMMETRIC MATRICES

Since the symmetric matrices play an important role in applications, we shall now pay some special attention to their eigenvalue problems.

As we have seen in Example 1-8, the eigenvalues and eigenvectors of a real matrix may, in general, be complex. However, for a real symmetric matrix, the eigenvalues must all be real. In fact, if we let $\bar{\mathbf{A}}$ be a real symmetric matrix and $\mathbf{u}$ be an eigenvector of $\bar{\mathbf{A}}$ corresponding to the eigenvalue λ, then

$$\bar{\mathbf{A}} \cdot \mathbf{u} = \lambda \mathbf{u} \tag{1.168}$$

Taking the complex conjugate of this equation and noting that $\bar{\mathbf{A}}$ is real, we have

$$\bar{\mathbf{A}} \cdot \mathbf{u}^* = \lambda^* \mathbf{u}^* \tag{1.169}$$

Dot-multiplying Eq. (1.168) by $\mathbf{u}^*$, Eq. (1.169) by $\mathbf{u}$, and subtracting, we obtain

$$\mathbf{u}^* \cdot \bar{\mathbf{A}} \cdot \mathbf{u} - \mathbf{u} \cdot \bar{\mathbf{A}} \cdot \mathbf{u}^* = (\lambda - \lambda^*)\mathbf{u} \cdot \mathbf{u}^* \tag{1.170}$$

since $\mathbf{u}^* \cdot \mathbf{u} = \mathbf{u} \cdot \mathbf{u}^*$. Recall that $\tilde{\bar{\mathbf{A}}} = \bar{\mathbf{A}}$. Furthermore, $\mathbf{u} \cdot \bar{\mathbf{A}} \cdot \mathbf{u}^*$ is a number, and the transpose of a number (1×1 matrix) is itself the number. Thus

$$\mathbf{u} \cdot \bar{\mathbf{A}} \cdot \mathbf{u}^* = (\mathbf{u} \cdot \bar{\mathbf{A}} \cdot \mathbf{u}^*)^T = \mathbf{u}^* \cdot \bar{\mathbf{A}} \cdot \mathbf{u}$$

and Eq. (1.170) becomes

$$(\lambda - \lambda^*)\mathbf{u} \cdot \mathbf{u}^* = 0$$

since $\mathbf{u} \cdot \mathbf{u}^* \neq 0$, $\lambda = \lambda^*$ and hence the eigenvalues are real.

The eigenvectors are determined by adj $(\bar{\mathbf{A}} - \lambda_i \bar{\mathbf{I}})$ if the characteristic matrix $\bar{\mathbf{A}} - \lambda_i \bar{\mathbf{I}}$ is planar, and by $\bar{\mathbf{A}} - \lambda_i \bar{\mathbf{I}}$ if it is linear. But for a real symmetric matrix, both $\bar{\mathbf{A}}$ and λ_i are real. This means that the eigenvectors of a real symmetric matrix may always be taken to be real.

Another property of a real symmetric matrix is that the eigenvectors corresponding to different eigenvalues are orthogonal; i.e., if $\mathbf{u}_i$ is an eigenvector corresponding to eigenvalue λ_i, and $\mathbf{u}_j$ is an eigenvector corresponding to eigenvalue λ_j $(\lambda_j \neq \lambda_i)$, then $\mathbf{u}_i \cdot \mathbf{u}_j = 0$. To prove this, we observe that, by assumption,

$$\bar{\mathbf{A}} \cdot \mathbf{u}_i = \lambda_i \mathbf{u}_i \qquad \text{and} \qquad \bar{\mathbf{A}} \cdot \mathbf{u}_j = \lambda_j \mathbf{u}_j \qquad \text{(no sum)}$$

Thus

$$\mathbf{u}_j \cdot \bar{\mathbf{A}} \cdot \mathbf{u}_i = \lambda_i \mathbf{u}_j \cdot \mathbf{u}_i \qquad \text{and} \qquad \mathbf{u}_i \cdot \bar{\mathbf{A}} \cdot \mathbf{u}_j = \lambda_j \mathbf{u}_i \cdot \mathbf{u}_j \qquad \text{(no sum)}$$

Subtracting and noting that $\mathbf{u}_j \cdot \bar{\mathbf{A}} \cdot \mathbf{u}_i = \mathbf{u}_i \cdot \bar{\mathbf{A}} \cdot \mathbf{u}_j$, we have

$$(\lambda_i - \lambda_j)\mathbf{u}_i \cdot \mathbf{u}_j = 0$$

But $\lambda_i \neq \lambda_j$; hence it follows that $\mathbf{u}_i \cdot \mathbf{u}_j = 0$. Since the eigenvectors of a real symmetric matrix are real, we can always divide the vector by its magnitude to obtain a unit vector. When the eigenvectors are of unit length, they are said to have been *normalized*. A set of orthogonal unit vectors is called *orthonormal*.

Since the characteristic equation (1.162) is a cubic equation in λ, the three real eigenvalues λ_1, λ_2, and λ_3 of a real symmetric matrix $\bar{\mathbf{A}}$ may or may not be distinct. Let us consider the following three cases:

Case 1 If all the three eigenvalues are distinct such that $\lambda_1 < \lambda_2 < \lambda_3$, the corresponding eigenvectors form an orthonormal set denoted by $\hat{\mathbf{e}}_1$, $\hat{\mathbf{e}}_2$, and $\hat{\mathbf{e}}_3$ where $\hat{}$ over a letter indicates a unit vector. Then

$$\bar{\mathbf{A}} \cdot \hat{\mathbf{e}}_i = \lambda_i \hat{\mathbf{e}}_i \qquad \text{(no sum)} \tag{1.171}$$

with

$$\hat{\mathbf{e}}_i \cdot \hat{\mathbf{e}}_j = \delta_{ij} \tag{1.172}$$

If $\hat{\mathbf{e}}_i$ is an eigenvector of $\bar{\mathbf{A}}$, it follows from Eq. (1.171) that $-\hat{\mathbf{e}}_i$ is also an eigenvector of $\bar{\mathbf{A}}$ corresponding to the same eigenvalue λ_i. Thus we may always choose $\hat{\mathbf{e}}_i$ to satisfy

$$\hat{\mathbf{e}}_1 \times \hat{\mathbf{e}}_2 = \hat{\mathbf{e}}_3 \tag{1.173}$$

that is, a right-handed system. In this case, Eq. (1.135) gives

$$\mathbf{b}_i = \mathbf{a}_i = \hat{\mathbf{e}}_i$$

and

$$\mathbf{m}_i = \bar{\mathbf{A}} \cdot \mathbf{b}_i = \lambda_i \hat{\mathbf{e}}_i \qquad \text{(no sum)}$$

The completeness relation (1.141) becomes

$$\bar{\mathbf{I}} = \hat{\mathbf{e}}_i \hat{\mathbf{e}}_i = \hat{\mathbf{e}}_1 \hat{\mathbf{e}}_1 + \hat{\mathbf{e}}_2 \hat{\mathbf{e}}_2 + \hat{\mathbf{e}}_3 \hat{\mathbf{e}}_3 \tag{1.174}$$

and the dyadic representation (1.143) of the matrix $\bar{\mathbf{A}}$ takes the form

$$\bar{\mathbf{A}} = \lambda_1 \hat{\mathbf{e}}_1 \hat{\mathbf{e}}_1 + \lambda_2 \hat{\mathbf{e}}_2 \hat{\mathbf{e}}_2 + \lambda_3 \hat{\mathbf{e}}_3 \hat{\mathbf{e}}_3 \tag{1.175}$$

Now, if we choose unit vectors as

$$\hat{\mathbf{e}}_1 = \begin{bmatrix} 1 \\ 0 \\ 0 \end{bmatrix} \quad \hat{\mathbf{e}}_2 = \begin{bmatrix} 0 \\ 1 \\ 0 \end{bmatrix} \quad \hat{\mathbf{e}}_3 = \begin{bmatrix} 0 \\ 0 \\ 1 \end{bmatrix} \tag{1.176}$$

and expand the dyads in accordance with Eq. (1.81), we obtain the explicit representation of the matrix $\bar{\mathbf{A}}$:

$$\bar{\mathbf{A}} = \begin{bmatrix} \lambda_1 & 0 & 0 \\ 0 & \lambda_2 & 0 \\ 0 & 0 & \lambda_3 \end{bmatrix} \tag{1.177}$$

That is, with respect to an orthonormal set of eigenvectors, the matrix takes a diagonal form and the diagonal elements are the eigenvalues of $\bar{\mathbf{A}}$.

Elimination of $\hat{\mathbf{e}}_2 \hat{\mathbf{e}}_2$ from Eqs. (1.174) and (1.175) yields

$$\bar{\mathbf{A}} = \lambda_2 \bar{\mathbf{I}} + (\lambda_3 - \lambda_2)\hat{\mathbf{e}}_3 \hat{\mathbf{e}}_3 - (\lambda_2 - \lambda_1)\hat{\mathbf{e}}_1 \hat{\mathbf{e}}_1 \tag{1.178}$$

Now defining the scalars

$$\gamma_3 = \sqrt{\frac{\lambda_3 - \lambda_2}{\lambda_3 - \lambda_1}} \qquad \gamma_2 = \sqrt{\frac{\lambda_2 - \lambda_1}{\lambda_3 - \lambda_1}}$$

and the unit vectors

$$\hat{\mathbf{m}} = \gamma_3 \hat{\mathbf{e}}_3 + \gamma_2 \hat{\mathbf{e}}_1$$
$$\hat{\mathbf{n}} = \gamma_3 \hat{\mathbf{e}}_3 - \gamma_2 \hat{\mathbf{e}}_1$$

where $\hat{\mathbf{m}}^2 = \hat{\mathbf{n}}^2 = 1$, we can write Eq. (1.178) in an alternative form:

$$\bar{\mathbf{A}} = \alpha \bar{\mathbf{I}} + \beta(\hat{\mathbf{m}}\hat{\mathbf{n}} + \hat{\mathbf{n}}\hat{\mathbf{m}}) \tag{1.179}$$

Here $\alpha = \lambda_2$ and $\beta = (\lambda_3 - \lambda_1)/2$. This form clearly shows that the matrix $\bar{\mathbf{A}}$ is characterized by two distinct directions $\hat{\mathbf{m}}$ and $\hat{\mathbf{n}}$. In applications, the vectors $\hat{\mathbf{m}}$ and $\hat{\mathbf{n}}$ usually have a definite geometrical or physical significance. A matrix having the form of Eq. (1.175) or Eq. (1.179) is called *biaxial.*

Case 2 If two eigenvalues are equal, for example, $\lambda_1 = \lambda_2 \neq \lambda_3$, then Eq. (1.175) becomes

$$\bar{\mathbf{A}} = \lambda_1(\hat{\mathbf{e}}_1 \hat{\mathbf{e}}_1 + \hat{\mathbf{e}}_2 \hat{\mathbf{e}}_2) + \lambda_3 \hat{\mathbf{e}}_3 \hat{\mathbf{e}}_3 \tag{1.180}$$

and the corresponding matrix representation (1.177) takes the form

$$\bar{\mathbf{A}} = \begin{bmatrix} \lambda_1 & 0 & 0 \\ 0 & \lambda_1 & 0 \\ 0 & 0 & \lambda_3 \end{bmatrix} \tag{1.181}$$

We note from Eq. (1.180) that any vector lying on the plane formed by vectors $\hat{\mathbf{e}}_1$ and $\hat{\mathbf{e}}_2$ is an eigenvector of $\bar{\mathbf{A}}$ with the corresponding eigenvalue λ_1. In other words, any vector on a plane which is perpendicular to $\hat{\mathbf{e}}_3$, is an eigenvector of $\bar{\mathbf{A}}$. Making use of the completeness relation (1.174), we may write Eq. (1.180) in an alternative form:

$$\bar{\mathbf{A}} = \alpha \bar{\mathbf{I}} + \gamma \hat{\mathbf{c}}\hat{\mathbf{c}} \tag{1.182}$$

where $\alpha = \lambda_1$, $\gamma = \lambda_3 - \lambda_1$, and $\hat{\mathbf{c}} = \hat{\mathbf{e}}_3$. Because of the unique direction $\hat{\mathbf{c}}$ which distinguishes it from all other directions that are perpendicular to $\hat{\mathbf{c}}$, the matrix $\bar{\mathbf{A}}$ is called *uniaxial.*

Case 3 Finally, if all the three eigenvalues of the matrix $\bar{\mathbf{A}}$ are equal, $\lambda_1 = \lambda_2 = \lambda_3 = \lambda$, from Eqs. (1.175) and (1.174), we have

$$\bar{\mathbf{A}} = \lambda \bar{\mathbf{I}} \tag{1.183}$$

That is, $\bar{\mathbf{A}}$ is a scalar matrix. In this case, all vectors are eigenvectors of the matrix $\bar{\mathbf{A}}$. The scalar matrix is called *isotropic.*

Example 1.9 Find a set of three orthonormal eigenvectors for the symmetric matrix

$$\bar{\mathbf{A}} = \begin{bmatrix} 3 & -1 & 0 \\ -1 & 2 & -1 \\ 0 & -1 & 3 \end{bmatrix}$$

Solution From $\bar{\mathbf{A}}$, we find

$$\bar{\mathbf{A}}^2 = \begin{bmatrix} 10 & -5 & 1 \\ -5 & 6 & -5 \\ 1 & -5 & 10 \end{bmatrix} \qquad \bar{\mathbf{A}}^3 = \begin{bmatrix} 35 & -21 & 8 \\ -21 & 22 & -21 \\ 8 & -21 & 35 \end{bmatrix}$$

Hence

$$\bar{\mathbf{A}}_t = 8 \qquad (\bar{\mathbf{A}}^2)_t = 26 \qquad (\bar{\mathbf{A}}^3)_t = 92$$

Substitution of the above into Eq. (1.163) yields

$$(\text{adj } \bar{\mathbf{A}})_t = 19 \qquad |\bar{\mathbf{A}}| = 12$$

The characteristic equation (1.162) becomes

$$\lambda^3 - 8\lambda^2 + 19\lambda - 12 = (\lambda - 1)(\lambda - 3)(\lambda - 4) = 0$$

and has roots $\lambda_1 = 3$, $\lambda_2 = 1$, and $\lambda_3 = 4$. The adjoint matrices of $\bar{\mathbf{A}}$ and $(\bar{\mathbf{A}} - \lambda\bar{\mathbf{I}})$ are

$$\text{adj } \bar{\mathbf{A}} = \bar{\mathbf{A}}^2 - \bar{\mathbf{A}}_t \bar{\mathbf{A}} + (\text{adj } \bar{\mathbf{A}})_t \bar{\mathbf{I}} = \begin{bmatrix} 5 & 3 & 1 \\ 3 & 9 & 3 \\ 1 & 3 & 5 \end{bmatrix}$$

and

$$\operatorname{adj}(\bar{\mathbf{A}} - \lambda\bar{\mathbf{I}}) = \operatorname{adj}\bar{\mathbf{A}} + (\bar{\mathbf{A}} - \bar{\mathbf{A}}_t\bar{\mathbf{I}})\lambda + \lambda^2\bar{\mathbf{I}}$$

$$= \begin{bmatrix} \lambda^2 - 5\lambda + 5 & 3 - \lambda & 1 \\ 3 - \lambda & (3 - \lambda)^2 & 3 - \lambda \\ 1 & 3 - \lambda & \lambda^2 - 5\lambda + 5 \end{bmatrix}$$

respectively. For $\lambda_1 = 3$, the adjoint matrix is

$$\operatorname{adj}(\bar{\mathbf{A}} - \lambda_1\bar{\mathbf{I}}) = \begin{bmatrix} -1 & 0 & 1 \\ 0 & 0 & 0 \\ 1 & 0 & -1 \end{bmatrix} = \begin{bmatrix} 1 \\ 0 \\ -1 \end{bmatrix} [-1, 0, 1]$$

For $\lambda_2 = 1$, the adjoint matrix is

$$\operatorname{adj}(\bar{\mathbf{A}} - \lambda_2\bar{\mathbf{I}}) = \begin{bmatrix} 1 & 2 & 1 \\ 2 & 4 & 2 \\ 1 & 2 & 1 \end{bmatrix} = \begin{bmatrix} 1 \\ 2 \\ 1 \end{bmatrix} [1, \quad 2, \quad 1]$$

For $\lambda_3 = 4$, the adjoint matrix is

$$\operatorname{adj}(\bar{\mathbf{A}} - \lambda_3\bar{\mathbf{I}}) = \begin{bmatrix} 1 & -1 & 1 \\ -1 & 1 & -1 \\ 1 & -1 & 1 \end{bmatrix} = \begin{bmatrix} 1 \\ -1 \\ 1 \end{bmatrix} [1, -1, 1]$$

Choosing vector $\mathbf{c}$ with components $c_1 = c_2 = 0$, and $c_3 = 1$ in Eq. (1.165) and normalizing the resulted vector, we obtain the normalized eigenvectors corresponding to the eigenvalues $\lambda_1 = 3$, $\lambda_2 = 1$, and $\lambda_3 = 4$:

$$\hat{\mathbf{u}}_1 = \frac{1}{\sqrt{2}} \begin{bmatrix} 1 \\ 0 \\ -1 \end{bmatrix} \qquad \hat{\mathbf{u}}_2 = \frac{1}{\sqrt{6}} \begin{bmatrix} 1 \\ 2 \\ 1 \end{bmatrix} \qquad \hat{\mathbf{u}}_3 = \frac{1}{\sqrt{3}} \begin{bmatrix} 1 \\ -1 \\ 1 \end{bmatrix}$$

Here we see that all the eigenvalues and eigenvectors are real as expected, and

$$\hat{\mathbf{u}}_i \cdot \hat{\mathbf{u}}_j = \delta_{ij}$$

$$\hat{\mathbf{u}}_1 \times \hat{\mathbf{u}}_2 = \hat{\mathbf{u}}_3$$

Therefore

$$\bar{\mathbf{I}} = \hat{\mathbf{u}}_1\hat{\mathbf{u}}_1 + \hat{\mathbf{u}}_2\hat{\mathbf{u}}_2 + \hat{\mathbf{u}}_3\hat{\mathbf{u}}_3$$

and

$$\bar{\mathbf{A}} = 3\hat{\mathbf{u}}_1\hat{\mathbf{u}}_1 + \hat{\mathbf{u}}_2\hat{\mathbf{u}}_2 + 4\hat{\mathbf{u}}_3\hat{\mathbf{u}}_3$$

or

$$\bar{\mathbf{A}} = 3\bar{\mathbf{I}} + \tfrac{3}{2}(\hat{\mathbf{m}}\hat{\mathbf{n}} + \hat{\mathbf{n}}\hat{\mathbf{m}})$$

where

$$\hat{\mathbf{m}} = \frac{1}{3} \begin{bmatrix} 2 \\ 1 \\ 2 \end{bmatrix} \qquad \hat{\mathbf{n}} = \begin{bmatrix} 0 \\ -1 \\ 0 \end{bmatrix}$$

as we may readily verify.

Example 1.10 Find a set of three orthonormal eigenvectors for the symmetric matrix

$$\bar{\mathbf{A}} = \begin{bmatrix} 0 & 1 & 1 \\ 1 & 0 & 1 \\ 1 & 1 & 0 \end{bmatrix}$$

SOLUTION From $\bar{\mathbf{A}}$, we have

$$\bar{\mathbf{A}}^2 = \begin{bmatrix} 2 & 1 & 1 \\ 1 & 2 & 1 \\ 1 & 1 & 2 \end{bmatrix} \qquad \bar{\mathbf{A}}^3 = \begin{bmatrix} 2 & 3 & 3 \\ 3 & 2 & 3 \\ 3 & 3 & 2 \end{bmatrix}$$

Hence

$$\bar{\mathbf{A}}_t = 0 \qquad (\bar{\mathbf{A}}^2)_t = 6 \qquad (\bar{\mathbf{A}}^3)_t = 6$$

Substitution of the above into Eq. (1.163) yields

$$(\text{adj } \bar{\mathbf{A}})_t = -3 \qquad |\bar{\mathbf{A}}| = 2$$

The characteristic equation (1.162) becomes

$$\lambda^3 - 3\lambda - 2 = (\lambda - 2)(\lambda + 1)^2 = 0$$

and has roots $\lambda_1 = 2$ and $\lambda_2 = \lambda_3 = -1$. The adjoint matrices of $\bar{\mathbf{A}}$ and $\bar{\mathbf{A}} - \lambda\bar{\mathbf{I}}$ are

$$\text{adj } \bar{\mathbf{A}} = \bar{\mathbf{A}}^2 - 3\bar{\mathbf{I}}$$

and

$$\text{adj } (\bar{\mathbf{A}} - \lambda\bar{\mathbf{I}}) = \lambda^2\bar{\mathbf{I}} + \lambda\bar{\mathbf{A}} + \bar{\mathbf{A}}^2 - 3\bar{\mathbf{I}}$$

respectively. For $\lambda_1 = 2$, the adjoint matrix is

$$\text{adj } (\bar{\mathbf{A}} - \lambda_1\bar{\mathbf{I}}) = \bar{\mathbf{I}} + 2\bar{\mathbf{A}} + \bar{\mathbf{A}}^2 = \begin{bmatrix} 3 & 3 & 3 \\ 3 & 3 & 3 \\ 3 & 3 & 3 \end{bmatrix} = \begin{bmatrix} 1 \\ 1 \\ 1 \end{bmatrix} [3, \ 3, \ 3]$$

Hence, the normalized eigenvector corresponding to eigenvalue $\lambda_1 = 2$ is

$$\hat{\mathbf{u}}_1 = \frac{1}{\sqrt{3}} \begin{bmatrix} 1 \\ 1 \\ 1 \end{bmatrix}$$

For $\lambda_2 = \lambda_3 = -1$, the adjoint matrix adj $(\bar{\mathbf{A}} - \lambda_2\bar{\mathbf{I}})$ is zero. In this case, the characteristic matrix of $\bar{\mathbf{A}}$ is linear; that is,

$$\bar{\mathbf{A}} - \lambda_2\bar{\mathbf{I}} = \begin{bmatrix} 1 & 1 & 1 \\ 1 & 1 & 1 \\ 1 & 1 & 1 \end{bmatrix} = \begin{bmatrix} 1 \\ 1 \\ 1 \end{bmatrix} [1, \ 1, \ 1] = \mathbf{ab}$$

Hence any vector in the plane perpendicular to $\mathbf{b} = [1,\ 1,\ 1]$ is an eigenvector. If we choose the unit vector

$$\hat{\mathbf{u}}_2 = \frac{1}{\sqrt{2}}\begin{bmatrix} 1 \\ -1 \\ 0 \end{bmatrix}$$

as the eigenvector corresponding to the eigenvalue $\lambda_2 = -1$, then the unit vector

$$\hat{\mathbf{u}}_3 = \hat{\mathbf{u}}_1 \times \hat{\mathbf{u}}_2 = \frac{1}{\sqrt{6}}\begin{bmatrix} 1 \\ 1 \\ -2 \end{bmatrix}$$

gives the eigenvector for $\lambda_3 = -1$. Alternatively, from $\bar{\mathbf{A}} - \lambda_2 \bar{\mathbf{I}} = \mathbf{ab}$, we see that $\mathbf{a}$ is an eigenvector of $\bar{\mathbf{A}}$ corresponding to the eigenvalue $\lambda_2 + \mathbf{a} \cdot \mathbf{b} = -1 + 3 = 2 = \lambda_1$. Again, the eigenvectors form an orthonormal set:

$$\hat{\mathbf{u}}_i \cdot \hat{\mathbf{u}}_j = \delta_{ij}$$

Therefore

$$\bar{\mathbf{I}} = \hat{\mathbf{u}}_1\hat{\mathbf{u}}_1 + \hat{\mathbf{u}}_2\hat{\mathbf{u}}_2 + \hat{\mathbf{u}}_3\hat{\mathbf{u}}_3$$

and

$$\bar{\mathbf{A}} = 2\hat{\mathbf{u}}_1\hat{\mathbf{u}}_1 - \hat{\mathbf{u}}_2\hat{\mathbf{u}}_2 - \hat{\mathbf{u}}_3\hat{\mathbf{u}}_3$$

or

$$\bar{\mathbf{A}} = -\bar{\mathbf{I}} + 3\hat{\mathbf{u}}_1\hat{\mathbf{u}}_1$$

as we may easily verify.

As an application of an orthonormal set of eigenvectors, we consider the problem of finding the solution of the inhomogeneous equation

$$(\bar{\mathbf{A}} - \lambda\bar{\mathbf{I}}) \cdot \mathbf{u} = \mathbf{f} \tag{1.184}$$

where $\bar{\mathbf{A}}$ and $\mathbf{f}$ are given and λ is an arbitrary parameter. Let $\hat{\mathbf{u}}_1$, $\hat{\mathbf{u}}_2$, and $\hat{\mathbf{u}}_3$ be the orthonormal set of eigenvectors belonging to the eigenvalues λ_1, λ_2, and λ_3 of the matrix $\bar{\mathbf{A}}$, namely,

$$\bar{\mathbf{A}} \cdot \hat{\mathbf{u}}_i = \lambda_i \hat{\mathbf{u}}_i \qquad \text{(no sum)}$$

where

$$\hat{\mathbf{u}}_i \cdot \hat{\mathbf{u}}_j = \delta_{ij} \tag{1.185}$$

We expand both the given vector $\mathbf{f}$ and the unknown vector $\mathbf{u}$ in terms of the orthonormal set

$$\mathbf{f} = f_i \hat{\mathbf{u}}_i = f_1\hat{\mathbf{u}}_1 + f_2\hat{\mathbf{u}}_2 + f_3\hat{\mathbf{u}}_3 \qquad f_i = \mathbf{f} \cdot \hat{\mathbf{u}}_i$$

and

$$\mathbf{u} = \alpha_i \hat{\mathbf{u}}_i = \alpha_1\hat{\mathbf{u}}_1 + \alpha_2\hat{\mathbf{u}}_2 + \alpha_3\hat{\mathbf{u}}_3$$

Substitution of the above into Eq. (1.184) gives

$$(\bar{\mathbf{A}} - \lambda\bar{\mathbf{I}}) \cdot (\alpha_i \hat{\mathbf{u}}_i) = (\lambda_i - \lambda)\alpha_i \hat{\mathbf{u}}_i = f_i \hat{\mathbf{u}}_i \tag{1.186}$$

Dot-multiplying both sides of Eq. (1.186) by $\hat{\mathbf{u}}_j$ and making use of the orthonormal condition (1.185), we obtain

$$(\lambda_i - \lambda)\alpha_i = f_i = \mathbf{f} \cdot \hat{\mathbf{u}}_i \tag{1.187}$$

Hence if $\lambda \neq \lambda_i$ for any $i = 1, 2, 3$, then

$$\mathbf{u} = \frac{(\mathbf{f} \cdot \hat{\mathbf{u}}_1)\hat{\mathbf{u}}_1}{\lambda_1 - \lambda} + \frac{(\mathbf{f} \cdot \hat{\mathbf{u}}_2)\hat{\mathbf{u}}_2}{\lambda_2 - \lambda} + \frac{(\mathbf{f} \cdot \hat{\mathbf{u}}_3)\hat{\mathbf{u}}_3}{\lambda_3 - \lambda} \tag{1.188}$$

On the other hand, if $\lambda = \lambda_j$ for some j, then Eq. (1.187) gives a contradictory equation when $i = j$ unless $f_j = 0$, in which case the corresponding equation is simply an identity. For example, if $\lambda = \lambda_3$, then Eq. (1.184) has no solution unless $f_3 = \mathbf{f} \cdot \hat{\mathbf{u}}_3 = 0$, in which case a solution is given by

$$\mathbf{u} = \frac{(\mathbf{f} \cdot \hat{\mathbf{u}}_1)\hat{\mathbf{u}}_1}{\lambda_1 - \lambda} + \frac{(\mathbf{f} \cdot \hat{\mathbf{u}}_2)\hat{\mathbf{u}}_2}{\lambda_2 - \lambda} + \alpha\hat{\mathbf{u}}_3 \tag{1.189}$$

where α is an arbitrary constant.

Alternatively, if λ is not an eigenvalue of $\bar{\mathbf{A}}$, the coefficient matrix $\bar{\mathbf{A}} - \lambda\bar{\mathbf{I}}$ is nonsingular and the inverse matrix of $\bar{\mathbf{A}} - \lambda\bar{\mathbf{I}}$ exists. Hence

$$\mathbf{u} = (\bar{\mathbf{A}} - \lambda\bar{\mathbf{I}})^{-1} \cdot \mathbf{f} \tag{1.190}$$

gives a unique solution to Eq. (1.184). In summary, we see that there are two different methods for solving Eq. (1.184). One method is to find the inverse matrix $(\bar{\mathbf{A}} - \lambda\bar{\mathbf{I}})^{-1}$. The other method is to expand the vectors in terms of the eigenvectors of $\bar{\mathbf{A}}$. The following example illustrates both techniques.

Example 1.11 Find the solution of the following inhomogeneous equation

$$(\bar{\mathbf{A}} - \lambda\bar{\mathbf{I}}) \cdot \mathbf{u} = \mathbf{f}$$

where

$$\bar{\mathbf{A}} = \begin{bmatrix} 3 & -1 & 0 \\ -1 & 2 & -1 \\ 0 & -1 & 3 \end{bmatrix} \qquad \mathbf{f} = \begin{bmatrix} 6 \\ -3 \\ -6 \end{bmatrix}$$

and λ is an arbitrary parameter.

Solution According to Example 1.9, the normalized eigenvectors belonging to the eigenvalues $\lambda_1 = 3$, $\lambda_2 = 1$, and $\lambda_3 = 4$ of the matrix $\bar{\mathbf{A}}$ are

$$\hat{\mathbf{u}}_1 = \frac{1}{\sqrt{2}}\begin{bmatrix} 1 \\ 0 \\ -1 \end{bmatrix} \qquad \hat{\mathbf{u}}_2 = \frac{1}{\sqrt{6}}\begin{bmatrix} 1 \\ 2 \\ 1 \end{bmatrix} \qquad \hat{\mathbf{u}}_3 = \frac{1}{\sqrt{3}}\begin{bmatrix} 1 \\ -1 \\ 1 \end{bmatrix}$$

Hence

$$f_1 = \mathbf{f} \cdot \hat{\mathbf{u}}_1 = \frac{12}{\sqrt{2}} \qquad f_2 = \mathbf{f} \cdot \hat{\mathbf{u}}_2 = -\sqrt{6} \qquad f_3 = \mathbf{f} \cdot \hat{\mathbf{u}}_3 = \sqrt{3}$$

If $\lambda \neq \lambda_i$, substitution of the above into Eq. (1.188) gives

$$\mathbf{u} = \frac{6}{3-\lambda}\begin{bmatrix} 1 \\ 0 \\ -1 \end{bmatrix} + \frac{1}{\lambda - 1}\begin{bmatrix} 1 \\ 2 \\ 1 \end{bmatrix} + \frac{1}{4-\lambda}\begin{bmatrix} 1 \\ -1 \\ 1 \end{bmatrix}$$

or

$$u_1 = \frac{6}{3-\lambda} + \frac{1}{\lambda - 1} + \frac{1}{4-\lambda}$$

$$u_2 = \frac{2}{\lambda - 1} + \frac{1}{\lambda - 4}$$

$$u_3 = \frac{6}{\lambda - 3} + \frac{1}{\lambda - 1} + \frac{1}{4-\lambda}$$

Alternatively, we find the inverse of the coefficient matrix

$$(\bar{\mathbf{A}} - \lambda\bar{\mathbf{I}})^{-1} = \frac{1}{(3-\lambda)(\lambda-1)(\lambda-4)}\begin{bmatrix} \lambda^2 - 5\lambda + 5 & 3-\lambda & 1 \\ 3-\lambda & (3-\lambda)^2 & 3-\lambda \\ 1 & 3-\lambda & \lambda^2 - 5\lambda + 5 \end{bmatrix}$$

Hence, according to Eq. (1.190), we obtain

$$u_1 = \frac{6\lambda^2 - 27\lambda + 15}{(3-\lambda)(\lambda-1)(\lambda-4)} = \frac{6}{3-\lambda} + \frac{1}{\lambda-1} + \frac{1}{4-\lambda}$$

$$u_2 = \frac{3(\lambda - 3)}{(\lambda-1)(\lambda-4)} = \frac{2}{\lambda-1} + \frac{1}{\lambda-4}$$

$$u_3 = \frac{-6\lambda^2 + 33\lambda - 33}{(3-\lambda)(\lambda-1)(\lambda-4)} = \frac{6}{\lambda-3} + \frac{1}{\lambda-1} + \frac{1}{4-\lambda}$$

which check with the results obtained by the method of eigenvector expansion.

1.9 QUADRATIC FORMS AND SURFACES

If $\mathbf{r} = [x_i]$ is a real vector and $\bar{\mathbf{A}} = [a_{ij}]$ is a real matrix, the expression

$$Q = \mathbf{r} \cdot \bar{\mathbf{A}} \cdot \mathbf{r} \leftrightarrow x_i a_{ij} x_j \tag{1.191}$$

is called a *quadratic form* in the three variables x_1, x_2, and x_3. The matrix $\bar{\mathbf{A}}$ is called the *coefficient matrix* of the quadratic form. Without loss of generality, we can assume that $\bar{\mathbf{A}}$ is symmetric, for

$$\begin{aligned} Q &= \mathbf{r} \cdot [\tfrac{1}{2}(\bar{\mathbf{A}} + \tilde{\bar{\mathbf{A}}}) + \tfrac{1}{2}(\bar{\mathbf{A}} - \tilde{\bar{\mathbf{A}}})] \cdot \mathbf{r} \\ &= \mathbf{r} \cdot [\tfrac{1}{2}(\bar{\mathbf{A}} + \tilde{\bar{\mathbf{A}}})] \cdot \mathbf{r} \end{aligned}$$

and $(\bar{\mathbf{A}} + \tilde{\bar{\mathbf{A}}})/2$ is symmetric.

Let us next examine the extremum (maximum or minimum) of the quadratic form Q over a unit sphere $\mathbf{r}^2 = \mathbf{r} \cdot \mathbf{r} = 1$. We employ the method of the Lagrange multiplier by constructing a new quadratic function:

$$F = \mathbf{r} \cdot \bar{\mathbf{A}} \cdot \mathbf{r} - \lambda(\mathbf{r}^2 - 1) \tag{1.192}$$

where λ is the *Lagrange multiplier*. To determine the *stationary points* of F, we note that

$$\begin{aligned} \frac{\partial F}{\partial x_i} &= \frac{\partial}{\partial x_i} [x_j a_{jk} x_k - \lambda(x_j x_j - 1)] \\ &= \delta_{ij} a_{jk} x_k + x_j a_{jk} \delta_{ik} - 2\lambda x_j \delta_{ij} \\ &= 2(a_{ik} x_k - \lambda x_i) \end{aligned}$$

or, in direct form,

$$\frac{\partial F}{\partial \mathbf{r}} = 2(\bar{\mathbf{A}} \cdot \mathbf{r} - \lambda \mathbf{r}) \tag{1.193}$$

where the *gradient* of a scalar function F is defined by

$$\frac{\partial F}{\partial \mathbf{r}} = \nabla F = \left[\frac{\partial F}{\partial x_i}\right] = \begin{bmatrix} \dfrac{\partial F}{\partial x_1} \\ \dfrac{\partial F}{\partial x_2} \\ \dfrac{\partial F}{\partial x_3} \end{bmatrix} \tag{1.194}$$

The condition $\partial F/\partial \mathbf{r} = \mathbf{0}$ of the stationary points yields the expression

$$\bar{\mathbf{A}} \cdot \mathbf{r} = \lambda \mathbf{r}$$

Dot-multiplying both sides of the above by $\mathbf{r}$ and noting that $\mathbf{r}^2 = 1$, we have

$$\mathbf{r} \cdot \bar{\mathbf{A}} \cdot \mathbf{r} = \lambda$$

That is, the extremum of the quadratic form $Q = \mathbf{r} \cdot \bar{\mathbf{A}} \cdot \mathbf{r}$, subject to the condition $\mathbf{r}^2 = 1$, occurs when the Lagrange multiplier λ is an eigenvalue of the matrix $\bar{\mathbf{A}}$ and $\mathbf{r}$ is the corresponding eigenvector. In other words, we may characterize the three real eigenvalues λ_i (such that $\lambda_1 < \lambda_2 < \lambda_3$) and the corresponding three real eigenvectors $\mathbf{r}_i$ of a real symmetric matrix $\bar{\mathbf{A}}$ by the following properties:

1. The largest eigenvalue λ_3 of $\bar{\mathbf{A}}$ is the maximum value of the quadratic form $Q = \mathbf{r} \cdot \bar{\mathbf{A}} \cdot \mathbf{r}$ on the unit sphere. The vector $\mathbf{r}_3$, drawn from the origin to the point on the unit sphere where this maximum is achieved, is the eigenvector corresponding to the eigenvalue λ_3.
2. The smallest eigenvalue λ_1 of $\bar{\mathbf{A}}$ is the minimum value of the quadratic form

$Q = \mathbf{r} \cdot \bar{\mathbf{A}} \cdot \mathbf{r}$ on the unit sphere. The unit vector $\mathbf{r}_1$ which gives this minimum is the eigenvector corresponding to λ_1.

3. The eigenvalue λ_2 is the maximum value of the quadratic form $Q = \mathbf{r} \cdot \bar{\mathbf{A}} \cdot \mathbf{r}$ for all $\mathbf{r}$ of unit length for which $\mathbf{r} \cdot \mathbf{r}_3 = 0$. The vector $\mathbf{r}_2$ for which the quadratic form assumes this maximum is the eigenvector belonging to λ_2.

The surface in three-dimensional space defined by the equation

$$\mathbf{r} \cdot \bar{\mathbf{A}} \cdot \mathbf{r} = 1 \tag{1.195}$$

where $\bar{\mathbf{A}}$ is a real symmetric matrix, is known as a *quadratic surface*. Substituting the dyadic representation (1.175) of the matrix $\bar{\mathbf{A}}$ into Eq. (1.195) and noting that $\mathbf{r} \cdot \hat{\mathbf{e}}_i = x_i$, we obtain

$$\lambda_1 x_1^2 + \lambda_2 x_2^2 + \lambda_3 x_3^2 = 1 \tag{1.196}$$

The following cases arise: (*a*) If two of the three eigenvalues are zero, Eq. (1.196) represents two parallel planes. (*b*) If one of the three eigenvalues is zero, Eq. (1.196) represents an elliptic or hyperbolic cylinder. (*c*) If one of the eigenvalues is negative, the surface represented by Eq. (1.196) is a hyperboloid of one sheet. In the special case where the two positive λ's are equal, the surface is a hyperboloid of revolution. (*d*) If two of the eigenvalues are negative, the surface is a hyperboloid of two sheets. Again, when the two negative λ's are equal, the surface is a hyperboloid of revolution. (*e*) Finally, if all the three eigenvalues are positive, Eq. (1.196) represents an ellipsoid. Comparing it with the standard ellipsoid equation

$$\frac{x_1^2}{a^2} + \frac{x_2^2}{b^2} + \frac{x_3^2}{c^2} = 1 \tag{1.197}$$

we find that the semiaxes of the ellipsoid are equal to the reciprocals of the square roots of the eigenvalues: $a = 1/\sqrt{\lambda_1}$, $b = 1/\sqrt{\lambda_2}$, and $c = 1/\sqrt{\lambda_3}$. A radius vector drawn from the origin to a point on the ellipsoid is called the *principal axis* of the ellipsoid if it is normal to the ellipsoid at the point. According to vector analysis, the vector $\nabla f = \partial f/\partial \mathbf{r}$ is normal to the surface $f(\mathbf{r}) = \text{const}$ at the point $\mathbf{r}$. Applying the definition (1.194) to the quadratic from $Q = \mathbf{r} \cdot \bar{\mathbf{A}} \cdot \mathbf{r}$, we find the vector normal to the ellipsoid $\mathbf{r} \cdot \bar{\mathbf{A}} \cdot \mathbf{r} = 1$ at $\mathbf{r}$:

$$\nabla Q = \frac{\partial Q}{\partial \mathbf{r}} = 2(\bar{\mathbf{A}} \cdot \mathbf{r}) \tag{1.198}$$

Hence the equation defining a principal axis $\mathbf{r}$ is

$$\bar{\mathbf{A}} \cdot \mathbf{r} = \lambda \mathbf{r}$$

In other words, the principal axes are along the directions of the eigenvectors of the matrix $\bar{\mathbf{A}}$. When two of the three eigenvalues are equal (uniaxial), the surface described by Eq. (1.196) is an ellipsoid of revolution about the axis of vector $\hat{\mathbf{c}}$, and when all three are equal (isotropic), it is a sphere.

1.10 COMPLEX VECTORS AND MATRICES

Since vectors and matrices occurring in our later applications contain complex components, we will consider some of their important properties here.

In general, operations with complex vectors and matrices follow the usual rules of vector and matrix algebras and of algebra of complex numbers. For example, the conjugate vector of $\mathbf{c} = \mathbf{a} + i\mathbf{b}$ is the vector

$$\mathbf{c}^* = \mathbf{a} - i\mathbf{b} \tag{1.199}$$

where **a** and **b** are two real three-dimensional vectors. Similarly, we have

$$\mathbf{c}^2 = \mathbf{c} \cdot \mathbf{c} = \mathbf{a}^2 - \mathbf{b}^2 + i2(\mathbf{a} \cdot \mathbf{b}) \tag{1.200}$$

In order to preserve the property that a nonzero vector has finite length, we define the *hermitian dot product* of two complex vectors **c** and **p** as $\mathbf{c}^* \cdot \mathbf{p}$. Hence

$$\mathbf{c}^* \cdot \mathbf{c} = \mathbf{a}^2 + \mathbf{b}^2 \tag{1.201}$$

is real and positive. The *length* or *magnitude of the complex vector* **c**, denoted by $|\mathbf{c}|$, is accordingly defined as the positive square root of the hermitian dot product of **c** with itself; i.e.,

$$|\mathbf{c}| = +\sqrt{\mathbf{c}^* \cdot \mathbf{c}} \tag{1.202}$$

This definition of $|\mathbf{c}|$ has the property that $|\mathbf{c}|$ is real, and $|\mathbf{c}| > 0$ if $\mathbf{c} \neq \mathbf{0}$. Also it reduces to the definition of length of a vector when **c** is real. For any complex vector **c**, the cross product of **c** with itself always vanishes:

$$\mathbf{c} \times \mathbf{c} = \mathbf{0} \tag{1.203}$$

But

$$\mathbf{c} \times \mathbf{c}^* = i2(\mathbf{b} \times \mathbf{a}) \tag{1.204}$$

is purely imaginary. Thus, if $\mathbf{c} \times \mathbf{c}^*$ is zero, either **a** or **b** is zero or they are parallel. On the other hand, if $\mathbf{c} \times \mathbf{c}^* \neq \mathbf{0}$, we may represent any given vector **p** as a linear combination of **c**, $\mathbf{c}^*$, and $(\mathbf{c} \times \mathbf{c}^*)$; that is,

$$\mathbf{p} = \alpha\mathbf{c} + \beta\mathbf{c}^* + \gamma(\mathbf{c} \times \mathbf{c}^*) \tag{1.205}$$

where

$$\alpha = \frac{(\mathbf{p} \times \mathbf{c}^*) \cdot (\mathbf{c} \times \mathbf{c}^*)}{(\mathbf{c} \times \mathbf{c}^*)^2} \qquad \beta = \frac{(\mathbf{p} \times \mathbf{c}) \cdot (\mathbf{c}^* \times \mathbf{c})}{(\mathbf{c} \times \mathbf{c}^*)^2} \qquad \gamma = \frac{\mathbf{p} \cdot (\mathbf{c} \times \mathbf{c}^*)}{(\mathbf{c} \times \mathbf{c}^*)^2} \tag{1.206}$$

To obtain α, we simply take the cross product of Eq. (1.205) with $\mathbf{c}^*$ and then dot the resulted equation with $(\mathbf{c} \times \mathbf{c}^*)$. Similarly, we can show that β and γ are as given in Eq. (1.206).

Next let us turn to the general case of complex matrices. When the two operations of transposition and complex conjugation are carried out one after another on a matrix $\bar{\mathbf{A}}$, the resulting matrix is called the *hermitian conjugate* of $\bar{\mathbf{A}}$ and is denoted by $\bar{\mathbf{A}}^+$; that is, $\bar{\mathbf{A}}^+ = \tilde{\bar{\mathbf{A}}}^*$. The operation of the hermitian conjugate has the following properties:

(*a*) $(\bar{\mathbf{A}}^+)^+ = \bar{\mathbf{A}}$

(*b*) $(\bar{\mathbf{A}} + \bar{\mathbf{B}})^+ = \bar{\mathbf{A}}^+ + \bar{\mathbf{B}}^+$

(*c*) $(\alpha\bar{\mathbf{A}})^+ = \alpha^*\bar{\mathbf{A}}^+$ (1.207)

(*d*) $(\bar{\mathbf{A}} \cdot \mathbf{B})^+ = \bar{\mathbf{B}}^+ \cdot \bar{\mathbf{A}}^+$

(*e*) $(\bar{\mathbf{A}}^{-1})^+ = (\bar{\mathbf{A}}^+)^{-1}$

The proofs of (*a*), (*b*), and (*c*) follow directly from the definition. Part (*d*) is an immediate consequence of Eq. (1.25). Finally, by taking the hermitian conjugate of the following identity,

$$\bar{\mathbf{A}}^{-1} \cdot \bar{\mathbf{A}} = \bar{\mathbf{A}} \cdot \bar{\mathbf{A}}^{-1} = \bar{\mathbf{I}}$$

and using the result of part (*d*), we obtain

$$\bar{\mathbf{A}}^+ \cdot (\bar{\mathbf{A}}^{-1})^+ = (\bar{\mathbf{A}}^{-1})^+ \cdot \bar{\mathbf{A}}^+ = \bar{\mathbf{I}}^+ = \bar{\mathbf{I}}$$

and hence, prove part (*e*).

A matrix $\bar{\mathbf{A}} = [a_{ij}]$ is called *hermitian* if

$$\bar{\mathbf{A}}^+ = \bar{\mathbf{A}} \leftrightarrow a_{ij}^* = a_{ji} \tag{1.208}$$

and is called *antihermitian* if

$$\bar{\mathbf{A}}^+ = -\bar{\mathbf{A}} \leftrightarrow a_{ij}^* = -a_{ji} \tag{1.209}$$

Clearly, the hermitian and antihermitian matrices are generalizations of real symmetric and antisymmetric matrices. Thus, many results true for real symmetric matrices may also be generalized to hermitian matrices. For instance, if $\bar{\mathbf{A}}$ is a hermitian matrix and $\mathbf{u}$ is any complex vector, the expression

$$H = \mathbf{u}^* \cdot \bar{\mathbf{A}} \cdot \mathbf{u} = u_i^* a_{ij} u_j \tag{1.210}$$

is known as a *hermitian form*, a generalization of quadratic form. Since H is a scalar, it is equal to its transpose. Hence

$$H^* = \mathbf{u} \cdot \bar{\mathbf{A}}^* \cdot \mathbf{u}^* = \mathbf{u}^* \cdot \tilde{\bar{\mathbf{A}}}^* \cdot \mathbf{u} = \mathbf{u}^* \cdot \bar{\mathbf{A}} \cdot \mathbf{u} = H$$

This proves that the value of a hermitian form is always real.

We now show that every hermitian matrix $\bar{\mathbf{A}}$ can be expressed as

$$\bar{\mathbf{A}} = \bar{\mathbf{S}} + i\bar{\mathbf{N}} \tag{1.211}$$

where $\bar{\mathbf{S}}$ is a real symmetric matrix and $\bar{\mathbf{N}}$ is a real antisymmetric matrix. Since an antisymmetric matrix can be represented by $\mathbf{u} \times \bar{\mathbf{I}}$, Eq. (1.211) may also be written as

$$\bar{\mathbf{A}} = \bar{\mathbf{S}} + i(\mathbf{u} \times \bar{\mathbf{I}}) \tag{1.212}$$

where $\mathbf{u}$ is a real vector. Indeed, by taking the hermitian conjugate of Eq. (1.211), we have

$$\begin{aligned}\bar{\mathbf{A}}^+ &= \bar{\mathbf{S}}^+ - i\bar{\mathbf{N}}^+ \\ &= \tilde{\bar{\mathbf{S}}} - i\tilde{\bar{\mathbf{N}}} \\ &= \bar{\mathbf{S}} + i\bar{\mathbf{N}} = \bar{\mathbf{A}}\end{aligned}$$

That is, $\bar{\mathbf{A}}$ is hermitian, which proves the result.

Similar to the case of a real symmetric matrix, the eigenvalues of a hermitian matrix are real and the eigenvectors corresponding to distinct eigenvalues are orthogonal in the sense that the hermitian dot product vanishes. In other words, every hermitian matrix possesses a complete set of orthonormal eigenvectors. In this case, the completeness relation becomes

$$\bar{\mathbf{I}} = \hat{\mathbf{u}}_i \hat{\mathbf{u}}_i^* = \hat{\mathbf{u}}_i^* \hat{\mathbf{u}}_i \tag{1.213}$$

where

$$\hat{\mathbf{u}}_i^* \cdot \hat{\mathbf{u}}_j = \delta_{ij} \tag{1.214}$$

and $\hat{\mathbf{u}}_1$, $\hat{\mathbf{u}}_2$, and $\hat{\mathbf{u}}_3$ are the orthonormal eigenvectors of the hermitian matrix $\bar{\mathbf{A}}$. The *dyadic decomposition* of the matrix $\bar{\mathbf{A}}$ takes the form

$$\bar{\mathbf{A}} = \lambda_1 \hat{\mathbf{u}}_1 \hat{\mathbf{u}}_1^* + \lambda_2 \hat{\mathbf{u}}_2 \hat{\mathbf{u}}_2^* + \lambda_3 \hat{\mathbf{u}}_3 \hat{\mathbf{u}}_3^* \tag{1.215}$$

where λ_1, λ_2, and λ_3 are the eigenvalues of $\bar{\mathbf{A}}$. Other forms of representation of the matrix $\bar{\mathbf{A}}$ can be derived similar to those of the symmetric matrix. We now consider the following example.

Example 1.12 Find the eigenvalues and the orthonormal set of eigenvectors for

$$\bar{\mathbf{A}} = a_\perp(\bar{\mathbf{I}} - \hat{\mathbf{b}}\hat{\mathbf{b}}) + ia_\times(\hat{\mathbf{b}} \times \bar{\mathbf{I}}) + a_\parallel \hat{\mathbf{b}}\hat{\mathbf{b}}$$

where $\hat{\mathbf{b}}$ is a unit vector and $a_\perp$, $a_\times$, and $a_\parallel$ are real scalars.

SOLUTION First, we note that $\bar{\mathbf{A}}$ is hermitian and has the form of Eq. (1.212) with $\bar{\mathbf{S}} = a_\perp \bar{\mathbf{I}} + (a_\parallel - a_\perp)\hat{\mathbf{b}}\hat{\mathbf{b}}$ and $\mathbf{u} = a_\times \hat{\mathbf{b}}$. Secondly, if we choose $\hat{\mathbf{b}} = [0, 0, 1]^T$, then

$$\hat{\mathbf{b}}\hat{\mathbf{b}} = \begin{bmatrix} 0 & 0 & 0 \\ 0 & 0 & 0 \\ 0 & 0 & 1 \end{bmatrix} \qquad \bar{\mathbf{I}} - \hat{\mathbf{b}}\hat{\mathbf{b}} = \begin{bmatrix} 1 & 0 & 0 \\ 0 & 1 & 0 \\ 0 & 0 & 0 \end{bmatrix} \qquad \hat{\mathbf{b}} \times \bar{\mathbf{I}} = \begin{bmatrix} 0 & -1 & 0 \\ 1 & 0 & 0 \\ 0 & 0 & 0 \end{bmatrix}$$

and $\bar{\mathbf{A}}$ takes the matrix representation

$$\bar{\mathbf{A}} = \begin{bmatrix} a_\perp & -ia_\times & 0 \\ ia_\times & a_\perp & 0 \\ 0 & 0 & a_\parallel \end{bmatrix}$$

Using the results of Example 1.6, we find the determinant and adjoint of $\bar{\mathbf{A}} - \lambda\bar{\mathbf{I}}$:

$$
\begin{aligned}
|\bar{\mathbf{A}} - \lambda\bar{\mathbf{I}}| &= |(a_\perp - \lambda)\bar{\mathbf{I}} + ia_\times(\hat{\mathbf{b}} \times \bar{\mathbf{I}}) + (a_\parallel - a_\perp)\hat{\mathbf{b}}\hat{\mathbf{b}}| \\
&= (a_\perp - \lambda)^3 + (a_\parallel - a_\perp)(a_\perp - \lambda)^2 - a_\times^2(a_\perp - \lambda) - a_\times^2(a_\parallel - a_\perp) \\
&= (a_\perp + a_\times - \lambda)(a_\perp - a_\times - \lambda)(a_\parallel - \lambda) \\
\operatorname{adj}(\bar{\mathbf{A}} - \lambda\bar{\mathbf{I}}) &= \operatorname{adj}[(a_\perp - \lambda)\bar{\mathbf{I}} + ia_\times(\hat{\mathbf{b}} \times \bar{\mathbf{I}}) + (a_\parallel - a_\perp)\hat{\mathbf{b}}\hat{\mathbf{b}}] \\
&= (a_\parallel - \lambda)[(a_\perp - \lambda)\bar{\mathbf{I}} - ia_\times(\hat{\mathbf{b}} \times \bar{\mathbf{I}})] + [(a_\perp - \lambda)(a_\perp - a_\parallel) - a_\times^2]\hat{\mathbf{b}}\hat{\mathbf{b}}
\end{aligned}
$$

The eigenvalues of $\bar{\mathbf{A}}$ are, therefore,

$$\lambda_1 = a_\perp + a_\times \qquad \lambda_2 = a_\perp - a_\times \qquad \lambda_3 = a_\parallel$$

and are real as they should be. The adjoints of $\bar{\mathbf{A}} - \lambda\bar{\mathbf{I}}$ corresponding to λ_1, λ_2, and λ_3 are

$$
\begin{aligned}
\operatorname{adj}(\bar{\mathbf{A}} - \lambda_1\bar{\mathbf{I}}) &= \alpha[\bar{\mathbf{I}} - \hat{\mathbf{b}}\hat{\mathbf{b}} + i(\hat{\mathbf{b}} \times \bar{\mathbf{I}})] \\
\operatorname{adj}(\bar{\mathbf{A}} - \lambda_2\bar{\mathbf{I}}) &= \beta[\bar{\mathbf{I}} - \hat{\mathbf{b}}\hat{\mathbf{b}} - i(\hat{\mathbf{b}} \times \bar{\mathbf{I}})] \\
\operatorname{adj}(\bar{\mathbf{A}} - \lambda_3\bar{\mathbf{I}}) &= \gamma\hat{\mathbf{b}}\hat{\mathbf{b}}
\end{aligned}
$$

where

$$
\begin{aligned}
\alpha &= a_\times(a_\times + a_\perp - a_\parallel) \\
\beta &= a_\times(a_\times - a_\perp + a_\parallel) \\
\gamma &= (a_\perp - a_\parallel)^2 - a_\times^2
\end{aligned}
$$

Now if we choose $\hat{\mathbf{b}} = [0, 0, 1]^T$, the matrix representations of the adjoints become

$$\operatorname{adj}(\bar{\mathbf{A}} - \lambda_1\bar{\mathbf{I}}) = \alpha\begin{bmatrix} 1 & -i & 0 \\ i & 1 & 0 \\ 0 & 0 & 0 \end{bmatrix} = \begin{bmatrix} 1 \\ i \\ 0 \end{bmatrix}[\alpha, -i\alpha, 0]$$

$$\operatorname{adj}(\bar{\mathbf{A}} - \lambda_2\bar{\mathbf{I}}) = \beta\begin{bmatrix} 1 & i & 0 \\ -i & 1 & 0 \\ 0 & 0 & 0 \end{bmatrix} = \begin{bmatrix} 1 \\ -i \\ 0 \end{bmatrix}[\beta, i\beta, 0]$$

$$\operatorname{adj}(\bar{\mathbf{A}} - \lambda_3\bar{\mathbf{I}}) = \gamma\begin{bmatrix} 0 & 0 & 0 \\ 0 & 0 & 0 \\ 0 & 0 & 1 \end{bmatrix} = \begin{bmatrix} 0 \\ 0 \\ 1 \end{bmatrix}[0, 0, \gamma]$$

Since an eigenvector of $\bar{\mathbf{A}}$ is parallel to the vector $[\operatorname{adj}(\bar{\mathbf{A}} - \lambda\bar{\mathbf{I}})] \cdot \mathbf{c}$, we obtain the normalized eigenvectors corresponding to the eigenvalues $\lambda_1 = a_\perp + a_\times$, $\lambda_2 = a_\perp - a_\times$, and $\lambda_3 = a_\parallel$:

$$\hat{\mathbf{u}}_1 = \frac{1}{\sqrt{2}}\begin{bmatrix} 1 \\ i \\ 0 \end{bmatrix} \qquad \hat{\mathbf{u}}_2 = \frac{1}{\sqrt{2}}\begin{bmatrix} 1 \\ -i \\ 0 \end{bmatrix} \qquad \hat{\mathbf{u}}_3 = \begin{bmatrix} 0 \\ 0 \\ 1 \end{bmatrix}$$

Here we note that

$$\hat{\mathbf{u}}_i^* \cdot \hat{\mathbf{u}}_j = \delta_{ij}$$

and

$$\hat{\mathbf{u}}_1\hat{\mathbf{u}}_1^* + \hat{\mathbf{u}}_2\hat{\mathbf{u}}_2^* + \hat{\mathbf{u}}_3\hat{\mathbf{u}}_3^*$$

$$= \frac{1}{2}\begin{bmatrix} 1 \\ i \\ 0 \end{bmatrix}[1, -i, 0] + \frac{1}{2}\begin{bmatrix} 1 \\ -i \\ 0 \end{bmatrix}[1, i, 0] + \begin{bmatrix} 0 \\ 0 \\ 1 \end{bmatrix}[0, 0, 1]$$

$$= \frac{1}{2}\begin{bmatrix} 1 & -i & 0 \\ i & 1 & 0 \\ 0 & 0 & 0 \end{bmatrix} + \frac{1}{2}\begin{bmatrix} 1 & i & 0 \\ -i & 1 & 0 \\ 0 & 0 & 0 \end{bmatrix} + \begin{bmatrix} 0 & 0 & 0 \\ 0 & 0 & 0 \\ 0 & 0 & 1 \end{bmatrix} = \bar{\mathbf{I}}$$

$$\lambda_1\hat{\mathbf{u}}_1\hat{\mathbf{u}}_1^* + \lambda_2\hat{\mathbf{u}}_2\hat{\mathbf{u}}_2^* + \lambda_3\hat{\mathbf{u}}_3\hat{\mathbf{u}}_3^*$$

$$= \frac{a_\perp + a_\times}{2}\begin{bmatrix} 1 & -i & 0 \\ i & 1 & 0 \\ 0 & 0 & 0 \end{bmatrix} + \frac{a_\perp - a_\times}{2}\begin{bmatrix} 1 & i & 0 \\ -i & 1 & 0 \\ 0 & 0 & 0 \end{bmatrix} + a_\parallel\begin{bmatrix} 0 & 0 & 0 \\ 0 & 0 & 0 \\ 0 & 0 & 1 \end{bmatrix}$$

$$= \bar{\mathbf{A}}$$

which verify the completeness relation and the dyadic decomposition of the matrix $\bar{\mathbf{A}}$.

PROBLEMS

1.1 The common vector operations in direct and index notations are summarized below. **u** and **v** are two arbitrary vector functions of $\mathbf{r} = [x_i]$; ϕ and Ψ are two scalar functions of **r**.

The dot product of two vectors:

$$\mathbf{u} \cdot \mathbf{v} \leftrightarrow u_i v_i$$

The dyadic product of two vectors:

$$\mathbf{u}\mathbf{v} \leftrightarrow u_i v_j$$

The cross product of two vectors:

$$\mathbf{u} \times \mathbf{v} \leftrightarrow \varepsilon_{ijk} u_j v_k$$

The del operator:

$$\nabla = \frac{\partial}{\partial \mathbf{r}} \leftrightarrow \frac{\partial}{\partial x_i} = \partial_i$$

The gradient of a scalar function:

$$\nabla \phi \leftrightarrow \frac{\partial \phi}{\partial x_i} = \partial_i \phi$$

The divergence of a vector function:

$$\nabla \cdot \mathbf{u} \leftrightarrow \frac{\partial u_i}{\partial x_i} = \partial_i u_i$$

The curl of a vector function:

$$\nabla \times \mathbf{u} \leftrightarrow \varepsilon_{ijk} \frac{\partial u_k}{\partial x_j} = \varepsilon_{ijk} \partial_j u_k$$

Use index notation to prove the following vector identities:

(*a*) $\nabla \times \nabla \phi = \mathbf{0}$

(*b*) $\nabla \cdot (\nabla \times \mathbf{u}) = 0$

(*c*) $\nabla(\phi \Psi) = \phi \nabla \Psi + \Psi \nabla \phi$

(*d*) $\nabla \cdot (\phi \mathbf{u}) = \mathbf{u} \cdot \nabla \phi + \phi \nabla \cdot \mathbf{u}$

(*e*) $\nabla \times (\phi \mathbf{u}) = \nabla \phi \times \mathbf{u} + \phi \nabla \times \mathbf{u}$

(*f*) $\nabla \cdot (\mathbf{u} \times \mathbf{v}) = \mathbf{v} \cdot (\nabla \times \mathbf{u}) - \mathbf{u} \cdot (\nabla \times \mathbf{v})$

(*g*) $\nabla \times (\nabla \times \mathbf{u}) = \nabla(\nabla \cdot \mathbf{u}) - \nabla^2 \mathbf{u}$

1.2 Let $\mathbf{R} = \mathbf{r} - \mathbf{r}'$ be the *distance vector* drawn from $\mathbf{r}'$ to $\mathbf{r}$ as shown in Fig. 1.1, and let $R = |\mathbf{R}|$ be its magnitude and $\hat{\mathbf{R}} = \mathbf{R}/R$ be the unit vector in the direction of $\mathbf{R}$. If $f(R)$ is any scalar function of the magnitude R and $\mathbf{a}$ is a constant vector. Show that:

(*a*) $\nabla f(R) = \dfrac{df(R)}{dR} \hat{\mathbf{R}}$

(*b*) $\nabla R^n = nR^{n-1} \hat{\mathbf{R}} \qquad \nabla \dfrac{1}{R} = -\dfrac{\hat{\mathbf{R}}}{R^2}$

(*c*) $\nabla \dfrac{e^{ikR}}{R} = \left(ik - \dfrac{1}{R}\right) \dfrac{e^{ikR}}{R} \hat{\mathbf{R}}$

(*d*) $\nabla \cdot \hat{\mathbf{R}} = \dfrac{2}{R} \qquad \nabla \cdot \mathbf{R} = 3 \qquad \nabla \cdot (R\mathbf{a}) = \mathbf{a} \cdot \hat{\mathbf{R}}$

(*e*) $\nabla \cdot [f(R)\hat{\mathbf{R}}] = \dfrac{df(R)}{dR} + \dfrac{2f(R)}{R}$

(*f*) $\nabla \cdot (R^n \hat{\mathbf{R}}) = (n+2)R^{n-1}$

(*g*) $\nabla \cdot \left(\dfrac{e^{ikR}}{R} \hat{\mathbf{R}}\right) = \left(ik + \dfrac{1}{R}\right) \dfrac{e^{ikR}}{R}$

(*h*) $(\mathbf{a} \cdot \nabla) R = \mathbf{a} \cdot \hat{\mathbf{R}} \qquad (\mathbf{a} \cdot \nabla) \mathbf{R} = \mathbf{a}$

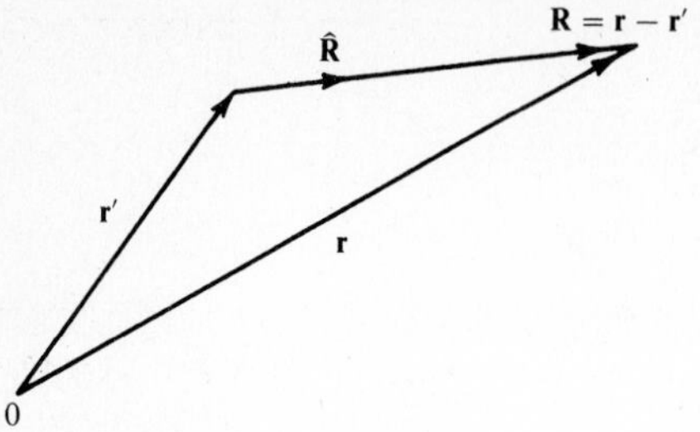

Figure 1.1 Distance vector $\mathbf{R} = \mathbf{r} - \mathbf{r}'$.

(*i*) $\nabla^2 f(R) = \dfrac{d^2 f(R)}{dR^2} + \dfrac{2}{R}\dfrac{df}{dR}$

(*j*) $\nabla^2 R^n = n(n+1)R^{n-2}$

(*k*) $\nabla \times [f(R)\hat{\mathbf{R}}] = 0 \qquad \nabla \times \mathbf{R} = 0 \qquad \nabla \times \hat{\mathbf{R}} = 0$

(*l*) $\nabla \times (R\mathbf{a}) = \hat{\mathbf{R}} \times \mathbf{a}$

(*m*) $\nabla \times (\mathbf{a} \times \mathbf{R}) = 2\mathbf{a}$

1.3 Show that

$$(\mathbf{u} \times \bar{\mathbf{I}}) \cdot (\mathbf{v} \times \bar{\mathbf{I}}) \cdot (\mathbf{w} \times \bar{\mathbf{I}}) = (\mathbf{u} \times \mathbf{w})\mathbf{v} - (\mathbf{w} \cdot \mathbf{v})(\mathbf{u} \times \bar{\mathbf{I}})$$
$$= \mathbf{v}(\mathbf{u} \times \mathbf{w}) - (\mathbf{v} \cdot \mathbf{u})(\mathbf{w} \times \bar{\mathbf{I}})$$

1.4 Show that

$$(\bar{\mathbf{A}} \cdot \mathbf{a}) \times (\bar{\mathbf{A}} \cdot \mathbf{b}) \cdot (\bar{\mathbf{A}} \cdot \mathbf{c}) = |\bar{\mathbf{A}}|\,(\mathbf{a} \times \mathbf{b} \cdot \mathbf{c})$$

and hence prove that

$$|\bar{\mathbf{A}} \cdot \bar{\mathbf{B}}| = |\bar{\mathbf{A}}|\,|\bar{\mathbf{B}}|$$

1.5 Let $\bar{\mathbf{A}} = \hat{\mathbf{a}} \times \bar{\mathbf{I}}$ and $\hat{\mathbf{a}}$ be a unit vector. Show that

$$\bar{\mathbf{A}}^2 = \hat{\mathbf{a}}\hat{\mathbf{a}} - \bar{\mathbf{I}} \qquad \bar{\mathbf{A}}^3 = -\bar{\mathbf{A}} \qquad \bar{\mathbf{A}}^4 = \bar{\mathbf{I}} - \hat{\mathbf{a}}\hat{\mathbf{a}} \qquad \bar{\mathbf{A}}^5 = \mathbf{A}$$

1.6 Show that

(*a*) $|\lambda\bar{\mathbf{I}} + \mathbf{al} + \mathbf{bm} + \mathbf{cn}| = \lambda^3 + (\mathbf{a} \cdot \mathbf{l} + \mathbf{b} \cdot \mathbf{m} + \mathbf{c} \cdot \mathbf{n})\lambda^2$

$+ [(\mathbf{m} \times \mathbf{n}) \cdot (\mathbf{b} \times \mathbf{c}) + (\mathbf{n} \times \mathbf{l}) \cdot (\mathbf{c} \times \mathbf{a}) + (\mathbf{l} \times \mathbf{m}) \cdot (\mathbf{a} \times \mathbf{b})]\lambda$

$+ (\mathbf{a} \cdot \mathbf{b} \times \mathbf{c})(\mathbf{l} \cdot \mathbf{m} \times \mathbf{n})$

(*b*) $\operatorname{adj}(\lambda\bar{\mathbf{I}} + \mathbf{al} + \mathbf{bm} + \mathbf{cn})$

$= \lambda[(\lambda + \mathbf{a} \cdot \mathbf{l} + \mathbf{b} \cdot \mathbf{m} + \mathbf{c} \cdot \mathbf{n})\bar{\mathbf{I}} - (\mathbf{al} + \mathbf{bm} + \mathbf{cn})]$

$+ (\mathbf{m} \times \mathbf{n})(\mathbf{b} \times \mathbf{c}) + (\mathbf{n} \times \mathbf{l})(\mathbf{c} \times \mathbf{a}) + (\mathbf{l} \times \mathbf{m})(\mathbf{a} \times \mathbf{b})$

1.7 Show that

(*a*) $|\lambda\bar{\mathbf{I}} + \mathbf{c} \times \bar{\mathbf{I}} + \mathbf{al} + \mathbf{bm}| = \lambda^3 + (\mathbf{a} \cdot \mathbf{l} + \mathbf{b} \cdot \mathbf{m})\lambda^2$

$+ [\mathbf{c}^2 + (\mathbf{l} \times \mathbf{m}) \cdot (\mathbf{a} \times \mathbf{b}) - \mathbf{a} \cdot (\mathbf{l} \times \mathbf{c}) - \mathbf{b} \cdot (\mathbf{m} \times \mathbf{c})]\lambda$

$+ (\mathbf{l} \cdot \mathbf{c})(\mathbf{a} \cdot \mathbf{c}) + (\mathbf{m} \cdot \mathbf{c})(\mathbf{b} \cdot \mathbf{c}) + (\mathbf{a} \cdot \mathbf{c})(\mathbf{l} \times \mathbf{m} \cdot \mathbf{b}) - (\mathbf{b} \cdot \mathbf{c})(\mathbf{l} \times \mathbf{m} \cdot \mathbf{a})$

(*b*) $\operatorname{adj}(\lambda\bar{\mathbf{I}} + \mathbf{c} \times \bar{\mathbf{I}} + \mathbf{al} + \mathbf{bm})$

$$= [\lambda^2 + (\mathbf{a}\cdot\mathbf{l} + \mathbf{b}\cdot\mathbf{m})\lambda + \mathbf{b}\times\mathbf{c}\cdot\mathbf{m} + \mathbf{a}\times\mathbf{c}\cdot\mathbf{l}]\bar{\mathbf{I}}$$
$$- (\lambda + \mathbf{a}\cdot\mathbf{l} + \mathbf{b}\cdot\mathbf{m})(\mathbf{c}\times\bar{\mathbf{I}}) - \lambda(\mathbf{al} + \mathbf{bm})$$
$$+ (\mathbf{l}\times\mathbf{m})(\mathbf{a}\times\mathbf{b}) + \mathbf{cc} + (\mathbf{c}\times\mathbf{a})\mathbf{l} + (\mathbf{c}\times\mathbf{b})\mathbf{m}$$
$$+ \mathbf{a}(\mathbf{l}\times\mathbf{c}) + \mathbf{b}(\mathbf{m}\times\mathbf{c})$$

1.8 If $\bar{\mathbf{A}}$ and $\bar{\mathbf{B}}$ are matrices and **u** and **v** are vectors, show that

(*a*) $(\bar{\mathbf{A}}\times\mathbf{u})^T = -\mathbf{u}\times\tilde{\bar{\mathbf{A}}}$ $\quad(\mathbf{u}\times\bar{\mathbf{A}})^T = -\tilde{\bar{\mathbf{A}}}\times\mathbf{u}$

(*b*) $\mathbf{u}\cdot(\bar{\mathbf{A}}\times\mathbf{v}) = (\mathbf{u}\cdot\bar{\mathbf{A}})\times\mathbf{v}$ $\quad(\mathbf{u}\times\bar{\mathbf{A}})\cdot\mathbf{v} = \mathbf{u}\times(\bar{\mathbf{A}}\cdot\mathbf{v})$

(*c*) $(\bar{\mathbf{A}}\times\mathbf{u})\times\mathbf{v} = (\bar{\mathbf{A}}\cdot\mathbf{v})\mathbf{u} - (\mathbf{u}\cdot\mathbf{v})\bar{\mathbf{A}}$

$\mathbf{u}\times(\mathbf{v}\times\bar{\mathbf{A}}) = \mathbf{v}(\mathbf{u}\cdot\bar{\mathbf{A}}) - (\mathbf{u}\cdot\mathbf{v})\bar{\mathbf{A}}$

(*d*) $\bar{\mathbf{A}}\times(\mathbf{u}\times\mathbf{v}) = (\bar{\mathbf{A}}\cdot\mathbf{v})\mathbf{u} - (\bar{\mathbf{A}}\cdot\mathbf{u})\mathbf{v}$

$(\mathbf{u}\times\mathbf{v})\times\bar{\mathbf{A}} = \mathbf{v}(\mathbf{u}\cdot\bar{\mathbf{A}}) - \mathbf{u}(\mathbf{v}\cdot\bar{\mathbf{A}})$

(*e*) $(\mathbf{u}\times\bar{\mathbf{A}})\cdot\bar{\mathbf{B}} = \mathbf{u}\times(\bar{\mathbf{A}}\cdot\bar{\mathbf{B}})$ $\quad\bar{\mathbf{A}}\cdot(\bar{\mathbf{B}}\times\mathbf{u}) = (\bar{\mathbf{A}}\cdot\bar{\mathbf{B}})\times\mathbf{u}$

(*f*) $(\bar{\mathbf{A}}\times\mathbf{u})\cdot\bar{\mathbf{B}} = \bar{\mathbf{A}}\cdot(\mathbf{u}\times\bar{\mathbf{B}})$

1.9 Let $\mathbf{r} = x_i\hat{\mathbf{e}}_i$, $r = |\mathbf{r}| = (x_1^2 + x_2^2 + x_3^2)^{1/2}$ and $\hat{\mathbf{r}} = \mathbf{r}/r$. Using the definitions of the gradient of a scalar function $f(\mathbf{r})$:

$$\nabla f = \frac{\partial f}{\partial \mathbf{r}} = \left[\frac{\partial f}{\partial x_i}\right]$$

and the gradient of a vector function $\mathbf{g}(\mathbf{r})$:

$$\nabla \mathbf{g} = \frac{\partial \mathbf{g}}{\partial \mathbf{r}} = \left[\frac{\partial g_i}{\partial x_j}\right]$$

show that

(*a*) $\nabla r = \hat{\mathbf{r}}$

(*b*) $\nabla(\mathbf{c}\cdot\mathbf{r}) = \mathbf{c}$

(*c*) $\nabla(\mathbf{r}\cdot\bar{\mathbf{C}}\cdot\mathbf{r}) = (\bar{\mathbf{C}} + \tilde{\bar{\mathbf{C}}})\cdot\mathbf{r}$

(*d*) $\nabla\mathbf{r} = \bar{\mathbf{I}}$

(*e*) $\nabla(\mathbf{r}\times\mathbf{c}) = \bar{\mathbf{I}}\times\mathbf{c}$

(*f*) $\nabla\hat{\mathbf{r}} = \dfrac{1}{r}(\bar{\mathbf{I}} - \hat{\mathbf{r}}\hat{\mathbf{r}})$

(*g*) $\nabla\nabla\dfrac{1}{r} = \dfrac{3\hat{\mathbf{r}}\hat{\mathbf{r}} - \bar{\mathbf{I}}}{r^3}$

(*h*) $\nabla(\bar{\mathbf{C}}\cdot\mathbf{r}) = \bar{\mathbf{C}}$

where $\bar{\mathbf{C}}$ is a constant matrix and **c** is a constant vector.

1.10 Let ϕ, **u**, **v**, and $\bar{\mathbf{A}} = [a_{ij}]$ be functions of **r**, and let **c** and $\bar{\mathbf{C}}$ be constant vector and matrix respectively. The divergence and the curl of a matrix are defined by

$$\nabla\cdot\bar{\mathbf{A}} \leftrightarrow \partial_i a_{ij}$$

and

$$\nabla \times \bar{\mathbf{A}} \leftrightarrow \varepsilon_{ijk}\,\partial_j\,a_{kl}$$

respectively. Show that

(a) $\nabla \cdot (\mathbf{uv}) = (\nabla \cdot \mathbf{u})\mathbf{v} + (\mathbf{u} \cdot \nabla)\mathbf{v}$

(b) $\nabla \cdot (\phi\bar{\mathbf{A}}) = \nabla\phi \cdot \bar{\mathbf{A}} + \phi\nabla \cdot \bar{\mathbf{A}}$

$\nabla \cdot (\phi\bar{\mathbf{I}}) = \nabla\phi$

(c) $\nabla \cdot (\mathbf{vu} - \mathbf{uv}) = \nabla \times (\mathbf{u} \times \mathbf{v})$

(d) $\nabla \cdot (\bar{\mathbf{A}} \cdot \mathbf{u}) = (\nabla \cdot \bar{\mathbf{A}}) \cdot \mathbf{u} + (\bar{\mathbf{A}} \cdot \nabla\mathbf{u})_t$

$\nabla^2(\mathbf{r} \cdot \bar{\mathbf{C}} \cdot \mathbf{r}) = 2\bar{\mathbf{C}}_t \qquad \nabla \cdot (\bar{\mathbf{C}} \cdot \mathbf{r}) = \bar{\mathbf{C}}_t$

(e) $\nabla \cdot (\mathbf{u} \times \bar{\mathbf{A}}) = (\nabla \times \mathbf{u}) \cdot \bar{\mathbf{A}} - \mathbf{u} \cdot (\nabla \times \bar{\mathbf{A}})$

$\nabla \cdot (\bar{\mathbf{I}} \times \mathbf{u}) = \nabla \times \mathbf{u}$

(f) $\nabla \cdot \nabla \times \bar{\mathbf{A}} = 0$

(g) $\nabla(\phi\mathbf{u}) = (\nabla\phi)\mathbf{u} + \phi\nabla\mathbf{u}$

(h) $\nabla(\mathbf{u} \cdot \mathbf{v}) = (\nabla\mathbf{u}) \cdot \mathbf{v} + (\nabla\mathbf{v}) \cdot \mathbf{u}$

(i) $\mathbf{u} \times (\nabla \times \mathbf{v}) = (\nabla\mathbf{v}) \cdot \mathbf{u} - (\mathbf{u} \cdot \nabla)\mathbf{v}$

(j) $(\mathbf{u} \times \nabla) \times \mathbf{v} = (\nabla\mathbf{v}) \cdot \mathbf{u} - \mathbf{u}(\nabla \cdot \mathbf{v})$

(k) $\nabla \times (\mathbf{uv}) = (\nabla \times \mathbf{u})\mathbf{v} - \mathbf{u} \times \nabla\mathbf{v}$

$\nabla \times (\nabla\mathbf{u}) = 0$

(l) $\nabla(\mathbf{u} \times \mathbf{v}) = (\nabla\mathbf{u}) \times \mathbf{v} - (\nabla\mathbf{v}) \times \mathbf{u}$

(m) $\nabla \times (\phi\bar{\mathbf{A}}) = \nabla\phi \times \bar{\mathbf{A}} + \phi\nabla \times \bar{\mathbf{A}}$

$\nabla \times (\phi\bar{\mathbf{I}}) = \nabla\phi \times \bar{\mathbf{I}}$

(n) $\nabla \times (\bar{\mathbf{A}} \cdot \mathbf{c}) = (\nabla \times \bar{\mathbf{A}}) \cdot \mathbf{c}$

(o) $\bar{\mathbf{C}} \cdot (\nabla \times \mathbf{u}) = \nabla \cdot (\mathbf{u} \times \bar{\mathbf{C}})$

1.11 Find the eigenvalues and the normalized eigenvectors of the following matrices:

(a)

$$\bar{\mathbf{A}} = \begin{bmatrix} 2 & -2 & 3 \\ 1 & 1 & 1 \\ 1 & 3 & -1 \end{bmatrix}$$

(b)

$$\bar{\mathbf{A}} = \begin{bmatrix} \cos\theta & -\sin\theta & 0 \\ \sin\theta & \cos\theta & 0 \\ 0 & 0 & 1 \end{bmatrix}$$

(c)

$$\bar{\mathbf{A}} = \begin{bmatrix} \alpha & \beta & \beta \\ \beta & \alpha & \beta \\ \beta & \beta & \alpha \end{bmatrix}$$

where α and $\beta \neq 0$ are scalars.

1.12 For a given biaxial matrix of the form

$$\bar{\mathbf{A}} = \alpha\bar{\mathbf{I}} + \beta(\hat{\mathbf{m}}\hat{\mathbf{n}} + \hat{\mathbf{n}}\hat{\mathbf{m}})$$

where α and β are scalars and $\hat{\mathbf{m}}$ and $\hat{\mathbf{n}}$ are unit vectors, show that

(*a*) $\bar{\mathbf{A}}_t = 3\alpha + 2\beta(\hat{\mathbf{m}} \cdot \hat{\mathbf{n}})$

(*b*) $|\bar{\mathbf{A}}| = \alpha[(\alpha + \beta\hat{\mathbf{m}} \cdot \hat{\mathbf{n}})^2 - \beta^2]$

(*c*) $\text{adj } \bar{\mathbf{A}} = \alpha[(\alpha + 2\beta\hat{\mathbf{m}} \cdot \hat{\mathbf{n}})\bar{\mathbf{I}} - \beta(\hat{\mathbf{m}}\hat{\mathbf{n}} + \hat{\mathbf{n}}\hat{\mathbf{m}})] - \beta^2(\hat{\mathbf{m}} \times \hat{\mathbf{n}})(\hat{\mathbf{m}} \times \hat{\mathbf{n}})$

(*d*) $(\text{adj } \bar{\mathbf{A}})_t = 3\alpha^2 + 4\alpha\beta(\hat{\mathbf{m}} \cdot \hat{\mathbf{n}}) - \beta^2(\hat{\mathbf{m}} \times \hat{\mathbf{n}})^2$

(*e*) the eigenvalues and the corresponding eigenvectors of $\bar{\mathbf{A}}$ are

$$\lambda_1 = \alpha \qquad \mathbf{u}_1 = \hat{\mathbf{m}} \times \hat{\mathbf{n}}$$

$$\lambda_2 = \alpha + \beta(1 + \hat{\mathbf{m}} \cdot \hat{\mathbf{n}}) \qquad \mathbf{u}_2 = \hat{\mathbf{m}} + \hat{\mathbf{n}}$$

$$\lambda_3 = \alpha - \beta(1 - \hat{\mathbf{m}} \cdot \hat{\mathbf{n}}) \qquad \mathbf{u}_3 = \hat{\mathbf{m}} - \hat{\mathbf{n}}$$

1.13 For a given uniaxial matrix of the form

$$\bar{\mathbf{A}} = \alpha\bar{\mathbf{I}} + \beta\hat{\mathbf{c}}\hat{\mathbf{c}}$$

where α and β are scalars and $\hat{\mathbf{c}}$ is a unit vector, show that

(*a*) $\bar{\mathbf{A}}_t = 3\alpha + \beta$

(*b*) $|\bar{\mathbf{A}}| = \alpha^2(\alpha + \beta)$

(*c*) $\text{adj } \bar{\mathbf{A}} = \alpha[(\alpha + \beta)\bar{\mathbf{I}} - \beta\hat{\mathbf{c}}\hat{\mathbf{c}}] = \alpha[\alpha\bar{\mathbf{I}} - \beta(\hat{\mathbf{c}} \times \bar{\mathbf{I}})^2]$

(*d*) $(\text{adj } \bar{\mathbf{A}})_t = \alpha(3\alpha + 2\beta)$

(*e*) the eigenvalues and the corresponding eigenvectors of $\bar{\mathbf{A}}$ are

$$\lambda_1 = \alpha \qquad \mathbf{u}_1 = \text{any vector perpendicular to } \hat{\mathbf{c}}$$

$$\lambda_3 = \alpha + \beta \qquad \mathbf{u}_3 = \hat{\mathbf{c}}$$

$$\lambda_2 = \alpha \qquad \mathbf{u}_2 = \hat{\mathbf{c}} \times \mathbf{u}_1$$

1.14 (*a*) Prove the *Cayley-Hamilton theorem* which states that every matrix $\bar{\mathbf{A}}$ satisfies its own characteristic equation. That is, if λ in the characteristic equation

$$\lambda^3 - \bar{\mathbf{A}}_t \lambda^2 + (\text{adj } \bar{\mathbf{A}})_t \lambda - |\bar{\mathbf{A}}| = 0$$

is replaced by the matrix $\bar{\mathbf{A}}$, then $\bar{\mathbf{A}}$ satisfies the polynomial equation

$$\bar{\mathbf{A}}^3 - \bar{\mathbf{A}}_t \bar{\mathbf{A}}^2 + (\text{adj } \bar{\mathbf{A}})_t \bar{\mathbf{A}} - |\bar{\mathbf{A}}|\, \bar{\mathbf{I}} = \bar{\mathbf{0}}$$

Hint: Use Eq. (1.64)

(*b*) Use the result of the Cayley-Hamilton theorem to obtain the determinant (1.57) and the adjoint (1.64) and its trace of the matrix $\bar{\mathbf{A}}$ in direct forms.

1.15 Let λ_1, λ_2, and λ_3 be the three distinct eigenvalues of the matrix $\bar{\mathbf{A}}$ and $\mathbf{u}_1$, $\mathbf{u}_2$, and $\mathbf{u}_3$ be the corresponding eigenvectors. Define the *projectors* by

$$\bar{\mathbf{P}}_1 = \frac{(\bar{\mathbf{A}} - \lambda_2\bar{\mathbf{I}}) \cdot (\bar{\mathbf{A}} - \lambda_3\bar{\mathbf{I}})}{(\lambda_1 - \lambda_2)(\lambda_1 - \lambda_3)}$$

$$\bar{\mathbf{P}}_2 = \frac{(\bar{\mathbf{A}} - \lambda_1\bar{\mathbf{I}}) \cdot (\bar{\mathbf{A}} - \lambda_3\bar{\mathbf{I}})}{(\lambda_2 - \lambda_1)(\lambda_2 - \lambda_3)}$$

$$\bar{\mathbf{P}}_3 = \frac{(\bar{\mathbf{A}} - \lambda_1\bar{\mathbf{I}}) \cdot (\bar{\mathbf{A}} - \lambda_2\bar{\mathbf{I}})}{(\lambda_3 - \lambda_1)(\lambda_3 - \lambda_2)}$$

and show that

(*a*) $\bar{\mathbf{P}}_i$ projects any three-dimensional vector in the direction of the eigenvector $\mathbf{u}_i$;

(*b*)

$$\bar{\mathbf{P}}_i^2 = \bar{\mathbf{P}}_i$$

$$\bar{\mathbf{P}}_i \cdot \bar{\mathbf{P}}_j = \bar{0}$$

$$\bar{\mathbf{P}}_1 + \bar{\mathbf{P}}_2 + \bar{\mathbf{P}}_3 = \bar{\mathbf{I}}$$

(*c*) if $f(\bar{\mathbf{A}})$ is any polynomial function of the matrix $\bar{\mathbf{A}}$, then

$$f(\bar{\mathbf{A}}) = f(\lambda_1)\bar{\mathbf{P}}_1 + f(\lambda_2)\bar{\mathbf{P}}_2 + f(\lambda_3)\bar{\mathbf{P}}_3$$

As some special cases:

If $f(\bar{\mathbf{A}}) = \bar{\mathbf{I}}$, then

$$\bar{\mathbf{I}} = \bar{\mathbf{P}}_1 + \bar{\mathbf{P}}_2 + \bar{\mathbf{P}}_3$$

If $f(\bar{\mathbf{A}}) = \bar{\mathbf{A}}$, then

$$\bar{\mathbf{A}} = \lambda_1 \bar{\mathbf{P}}_1 + \lambda_2 \bar{\mathbf{P}}_2 + \lambda_3 \bar{\mathbf{P}}_3$$

If $f(\bar{\mathbf{A}}) = \operatorname{adj} \bar{\mathbf{A}}$, then

$$\operatorname{adj} \bar{\mathbf{A}} = \lambda_2 \lambda_3 \bar{\mathbf{P}}_1 + \lambda_1 \lambda_3 \bar{\mathbf{P}}_2 + \lambda_1 \lambda_2 \bar{\mathbf{P}}_3$$

1.16 Show that:

(*a*) the diagonal elements, the trace, and the determinant of a hermitian matrix are real.

(*b*) the adjoint and the inverse of a hermitian matrix are hermitian.

(*c*) the diagonal elements, the trace, and the determinant of an antihermitian matrix are purely imaginary.

1.17 Show that any complex matrix $\bar{\mathbf{A}}$ can always be expressed uniquely as:

(*a*) the sum of a real and a purely imaginary matrix;

(*b*) the sum of a symmetric and an antisymmetric matrix;

(*c*) the sum of a hermitian and an antihermitian matrix.

1.18 Show that:

(*a*) $i\bar{\mathbf{A}}$ is antihermitian if $\bar{\mathbf{A}}$ is hermitian, and $i\bar{\mathbf{A}}$ is hermitian if $\bar{\mathbf{A}}$ is antihermitian;

(*b*) if $\bar{\mathbf{A}}$ and $\bar{\mathbf{B}}$ are hermitian matrices, then the matrices $\bar{\mathbf{A}} \pm \bar{\mathbf{B}}$ and $\bar{\mathbf{A}} \cdot \bar{\mathbf{B}} + \bar{\mathbf{B}} \cdot \bar{\mathbf{A}}$ are also hermitian; the product $\bar{\mathbf{A}} \cdot \bar{\mathbf{B}}$ is hermitian if and only if $\bar{\mathbf{A}}$ and $\bar{\mathbf{B}}$ commute.

1.19 Consider the *general eigenvalue problem*:

$$\bar{\mathbf{A}} \cdot \mathbf{a} = \lambda \bar{\mathbf{B}} \cdot \mathbf{a}$$

where $\bar{\mathbf{A}}$ and $\bar{\mathbf{B}}$ are either hermitian or real and symmetric matrices. Let $\mathbf{a}_1$, $\mathbf{a}_2$, and $\mathbf{a}_3$ be the eigenvectors corresponding, respectively, to the distinct eigenvalues λ_1, λ_2, and λ_3. Show that

(*a*) $\lambda = 0$ is an eigenvalue if and only if $|\bar{\mathbf{A}}| = 0$;

(*b*) if $\bar{\mathbf{B}}$ is definite (that is, $\mathbf{a}^* \cdot \bar{\mathbf{B}} \cdot \mathbf{a} \neq 0$), then the eigenvalues are real;

(*c*) if $\bar{\mathbf{A}}$ and $\bar{\mathbf{B}}$ are both positive-definite (that is, $\mathbf{a}^* \cdot \bar{\mathbf{A}} \cdot \mathbf{a} > 0$, $\mathbf{a}^* \cdot \bar{\mathbf{B}} \cdot \mathbf{a} > 0$ for all $\mathbf{a} \neq \mathbf{0}$) or both negative-definite (that is, $\mathbf{a}^* \cdot \bar{\mathbf{A}} \cdot \mathbf{a} < 0$, $\mathbf{a}^* \cdot \bar{\mathbf{B}} \cdot \mathbf{a} < 0$ for all $\mathbf{a} \neq \mathbf{0}$), then the eigenvalues are all positive;

(*d*) the two vectors $\mathbf{a}_i$ and $\mathbf{c}_j$ are orthogonal in the sense that $\mathbf{a}_i^* \cdot \mathbf{c}_j = N_i^2 \delta_{ij}$ where $\mathbf{c}_j = \bar{\mathbf{B}} \cdot \mathbf{a}_j$ and $N_i^2 = \mathbf{a}_i^* \cdot \mathbf{c}_i$ (no sum);

(*e*) any two eigenvectors $\mathbf{a}_i$ and $\mathbf{a}_j$ are orthogonal relative to $\bar{\mathbf{A}}$, that is, $\mathbf{a}_i^* \cdot \bar{\mathbf{A}} \cdot \mathbf{a}_j = \lambda_i N_i^2 \delta_{ij}$ (no sum);

(*f*) the completeness relation is

$$\bar{\mathbf{I}} = \frac{\mathbf{a}_1 \mathbf{c}_1^*}{N_1^2} + \frac{\mathbf{a}_2 \mathbf{c}_2^*}{N_2^2} + \frac{\mathbf{a}_3 \mathbf{c}_3^*}{N_3^2}$$

$$= \frac{\mathbf{c}_1 \mathbf{a}_1^*}{N_1^2} + \frac{\mathbf{c}_2 \mathbf{a}_2^*}{N_2^2} + \frac{\mathbf{c}_3 \mathbf{a}_3^*}{N_3^2}$$

(*g*) if $\lambda \neq \lambda_i$ for $i = 1, 2, 3$, then

$$(\bar{\mathbf{A}} - \lambda\bar{\mathbf{B}})^{-1} = \frac{\mathbf{a}_1\,\mathbf{a}_1^*}{(\lambda_1 - \lambda)N_1^2} + \frac{\mathbf{a}_2\,\mathbf{a}_2^*}{(\lambda_2 - \lambda)N_2^2} + \frac{\mathbf{a}_3\,\mathbf{a}_3^*}{(\lambda_3 - \lambda)N_3^2}$$

1.20 Consider an arbitrary vector $\mathbf{u}$, which is to undergo a rotation through the angle θ about an axis along the unit vector $\hat{\mathbf{c}}$ as shown in Fig. 1.2. Let $\mathbf{u}'$ denote the resulting vector. Show that

$$\mathbf{u}' = \bar{\mathbf{R}}(\theta) \cdot \mathbf{u}$$

where

$$\bar{\mathbf{R}}(\theta) = \cos\theta\,\bar{\mathbf{I}} + (1 - \cos\theta)\hat{\mathbf{c}}\hat{\mathbf{c}} + \sin\theta(\hat{\mathbf{c}} \times \bar{\mathbf{I}})$$

Next, verify the following properties:

(*a*) $\bar{\mathbf{R}}(0) = \bar{\mathbf{I}}$

(*b*) $|\bar{\mathbf{R}}(\theta)| = 1$

(*c*) $\bar{\mathbf{R}}(\theta) \cdot \hat{\mathbf{c}} = \hat{\mathbf{c}}$

(*d*) $\bar{\mathbf{R}}(\theta) \cdot \tilde{\bar{\mathbf{R}}}(\theta) = \tilde{\bar{\mathbf{R}}}(\theta) \cdot \bar{\mathbf{R}}(\theta) = \bar{\mathbf{I}}$, hence $\bar{\mathbf{R}}^{-1}(\theta) = \tilde{\bar{\mathbf{R}}}(\theta) = \bar{\mathbf{R}}(-\theta)$ that is, $\bar{\mathbf{R}}(\theta)$ is orthogonal

(*e*) $\bar{\mathbf{R}}(\theta) \cdot \bar{\mathbf{R}}(\phi) = \bar{\mathbf{R}}(\theta + \phi)$

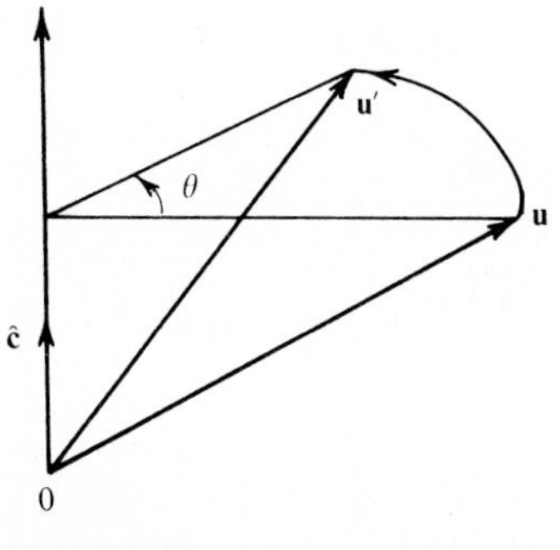

Figure 1.2 Rotation of the vector $\mathbf{u}$ about the axis $\hat{\mathbf{c}}$, through the angle θ, to yield the new vector $\mathbf{u}'$.

1.21 Let $(\hat{\mathbf{x}}, \hat{\mathbf{y}}, \hat{\mathbf{z}})$, $(\hat{\boldsymbol{\rho}}, \hat{\boldsymbol{\phi}}, \hat{\mathbf{z}})$, and $(\hat{\mathbf{r}}, \hat{\boldsymbol{\theta}}, \hat{\boldsymbol{\phi}})$ be the unit vectors of the rectangular (or cartesian), cylindrical, and spherical coordinate systems respectively. Show that

$$\hat{\boldsymbol{\rho}} = \cos\phi\,\hat{\mathbf{x}} + \sin\phi\,\hat{\mathbf{y}}$$

$$\hat{\boldsymbol{\phi}} = -\sin\phi\,\hat{\mathbf{x}} + \cos\phi\,\hat{\mathbf{y}}$$

$$\hat{\mathbf{z}} = \hat{\mathbf{z}}$$

and

$$\hat{\mathbf{r}} = \sin\theta\cos\phi\,\hat{\mathbf{x}} + \sin\theta\sin\phi\,\hat{\mathbf{y}} + \cos\theta\,\hat{\mathbf{z}}$$

$$\hat{\boldsymbol{\theta}} = \cos\theta\cos\phi\,\hat{\mathbf{x}} + \cos\theta\sin\phi\,\hat{\mathbf{y}} - \sin\theta\,\hat{\mathbf{z}}$$

$$\hat{\boldsymbol{\phi}} = -\sin\phi\,\hat{\mathbf{x}} + \cos\phi\,\hat{\mathbf{y}}$$

where θ and ϕ are the angles as shown in Fig. 1.3.

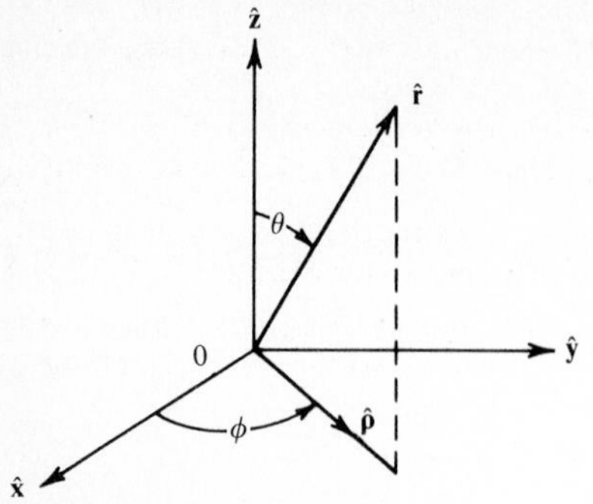

Figure 1.3 Orientations of unit vectors $\hat{\boldsymbol{\rho}}$ and $\hat{\mathbf{r}}$.

1.22 With respect to the rectangular, cylindrical, and spherical coordinate systems, a vector **u** may be represented as

$$\begin{aligned}\mathbf{u} &= u_x\hat{\mathbf{x}} + u_y\hat{\mathbf{y}} + u_z\hat{\mathbf{z}} \\ &= u_\rho\hat{\boldsymbol{\rho}} + u_\phi\hat{\boldsymbol{\phi}} + u_z\hat{\mathbf{z}} \\ &= u_r\hat{\mathbf{r}} + u_\theta\hat{\boldsymbol{\theta}} + u_\phi\hat{\boldsymbol{\phi}}\end{aligned}$$

Show that the transformation of components from rectangular to cylindrical coordinates is given by

$$\begin{bmatrix} u_\rho \\ u_\phi \\ u_z \end{bmatrix} = \begin{bmatrix} \cos\phi & \sin\phi & 0 \\ -\sin\phi & \cos\phi & 0 \\ 0 & 0 & 1 \end{bmatrix} \begin{bmatrix} u_x \\ u_y \\ u_z \end{bmatrix}$$

and that the transformation from rectangular to spherical is

$$\begin{bmatrix} u_r \\ u_\theta \\ u_\phi \end{bmatrix} = \begin{bmatrix} \sin\theta\cos\phi & \sin\theta\sin\phi & \cos\theta \\ \cos\theta\cos\phi & \cos\theta\sin\phi & -\sin\theta \\ -\sin\phi & \cos\phi & 0 \end{bmatrix} \begin{bmatrix} u_x \\ u_y \\ u_z \end{bmatrix}$$

REFERENCES

Bellman, R.: *Introduction to Matrix Analysis*, 2d ed., McGraw-Hill Book Company, New York, 1970.

Bowman, F.: *An Introduction to Determinants and Matrices*, The English Universities Press Ltd., London, 1962.

Bradbury, T. C.: *Theoretical Mechanics*, John Wiley & Sons, Inc., New York, 1968, chaps. 1, 2, and 3.

Brand, L.: *Vector and Tensor Analysis*, John Wiley & Sons, Inc., New York, 1947.

Chorlton, F.: *Vector and Tensor Methods*, Halsted Press, John Wiley & Sons, Inc., New York, 1976, chaps. 12, 13.

Coburn, N.: *Vector and Tensor Analysis*, Dover Publications, Inc., New York, 1970.

Gibbs, J. W., and E. B. Wilson: *Vector Analysis*, Dover Publications, Inc., New York, 1960.

Goodbody, A. M.: *Cartesian Tensors: With Applications to Mechanics, Fluid Mechanics and Elasticity*, Halsted Press, John Wiley & Sons, Inc., New York, 1982.

Heading, J.: *Matrix Theory for Physicists*, Longmans, Green and Co., London, 1958.

Jaunzemis, W.: *Continuum Mechanics*, The Macmillan Co., New York, 1967, chap. 1.

Joshi, A. W.: *Matrices and Tensors in Physics*, Halsted Press, John Wiley & Sons, Inc., New York, 1975.

Luehr, C. P., and M. Rosenbaum: "Intrinsic Vector and Tensor Techniques in Minkowski Space with Applications to Special Relativity," *J. Math. Phys.*, vol. 9, no. 2, 1968, pp. 284–298.

McConnell, A. J.: *Application of Tensor Analysis*, Dover Publications, Inc., New York, 1957.

Maxwell, E. A.: *Coordinate Geometry with Vectors and Tensors*, Oxford University Press, London, 1958.

Milne, E. A.: *Vectorial Mechanics*, Interscience Publishers Inc., New York, 1948, pt. 1.

Portis, A. M.: *Electromagnetic Fields: Sources and Media*, John Wiley & Sons, Inc., New York, 1978, app. C.

Spencer, A. J. M.: "Theory of Invariants," in *Continuum Physics*, A. Cemal Eringen (ed.), Academic Press, New York, 1971, pp. 239–353.

Van Bladel, J.: *Electromagnetic Fields*, McGraw-Hill Book Company, New York, 1964, apps. 1 and 3.

Weatherburn, C. E.: *Advanced Vector Analysis with Applications to Mathematical Physics*, G. Bell & Sons, Ltd., London, 1957.

Wills, A. P.: *Vector Analysis with an Introduction to Tensor Analysis*, Dover Publications, Inc., New York, 1958.

Wylie, C. R.: *Advanced Engineering Mathematics*, 4th ed., McGraw-Hill Book Company, New York, 1975, chaps. 10 and 11.

CHAPTER

TWO

BASIC EQUATIONS OF ELECTRODYNAMICS

It has now been well established through experiments that the propagation and excitation of electromagnetic waves are governed mathematically by the Maxwell equations. We assume that the reader is familiar with the more elementary aspects of the electromagnetic theory.† In this chapter, we shall review the Maxwell equations and some of their consequences in both time and frequency domains. We shall further derive the transformation formulas for sources and field vectors to moving systems. The remaining chapters will be devoted to applications of the theory to problems concerning the propagation and excitation of electromagnetic waves in various media.

2.1 MAXWELL'S EQUATIONS

In the International System (SI) of units, which will be used throughout this book, the *Maxwell equations* in differential forms are:

$$\nabla \times \mathscr{E} = -\frac{\partial \mathscr{B}}{\partial t} \qquad \text{(Faraday's law of induction)} \tag{2.1}$$

$$\nabla \times \mathscr{H} = \frac{\partial \mathscr{D}}{\partial t} + \mathscr{J} \qquad \text{(generalized Ampere's law)} \tag{2.2}$$

$$\nabla \cdot \mathscr{B} = 0 \qquad \text{(Gauss' law for magnetic field)} \tag{2.3}$$

$$\nabla \cdot \mathscr{D} = \rho \qquad \text{(Gauss' law for electric field)} \tag{2.4}$$

† For typical introductory and intermediate texts on electromagnetic theory, see references and bibliography.

where $\mathscr{E}$ = electric field intensity (vector), volts/meter (V/m)
$\mathscr{H}$ = magnetic field intensity (vector), amperes/meter (A/m)
$\mathscr{D}$ = electric flux density (vector), coulombs/meter^2 (C/m^2)
$\mathscr{B}$ = magnetic flux density (vector), webers/meter^2 (Wb/m^2)
$\mathscr{J}$ = electric current density (vector), amperes/meter^2 (A/m^2)
ρ = electric charge density (scalar), coulombs/meter^3 (C/m^3)

All the above quantities, in the general case, are arbitrary real functions of space, i.e., the position vector **r** (meters) and time t (seconds). $\mathscr{J}$ includes the conduction current $\mathscr{J}_c$, which varies with the field, and the externally applied current $\mathscr{J}_{ex}$, which is independent of the field.

Taking the divergence of Eq. (2.2), noting that $\nabla \cdot (\nabla \times \mathscr{H}) = 0$ and using Eq. (2.4), we obtain the equation of continuity

$$\nabla \cdot \mathscr{J} = -\frac{\partial \rho}{\partial t} \tag{2.5}$$

It expresses, in differential form, the *law of conservation of an electric charge.*

The equation of continuity represents a general experimental result and may be considered as a basic natural law. Including Eq. (2.5), we may show that not all Eqs. (2.1) to (2.4) are independent. For example, Eqs. (2.3) and (2.4) can be obtained from Eqs. (2.1), (2.2), and (2.5). In fact, by taking the divergence of Eq. (2.1), we have

$$\frac{\partial}{\partial t}(\nabla \cdot \mathscr{B}) = 0$$

It follows that at any point in space, the divergence of the magnetic flux density must be a constant in time. If this constant had a value other than zero at any point in space, we would have to admit the presence of a free magnetic "charge" at that point forever. The fact that not even one such charge has ever been found leads us to reject this possibility on physical grounds. Thus we conclude that

$$\nabla \cdot \mathscr{B} = 0$$

which coincides with Eq. (2.3). Similarly, by taking the divergence of Eq. (2.2) and using the result of Eq. (2.5), we obtain Eq. (2.4). Therefore, we may choose as basic equations either the set of Eqs. (2.1), (2.2), and (2.5) or the set of Eqs. (2.1) and (2.2), supplemented by Eqs. (2.3) and (2.4) as initial conditions.

For reasons of mathematical generality and symmetry, we may introduce the fictitious magnetic charge density ρ_m and the magnetic current density $\mathscr{J}_m$. They satisfy the equation of continuity

$$\nabla \cdot \mathscr{J}_m = -\frac{\partial \rho_m}{\partial t} \tag{2.6}$$

Thus the *generalized Maxwell equations* take the form

$$\nabla \times \mathscr{E} = -\frac{\partial \mathscr{B}}{\partial t} - \mathscr{J}_m \tag{2.7}$$

$$\nabla \times \mathscr{H} = \frac{\partial \mathscr{D}}{\partial t} + \mathscr{J} \tag{2.8}$$

$$\nabla \cdot \mathscr{B} = \rho_m \tag{2.9}$$

$$\nabla \cdot \mathscr{D} = \rho \tag{2.10}$$

Here we should note that the free magnetic charge does not exist in reality.

Integrating Eqs. (2.1) and (2.2) respectively over a surface S, bounded by the curve C, and using Stokes' theorem of vector analysis, we obtain the Maxwell equations in integral forms:

$$\oint_c \mathscr{E} \cdot d\mathbf{l} = -\frac{d}{dt}\int_s \mathscr{B} \cdot d\mathbf{s} \tag{2.11}$$

$$\oint_c \mathscr{H} \cdot d\mathbf{l} = \frac{d}{dt}\int_s \mathscr{D} \cdot d\mathbf{s} + \int_s \mathscr{J} \cdot d\mathbf{s} \tag{2.12}$$

where $d\mathbf{s}$ is an infinitesimal vector area on the surface S and $d\mathbf{l}$ is a differential vector length. The circle on the left-hand side indicates the closed-line integration. We employ the usual convention that $d\mathbf{l}$ encircles $d\mathbf{s}$ according to the right-hand rule as shown in Fig. 2.1. Similarly, integrating Eqs. (2.3), (2.4), and (2.5) over a volume V, bounded by the surface S, and using the divergence theorem, we obtain the remaining equations

$$\oint_s \mathscr{D} \cdot d\mathbf{s} = \int_V \rho \, d^3r \tag{2.13}$$

$$\oint_s \mathscr{B} \cdot d\mathbf{s} = 0 \tag{2.14}$$

and

$$\oint_s \mathscr{J} \; d\mathbf{s} = -\frac{d}{dt}\int_V \rho \, d^3r \tag{2.15}$$

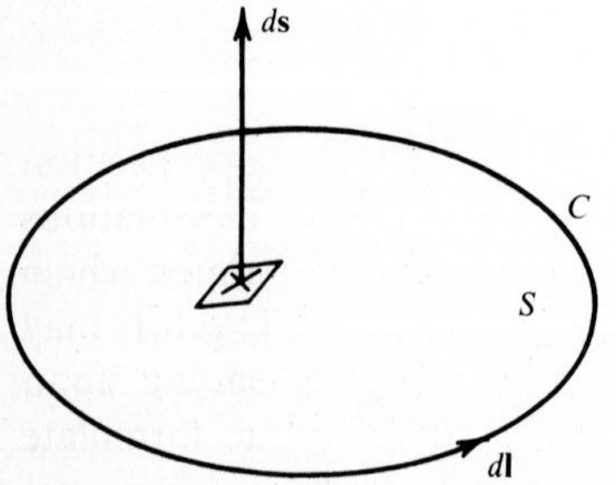

Figure 2.1 Relation of vector area and the direction of traversal of $d\mathbf{l}$.

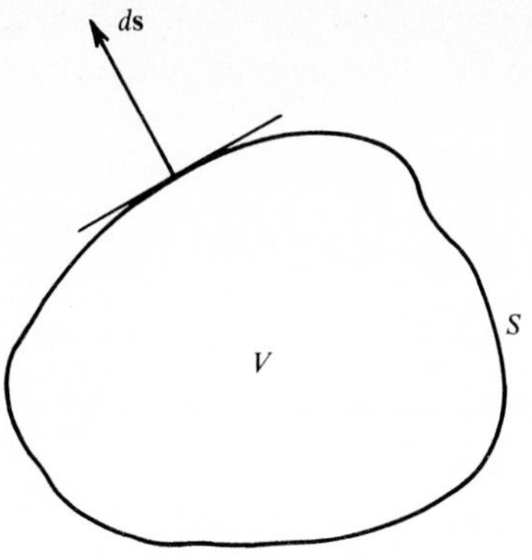

Figure 2.2 A volume V bounded by a closed surface S. The vector area $d\mathbf{s}$ is positive if it points away from V.

where the circle on the left-hand side denotes the closed-surface integration and d^3r is a three-dimensional volume element. We use the convention that $d\mathbf{s}$ is positive if it points outward from a closed surface S as shown in Fig. 2.2.

The differential forms of the Maxwell equations are valid at every point in space provided that the field vectors are well-behaved at the point, that is, they must be continuous and have continuous derivatives. On the other hand, the integral forms of the Maxwell equations are valid over a region, thus they are useful in deriving the boundary conditions at the interface.

2.2 CONSTITUTIVE RELATIONS

The vector equations (2.1), (2.2) and the scalar equations (2.3), (2.4) are equivalent to eight scalar equations for twelve scalar unknowns (three components for every field vector $\mathscr{E}$, $\mathscr{D}$, $\mathscr{H}$, $\mathscr{B}$), even if we consider the charge density ρ and the current density $\mathscr{J}$ as given quantities. To allow a unique determination of the field vectors, Maxwell's equations must be supplemented by relations which describe the behavior of the medium under the influence of the field. These subsidiary relations are called *constitutive relations* which can be established by experimentation or deduced from atomic theory. The constitutive relations may, in general, be written in the following functional forms

$$\begin{aligned} \mathscr{D} &= \mathscr{D}(\mathscr{E}, \mathscr{H}) \\ \mathscr{B} &= \mathscr{B}(\mathscr{E}, \mathscr{H}) \\ \mathscr{J}_c &= \mathscr{J}_c(\mathscr{E}, \mathscr{H}) \end{aligned} \tag{2.16}$$

If the field vectors are linearly related so that the principle of superposition applies, the medium is said to be *linear*. A combination of Maxwell's equations and linear constitutive relations forms the basis of linear electrodynamics, which is the main concern of this book. The linear relations among the field vectors may be either algebraic or involve differentiation and integration, depending upon the medium under consideration. In general, they are rather difficult to formulate in the time domain. Explicit forms of the constitutive relations for various media will be undertaken in the later chapters. In free space, the constitutive relations take the simpliest forms:

$$\mathscr{D} = \varepsilon_0 \mathscr{E}$$
$$\mathscr{B} = \mu_0 \mathscr{H} \tag{2.17}$$

where the permittivity ε_0 and the permeability μ_0 of free space are given respectively by

$$\varepsilon_0 = 8.854 \times 10^{-12} \cong \frac{1}{36\pi} \times 10^{-9} \text{ F/m}$$

and
$$\mu_0 = 4\pi \times 10^{-7} \text{ H/m} \tag{2.18}$$

in the SI system of units.

A medium is said to be *isotropic* if the electrical and magnetic properties at a given point are independent of the direction of the field at the point. For example, in *simple matter*,

$$\mathscr{D} = \varepsilon_0 \varepsilon \mathscr{E}$$
$$\mathscr{B} = \mu_0 \mu \mathscr{H} \tag{2.19}$$
$$\mathscr{J}_c = \sigma \mathscr{E}$$

where the dielectric constant ε, the relative permeability μ, and the conductivity σ are scalars. If ε, μ, and σ have the same values at every point in space, the medium is said to be *homogeneous*. On the other hand, if the electrical and magnetic properties of a medium depend upon the directions of field vectors, the medium is called *anisotropic*. For example,

$$\mathscr{D} = \varepsilon_0 \bar{\boldsymbol{\varepsilon}} \cdot \mathscr{E}$$
$$\mathscr{B} = \mu_0 \bar{\boldsymbol{\mu}} \cdot \mathscr{H} \tag{2.20}$$

where $\bar{\boldsymbol{\varepsilon}}$ and $\bar{\boldsymbol{\mu}}$ are respectively the dielectric and permeability tensors.† In other cases, both $\mathscr{D}$ and $\mathscr{B}$ may depend on $\mathscr{E}$ and $\mathscr{H}$:

$$\mathscr{D} = \varepsilon_0 \bar{\boldsymbol{\varepsilon}} \cdot \mathscr{E} + \bar{\boldsymbol{\xi}} \cdot \mathscr{H}$$
$$\mathscr{B} = \bar{\boldsymbol{\zeta}} \cdot \mathscr{E} + \mu_0 \bar{\boldsymbol{\mu}} \cdot \mathscr{H} \tag{2.21}$$

where $\bar{\boldsymbol{\xi}}$ and $\bar{\boldsymbol{\zeta}}$ are referred to as *magnetoelectric tensors*. The constitutive relations (2.19) to (2.21) which connect the field vectors at the same instant t are valid for slowly time-varying electromagnetic fields. For a rapidly varying field, the state of the medium depends not only on the field at present time t but also on the previous times. Thus for a linear, time-invariant medium, we may write the

† For a definition of a tensor, see Appendix A.

constitutive relations (2.20) as

$$\mathscr{D}(t) = \varepsilon_0 \int_{-\infty}^{t} \bar{\bar{\varepsilon}}(t - t') \cdot \mathscr{E}(t')\, dt'$$

$$\mathscr{B}(t) = \mu_0 \int_{-\infty}^{t} \bar{\bar{\mu}}(t - t') \cdot \mathscr{H}(t')\, dt' \tag{2.22}$$

which take into account the influence of the prior history on the electromagnetic properties of the medium. Since the field vectors $\mathscr{D}(t)$ and $\mathscr{E}(t)$ are real, $\bar{\bar{\varepsilon}}(t - t')$ in Eq. (2.22) must be real; so must $\bar{\bar{\mu}}(t - t')$ also be real. The medium characterized by the constitutive relations (2.22) is called a time- or frequency-dispersive medium or simply a *dispersive medium.*

2.3 BOUNDARY CONDITIONS

The Maxwell equations (2.1) to (2.4) are valid in regions where the physical properties of the medium vary continuously. However, across any surface which bounds one medium from another the constitutive parameters such as μ, ε, or σ may change abruptly, and we may expect corresponding changes in the field vectors. In order to continue the solution of Maxwell's equations from one region to another so that the resulting solution is unique and valid everywhere, we need boundary conditions to impose on the field vectors at the interface. They can be determined from the integral form of the Maxwell equations (2.11) to (2.14).

Consider two different media separated by an interface S as shown in Fig. 2.3. Let $\hat{\mathbf{q}}$ be a unit vector normal to the interface, pointing from medium 1 toward medium 2. Let us take the surface integral in Eq. (2.14) over the closed surface of a small cylinder. The contribution from the curved surface of the

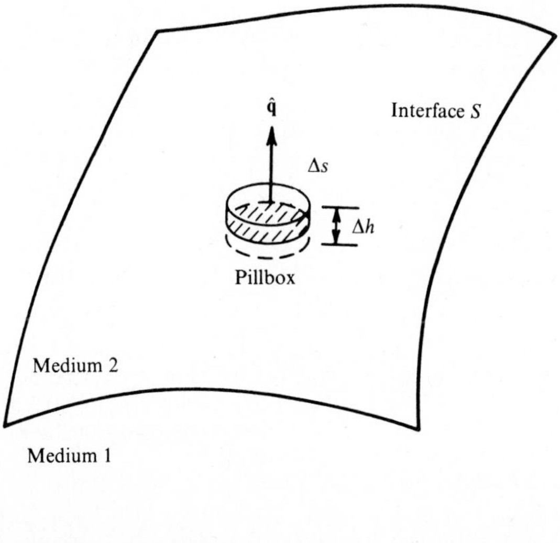

Figure 2.3 Pillbox used in obtaining boundary conditions for $\mathscr{B}$ and $\mathscr{D}$.

cylinder is directly proportional to Δh, and, for a sufficiently small cross-sectional area Δs of the cylinder, $\mathscr{B}$ may be considered constant over each end. In the limit as $\Delta h \to 0$, the ends of the cylinder lie just on either side of S and the contribution from the curved surface becomes vanishingly small. Thus, we obtain

$$(\mathscr{B}_2 - \mathscr{B}_1) \cdot \hat{\mathbf{q}} = 0 \tag{2.23}$$

which states that the normal component of $\mathscr{B}$ across a boundary surface of two media must be continuous.

Similarly, carrying out the integral in Eq. (2.13) over the cylinder in Fig. 2.3, we obtain

$$(\mathscr{D}_2 - \mathscr{D}_1) \cdot \hat{\mathbf{q}} = \lim_{\Delta h \to 0} \rho \, \Delta h = \rho_s \tag{2.24}$$

where ρ_s is the surface charge density measured in coulombs per square meter. Equation (2.24) states that in the presence of a layer of surface charge density ρ_s on S, the normal component of $\mathscr{D}$ changes abruptly across the interface, and the amount of discontinuity is equal to the surface charge density. On the other hand, if there is no surface charge ($\rho_s = 0$) on S as in the case of two different dielectrics, the normal component of $\mathscr{D}$ must be continuous.

Turning next to the behavior of the tangential components, we replace the cylinder in Fig. 2.3 by a small rectangular loop of area ΔA bounded by sides of length Δl, parallel to the interface S, and ends of length Δh, perpendicular to S as shown in Fig. 2.4. The surface integral on the right-hand side of Eq. (2.11) vanishes when $\Delta h \to 0$ as do the contributions to the line integral from the segments of the ends. Thus we have

$$(\mathscr{E}_2 - \mathscr{E}_1) \cdot \hat{\mathbf{t}} = 0 \tag{2.25}$$

where $\hat{\mathbf{t}}$ is the unit vector tangent to the interface. Alternatively, Eq. (2.25) may also be expressed in terms of the unit normal vector $\hat{\mathbf{q}}$ as

$$\hat{\mathbf{q}} \times (\mathscr{E}_2 - \mathscr{E}_1) = \mathbf{0} \tag{2.26}$$

which states that the tangential component of $\mathscr{E}$ across a boundary surface of two media must be continuous.

Similarly, we can show that Eq. (2.12) leads to the relation

$$\hat{\mathbf{q}} \times (\mathscr{H}_2 - \mathscr{H}_1) = \lim_{\Delta h \to 0} \mathscr{J} \, \Delta h = \mathscr{J}_s \tag{2.27}$$

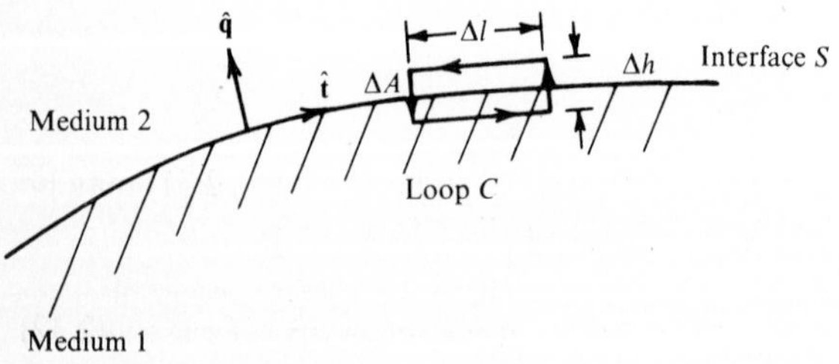

Figure 2.4 Rectangular loop used in obtaining boundary conditions for $\mathscr{E}$ and $\mathscr{H}$.

where $\mathscr{J}_S$ is the surface current density measured in amperes per meter. Equation (2.27) states that the tangential component of $\mathscr{H}$ across any boundary surface of two media is discontinuous. The amount of discontinuity is equal to the surface current density. If the current density $\mathscr{J}$ is finite, as it must be in any medium of finite conductivity, then the limit on the right-hand side of Eq. (2.27) vanishes. Hence we have

$$\hat{\mathbf{q}} \times (\mathscr{H}_2 - \mathscr{H}_1) = \mathbf{0} \tag{2.28}$$

i.e., if there is no surface current ($\mathscr{J}_s = \mathbf{0}$) on S, as in the case of two different dielectrics, the tangential component of $\mathscr{H}$ must be continuous.

As a special case, if medium 1 is a perfect conductor ($\sigma_1 \to \infty$) and medium 2 is a perfect dielectric ($\sigma_2 = 0$), then all the field vectors in medium 1 vanish identically and the boundary conditions become

$$\begin{aligned} \hat{\mathbf{q}} \cdot \mathscr{B}_2 &= 0 \qquad & \hat{\mathbf{q}} \cdot \mathscr{D}_2 &= \rho_s \\ \hat{\mathbf{q}} \times \mathscr{E}_2 &= \mathbf{0} \qquad & \hat{\mathbf{q}} \times \mathscr{H}_2 &= \mathscr{J}_s \end{aligned} \tag{2.29}$$

In practice, since the field vectors are related by the Maxwell equations, the boundary condition (2.26) implies Eq. (2.23) and the boundary condition (2.27) implies Eq. (2.24). In other words, satisfaction of the boundary conditions on the tangential components of the electric and magnetic fields automatically ensures the satisfaction of the boundary conditions on the normal components of the magnetic and electric flux densities respectively (see Prob. 2.6). This simplifies the analytical construction of a solution to the Maxwell equations.

Finally, when solving for fields in an unbounded region of space, the behavior of the field at infinity must also be specified. This boundary condition is called a *radiation condition* and requires that the energy flow at infinity shall be directed away from the source.

2.4 ENERGY AND POWER

From the Maxwell equations, we may formulate the law of conservation of energy for the electromagnetic fields. Dotting both sides of Eqs. (2.1) and (2.2) by $\mathscr{H}$ and $-\mathscr{E}$ respectively, then adding, and using the vector identity:

$$\mathbf{b} \cdot (\nabla \times \mathbf{a}) - \mathbf{a} \cdot (\nabla \times \mathbf{b}) = \nabla \cdot (\mathbf{a} \times \mathbf{b}) \tag{2.30}$$

we obtain the following relation

$$\mathscr{E} \cdot \frac{\partial \mathscr{D}}{\partial t} + \mathscr{H} \cdot \frac{\partial \mathscr{B}}{\partial t} = -\nabla \cdot (\mathscr{E} \times \mathscr{H}) - \mathscr{E} \cdot \mathscr{J}_c \tag{2.31}$$

Now, if $\mathscr{D}$, $\mathscr{B}$, $\mathscr{E}$, and $\mathscr{H}$ are so related that the left-hand side of Eq. (2.31) may be expressed as the time derivative of a scalar function W, i.e., if

$$\mathscr{E} \cdot \frac{\partial \mathscr{D}}{\partial t} + \mathscr{H} \cdot \frac{\partial \mathscr{B}}{\partial t} = \frac{\partial W}{\partial t} \tag{2.32}$$

then Eq. (2.31) may be written as

$$\frac{\partial W}{\partial t} = -\nabla \cdot \mathscr{P} - Q \tag{2.33}$$

In Eq. (2.33),

$$\mathscr{P} = \mathscr{E} \times \mathscr{H} \tag{2.34}$$

is the *Poynting vector* measured in watts per square meter and represents the amount of power crossing a unit area in the direction perpendicular to both $\mathscr{E}$ and $\mathscr{H}$, and

$$Q = \mathscr{E} \cdot \mathscr{J}_c \tag{2.35}$$

is the joule heat which represents the energy dissipated in a medium per unit volume per second. Equation (2.33) has the form of a conservation law and is known as the differential form of the *Poynting theorem.* In fact, integrating Eq. (2.33) over an arbitrary volume V, bounded by the closed surface S, and using the divergence theorem of vector analysis, we find the integral form of the Poynting theorem:

$$-\frac{d}{dt}\int_V W\, d^3r = \oint_S \mathscr{P} \cdot d\mathbf{s} + \int_V Q\, d^3r \tag{2.36}$$

Since each term in Eq. (2.36) has the dimension of watts, the Poynting theorem is a power theorem. The time rate of decrease of the volume integral on the left-hand side of Eq. (2.36) is equal to the sum of the power flux density vector $\mathscr{P}$ crossing the surface S and the dissipation of heat inside the volume V. Therefore, we may interpret W as the electromagnetic energy density measured in joules per cubic meter. Let us assume that in the remote past, there is no stored energy because the fields are zero, i.e., for $t \to -\infty$, $\mathscr{E}(-\infty) = \mathscr{H}(-\infty) = 0$, $W(-\infty) = 0$. Then from Eq. (2.32), we obtain the total instantaneous energy density

$$W = W_e + W_m \tag{2.37}$$

where

$$W_e = \int \mathscr{E} \cdot \frac{\partial \mathscr{D}}{\partial t}\, dt \tag{2.38}$$

is the instantaneous *electric energy density* and

$$W_m = \int \mathscr{H} \cdot \frac{\partial \mathscr{B}}{\partial t}\, dt \tag{2.39}$$

is the instantaneous *magnetic energy density.*

As an example, let us consider an isotropic, nondispersive, lossless medium. In this case, the constitutive relations are

$$\begin{aligned} \mathscr{D} &= \varepsilon_0\, \varepsilon \mathscr{E} \\ \mathscr{B} &= \mu_0\, \mu \mathscr{H} \end{aligned} \tag{2.40}$$

where ε and μ are constants. Substituting Eq. (2.40) into Eqs. (2.38) and (2.39) and taking into account the identities

$$\mathscr{E} \cdot \frac{\partial \mathscr{D}}{\partial t} = \frac{1}{2} \frac{\partial}{\partial t} (\mathscr{E} \cdot \mathscr{D})$$

$$\mathscr{H} \cdot \frac{\partial \mathscr{B}}{\partial t} = \frac{1}{2} \frac{\partial}{\partial t} (\mathscr{H} \cdot \mathscr{B})$$

we find

$$\begin{aligned} W_e &= \tfrac{1}{2} \varepsilon_0 \, \varepsilon \mathscr{E}^2 = \tfrac{1}{2} \mathscr{E} \cdot \mathscr{D} \\ W_m &= \tfrac{1}{2} \mu_0 \, \mu \mathscr{H}^2 = \tfrac{1}{2} \mathscr{H} \cdot \mathscr{B} \end{aligned} \tag{2.41}$$

2.5 MONOCHROMATIC FIELDS

Since the Maxwell equations together with the linear constitutive relations form a linear system of partial differential equations, no generality is lost by considering the sinusoidal time-varying fields alone. In fact, according to Fourier analysis, an arbitrary time-varying function $f(t)$ can always be represented as a superposition of sinusoidal functions, i.e.,

$$f(t) = \frac{1}{2\pi} \int_{-\infty}^{+\infty} f(\omega) e^{-i\omega t} \, d\omega \tag{2.42}$$

Here we have used the same symbol with a different argument to distinguish the function and its Fourier transform. If a field vector varies sinusoidally in time with a single angular frequency ω, the field is said to be *monochromatic*. To transform Maxwell's equations of a monochromatic field from time to frequency domain, we choose the time-dependent factor $e^{-i\omega t}$ and adopt the convention

$$\mathscr{E} = \text{Re}\,(\mathbf{E} e^{-i\omega t}) \tag{2.43}$$

where the time domain $\mathscr{E}$ is a real vector function of the position vector $\mathbf{r}$ and time t, and the frequency domain $\mathbf{E}$ is a complex vector whose cartesian components are complex functions of $\mathbf{r}$. Re is a shorthand for the "real part of," $\omega = 2\pi f$ is the angular frequency and f is the frequency. We note that for a real linear operator $\mathscr{L}$ such as $\partial/\partial t, \mathbf{\nabla}$, $\int dt$, and multiplicaton by a real function, etc., the following rule is valid

$$\mathscr{L}(\text{Re}\, g) = \text{Re}\,(\mathscr{L} g) \tag{2.44}$$

for any complex function g. Substituting Eq. (2.43) and similar expressions for $\mathscr{B}$, $\mathscr{H}$, $\mathscr{D}$, $\mathscr{J}$, ρ into Eqs. (2.1) to (2.4), we obtain, after omitting the real part sign and the factor $e^{-i\omega t}$, the complex Maxwell equations in frequency domain

$$\begin{aligned} \nabla \times \mathbf{E} &= i\omega \mathbf{B} \\ \nabla \times \mathbf{H} &= -i\omega \mathbf{D} + \mathbf{J} \\ \nabla \cdot \mathbf{B} &= 0 \\ \nabla \cdot \mathbf{D} &= \rho \end{aligned} \tag{2.45}$$

and from Eq. (2.5), the equation of continuity

$$\nabla \cdot \mathbf{J} = i\omega\rho \tag{2.46}$$

The constitutive relations for nondispersive media remain valid in the frequency domain. For a dispersive medium, we substitute the monochromatic field of the form (2.43) into Eq. (2.22) and obtain, in the frequency domain,

$$\begin{aligned} \mathbf{D} &= \varepsilon_0 \bar{\boldsymbol{\varepsilon}}(\omega) \cdot \mathbf{E} \\ \mathbf{B} &= \mu_0 \bar{\boldsymbol{\mu}}(\omega) \cdot \mathbf{H} \end{aligned} \tag{2.47}$$

where

$$\begin{aligned} \bar{\boldsymbol{\varepsilon}}(\omega) &= \int_0^\infty \bar{\boldsymbol{\varepsilon}}(t) e^{i\omega t}\, dt \\ \bar{\boldsymbol{\mu}}(\omega) &= \int_0^\infty \bar{\boldsymbol{\mu}}(t) e^{i\omega t}\, dt \end{aligned} \tag{2.48}$$

are complex functions of ω, that is, the medium is dispersive. The distinct advantage of carrying out discussion in the frequency rather than in the time domain is that this permits us to reduce the integral relations in Eq. (2.22) to the algebraic relations (2.47). From the fact that $\bar{\boldsymbol{\varepsilon}}(t)$ and $\bar{\boldsymbol{\mu}}(t)$ are real, we conclude from Eq. (2.48) that for real ω

$$\begin{aligned} \bar{\boldsymbol{\varepsilon}}^*(\omega) &= \bar{\boldsymbol{\varepsilon}}(-\omega) \\ \bar{\boldsymbol{\mu}}^*(\omega) &= \bar{\boldsymbol{\mu}}(-\omega) \end{aligned} \tag{2.49}$$

In terms of the real and imaginary parts of $\bar{\boldsymbol{\varepsilon}}(\omega)$ and $\bar{\boldsymbol{\mu}}(\omega)$, we can rewrite Eq. (2.49) as

$$\begin{aligned} \operatorname{Re} \bar{\boldsymbol{\varepsilon}}(\omega) &= \operatorname{Re} \bar{\boldsymbol{\varepsilon}}(-\omega) \\ \operatorname{Re} \bar{\boldsymbol{\mu}}(\omega) &= \operatorname{Re} \bar{\boldsymbol{\mu}}(-\omega) \\ \operatorname{Im} \bar{\boldsymbol{\varepsilon}}(\omega) &= -\operatorname{Im} \bar{\boldsymbol{\varepsilon}}(-\omega) \\ \operatorname{Im} \bar{\boldsymbol{\mu}}(\omega) &= -\operatorname{Im} \bar{\boldsymbol{\mu}}(-\omega) \end{aligned} \tag{2.50}$$

Thus the real parts of $\bar{\boldsymbol{\varepsilon}}(\omega)$ and $\bar{\boldsymbol{\mu}}(\omega)$ are even functions of ω and the imaginary parts of $\bar{\boldsymbol{\varepsilon}}(\omega)$ and $\bar{\boldsymbol{\mu}}(\omega)$ are odd functions of ω.

A lossy medium can either be characterized by a conductivity tensor $\bar{\boldsymbol{\sigma}}$ or a *complex dielectric tensor* $\bar{\boldsymbol{\varepsilon}}$. In fact, from a knowledge of the conduction current

$\mathbf{J}_c = \bar{\boldsymbol{\sigma}} \cdot \mathbf{E}$, the complex dielectric tensor can be found by noting that the total current density is the sum of $\mathbf{J}_c$ and the free-space displacement current density $-i\omega\varepsilon_0 \mathbf{E}$. By regarding this total current density as a displacement current in a dielectric medium, we obtain

$$-i\omega\varepsilon_0 \bar{\boldsymbol{\varepsilon}} \cdot \mathbf{E} = \bar{\boldsymbol{\sigma}} \cdot \mathbf{E} - i\omega\varepsilon_0 \mathbf{E} \tag{2.51}$$

Thus it follows that the complex dielectric tensor $\bar{\boldsymbol{\varepsilon}}$ is related to the conductivity tensor $\bar{\boldsymbol{\sigma}}$ by

$$\bar{\boldsymbol{\varepsilon}} = \bar{\mathbf{I}} + i \frac{1}{\omega\varepsilon_0} \bar{\boldsymbol{\sigma}} \tag{2.52}$$

For monochromatic fields, the *instantaneous Poynting vector* $\mathscr{P}$, defined by Eq. (2.34), may be written as

$$\begin{aligned} \mathscr{P} &= \mathscr{E} \times \mathscr{H} \\ &= \tfrac{1}{2} \operatorname{Re} (\mathbf{E} \times \mathbf{H}^*) + \tfrac{1}{2} \operatorname{Re} [(\mathbf{E} \times \mathbf{H}) e^{-i2\omega t}] \end{aligned} \tag{2.53}$$

The *time-averaged Poynting vector* denoted by $\langle \mathscr{P} \rangle$, which is generally of much more significance in application, is defined by

$$\langle \mathscr{P} \rangle = \frac{1}{T} \int_0^T \mathscr{P} \, dt \tag{2.54}$$

where $T = 2\pi/\omega$ is the period. Substituting Eq. (2.53) into Eq. (2.54) and noting that the integration of $e^{-i2\omega t}$ over a period vanishes, we obtain

$$\langle \mathscr{P} \rangle = \tfrac{1}{2} \operatorname{Re} (\mathbf{E} \times \mathbf{H}^*) \tag{2.55}$$

Since in the course of problem solving, field solutions are frequently obtained in the frequency domain, Eq. (2.55) suggests that we define a *complex Poynting vector*

$$\mathbf{P} = \tfrac{1}{2}(\mathbf{E} \times \mathbf{H}^*) \tag{2.56}$$

the real part of which gives the time average Poynting vector. Similarly we may find, for example, from Eq. (2.41) the time-averaged electric and magnetic energy densities

$$\langle W_e \rangle = \tfrac{1}{4}\varepsilon_0 \, \varepsilon \mathbf{E} \cdot \mathbf{E}^* \tag{2.57}$$

and

$$\langle W_m \rangle = \tfrac{1}{4}\mu_0 \, \mu \mathbf{H} \cdot \mathbf{H}^* \tag{2.58}$$

respectively for an isotropic, nondispersive medium.

In a source-free region (that is, $\mathbf{J} = \mathbf{0}$, $\rho = 0$), the difference of the dot product of $\mathbf{H}^*$ by $\nabla \times \mathbf{E} = i\omega\mathbf{B}$ and the dot product of $\mathbf{E}$ by $\nabla \times \mathbf{H}^* = i\omega\mathbf{D}^*$, gives

$$\mathbf{H}^* \cdot \nabla \times \mathbf{E} - \mathbf{E} \cdot \nabla \times \mathbf{H}^* = i\omega(\mathbf{H}^* \cdot \mathbf{B} - \mathbf{E} \cdot \mathbf{D}^*) \tag{2.59}$$

Using the vector identity

$$\nabla \cdot (\mathbf{E} \times \mathbf{H}^*) = \mathbf{H}^* \cdot \nabla \times \mathbf{E} - \mathbf{E} \cdot \nabla \times \mathbf{H}^* \tag{2.60}$$

and multiplying both sides of Eq. (2.59) by $-\frac{1}{2}$, we obtain the *complex Poynting theorem* in the frequency domain:

$$-\nabla \cdot \mathbf{P} = \frac{i\omega}{2}(\mathbf{E} \cdot \mathbf{D}^* - \mathbf{H}^* \cdot \mathbf{B}) \tag{2.61}$$

The real part of Eq. (2.61), that is,

$$-\nabla \cdot \langle \mathscr{P} \rangle = \text{Re}\left[\frac{i\omega}{2}(\mathbf{E} \cdot \mathbf{D}^* - \mathbf{H}^* \cdot \mathbf{B})\right] \tag{2.62}$$

expresses the conservation of time-averaged power, the term on the left representing the total time-averaged power entering a point and the terms on the right indicating the average power dissipated as heat loss per unit volume at that point. Since the dissipation of energy as heat is always positive, we conclude from Eq. (2.62) that a medium is lossy if

$$-\nabla \cdot \langle \mathscr{P} \rangle = -\text{Re}\,(\nabla \cdot \mathbf{P}) > 0 \tag{2.63}$$

and is lossless if $-\nabla \cdot \langle \mathscr{P} \rangle = 0$ or

$$\text{Re}\left[\frac{i\omega}{2}(\mathbf{E} \cdot \mathbf{D}^* - \mathbf{H}^* \cdot \mathbf{B})\right] = 0 \tag{2.64}$$

To illustrate the implications of Eqs. (2.63) and (2.64) on the constitutive parameters, let us consider an isotropic, lossy, nondispersive medium. In this case, the constitutive relations are

$$\begin{aligned} \mathbf{D} &= \varepsilon_0 \varepsilon \mathbf{E} \\ \mathbf{B} &= \mu_0 \mu \mathbf{H} \end{aligned} \tag{2.65}$$

where

$$\begin{aligned} \varepsilon &= \varepsilon_r + i\varepsilon_i \\ \mu &= \mu_r + i\mu_i \end{aligned} \tag{2.66}$$

are complex constants. Substituting Eq. (2.65) into the right-hand side of Eq. (2.62), we obtain

$$\text{Re}\left[\frac{i\omega}{2}(\mathbf{E} \cdot \mathbf{D}^* - \mathbf{H}^* \cdot \mathbf{B})\right] = \frac{\omega}{2}(\varepsilon_0 \varepsilon_i \mathbf{E} \cdot \mathbf{E}^* + \mu_0 \mu_i \mathbf{H} \cdot \mathbf{H}^*) \tag{2.67}$$

Since

$$\begin{aligned} \langle \mathscr{E}^2 \rangle &= \tfrac{1}{2}(\mathbf{E} \cdot \mathbf{E}^*) \\ \langle \mathscr{H}^2 \rangle &= \tfrac{1}{2}(\mathbf{H} \cdot \mathbf{H}^*) \end{aligned} \tag{2.68}$$

thus Eq. (2.62) becomes

$$-\nabla \cdot \langle \mathscr{P} \rangle = \omega\varepsilon_0 \varepsilon_i \langle \mathscr{E}^2 \rangle + \omega\mu_0 \mu_i \langle \mathscr{H}^2 \rangle \tag{2.69}$$

i.e., the electric and magnetic losses are determined by the imaginary parts of ε and μ respectively. It follows from Eq. (2.63) that for a lossy medium

$$\varepsilon_i > 0 \qquad \mu_i > 0 \tag{2.70}$$

However, the signs for the real parts of ε and μ are subject to no physical restriction. From Eqs. (2.64) and (2.69), it is also evident that for a lossless medium, $\varepsilon_i = 0$, $\mu_i = 0$, that is, ε and μ must be real. As a second example, let us consider media characterized by the constitutive relations

$$\begin{aligned} \mathbf{D} &= \varepsilon_0 \bar{\boldsymbol{\varepsilon}} \cdot \mathbf{E} + \bar{\boldsymbol{\xi}} \cdot \mathbf{H} \\ \mathbf{B} &= \bar{\boldsymbol{\zeta}} \cdot \mathbf{E} + \mu_0 \bar{\boldsymbol{\mu}} \cdot \mathbf{H} \end{aligned} \tag{2.71}$$

where $\bar{\boldsymbol{\varepsilon}}$, $\bar{\boldsymbol{\xi}}$, $\bar{\boldsymbol{\zeta}}$, and $\bar{\boldsymbol{\mu}}$ are complex-constant tensors. Substituting Eq. (2.71) into the right-hand side of Eq. (2.62), after some simplification, we obtain

$$\begin{aligned} \operatorname{Re}\left[\frac{i\omega}{2}(\mathbf{E}\cdot\mathbf{D}^* - \mathbf{H}^*\cdot\mathbf{B})\right] = \frac{i\omega}{4}[&\varepsilon_0 \mathbf{E}^* \cdot (\bar{\boldsymbol{\varepsilon}}^+ - \bar{\boldsymbol{\varepsilon}}) \cdot \mathbf{E} \\ &+ \mu_0 \mathbf{H}^* \cdot (\bar{\boldsymbol{\mu}}^+ - \bar{\boldsymbol{\mu}}) \cdot \mathbf{H} + \mathbf{H}^* \cdot (\bar{\boldsymbol{\xi}}^+ - \bar{\boldsymbol{\zeta}}) \cdot \mathbf{E} \\ &+ \mathbf{E}^* \cdot (\bar{\boldsymbol{\zeta}}^+ - \bar{\boldsymbol{\xi}}) \cdot \mathbf{H}] \end{aligned} \tag{2.72}$$

Since $\mathbf{E}$ and $\mathbf{H}$ are two arbitrary complex vectors, the lossless condition (2.64) requires that

$$\begin{aligned} \bar{\boldsymbol{\varepsilon}}^+ &= \bar{\boldsymbol{\varepsilon}} \\ \bar{\boldsymbol{\mu}}^+ &= \bar{\boldsymbol{\mu}} \\ \bar{\boldsymbol{\zeta}}^+ &= \bar{\boldsymbol{\xi}} \end{aligned} \tag{2.73}$$

In other words, to be consistent with the law of conservation of power, the constitutive tensors of a lossless medium characterized by the constitutive relation (2.71) can not be chosen arbitrarily; instead, $\bar{\boldsymbol{\varepsilon}}$ and $\bar{\boldsymbol{\mu}}$ must be hermitian and $\bar{\boldsymbol{\zeta}}^+ = \bar{\boldsymbol{\xi}}$.

2.6 POYNTING'S THEOREM FOR QUASI-MONOCHROMATIC FIELDS

In Sec. 2.4 we defined the energy density as the energy absorbed by the medium during the buildup of the electromagnetic field. This definition is not directly applicable to monochromatic fields, because they are not zero for $t = -\infty$. To avoid this difficulty we consider the fields with complex frequency

$$s = \omega - i\nu$$

where ω and ν are real positive numbers and ν is small such that $\nu \ll \omega$. In this case, we have *quasi-monochromatic fields*:

$$\begin{aligned}\mathscr{E} &= \text{Re}\,(\mathbf{E}e^{-ist}) = \text{Re}\,[\mathbf{E}(t)e^{-i\omega t}]\\ \mathscr{H} &= \text{Re}\,(\mathbf{H}e^{-ist}) = \text{Re}\,[\mathbf{H}(t)e^{-i\omega t}]\end{aligned} \tag{2.74}$$

where the quasi-monochromatic nature of the fields are manifested by the fact that the functions $\mathbf{E}(t) = \mathbf{E}e^{-\nu t}$ and $\mathbf{H}(t) = \mathbf{H}e^{-\nu t}$ vary very slowly over the *quasi-period* $T = 2\pi/\omega$. With Eq. (2.74), we obtain the instantaneous Poynting vector

$$\begin{aligned}\mathscr{P} &= \mathscr{E} \times \mathscr{H}\\ &= \tfrac{1}{4}[\mathbf{E}(t) \times \mathbf{H}^*(t) + \mathbf{E}^*(t) \times \mathbf{H}(t)]\\ &\quad + \tfrac{1}{4}[\mathbf{E}(t) \times \mathbf{H}(t)e^{-i2\omega t} + \mathbf{E}^*(t) \times \mathbf{H}^*(t)e^{i2\omega t}]\end{aligned} \tag{2.75}$$

Next, we carry out average of Eq. (2.75) over high-frequency ω, denoted by the same symbol $\langle\ \rangle$ as in Eq. (2.54). Over the quasi-period $T = 2\pi/\omega$, $\mathbf{E}(t)$ and $\mathbf{H}(t)$ are slowly varying functions and thus may be considered as constants over T. This averaging is equivalent to neglecting the terms containing factors $e^{\pm i2\omega t}$. We then obtain the following

$$\langle\mathscr{P}\rangle = \tfrac{1}{4}(\mathbf{E} \times \mathbf{H}^* + \mathbf{E}^* \times \mathbf{H})e^{-2\nu t} \tag{2.76}$$

It is noted that this average is itself a slowly varying function of t. If we consider dispersive, lossless media characterized by the constitutive relations (2.22), for a time variation of e^{-ist} the Maxwell equations and the constitutive relations become

$$\begin{aligned}\nabla \times \mathbf{E} &= is\mathbf{B}\\ \nabla \times \mathbf{H} &= -is\mathbf{D}\end{aligned} \tag{2.77}$$

and

$$\begin{aligned}\mathbf{D} &= \varepsilon_0\,\bar{\boldsymbol{\varepsilon}}(s) \cdot \mathbf{E}\\ \mathbf{B} &= \mu_0\,\bar{\boldsymbol{\mu}}(s) \cdot \mathbf{H}\end{aligned} \tag{2.78}$$

respectively. Substituting Eq. (2.78) into Eq. (2.77) and then into the divergence of Eq. (2.76), we obtain

$$\begin{aligned}\nabla \cdot \langle\mathscr{P}\rangle = \frac{ie^{-2\nu t}}{4}\{&\varepsilon_0\,\mathbf{E}^* \cdot [s\bar{\boldsymbol{\varepsilon}}(s) - s^*\bar{\boldsymbol{\varepsilon}}^+(s)] \cdot \mathbf{E}\\ &+ \mu_0\,\mathbf{H}^* \cdot [s\bar{\boldsymbol{\mu}}(s) - s^*\bar{\boldsymbol{\mu}}^+(s)] \cdot \mathbf{H}\}\end{aligned} \tag{2.79}$$

For sufficiently small ν, we expand $s\bar{\boldsymbol{\varepsilon}}(s)$ in the Taylor series about ω, and neglect the second and higher order terms in ν; then

$$s\bar{\boldsymbol{\varepsilon}}(s) \cong \omega\bar{\boldsymbol{\varepsilon}}(\omega) - i\nu\,\frac{\partial[\omega\bar{\boldsymbol{\varepsilon}}(\omega)]}{\partial\omega}$$

and

$$s^*\bar{\varepsilon}^+(s) \cong \omega\bar{\varepsilon}^+(\omega) + i\nu \frac{\partial[\omega\bar{\varepsilon}^+(\omega)]}{\partial\omega}$$

Hence

$$s\bar{\varepsilon}(s) - s^*\bar{\varepsilon}^+(s) \cong -i2\nu \frac{\partial[\omega\bar{\varepsilon}(\omega)]}{\partial\omega} \tag{2.80}$$

Here we have used the condition that $\bar{\varepsilon}(\omega)$ is hermitian [that is, $\bar{\varepsilon}^+(\omega) = \bar{\varepsilon}(\omega)$], for a lossless medium. Similarly, we obtain

$$s\bar{\mu}(s) - s^*\bar{\mu}^+(s) \cong -i2\nu \frac{\partial[\omega\bar{\mu}(\omega)]}{\partial\omega} \tag{2.81}$$

Substituting Eqs. (2.80) and (2.81) into Eq. (2.79), we obtain the first order in ν

$$\nabla \cdot \langle \mathscr{P} \rangle = 2\nu\left\{\frac{\varepsilon_0}{4}\mathbf{E}^* \cdot \frac{\partial[\omega\bar{\varepsilon}(\omega)]}{\partial\omega} \cdot \mathbf{E} + \frac{\mu_0}{4}\mathbf{H}^* \cdot \frac{\partial[\omega\bar{\mu}(\omega)]}{\partial\omega} \cdot \mathbf{H}\right\}e^{-2\nu t} \tag{2.82}$$

But

$$\frac{\partial e^{-2\nu t}}{\partial t} = -2\nu e^{-2\nu t} \tag{2.83}$$

Thus, we may write Eq. (2.82) as

$$-\nabla \cdot \langle \mathscr{P} \rangle = \frac{\partial \langle W \rangle}{\partial t} \tag{2.84}$$

which is the *Poynting theorem for the quasi-monochromatic fields*, where $\langle \mathscr{P} \rangle$ is the Poynting vector averaged over the quasi-period $T = 2\pi/\omega$ and is given by Eq. (2.76), and where

$$\langle W \rangle = \langle W_e \rangle + \langle W_m \rangle \tag{2.85}$$

is the total average energy density for a lossless, dispersive anisotropic medium. In Eq. (2.85),

$$\langle W_e \rangle = \frac{\varepsilon_0}{4}\mathbf{E}^*(t) \cdot \frac{\partial[\omega\bar{\varepsilon}(\omega)]}{\partial\omega} \cdot \mathbf{E}(t) \tag{2.86}$$

and

$$\langle W_m \rangle = \frac{\mu_0}{4}\mathbf{H}^*(t) \cdot \frac{\partial[\omega\bar{\mu}(\omega)]}{\partial\omega} \cdot \mathbf{H}(t) \tag{2.87}$$

are, respectively, the average *electric* and *magnetic energy densities*. Since the stored energy is positive, the tensors $\partial[\omega\bar{\varepsilon}(\omega)]/\partial\omega$ and $\partial[\omega\bar{\mu}(\omega)]/\partial\omega$ must be positive definite. For a nondispersive, isotropic medium: $\bar{\boldsymbol{\varepsilon}}(\omega) = \varepsilon\bar{\mathbf{I}}$ and $\bar{\boldsymbol{\mu}}(\omega) = \mu\bar{\mathbf{I}}$ where ε and μ are constants and independent of ω, Eqs. (2.86) and (2.87) reduce to Eqs. (2.57) and (2.58) respectively.

Since the Poynting vector and the energy density have the dimensions of watt per square meter and joule per cubic meter respectively, their ratio gives a quantity with dimension of velocity. This velocity represents the *velocity of energy transport* and is denoted by

$$\mathbf{v}_E = \frac{\langle \mathscr{P} \rangle}{\langle W \rangle} \tag{2.88}$$

which never exceeds the velocity of light.

2.7 THE LORENTZ TRANSFORMATION

Thus far we have reviewed the Maxwell equations for stationary media and some of their direct consequences. In this and the next sections, we shall discuss briefly the part of the special theory of relativity which has application in the electrodynamics of moving media.

The special theory of relativity, as it was developed by Einstein, rests on two postulates:

1. The principle of relativity according to which the Maxwell equations must have the same mathematical form in all *inertial systems*—coordinate systems that are in uniform relative motion
2. The principle of the constancy of the velocity of light according to which the velocity of light in free space has the same value c in all inertial systems

Both postulates can be summarized by the statement that the Maxwell equations must be *covariant* under the Lorentz transformation. In other words, if we write Maxwell's equations in an inertial system Σ, after the space-time coordinates are transformed according to the Lorentz transformation from Σ to Σ', which is moving at a uniform velocity with respect to Σ, the field vectors, the charge and current densities, must transform in such a way that the transformed equations in Σ' have the same mathematical appearance as the original equations in Σ.

To derive the Lorentz transformation, let us consider an inertial system Σ' which moves with a uniform velocity $\mathbf{v}$ relative to another inertial system Σ. Both Σ' and Σ have the same orientation. An event characterized by the coordinate $(\mathbf{r}, t)$ in Σ has the coordinate $(\mathbf{r}', t')$ in Σ'. At time $t = t' = 0$, the origins of Σ and Σ' coincide, and a light pulse is emitted by a source from the coincident origins. We are looking for a transformation between $(\mathbf{r}, t)$ and $(\mathbf{r}', t')$ such that the spherical wavefront

$$\mathbf{r}'^2 - c^2 t'^2 = 0 \tag{2.89}$$

in Σ' corresponds to the spherical wavefront

$$\mathbf{r}^2 - c^2t^2 = 0 \tag{2.90}$$

in Σ. Notice that the same value c was used for the velocity of light in both Σ and Σ' so as to be in accord with the second postulate of the principle of the constancy of the velocity of light. The transformation must be linear in order to give a one-to-one correspondence between the events in Σ and Σ'. In addition, it must also be reduced to the *galilean transformation* of classical mechanics

$$\begin{aligned}\mathbf{r}' &= \mathbf{r} - \mathbf{v}t \\ t' &= t\end{aligned} \tag{2.91}$$

when $v \ll c$.

Let us consider a linear transformation between $(\mathbf{r}', t')$ and $(\mathbf{r}, t)$ of the following form:

$$\begin{aligned}\mathbf{r}' &= \bar{\mathbf{L}} \cdot \mathbf{r} - \gamma\mathbf{v}t \\ t' &= \gamma(t - \mathbf{m} \cdot \mathbf{r})\end{aligned} \tag{2.92}$$

where the scalar γ, the vector $\mathbf{m}$, and the dyadic $\bar{\mathbf{L}}$, that depend only on c and $\mathbf{v}$, are unknowns to be determined. The same scalar γ appears in both equations because the origin of the system Σ must move with the velocity $-\mathbf{v}$ as seen from the system Σ'. Moreover, $\bar{\mathbf{L}}$ is a symmetric dyadic as it is a function of $\mathbf{v}$ alone.

From Eqs. (2.89) and (2.90) the desired transformation (2.92) must be such that

$$\mathbf{r}'^2 - c^2t'^2 = \mathbf{r}^2 - c^2t^2 \tag{2.93}$$

Substituting Eq. (2.92) into the left-hand side of Eq. (2.93), we obtain

$$\begin{aligned}\mathbf{r}'^2 - c^2t'^2 &= \mathbf{r} \cdot (\bar{\mathbf{L}}^2 - c^2\gamma^2\mathbf{mm}) \cdot \mathbf{r} \\ &\quad - 2t\gamma(\mathbf{v} \cdot \bar{\mathbf{L}} - c^2\gamma\mathbf{m}) \cdot \mathbf{r} - \gamma^2\left(1 - \frac{v^2}{c^2}\right)c^2t^2 \\ &= \mathbf{r} \cdot \bar{\mathbf{I}} \cdot \mathbf{r} - c^2t^2\end{aligned} \tag{2.94}$$

Since $\mathbf{r}$ and t are arbitrary, Eq. (2.94) holds, if

$$\bar{\mathbf{L}}^2 - c^2\gamma^2\mathbf{mm} = \bar{\mathbf{I}} \tag{2.95}$$

$$\mathbf{v} \cdot \bar{\mathbf{L}} - c^2\gamma\mathbf{m} = \mathbf{0} \tag{2.96}$$

$$\gamma^2\left(1 - \frac{v^2}{c^2}\right) = 1 \tag{2.97}$$

From Eq. (2.97), we get

$$\gamma = \frac{1}{\sqrt{1 - \beta^2}} \tag{2.98}$$

where $\beta = v/c$. In order that Eq. (2.92) be consistent with Eq. (2.91), we choose the positive square root so that γ approaches 1 when β approaches 0. Eliminating $\mathbf{m}$ from Eqs. (2.95) and (2.96), we have

$$\bar{\mathbf{L}} \cdot (\bar{\mathbf{I}} - \beta^2 \hat{\mathbf{v}}\hat{\mathbf{v}}) \cdot \bar{\mathbf{L}} = \bar{\mathbf{I}} \tag{2.99}$$

where $\hat{\mathbf{v}}$ is the unit vector in the direction of $\mathbf{v}$. Multiplication of Eq. (2.99) by $\bar{\mathbf{L}}^{-1}$ from the left and by $\bar{\mathbf{L}}$ from the right gives

$$(\bar{\mathbf{I}} - \beta^2 \hat{\mathbf{v}}\hat{\mathbf{v}}) \cdot \bar{\mathbf{L}}^2 = \bar{\mathbf{I}}$$

or

$$\bar{\mathbf{L}}^2 = (\bar{\mathbf{I}} - \beta^2 \bar{\mathbf{v}}\bar{\mathbf{v}})^{-1} \tag{2.100}$$

But from Eq. (1.126), we have

$$\begin{aligned} |\bar{\mathbf{I}} - \beta^2 \hat{\mathbf{v}}\hat{\mathbf{v}}| &= 1 - \beta^2 \\ \operatorname{adj} (\bar{\mathbf{I}} - \beta^2 \hat{\mathbf{v}}\hat{\mathbf{v}}) &= (1 - \beta^2)\bar{\mathbf{I}} + \beta^2 \hat{\mathbf{v}}\hat{\mathbf{v}} \end{aligned} \tag{2.101}$$

Thus Eq. (2.100) becomes

$$\begin{aligned} \bar{\mathbf{L}}^2 &= \bar{\mathbf{I}} + \gamma^2 \beta^2 \hat{\mathbf{v}}\hat{\mathbf{v}} \\ &= \bar{\mathbf{I}} + (\gamma^2 - 1)\hat{\mathbf{v}}\hat{\mathbf{v}} \\ &= [\bar{\mathbf{I}} + (\gamma - 1)\hat{\mathbf{v}}\hat{\mathbf{v}}]^2 \end{aligned} \tag{2.102}$$

consequently

$$\bar{\mathbf{L}} = \bar{\mathbf{I}} + (\gamma - 1)\hat{\mathbf{v}}\hat{\mathbf{v}} \tag{2.103}$$

It is also interesting to note, from Eq. (1.126) that

$$\begin{aligned} |\bar{\mathbf{L}}| &= \gamma \\ \operatorname{adj} \bar{\mathbf{L}} &= \gamma \bar{\mathbf{I}} + (1 - \gamma)\hat{\mathbf{v}}\hat{\mathbf{v}} \end{aligned} \tag{2.104}$$

hence

$$\bar{\mathbf{L}}^{-1} = \bar{\mathbf{I}} - (1 - \gamma^{-1})\hat{\mathbf{v}}\hat{\mathbf{v}} = \bar{\mathbf{L}} - \beta^2 \gamma \hat{\mathbf{v}}\hat{\mathbf{v}} \tag{2.105}$$

Substituting Eq. (2.103) back into Eq. (2.96), we find

$$\mathbf{m} = \frac{\mathbf{v} \cdot \bar{\mathbf{L}}}{\gamma c^2} = \frac{\mathbf{v}}{c^2} \tag{2.106}$$

The linear transformation (2.92) with $\bar{\mathbf{L}}$ and $\mathbf{m}$ given by Eqs. (2.103) and (2.106), respectively, is known as the *Lorentz transformation.* For easy reference, we repeat it below

$$\begin{aligned} \mathbf{r}' &= \bar{\mathbf{L}} \cdot \mathbf{r} - \gamma \mathbf{v} t \\ t' &= \gamma\left(t - \frac{\mathbf{v} \cdot \mathbf{r}}{c^2}\right) \end{aligned} \tag{2.107}$$

same mathematical form in all inertial systems. Thus, in the unprimed system Σ, the Maxwell equations are

$$\nabla \times \mathscr{H} - \frac{\partial \mathscr{D}}{\partial t} = \mathscr{J} \tag{2.119}$$

$$\nabla \cdot \mathscr{D} = \rho \tag{2.120}$$

$$\nabla \times \mathscr{E} + \frac{\partial \mathscr{B}}{\partial t} = \mathbf{0} \tag{2.121}$$

$$\nabla \cdot \mathscr{B} = 0 \tag{2.122}$$

and the equation of continuity is

$$\nabla \cdot \mathscr{J} = -\frac{\partial \rho}{\partial t} \tag{2.123}$$

where $\mathscr{E}$, $\mathscr{D}$, $\mathscr{H}$, $\mathscr{B}$, $\mathscr{J}$, and ρ are functions of the unprimed variables $\mathbf{r}$, t. In the primed system Σ', which moves with a uniform velocity $\mathbf{v}$ with respect to Σ, they are

$$\nabla' \times \mathscr{H}' - \frac{\partial \mathscr{D}'}{\partial t'} = \mathscr{J}' \tag{2.124}$$

$$\nabla' \cdot \mathscr{D}' = \rho' \tag{2.125}$$

$$\nabla' \times \mathscr{E}' + \frac{\partial \mathscr{B}'}{\partial t'} = \mathbf{0} \tag{2.126}$$

$$\nabla' \cdot \mathscr{B}' = 0 \tag{2.127}$$

and

$$\nabla' \cdot \mathscr{J}' = -\frac{\partial \rho'}{\partial t'} \tag{2.128}$$

where $\mathscr{E}'$, $\mathscr{D}'$, $\mathscr{H}'$, $\mathscr{B}'$, $\mathscr{J}'$, and ρ' are functions of the primed variables $\mathbf{r}'$, t'. Since $\mathbf{r}$, t are related to $\mathbf{r}'$, t' by the Lorentz transformation, we may thus consider the sources and field vectors in Σ as functions of $\mathbf{r}'$, t'. To derive transformation formulas for sources $\mathscr{J}$ and ρ, we substitute Eqs. (2.114) and (2.118) into Eq. (2.123). Noting that $\bar{\mathbf{L}}$ is symmetric and $\mathbf{v}$ is a constant vector, we obtain

$$\nabla' \cdot (\bar{\mathbf{L}} \cdot \mathscr{J} - \gamma \mathbf{v} \rho) = -\frac{\partial}{\partial t'}\left(\gamma \rho - \frac{\gamma}{c^2}\mathbf{v} \cdot \mathscr{J}\right) \tag{2.129}$$

Equations (2.128) and (2.129) have the same mathematical form showing that they are Lorentz covariant, provided that the current and charge densities transform according to

$$\mathscr{J}' = \bar{\mathbf{L}} \cdot \mathscr{J} - \gamma \mathbf{v} \rho \tag{2.130}$$

$$\rho' = \gamma\left(\rho - \frac{\mathbf{v} \cdot \mathscr{J}}{c^2}\right) \tag{2.131}$$

Here we note that $\mathscr{J}$ and ρ transform in the same way as $\mathbf{r}$ and t; that is, the mathematical forms of Eqs. (2.130) and (2.131) are the same as that of the Lorentz transformation (2.107).

To determine the transformations for the field vectors $\mathscr{H}$ and $\mathscr{D}$, we substitute $\mathscr{J}$ of Eq. (2.119) and ρ of Eq. (2.120) into the right-hand side of Eq. (2.130), use the transformations (2.114) and (2.118), and obtain

$$\bar{\mathbf{L}} \cdot [(\bar{\mathbf{L}} \cdot \nabla') \times \mathscr{H}] + \gamma\{\bar{\mathbf{L}} \cdot (\mathbf{v} \cdot \nabla')\mathscr{D} - \mathbf{v}[(\bar{\mathbf{L}} \cdot \nabla') \cdot \mathscr{D}]\}$$
$$+ \frac{\partial}{\partial t'}\left[-\frac{\gamma\beta}{c}\bar{\mathbf{L}} \cdot (\hat{\mathbf{v}} \times \mathscr{H}) - \gamma(\bar{\mathbf{L}} \cdot \mathscr{D} - \gamma\beta^2\hat{\mathbf{v}}\hat{\mathbf{v}} \cdot \mathscr{D})\right] = \mathscr{J}' \quad (2.132)$$

But

$$\begin{aligned} \bar{\mathbf{L}} \cdot [(\bar{\mathbf{L}} \cdot \nabla') \times \mathscr{H}] &= \nabla' \times (\gamma\bar{\mathbf{L}}^{-1} \cdot \mathscr{H}) \\ \bar{\mathbf{L}} \cdot (\mathbf{v} \cdot \nabla')\mathscr{D} - \mathbf{v}[(\bar{\mathbf{L}} \cdot \nabla') \cdot \mathscr{D}] &= -\nabla' \times (\mathbf{v} \times \mathscr{D}) \\ \bar{\mathbf{L}} \cdot (\hat{\mathbf{v}} \times \mathscr{H}) &= \hat{\mathbf{v}} \times \mathscr{H} \\ \bar{\mathbf{L}} \cdot \mathscr{D} - \gamma\beta^2\hat{\mathbf{v}}\hat{\mathbf{v}} \cdot \mathscr{D} &= \bar{\mathbf{L}}^{-1} \cdot \mathscr{D} \end{aligned} \quad (2.133)$$

where $\bar{\mathbf{L}}^{-1}$ is given by Eq. (2.105). Hence Eq. (2.132) becomes

$$\nabla' \times [\gamma(\bar{\mathbf{L}}^{-1} \cdot \mathscr{H} - \mathbf{v} \times \mathscr{D})] - \frac{\partial}{\partial t'}\left[\gamma\left(\bar{\mathbf{L}}^{-1} \cdot \mathscr{D} + \frac{1}{c^2}\mathbf{v} \times \mathscr{H}\right)\right] = \mathscr{J}' \quad (2.134)$$

Equations (2.124) and (2.134) have the same mathematical form provided that the field vectors $\mathscr{H}$ and $\mathscr{D}$ transform according to

$$\begin{aligned} \mathscr{H}' &= \gamma(\bar{\mathbf{L}}^{-1} \cdot \mathscr{H} - \mathbf{v} \times \mathscr{D}) \\ \mathscr{D}' &= \gamma\left(\bar{\mathbf{L}}^{-1} \cdot \mathscr{D} + \frac{\mathbf{v} \times \mathscr{H}}{c^2}\right) \end{aligned} \quad (2.135)$$

Similarly, the transformations for $\mathscr{E}$ and $\mathscr{B}$ can be obtained from Eqs. (2.121) and (2.126). To this end, we dot-multiply Eq. (2.121) from the left by $\bar{\mathbf{L}}$ and add to it the product of $\gamma\mathbf{v}$ and Eq. (2.122). Using again the results of Eqs. (2.114) and (2.118), we obtain

$$\bar{\mathbf{L}} \cdot [(\bar{\mathbf{L}} \cdot \nabla') \times \mathscr{E}] - \gamma\{\bar{\mathbf{L}} \cdot (\mathbf{v} \cdot \nabla')\mathscr{B} - \mathbf{v}[(\bar{\mathbf{L}} \cdot \nabla') \cdot \mathscr{B}]\}$$
$$+ \frac{\partial}{\partial t'}\left[-\frac{\gamma\beta}{c}\bar{\mathbf{L}} \cdot (\hat{\mathbf{v}} \times \mathscr{E}) + \gamma(\bar{\mathbf{L}} \cdot \mathscr{B} - \gamma\beta^2\hat{\mathbf{v}}\hat{\mathbf{v}} \cdot \mathscr{B})\right] = \mathbf{0} \quad (2.136)$$

Making use of the formulas similar to Eq. (2.133), we may write Eq. (2.136) as

$$\nabla' \times [\gamma(\bar{\mathbf{L}}^{-1} \cdot \mathscr{E} + \mathbf{v} \times \mathscr{B})] + \frac{\partial}{\partial t'}\left[\gamma\left(\bar{\mathbf{L}}^{-1} \cdot \mathscr{B} - \frac{\mathbf{v} \times \mathscr{E}}{c^2}\right)\right] = \mathbf{0} \quad (2.137)$$

Equations (2.137) and (2.126) have the same mathematical form, and the two are exactly the same if

$$\begin{aligned} \mathscr{E}' &= \gamma(\bar{\mathbf{L}}^{-1} \cdot \mathscr{E} + \mathbf{v} \times \mathscr{B}) \\ \mathscr{B}' &= \gamma\left(\bar{\mathbf{L}}^{-1} \cdot \mathscr{B} - \frac{\mathbf{v} \times \mathscr{E}}{c^2}\right) \end{aligned} \tag{2.138}$$

This completes the derivations of transformation formulas for sources and field vectors in two systems moving with a relative uniform velocity. The inverse transformations can again be obtained either by solving the unprimed quantities in Eqs. (2.130), (2.131), (2.135), and (2.138) directly in terms of the primed quantities or by simply interchanging the primed and unprimed quantities and replacing $\mathbf{v}$ by $-\mathbf{v}$. In either case, we obtain

$$\begin{aligned} \mathscr{J} &= \bar{\mathbf{L}} \cdot \mathscr{J}' + \gamma \mathbf{v} \rho' \\ \rho &= \gamma\left(\rho' + \frac{\mathbf{v} \cdot \mathscr{J}'}{c^2}\right) \\ \mathscr{H} &= \gamma(\bar{\mathbf{L}}^{-1} \cdot \mathscr{H}' + \mathbf{v} \times \mathscr{D}') \\ \mathscr{D} &= \gamma\left(\bar{\mathbf{L}}^{-1} \cdot \mathscr{D}' - \frac{\mathbf{v} \times \mathscr{H}'}{c^2}\right) \\ \mathscr{E} &= \gamma(\bar{\mathbf{L}}^{-1} \cdot \mathscr{E}' - \mathbf{v} \times \mathscr{B}') \\ \mathscr{B} &= \gamma\left(\bar{\mathbf{L}}^{-1} \cdot \mathscr{B}' + \frac{\mathbf{v} \times \mathscr{E}'}{c^2}\right) \end{aligned} \tag{2.139}$$

In summary, we have shown that the Maxwell equations are Lorentz covariant, provided that the sources and field vectors are transformed according to Eqs. (2.130), (2.131), (2.135), and (2.138).

PROBLEMS

2.1 Starting with Maxwell's equations, derive the Coulomb law of force between two stationary point electric charges Q_0 and Q:

$$\mathscr{F} = Q_0 \mathscr{E} = \frac{Q_0 Q}{4\pi\varepsilon_0 \varepsilon R^2} \hat{\mathbf{R}}$$

2.2 A plane $y = 0$ separates the region of free space ($y < 0$) from the region occupied by a perfect conductor ($y > 0$). A plane wave is traveling in the $+z$ direction with its magnetic field intensity given by

$$\mathscr{H}(z, t) = H_0 f\left(t - \frac{z}{v}\right) \hat{\mathbf{x}}$$

where f is an arbitrary function of $t - \dfrac{z}{v}$ and H_0 is a constant.

(*a*) In which region of the space does the wave exist?
(*b*) Find the value of v.
(*c*) Find the electric field intensity.

(*d*) Find the surface charge and current densities.
(*e*) Find the Poynting vector.

2.3 At the time $t = 0$ there is a charge distribution ρ_0 in a medium of nonzero conductivity σ and dielectric constant ε. Determine how this charge distribution changes with time.

2.4 (*a*) Show that for a linear, time-invariant dielectric medium which satisfies the causality requirement: the response follows the stimulus, the real part $\varepsilon_1(\omega)$ and the imaginary part $\varepsilon_2(\omega)$ of the complex dielectric constant $\varepsilon(\omega)$ are related by

$$\varepsilon_1(\omega) = 1 + \frac{2}{\pi} P \int_0^\infty \frac{\varepsilon_2(\omega')}{\omega'^2 - \omega^2} \omega' \, d\omega'$$

and

$$\varepsilon_2(\omega) = -\frac{2\omega}{\pi} P \int_0^\infty \frac{\varepsilon_1(\omega')}{\omega'^2 - \omega^2} \, d\omega'$$

where P denotes the Cauchy principal value. These are known as the *Kramers-Kronig relations*. *Hint*: assume that $\varepsilon(\omega)$ is an analytical function of ω and carry out the integral

$$\oint \frac{\varepsilon(\omega')}{\omega' - \omega} \, d\omega'$$

over the contour which is made up of the real axis indented slightly into the upper half-plane near $\omega' = \omega$, and a large semicircle of radius R that closes the contour in the upper half-plane. (*b*) Determine the real part $\varepsilon_1(\omega)$ of the dielectric constant, if the imaginary part is given by

$$\varepsilon_2(\omega) = \frac{(a-1)\, b\omega}{1 + b^2\omega^2}$$

where a and b are constants.

2.5 When a current-carrying conductor is placed in a uniform magnetic field $\mathscr{B}_0$, an electric field is developed in the direction perpendicular to both the current and the applied magnetic field. This phenomenon is known as the *Hall effect*. (*a*) Using an atomic model, justify the following "generalized" Ohm's law:

$$\mathscr{J} + \mu_H \mathscr{B}_0 \times \mathscr{J} = \sigma \mathscr{E}$$

where

$$\mu_H = \frac{q\tau}{m} = \frac{\sigma}{nq} = R_H \sigma$$

is called the *Hall mobility*. m is the mass of charge q, n is the number density, τ is the mean collision time and R_H is the *Hall coefficient*. (*b*) Show that the result of (*a*) may be expressed as

$$\mathscr{J} = \bar{\boldsymbol{\sigma}} \cdot \mathscr{E}$$

where the conductivity tensor takes the form

$$\bar{\boldsymbol{\sigma}} = \sigma_\perp(\bar{\mathbf{I}} - \hat{\mathbf{b}}\hat{\mathbf{b}}) - \sigma_H(\hat{\mathbf{b}} \times \bar{\mathbf{I}}) + \sigma_\parallel \hat{\mathbf{b}}\hat{\mathbf{b}}$$

$\hat{\mathbf{b}}$ is the unit vector in the direction of $\mathscr{B}_0$ and

$$\sigma_\parallel = \sigma = \text{longitudinal conductivity}$$

$$\sigma_\perp = \frac{\sigma}{1 + \mu_H^2 \mathscr{B}_0^2} = \text{transverse conductivity}$$

$$\sigma_H = \frac{\mu_H \mathscr{B}_0 \sigma}{1 + \mu_H^2 \mathscr{B}_0^2} = \text{Hall conductivity}$$

2.6 Show that for a time-varying field, the boundary condition (2.26) implies (2.23) and the boundary condition (2.27) implies (2.24).

2.7 For media characterized by the constitutive relations

$$\mathscr{D} = \varepsilon_0 \bar{\varepsilon} \cdot \mathscr{E} + \bar{\xi} \cdot \mathscr{H}$$

$$\mathscr{B} = \bar{\zeta} \cdot \mathscr{E} + \mu_0 \bar{\mu} \cdot \mathscr{H}$$

where $\bar{\varepsilon}$, $\bar{\mu}$, $\bar{\zeta}$, $\bar{\xi}$, are constant tensors, show that the energy density (2.37) is given by

$$W = \frac{1}{2}(\mathscr{E} \cdot \mathscr{D} + \mathscr{H} \cdot \mathscr{B})$$

provided that $\bar{\mu}$ and $\bar{\varepsilon}$ are symmetric and that $\tilde{\bar{\zeta}} = \bar{\xi}$.

2.8 Consider two sets of monochromatic sources $\mathbf{J}_1$ and $\mathbf{J}_2$ of the same frequency, existing in an anisotropic medium characterized by the symmetric dielectric tensor $\bar{\varepsilon}$. Denote the fields produced by the source $\mathbf{J}_1$ alone by $\mathbf{E}_1$, $\mathbf{H}_1$, and the fields produced by the source $\mathbf{J}_2$ alone by $\mathbf{E}_2$, $\mathbf{H}_2$. Show that the sources and the electric fields satisfy the following *reciprocity relation*:

$$\int_V \mathbf{E}_2 \cdot \mathbf{J}_1 \, d^3r = \int_V \mathbf{E}_1 \cdot \mathbf{J}_2 \, d^3r$$

This relation is also valid in the special case of isotropic media.

2.9 Consider a given current distribution $\mathbf{J}$ in a region of space V filled with a linear, isotropic conducting medium. Let the tangential component of either $\mathbf{E}$, or $\mathbf{H}$, at each point on the bounding surface S be specified. Suppose there are two different fields $(\mathbf{E}_1, \mathbf{H}_1)$ and $(\mathbf{E}_2, \mathbf{H}_2)$ which are excited by the source $\mathbf{J}$ in V and satisfy the given boundary conditions on S. (*a*) Show that the difference fields $\mathbf{E}_3 = \mathbf{E}_2 - \mathbf{E}_1$ and $\mathbf{H}_3 = \mathbf{H}_2 - \mathbf{H}_1$ satisfy the source-free Maxwell equations at points in V and homogeneous boundary conditions on S. (*b*) Show that if $\sigma \neq 0$, $\mathbf{E}_3 = \mathbf{H}_3 = \mathbf{0}$ everywhere in V. That is, a field in a lossy region is uniquely determined by the sources within the region and the tangential component of $\mathbf{E}$ or $\mathbf{H}$ on S. The field in a lossless medium may be regarded as the limit of the corresponding field in a lossy medium when $\sigma \to 0$. Hence *uniqueness theorem* also applies to lossless media.

2.10 The determination of electromagnetic fields in a homogeneous, isotropic medium is often facilitated by the use of auxiliary functions known as *potentials*. According to Eq. (2.3) the vector $\mathscr{B}$ is divergenceless, and thus we can represent it as the curl of a *vector potential* $\mathscr{A}$:

$$\mathscr{B} = \nabla \times \mathscr{A}$$

Hence, from Eq. (2.1) it follows that

$$\mathscr{E} = -\frac{\partial \mathscr{A}}{\partial t} - \nabla \phi$$

where ϕ is called a *scalar potential*. (*a*) Show that the vector and scalar potentials satisfy the inhomogeneous wave equations

$$\nabla^2 \mathscr{A} - \frac{\mu\varepsilon}{c^2}\frac{\partial^2 \mathscr{A}}{\partial t^2} = -\mu_0 \mu \mathscr{J}$$

$$\nabla^2 \phi - \frac{\mu\varepsilon}{c^2}\frac{\partial^2 \phi}{\partial t^2} = -\frac{\rho}{\varepsilon_0 \varepsilon}$$

provided that they are related by the *Lorentz condition*

$$\nabla \cdot \mathscr{A} + \frac{\mu\varepsilon}{c^2}\frac{\partial \phi}{\partial t} = 0$$

where $c = 1/\sqrt{\mu_0 \varepsilon_0}$ is the velocity of light in free space.

(*b*) Show that the Lorentz condition implies the equation of continuity (2.5). (*c*) If we choose the vector potential to be divergenceless, that is, $\nabla \cdot \mathscr{A} = 0$ (Coulomb condition), find the equations satisfied by $\mathscr{A}$ and ϕ.

2.11 An electromagnetic wave can carry energy. It can also exert stress which may be a pressure, a tension, or a shear in a material medium. Let us consider a monochromatic plane wave propagating in a homogeneous, isotropic medium.

(*a*) Taking the cross product of Eq. (2.1) with $\mathscr{D}$ and Eq. (2.2) with $\mathscr{B}$ and then adding, show that

$$(\nabla \times \mathscr{E}) \times \mathscr{D} + (\nabla \times \mathscr{H}) \times \mathscr{B} = \mathscr{J} \times \mathscr{B} + \frac{\partial}{\partial t}(\mathscr{D} \times \mathscr{B})$$

(*b*) Making use of the identities in Prob. 1.10, show that

$$\nabla \cdot \bar{\mathscr{T}} = (\nabla \times \mathscr{E}) \times \mathscr{D} + (\nabla \times \mathscr{H}) \times \mathscr{B} + \mathscr{E}(\nabla \cdot \mathscr{D}) + \mathscr{H}(\nabla \cdot \mathscr{B})$$

where
$$\bar{\mathscr{T}} = \mathscr{D}\mathscr{E} + \mathscr{B}\mathscr{H} - \tfrac{1}{2}(\mathscr{D} \cdot \mathscr{E} + \mathscr{B} \cdot \mathscr{H})\bar{\mathbf{I}}$$

is the *Maxwell stress tensor* which has the dimensions of force per unit area.

(*c*) Making use of the two divergence Maxwell equations, show that

$$\mathbf{f} = \nabla \cdot \bar{\mathscr{T}} - \frac{\mu\varepsilon}{c^2}\frac{\partial}{\partial t}(\mathscr{E} \times \mathscr{H})$$

where
$$\mathbf{f} = \rho\mathscr{E} + \mathscr{J} \times \mathscr{B}$$

is the Lorentz force density which has the dimensions of force per unit volume.

(*d*) Integrating the result of part (*c*) over a volume V bounded by a surface S, show that the total force $\mathbf{F}$ is

$$\mathbf{F} = \oint_s \left[\varepsilon_0 \varepsilon\, \mathscr{E}(\mathscr{E} \cdot \hat{\mathbf{n}}) - \frac{\varepsilon_0 \varepsilon}{2}\mathscr{E}^2 \hat{\mathbf{n}}\right] ds$$
$$+ \oint_s \left[\mu_0 \mu\, \mathscr{H}(\mathscr{H} \cdot \hat{\mathbf{n}}) - \frac{\mu_0 \mu}{2}\mathscr{H}^2 \hat{\mathbf{n}}\right] ds - \frac{\mu\varepsilon}{c^2}\int_V \frac{\partial}{\partial t}(\mathscr{E} \times \mathscr{H})\, d^3r$$

where $\hat{\mathbf{n}}$ is a unit vector normal to ds.

2.12 Show that each term of the Maxwell stress tensor defined in Prob. 2.11*b* has the dimensions of force per unit area or energy per unit volume.

2.13 Consider two inertial frames Σ and Σ' with Σ' moving with uniform velocity v along the common positive x axis relative to Σ. Assume that the Lorentz transformations are linear transformations of the form

$$x' = Ax + Bt$$
$$t' = Cx + Dt$$

Since the origin of Σ' corresponding to $x' = 0$ moves with velocity v, we have $-B/A = v$ so that $x' = A(x - vt)$. Substituting x' and t' in the identity $x^2 - c^2 t^2 = x'^2 - c^2 t'^2$ and equating coefficients of x^2, t^2, and xt, show that the resulting transformations for x' and t' are the Lorentz transformations.

2.14 Show that when the operator $\bar{\mathbf{L}}$ defined in Eq. (2.103) operates on a vector $\mathbf{a} = \mathbf{a}_\perp + \mathbf{a}_\parallel$, it increases the component of $\mathbf{a}$ parallel to $\hat{\mathbf{v}}$ by γ times and keeps the component perpendicular to $\hat{\mathbf{v}}$ unchanged:

$$\bar{\mathbf{L}} \cdot \mathbf{a} = \gamma \mathbf{a}_\parallel + \mathbf{a}_\perp$$

Hence $\bar{\mathbf{L}}$ is referred to as a *longitudinal stretcher operator*. On the other hand, show that

$$(\gamma \bar{\mathbf{L}}^{-1}) \cdot \mathbf{a} = \gamma \mathbf{a}_\perp + \mathbf{a}_\parallel$$

Hence $\gamma \bar{\mathbf{L}}^{-1}$ is a *transerve stretcher operator*.

2.15 (*a*) Show that the transformation laws (2.135) and (2.138) can be decomposed into components parallel and perpendicular to the direction of motion:

$$\mathscr{E}' = (\mathscr{E} \cdot \hat{\mathbf{v}})\hat{\mathbf{v}} + \gamma[(\bar{\mathbf{I}} - \hat{\mathbf{v}}\hat{\mathbf{v}}) \cdot \mathscr{E} + \mathbf{v} \times \mathscr{B}]$$

$$\mathscr{H}' = (\mathscr{H} \cdot \hat{\mathbf{v}})\hat{\mathbf{v}} + \gamma[(\bar{\mathbf{I}} - \hat{\mathbf{v}}\hat{\mathbf{v}}) \cdot \mathscr{H} - \mathbf{v} \times \mathscr{D}]$$

$$\mathscr{D}' = (\mathscr{D} \cdot \hat{\mathbf{v}})\hat{\mathbf{v}} + \gamma\left[(\bar{\mathbf{I}} - \hat{\mathbf{v}}\hat{\mathbf{v}}) \cdot \mathscr{D} + \frac{\mathbf{v} \times \mathscr{H}}{c^2}\right]$$

$$\mathscr{B}' = (\mathscr{B} \cdot \hat{\mathbf{v}})\hat{\mathbf{v}} + \gamma\left[(\bar{\mathbf{I}} - \hat{\mathbf{v}}\hat{\mathbf{v}}) \cdot \mathscr{B} - \frac{\mathbf{v} \times \mathscr{E}}{c^2}\right]$$

Thus, the field components parallel to the direction of motion remain unchanged. However, the electric and magnetic fields perpendicular to the direction of motion are interrelated as indicated in the terms inside the square brackets.

(*b*) Considering a special case $\hat{\mathbf{v}} = \hat{\mathbf{x}}$, show that the transformation laws reduce to

$$\mathscr{E}'_x = \mathscr{E}_x \qquad \mathscr{H}'_x = \mathscr{H}_x$$

$$\mathscr{E}'_y = \gamma(\mathscr{E}_y - v\mathscr{B}_z) \qquad \mathscr{H}'_y = \gamma(\mathscr{H}_y + v\mathscr{D}_z)$$

$$\mathscr{E}'_z = \gamma(\mathscr{E}_z + v\mathscr{B}_y) \qquad \mathscr{H}'_z = \gamma(\mathscr{H}_z - v\mathscr{D}_y)$$

$$\mathscr{D}'_x = \mathscr{D}_x \qquad \mathscr{B}'_x = \mathscr{B}_x$$

$$\mathscr{D}'_y = \gamma\left(\mathscr{D}_y - \frac{v\mathscr{H}_z}{c^2}\right) \qquad \mathscr{B}'_y = \gamma\left(\mathscr{B}_y + \frac{v\mathscr{E}_z}{c^2}\right)$$

$$\mathscr{D}'_z = \gamma\left(\mathscr{D}_z + \frac{v\mathscr{H}_y}{c^2}\right) \qquad \mathscr{B}'_z = \gamma\left(\mathscr{B}_z - \frac{v\mathscr{E}_y}{c^2}\right)$$

2.16 An observer finds that a certain region of space contains a static electric field $\mathscr{E} = 20\hat{\mathbf{z}}$ V/m. How does this field appear to an observer moving through the region in the $+y$ direction with velocity of (*a*) 30 m/s, (*b*) 3×10^7 m/s, and (*c*) 3×10^8 m/s?

2.17 Repeat Prob. 2.16 for the case where the stationary observer finds that the region contains only a static magnetic field $\mathbf{B} = 30 \times 10^{-9}\hat{\mathbf{x}}$ Wb/m^2.

2.18 Show that the electric field at a point $\mathbf{r}$ in free space due to a point charge of magnitude Q at $\mathbf{r}_0$ moving with uniform velocity $\mathbf{v}$ is given by

$$\mathscr{E} = \frac{Q}{4\pi\varepsilon_0 R^2} \frac{1 - \beta^2}{(1 - \beta^2 \sin^2 \theta)^{3/2}} \hat{\mathbf{R}}$$

where $\mathbf{R}$ is the distance vector from $\mathbf{r}_0$ to $\mathbf{r}$ and θ is the angle between $\mathbf{R}$ and $\mathbf{v}$. Note that when $v = 0$ this reduces to the result as given in Prob. 2.1 for a static charge. An examination of the above result also reveals that the electric field due to a moving charge is increased in the direction perpendicular to the motion, and is decreased along the direction of motion. But according to Prob. 2.15, the components of the electric field parallel to $\mathbf{v}$ must be equal: $\mathscr{E}'_\parallel = \mathscr{E}_\parallel = (\mathscr{E} \cdot \hat{\mathbf{v}})\hat{\mathbf{v}}$. How is this reconciled with the fact that the electric field is decreased along the direction of motion?

2.19 Show that the magnetic field in free space due to a charge Q moving with uniform velocity $\mathbf{v}$ is given by

$$\mathscr{B} = \frac{\mu_0}{4\pi} \frac{Q(\mathbf{v} \times \hat{\mathbf{R}})}{R^2} \frac{1 - \beta^2}{(1 - \beta^2 \sin^2 \theta)^{3/2}}$$

$$= \frac{1}{c^2} \mathbf{v} \times \mathscr{E}$$

where $\mathscr{E}$ is given in Prob. 2.18. Note that when $v \ll c$ this reduces to the Biot-Savart law:

$$d\mathscr{B} = \frac{\mu_0}{4\pi} \frac{\mathscr{J} \times \hat{\mathbf{R}}}{R^2} \, d^3r$$

for a steady current density $\mathscr{J}$.

2.20 (*a*) Show that the Lorentz transformation can be written as

$$x'_\alpha = a_{\alpha\beta} x_\beta \qquad \alpha = 1, 2, 3, 4$$

where summation over the repeated Greek index β from 1 to 4 is implied, and $[x_\beta] = (x_1, x_2, x_3, x_4) = (\mathbf{r}, ict)$, $[x'_\alpha] = (\mathbf{r}', ict')$ and

$$[a_{\alpha\beta}] = \begin{bmatrix} \bar{\mathbf{I}} + (\gamma - 1)\hat{\mathbf{v}}\hat{\mathbf{v}} & i\gamma \dfrac{\mathbf{v}}{c} \\ -i\gamma \dfrac{\tilde{\mathbf{v}}}{c} & \gamma \end{bmatrix}$$

$$= \begin{bmatrix} 1 + (\gamma - 1)\dfrac{v_x^2}{v^2} & (\gamma - 1)\dfrac{v_x v_y}{v^2} & (\gamma - 1)\dfrac{v_x v_z}{v^2} & i\gamma \dfrac{v_x}{c} \\ (\gamma - 1)\dfrac{v_y v_x}{v^2} & 1 + (\gamma - 1)\dfrac{v_y^2}{v^2} & (\gamma - 1)\dfrac{v_y v_z}{v^2} & i\gamma \dfrac{v_y}{c} \\ (\gamma - 1)\dfrac{v_z v_x}{v^2} & (\gamma - 1)\dfrac{v_z v_y}{v^2} & 1 + (\gamma - 1)\dfrac{v_z^2}{v^2} & i\gamma \dfrac{v_z}{c} \\ -i\gamma \dfrac{v_x}{c} & -i\gamma \dfrac{v_y}{c} & -i\gamma \dfrac{v_z}{c} & \gamma \end{bmatrix}$$

(*b*) Show that the transformation matrix is orthogonal, that is,

$$a_{\alpha\beta} a_{\alpha\nu} = a_{\beta\alpha} a_{\nu\alpha} = \delta_{\beta\nu}$$

(*c*) Assuming that the space-time coordinates undergo the Lorentz transformation, we define a 4-vector as a set of four quantities w_α ($\alpha = 1, 2, 3, 4$) that transform like the coordinates (see Appendix A):

$$w'_\alpha = a_{\alpha\beta} w_\beta$$

If $\mathscr{J}$, ρ, $\mathscr{A}$, and ϕ are the current density vector, the charge density, the magnetic vector potential, and the scalar potential, show that

$$[s_\alpha] = \begin{bmatrix} \mathscr{J} \\ ic\rho \end{bmatrix}, \; [\Phi_\alpha] = \begin{bmatrix} \mathscr{A} \\ \dfrac{i}{c}\phi \end{bmatrix} \qquad [u_\alpha] = \begin{bmatrix} \gamma \mathbf{u} \\ i\gamma c \end{bmatrix} \qquad \left[\frac{\partial}{\partial x_\alpha}\right] = \begin{bmatrix} \nabla \\ \dfrac{1}{ic}\dfrac{\partial}{\partial t} \end{bmatrix}$$

are 4-vectors.

(*d*) Show that the operator

$$\nabla^2 - \frac{1}{c^2}\frac{\partial^2}{\partial t^2}$$

is invariant in mathematical form under the Lorentz transformation.

2.21 (*a*) Show that in the rest frame Σ' of an isotropic medium, the inhomogeneous wave equations satisfied by the vector and scalar potentials (see Prob. 2.10) may be combined to yield:

$$[-1 - (\mu\varepsilon - 1)\delta_{4\alpha}]\left(\nabla'^2 - \frac{\mu\varepsilon}{c^2}\frac{\partial^2}{\partial t'^2}\right)\Phi'_\alpha = \mu_0 \mu s'_\alpha$$

where $\mu_0 \mu$ and $\varepsilon_0 \varepsilon$ are the permeability and the permittivity of the medium with respect to Σ', $\delta_{4\alpha}$ is the Kronecker delta.

(*b*) Show that the differential operator

$$\nabla'^2 - \frac{\mu\varepsilon}{c^2}\frac{\partial^2}{\partial t'^2}$$

in Σ' transforms to the differential operator

$$\mathscr{L} = \nabla^2 - \frac{1}{c^2}\frac{\partial^2}{\partial t^2} - \frac{(\mu\varepsilon - 1)\gamma^2}{c^2}\left(\frac{\partial}{\partial t} + \mathbf{v}\cdot\nabla\right)^2$$

in Σ.

(*c*) Using the results of Prob. 2.20, show that

$$a_{4\nu} a_{4\beta} s_\beta = -\frac{1}{c^2} u_\nu s_\beta u_\beta$$

(*d*) Transforming 4-potential vector Φ'_α and 4-current vector s'_α from Σ' to Σ, with respect to which the medium is moving at velocity $\mathbf{v}$, show that

$$a_{4\beta} \mathscr{L} \mathbf{\Phi}_\beta = -\frac{\mu_0}{\varepsilon} a_{4\beta} S_\beta$$

(*e*) Finally, making use of the orthogonality relation given in Prob. 2.20*b*, show that the wave equations for a moving medium become

$$\mathscr{L}\mathscr{A} = -\mu_0 \mu \mathscr{J} - \frac{(\mu\varepsilon - 1)\gamma^2}{\varepsilon_0 \varepsilon c^4}(\mathbf{v}\cdot\mathscr{J} - c^2\rho)\mathbf{v}$$

$$\mathscr{L}\phi = -\mu_0 \mu c^2 \rho - \frac{(\mu\varepsilon - 1)\gamma^2}{\varepsilon_0 \varepsilon c^2}(\mathbf{v}\cdot\mathscr{J} - c^2\rho)$$

2.22 By direct substitution, show that the following quantities are invariants under the Lorentz transformation: (*a*) $\mathscr{B}\cdot\mathscr{E}$; (*b*) $\mathscr{H}\cdot\mathscr{D}$; (*c*) $\mathscr{E}^2 - c^2\mathscr{B}^2$; (*d*) $\mathscr{H}^2 - c^2\mathscr{D}^2$; (*e*) $\mathscr{B}\cdot\mathscr{H} - \mathscr{E}\cdot\mathscr{D}$. *Hint*: use the transformation formulas for field vectors, show that $\mathscr{B}'\cdot\mathscr{E}' = \mathscr{B}\cdot\mathscr{E}$, etc.

2.23 Using the results of the previous problem, show that: (*a*) if $\mathscr{E} > c\mathscr{B}$ in one inertial frame, then $\mathscr{E} > c\mathscr{B}$ in any other inertial frame; (*b*) if $\mathscr{E}$ and $\mathscr{B}$ are perpendicular in one inertial frame, then they are perpendicular in any other inertial frame; (*c*) if the angle between $\mathscr{E}$ and $\mathscr{B}$ is acute (or obtuse) in one inertial frame, it is acute (or obtuse) in any other inertial frame.

2.24 It is known that the electric and magnetic fields in an inertial frame Σ are mutually perpendicular ($\mathscr{E} \perp \mathscr{H}$). Find the velocity of the frame Σ' moving relative to Σ, in which only the electric or only the magnetic field is present. Does this problem always have a solution, and if so, is it unique?

2.25 Derive the results of Eq. (2.139) by directly solving the unprimed quantities in Eqs. (2.130), (2.131), (2.135), and (2.138) in terms of the primed quantities.

REFERENCES

Electromagnetic theory†

Jordan, E. C., and K. G. Balmain: *Electromagnetic Waves and Radiating Systems*, 2d ed., Prentice-Hall, Inc., Englewood Cliffs, N.J., 1968.

Kraus, J. D., and K. R. Carver: *Electromagnetics*, 2d ed., McGraw-Hill Book Company, New York, 1973.

† For more references, see bibliography.

Lorrain, P., and D. C. Corson: *Electromagnetic Fields and Waves*, 2d ed., W. H. Freeman and Company, San Francisco, 1970.

Papas, C. H.: *Theory of Electromagnetic Wave Propagation*, McGraw-Hill Book Company, New York, 1965.

Paris, D. T., and F. K. Hurd: *Basic Electromagnetic Theory*, McGraw-Hill Book Company, New York, 1969.

Plonsey, R., and R. E. Collin: *Principles and Applications of Electromagnetic Fields*, McGraw-Hill Book Company, New York, 1961.

Ramo, S., J. R. Whinnery, and T. Van Duzer: *Fields and Waves in Communication Electronics*, John Wiley & Sons, Inc., New York, 1965.

Sommerfeld, A.: *Electrodynamics*, Academic Press, New York, 1949.

Stratton, J. A.: *Electromagnetic Theory*, McGraw-Hill Book Company, New York, 1941.

Energy density in dispersive media

Felsen, L. B., and N. Marcuvitz: *Radiation and Scattering of Waves*, Prentice-Hall, Inc., Englewood Cliffs, N.J., 1973, pp. 78–86.

Ginzburg, V. L.: *The Propagation of Electromagnetic Waves in Plasmas*, Pergamon Press, New York, 1964, appendixes A and B.

Kurss, H.: "Dispersion Relations, Stored Energy and Group Velocity for Anisotropic Electromagnetic Media," *Quart. Appl. Math.*, vol. 26, 1968, pp. 373–387.

Landau, L. D., and E. M. Lifshitz: *Electrodynamics of Continuous Media*, Pergamon Press, New York, 1960, pp. 253–256.

Stern, F.: "Elementary Theory of the Optical Properties of Solids," in F. Seitz and D. Turnbull (eds.), *Solid State Physics, Advances in Research and Applications*, Academic Press, New York, 1963, vol. 15, pp. 299–408.

Tonning, A.: "Energy Density in Continuous Electromagnetic Media," *IRE Trans. Antennas Propagation*, vol. AP-8, 1960, pp. 428–434.

Electromagnetism and relativity

Cullwick, E. G.: *Electromagnetism and Relativity*, John Wiley & Sons, Inc., New York, 1959.

Dunn, D. A.: *Models of Particles and Moving Media*, Academic Press, New York, 1971.

Heading, J.: *Electromagnetic Theory and Special Relativity*, University Tutorial Press, Ltd., London, 1964.

Hughes, W. F., and F. J. Young: *The Electromagnetodynamics of Fluids*, John Wiley & Sons, Inc., New York, 1966.

Møller, C.: *The Theory of Relativity*, Oxford University Press, London, 1966.

Pauli, W.: *Theory of Relativity*, Pergamon Press, New York, 1958.

Rosser, W. G. V.: *An Introduction to the Theory of Relativity*, Butterworths, London, 1964.

Ugarov, V. A.: *Special Theory of Relativity*, MIR Publishers, Moscow, 1979.

CHAPTER

THREE

SOME GENERAL PROPERTIES OF PLANE WAVES

In the previous chapter we reviewed the basic equations of electrodynamics. Now we shall examine plane wave solutions of the Maxwell equations in a source-free region. The plane waves are not only the simplest solutions of the Maxwell equations but they are also good approximations to many real wave problems. For instance, the waves at large distances from a transmitting antenna or from diffracting objects are well represented by plane waves. Furthermore, the Fourier integral enables us to represent more complicated waves as a superposition of plane waves; so in this sense the plane waves are the basic building blocks for all wave problems. Even in their own right, the basic concepts of plane wave propagation, reflection, and refraction have extensive applications in optics and other wave problems.

3.1 MONOCHROMATIC PLANE WAVES

Let us consider a monochromatic field whose variation in space is also sinusoidal. In this case, the instantaneous vector $\mathscr{E}$ takes the form

$$\mathscr{E} = \text{Re}\,(\mathbf{E}e^{-i\omega t}) \tag{3.1}$$

where the complex vector $\mathbf{E}$ depends on the position vector $\mathbf{r}$ and is given by

$$\mathbf{E} = \mathbf{E}_0\, e^{i\mathbf{k}\cdot\mathbf{r}} \tag{3.2}$$

In Eq. (3.2), $\mathbf{E}_0$ is a complex-constant amplitude vector, independent of $\mathbf{r}$, and the constant vector $\mathbf{k}$ is called the *wave vector*. In this chapter, we shall assume that $\mathbf{k} = k\hat{\mathbf{k}}$ is real and call k the *wave number* and the unit vector $\hat{\mathbf{k}}$ the *wave normal*.

Expression (3.1) is a constant vector when the phase

$$\phi = \mathbf{k} \cdot \mathbf{r} - \omega t \tag{3.3}$$

is a constant. Hence, we have

$$\mathbf{r} \cdot \hat{\mathbf{k}} = \zeta = \frac{c_1 + \omega t}{k} \tag{3.4}$$

which is an equation of a plane in the three-dimensional space. That is, at a given instant t, the surface of constant phase defined by $\phi = c_1$ is a plane which is perpendicular to the unit vector $\hat{\mathbf{k}}$ and is located at a distance ζ away from the origin as shown in Fig. 3.1. As t increases, the value of ζ must also increase to maintain $\phi = c_1$, and the plane defined by Eq. (3.4) will move in the direction of $\hat{\mathbf{k}}$. For this reason, the field is called a *plane wave*. It is also a uniform plane wave because on the plane of constant phase, the amplitude of the wave is also a constant. Differentiating Eq. (3.4) with respect to t, we obtain the velocity with which a plane of constant phase travels in the direction of $\hat{\mathbf{k}}$:

$$v_p = \frac{\partial \zeta}{dt} = \frac{\omega}{k} \tag{3.5}$$

and the vector

$$\mathbf{v}_p = v_p \hat{\mathbf{k}} = \frac{\omega}{k^2} \mathbf{k} \tag{3.6}$$

is called the *phase velocity vector* of the wave. The magnitude v_p gives the phase velocity of the wave and $\hat{\mathbf{k}}$ gives the direction of the wave propagation. It should be noted that the phase velocity represents the velocity of propagation of a plane of constant phase and does not necessarily coincide with the velocity of energy propagation.

The distance between two planes of constant phase whose phase difference is 2π radians is called the *wavelength* λ of the wave. For two planes of constant phase defined by the equations

$$\begin{aligned} \phi_1 &= \mathbf{k} \cdot \mathbf{r}_1 - \omega t \\ \phi_2 &= \mathbf{k} \cdot \mathbf{r}_2 - \omega t \end{aligned} \tag{3.7}$$

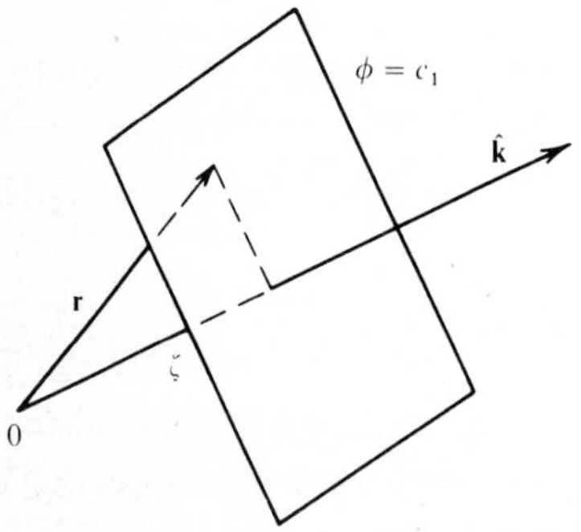

Figure 3.1 Uniform plane wave propagating in the direction $\hat{\mathbf{k}}$.

where $\mathbf{r}_1$ and $\mathbf{r}_2$ are the position vectors for points on the first and second planes of constant phase respectively (see Fig. 3.2), we obtain

$$\phi_2 - \phi_1 = 2\pi = \mathbf{k} \cdot (\mathbf{r}_2 - \mathbf{r}_1) = k\lambda$$

Hence

$$\lambda = \frac{2\pi}{k} \tag{3.8}$$

Similarly, the time T in which the phase ϕ changes by 2π radians, is called a *period* and is given by

$$T = \frac{2\pi}{\omega} = \frac{1}{f} \tag{3.9}$$

Sometimes, it is more convenient to use the *refractive index vector* $\mathbf{n}$ defined by

$$\mathbf{n} = \frac{c}{\omega}\mathbf{k} \tag{3.10}$$

The magnitude

$$n = \frac{ck}{\omega} \tag{3.11}$$

of the vector $\mathbf{n}$ is called the *index of refraction* and the direction of $\mathbf{n}$ coincides with the wave normal $\hat{\mathbf{k}}$. In terms of $\mathbf{n}$, the phase ϕ of the wave may be written as

$$\phi = \omega\left(\frac{1}{c}\mathbf{n} \cdot \mathbf{r} - t\right) \tag{3.12}$$

So far we have considered the monochromatic plane wave of the form (3.1). We shall next examine conditions under which the monochromatic wave satisfies the Maxwell equations. Since the field vectors are linearly related by the Maxwell equations, they must all have the same space variation $e^{i\mathbf{k}\cdot\mathbf{r}}$ as $\mathbf{E}$. Also

$$\nabla e^{i\mathbf{k}\cdot\mathbf{r}} = i\mathbf{k}e^{i\mathbf{k}\cdot\mathbf{r}} \tag{3.13}$$

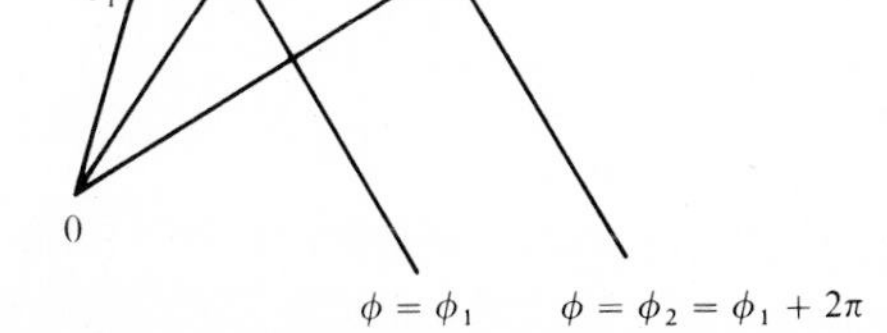

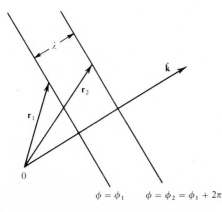

Figure 3.2 Two planes of constant phase differing in phase by 2π radians.

and

$$\begin{aligned}\nabla \times \mathbf{E} &= \nabla \times (\mathbf{E}_0 e^{i\mathbf{k}\cdot\mathbf{r}}) \\ &= e^{i\mathbf{k}\cdot\mathbf{r}}(\nabla \times \mathbf{E}_0) + (\nabla e^{i\mathbf{k}\cdot\mathbf{r}}) \times \mathbf{E}_0 \\ &= i\mathbf{k} \times \mathbf{E} \qquad (3.14)\end{aligned}$$

because $\mathbf{E}_0$ is independent of $\mathbf{r}$, $\nabla \times \mathbf{E}_0 = \mathbf{0}$. Now, substituting Eq. (3.2) and similar expressions for $\mathbf{D}$, $\mathbf{H}$, and $\mathbf{B}$ into Maxwell's equations (2.45) in a source-free region (that is, $\mathbf{J} = \mathbf{0}$, $\rho = 0$), and making use of Eqs. (3.13) and (3.14), we obtain

$$\omega \mathbf{B}_0 = \mathbf{k} \times \mathbf{E}_0 \qquad (3.15)$$

$$\omega \mathbf{D}_0 = -\mathbf{k} \times \mathbf{H}_0 \qquad (3.16)$$

and

$$\mathbf{k} \cdot \mathbf{D}_0 = \mathbf{k} \cdot \mathbf{B}_0 = 0 \qquad (3.17)$$

That is, to insure that the monochromatic plane wave of the form (3.2) is a solution of the Maxwell equations, the complex constant-amplitude vectors $\mathbf{E}_0$, $\mathbf{D}_0$, $\mathbf{H}_0$, and $\mathbf{B}_0$ (or $\mathbf{E}$, $\mathbf{D}$, $\mathbf{H}$, and $\mathbf{B}$) must be related by Eqs. (3.15) to (3.17). In other words, not all of them can be chosen arbitrarily.

To be more specific, let us consider a lossless, isotropic, nondispersive medium. In this case, the constitutive relations are

$$\begin{aligned}\mathbf{D}_0 &= \varepsilon_0 \varepsilon \mathbf{E}_0 \\ \mathbf{B}_0 &= \mu_0 \mu \mathbf{H}_0\end{aligned} \qquad (3.18)$$

where μ and ε are real constants. Substituting Eq. (3.18) into Eqs. (3.15) and (3.16), after eliminating either $\mathbf{H}_0$ or $\mathbf{E}_0$ from the resulting equations, we obtain

$$(n^2 - \mu\varepsilon)\mathbf{E}_0 = \mathbf{0} \qquad (3.19)$$

or

$$(n^2 - \mu\varepsilon)\mathbf{H}_0 = \mathbf{0} \qquad (3.20)$$

The homogeneous equation (3.19) [or (3.20)] has a nonzero solution $\mathbf{E}_0$ (or $\mathbf{H}_0$) only if the coefficient vanishes, i.e., if

$$n^2 = \mu\varepsilon \qquad (3.21)$$

or

$$k^2 = k_0^2 \mu\varepsilon = \frac{\omega^2 \mu\varepsilon}{c^2} \qquad (3.22)$$

where $k_0 = \omega\sqrt{\mu_0 \varepsilon_0} = \omega/c$ is the wave number in free space. Equation (3.22) which relates the wave number k and the frequency ω is called the *dispersion equation*. In this case, we find

$$\mathbf{H}_0 = \frac{1}{\eta}(\hat{\mathbf{k}} \times \mathbf{E}_0) \qquad (3.23)$$

and

$$\mathbf{E}_0 = -\eta(\hat{\mathbf{k}} \times \mathbf{H}_0) \qquad (3.24)$$

where

$$\eta = \sqrt{\frac{\mu_0 \mu}{\varepsilon_0 \varepsilon}} \qquad (3.25)$$

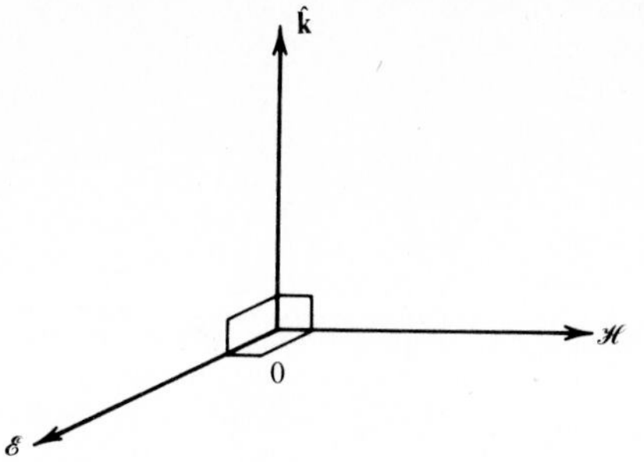

Figure 3.3 Relative orientations of $\mathscr{E}$, $\mathscr{H}$, and $\hat{\mathbf{k}}$.

is the *intrinsic impedance* of the medium. To interpret Eqs. (3.23) and (3.24), we obtain the relations in the time domain:

$$\mathscr{H} = \frac{1}{\eta}(\hat{\mathbf{k}} \times \mathscr{E}) \tag{3.26}$$

and

$$\mathscr{E} = -\eta(\hat{\mathbf{k}} \times \mathscr{H}) \tag{3.27}$$

from which it follows that

$$\hat{\mathbf{k}} \cdot \mathscr{H} = \hat{\mathbf{k}} \cdot \mathscr{E} = \mathscr{E} \cdot \mathscr{H} = 0 \tag{3.28}$$

In words, the monochromatic plane wave in a lossless, isotropic medium is a *transverse electromagnetic wave*, i.e., the electric and magnetic field vectors lie in the plane (the plane of constant phase) normal to the direction of wave propagation. In other words, $\mathscr{E}$, $\mathscr{H}$, and $\hat{\mathbf{k}}$ form a right-handed orthogonal triad of vectors as shown in Fig. 3.3. Hence, in the given medium, we may specify ω, $\hat{\mathbf{k}}$, and $\mathscr{E}$ arbitrarily with $\hat{\mathbf{k}} \cdot \mathscr{E} = 0$. Then $\mathscr{H}$ and k follow from Eqs. (3.26) and (3.22) respectively.

3.2 STATES OF POLARIZATION

In this section, we shall examine the states of polarization of a monochromatic plane wave. By polarization of a wave, we mean the curve traced out by the tip of the instantaneous electric field intensity $\mathscr{E}$ of Eq. (3.1) at a fixed point in space as time t varies. To be specific, let us write the complex cartesian components of the constant vector $\mathbf{E}_0$ as

$$\begin{aligned} E_{0x} &= E_{1x} + iE_{2x} \\ E_{0y} &= E_{1y} + iE_{2y} \\ E_{0z} &= E_{1z} + iE_{2z} \end{aligned} \tag{3.29}$$

The real quantities E_{1x}, E_{1y}, E_{1z}, and E_{2x}, E_{2y}, E_{2z} are the cartesian components of two real vectors $\mathbf{E}_1$ and $\mathbf{E}_2$ respectively. In other words, we may write the complex vector $\mathbf{E}_0$ as

$$\mathbf{E}_0 = \mathbf{E}_1 + i\mathbf{E}_2 \tag{3.30}$$

where $\mathbf{E}_1 = E_{1x}\hat{\mathbf{x}} + E_{1y}\hat{\mathbf{y}} + E_{1z}\hat{\mathbf{z}}$

$\mathbf{E}_2 = E_{2x}\hat{\mathbf{x}} + E_{2y}\hat{\mathbf{y}} + E_{2z}\hat{\mathbf{z}}$

and $\hat{\mathbf{x}}$, $\hat{\mathbf{y}}$, and $\hat{\mathbf{z}}$ are unit vectors along the x, y, and z axes, respectively. Substituting Eq. (3.30) into Eq. (3.1), we obtain the instantaneous electric field intensity

$$\begin{aligned}\mathscr{E} &= \operatorname{Re}(\mathbf{E}_0 e^{i\phi}) \\ &= \cos\phi\, \mathbf{E}_1 - \sin\phi\, \mathbf{E}_2 \end{aligned} \tag{3.31}$$

According to this representation, we see that $\mathscr{E}$ lies on the plane formed by vectors $\mathbf{E}_1$ and $\mathbf{E}_2$, but its direction is constantly changing from $\mathbf{E}_1$ at $\phi = 0$ to $-\mathbf{E}_2$ at $\phi = \pi/2$ to $-\mathbf{E}_1$ at $\phi = \pi$ to $\mathbf{E}_2$ at $\phi = 3\pi/2$ and then back to $\mathbf{E}_1$ at $\phi = 2\pi$. For any other ϕ, $\mathscr{E}$ in Eq. (3.31) is simply the vector sum of the two sinusoidally varying components along $\mathbf{E}_1$ and $\mathbf{E}_2$. Since $d\mathscr{E}/d\phi$ at each ϕ is readily computed, the curve traced out by the tip of $\mathscr{E}$ can thus be sketched. To show that this curve is an ellipse, we take the cross product of Eq. (3.31) first with respect to $\mathbf{E}_1$ and then with respect to $\mathbf{E}_2$:

$$\mathscr{E} \times \mathbf{E}_1 = (\mathbf{E}_1 \times \mathbf{E}_2) \sin\phi$$

$$\mathscr{E} \times \mathbf{E}_2 = (\mathbf{E}_1 \times \mathbf{E}_2) \cos\phi$$

Next, taking the dot product of each of the above equations with itself and then adding, we eliminate the parameter ϕ and obtain the equation for the polarization curve

$$(\mathscr{E} \times \mathbf{E}_1)^2 + (\mathscr{E} \times \mathbf{E}_2)^2 = (\mathbf{E}_1 \times \mathbf{E}_2)^2 \tag{3.32}$$

According to Eq. (3.31), we also have the square of the amplitude

$$|\mathscr{E}|^2 = \mathscr{E}^2 = \tfrac{1}{2}(\mathbf{E}_1^2 + \mathbf{E}_2^2) + \tfrac{1}{2}(\mathbf{E}_1^2 - \mathbf{E}_2^2)\cos 2\phi - \mathbf{E}_1 \cdot \mathbf{E}_2 \sin 2\phi \tag{3.33}$$

which is finite for all ϕ. It is thus clear that the plane quadric curve represented by Eq. (3.32) is an ellipse as illustrated in Fig. 3.4. In this general case, the wave is said to be *elliptically polarized*, and the curve (3.32) gives the polarization ellipse.

Some special cases of polarization are of interest, and we discuss them below.

Case 1 If $\mathbf{E}_1$ and $\mathbf{E}_2$ are mutually perpendicular but not equal in magnitude, i.e., if

$$\mathbf{E}_1 \cdot \mathbf{E}_2 = 0 \qquad \text{or} \qquad \mathbf{E}_0^{*2} = \mathbf{E}_0^2$$

and

$$|\mathbf{E}_1| \neq |\mathbf{E}_2| \tag{3.34}$$

the tip of $\mathscr{E}$ traces out a canonical ellipse. In this case, $\mathbf{E}_1$ and $\mathbf{E}_2$ coincide with the major and minor semiaxes of the polarization ellipse respectively. In fact, if

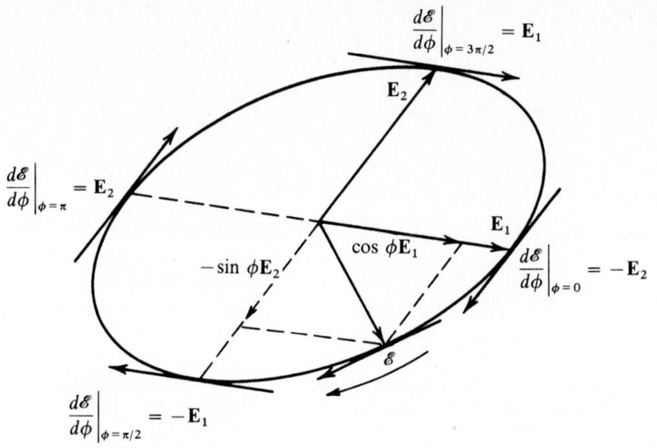

Figure 3.4 Polarization ellipse of a monochromatic field.

we choose $\mathbf{E}_1$ along $0x$ axis and $\mathbf{E}_2$ along $0y$ axis, then

$$(\mathscr{E} \times \mathbf{E}_1)^2 = \mathbf{E}_1^2 \mathscr{E}_y^2$$

$$(\mathscr{E} \times \mathbf{E}_2)^2 = \mathbf{E}_2^2 \mathscr{E}_x^2$$

Also, according to Eq. (3.34)

$$(\mathbf{E}_1 \times \mathbf{E}_2)^2 = \mathbf{E}_1^2 \mathbf{E}_2^2$$

Substituting the above into Eq. (3.32), we obtain the equation of canonical ellipse

$$\frac{\mathscr{E}_x^2}{\mathbf{E}_1^2} + \frac{\mathscr{E}_y^2}{\mathbf{E}_2^2} = 1 \tag{3.35}$$

The polarization ellipse is shown in Fig. 3.5.

Case 2 The polarization ellipse degenerates into a circle if the magnitude of $\mathscr{E}$ is independent of ϕ. From Eq. (3.33), $|\mathscr{E}|^2$ is a constant if the coefficients of cos

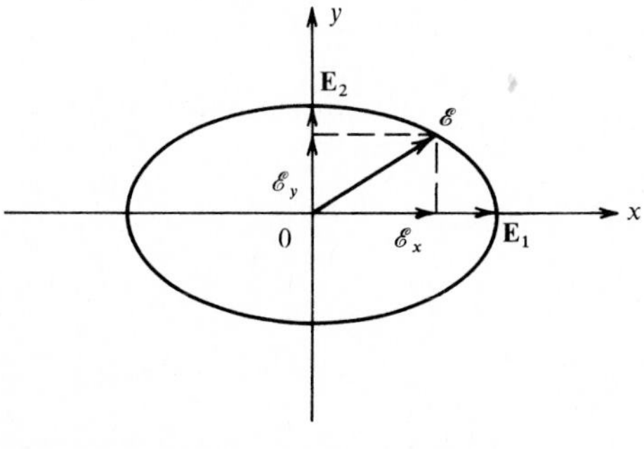

Figure 3.5 Canonical ellipse.

2ϕ and $\sin 2\phi$ vanish, i.e., if

$$\begin{aligned} \mathbf{E}_1 \cdot \mathbf{E}_2 &= 0 \\ \mathbf{E}_1^2 - \mathbf{E}_2^2 &= 0 \end{aligned} \tag{3.36}$$

or $\mathbf{E}_1 \perp \mathbf{E}_2$ and $|\mathbf{E}_1| = |\mathbf{E}_2|$. In this case, the wave is said to be *circularly polarized* and the radius of the polarization circle is $|\mathscr{E}| = |\mathbf{E}_1| = |\mathbf{E}_2|$ as illustrated in Fig. 3.6. Since $\mathbf{E}_0 = \mathbf{E}_1 + i\mathbf{E}_2$ and $\mathbf{E} = \mathbf{E}_0\, e^{i\mathbf{k}\cdot\mathbf{r}}$, the condition (3.36) for circular polarization may also be expressed as

$$\mathbf{E}_0^2 = 0 \tag{3.37}$$

or

$$\mathbf{E}^2 = 0 \tag{3.38}$$

Case 3 A wave is said to be *linearly polarized* if the direction of $\mathscr{E}$ remains unchanged but oscillates back and forth with a sinusoidally varying magnitude. In other words, the change in $\mathscr{E}$ with respect to ϕ is always parallel to $\mathscr{E}$. This condition may be written as

$$\mathscr{E} \times \frac{d\mathscr{E}}{d\phi} = \mathbf{0} \tag{3.39}$$

But from Eq. (3.31), we have

$$\frac{d\mathscr{E}}{d\phi} = -\sin\phi\mathbf{E}_1 - \cos\phi\,\mathbf{E}_2$$

Hence

$$\mathscr{E} \times \frac{d\mathscr{E}}{d\phi} = -(\mathbf{E}_1 \times \mathbf{E}_2) \tag{3.40}$$

and the condition (3.39) for linear polarization becomes

$$\mathbf{E}_1 \times \mathbf{E}_2 = \mathbf{0} \tag{3.41}$$

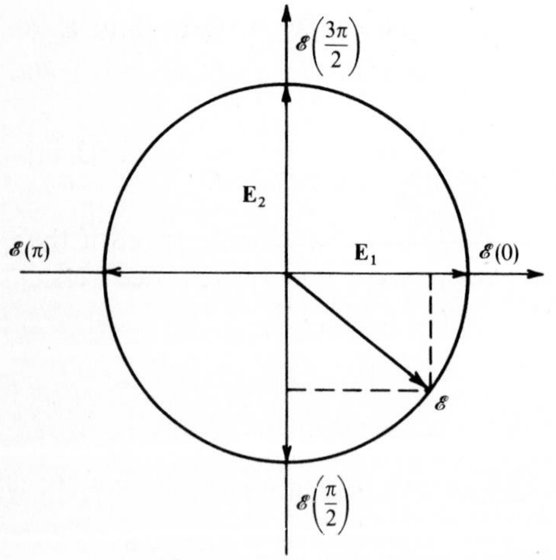

Figure 3.6 Circular polarization.

Thus, linear polarization results if either $\mathbf{E}_1$ or $\mathbf{E}_2$ is zero or if $\mathbf{E}_1$ is parallel or antiparallel to $\mathbf{E}_2$. On the other hand, substituting condition (3.41) into Eq. (3.32), we see that the sum of two positive terms must be zero which is possible only if each term vanishes, i.e.,

$$(\mathscr{E} \times \mathbf{E}_1)^2 = (\mathscr{E} \times \mathbf{E}_2)^2 = 0 \tag{3.42}$$

Hence for linear polarization, $\mathscr{E}$ must be parallel to $\mathbf{E}_1$ and $\mathbf{E}_2$. The case of $\mathbf{E}_2 = \mathbf{0}$ is shown in Fig. 3.7. Alternatively, since

$$\mathbf{E}_0 \times \mathbf{E}_0^* = i2(\mathbf{E}_2 \times \mathbf{E}_1) \tag{3.43}$$

condition (3.41) shows that linear polarization results whenever

$$\mathbf{E}_0 \times \mathbf{E}_0^* = \mathbf{0} \tag{3.44}$$

or

$$\mathbf{E} \times \mathbf{E}^* = \mathbf{0} \tag{3.45}$$

In general, if $\mathbf{E} \times \mathbf{E}^* \neq \mathbf{0}$ and $\mathbf{E}^2 \neq 0$, the wave is elliptically polarized.

In addition to the shapes of the polarization curves, there are two different senses of rotation in which the tip of $\mathscr{E}$ is described. In isotropic media, the electric field intensity $\mathscr{E}$ is always perpendicular to the direction of wave propagation $\hat{\mathbf{k}}$. We may thus define the right- or the left-handed elliptically (or circularly) polarized wave according to the direction in which the tip of $\mathscr{E}$ rotates with respect to $\hat{\mathbf{k}}$. Since both $\mathscr{E}$ and $d\mathscr{E}/d\phi$ lie on the same plane formed by the vectors $\mathbf{E}_1$ and $\mathbf{E}_2$, the cross product of $\mathscr{E}$ and $d\mathscr{E}/d\phi$ is either parallel or antiparallel to $\hat{\mathbf{k}}$. If $\hat{\mathbf{k}}$ coincides with the direction of $\mathscr{E} \times d\mathscr{E}/d\phi$, by definition, the wave is said to be left-handed elliptically (or circularly) polarized and vice versa. Consequently, the senses of rotation may be formulated as

$$\begin{aligned} &\hat{\mathbf{k}} \cdot \left(\mathscr{E} \times \frac{d\mathscr{E}}{d\phi}\right) > 0 \qquad \text{left-handed polarized wave} \\ &\hat{\mathbf{k}} \cdot \left(\mathscr{E} \times \frac{d\mathscr{E}}{d\phi}\right) < 0 \qquad \text{right-handed polarized wave} \end{aligned} \tag{3.46}$$

Using Eq. (3.40), we may state the above conditions in terms of $\mathbf{E}_1$ and $\mathbf{E}_2$; namely,

$$\begin{aligned} &\hat{\mathbf{k}} \cdot (\mathbf{E}_1 \times \mathbf{E}_2) > 0 \qquad \text{right-handed polarized wave} \\ &\hat{\mathbf{k}} \cdot (\mathbf{E}_1 \times \mathbf{E}_2) < 0 \qquad \text{left-handed polarized wave} \end{aligned} \tag{3.47}$$

In words, the criteria (3.47) state that the wave is right-handed polarized if the three real vectors $\mathbf{E}_1$, $\mathbf{E}_2$, and $\hat{\mathbf{k}}$ form a right-handed system and vice versa. Alternately, according to Eq. (3.43), we may write the criteria (3.47) as

$$\begin{aligned} &i\hat{\mathbf{k}} \cdot (\mathbf{E}_0 \times \mathbf{E}_0^*) > 0 \qquad \text{right-handed polarized wave} \\ &i\hat{\mathbf{k}} \cdot (\mathbf{E}_0 \times \mathbf{E}_0^*) < 0 \qquad \text{left-handed polarized wave} \end{aligned} \tag{3.48}$$

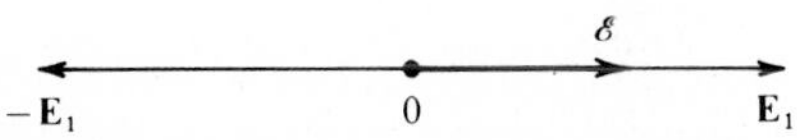

Figure 3.7 Linear polarization when $\mathbf{E}_2 = \mathbf{0}$.

In the case of circular polarization, we may combine conditions (3.36) and (3.47) to give

$$\begin{aligned} \mathbf{E}_2 &= \hat{\mathbf{k}} \times \mathbf{E}_1 \qquad \text{right-handed circularly polarized wave} \\ \mathbf{E}_2 &= -\hat{\mathbf{k}} \times \mathbf{E}_1 \qquad \text{left-handed circularly polarized wave} \end{aligned} \tag{3.49}$$

It is an easy matter to show that Eq. (3.49) satisfies both Eqs. (3.36) and (3.47). Moreover, condition (3.37) for circular polarization may be viewed as the condition for $\mathbf{E}_0$ to be orthogonal to itself. Since $\hat{\mathbf{k}} \cdot \mathbf{E}_0 = 0$, vector $\mathbf{E}_0$ being orthogonal to both $\hat{\mathbf{k}}$ and $\mathbf{E}_0$ may thus be represented as

$$\mathbf{E}_0 = C(\hat{\mathbf{k}} \times \mathbf{E}_0) \tag{3.50}$$

To determine the constant C, we take the cross product of Eq. (3.50) with $\hat{\mathbf{k}}$ and obtain

$$\hat{\mathbf{k}} \times \mathbf{E}_0 = -C\mathbf{E}_0 \tag{3.51}$$

A comparison of Eq. (3.50) with Eq. (3.51) gives $C^2 = -1$ or $C = \pm i$. Thus

$$\mathbf{E}_0 = \pm i(\hat{\mathbf{k}} \times \mathbf{E}_0) \tag{3.52}$$

Next, taking the dot product of Eq. (3.52) with $\mathbf{E}_0^*$ and interchanging the position of the dot and cross in the scalar triple product, we obtain

$$i\hat{\mathbf{k}} \cdot (\mathbf{E}_0 \times \mathbf{E}_0^*) = \pm |\mathbf{E}_0|^2 \tag{3.53}$$

Comparing the above with Eq. (3.48), we finally conclude that

$$\begin{aligned} \mathbf{E}_0 &= i(\hat{\mathbf{k}} \times \mathbf{E}_0) \qquad \text{right-handed circularly polarized wave} \\ \mathbf{E}_0 &= -i(\hat{\mathbf{k}} \times \mathbf{E}_0) \qquad \text{left-handed circularly polarized wave} \end{aligned} \tag{3.54}$$

3.3 DETERMINATION OF POLARIZATION ELLIPSE

For given real vectors $\mathbf{E}_1$ and $\mathbf{E}_2$, let us now determine the magnitudes and directions of the major and minor axes of the polarization ellipse. Since at a given point in space, the instantaneous vector $\mathscr{E}$ lies on the plane formed by vectors $\mathbf{E}_1$ and $\mathbf{E}_2$, we may thus write $\mathscr{E}$ as a linear combination of them, namely

$$\mathscr{E} = \alpha\mathbf{E}_1 + \beta\mathbf{E}_2 \tag{3.55}$$

where α and β are two arbitrary constants. Substituting Eq. (3.55) into the equation of polarization ellipse Eq. (3.32), we obtain a condition relating α and β:

$$\alpha^2 + \beta^2 = 1 \tag{3.56}$$

Since the major and minor axes of the polarization ellipse correspond to the extremum (maximum or minimum) of $\mathscr{E}^2$ subject to the constraint (3.56), we may employ the method of the Lagrange multiplier. Consider the problem of determining the extremum of the function

$$f(\alpha, \beta) = \mathscr{E}^2 - \lambda(\alpha^2 + \beta^2 - 1)$$
$$= \alpha^2 \mathbf{E}_1^2 + \beta^2 \mathbf{E}_2^2 + 2\alpha\beta \mathbf{E}_1 \cdot \mathbf{E}_2 - \lambda(\alpha^2 + \beta^2 - 1) \tag{3.57}$$

where λ is a parameter to be determined and is called a Lagrange multiplier. The conditions $\partial f/\partial \alpha = \partial f/\partial \beta = 0$ yield the two linear equations in α and β

$$\begin{aligned} (\mathbf{E}_1^2 - \lambda)\alpha + (\mathbf{E}_1 \cdot \mathbf{E}_2)\beta &= 0 \\ (\mathbf{E}_1 \cdot \mathbf{E}_2)\alpha + (\mathbf{E}_2^2 - \lambda)\beta &= 0 \end{aligned} \tag{3.58}$$

Upon multiplying the first equation by α, the second by β, then adding and using Eq. (3.56), we obtain

$$\lambda = \mathscr{E}_{\text{ext}}^2 \tag{3.59}$$

That is, the Lagrange multiplier λ yields the extremum of $\mathscr{E}^2$. On the other hand, the condition for the existence of nonzero α and β in Eq. (3.58) is that the determinant of the coefficient matrix must vanish:

$$\begin{vmatrix} \mathbf{E}_1^2 - \lambda & \mathbf{E}_1 \cdot \mathbf{E}_2 \\ \mathbf{E}_1 \cdot \mathbf{E}_2 & \mathbf{E}_2^2 - \lambda \end{vmatrix} = 0 \tag{3.60}$$

or

$$\lambda^2 - (\mathbf{E}_1^2 + \mathbf{E}_2^2)\lambda + (\mathbf{E}_1 \times \mathbf{E}_2)^2 = 0 \tag{3.61}$$

Hence the roots of Eq. (3.61) are

$$\lambda_1 = \mathscr{E}_{\max}^2 = \tfrac{1}{2}\left[(\mathbf{E}_1^2 + \mathbf{E}_2^2) + \sqrt{(\mathbf{E}_1^2 + \mathbf{E}_2^2)^2 - 4(\mathbf{E}_1 \times \mathbf{E}_2)^2}\right] \tag{3.62}$$

and

$$\lambda_2 = \mathscr{E}_{\min}^2 = \tfrac{1}{2}\left[(\mathbf{E}_1^2 + \mathbf{E}_2^2) - \sqrt{(\mathbf{E}_1^2 + \mathbf{E}_2^2)^2 - 4(\mathbf{E}_1 \times \mathbf{E}_2)^2}\right] \tag{3.63}$$

Since the discriminant

$$(\mathbf{E}_1^2 + \mathbf{E}_2^2)^2 - 4(\mathbf{E}_1 \times \mathbf{E}_2)^2 = (\mathbf{E}_1^2 - \mathbf{E}_2^2)^2 + 4(\mathbf{E}_1 \cdot \mathbf{E}_2)^2 \tag{3.64}$$

is clearly nonnegative and less than $(\mathbf{E}_1^2 + \mathbf{E}_2^2)^2$, the two roots λ_1 and λ_2 of Eq. (3.61) are always real and nonnegative. Also from Eqs. (3.62) and (3.63) [or by the roots and coefficients relations in Eq. (3.61)], we have

$$\begin{aligned} \lambda_1 + \lambda_2 &= \mathbf{E}_1^2 + \mathbf{E}_2^2 \\ \lambda_1 \lambda_2 &= (\mathbf{E}_1 \times \mathbf{E}_2)^2 \end{aligned} \tag{3.65}$$

and from Eq. (3.58)

$$\frac{\alpha}{\beta} = \frac{\lambda - \mathbf{E}_2^2}{\mathbf{E}_1 \cdot \mathbf{E}_2} = \frac{\mathbf{E}_1 \cdot \mathbf{E}_2}{\lambda - \mathbf{E}_1^2} \tag{3.66}$$

where λ represents either λ_1 or λ_2. Substituting Eq. (3.66) into Eq. (3.55), we find the directions of the major and minor axes of the polarization ellipse

$$\mathscr{E}_{max} \parallel [\mathbf{E}_2 \times (\mathbf{E}_1 \times \mathbf{E}_2) - \lambda_1 \mathbf{E}_1]$$

and

$$\mathscr{E}_{min} \parallel [\mathbf{E}_1 \times (\mathbf{E}_2 \times \mathbf{E}_1) - \lambda_2 \mathbf{E}_2] \tag{3.67}$$

respectively. The general formulas (3.62) and (3.63) include both the circular and linear polarizations as special cases. In fact, according to condition (3.36) for a circular polarization, the discriminant (3.64) vanishes and the radius of the polarization circle is

$$|\mathscr{E}| = |\mathbf{E}_1| = |\mathbf{E}_2|$$

For a linear polarization, from condition (3.41) we have $|\mathscr{E}_{min}| = 0$, thus the tip of $\mathscr{E}$ oscillates along the direction of $\mathscr{E}_{max}$ with maximum value

$$|\mathscr{E}_{max}| = \sqrt{\mathbf{E}_1^2 + \mathbf{E}_2^2} = \sqrt{\mathbf{E} \cdot \mathbf{E}^*} = |\mathbf{E}| \tag{3.68}$$

The above formulation provides a simple geometrical interpretation and method of construction of the polarization ellipse. To see these, let $\alpha < 1$ be given, the tip of $\mathscr{E}$ in Eq. (3.55) as β varies, traces out a straight line which is the sum of vector $\alpha\mathbf{E}_1$ in the direction of $\mathbf{E}_1$ and a vector parallel to $\mathbf{E}_2$. The intersection of this straight line with the polarization ellipse (3.32) determines the value of β according to Eq. (3.56); i.e.,

$$\beta = \pm\sqrt{1-\alpha^2}$$

We may thus write Eq. (3.55) as

$$\mathscr{E} = \alpha\mathbf{E}_1 \pm \sqrt{1-\alpha^2}\,\mathbf{E}_2 \tag{3.69}$$

which shows that the points of intersections are located at equal distance $\sqrt{1-\alpha^2}\,|\mathbf{E}_2|$ away from the point $\alpha\mathbf{E}_1$ along the line parallel to $\mathbf{E}_2$ as illustrated in Fig. 3.8. In other words, all chords of the ellipse parallel to $\mathbf{E}_2$ are bisected by $\mathbf{E}_1$, and, similarly, all chords parallel to $\mathbf{E}_1$ are bisected by $\mathbf{E}_2$. Hence, vectors $\mathbf{E}_1$

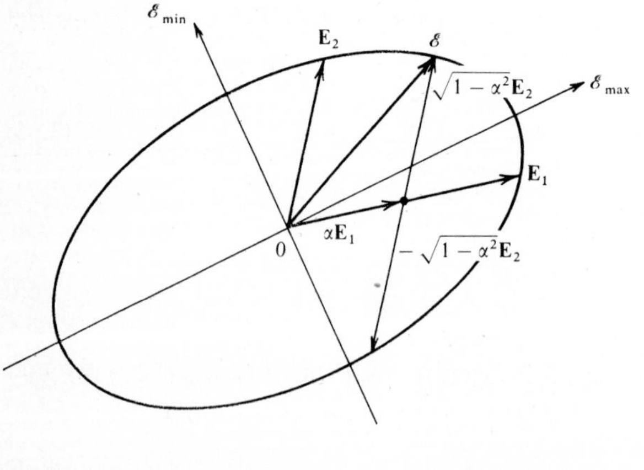

Figure 3.8 Determination of polarization ellipse when $\mathbf{E}_1$ and $\mathbf{E}_2$ are given.

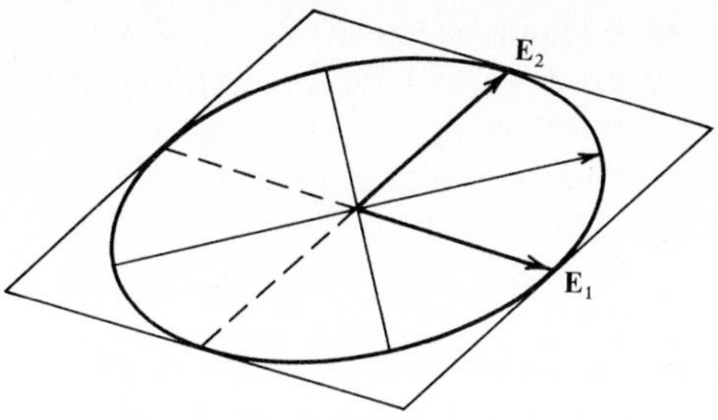

Figure 3.9 Location of major semiaxis when the angle between $\mathbf{E}_1$ and $\mathbf{E}_2$ is acute.

and $\mathbf{E}_2$ are the *conjugate semidiameters* of the ellipse (3.32). It is therefore clear that the polarization ellipse is completely and uniquely determined by the vectors $\mathbf{E}_1$ and $\mathbf{E}_2$. The reverse, however, is not true.

Based on the above argument, we may easily construct the polarization ellipse. Since the tangent to the ellipse at the end point of any given diameter is parallel to another diameter, the ellipse is thus bounded by the parallelogram with the end points of $\mathbf{E}_1$ and $\mathbf{E}_2$ at the middle of each side (see Figs. 3.9 and 3.10). The major semiaxis of the ellipse lies closer to the larger vectors of $\mathbf{E}_1$ and $\mathbf{E}_2$. Also, if the angle between vectors $\mathbf{E}_1$ and $\mathbf{E}_2$ is acute, the major semiaxis lies between them (see Fig. 3.9). On the other hand, if the angle is obtuse, then the minor semiaxis lies between vectors $\mathbf{E}_1$ and $\mathbf{E}_2$ (see Fig. 3.10).

3.4 SUPERPOSITION OF WAVES OF DIFFERENT POLARIZATIONS

In studying wave phenomena, we often have to superpose two or more waves and examine how the specific properties of each constituent wave influence the ultimate form of the resultant wave. Since we consider only linear media in this book, the principle of superposition applies: the sum of two monochromatic plane waves each of which is a solution of Maxwell's equations yields another wave satisfying the same equations.

Let us consider two monochromatic plane waves propagating in the same

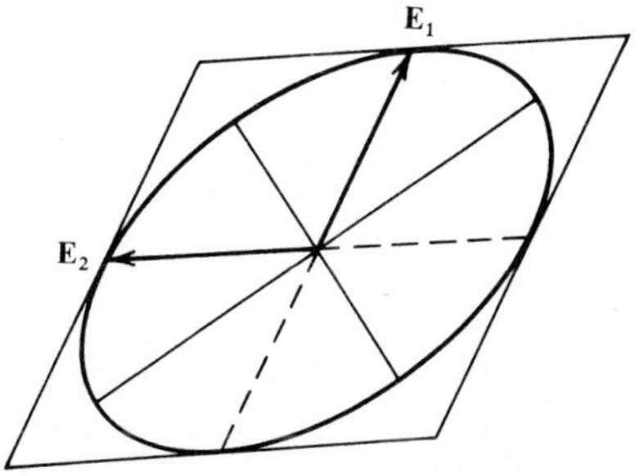

Figure 3.10 Location of minor semiaxis when the angle between $\mathbf{E}_1$ and $\mathbf{E}_2$ is obtuse.

direction with the same phase velocity. As a result of linear superposition, the resultant wave is again a monochromatic plane wave having the same phase velocity and direction of propagation. In other words, the phase $\phi = \mathbf{k} \cdot \mathbf{r} - \omega t$ in Eq. (3.1) for all such waves are identical but their complex amplitude vectors $\mathbf{E}_0$ are different, i.e., we superpose waves with different polarizations and intensities. From the mathematical point of view, the problem reduces to the addition of complex vectors in the same plane.

We now wish to establish that a wave of any (elliptical, circular, or linear) polarization may always be considered as a sum of two linearly polarized waves with arbitrary directions of oscillations. The simplest way to convince ourselves is to treat the problem geometrically. For a given wave of any polarization $\mathbf{E}_0 = \mathbf{E}_1 + i\mathbf{E}_2$, we choose two arbitrary directions OA and OB which lie on the same plane formed by the real vectors $\mathbf{E}_1$ and $\mathbf{E}_2$. Decomposing both vectors $\mathbf{E}_1$ and $\mathbf{E}_2$ in these directions (see Fig. 3.11), we have

$$\begin{aligned} \mathbf{E}_0 &= \mathbf{E}_1^{(1)} + \mathbf{E}_1^{(2)} + i(\mathbf{E}_2^{(1)} + \mathbf{E}_2^{(2)}) \\ &= \mathbf{E}_0^{(1)} + \mathbf{E}_0^{(2)} \end{aligned} \tag{3.70}$$

where

$$\begin{aligned} \mathbf{E}_0^{(1)} &= \mathbf{E}_1^{(1)} + i\mathbf{E}_2^{(1)} \\ \mathbf{E}_0^{(2)} &= \mathbf{E}_1^{(2)} + i\mathbf{E}_2^{(2)} \end{aligned} \tag{3.71}$$

Since

$$\mathbf{E}_1^{(1)} \parallel \mathbf{E}_2^{(1)} \parallel OA$$

$$\mathbf{E}_1^{(2)} \parallel \mathbf{E}_2^{(2)} \parallel OB$$

it is thus clear that the complex vectors $\mathbf{E}_0^{(1)}$ and $\mathbf{E}_0^{(2)}$ determine two linearly polarized waves. A special case of such a decomposition is when $OA \parallel \mathbf{E}_1$ and $OB \parallel \mathbf{E}_2$; that is, for a wave of any polarization $\mathbf{E}_0$, we can always represent it as a sum of two linearly polarized waves with $\mathbf{E}_0^{(1)} = \mathbf{E}_1$ and $\mathbf{E}_0^{(2)} = i\mathbf{E}_2$.

In practice, the most common decomposition of a wave of any polarization is that into two linearly polarized waves in mutually perpendicular directions ($OA \perp OB$). However, it should be emphasized once more that the directions of

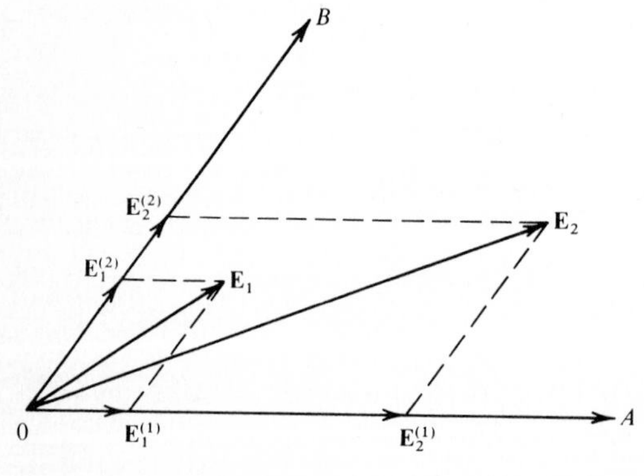

Figure 3.11 Frequency domain diagram: decomposition of elliptical polarization into two linear polarizations.

oscillations of two linearly polarized waves into which a given wave is decomposed, may be completely arbitrary.

As an example, let us take two linearly polarized waves

$$\mathbf{E}_0^{(1)} = \mathbf{E}_1^{(1)} + i\mathbf{E}_2^{(1)}$$

and

$$\mathbf{E}_0^{(2)} = \mathbf{E}_1^{(2)} + i\mathbf{E}_2^{(2)} \tag{3.72}$$

where

$$\mathbf{E}_2^{(1)} = \alpha_1 \mathbf{E}_1^{(1)}$$

$$\mathbf{E}_2^{(2)} = \alpha_2 \mathbf{E}_1^{(2)} \tag{3.73}$$

The resultant wave $\mathbf{E}_0 = \mathbf{E}_1 + i\mathbf{E}_2$ with

$$\mathbf{E}_1 = \mathbf{E}_1^{(1)} + \mathbf{E}_1^{(2)}$$

$$\mathbf{E}_2 = \mathbf{E}_2^{(1)} + \mathbf{E}_2^{(2)} \tag{3.74}$$

will also be linearly polarized if

$$\mathbf{E}_2 = \alpha \mathbf{E}_1 \tag{3.75}$$

Substituting Eqs. (3.73) and (3.74) into Eq. (3.75), we obtain

$$(\alpha_1 - \alpha)\mathbf{E}_1^{(1)} = (\alpha - \alpha_2)\mathbf{E}_1^{(2)}$$

which is possible if either

$$\mathbf{E}_1^{(1)} \parallel \mathbf{E}_1^{(2)} \parallel \mathbf{E}_2^{(1)} \parallel \mathbf{E}_2^{(2)}$$

or

$$\alpha_1 = \alpha_2 = \alpha$$

In other words, the superposition of two linearly polarized waves gives rise to a linearly polarized wave provided that either all vectors $\mathbf{E}_1^{(1)}$, $\mathbf{E}_1^{(2)}$, $\mathbf{E}_2^{(1)}$, $\mathbf{E}_2^{(2)}$ are parallel to each other or the ratios of the magnitudes of the imaginary part to the real part: α_1, α_2, α all coincide.

We now turn to the case of circular polarization and verify the following: let $\mathbf{E}_0^{(1)}$ and $\mathbf{E}_0^{(2)}$ be two complex-amplitude vectors which lie on the same plane and which characterize two circularly polarized waves, then

$$\mathbf{E}_0^{(1)} \cdot \mathbf{E}_0^{(2)} = \mathbf{E}_0^{(1)*} \cdot \mathbf{E}_0^{(2)*} = 0 \tag{3.76}$$

$$\mathbf{E}_0^{(1)} \times \mathbf{E}_0^{(2)} = \mathbf{E}_0^{(1)*} \times \mathbf{E}_0^{(2)*} = \mathbf{0} \tag{3.77}$$

if the two waves have the same senses of rotation, and

$$\mathbf{E}_0^{(1)} \cdot \mathbf{E}_0^{(2)*} = \mathbf{E}_0^{(1)*} \cdot \mathbf{E}_0^{(2)} = 0 \tag{3.78}$$

$$\mathbf{E}_0^{(1)} \times \mathbf{E}_0^{(2)*} = \mathbf{E}_0^{(1)*} \times \mathbf{E}_0^{(2)} = \mathbf{0} \tag{3.79}$$

if the senses of rotation are opposite. In fact, according to Eq. (3.54), the two circularly polarized waves $\mathbf{E}_0^{(1)}$ and $\mathbf{E}_0^{(2)}$ satisfy

$$\mathbf{E}_0^{(1)} = \pm i(\hat{\mathbf{k}} \times \mathbf{E}_0^{(1)}) \tag{3.80}$$

and

$$\mathbf{E}_0^{(2)} = \pm i(\hat{\mathbf{k}} \times \mathbf{E}_0^{(2)})$$

where either the upper or the lower signs must be used. To show Eq. (3.76), we note that $\hat{\mathbf{k}} \cdot \mathbf{E}_0^{(1)} = \hat{\mathbf{k}} \cdot \mathbf{E}_0^{(2)} = 0$, hence

$$\begin{aligned} \mathbf{E}_0^{(1)} \cdot \mathbf{E}_0^{(2)} &= [\pm i(\hat{\mathbf{k}} \times \mathbf{E}_0^{(1)})] \cdot [\pm i(\hat{\mathbf{k}} \times \mathbf{E}_0^{(2)})] \\ &= -(\hat{\mathbf{k}} \times \mathbf{E}_0^{(1)}) \cdot (\hat{\mathbf{k}} \times \mathbf{E}_0^{(2)}) \\ &= -\mathbf{E}_0^{(1)} \cdot \mathbf{E}_0^{(2)} \end{aligned}$$

and Eq. (3.76) follows. Similarly, we have

$$\begin{aligned} \mathbf{E}_0^{(1)} \times \mathbf{E}_0^{(2)} &= \mathbf{E}_0^{(1)} \times [\pm i(\hat{\mathbf{k}} \times \mathbf{E}_0^{(2)})] \\ &= \pm i[(\mathbf{E}_0^{(1)} \cdot \mathbf{E}_0^{(2)})\hat{\mathbf{k}} - (\hat{\mathbf{k}} \cdot \mathbf{E}_0^{(1)})\mathbf{E}_0^{(2)}] \\ &= \mathbf{0} \end{aligned}$$

which is Eq. (3.77). In the case of two oppositely rotated circularly polarized waves, Eqs. (3.78) and (3.79) can be proved similarly by considering, for instance, $\mathbf{E}_0^{(1)} = \pm i(\hat{\mathbf{k}} \times \mathbf{E}_C^{(1)})$ and $\mathbf{E}_0^{(2)} = \mp i(\hat{\mathbf{k}} \times \mathbf{E}_0^{(2)})$.

With the above results, we may proceed to establish that the sum of two circularly polarized waves of the same sense of rotation gives rise to another circularly polarized wave of the same sense of rotation. In fact, using Eq. (3.80), we obtain

$$\begin{aligned} \mathbf{E}_0 &= \mathbf{E}_0^{(1)} + \mathbf{E}_0^{(2)} \\ &= \pm i\hat{\mathbf{k}} \times (\mathbf{E}_0^{(1)} + \mathbf{E}_0^{(2)}) \\ &= \pm i(\hat{\mathbf{k}} \times \mathbf{E}_0) \end{aligned}$$

i.e., a circularly polarized wave having the same sense of rotation as constituent waves. The same conclusion may be reached by squaring $\mathbf{E}_0$ and using Eqs. (3.37) and (3.76). We may also interpret the result geometrically either in the time or in the frequency domain. In the time domain, the vector sum of two instantaneous circularly polarized vectors $\mathscr{E}^{(1)}$ and $\mathscr{E}^{(2)}$, which have the same sense of rotation and angular velocity, gives a circularly polarized vector $\mathscr{E}$ having the same properties as shown in Fig. 3.12. In the frequency domain, the real and imaginary

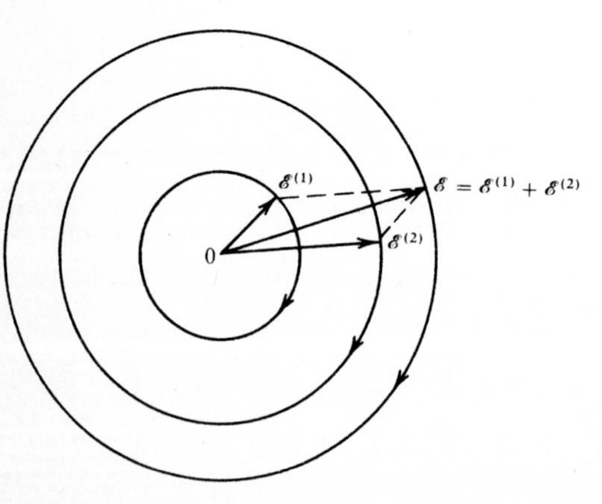

Figure 3.12 Time domain diagram: sum of two circular polarizations of the same sense of rotation.

parts of the two complex vectors $\mathbf{E}_0^{(1)} = \mathbf{E}_1^{(1)} + i\mathbf{E}_2^{(1)}$ and $\mathbf{E}_0^{(2)} = \mathbf{E}_1^{(2)} + i\mathbf{E}_2^{(2)}$ of two circularly polarized waves must satisfy condition (3.36), that is

$$\mathbf{E}_1^{(1)} \perp \mathbf{E}_2^{(1)} \qquad |\mathbf{E}_1^{(1)}| = |\mathbf{E}_2^{(1)}|$$

and

$$\mathbf{E}_1^{(2)} \perp \mathbf{E}_2^{(2)} \qquad |\mathbf{E}_1^{(2)}| = |\mathbf{E}_2^{(2)}|$$

After combining the real parts $\mathbf{E}_1^{(1)}, \mathbf{E}_1^{(2)}$ and the imaginary parts $\mathbf{E}_2^{(1)}, \mathbf{E}_2^{(2)}$ separately, we obtain a pair of real, mutually perpendicular vectors $\mathbf{E}_1 = \mathbf{E}_1^{(1)} + \mathbf{E}_1^{(2)}$ and $\mathbf{E}_2 = \mathbf{E}_2^{(1)} + \mathbf{E}_2^{(2)}$ of equal magnitudes. In other words, $\mathbf{E}_1$ and $\mathbf{E}_2$ also satisfy Eq. (3.36) and thus represent a circularly polarized wave (see Fig. 3.13).

On the other hand, if $\mathbf{E}_0^{(1)}$ and $\mathbf{E}_0^{(2)}$ are two circularly polarized waves of opposite senses of rotation, then the sum $\mathbf{E}_0^{(1)} + \mathbf{E}_0^{(2)} = \mathbf{E}_0$ may give a wave of any polarization including circular polarization. In fact, according to Eq. (3.37): $\mathbf{E}_0^{(1)2} = \mathbf{E}_0^{(2)2} = 0$, thus

$$\mathbf{E}_0^2 = 2\mathbf{E}_0^{(1)} \cdot \mathbf{E}_0^{(2)} \neq \mathbf{0}$$

Otherwise, waves $\mathbf{E}_0^{(1)}$ and $\mathbf{E}_0^{(2)}$ would have the same senses of rotation [cf. Eq. (3.76)]. In addition, using Eq. (3.78), we find the magnitude square of $\mathbf{E}_0$:

$$|\mathbf{E}_0|^2 = |\mathbf{E}_0^{(1)}|^2 + |\mathbf{E}_0^{(2)}|^2 \tag{3.81}$$

Also, from Eq. (3.79), we have

$$\mathbf{E}_0 \times \mathbf{E}_0^* = \mathbf{E}_0^{(1)} \times \mathbf{E}_0^{(1)*} + \mathbf{E}_0^{(2)} \times \mathbf{E}_0^{(2)*} \tag{3.82}$$

and thus the magnitude square

$$|\mathbf{E}_0 \times \mathbf{E}_0^*|^2 = |\mathbf{E}_0^{(1)}|^4 + |\mathbf{E}_0^{(2)}|^4 - 2(\mathbf{E}_0^{(1)} \cdot \mathbf{E}_0^{(2)})(\mathbf{E}_0^{(1)} \cdot \mathbf{E}_0^{(2)})^* \tag{3.83}$$

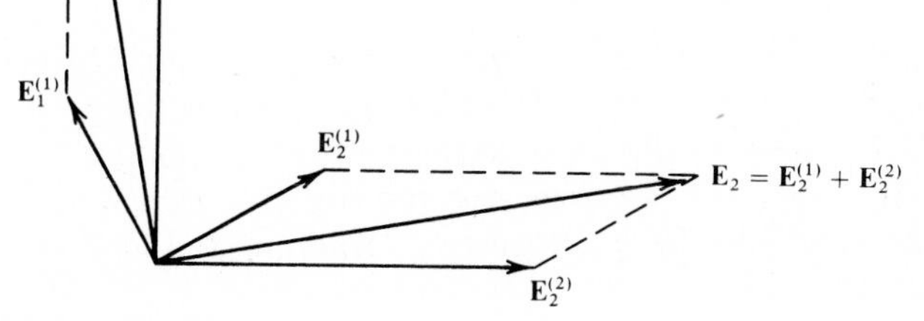

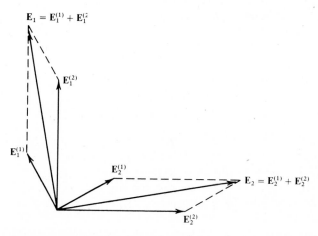

Figure 3.13 Frequency domain diagram: sum of two circular polarizations of the same sense of rotation.

Here we have used the condition (3.78). Furthermore, according to Eq. (3.79)

$$|\mathbf{E}_0^{(1)} \times \mathbf{E}_0^{(2)*}|^2 = (\mathbf{E}_0^{(1)} \times \mathbf{E}_0^{(2)*}) \cdot (\mathbf{E}_0^{(1)*} \times \mathbf{E}_0^{(2)}) = 0$$
$$= |\mathbf{E}_0^{(1)}|^2 |\mathbf{E}_0^{(2)}|^2 - (\mathbf{E}_0^{(1)} \cdot \mathbf{E}_0^{(2)})(\mathbf{E}_0^{(1)} \cdot \mathbf{E}_0^{(2)})^*$$

Hence Eq. (3.83) becomes

$$|\mathbf{E}_0 \times \mathbf{E}_0^*|^2 = (|\mathbf{E}_0^{(1)}|^2 - |\mathbf{E}_0^{(2)}|^2)^2$$

or

$$|\mathbf{E}_0 \times \mathbf{E}_0^*| = |\mathbf{E}_0^{(1)}|^2 - |\mathbf{E}_0^{(2)}|^2 \tag{3.84}$$

when $|\mathbf{E}_0^{(1)}| \geq |\mathbf{E}_0^{(2)}|$. We thus justify the statement.

Next, let us show that a given wave $\mathbf{E}_0 = \mathbf{E}_1 + i\mathbf{E}_2$ of any polarization may always be uniquely decomposed into a sum of two circular polarizations $\mathbf{E}_0^{(1)}$ and $\mathbf{E}_0^{(2)}$ of opposite senses of rotation. In fact, according to Eq. (3.54)

$$\begin{aligned} \mathbf{E}_0^{(1)} &= i(\hat{\mathbf{k}} \times \mathbf{E}_0^{(1)}) \\ \mathbf{E}_0^{(2)} &= -i(\hat{\mathbf{k}} \times \mathbf{E}_0^{(2)}) \end{aligned} \tag{3.85}$$

Next, taking the cross product of the sum of these expressions with $i\hat{\mathbf{k}}$, we obtain

$$\mathbf{E}_0^{(1)} - \mathbf{E}_0^{(2)} = i(\hat{\mathbf{k}} \times \mathbf{E}_0) \tag{3.86}$$

which together with $\mathbf{E}_0^{(1)} + \mathbf{E}_0^{(2)} = \mathbf{E}_0$ gives

$$\begin{aligned} \mathbf{E}_0^{(1)} &= \tfrac{1}{2}[\mathbf{E}_0 + i(\hat{\mathbf{k}} \times \mathbf{E}_0)] \\ \mathbf{E}_0^{(2)} &= \tfrac{1}{2}[\mathbf{E}_0 - i(\hat{\mathbf{k}} \times \mathbf{E}_0)] \end{aligned} \tag{3.87}$$

Moreover, by substituting $\mathbf{E}_0 = \mathbf{E}_1 + i\mathbf{E}_2$ into Eq. (3.87) and letting

$$\begin{aligned} \mathbf{E}_0^{(1)} &= \mathbf{E}_1^{(1)} + i\mathbf{E}_2^{(1)} \\ \mathbf{E}_0^{(2)} &= \mathbf{E}_1^{(2)} + i\mathbf{E}_2^{(2)} \end{aligned} \tag{3.88}$$

we obtain the real parts

$$\begin{aligned} \mathbf{E}_1^{(1)} &= \tfrac{1}{2}(\mathbf{E}_1 - \hat{\mathbf{k}} \times \mathbf{E}_2) \\ \mathbf{E}_1^{(2)} &= \tfrac{1}{2}(\mathbf{E}_1 + \hat{\mathbf{k}} \times \mathbf{E}_2) \end{aligned} \tag{3.89}$$

and the imaginary parts

$$\begin{aligned} \mathbf{E}_2^{(1)} &= \hat{\mathbf{k}} \times \mathbf{E}_1^{(1)} = \tfrac{1}{2}(\mathbf{E}_2 + \hat{\mathbf{k}} \times \mathbf{E}_1) \\ \mathbf{E}_2^{(2)} &= -\hat{\mathbf{k}} \times \mathbf{E}_1^{(2)} = \tfrac{1}{2}(\mathbf{E}_2 - \hat{\mathbf{k}} \times \mathbf{E}_1) \end{aligned} \tag{3.90}$$

In summary, Eq. (3.87) or Eqs. (3.89) and (3.90) provide means of determining the two circular polarizations of opposite senses of rotation in terms of the given polarization $\mathbf{E}_0 = \mathbf{E}_1 + i\mathbf{E}_2$. To show that this decomposition is unique, we assume that there were two possible different decompositions: $\mathbf{E}_0 = \mathbf{E}_0^{(1)} + \mathbf{E}_0^{(2)} = \mathbf{E}_0^{(1)\prime} + \mathbf{E}_0^{(2)\prime}$ where $\mathbf{E}_0^{(1)}$ and $\mathbf{E}_0^{(1)\prime}$ have one sense of rotation and $\mathbf{E}_0^{(2)}$ and $\mathbf{E}_0^{(2)\prime}$ have another, then

$$(\mathbf{E}_0^{(1)} - \mathbf{E}_0^{(1)\prime}) + (\mathbf{E}_0^{(2)} - \mathbf{E}_0^{(2)\prime}) = \mathbf{0}$$

But the two parentheses in the equation represent two circular polarizations of opposite senses of rotation, thus according to Eq. (3.81), we must have

$$|\mathbf{E}_0^{(1)} - \mathbf{E}_0^{(1)\prime}|^2 + |\mathbf{E}_0^{(2)} - \mathbf{E}_0^{(2)\prime}|^2 = 0$$

which is possible only if $\mathbf{E}_0^{(1)} = \mathbf{E}_0^{(1)\prime}$ and $\mathbf{E}_0^{(2)} = \mathbf{E}_0^{(2)\prime}$, that is, the decomposition is unique. When the wave to be decomposed is itself a circularly polarized wave, we may still use Eq. (3.87) except in this case one of the waves $\mathbf{E}_0^{(1)}$ and $\mathbf{E}_0^{(2)}$ must be zero and the other yields $\mathbf{E}_0$.

By similar procedure, we may study various other decompositions. For instance, the decomposition of a given wave into a linear and a circular (see Prob. 3.12), or into two elliptical, or into one elliptical and one circular polarization, etc.

The coordinate-free description of wave polarizations discussed in this section has the advantage of generality and, at the same time, provides a simple means in reaching various relations concerning the superposition of waves of different polarizations.

3.5 LAWS OF REFLECTION AND REFRACTION

So far we have considered only plane wave propagation in an unbounded medium. In practice, however, the medium has finite dimension; thus a more interesting problem is the effects of boundary surface—surface due to an abrupt change in physical properties, on the wave propagation. Suppose that a monochromatic plane wave, propagating in the first medium is incident upon a plane interface of discontinuity marking the boundary of the second medium of different electrical and magnetic properties. The incident wave splits into transmitted waves proceeding into the second medium and reflected waves propagating back into the first medium. The existence of these waves is necessary in order to satisfy the boundary conditions at the interface. In a general problem, the amplitude, polarization, direction of propagation and frequency of the incident wave are given, and we seek these characteristics of the reflected and transmitted waves. In this section, we shall investigate the general laws of reflection and refraction and leave the relations connecting the amplitudes to the following chapters.

According to Sec. 2.3, the boundary conditions are linear relations. We may thus write them in frequency domain as

$$\mathbf{E}_{\mathrm{I}} \times \hat{\mathbf{q}} = \mathbf{E}_{\mathrm{II}} \times \hat{\mathbf{q}} \tag{3.91}$$

$$\mathbf{H}_{\mathrm{I}} \times \hat{\mathbf{q}} = \mathbf{H}_{\mathrm{II}} \times \hat{\mathbf{q}} \tag{3.92}$$

$$\mathbf{B}_{\mathrm{I}} \cdot \hat{\mathbf{q}} = \mathbf{B}_{\mathrm{II}} \cdot \hat{\mathbf{q}} \tag{3.93}$$

$$\mathbf{D}_{\mathrm{I}} \cdot \hat{\mathbf{q}} = \mathbf{D}_{\mathrm{II}} \cdot \hat{\mathbf{q}} \tag{3.94}$$

where $\hat{\mathbf{q}}$ is a unit vector normal to the interface pointing from the first to the second medium. Here we have assumed that there are no surface charge and current densities on the interface. The subscripts I and II denote the total complex field vectors in the first and second medium respectively. However, we should note that the boundary conditions (3.91) to (3.94) hold only for points on the interface. Let 0 be a fixed origin which, for convenience, is assumed to be located on the interface. If $\mathbf{r}$ is the position vector drawn from 0 to any point in space, then the vector

$$\hat{\mathbf{q}} \times \mathbf{r} = \mathbf{r}_p \tag{3.95}$$

defines the position vector of a point on the interface. In other words, the boundary conditions (3.91) to (3.94) hold only for points $\mathbf{r}_p$.

Now let us denote the electric field intensity of the incident wave which exists in medium **I**, by

$$\mathbf{E}_i = \mathbf{E}_{0i} \exp(i\mathbf{k}_i \cdot \mathbf{r})$$

where $\mathbf{E}_{0i}$ and $\mathbf{k}_i$ are respectively the complex constant-amplitude and wave vectors of the incident wave. Similarly, the reflected waves which also exist in medium I are of the form

$$\mathbf{E}_r^{(1)} = \mathbf{E}_{0r}^{(1)} \exp(i\mathbf{k}_r^{(1)} \cdot \mathbf{r})$$
$$\mathbf{E}_r^{(2)} = \mathbf{E}_{0r}^{(2)} \exp(i\mathbf{k}_r^{(2)} \cdot \mathbf{r})$$
$$\mathbf{E}_r^{(3)} = \mathbf{E}_{0r}^{(3)} \exp(i\mathbf{k}_r^{(3)} \cdot \mathbf{r})$$
$$\cdots\cdots\cdots\cdots\cdots\cdots\cdots\cdots$$

where $\mathbf{k}_r^{(1)}, \mathbf{k}_r^{(2)}, \mathbf{k}_r^{(3)}, \ldots$ are the wave vectors of the first, second, third, ... reflected waves, respectively, and $\mathbf{E}_{0r}^{(1)}, \mathbf{E}_{0r}^{(2)}, \mathbf{E}_{0r}^{(3)}, \ldots$ are the corresponding complex constant-amplitude vectors, all as yet undetermined. The transmitted waves which exist in medium II are of the forms†

$$\mathbf{E}_t^{(1)} = \mathbf{E}_{0t}^{(1)} \exp(i\mathbf{k}_t^{(1)} \cdot \mathbf{r})$$
$$\mathbf{E}_t^{(2)} = \mathbf{E}_{0t}^{(2)} \exp(i\mathbf{k}_t^{(2)} \cdot \mathbf{r})$$
$$\mathbf{E}_t^{(3)} = \mathbf{E}_{0t}^{(3)} \exp(i\mathbf{k}_t^{(3)} \cdot \mathbf{r})$$
$$\cdots\cdots\cdots\cdots\cdots\cdots\cdots\cdots$$

where $\mathbf{k}_t^{(1)}, \mathbf{k}_t^{(2)}, \mathbf{k}_t^{(3)}, \ldots$ are respectively, the wave vectors of the first, second, third, ... transmitted waves, and $\mathbf{E}_{0t}^{(1)}, \mathbf{E}_{0t}^{(2)}$, and $\mathbf{E}_{0t}^{(3)}, \ldots$ are the corresponding complex constant-amplitude vectors. The total number of the reflected (or transmitted) waves depends upon the electrical and magnetic properties of medium **I** (or medium **II**) and will be discussed in detail in later chapters. The other field vectors of the waves can be obtained from the Maxwell equations and the constitutive relations.

Substituting the electric field intensities of the incident, reflected, and trans-

† Subscript t of a vector indicating the transmitted wave should not be confused with the subscript t of a matrix indicating the trace of the matrix.

mitted waves into the boundary condition (3.91), we have

$$(\mathbf{E}_{0i} \times \hat{\mathbf{q}}) \exp(i\mathbf{k}_i \cdot \mathbf{r}) + (\mathbf{E}_{0r}^{(1)} \times \hat{\mathbf{q}}) \exp(i\mathbf{k}_r^{(1)} \cdot \mathbf{r}) + (\mathbf{E}_{0r}^{(2)} \times \hat{\mathbf{q}}) \exp(i\mathbf{k}_r^{(2)} \cdot \mathbf{r})$$
$$+ \cdots - (\mathbf{E}_{0t}^{(1)} \times \hat{\mathbf{q}}) \exp(i\mathbf{k}_t^{(1)} \cdot \mathbf{r}) - (\mathbf{E}_{0t}^{(2)} \times \hat{\mathbf{q}}) \exp(i\mathbf{k}_t^{(2)} \cdot \mathbf{r}) \cdots = \mathbf{0} \quad (3.96)$$

which must be true for all the points $\mathbf{r}_p$ on the interface. Since the vector coefficients of the exponential factors $\exp(i\mathbf{k} \cdot \mathbf{r})$, $\exp(i\mathbf{k}_r^{(1)} \cdot \mathbf{r})$, ... are constant vectors, hence Eq. (3.96) represents linear relations among the exponential functions. As it is well known that the exponential functions of different exponents are linearly independent, Eq. (3.96) holds either all the constant vector coefficients vanish or all the exponents $\mathbf{k}_i \cdot \mathbf{r}$, $\mathbf{k}_r^{(1)} \cdot \mathbf{r}$, $\mathbf{k}_r^{(2)} \cdot \mathbf{r}$, ..., $\mathbf{k}_t^{(1)} \cdot \mathbf{r}$, ... are equal at all points on the interface. The first case is impossible, since $\mathbf{E}_{0i}$ of the incident wave can be chosen arbitrarily. Hence we have

$$\mathbf{k}_i \cdot \mathbf{r}_p = \mathbf{k}_r^{(1)} \cdot \mathbf{r}_p = \mathbf{k}_r^{(2)} \cdot \mathbf{r}_p = \mathbf{k}_r^{(3)} \cdot \mathbf{r}_p = \cdots$$
$$= \mathbf{k}_t^{(1)} \cdot \mathbf{r}_p = \mathbf{k}_t^{(2)} \cdot \mathbf{r}_p = \mathbf{k}_t^{(3)} \cdot \mathbf{r}_p = \cdots \quad (3.97)$$

for all $\mathbf{r}_p$ on the interface. Equation (3.97) may also be written as

$$\begin{aligned}
(\mathbf{k}_r^{(1)} - \mathbf{k}_i) \cdot \mathbf{r}_p &= 0 \\
(\mathbf{k}_r^{(2)} - \mathbf{k}_i) \cdot \mathbf{r}_p &= 0 \\
(\mathbf{k}_r^{(3)} - \mathbf{k}_i) \cdot \mathbf{r}_p &= 0 \\
\cdots\cdots&\cdots\cdots \\
(\mathbf{k}_t^{(1)} - \mathbf{k}_i) \cdot \mathbf{r}_p &= 0 \\
(\mathbf{k}_t^{(2)} - \mathbf{k}_i) \cdot \mathbf{r}_p &= 0 \\
\cdots\cdots&\cdots\cdots
\end{aligned} \quad (3.98)$$

Now, replacing $\mathbf{r}_p$ in Eq. (3.98) by $\hat{\mathbf{q}} \times \mathbf{r}$, interchanging the positions of the dot and cross in the scalar triple product, and noting that $\mathbf{r}$ is arbitrary, we obtain

$$(\mathbf{k}_r^{(1)} - \mathbf{k}_i) \times \hat{\mathbf{q}} = (\mathbf{k}_r^{(2)} - \mathbf{k}_i) \times \hat{\mathbf{q}} = \cdots$$
$$= (\mathbf{k}_t^{(1)} - \mathbf{k}_i) \times \hat{\mathbf{q}} = (\mathbf{k}_t^{(2)} - \mathbf{k}_i) \times \hat{\mathbf{q}} = \cdots = \mathbf{0} \quad (3.99)$$

or finally

$$\mathbf{k}_i \times \hat{\mathbf{q}} = \mathbf{k}_r^{(1)} \times \hat{\mathbf{q}} = \mathbf{k}_r^{(2)} \times \hat{\mathbf{q}} = \cdots$$
$$= \mathbf{k}_t^{(1)} \times \hat{\mathbf{q}} = \mathbf{k}_t^{(2)} \times \hat{\mathbf{q}} = \cdots \quad (3.100)$$

which expresses the general laws of reflection and refraction in vector forms. They must be satisfied by the wave vectors of the incident, reflected, and transmitted waves on the interface of any two linear, homogeneous media. In other words, Eq. (3.100) states that across the interface of any two linear, homogeneous media, the tangential components of the wave vectors of the incident, reflected, and transmitted waves, must all be equal. Now denoting the constant vector

$$\mathbf{k}_i \times \hat{\mathbf{q}} = \mathbf{a} \quad (3.101)$$

and assuming that it is real, we see, from Eq. (3.100) that the vectors $\hat{\mathbf{q}}$, $\mathbf{k}_i$, $\mathbf{k}_r^{(1)}$, $\mathbf{k}_r^{(2)}$, $\mathbf{k}_r^{(3)}$, ..., $\mathbf{k}_t^{(1)}$, $\mathbf{k}_t^{(2)}$, $\mathbf{k}_t^{(3)}$, ..., are perpendicular to the same vector $\mathbf{a}$ and, consequently, they all lie on the same plane. The plane formed by the vectors $\hat{\mathbf{q}}$ and $\mathbf{k}_i$ is called the *plane of incidence* and is defined by the equation

$$\mathbf{a} \cdot \mathbf{r} = 0 \tag{3.102}$$

where $\mathbf{r}$ is a position vector drawn from a fixed origin located on the plane of incidence. We thus conclude that the vector $\hat{\mathbf{q}}$ and the wave vectors of the incident, reflected, and transmitted waves all lie on the plane of incidence.

By taking the magnitudes of the cross products in Eq. (3.100), we obtain the more familiar form of the laws of reflection and refraction:

$$\begin{aligned} |\mathbf{a}| &= |\mathbf{k}_i \times \hat{\mathbf{q}}| = |\mathbf{k}_r^{(1)} \times \hat{\mathbf{q}}| = |\mathbf{k}_r^{(2)} \times \hat{\mathbf{q}}| = \cdots \\ &= |\mathbf{k}_t^{(1)} \times \hat{\mathbf{q}}| = |\mathbf{k}_t^{(2)} \times \hat{\mathbf{q}}| = \cdots \end{aligned} \tag{3.103}$$

or, in terms of angles

$$\begin{aligned} k_i \sin \theta_i &= k_r^{(1)} \sin \theta_r^{(1)} = k_r^{(2)} \sin \theta_r^{(2)} = \cdots \\ &= k_t^{(1)} \sin \theta_t^{(1)} = k_t^{(2)} \sin \theta_t^{(2)} = \cdots \end{aligned} \tag{3.104}$$

where k_i is the wave number of the incident wave, and θ_i is the angle between the vectors $\mathbf{k}_i$ and $\hat{\mathbf{q}}$ and is called the *angle of incidence*. Likewise, $k_r^{(1)}$, $k_r^{(2)}$, $k_r^{(3)}$, ... are the wave numbers of the reflected waves and $\theta_r^{(1)}$, $\theta_r^{(2)}$, ... are the corresponding angles of reflection formed, respectively, by the vectors $\mathbf{k}_r^{(1)}$, $\mathbf{k}_r^{(2)}$, $\mathbf{k}_r^{(3)}$, ... with $\hat{\mathbf{q}}$, while $k_t^{(1)}$, $k_t^{(2)}$, $k_t^{(3)}$, ... are the wave numbers of the transmitted waves and $\theta_t^{(1)}$, $\theta_t^{(2)}$, ... are the corresponding angles of transmission (or angles of refraction) formed, respectively, by the vectors $\mathbf{k}_t^{(1)}$, $\mathbf{k}_t^{(2)}$, $\mathbf{k}_t^{(3)}$, ... with $\hat{\mathbf{q}}$.

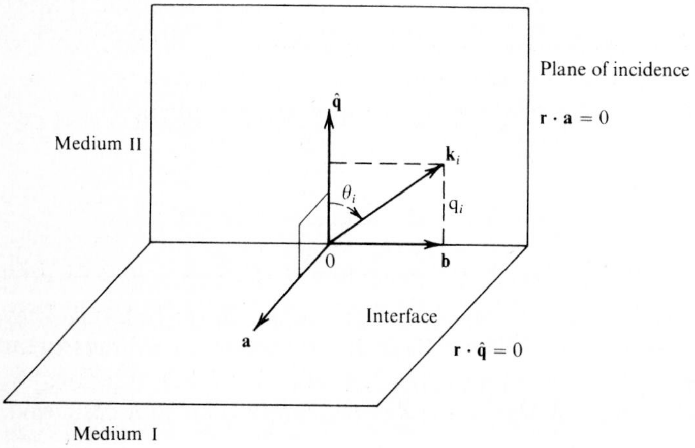

Figure 3.14 Orientations of interface and the plane of incidence.

To provide a simple geometrical construction of the wave vectors, we form the cross product of $\hat{\mathbf{q}}$ with Eq. (3.100) and then expand the vector triple products:

$$\begin{aligned}
\hat{\mathbf{q}} \times \mathbf{a} &= \mathbf{k}_i - (\mathbf{k}_i \cdot \hat{\mathbf{q}})\hat{\mathbf{q}} \\
&= \mathbf{k}_r^{(1)} - (\mathbf{k}_r^{(1)} \cdot \hat{\mathbf{q}})\hat{\mathbf{q}} \\
&= \mathbf{k}_r^{(2)} - (\mathbf{k}_r^{(2)} \cdot \hat{\mathbf{q}})\hat{\mathbf{q}} \\
&\cdots\cdots\cdots\cdots \\
&= \mathbf{k}_t^{(1)} - (\mathbf{k}_t^{(1)} \cdot \hat{\mathbf{q}})\hat{\mathbf{q}} \\
&= \mathbf{k}_t^{(2)} - (\mathbf{k}_t^{(2)} \cdot \hat{\mathbf{q}})\hat{\mathbf{q}} \\
&\cdots\cdots\cdots\cdots
\end{aligned}$$

or

$$\begin{array}{ll}
\mathbf{k}_i = \mathbf{b} + q_i \hat{\mathbf{q}} & q_i = \mathbf{k}_i \cdot \hat{\mathbf{q}} \\
\mathbf{k}_r^{(1)} = \mathbf{b} + q_r^{(1)} \hat{\mathbf{q}} & q_r^{(1)} = \mathbf{k}_r^{(1)} \cdot \hat{\mathbf{q}} \\
\mathbf{k}_r^{(2)} = \mathbf{b} + q_r^{(2)} \hat{\mathbf{q}} & q_r^{(2)} = \mathbf{k}_r^{(2)} \cdot \hat{\mathbf{q}} \\
\cdots\cdots\cdots\cdots & \cdots\cdots\cdots\cdots \\
\mathbf{k}_t^{(1)} = \mathbf{b} + q_t^{(1)} \hat{\mathbf{q}} & q_t^{(1)} = \mathbf{k}_t^{(1)} \cdot \hat{\mathbf{q}} \\
\mathbf{k}_t^{(2)} = \mathbf{b} + q_t^{(2)} \hat{\mathbf{q}} & q_t^{(2)} = \mathbf{k}_t^{(2)} \cdot \hat{\mathbf{q}} \\
\cdots\cdots\cdots\cdots & \cdots\cdots\cdots\cdots
\end{array} \tag{3.105}$$

where

$$\mathbf{b} = \hat{\mathbf{q}} \times \mathbf{a} \tag{3.106}$$

In summary, Eq. (3.105) states that (1) all the wave vectors of the incident, reflected, and transmitted waves lie on the plane of incidence which is defined by vectors $\mathbf{b}$ and $\hat{\mathbf{q}}$ (or $\mathbf{k}_i$ and $\hat{\mathbf{q}}$) (see Fig. 3.14); (2) the projections of the wave vectors $\mathbf{k}_i$, $\mathbf{k}_r^{(1)}$, $\mathbf{k}_r^{(2)}$, ... on the interface are equal to the same constant vector $\mathbf{b}(|\mathbf{b}| = |\mathbf{a}|)$, an alternative form of the laws of reflection and refraction; and (3) the projections of $\mathbf{k}_i$, $\mathbf{k}_r^{(1)}$, $\mathbf{k}_r^{(2)}$, ... on the unit normal to the interface are, respectively, q_i, $q_r^{(1)}$, $q_r^{(2)}$, ... Based on these, we may formulate the following general rule: letting 0 be a fixed origin located on the interface, the tips of all the wave vectors of the incident, reflected, and transmitted waves drawn from 0 must lie on a straight-line which passes through the end point of vector $\mathbf{b}$ and is parallel to $\hat{\mathbf{q}}$ (see Fig. 3.15).

According to Eq. (3.105), we may derive some general relations connecting any three wave vectors. For instance, if $\mathbf{k}_\alpha$ represents any one wave vector in Eq. (3.105), then

$$\mathbf{k}_\alpha \times \mathbf{k}_\beta + \mathbf{k}_\beta \times \mathbf{k}_\gamma + \mathbf{k}_\gamma \times \mathbf{k}_\alpha = 0 \tag{3.107}$$

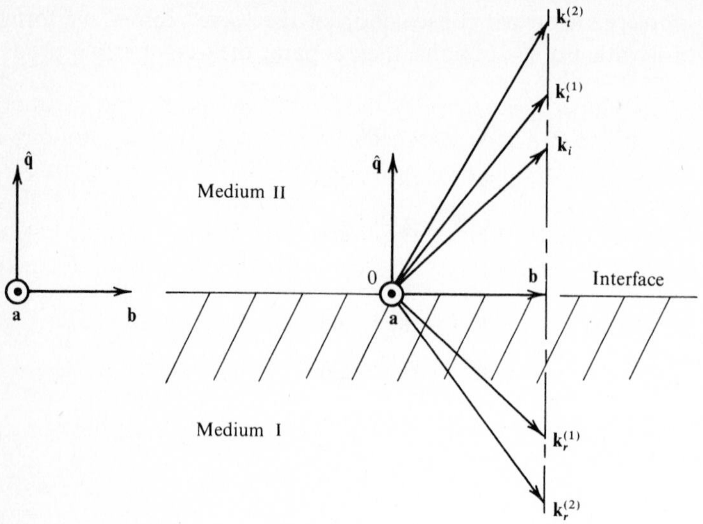

Figure 3.15 Geometrical construction of wave vectors.

This can be proved by noting that

$$\begin{aligned}\mathbf{k}_\alpha \times \mathbf{k}_\beta &= (\mathbf{b} + q_\alpha \hat{\mathbf{q}}) \times (\mathbf{b} + q_\beta \hat{\mathbf{q}}) \\ &= q_\alpha(\hat{\mathbf{q}} \times \mathbf{b}) + q_\beta(\mathbf{b} \times \hat{\mathbf{q}}) \\ &= (q_\beta - q_\alpha)\mathbf{a} \qquad (3.108)\end{aligned}$$

Substituting Eq. (3.108) and similar expressions into the left-hand side of Eq. (3.107), we establish the identity. A similar proof gives the following identities:

$$q_\alpha(\mathbf{k}_\beta \times \mathbf{k}_\gamma) + q_\beta(\mathbf{k}_\gamma \times \mathbf{k}_\alpha) + q_\gamma(\mathbf{k}_\alpha \times \mathbf{k}_\beta) = \mathbf{0} \qquad (3.109)$$

and

$$(\mathbf{k}_\alpha \times \mathbf{k}_\beta)\mathbf{k}_\gamma + (\mathbf{k}_\beta \times \mathbf{k}_\gamma)\mathbf{k}_\alpha + (\mathbf{k}_\gamma \times \mathbf{k}_\alpha)\mathbf{k}_\beta = \bar{\mathbf{0}} \qquad (3.110)$$

Equation (3.110) is a dyadic equation and follows from the fact that $q_\alpha = (\mathbf{k}_\alpha - \mathbf{b}) \cdot \hat{\mathbf{q}}$, Eqs. (3.109) and (3.107).

As an interesting application of the laws of reflection and refraction (3.100), we can easily show that the boundary conditions for the normal components of **B** and **D** follow from the boundary conditions for the tangential components of **E** and **H** and the Maxwell equations. For example, we may prove that Eq. (3.93) follows from Eq. (3.91) and the Maxwell equation $\omega\mathbf{B} = \mathbf{k} \times \mathbf{E}$. Indeed, the total **B** fields in medium I and II are

$$\begin{aligned}\mathbf{B}_\mathrm{I} &= \mathbf{B}_i + \sum_\alpha \mathbf{B}_r^{(\alpha)} \\ &= \frac{1}{\omega}\left[(\mathbf{k}_i \times \mathbf{E}_i) + \sum_\alpha (\mathbf{k}_r^{(\alpha)} \times \mathbf{E}_r^{(\alpha)})\right]\end{aligned}$$

and
$$\mathbf{B}_{\mathrm{II}} = \sum_{\beta} \mathbf{B}_t^{(\beta)} = \frac{1}{\omega} \sum_{\beta} (\mathbf{k}_t^{(\beta)} \times \mathbf{E}_t^{(\beta)})$$

respectively; where the superscript α denotes the αth reflected wave and the superscript β denotes the βth transmitted wave, and where the summation over α means the sum over all the reflected waves in medium I and the summation over β means the sum over all the transmitted waves in medium II. According to Eqs. (3.100), (3.101), and the property of scalar triple product that the position of dot and cross may be interchanged without altering the value of the product, we write

$$\begin{aligned}
\omega(\mathbf{B}_{\mathrm{I}} - \mathbf{B}_{\mathrm{II}}) \cdot \hat{\mathbf{q}} &= (\mathbf{k}_i \times \mathbf{E}_i + \sum_{\alpha} \mathbf{k}_r^{(\alpha)} \times \mathbf{E}_r^{(\alpha)} - \sum_{\beta} \mathbf{k}_t^{(\beta)} \times \mathbf{E}_t^{(\beta)}) \cdot \hat{\mathbf{q}} \\
&= -\mathbf{E}_i \cdot (\mathbf{k}_i \times \hat{\mathbf{q}}) - \sum_{\alpha} \mathbf{E}_r^{(\alpha)} \cdot (\mathbf{k}_r^{(\alpha)} \times \hat{\mathbf{q}}) + \sum_{\beta} \mathbf{E}_t^{(\beta)} \cdot (\mathbf{k}_t^{(\beta)} \times \hat{\mathbf{q}}) \\
&= -\left(\mathbf{E}_i + \sum_{\alpha} \mathbf{E}_r^{(\alpha)} - \sum_{\beta} \mathbf{E}_t^{(\beta)}\right) \cdot \mathbf{a} \\
&= \mathbf{k}_i \cdot [(\mathbf{E}_{\mathrm{I}} - \mathbf{E}_{\mathrm{II}}) \times \hat{\mathbf{q}}]
\end{aligned}$$

which vanishes if $\mathbf{E}_{\mathrm{I}}$ and $\mathbf{E}_{\mathrm{II}}$ satisfy Eq. (3.91). In other words, Eq. (3.93) follows from Eq. (3.91). Similarly we can show that Eq. (3.94) follows from Eq. (3.92) and Maxwell's equation $\omega\mathbf{D} = -\mathbf{k} \times \mathbf{H}$.

3.6 GROUP VELOCITY

In previous sections, we have considered the monochromatic plane wave of the form

$$\mathbf{E}(\mathbf{r}, t) = \mathbf{E}_0 \, e^{i(\mathbf{k}\cdot\mathbf{r} - \omega t)}$$

where the frequency ω is related to the wave vector $\mathbf{k}$ by the dispersion equation $\omega = \omega(\mathbf{k})$ as given in Eq. (3.22) for an isotropic medium. Monochromatic plane waves which are infinite in time duration and spatial extent are idealizations never realized in practice. The surfaces of constant phase defined by $\mathbf{k} \cdot \mathbf{r} - \omega t =$ const propagate with phase velocity $\mathbf{v}_p = \omega\mathbf{k}/k^2$ which, in general, does not represent the velocity of energy transport, and may thus assume values greater or less than the velocity of light.

However, waves encountered in practice are bounded either in time, such as waves are turned on for a finite time duration, or in space or both. These waves are called *packets*. According to the Fourier integral, we may represent a wave packet as a superposition of a group of monochromatic plane waves:

$$\mathbf{E}(\mathbf{r}, t) = \frac{1}{(2\pi)^3} \int_{-\infty}^{+\infty} \mathbf{E}(\mathbf{k}) e^{i(\mathbf{k}\cdot\mathbf{r} - \omega t)} \, d^3k \qquad (3.111)$$

where $\omega = \omega(\mathbf{k})$ and $d^3k = dk_x\, dk_y\, dk_z$ is the three-dimensional volume element in $\mathbf{k}$-space. The Fourier amplitude function $\mathbf{E}(\mathbf{k})$ has a sharp peak at $\mathbf{k}_0$ and is negligible outside a small region. Within this region, $\omega(\mathbf{k})$ will deviate but slightly from $\omega_0 = \omega(\mathbf{k}_0)$ and may thus be represented by the first two terms of a Taylor series expansion:

$$\omega(\mathbf{k}) = \omega(\mathbf{k}_0) + \left(\frac{\partial \omega}{\partial \mathbf{k}}\right)_{\mathbf{k}_0} \cdot (\mathbf{k} - \mathbf{k}_0) + \cdots \tag{3.112}$$

where $\partial\omega/\partial\mathbf{k}$ designates the gradient of ω in $\mathbf{k}$-space and is given by

$$\frac{\partial \omega}{\partial \mathbf{k}} = \frac{\partial \omega}{\partial k_x}\,\hat{\mathbf{x}} + \frac{\partial \omega}{\partial k_y}\,\hat{\mathbf{y}} + \frac{\partial \omega}{\partial k_z}\,\hat{\mathbf{z}} \tag{3.113}$$

in a cartesian coordinate system. Substituting Eq. (3.112) into Eq. (3.111), we obtain

$$\mathbf{E}(\mathbf{r}, t) \cong \mathbf{A}\left[\mathbf{r} - \left(\frac{\partial \omega}{\partial \mathbf{k}}\right)_{\mathbf{k}_0} t\right] \exp i(\mathbf{k}_0 \cdot \mathbf{r} - \omega_0 t) \tag{3.114}$$

which has the form of a monochromatic plane wave $\exp[i(\mathbf{k}_0 \cdot \mathbf{r} - \omega_0 t)]$ modulated by an amplitude function

$$\mathbf{A}\left[\mathbf{r} - \left(\frac{\partial \omega}{\partial \mathbf{k}}\right)_{\mathbf{k}_0} t\right] = \frac{1}{(2\pi)^3} \int_{-\infty}^{+\infty} \mathbf{E}(\mathbf{k}) \exp\left\{i(\mathbf{k} - \mathbf{k}_0) \cdot \left[\mathbf{r} - \left(\frac{\partial \omega}{\partial \mathbf{k}}\right)_{\mathbf{k}_0} t\right]\right\} d^3k \tag{3.115}$$

The amplitude vector $\mathbf{A}[\mathbf{r} - (\partial\omega/\partial\mathbf{k})_{\mathbf{k}_0} t]$ is constant over surfaces

$$\mathbf{r} - \left(\frac{\partial \omega}{\partial \mathbf{k}}\right)_{\mathbf{k}_0} \mathrm{t} = \text{const} \tag{3.116}$$

from which we conclude that the peak of the wave packet is propagated with the group velocity

$$\mathbf{v}_g = \frac{\partial \omega}{\partial \mathbf{k}} \tag{3.117}$$

where the value of $\partial\omega/\partial\mathbf{k}$ is evaluated at $\mathbf{k} = \mathbf{k}_0$. In other words, the group velocity is defined as the velocity with which the whole wave packet moves without distortion. In an isotropic medium, ω is a function of the magnitude of the wave vector $\mathbf{k}$ only, that is, $\omega = \omega(k)$ where $k = \sqrt{k_x^2 + k_y^2 + k_z^2}$. Hence

$$\mathbf{v}_g = \frac{\partial \omega}{\partial \mathbf{k}} = \frac{d\omega}{dk}\,\hat{\mathbf{k}} \tag{3.118}$$

that is, the group velocity is parallel to the wave vector $\mathbf{k}$.

The group velocity defined in Eq. (3.117) can be interpreted geometrically. According to vector analysis, the gradient of a scalar function $\omega(\mathbf{k})$ is a vector whose direction is normal to the surface $\omega(\mathbf{k}) = \text{const}$. This surface is called the *wave vector surface*, or the *dispersion surface*, since for each value of ω, the dispersion equation describes a surface in **k**-space. Consequently, the given values of ω and **k** determine a point on this surface and the group velocity has the direction of the normal to the wave vector surface at the given point as illustrated in Fig. 3.16.

Finally, we shall show that in a lossless, frequency-dispersive medium, when the values of ω and **k** are real (i.e., *transparent medium*), the group velocity is equal to the velocity of energy transport. For this purpose, we repeat the Maxwell equations (3.15) and (3.16) for anisotropic media

$$\mathbf{k} \times \mathbf{E}_0 = \omega\mu_0\,\bar{\boldsymbol{\mu}} \cdot \mathbf{H}_0 \tag{3.119}$$

$$\mathbf{k} \times \mathbf{H}_0 = -\omega\varepsilon_0\,\bar{\boldsymbol{\varepsilon}} \cdot \mathbf{E}_0 \tag{3.120}$$

and their complex conjugate

$$\begin{aligned} \mathbf{k} \times \mathbf{E}_0^* &= \omega\mu_0\,\bar{\boldsymbol{\mu}}^* \cdot \mathbf{H}_0^* \\ \mathbf{k} \times \mathbf{H}_0^* &= -\omega\varepsilon_0\,\bar{\boldsymbol{\varepsilon}}^* \cdot \mathbf{E}_0^* \end{aligned} \tag{3.121}$$

where $\mathbf{E}_0$ and $\mathbf{H}_0$ are functions of **k** and $\omega(\mathbf{k})$, $\bar{\boldsymbol{\varepsilon}}$ and $\bar{\boldsymbol{\mu}}$ are functions of $\omega(\mathbf{k})$ only. Equations (3.119) and (3.120) are valid for either the monochromatic plane wave or each Fourier component of the wave packet. Taking the dot products of Eqs. (3.119) and (3.120) with $\mathbf{H}_0^*$ and $\mathbf{E}_0^*$, respectively, and then subtracting, we obtain

$$\mu_0\,\mathbf{H}_0^* \cdot \omega\bar{\boldsymbol{\mu}} \cdot \mathbf{H}_0 + \varepsilon_0\,\mathbf{E}_0^* \cdot \omega\bar{\boldsymbol{\varepsilon}} \cdot \mathbf{E}_0 = \mathbf{k} \cdot (\mathbf{E}_0 \times \mathbf{H}_0^* + \mathbf{E}_0^* \times \mathbf{H}_0) \tag{3.122}$$

Next, differentiating Eq. (3.122) with respect to k_α ($\alpha = x, y, z$), and noting that

$$\frac{\partial \mathbf{k}}{\partial k_\alpha} = \sum_{\beta=1}^{3} \frac{\partial k_\beta}{\partial k_\alpha}\,\hat{\mathbf{x}}_\beta = \sum_{\beta=1}^{3} \delta_{\alpha\beta}\,\hat{\mathbf{x}}_\beta = \hat{\mathbf{x}}_\alpha \tag{3.123}$$

we have

$$\begin{aligned} &\left[\mathbf{H}_0^* \cdot \frac{\partial(\omega\mu_0\,\bar{\boldsymbol{\mu}})}{\partial\omega} \cdot \mathbf{H}_0 + \mathbf{E}_0^* \cdot \frac{\partial(\omega\varepsilon_0\,\bar{\boldsymbol{\varepsilon}})}{\partial\omega} \cdot \mathbf{E}_0\right]\frac{\partial\omega}{\partial k_\alpha} \\ &\qquad = \hat{\mathbf{x}}_\alpha \cdot (\mathbf{E}_0 \times \mathbf{H}_0^* + \mathbf{E}_0^* \times \mathbf{H}_0) + \frac{\partial \mathbf{H}_0^*}{\partial k_\alpha} \cdot (\mathbf{k} \times \mathbf{E}_0 - \omega\mu_0\,\bar{\boldsymbol{\mu}} \cdot \mathbf{H}_0) \\ &\qquad\quad + \frac{\partial \mathbf{H}_0}{\partial k_\alpha} \cdot (\mathbf{k} \times \mathbf{E}_0^* - \omega\mu_0\,\tilde{\bar{\boldsymbol{\mu}}} \cdot \mathbf{H}_0^*) - \frac{\partial \mathbf{E}_0^*}{\partial k_\alpha} \cdot (\mathbf{k} \times \mathbf{H}_0 + \omega\varepsilon_0\,\bar{\boldsymbol{\varepsilon}} \cdot \mathbf{E}_0) \\ &\qquad\quad - \frac{\partial \mathbf{E}_0}{\partial k_\alpha} \cdot (\mathbf{k} \times \mathbf{H}_0^* + \omega\varepsilon_0\,\tilde{\bar{\boldsymbol{\varepsilon}}} \cdot \mathbf{E}^*) \end{aligned} \tag{3.124}$$

For a lossless medium, $\bar{\boldsymbol{\varepsilon}}$ and $\bar{\boldsymbol{\mu}}$ are hermitian: $\tilde{\bar{\boldsymbol{\varepsilon}}} = \bar{\boldsymbol{\varepsilon}}^*$, $\tilde{\bar{\boldsymbol{\mu}}} = \bar{\boldsymbol{\mu}}^*$; thus, according to Eqs. (3.119) to (3.121), the coefficients of the derivatives of the amplitude vectors vanish and Eq. (3.124) becomes

$$\langle W \rangle \frac{\partial\omega}{\partial k_\alpha} = \langle \mathscr{P} \rangle \cdot \hat{\mathbf{x}}_\alpha \tag{3.125}$$

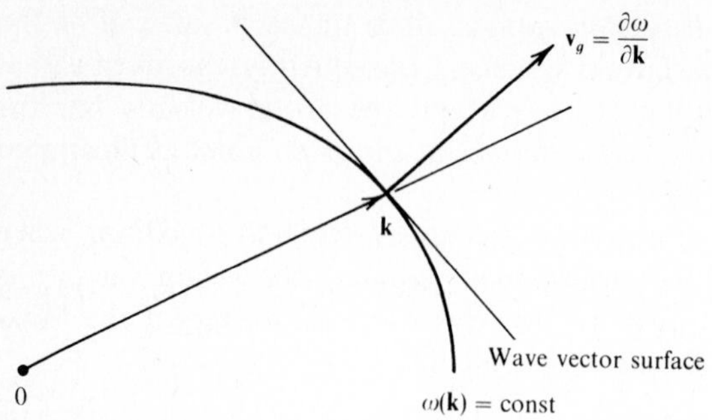

Figure 3.16 Wave vector surface and its relation to the group velocity.

where, according to Eq. (2.85)

$$\langle W \rangle = \frac{1}{4}\left[\mathbf{H}_0^* \cdot \frac{\partial(\omega\mu_0 \bar{\boldsymbol{\mu}})}{\partial\omega} \cdot \mathbf{H}_0 + \mathbf{E}_0^* \cdot \frac{\partial(\omega\varepsilon_0 \bar{\boldsymbol{\varepsilon}})}{\partial\omega} \cdot \mathbf{E}_0\right] \tag{3.126}$$

is the time-averaged energy density and

$$\langle \mathscr{P} \rangle = \tfrac{1}{4}(\mathbf{E}_0 \times \mathbf{H}_0^* + \mathbf{E}_0^* \times \mathbf{H}_0) \tag{3.127}$$

is the time-averaged Poynting vector. Combining the three components, we obtain Eq. (3.125) in vector form:

$$\frac{\partial \omega}{\partial \mathbf{k}} = \frac{\langle \mathscr{P} \rangle}{\langle W \rangle} \tag{3.128}$$

A comparison of Eqs. (2.88), (3.117), and (3.128) shows that $\mathbf{v}_g = \mathbf{v}_E$: the group velocity coincides with the velocity of energy transport.

For an electrically anisotropic (that is, $\bar{\boldsymbol{\mu}} = \mu\bar{\mathbf{I}}$) and frequency-dispersive medium, we can rewrite the time-averaged Poynting vector (3.127) with the aid of Maxwell's equation (3.119) as

$$\langle \mathscr{P} \rangle = \frac{1}{4\omega\mu_0\mu}[\mathbf{E}_0 \times (\mathbf{k} \times \mathbf{E}_0^*) + \mathbf{E}_0^* \times (\mathbf{k} \times \mathbf{E}_0)]$$

therefore,

$$\mathbf{k} \cdot \langle \mathscr{P} \rangle = \frac{k^2}{2\omega\mu_0\mu}[\mathbf{E}_0 \cdot \mathbf{E}_0^* - (\mathbf{E}_0 \cdot \hat{\mathbf{k}})(\mathbf{E}_0^* \cdot \hat{\mathbf{k}})] \geq 0 \tag{3.129}$$

The last inequality follows from the fact that the magnitude of any complex vector is nonnegative, i.e.,

$$[\mathbf{E}_0 - (\mathbf{E}_0 \cdot \hat{\mathbf{k}})\hat{\mathbf{k}}] \cdot [\mathbf{E}_0 - (\mathbf{E}_0 \cdot \hat{\mathbf{k}})\hat{\mathbf{k}}]^* \geq 0$$

Since the group velocity is in the same direction as $\langle \mathscr{P} \rangle$, Eq. (3.129) implies that the angle between the directions of the group velocity $\mathbf{v}_g$ (or the direction of energy transport) and the wave vector $\mathbf{k}$ (the direction of phase velocity $\mathbf{v}_p$) is acute.

PROBLEMS

3.1 A *gyrational medium* is characterized by the constitutive relations

$$\mathbf{D} = \varepsilon_0 \varepsilon \mathbf{E} + \xi \mathbf{H}$$

$$\mathbf{B} = \zeta \mathbf{E} + \mu_0 \mu \mathbf{H}$$

(*a*) Denoting any one of the field vectors **E**, **H**, **B**, and **D** by **F**, and using Maxwell's equations and the above constitutive relations, show that

$$\nabla^2 \mathbf{F} + \alpha \nabla \times \mathbf{F} + k_g^2 \mathbf{F} = \mathbf{0}$$

where

$$\alpha = i\omega(\zeta - \xi)$$

$$k_g^2 = \omega^2(\mu_0 \varepsilon_0 \mu\varepsilon - \zeta\xi)$$

(*b*) Introducing a scalar potential ϕ and a vector potential **A** such that

$$\mathbf{B} = \nabla \times \mathbf{A}$$

$$\mathbf{E} = i\omega \mathbf{A} - \nabla\phi$$

show that **A** and ϕ satisfy

$$\nabla^2 \mathbf{A} + \alpha \nabla \times \mathbf{A} + k_g^2 \mathbf{A} = \mathbf{0}$$

and

$$\nabla^2 \phi + k_g^2 \phi = 0$$

respectively, provided that they are related by

$$\nabla \cdot \mathbf{A} + i\omega(\zeta\xi - \mu_0 \varepsilon_0 \mu\varepsilon)\phi = 0$$

(*c*) If the gyrational medium is lossless, show that μ and ε are real, and $\xi = \zeta^*$. Hence α and k_g^2 are also real.

3.2 A monochromatic plane wave propagates in a gyrational medium with the constitutive relations given in Prob. 3.1. (*a*) Show that the field vectors **E**, **H**, **B**, and **D** are all perpendicular to the direction of wave propagation **k**. In other words, it is a TEM wave.

(*b*) Show that the dispersion equation in such a medium is

$$k^2 = \tfrac{1}{2}[2k_g^2 + \alpha^2 \pm \sqrt{(2k_g^2 + \alpha^2)^2 - 4k_g^4}]$$

Thus, if $\alpha \neq 0$, k^2 has two possible values and two waves can be propagated in the medium with two different phase velocities.

3.3 The constitutive relations for monochromatic plane waves in optically active crystals of the cubic system take the following form:

$$\mathbf{D} = \varepsilon_0 \varepsilon \mathbf{E} + ip(\mathbf{k} \times \mathbf{E})$$

$$\mathbf{B} = \mu_0 \mu \mathbf{H} + iq(\mathbf{k} \times \mathbf{H})$$

where p and q represent respectively the electric and magnetic activities of the medium.

(*a*) Show that the dispersion equation in such a medium is

$$(1 - \omega^2 pq)k^2 \mp \omega^2(q\varepsilon_0 \varepsilon + p\mu_0 \mu)k - k_0^2 \mu\varepsilon = 0$$

or

$$k_\pm = \frac{\sqrt{\omega^4(q\varepsilon_0 \varepsilon - p\mu_0 \mu)^2 + 4k_0^2 \mu\varepsilon} \pm \omega^2(q\varepsilon_0 \varepsilon + p\mu_0 \mu)}{2(1 - \omega^2 pq)}$$

(*b*) Show that the electric and magnetic fields may be written as

$$\mathbf{E} = \pm i(\hat{\mathbf{k}} \times \mathbf{E})$$

$$\mathbf{H} = \pm i(\hat{\mathbf{k}} \times \mathbf{H})$$

That is, both **E** and **H** are circularly polarized. The right- and left-handed circularly polarized waves are propagated in the medium with phase velocities ω/k_+ and ω/k_- respectively.

3.4 A monochromatic plane wave propagating in free space and in the z direction is described by the complex amplitude vector $\mathbf{E}_0 = \mathbf{E}_1 + i\mathbf{E}_2$. If $\mathbf{E}_1$ and $\mathbf{E}_2$ have the directions and magnitudes as shown in Fig. 3.17 at $t = 0$, $z = 0$, find the instantaneous field vectors $\mathscr{E}(z, t)$ and $\mathscr{H}(z, t)$. What is the polarization of the wave?

3.5 For a monochromatic plane wave, show that the areal velocity (the area swept out by $\mathscr{E}$ per unit time) is constant for an elliptical polarization.

3.6 (*a*) Show that Eq. (3.33) can be written in the form

$$|\mathscr{E}|^2 = a + b \cos (2\phi + \theta)$$

where

$$a = \tfrac{1}{2}(\mathbf{E}_1^2 + \mathbf{E}_2^2)$$

$$b^2 = \tfrac{1}{4}(\mathbf{E}_1^2 + \mathbf{E}_2^2)^2 - \mathbf{E}_1^2\mathbf{E}_2^2 \sin \xi$$

$$\tan \theta = \frac{2|\mathbf{E}_1|\,|\mathbf{E}_2| \cos \xi}{\mathbf{E}_1^2 - \mathbf{E}_2^2}$$

and ξ is the space angle between vectors $\mathbf{E}_1$ and $\mathbf{E}_2$. (*b*) Find the values of $|\mathbf{E}_1|/|\mathbf{E}_2|$ and ξ such that the ratio $(b/a)^2$ in part (*a*) is a maximum or a minimum. (*c*) Discuss and sketch the polarization of $\mathscr{E}$ in the two cases.

3.7 Show that a wave is circularly polarized if

$$|\mathbf{E}_0|^2 = |\mathbf{E}_0 \times \mathbf{E}_0^*|$$

and is linearly polarized if

$$|\mathbf{E}_0|^2 = |\mathbf{E}_0^2|$$

3.8 In application, it is more convenient to describe the polarization of a wave in terms of the complex amplitude vector $\mathbf{E}_0 = \mathbf{E}_1 + i\mathbf{E}_2$. To this end, show that (*a*) the polarization ellipse (3.32) may be expressed as

$$|\mathscr{E} \times \mathbf{E}_0|^2 = \tfrac{1}{4}|\mathbf{E}_0 \times \mathbf{E}_0^*|^2$$

(*b*) The major semiaxis $\mathbf{a}_0$ and the minor semiaxis $\mathbf{b}_0$ of the polarization ellipse are the real and

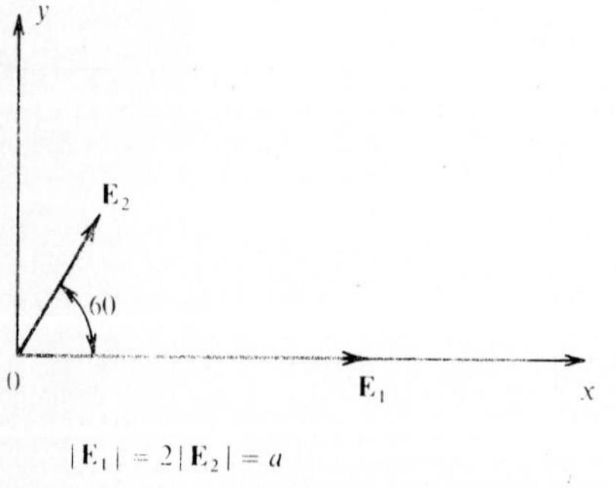

Figure 3.17 Orientations of real vectors $\mathbf{E}_1$ and $\mathbf{E}_2$.

imaginary parts of the complex vector

$$\mathbf{E}_{0_r} = \sqrt{\frac{|\mathbf{E}_0^2|}{\mathbf{E}_0^2}}\,\mathbf{E}_0$$

respectively, that is, $\mathbf{E}_{0_r} = \mathbf{a}_0 + i\mathbf{b}_0$ where

$$\mathbf{a}_0 = \frac{1}{2}\left(\sqrt{\frac{|\mathbf{E}_0^2|}{\mathbf{E}_0^2}}\,\mathbf{E}_0 + \sqrt{\frac{|\mathbf{E}_0^2|}{\mathbf{E}_0^{*2}}}\,\mathbf{E}_0^*\right) = \mathscr{E}_{\max}$$

$$\mathbf{b}_0 = \frac{1}{2i}\left(\sqrt{\frac{|\mathbf{E}_0^2|}{\mathbf{E}_0^2}}\,\mathbf{E}_0 - \sqrt{\frac{|\mathbf{E}_0^2|}{\mathbf{E}_0^{*2}}}\,\mathbf{E}_0^*\right) = \mathscr{E}_{\min}$$

and

$$\mathbf{a}_0 \cdot \mathbf{b}_0 = 0$$

3.9 Given two complex amplitude vectors $\mathbf{E}_0^{(1)}$ and $\mathbf{E}_0^{(2)}$ such that they lie on the same plane (i.e., for a real unit vector $\hat{\mathbf{k}}$, $\hat{\mathbf{k}} \cdot \mathbf{E}_0^{(1)} = \hat{\mathbf{k}} \cdot \mathbf{E}_0^{(2)} = 0$) and $\mathbf{E}_0^{(1)} \cdot \mathbf{E}_0^{(2)} = 0$, show that the instantaneous vectors $\mathscr{E}^{(1)}$ and $\mathscr{E}^{(2)}$ are mutually perpendicular at all instants, and the polarization curves described by them have the same shape (eccentricity) and sense of rotation.

3.10 Let

$$\gamma = \frac{|\mathbf{E}_0^2|}{|\mathbf{E}_0|^2}$$

show that the shapes of the polarization curves are determined by the values of γ according to the following:

$\gamma = 0$	a circle
$0 < \gamma < 1$	an ellipse
$\gamma = 1$	a straight line

3.11 Similar to Figs. 3.12 and 3.13, prove and construct geometrically both in time and frequency domains that the sum of two circularly polarized waves of opposite senses of rotation and (*a*) of unequal magnitudes, is an elliptically polarized wave whose sense of rotation is the same as that of the circularly polarized wave of larger amplitude; (*b*) of equal magnitudes, is a linearly polarized wave.

3.12 Show that a wave of any polarization can be decomposed into infinite sets of the sum of a linear polarization and a circular polarization of any sense of rotation. However, if the direction of oscillation of the linear polarization and the sense of rotation of the circular polarization are given, then the decomposition is unique.

3.13 As an alternative form to Eq. (3.105), show that the wave vectors of the reflected and transmitted waves can be expressed as

$$\mathbf{k}_r^{(1)} = \mathbf{k}_i + \alpha_r^{(1)}\hat{\mathbf{q}}$$

$$\mathbf{k}_r^{(2)} = \mathbf{k}_i + \alpha_r^{(2)}\hat{\mathbf{q}}$$

....

$$\mathbf{k}_t^{(1)} = \mathbf{k}_i + \alpha_t^{(1)}\hat{\mathbf{q}}$$

$$\mathbf{k}_t^{(2)} = \mathbf{k}_i + \alpha_t^{(2)}\hat{\mathbf{q}}$$

....

where $\alpha_r^{(1)}, \alpha_r^{(2)}, \ldots, \alpha_t^{(1)}, \alpha_t^{(2)}, \ldots$ are some constants. Next give a geometrical interpretation of the above relations.

3.14 As a simple example of group velocity, let us consider two waves of equal real amplitudes and with almost the same wave numbers and frequencies. Their sum has the form

$$\Psi(z, t) = \Psi_0 \cos(k_1 z - \omega_1 t) + \Psi_0 \cos(k_2 z - \omega_2 t)$$

where

$$\omega_1 = \omega_0 + \Delta\omega \qquad \omega_2 = \omega_0 - \Delta\omega$$

$$k_1 = k_0 + \Delta k \qquad k_2 = k_0 - \Delta k$$

show that

$$\Psi(z, t) = \Psi_m \cos(k_0 z - \omega_0 t)$$

Hence the envelope

$$\Psi_m = 2\Psi_0 \cos(\Delta k z - \Delta\omega t)$$

travels at the group velocity $v_g = d\omega/dk$ while the carrier $\cos(k_0 z - \omega_0 t)$ moves at the average phase velocity $v_p = \omega_0/k_0$.

3.15 Show that the group velocity v_g and the phase velocity v_p are related by

$$v_g = v_p + k\frac{dv_p}{dk}$$

$$= v_p - \lambda\frac{dv_p}{d\lambda}$$

where λ is the wavelength of the wave. Clearly, if the phase velocity is independent of λ, v_g and v_p are identical, and the medium is nondispersive. On the other hand, in a dispersive medium, two possibilities occur: normal dispersion, $dv_p/d\lambda > 0$, $v_g < v_p$, and anomalous dispersion, $dv_p/d\lambda < 0$, $v_g > v_p$.

3.16 Consider a medium whose dielectric constant is given by

$$\varepsilon(\omega) = 1 + \frac{\omega_p^2}{\omega_0^2 - \omega^2}$$

where ω_p^2 and ω_0^2 are constants. Assuming $\mu = 1$, show that the phase and group velocities for the case of large frequency ($\omega \gg \omega_0$) are given by

$$v_p = c\left(1 + \frac{\omega_p^2}{2\omega^2}\right) > c$$

$$v_g = c\left(1 - \frac{\omega_p^2}{2\omega^2}\right) < c$$

and for small frequency ($\omega \ll \omega_0$)

$$v_p = \frac{c}{\sqrt{\varepsilon_n}}\left(1 - \frac{\omega_p^2\omega^2}{2\varepsilon_n\omega_0^4}\right) < c$$

$$v_g = \frac{c}{\sqrt{\varepsilon_n}}\left(1 - \frac{3\omega_p^2\omega^2}{2\varepsilon_n\omega_0^4}\right) < c$$

where

$$\varepsilon_n = \varepsilon(0) = 1 + \frac{\omega_p^2}{\omega_0^2}$$

3.17 In a lossless, dispersive, anisotropic medium, if the index of refraction n depends on the angular frequency ω and the wave normal $\hat{\mathbf{k}}$, then

$$\omega(\mathbf{k}) = \frac{ck(\mathbf{k})}{n(\omega, \hat{\mathbf{k}})}$$

Using the definition (3.117) of the group velocity, show that

$$\mathbf{v}_g = \frac{\partial \omega}{\partial \mathbf{k}} = \frac{\mathbf{v}_p - \dfrac{v_p}{n}\dfrac{\partial n}{\partial \hat{\mathbf{k}}}}{1 + \dfrac{\omega}{n}\dfrac{\partial n}{\partial \omega}}$$

$$= \frac{v_p\left(\mathbf{n} - \dfrac{\partial n}{\partial \hat{\mathbf{k}}}\right)}{\dfrac{\partial(\omega n)}{\partial \omega}}$$

where $\partial/\partial\hat{\mathbf{k}}$ denotes the gradient operator with respect to the components of $\hat{\mathbf{k}}$. Thus as a result of frequency dispersion and anisotropy of the medium, the group velocity in general differs from the phase velocity $\mathbf{v}_p = v_p \hat{\mathbf{k}}$. The denominator $\partial(\omega n)/\partial\omega$ indicates the effect of frequency dispersion on the group velocity. The term proportional to $\partial n/\partial\hat{\mathbf{k}}$ represents the dependence of group velocity on the anisotropy of the medium.

3.18 In Prob. 3.17, let the magnitude of the group velocity be v_g and its direction be the unit vector $\hat{\mathbf{t}}$:

$$\mathbf{v}_g = v_g \hat{\mathbf{t}}$$

Show that

$$v_g = \frac{v_p \sec\theta}{1 + \dfrac{\omega}{n}\dfrac{\partial n}{\partial\omega}}$$

$$\hat{\mathbf{t}} = \cos\theta\left(\hat{\mathbf{k}} - \frac{1}{n}\frac{\partial n}{\partial\hat{\mathbf{k}}}\right)$$

$$\cos\theta = \frac{1}{\left(1 + \dfrac{1}{n^2}\dfrac{\partial n}{\partial\hat{\mathbf{k}}} \cdot \dfrac{\partial n}{\partial\hat{\mathbf{k}}}\right)^{1/2}}$$

where θ is the angle between the group velocity and the wave vector $\mathbf{k}$.

REFERENCES

Adler, R. B., L. J. Chu, and R. M. Fano: *Electromagnetic Energy Transmission and Radiation*, John Wiley & Sons, Inc., New York, 1960, chap. 1.

Azzam, R. M. A., and N. M. Bashara: *Ellipsometry and Polarized Light*, North-Holland Publishing Co., New York, 1977, chap. 1.

Brillouin, L.: *Wave Propagation and Group Velocity*, Academic Press, New York, 1960.

Kelso, J. M.: *Radio Ray Propagation in the Ionosphere*, McGraw-Hill Book Company, New York, 1964, chap. 2.

Kraus, J. D., and K. R. Carver: *Electromagnetics*, 2d ed., McGraw-Hill Book Company, New York, 1973, chap. 11.

Lighthill, M. J.: "Group Velocity," *J. Inst. Math. Applics.*, vol. 1, 1965, pp. 1–28.

Nelson, D. F.: "Group Velocity in Crystal Optics," *Am. J. Phys.*, vol. 45, no. 12, 1977, pp. 1187–1190.

Paris, D. T., and F. K. Hurd: *Basic Electromagnetic Theory*, McGraw-Hill Book Company, New York, 1969, chap. 7.

Stone, J. M.: *Radiation and Optics*, McGraw-Hill Book Company, New York, 1963, chap. 16.

CHAPTER

FOUR

PLANE WAVES IN ISOTROPIC MEDIA

We shall now undertake a more systematic and detailed study of the basic aspects of electromagnetic plane wave propagation in lossless and lossy isotropic media. We shall first review waves in an unbounded isotropic medium and then examine problems of reflection and refraction of waves from a plane interface which separates the two media. However, we shall not deal with sources of electromagnetic waves until Chap. 9.

4.1 UNIFORM PLANE WAVES IN LOSSLESS ISOTROPIC MEDIA

As shown in Sec. 3-1, in a homogeneous, lossless, isotropic medium characterized by the real constant parameters ε and μ, the uniform plane wave solution of the forms

$$\mathbf{E} = \mathbf{E}_0 \, e^{i\mathbf{k}\cdot\mathbf{r}} \tag{4.1}$$

$$\mathbf{H} = \mathbf{H}_0 \, e^{i\mathbf{k}\cdot\mathbf{r}} \tag{4.2}$$

satisfies the Maxwell equations. Here the wave number k must be related to the angular frequency ω by the dispersion equation:

$$k = \frac{\omega}{c} \sqrt{\mu\varepsilon} \tag{4.3}$$

and the complex constant-amplitude vectors by

$$\mathbf{H}_0 = \frac{1}{\eta} (\hat{\mathbf{k}} \times \mathbf{E}_0) \tag{4.4}$$

and
$$\mathbf{E}_0 = -\eta\,(\hat{\mathbf{k}} \times \mathbf{H}_0) \tag{4.5}$$

where
$$\eta = \sqrt{\frac{\mu_0\,\mu}{\varepsilon_0\,\varepsilon}} \tag{4.6}$$

is the intrinsic wave impedance, both k and η are real. The planes of constant phase are normal to $\hat{\mathbf{k}}$ and travel with the phase velocity

$$\mathbf{v}_p = \frac{\omega}{k}\,\hat{\mathbf{k}} = \frac{c}{\sqrt{\mu\varepsilon}}\,\hat{\mathbf{k}}$$

Over the plane of constant phase, the amplitude vector of the wave is also constant, that is, a uniform plane wave. By taking the real parts on both sides of Eqs. (4.4) and (4.5) according to Eq. (3.1), we obtain the relations in time domain:

$$\mathscr{H} = \frac{1}{\eta}\,(\hat{\mathbf{k}} \times \mathscr{E}) \tag{4.7}$$

and
$$\mathscr{E} = -\,\eta(\hat{\mathbf{k}} \times \mathscr{H}) \tag{4.8}$$

This implies that the vectors $\mathscr{E}$ and $\mathscr{H}$ of a uniform plane wave are always perpendicular to the direction of wave propagation $\hat{\mathbf{k}}$ and to each other as illustrated in Fig. 3.3. Since the planes of constant phase are also perpendicular to $\hat{\mathbf{k}}$, we conclude that $\mathscr{E}$ and $\mathscr{H}$ lie on the planes of constant phase, and the tips of $\mathscr{E}$ and $\mathscr{H}$ trace out curves of identical forms, rotate in the same direction but are oriented mutually perpendicular to each other.

The instantaneous and time-averaged electric and magnetic energy densities carried by the wave are given by Eqs. (2.41), (2.57), and (2.58). For convenience, we repeat them here:

$$W_e = \tfrac{1}{2}\varepsilon_0\,\varepsilon\,\mathscr{E}^2 \tag{4.9}$$

$$W_m = \tfrac{1}{2}\mu_0\,\mu\,\mathscr{H}^2 \tag{4.10}$$

and
$$\langle W_e\rangle = \tfrac{1}{4}\varepsilon_0\,\varepsilon\,|\mathbf{E}_0|^2 \tag{4.11}$$

$$\langle W_m\rangle = \tfrac{1}{4}\mu_0\,\mu\,|\mathbf{H}_0|^2 \tag{4.12}$$

From Maxwell's equations (4.7) and (4.8), we can easily show that the instantaneous electric and magnetic energy densities, at any instant, are equal to each other. In fact

$$\begin{aligned} W_e &= \tfrac{1}{2}\varepsilon_0\,\varepsilon\,\mathscr{E}\cdot\mathscr{E} \\ &= -\tfrac{1}{2}\varepsilon_0\,\varepsilon\eta\,(\hat{\mathbf{k}} \times \mathscr{H})\cdot\mathscr{E} \\ &= \tfrac{1}{2}\varepsilon_0\,\varepsilon\eta\,\mathscr{H}\cdot(\hat{\mathbf{k}} \times \mathscr{E}) \\ &= \tfrac{1}{2}\varepsilon_0\,\varepsilon\eta^2\mathscr{H}^2 = W_m \end{aligned} \tag{4.13}$$

Thus at any instant, the total instantaneous energy density is

$$\begin{aligned} W &= W_e + W_m \\ &= \varepsilon_0\,\varepsilon\mathscr{E}^2 = \mu_0\,\mu\mathscr{H}^2 \end{aligned} \tag{4.14}$$

By taking the time average on both sides of Eqs. (4.13) and (4.14), we obtain

$$\langle W_e \rangle = \langle W_m \rangle \tag{4.15}$$

and

$$\langle W \rangle = \tfrac{1}{2}\varepsilon_0 \varepsilon \, |\mathbf{E}_0|^2$$
$$= \tfrac{1}{2}\mu_0 \mu \, |\mathbf{H}_0|^2 \tag{4.16}$$

The instantaneous Poynting vector, defined by Eq. (2.34), is

$$\mathscr{P} = \mathscr{E} \times \mathscr{H} \tag{4.17}$$

Replacing $\mathscr{H}$ according to Eq. (4.7) and noting that $\hat{\mathbf{k}} \cdot \mathscr{E} = 0$, we have

$$\mathscr{P} = \frac{1}{\eta} \mathscr{E}^2 \hat{\mathbf{k}} \tag{4.18}$$

Also the time-averaged Poynting vector for a uniform plane wave in lossless media is

$$\langle \mathscr{P} \rangle = \tfrac{1}{2}\mathrm{Re}\,(\mathbf{E}_0 \times \mathbf{H}_0^*) \tag{4.19}$$

or

$$\langle \mathscr{P} \rangle = \frac{1}{2\eta} |\mathbf{E}_0|^2 \hat{\mathbf{k}} \tag{4.20}$$

Since an elliptically polarized wave can be expressed as a sum of two circularly polarized waves of opposite senses of rotation, we can show, with the aid of Eq. (3.81), that the time-averaged energy density (4.16) of an elliptically polarized wave is equal to the sum of the time-averaged energy densities of two circularly polarized waves of opposite senses of rotation; i.e.,

$$\langle W \rangle = \langle W^{(1)} \rangle + \langle W^{(2)} \rangle \tag{4.21}$$

where $\langle W \rangle$ given by Eq. (4.16) is the time-averaged energy density of the elliptically polarized wave and

$$\langle W^{(1)} \rangle = \frac{\varepsilon_0 \varepsilon}{2} |\mathbf{E}_0^{(1)}|^2$$
$$\langle W^{(2)} \rangle = \frac{\varepsilon_0 \varepsilon}{2} |\mathbf{E}_0^{(2)}|^2 \tag{4.22}$$

are the time-averaged energy densities of two circularly polarized waves into which $\mathbf{E}_0$ is decomposed: $\mathbf{E}_0 = \mathbf{E}_0^{(1)} + \mathbf{E}_0^{(2)}$. However, it should be noted that the instantaneous energy density

$$W = \frac{\varepsilon_0 \varepsilon}{2} |\mathbf{E}_0^{(1)} e^{i\phi} + \mathbf{E}_0^{(2)*} e^{-i\phi}|^2 \tag{4.23}$$

is not the simple sum of those pertinent to each polarization alone. A similar result is valid for the time-averaged Poynting vector:

$$\langle \mathscr{P} \rangle = \langle \mathscr{P}^{(1)} \rangle + \langle \mathscr{P}^{(2)} \rangle \tag{4.24}$$

where $\langle \mathscr{P} \rangle$ given by Eq. (4.20) is the time-averaged Poynting vector of the elliptically polarized wave and

$$\langle \mathscr{P}^{(1)} \rangle = \frac{1}{2\eta} |\mathbf{E}_0^{(1)}|^2 \hat{\mathbf{k}}$$
$$\langle \mathscr{P}^{(2)} \rangle = \frac{1}{2\eta} |\mathbf{E}_0^{(2)}|^2 \hat{\mathbf{k}} \tag{4.25}$$

are the time-averaged Poynting vectors of the two circularly polarized waves. On the other hand, if we decompose a wave of arbitrary polarization into two linear polarizations oscillating in mutually perpendicular directions, both the energy density (instantaneous and time-averaged) and the Poynting vector of the wave are equal to the sum of the energy densities and Poynting vectors of individual linearly polarized waves (see Prob. 4.7).

A comparison of Eqs. (4.20) and (4.16) shows that

$$\langle \mathscr{P} \rangle = \langle W \rangle \mathbf{v}_p \tag{4.26}$$

Hence, we may interpret the time-averaged Poynting vector of a uniform plane wave as the transfer of the time-averaged energy density through the medium with the velocity $\mathbf{v}_p$. According to Eqs. (2.88) and (3.118), the velocity of energy transport and the group velocity are given by

$$\mathbf{v}_E = \frac{\langle \mathscr{P} \rangle}{\langle W \rangle} = \frac{c}{\sqrt{\mu\varepsilon}} \hat{\mathbf{k}} = \mathbf{v}_p \tag{4.27}$$

and

$$\mathbf{v}_g = \frac{d\omega}{dk} \hat{\mathbf{k}} = \frac{c}{\sqrt{\mu\varepsilon}} \hat{\mathbf{k}} \tag{4.28}$$

respectively. Thus for a uniform plane wave in a lossless, nondispersive isotropic medium, the velocity of energy transport coincides with the group velocity and the phase velocity with which the planes of constant phase propagate through the medium.

4.2 NONUNIFORM PLANE WAVES IN LOSSLESS ISOTROPIC MEDIA

In the last section, we discussed the uniform plane waves in lossless isotropic media and considered only the case of a real wave vector **k**. However, the Maxwell equations in such media can admit more general solutions than the uniform plane waves. In fact, the plane waves of the forms (4.1) and (4.2) may also satisfy Maxwell's equations for a lossless isotropic media when the wave vector **k** is complex:

$$\mathbf{k} = \mathbf{k}_1 + i\mathbf{k}_2 \tag{4.29}$$

Here

$$\mathbf{k}_1 = k_{1x}\hat{\mathbf{x}} + k_{1y}\hat{\mathbf{y}} + k_{1z}\hat{\mathbf{z}} = k_1\hat{\mathbf{k}}_1$$

and

$$\mathbf{k}_2 = k_{2x}\hat{\mathbf{x}} + k_{2y}\hat{\mathbf{y}} + k_{2z}\hat{\mathbf{z}} = k_2\hat{\mathbf{k}}_2 \tag{4.30}$$

are two real space vectors. We shall call $\mathbf{k}_1$ the *phase vector* and $\mathbf{k}_2$ the *attenuation vector*. Substitution of Eqs. (4.1) and (4.2) with the complex wave vector $\mathbf{k}$ into the source-free Maxwell's equations, yields [cf. Eqs. (3.15) and (3.16)]

$$\mathbf{k} \times \mathbf{E}_0 = \omega\mu_0 \mu \mathbf{H}_0 \tag{4.31}$$

$$\mathbf{k} \times \mathbf{H}_0 = -\omega\varepsilon_0 \varepsilon \mathbf{E}_0 \tag{4.32}$$

Next, dot-premultiplying Eq. (4.31) by $\mathbf{k}$ and noting that the cross product of the complex vector $\mathbf{k}$ with itself is zero, we obtain

$$\omega\mu_0 \mu \mathbf{k} \cdot \mathbf{H}_0 = \mathbf{k} \cdot (\mathbf{k} \times \mathbf{E}_0)$$

$$= (\mathbf{k} \times \mathbf{k}) \cdot \mathbf{E}_0 = 0$$

or

$$\mathbf{k} \cdot \mathbf{H}_0 = 0 \tag{4.33}$$

Similarly, from Eq. (4.32)

$$\mathbf{k} \cdot \mathbf{E}_0 = 0 \tag{4.34}$$

Since $\mathbf{k}$, $\mathbf{H}_0$, and $\mathbf{E}_0$ are complex vectors, Eqs. (4.33) and (4.34) do not imply perpendicularity of vectors in space.

Eliminating $\mathbf{H}_0$ from Eqs. (4.31) and (4.32), and using the result of Eq. (4.34), we obtain

$$(\mathbf{k}^2 - k_0^2\mu\varepsilon)\mathbf{E}_0 = \mathbf{0} \tag{4.35}$$

where $k_0 = \omega\sqrt{\mu_0 \varepsilon_0}$. For a nonzero solution $\mathbf{E}_0$ to exist, the complex wave vector $\mathbf{k}$ must satisfy the dispersion equation

$$\mathbf{k}^2 = k_0^2 \mu\varepsilon \tag{4.36}$$

or

$$(k_1^2 - k_2^2) + i2\mathbf{k}_1 \cdot \mathbf{k}_2 = k_0^2 \mu\varepsilon$$

Equating the real and imaginary parts separately, we have

$$k_1^2 - k_2^2 = k_0^2 \mu\varepsilon \tag{4.37}$$

$$\mathbf{k}_1 \cdot \mathbf{k}_2 = 0 \tag{4.38}$$

Equations (4.36), (4.31), (4.32), (4.33), and (4.34) summarize the requirements imposed on the plane wave of the forms (4.1) and (4.2) by the Maxwell equations. If and when these conditions are satisfied, the plane waves will be acceptable solutions to the Maxwell equations in lossless isotropic media. To understand Eq. (4.38), we recall the instantaneous electric field intensity

$$\mathscr{E} = \text{Re}\,(\mathbf{E}_0\, e^{i(\mathbf{k}\cdot\mathbf{r} - \omega t)})$$

$$= \text{Re}\,[(\mathbf{E}_0\, e^{-\mathbf{k}_2\cdot\mathbf{r}})e^{i\phi_1}]$$

$$= \tfrac{1}{2}e^{-\mathbf{k}_2\cdot\mathbf{r}}(\mathbf{E}_0\, e^{i\phi_1} + \mathbf{E}_0^* e^{-i\phi_1}) \tag{4.39}$$

where

$$\phi_1 = \mathbf{k}_1 \cdot \mathbf{r} - \omega t \tag{4.40}$$

is the phase of the wave. A surface of constant phase at a given instant defined by $\mathbf{r} \cdot \hat{\mathbf{k}}_1 =$ constant is a plane normal to $\hat{\mathbf{k}}_1$. We thus call k_1 the *phase constant* and

$\hat{\mathbf{k}}_1$ the *unit wave normal.* From Eq. (4.40), we see that the plane of constant phase moves with the phase velocity

$$\mathbf{v}_p = \frac{\omega}{k_1}\hat{\mathbf{k}}_1 \tag{4.41}$$

The amplitude vector of the wave, on the other hand, varies as $\mathbf{E}_0 e^{-\mathbf{k}_2 \cdot \mathbf{r}}$. Accordingly, a surface of constant amplitude, defined by $\mathbf{r} \cdot \hat{\mathbf{k}}_2 =$ constant, is a plane normal to $\hat{\mathbf{k}}_2$. We thus call k_2 the *attenuation constant* and $\hat{\mathbf{k}}_2$ the *unit amplitude normal.* The field, defined by Eq. (4.39), is again a plane wave because its surfaces of constant phase are planes. But it is not a uniform plane wave, since the amplitude vector of the wave varies with position over the planes of constant phase. The plane wave characterized by the complex wave vector with $\hat{\mathbf{k}}_1 \neq \hat{\mathbf{k}}_2$ is called the *nonuniform plane wave.* Thus, for a nonuniform plane wave, the simultaneous constancy of amplitude and phase can occur only in the direction of the intersection of planes of constant phase and amplitude, i.e., in the direction of $\hat{\mathbf{k}}_1 \times \hat{\mathbf{k}}_2$.

The condition stated in Eq. (4.38) is that the real vectors $\mathbf{k}_1$ and $\mathbf{k}_2$ are perpendicular in space (if $k_1, k_2 \neq 0$). Moreover, as long as the frequency is not zero, $\omega\sqrt{\mu_0 \varepsilon_0 \mu\varepsilon} > 0$, and Eq. (4.37) guarantees that $k_1 > k_2$. Specifically

$$k_1^2 = k_2^2 + \omega^2 \mu_0 \varepsilon_0 \mu\varepsilon \geq \omega^2 \mu_0 \varepsilon_0 \mu\varepsilon > 0 \tag{4.42}$$

for $\omega \neq 0$. In other words, the phase velocity with which a nonuniform plane wave is propagated is less than the phase velocity with which a uniform plane wave is propagated in the same medium. In general, if $\mathbf{k}_2 \neq \mathbf{0}$, the field of a nonuniform plane wave (4.39) has planes of constant amplitude (normal to $\hat{\mathbf{k}}_2$) and planes of constant phase (normal to $\hat{\mathbf{k}}_1$) that are mutually perpendicular as illustrated in Fig. 4.1. We note that Eq. (4.37) does not uniquely determine k_1, k_2 and the expression (4.39) is physically meaningful only for the half-space such

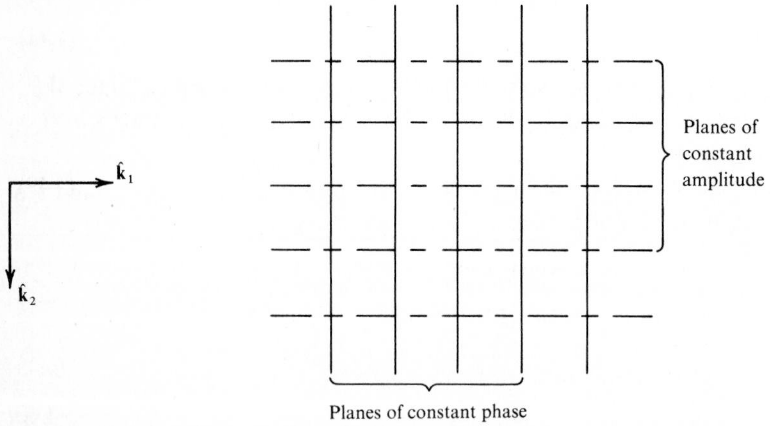

Figure 4.1 Planes of constant phase and of constant amplitude in a lossless medium.

that $\mathbf{k}_2 \cdot \mathbf{r} > 0$. Thus we shall find important applications of nonuniform plane waves in boundary-value problems such as the phenomenon of total internal reflection and the reflection of waves from conducting media.

The instantaneous electric and magnetic energy densities given by Eqs. (4.9) and (4.10) respectively are still valid for nonuniform plane waves. Substituting $\mathscr{E}$ of Eq. (4.39) and a similar form for $\mathscr{H}$ into Eqs. (4.9) and (4.10), and separating the time-dependent terms from the time-independent term, we obtain

$$\begin{aligned} W_e &= W_e^{(a)} + W_e^{(d)} \\ W_m &= W_m^{(a)} + W_m^{(d)} \end{aligned} \tag{4.43}$$

where the time-independent parts of the electric and magnetic energy densities denoted by the superscript (d) (dc part) are

$$\begin{aligned} W_e^{(d)} &= \tfrac{1}{4}\varepsilon_0 \varepsilon e^{-2\mathbf{k}_2\cdot\mathbf{r}} |\mathbf{E}_0|^2 \\ W_m^{(d)} &= \tfrac{1}{4}\mu_0 \mu e^{-2\mathbf{k}_2\cdot\mathbf{r}} |\mathbf{H}_0|^2 \end{aligned} \tag{4.44}$$

and the time-dependent parts denoted by the superscript (a) (ac part) are

$$\begin{aligned} W_e^{(a)} &= \tfrac{1}{8}\varepsilon_0 \varepsilon e^{-2\mathbf{k}_2\cdot\mathbf{r}}(\mathbf{E}_0^2 e^{i2\phi_1} + \mathbf{E}_0^{*2} e^{-i2\phi_1}) \\ W_m^{(a)} &= \tfrac{1}{8}\mu_0 \mu e^{-2\mathbf{k}_2\cdot\mathbf{r}}(\mathbf{H}_0^2 e^{i2\phi_1} + \mathbf{H}_0^{*2} e^{-i2\phi_1}) \end{aligned} \tag{4.45}$$

We now show that

$$W_e^{(a)} = W_m^{(a)} \tag{4.46}$$

In fact, from Maxwell's equations (4.31) and (4.32), we have

$$\begin{aligned} \varepsilon_0 \varepsilon \mathbf{E}_0^2 &= -\frac{1}{\omega}(\mathbf{k} \times \mathbf{H}_0) \cdot \mathbf{E}_0 \\ &= \frac{1}{\omega}\mathbf{H}_0 \cdot (\mathbf{k} \times \mathbf{E}_0) = \mu_0 \mu \mathbf{H}_0^2 \end{aligned} \tag{4.47}$$

and the complex conjugate of Eq. (4.47)

$$\varepsilon_0 \varepsilon \mathbf{E}_0^{*2} = \mu_0 \mu \mathbf{H}_0^{*2} \tag{4.48}$$

Substituting Eqs. (4.47) and (4.48) into $W_e^{(a)}$, we establish the result. Since $W_e^{(a)}$ and $W_m^{(a)}$ vary periodically in time with frequency 2ω, their time averages over a period must vanish, i.e.,

$$\langle W_e^{(a)} \rangle = \langle W_m^{(a)} \rangle = 0 \tag{4.49}$$

On the other hand, $W_e^{(d)}$ and $W_m^{(d)}$ do not depend on t, thus

$$\begin{aligned} \langle W_e^{(d)} \rangle &= W_e^{(d)} \\ \langle W_m^{(d)} \rangle &= W_m^{(d)} \end{aligned}$$

and consequently the time average of Eq. (4.43) gives

$$\begin{aligned} \langle W_e \rangle &= W_e^{(d)} \\ \langle W_m \rangle &= W_m^{(d)} \end{aligned} \tag{4.50}$$

Dot-multiplying the complex conjugate of Eq. (4.31) by $\mathbf{H}_0$ and Eq. (4.32) by $\mathbf{E}_0^*$, we obtain

$$\mu_0 \mu |\mathbf{H}_0|^2 = \frac{1}{\omega} \mathbf{k}^* \cdot (\mathbf{E}_0^* \times \mathbf{H}_0) \tag{4.51}$$

$$\varepsilon_0 \varepsilon |\mathbf{E}_0|^2 = \frac{1}{\omega} \mathbf{k} \cdot (\mathbf{E}_0^* \times \mathbf{H}_0) \tag{4.52}$$

In the case of a nonuniform plane wave, $\mathbf{k}^*$ is different from $\mathbf{k}$, thus Eq. (4.51) is not equal to Eq. (4.52) which in turn implies that $W_e^{(d)}$ is different from $W_m^{(d)}$. In other words, unlike the case of uniform plane wave, the instantaneous and time-averaged electric and magnetic energy densities carried by the nonuniform plane wave are not equal. The total instantaneous energy density carried by the wave is

$$\begin{aligned} W &= W_e + W_m \\ &= W^{(a)} + W^{(d)} \end{aligned} \tag{4.53}$$

where

$$\begin{aligned} W^{(a)} &= W_e^{(a)} + W_m^{(a)} \\ &= \tfrac{1}{4}\varepsilon_0 \varepsilon e^{-2\mathbf{k}_2 \cdot \mathbf{r}} (\mathbf{E}_0^2 e^{i2\phi_1} + \mathbf{E}_0^{*2} e^{-i2\phi_1}) \\ &= \tfrac{1}{4}\mu_0 \mu e^{-2\mathbf{k}_2 \cdot \mathbf{r}} (\mathbf{H}_0^2 e^{i2\phi_1} + \mathbf{H}_0^{*2} e^{-i2\phi_1}) \end{aligned} \tag{4.54}$$

and

$$\begin{aligned} W^{(d)} &= W_e^{(d)} + W_m^{(d)} \\ &= \tfrac{1}{4} e^{-2\mathbf{k}_2 \cdot \mathbf{r}} (\varepsilon_0 \varepsilon |\mathbf{E}_0|^2 + \mu_0 \mu |\mathbf{H}_0|^2) \end{aligned} \tag{4.55}$$

Hence

$$\langle W^{(a)} \rangle = 0$$

and

$$\langle W \rangle = W^{(d)}$$

In order to express $W^{(d)}$ in terms of $\mathbf{E}_0 = \mathbf{E}_1 + i\mathbf{E}_2$ alone, we use Eq. (4.31) and write

$$\begin{aligned} \mu_0 \mu |\mathbf{H}_0|^2 &= \frac{1}{\omega^2 \mu_0 \mu} (\mathbf{k} \times \mathbf{E}_0) \cdot (\mathbf{k}^* \times \mathbf{E}_0^*) \\ &= \frac{1}{\omega^2 \mu_0 \mu} [|\mathbf{k}|^2 |\mathbf{E}_0|^2 - (\mathbf{k} \cdot \mathbf{E}_0^*)(\mathbf{k} \cdot \mathbf{E}_0^*)^*] \end{aligned} \tag{4.56}$$

Since $\mathbf{k} \cdot \mathbf{E}_0 = 0$, thus

$$(\mathbf{k} \times \mathbf{k}^*) \cdot (\mathbf{E}_0 \times \mathbf{E}_0^*) = -(\mathbf{k} \cdot \mathbf{E}_0^*)(\mathbf{k} \cdot \mathbf{E}_0^*)^*$$

and $W^{(d)}$ becomes

$$\begin{aligned} W^{(d)} &= \frac{e^{-2\mathbf{k}_2 \cdot \mathbf{r}}}{4\omega^2 \mu_0 \mu} [(k_0^2 \mu\varepsilon + |\mathbf{k}|^2)|\mathbf{E}_0|^2 + (\mathbf{k} \times \mathbf{k}^*) \cdot (\mathbf{E}_0 \times \mathbf{E}_0^*)] \\ &= \frac{e^{-2\mathbf{k}_2 \cdot \mathbf{r}}}{4\omega^2 \mu_0 \mu} [(k_0^2 \mu\varepsilon + |\mathbf{k}|^2)|\mathbf{E}_0|^2 - 4(\mathbf{k}_1 \times \mathbf{k}_2) \cdot (\mathbf{E}_1 \times \mathbf{E}_2)] \end{aligned} \tag{4.57}$$

Using the dispersion equation (4.37) and (4.38) for a lossless medium, we find

$$k_0^2 \mu\varepsilon + |\mathbf{k}|^2 = 2k_1^2$$

Hence Eq. (4.57) reduces to

$$W^{(d)} = \frac{e^{-2\mathbf{k}_2\cdot\mathbf{r}}}{2\omega^2\mu_0\mu}[k_1^2(\mathbf{E}_1^2 + \mathbf{E}_2^2) - 2(\mathbf{k}_1 \times \mathbf{k}_2)\cdot(\mathbf{E}_1 \times \mathbf{E}_2)] \tag{4.58}$$

Similarly, we may write Eq. (4.54) as

$$W^{(a)} = \tfrac{1}{2}\varepsilon_0\varepsilon e^{-2\mathbf{k}_2\cdot\mathbf{r}}[(\mathbf{E}_1^2 - \mathbf{E}_2^2)\cos 2\phi_1 - 2(\mathbf{E}_1\cdot\mathbf{E}_2)\sin 2\phi_1] \tag{4.59}$$

Turning now to the instantaneous Poynting vector, we again separate Eq. (4.17) into the time-dependent part $\mathscr{P}^{(a)}$ and the time-independent part $\mathscr{P}^{(d)}$, that is,

$$\begin{aligned}\mathscr{P} &= \mathscr{E} \times \mathscr{H} \\ &= \mathscr{P}^{(a)} + \mathscr{P}^{(d)}\end{aligned} \tag{4.60}$$

where

$$\mathscr{P}^{(a)} = \tfrac{1}{2}e^{-2\mathbf{k}_2\cdot\mathbf{r}}\,\mathrm{Re}\,(\mathbf{E}_0 \times \mathbf{H}_0 e^{i2\phi_1}) \tag{4.61}$$

$$\mathscr{P}^{(d)} = \tfrac{1}{2}e^{-2\mathbf{k}_2\cdot\mathbf{r}}\,\mathrm{Re}\,(\mathbf{E}_0 \times \mathbf{H}_0^*) \tag{4.62}$$

Substituting $\mathbf{H}_0$ from Eq. (4.31) into Eq. (4.62) and noting from Eq. (4.34) that $\mathbf{k}\cdot\mathbf{E}_0 = 0$, hence

$$(\mathbf{k} - \mathbf{k}^*) \times (\mathbf{E}_0 \times \mathbf{E}_0^*) = (\mathbf{k}\cdot\mathbf{E}_0^*)\mathbf{E}_0 + (\mathbf{k}^*\cdot\mathbf{E}_0)\mathbf{E}_0^*$$

We may express $\mathscr{P}^{(d)}$ in terms of $\mathbf{E}_0$ alone:

$$\begin{aligned}\mathscr{P}^{(d)} &= \tfrac{1}{4}e^{-2\mathbf{k}_2\cdot\mathbf{r}}(\mathbf{E}_0 \times \mathbf{H}_0^* + \mathbf{E}_0^* \times \mathbf{H}_0) \\ &= \frac{e^{-2\mathbf{k}_2\cdot\mathbf{r}}}{4\omega\mu_0\mu}[\mathbf{E}_0 \times (\mathbf{k}^* \times \mathbf{E}_0^*) + \mathbf{E}_0^* \times (\mathbf{k} \times \mathbf{E}_0)] \\ &= \frac{e^{-2\mathbf{k}_2\cdot\mathbf{r}}}{4\omega\mu_0\mu}[|\mathbf{E}_0|^2(\mathbf{k} + \mathbf{k}^*) - (\mathbf{k} - \mathbf{k}^*) \times (\mathbf{E}_0 \times \mathbf{E}_0^*)]\end{aligned} \tag{4.63}$$

or

$$\begin{aligned}\mathscr{P}^{(d)} &= \frac{e^{-2\mathbf{k}_2\cdot\mathbf{r}}}{2\omega\mu_0\mu}[|\mathbf{E}_0|^2\mathbf{k}_1 - \mathbf{k}_2 \times i(\mathbf{E}_0 \times \mathbf{E}_0^*)] \\ &= \frac{e^{-2\mathbf{k}_2\cdot\mathbf{r}}}{2\omega\mu_0\mu}[(\mathbf{E}_1^2 + \mathbf{E}_2^2)\mathbf{k}_1 - 2\mathbf{k}_2 \times (\mathbf{E}_1 \times \mathbf{E}_2)]\end{aligned} \tag{4.64}$$

Similarly

$$\begin{aligned}\mathscr{P}^{(a)} &= \frac{e^{-2\mathbf{k}_2\cdot\mathbf{r}}}{2\omega\mu_0\mu}\,\mathrm{Re}\,(\mathbf{E}_0^2 e^{i2\phi_1}\mathbf{k}) \\ &= \frac{e^{-2\mathbf{k}_2\cdot\mathbf{r}}}{2\omega\mu_0\mu}\{[(\mathbf{E}_1^2 - \mathbf{E}_2^2)\cos 2\phi_1 - 2(\mathbf{E}_1\cdot\mathbf{E}_2)\sin 2\phi_1]\mathbf{k}_1 \\ &\quad - [(\mathbf{E}_1^2 - \mathbf{E}_2^2)\sin 2\phi_1 + 2(\mathbf{E}_1\cdot\mathbf{E}_2)\cos 2\phi_1]\mathbf{k}_2\}\end{aligned} \tag{4.65}$$

Therefore

$$\langle \mathscr{P}^{(a)} \rangle = \mathbf{0} \tag{4.66}$$

and

$$\langle \mathscr{P} \rangle = \mathscr{P}^{(d)} \tag{4.67}$$

It is interesting to note from Eq. (4.64) that the direction of the time-averaged Poynting vector depends on the vector $i(\mathbf{E}_0 \times \mathbf{E}_0^*)$, which in turn depends on the polarization of the wave, as such was not the case with uniform plane wave [cf. Eq. (4.20)].

From Eq. (4.64) and the dispersion equation (4.38), we can easily show that

$$\mathscr{P}^{(d)} \cdot \hat{\mathbf{k}}_2 = 0$$

or

$$\langle \mathscr{P} \rangle \cdot \hat{\mathbf{k}}_2 = 0 \tag{4.68}$$

that is, the time-averaged Poynting vector in a lossless isotropic medium is always perpendicular to $\hat{\mathbf{k}}_2$. Furthermore, comparing Eq. (4.58) with the dot product of $\mathbf{k}_1$ with Eq. (4.64), we find that

$$W^{(d)} = \frac{1}{\omega} \mathscr{P}^{(d)} \cdot \mathbf{k}_1 \tag{4.69}$$

or

$$\langle \mathscr{P} \rangle \cdot \hat{\mathbf{k}}_1 = \langle W \rangle v_p \tag{4.70}$$

where $v_p = \omega/k_1$ is the phase velocity. Equation (4.70) shows that the projection of the time-averaged Poynting vector of a nonuniform plane wave in the direction of $\hat{\mathbf{k}}_1$ is equal to the product of the time-averaged energy density and the phase velocity as in the case of uniform plane wave [cf. Eq. (4.26)]. We may also write Eq. (4.70) as

$$\frac{\langle \mathscr{P} \rangle}{\langle W \rangle} \cdot \hat{\mathbf{k}}_1 = \mathbf{v}_E \cdot \hat{\mathbf{k}}_1 = v_p \tag{4.71}$$

that is, for a nonuniform plane wave in a lossless isotropic medium, the projection of the velocity of energy transport in the direction of $\hat{\mathbf{k}}_1$ gives the phase velocity of the wave. On the other hand, comparing Eq. (4.59) with the dot product of $\hat{\mathbf{k}}_1$ with Eq. (4.65) and noting that $\hat{\mathbf{k}}_1 \cdot \hat{\mathbf{k}}_2 = 0$ for a lossless medium, we may express $W^{(a)}$ in terms of $\mathscr{P}^{(a)}$:

$$W^{(a)} = \frac{\omega \mu_0 \varepsilon_0 \mu\varepsilon}{k_1} \mathscr{P}^{(a)} \cdot \hat{\mathbf{k}}_1 \tag{4.72}$$

Hence, at a fixed point in space and for a general elliptical polarization, we may conclude that (1) $W^{(a)}$ and both the magnitude and direction of $\mathscr{P}^{(a)}$ vary periodically in time with frequency 2ω, and (2) $\mathscr{P}^{(d)}$ and $W^{(d)}$ are constants, independent of time. However, they do depend on the polarization of the wave, (3). In the case of uniform plane waves, vectors $\langle \mathscr{P} \rangle$, $\mathbf{v}_E$, and $\mathbf{v}_p$ point in the same fixed direction $\hat{\mathbf{k}}$. In contrast, as is evident from Eq. (4.64), the time-averaged Poynting vector and thus the velocity of energy transport of the nonuniform plane waves do not coincide in direction with the phase velocity defined in Eq. (4.41), (4). For a circular polarization: $\mathbf{E}_0^2 = 0$, thus $\mathscr{P}^{(a)} = \mathbf{0}$ and $W^{(a)} = 0$, the instantaneous Poynting vector and energy density are equal to the time-averaged values. In other words, $\mathscr{P}$ amd W are constants, and finally, (5), for a linear polarization: $\mathbf{E}_0 \times \mathbf{E}_0^* = 0$, $\langle \mathscr{P} \rangle$ becomes

$$\langle \mathscr{P} \rangle = \mathscr{P}^{(d)} = \frac{|\mathbf{E}_0|^2 e^{-2\mathbf{k}_2 \cdot \mathbf{r}}}{2\omega\mu_0 \mu} \mathbf{k}_1 \tag{4.73}$$

which means that the time-averaged energy propagates in the direction of wave normal $\hat{\mathbf{k}}_1$. The time averaged energy density is

$$\begin{aligned} \langle W \rangle &= \frac{1}{\omega} \mathscr{P}^{(d)} \cdot \mathbf{k}_1 \\ &= \frac{|\mathbf{E}_0|^2 e^{-2\mathbf{k}_2 \cdot \mathbf{r}}}{2\omega^2 \mu_0 \mu} k_1^2 \end{aligned} \tag{4.74}$$

In this case, we can again express $\langle \mathscr{P} \rangle$ in the form

$$\langle \mathscr{P} \rangle = \langle W \rangle \mathbf{v}_p \tag{4.75}$$

where the phase velocity $\mathbf{v}_p$ is given by Eq. (4.41).

4.3 UNIFORM AND NONUNIFORM PLANE WAVES IN LOSSY ISOTROPIC MEDIA

The presence of conductivity in a medium modifies somewhat the nature of electromagnetic wave propagation. Again, both uniform and nonuniform plane waves can arise. However, in a lossy medium, the electric field produces the conduction current, which in turn gives rise to the Joule heat loss. This loss decreases the energy carried by the wave. We thus expect the amplitude of the wave to decrease (a damped wave) as it travels through the medium. The constitutive relations for a homogeneous isotropic medium with a finite conductivity take the form

$$\mathbf{D} = \varepsilon_0 \varepsilon \mathbf{E} \tag{4.76}$$

$$\mathbf{B} = \mu_0 \mu \mathbf{H} \tag{4.77}$$

$$\mathbf{J}_c = \sigma \mathbf{E} \tag{4.78}$$

Substituting the above into the source-free Maxwell's equations (2.45) in frequency domain, we obtain

$$\nabla \times \mathbf{E} = i\omega\mu_0 \mu \mathbf{H} \tag{4.79}$$

$$\nabla \times \mathbf{H} = -i\omega\varepsilon_0 \varepsilon_c \mathbf{E} \tag{4.80}$$

where

$$\varepsilon_c = \varepsilon(1 + i\tau) \tag{4.81}$$

is the complex dielectric constant and

$$\tau = \frac{\sigma}{\omega\varepsilon_0 \varepsilon} \tag{4.82}$$

is called the *loss tangent* of the medium, which is the ratio of conduction current to the displacement current in a lossy medium. We search for a solution to Eqs. (4.79) and (4.80) of the form

$$\mathbf{E} = \mathbf{E}_0 \, e^{i\mathbf{k}\cdot\mathbf{r}} \tag{4.83}$$

$$\mathbf{H} = \mathbf{H}_0 \, e^{i\mathbf{k}\cdot\mathbf{r}} \tag{4.84}$$

which are identical with Eqs. (4.1) and (4.2) except that **k** now is complex.

Substitution of Eqs. (4.83) and (4.84) into Eqs. (4.79) and (4.80) yields

$$\mathbf{k} \times \mathbf{E}_0 = \omega\mu_0 \, \mu \mathbf{H}_0 \tag{4.85}$$

$$\mathbf{k} \times \mathbf{H}_0 = -\omega\varepsilon_0 \, \varepsilon_c \, \mathbf{E}_0 \tag{4.86}$$

which imply respectively

$$\mathbf{k} \cdot \mathbf{H}_0 = 0 \tag{4.87}$$

$$\mathbf{k} \cdot \mathbf{E}_0 = 0 \tag{4.88}$$

Since Eq. (4.86) has the same mathematical form as Eq. (4.32) for the lossless media except replacing ε by ε_c, by analogy, we may immediately write the dispersion equation (4.36) for a lossy medium as

$$\mathbf{k}^2 = k_0^2 \, \mu\varepsilon_c \tag{4.89}$$

Consequently, the phase vector $\mathbf{k}_1$ and the attenuation vector $\mathbf{k}_2$ of the complex wave vector $\mathbf{k} = \mathbf{k}_1 + i\mathbf{k}_2$ satisfy

$$\mathbf{k}_1^2 - \mathbf{k}_2^2 = k_0^2 \, \mu\varepsilon \tag{4.90}$$

$$\mathbf{k}_1 \cdot \mathbf{k}_2 = k_1 k_2 \cos \zeta = \tfrac{1}{2} \, k_0^2 \, \mu\varepsilon\tau \tag{4.91}$$

in which ζ is the space angle between $\mathbf{k}_1$ and $\mathbf{k}_2$.

From Eqs. (4.90) and (4.91), it is apparent that as long as ω and σ are nonzero, neither k_2 nor k_1 can be zero. In other words, if there is loss, the fields must suffer attenuation as well as phase shift in the steady state. Moreover, under the same conditions, $\cos \zeta \neq 0$; the nonuniform plane waves in lossy media are unlike those in lossless media (cf. Sec. 4.2), since here the planes of constant phase and the planes of constant amplitude cannot be mutually perpendicular. Furthermore, k_1 and k_2, being the magnitudes of two real vectors, must not be negative. It follows from the nonnegative character of μ and σ in Eq. (4.91) that

$$0 < \cos \zeta \leq 1 \qquad \text{if } \omega > 0$$

or

$$0 \leq \zeta < \frac{\pi}{2} \qquad \text{if } \omega > 0 \tag{4.92}$$

as illustrated in Fig. 4.2. Hence in a lossy medium, the range of ζ defined in Eq. (4.92) allows us to divide the study of exponential solution into the following two cases: (1) $\cos \zeta = 1$ (or $\zeta = 0$); and (2) $0 < \cos \zeta < 1$ (or $0 < \zeta < \pi/2$). The first case leads to uniform damped waves whereas the second case leads to nonuniform ones.

Case 1. Uniform damped waves When the planes of constant phase coincide with the planes of constant amplitude, $\zeta = 0$ or $\mathbf{k}_1 \parallel \mathbf{k}_2$, we have

$$\mathbf{k} \times \mathbf{k}^* = i2(\mathbf{k}_2 \times \mathbf{k}_1) = \mathbf{0} \tag{4.93}$$

and the plane wave is uniform (see Prob. 4.2). In this case, we may write

$$\mathbf{k}_1 = k_1 \hat{\mathbf{k}} \tag{4.94}$$

$$\mathbf{k}_2 = k_2 \hat{\mathbf{k}} \tag{4.95}$$

The wave vector becomes

$$\mathbf{k} = (k_1 + ik_2)\hat{\mathbf{k}} = k_c \hat{\mathbf{k}} \tag{4.96}$$

where

$$k_c = k_1 + ik_2 = k_0\sqrt{\mu\varepsilon_c} \tag{4.97}$$

is the complex wave number. Substituting Eq. (4.96) into the Maxwell equations (4.85) and (4.86), we obtain

$$\mathbf{H}_0 = \frac{1}{\eta_c}(\hat{\mathbf{k}} \times \mathbf{E}_0) \tag{4.98}$$

$$\mathbf{E}_0 = -\eta_c(\hat{\mathbf{k}} \times \mathbf{H}_0) \tag{4.99}$$

where

$$\eta_c = \frac{\omega\mu_0\mu}{k_c} = \sqrt{\frac{\mu_0\mu}{\varepsilon_0\varepsilon(1 + i\tau)}} \tag{4.100}$$

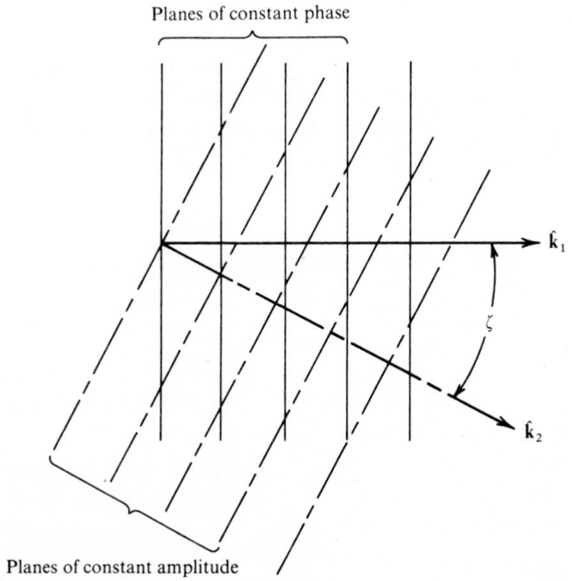

Figure 4.2 Planes of constant phase and of constant amplitude in a lossy medium.

is the *complex intrinsic wave impedance* of the medium. Substituting

$$\mathbf{E}_0 = \mathbf{E}_1 + i\mathbf{E}_2$$
$$\mathbf{H}_0 = \mathbf{H}_1 + i\mathbf{H}_2 \tag{4.101}$$

into Eqs. (4.87) and (4.88), we find that all four real vectors $\mathbf{E}_1$, $\mathbf{E}_2$, $\mathbf{H}_1$, and $\mathbf{H}_2$ lie on the same plane perpendicular to $\hat{\mathbf{k}}$. The instantaneous field vectors according to Eq. (4.39) are

$$\begin{aligned}\mathscr{E} &= e^{-\mathbf{k}_2\cdot\mathbf{r}}\,\mathrm{Re}\,(\mathbf{E}_0\, e^{i\phi_1}) \\ &= e^{-\mathbf{k}_2\cdot\mathbf{r}}(\mathbf{E}_1 \cos\phi_1 - \mathbf{E}_2 \sin\phi_1)\end{aligned} \tag{4.102}$$

and

$$\begin{aligned}\mathscr{H} &= e^{-\mathbf{k}_2\cdot\mathbf{r}}\,\mathrm{Re}\,(\mathbf{H}_0\, e^{i\phi_1}) \\ &= e^{-\mathbf{k}_2\cdot\mathbf{r}}\,\mathrm{Re}\left(\frac{\hat{\mathbf{k}} \times \mathbf{E}_0}{\eta_c}\, e^{i\phi_1}\right) \\ &= \frac{e^{-\mathbf{k}_2\cdot\mathbf{r}}}{|\eta_c|}\,[(\hat{\mathbf{k}} \times \mathbf{E}_1)\cos(\phi_1 - \xi) - (\hat{\mathbf{k}} \times \mathbf{E}_2)\sin(\phi_1 - \xi)]\end{aligned} \tag{4.103}$$

where we have written the complex wave impedance η_c in the polar form

$$\eta_c = |\eta_c| e^{i\xi} \tag{4.104}$$

The phase

$$\phi_1 = \mathbf{k}_1 \cdot \mathbf{r} - \omega t \tag{4.105}$$

of the wave determines the surfaces of constant phase which are planes perpendicular to $\hat{\mathbf{k}}$ and travel with the phase velocity

$$\mathbf{v}_p = \frac{\omega}{k_1}\,\hat{\mathbf{k}} \tag{4.106}$$

The amplitude of the wave decreases according to $e^{-\mathbf{k}_2\cdot\mathbf{r}}$ as the wave propagates through the lossy media. In a distance $\delta = \mathbf{r} \cdot \hat{\mathbf{k}}$ along the direction $\hat{\mathbf{k}}$ such that

$$k_2\,\delta = 1$$

the amplitude of a wave is found to decrease by a factor $1/e$. The distance

$$\delta = \frac{1}{k_2}$$

is known as the *skin depth* of a wave in a conducting medium. It is the distance that a plane wave is normally considered to penetrate the medium.

According to Eqs. (4.102) and (4.103), we conclude that at a fixed point in space, the ellipses traced out by $\mathscr{E}$ and $\mathscr{H}$ on the plane of constant phase have the same shape (more precisely eccentricity), sense of rotation, but their major (or minor) axes are mutually perpendicular. In this respect, the uniform damped plane waves in lossy media have much in common with nondamped plane waves in lossless media.

However, it should be noted that there is a difference between the two cases. Taking the dot product of $\mathscr{E}$ in Eq. (4.102) with $\mathscr{H}$ in Eq. (4.103), we have

$$\mathscr{E} \cdot \mathscr{H} = \frac{e^{-2\mathbf{k}_2 \cdot \mathbf{r}}}{|\eta_c|} [(\mathbf{E}_1 \times \mathbf{E}_2 \cdot \hat{\mathbf{k}}) \cos \phi_1 \sin (\phi_1 - \xi) - (\mathbf{E}_1 \times \mathbf{E}_2 \cdot \hat{\mathbf{k}}) \sin \phi_1 \cos (\phi_1 - \xi)] \tag{4.107}$$

which shows that in a lossless medium (that is, $\sigma = 0$ implies $\xi = 0$), the instantaneous electric and magnetic field vectors of a uniform plane wave are always mutually perpendicular. Evidently, the statement is generally not true for uniform plane waves in a lossy medium.

There is a special case worth mentioning: $\mathbf{E}_0$ is linearly polarized; $\mathbf{E}_1 \times \mathbf{E}_2 = 0$. In this case, the right-hand side of Eq. (4.107) vanishes identically. Thus, the instantaneous electric and magnetic field vectors of a linearly polarized uniform plane wave are mutually perpendicular at all times, whether or not the medium is lossy.

To determine the phase constant k_1 and the attenuation constant k_2, we solve

$$k_1^2 - k_2^2 = k_0^2 \mu\varepsilon \tag{4.108}$$

$$k_1 k_2 = \tfrac{1}{2} k_0^2 \mu\varepsilon\tau \tag{4.109}$$

by squaring Eqs. (4.108) and (4.109) and then adding to give

$$k_1^2 + k_2^2 = \pm k_0^2 \mu\varepsilon \sqrt{1 + \tau^2} \tag{4.110}$$

Summing Eqs. (4.108) and (4.110) and noting that k_1 (or k_2) must be real, k_1^2 must be positive; thus rejecting the negative square root in Eq. (4.110), we obtain the unique solution

$$k_1 = k_0 \sqrt{\mu\varepsilon} \left[\tfrac{1}{2}(\sqrt{1 + \tau^2} + 1)\right]^{1/2} \geq k_0 \sqrt{\mu\varepsilon} \tag{4.111}$$

Subtraction of Eqs. (4.110) and (4.108) gives

$$k_2 = k_0 \sqrt{\mu\varepsilon} \left[\tfrac{1}{2}(\sqrt{1 + \tau^2} - 1)\right]^{1/2} \tag{4.112}$$

We note that $k_c = k_1 + ik_2$ depends only on two parameters: $k_0\sqrt{\mu\varepsilon}$, which is the phase constant of a uniform plane wave in a lossless medium; and τ, the loss tangent. Two limiting cases: $\tau \ll 1$ and $\tau \gg 1$ are particularly interesting. When $\tau \ll 1$, a medium has small loss or the displacement current at the frequency in question is large compared to the conduction current. This is the case for poor conductors or magnetic media. The binomial expansions of Eqs. (4.111), (4.112), and (4.100) yield

$$k_1 \cong k_0\sqrt{\mu\varepsilon}\,(1 + \tfrac{1}{8}\tau^2 + \cdots) \cong k_0\sqrt{\mu\varepsilon}$$

$$k_2 \cong \tfrac{1}{2}k_0\,\tau\sqrt{\mu\varepsilon}\,(1 - \tfrac{1}{8}\tau^2 + \cdots) \simeq \tfrac{1}{2}\sigma\sqrt{\frac{\mu_0\,\mu}{\varepsilon_0\,\varepsilon}} \tag{4.113}$$

$$\eta_c \cong \sqrt{\frac{\mu_0\,\mu}{\varepsilon_0\,\varepsilon}}\,(1 - \tfrac{3}{8}\tau^2 - \tfrac{1}{2}i\tau + \cdots) \cong \sqrt{\frac{\mu_0\,\mu}{\varepsilon_0\,\varepsilon}}$$

On the other hand, when $\tau \gg 1$ a medium has large loss as in good conductors for which the conduction current exceeds greatly the displacement current. In this case, the binomial expansions of Eqs. (4.111), (4.112), and (4.100) lead to the results:

$$k_1 \cong k_0\sqrt{\frac{\mu\varepsilon\tau}{2}}\,(1 + \tfrac{1}{2}\tau^{-1} + \cdots) \cong \sqrt{\frac{\omega\mu_0\,\mu\sigma}{2}}$$

$$k_2 \cong k_0\sqrt{\frac{\mu\varepsilon\tau}{2}}\,(1 - \tfrac{1}{2}\tau^{-1} + \cdots) \cong \sqrt{\frac{\omega\mu_0\,\mu\sigma}{2}} \tag{4.114}$$

$$\eta_c \cong \sqrt{\frac{\omega\mu_0\,\mu}{2\sigma}}\,(1 - i)$$

In a nondispersive medium, k_1, k_2, and η_c vary directly as $\sqrt{\omega}$, and the skin depth is

$$\delta \cong \sqrt{\frac{2}{\omega\mu_0\,\mu\sigma}} \tag{4.115}$$

We next examine the energy densities of the damped uniform plane waves in a lossy isotropic medium. Expressions (4.44) and (4.50) remain valid. Thus, the time-averaged electric and magnetic energy densities may be written as

$$\langle W_e\rangle = \tfrac{1}{4}\varepsilon_0\,\varepsilon e^{-2\mathbf{k}_2\cdot\mathbf{r}}|\,\mathbf{E}_0\,|^2 \tag{4.116}$$

$$\langle W_m\rangle = \tfrac{1}{4}\mu_0\,\mu e^{-2\mathbf{k}_2\cdot\mathbf{r}}|\,\mathbf{H}_0\,|^2 \tag{4.117}$$

Now dot-multiplying Eq. (4.98) by its complex conjugate, we find

$$\mathbf{H}_0\cdot\mathbf{H}_0^* = \frac{1}{\eta_c\,\eta_c^*}\,(\hat{\mathbf{k}}\times\mathbf{E}_0)\cdot(\hat{\mathbf{k}}\times\mathbf{E}_0^*)$$

or

$$\mu_0\,\mu\,|\,\mathbf{H}_0\,|^2 = \sqrt{(1+\tau^2)}\,\varepsilon_0\,\varepsilon\,|\,\mathbf{E}_0\,|^2 \tag{4.118}$$

Thus, the time-averaged electric and magnetic energy densities in a lossy medium are related by

$$\langle W_m\rangle = \sqrt{(1+\tau^2)}\,\langle W_e\rangle \tag{4.119}$$

Consequently, within a good conductor, the time-averaged magnetic energy density in a plane wave far exceeds the time-averaged electric energy density. In other words, in a highly conducting medium, such as metal, most of the energy carried by the wave is stored in the magnetic field. This is due to the fact that the electric energy is lost during the generation of the conduction current which, in turn, gives rise to the Joule heat. In the two limiting cases discussed earlier, we have

$$\langle W_m \rangle = (1 + \tfrac{1}{2}\tau^2)\langle W_e \rangle \tag{4.120}$$

when $\tau \ll 1$ and

$$\langle W_m \rangle = \tau \langle W_e \rangle \tag{4.121}$$

when $\tau \gg 1$.

The expressions (4.60) to (4.67) for the Poynting vector remain valid in a lossy medium. However, for a uniform plane wave, we have

$$\hat{\mathbf{k}} \cdot \mathbf{E}_0 = \hat{\mathbf{k}} \cdot \mathbf{E}_0^* = 0$$

Thus, from Eqs. (4.64) and (4.67) the time-averaged Poynting vector becomes

$$\langle \mathscr{P} \rangle = \frac{k_1 e^{-2\mathbf{k}_2 \cdot \mathbf{r}}}{2\omega\mu_0\mu} |\mathbf{E}_0|^2 \hat{\mathbf{k}} \tag{4.122}$$

The total time-averaged energy density is

$$\langle W \rangle = \langle W_e \rangle + \langle W_m \rangle \tag{4.123}$$

Using Eqs. (4.119), (4.116), and (4.111), we may write Eq. (4.123) as

$$\langle W \rangle = \frac{k_1^2 e^{-2\mathbf{k}_2 \cdot \mathbf{r}}}{2\omega^2\mu_0\mu} |\mathbf{E}_0|^2 \tag{4.124}$$

Now, comparing Eq. (4.122) with Eq. (4.124), we obtain

$$\begin{aligned} \langle \mathscr{P} \rangle &= \langle W \rangle \frac{\omega}{k_1} \hat{\mathbf{k}} \\ &= \langle W \rangle \mathbf{v}_p \end{aligned} \tag{4.125}$$

where $\mathbf{v}_p$ defined in Eq. (4.106) is the phase velocity of a uniform plane wave in lossy medium, or

$$\mathbf{v}_E = \frac{\langle \mathscr{P} \rangle}{\langle W \rangle} = \mathbf{v}_p \tag{4.126}$$

Here again we see that the velocity of energy transport of a uniform plane wave in lossy media is equal to the phase velocity of the wave as in the case of lossless media.

As in the case of uniform undamped plane waves, we may also decompose a damped uniform plane wave of any polarization into linearly or circularly polarized components. The direct additivity of energy densities and Poynting vectors of the polarized components as discussed in Sec. 4.1 can easily be shown as applicable here.

Case 2. Nonuniform damped waves We consider next the case when the planes of constant phase do not coincide with the planes of constant amplitude: $0 < \zeta < \pi/2$. Hence

$$\mathbf{k} \times \mathbf{k}^* = i2(\mathbf{k}_2 \times \mathbf{k}_1) \neq \mathbf{0} \tag{4.127}$$

and, in this case, the plane wave is nonuniform (see Prob. 4.2). The phase constant k_1 and the attenuation constant k_2 can be expressed in terms of the known parameters of the medium (μ, ε, σ) along with the unknown parameter ζ. Comparison of Eq. (4.91) with Eq. (4.109) shows that they would be identical if τ in the latter were replaced by $\tau/\cos\zeta$. Accordingly, the solution to Eqs. (4.90) and (4.91) can be written down immediately from Eqs. (4.111) and (4.112) with this replacement. Thus

$$k_1 = k_0\sqrt{\mu\varepsilon}\left\{\frac{1}{2}\left[\sqrt{1+\left(\frac{\tau}{\cos\zeta}\right)^2}+1\right]\right\}^{1/2} \tag{4.128}$$

$$k_2 = k_0\sqrt{\mu\varepsilon}\left\{\frac{1}{2}\left[\sqrt{1+\left(\frac{\tau}{\cos\zeta}\right)^2}-1\right]\right\}^{1/2} \tag{4.129}$$

In general, since Maxwell's equations have not completely fixed ζ, its value must be determined finally by boundary conditions. However, keep in mind that Eq. (4.92) provides some restrictions from the field equations.

Interestingly, we can show that for nonuniform plane waves in lossless as well as in lossy media, simultaneous linear polarization for both $\mathbf{E}_0$ and $\mathbf{H}_0$ is not possible. Indeed, the cross product of Eqs. (4.86) and (4.85) gives

$$\mathbf{k} = c_1(\mathbf{E}_0 \times \mathbf{H}_0) \tag{4.130}$$

where

$$c_1 = \frac{k_0^2\,\mu\varepsilon_c}{\mathbf{k}\cdot(\mathbf{E}_0 \times \mathbf{H}_0)} \tag{4.131}$$

Now assume that a nonuniform plane wave has both its $\mathbf{E}_0$ and $\mathbf{H}_0$ linearly polarized, i.e., according to Eq. (3.44)

$$\mathbf{E}_0 \times \mathbf{E}_0^* = \mathbf{0}$$

or

$$\mathbf{E}_0^* = c_2\,\mathbf{E}_0 \tag{4.132}$$

and

$$\mathbf{H}_0 \times \mathbf{H}_0^* = \mathbf{0}$$

or

$$\mathbf{H}_0^* = c_3\,\mathbf{H}_0 \tag{4.133}$$

where c_2 and c_3 are two arbitrary complex constants. Substituting Eqs. (4.132) and (4.133) into the complex conjugate of Eq. (4.130), we find

$$\begin{aligned}\mathbf{k}^* &= c_1^* c_2^* c_3^*(\mathbf{E}_0 \times \mathbf{H}_0)\\ &= c_4\,\mathbf{k}\end{aligned} \tag{4.134}$$

where

$$c_4 = c_1^* c_2^* c_3^*$$

therefore $$\mathbf{k} \times \mathbf{k}^* = c_4(\mathbf{k} \times \mathbf{k}) = \mathbf{0}$$

which means a uniform plane wave, thus a contradiction to the assumption. We thus proved the statement. However, a circular polarization for the field vector $\mathbf{E}_0$ always accompanies a circular polarization for $\mathbf{H}_0$. To show this, we note again from Eqs. (4.85) and (4.86) that

$$\begin{aligned}\mu_0 \mu \mathbf{H}_0^2 &= \frac{1}{\omega}(\mathbf{k} \times \mathbf{E}_0) \cdot \mathbf{H}_0 \\ &= -\frac{1}{\omega}(\mathbf{k} \times \mathbf{H}_0) \cdot \mathbf{E}_0 \\ &= \varepsilon_0 \varepsilon_c \mathbf{E}_0^2 \end{aligned} \tag{4.135}$$

Thus, a circularly polarized $\mathbf{E}_0$ (that is, $\mathbf{E}_0^2 = 0$) implies a circularly polarized $\mathbf{H}_0$ (that is, $\mathbf{H}_0^2 = 0$) and vice versa. The time-averaged energy density (4.57) and Poynting vector (4.63) of a nonuniform plane wave in lossless media remain valid in lossy media. We repeat them here

$$\langle W \rangle = \frac{e^{-2\mathbf{k}_2 \cdot \mathbf{r}}}{4\omega^2 \mu_0 \mu}[(k_0^2 \mu\varepsilon + |\mathbf{k}|^2)|\mathbf{E}_0|^2 - |\mathbf{k} \cdot \mathbf{E}_0^*|^2] \tag{4.136}$$

$$\langle \mathscr{P} \rangle = \frac{e^{-2\mathbf{k}_2 \cdot \mathbf{r}}}{4\omega \mu_0 \mu}[|\mathbf{E}_0|^2(\mathbf{k} + \mathbf{k}^*) - (\mathbf{k} - \mathbf{k}^*) \times (\mathbf{E}_0 \times \mathbf{E}_0^*)] \tag{4.137}$$

and note that $\langle \mathscr{P} \rangle$ is neither proportional to $\mathbf{k}$ nor lies on the same plane with $\mathbf{k}$ (i.e., the plane formed by the two real vectors $\mathbf{k}_1$ and $\mathbf{k}_2$), since

$$\langle \mathscr{P} \rangle \cdot (\mathbf{k} \times \mathbf{k}^*) \neq 0$$

Dot-multiplying $\langle \mathscr{P} \rangle$ by the real vector

$$\mathbf{k}_1 = \operatorname{Re} \mathbf{k} = \tfrac{1}{2}(\mathbf{k} + \mathbf{k}^*)$$

we obtain

$$\langle \mathscr{P} \rangle \cdot \mathbf{k}_1 = \frac{e^{-2\mathbf{k}_2 \cdot \mathbf{r}}}{8\omega \mu_0 \mu}[|\mathbf{E}_0|^2(\mathbf{k} + \mathbf{k}^*)^2 - 2|\mathbf{k} \cdot \mathbf{E}_0^*|^2] \tag{4.138}$$

But, according to the dispersion equation (4.89)

$$\begin{aligned}(\mathbf{k} + \mathbf{k}^*)^2 &= 2|\mathbf{k}|^2 + k_0^2 \mu(\varepsilon_c + \varepsilon_c^*) \\ &= 2(|\mathbf{k}|^2 + k_0^2 \mu\varepsilon)\end{aligned} \tag{4.139}$$

thus, once again, we arrive at the relation

$$\begin{aligned}\langle \mathscr{P} \rangle \cdot \mathbf{k}_1 &= \frac{e^{-2\mathbf{k}_2 \cdot \mathbf{r}}}{4\omega \mu_0 \mu}[(k_0^2 \mu\varepsilon + |\mathbf{k}|^2)|\mathbf{E}_0|^2 - |\mathbf{k} \cdot \mathbf{E}_0^*|^2] \\ &= \omega \langle W \rangle\end{aligned}$$

or $$\langle \mathscr{P} \rangle \cdot \hat{\mathbf{k}}_1 = \langle W \rangle v_p \tag{4.140}$$

where

$$v_p = \frac{\omega}{k_1}$$

is the phase velocity of the wave.

4.4 REFLECTION AND TRANSMISSION OF WAVES AT PLANE INTERFACE

Thus far we have considered the propagation of plane electromagnetic waves in unbounded isotropic media. Beginning with this section, we shall examine the effects of a discontinuity in the medium on propagation and shall restrict ourselves to two homogeneous isotropic media separated by an infinite plane interface as illustrated in Fig. 3.14. As was shown in Sec. 3.5, in order to obtain continuity of the tangential components of the electric and magnetic field intensities at the interface, some valid relation must exist among the field vectors of the incident, the reflected, and the transmitted waves for all points on the interface. Such a relation is possible if (1) all the complex field vectors are identical functions of position on the interface, and (2) there exist certain relations among the amplitude vectors of the incident, the reflected, and the transmitted waves at the interface. Condition (1) results in the laws of reflection and refraction and was discussed in Sec. 3.5 for the general case. Condition (2) leads to the Fresnel equations and will be derived in the next three sections. Here we shall determine wave vectors in each medium and consider some general aspects of decomposition of amplitude vectors.

As was shown in Sec. 3.5, condition (1) implies that for a given wave vector $\mathbf{k}_i$ of the incident wave, the wave vectors $\mathbf{k}_r^{(1)}, \mathbf{k}_r^{(2)}, \mathbf{k}_r^{(3)}, \ldots$ of the reflected waves and $\mathbf{k}_t^{(1)}, \mathbf{k}_t^{(2)}, \mathbf{k}_t^{(3)}, \cdots$ of the transmitted waves, satisfy the laws of reflection and refraction (3.100) or (3.105). What remain to be determined are: the total number of reflected waves, the total number of transmitted waves, and the parameters q_α in the expression (3.105).

The answers to the above questions depend upon the electric and magnetic properties of the media. In order for waves to exist in a lossless isotropic medium 1 or 2 (see Fig. 4.3), the wave vectors of the reflected waves in medium 1 must satisfy the dispersion equation

$$\mathbf{k}_r^2 = k_0^2 \mu_1 \varepsilon_1 \tag{4.141}$$

and those of the transmitted waves in medium 2 must satisfy the dispersion equation

$$\mathbf{k}_t^2 = k_0^2 \mu_2 \varepsilon_2 \tag{4.142}$$

These equations provide additional conditions under which the parameters q_α and the number of reflected and transmitted waves can be determined.

From Eq. (3.105), we may write

$$\mathbf{k}_r = \mathbf{b} + q_r \hat{\mathbf{q}} \tag{4.143}$$

where
$$\mathbf{b} = \hat{\mathbf{q}} \times \mathbf{a}$$
$$q_r = \mathbf{k}_r \cdot \hat{\mathbf{q}} \tag{4.144}$$

By taking the dot product of Eq. (4.143) with itself, we obtain

$$\mathbf{k}_r^2 = \mathbf{b}^2 + q_r^2 = \mathbf{a}^2 + q_r^2$$

or
$$q_r^2 = \mathbf{k}_r^2 - \mathbf{a}^2 \tag{4.145}$$

Similarly

$$q_i^2 = \mathbf{k}_i^2 - \mathbf{a}^2 \tag{4.146}$$

Since both the reflected and the incident waves are in medium 1, they satisfy the same dispersion equation (4.141). Thus Eqs. (4.145) and (4.146) give

$$q_r^2 = q_i^2$$

or
$$q_r = \pm q_i \tag{4.147}$$

That is, two waves in general can exist in isotropic medium 1. The solution $q_r = +q_i$ leads to $\mathbf{k}_r = \mathbf{k}_i$ which is the incident wave. The other solution

$$q_r = -q_i \tag{4.148}$$

or
$$(\mathbf{k}_i + \mathbf{k}_r) \cdot \hat{\mathbf{q}} = 0$$

leads to

$$(\hat{\mathbf{k}}_r \cdot \hat{\mathbf{q}}) = -(\hat{\mathbf{k}}_i \cdot \hat{\mathbf{q}}) \tag{4.149}$$

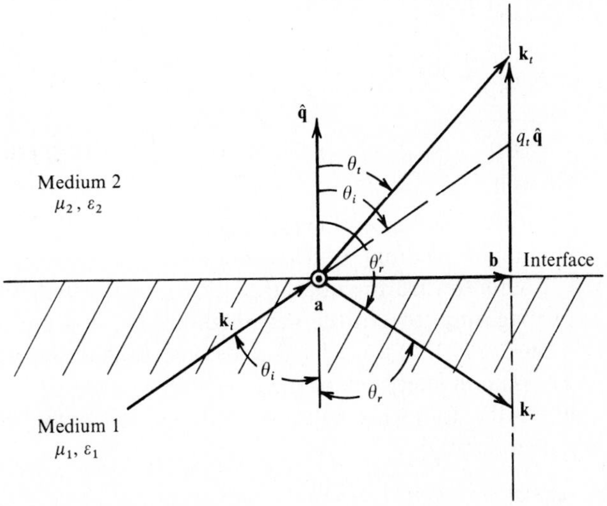

Figure 4.3 Orientation of the wave vectors with respect to the normal to the interface of two isotropic media.

which is equivalent to (see Fig. 4.3)

$$\cos\theta'_r = -\cos\theta_i$$

Meanwhile, according to Eq. (3.104)

$$\sin\theta'_r = \sin\theta_i$$

Hence

$$\theta_i = \pi - \theta'_r \tag{4.150}$$

In other words, the angle of incidence θ_i is equal to the angle of reflection θ_r. Here the angle $\pi - \theta'_r$ is taken to be the angle of reflection (see Fig. 4.3). Similarly, substituting the wave vector

$$\mathbf{k}_t = \mathbf{b} + q_t \hat{\mathbf{q}} \tag{4.151}$$

of the transmitted wave into Eq. (4.142), we obtain

$$\mathbf{k}_t^2 = \mathbf{a}^2 + q_t^2 = k_0^2 \mu_2 \varepsilon_2$$

and hence

$$q_t = \pm\sqrt{\mathbf{k}_t^2 - \mathbf{a}^2}$$

But

$$\mathbf{a}^2 = (\mathbf{k}_i \times \hat{\mathbf{q}})^2 = k_i^2 \sin^2\theta_i$$

Thus

$$q_t = \pm\sqrt{k_t^2 - k_i^2 \sin^2\theta_i} \tag{4.152}$$

For a real q_t, the following condition must be fulfilled

$$\frac{k_t^2}{k_i^2} \geq \sin^2\theta_i \tag{4.153}$$

or

$$\sqrt{\frac{\mu_2\varepsilon_2}{\mu_1\varepsilon_1}} \geq \sin\theta_i \tag{4.154}$$

In this case, q_t can either be positive or negative which again signifies that in general two waves can propagate in medium 2. However, the negative sign implies that the wave vector $\mathbf{k}_t$ points toward the interface (see Fig. 4.3) which, in turn, means that the wave is propagating toward the first medium, i.e., it is not a transmitted wave. Therefore, q_t in Eq. (4.152) must be positive and thus the wave vector $\mathbf{k}_t$ of the transmitted wave is completely determined.

From Eq. (3.104), we obtain the following relation between the angle of incidence θ_i and the angle of transmission θ_t

$$\frac{\sin\theta_i}{\sin\theta_t} = \frac{k_t}{k_i} = \frac{n_t}{n_i} = n_{21} \tag{4.155}$$

where n_t and n_i are the indices of refraction of the second and first medium respectively. n_{21} is called the *relative index of refraction* of the second medium with respect to the first. Equation (4.155) is known as the Snell law of refraction. When $n_t > n_i$ (or $n_{21} > 1$), we say that the second medium is optically denser than the first. In this case

$$\sin \theta_t = \frac{1}{n_{21}} \sin \theta_i < \sin \theta_i$$

so that there exists a real angle of transmission θ_t for every angle of incidence and $\theta_t < \theta_i$. On the other hand, if the second medium is optically less dense than the first, i.e., if $n_{21} < 1$, we can obtain a real angle θ_t only for those angles of incidence for which $\sin \theta_i < n_{21}$. For larger values of θ_i, total reflection occurs, which will be discussed in Sec. 4.9.

From the above consideration, we may conclude that when a given uniform plane wave is incident on the interface of two lossless isotropic media, there exists only one reflected wave and one transmitted wave. The wave vectors are determined analytically by

$$\mathbf{k}_r = \mathbf{b} - (\mathbf{k}_i \cdot \hat{\mathbf{q}})\hat{\mathbf{q}} \tag{4.156}$$

and

$$\mathbf{k}_t = \mathbf{b} + \sqrt{k_0^2 \mu_2 \epsilon_2 - \mathbf{a}^2}\,\hat{\mathbf{q}} \tag{4.157}$$

On the other hand, the wave vectors may also be constructed geometrically. According to Eq. (3.99), we may write

$$\begin{aligned} \mathbf{k}_r &= \mathbf{k}_i + \alpha_r \hat{\mathbf{q}} \\ \mathbf{k}_t &= \mathbf{k}_i + \alpha_t \hat{\mathbf{q}} \end{aligned} \tag{4.158}$$

where α_r and α_t are two arbitrary constants. Equation (4.158) means that the tips of the wave vectors $\mathbf{k}_r$ and $\mathbf{k}_t$, drawn from a fixed origin 0 on the interface, must lie on a straight line which passes through the tip of the given wave vector $\mathbf{k}_i$ and is parallel to $\hat{\mathbf{q}}$ (see Fig. 4.4). Since vector $\mathbf{k}_r$ and $\mathbf{k}_t$ must also satisfy respectively the dispersion equations (4.141) and (4.142), and geometrically they represent two spheres with radii k_r and k_t. In the present case, they are only two semicircles: one defined by Eq. (4.141) in the first medium (below the interface) and the other defined by Eq. (4.142) in the second medium (above the interface). As a result, to find $\mathbf{k}_r$ and $\mathbf{k}_t$ is actually to determine the points of intersections of the straight line (4.158) with the lower and upper semicircles as illustrated in Fig. 4.4.

We shall now turn to condition (2) mentioned earlier in this section. Considering the incident wave to be known, we must find the amplitude vectors of the reflected and transmitted waves that will ensure continuity of the tangential components of $\mathbf{E}$ and of $\mathbf{H}$ at the interface. For this purpose, we write the uniform incident wave

$$\begin{aligned} \mathbf{E}_i &= \mathbf{E}_{0i} e^{i\mathbf{k}_i \cdot \mathbf{r}} \\ \mathbf{H}_i &= \mathbf{H}_{0i} e^{i\mathbf{k}_i \cdot \mathbf{r}} = \frac{1}{\eta_i} (\hat{\mathbf{k}}_i \times \mathbf{E}_i) \end{aligned} \tag{4.159}$$

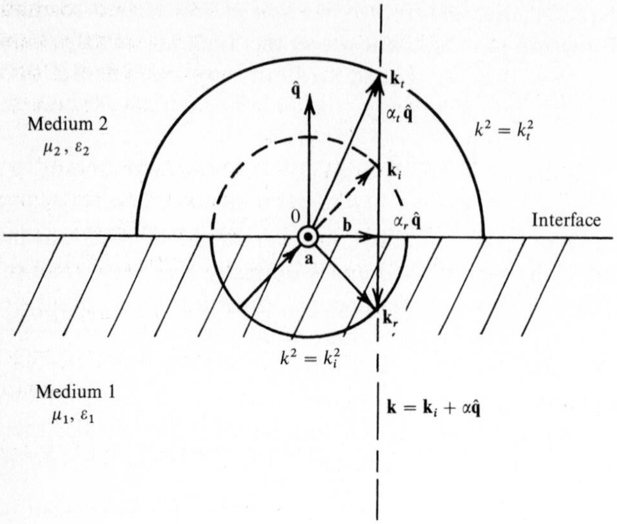

Figure 4.4 Geometrical determination of wave vectors in two isotropic media. The vector **a** points out of the paper.

the reflected wave

$$\begin{aligned} \mathbf{E}_r &= \mathbf{E}_{0r}\, e^{i\mathbf{k}_r \cdot \mathbf{r}} \\ \mathbf{H}_r &= \mathbf{H}_{0r}\, e^{i\mathbf{k}_r \cdot \mathbf{r}} = \frac{1}{\eta_i} (\hat{\mathbf{k}}_r \times \mathbf{E}_r) \end{aligned} \tag{4.160}$$

and the transmitted wave

$$\begin{aligned} \mathbf{E}_t &= \mathbf{E}_{0t}\, e^{i\mathbf{k}_t \cdot \mathbf{r}} \\ \mathbf{H}_t &= \mathbf{H}_{0t}\, e^{i\mathbf{k}_t \cdot \mathbf{r}} = \frac{1}{\eta_t} (\hat{\mathbf{k}}_t \times \mathbf{E}_t) \end{aligned} \tag{4.161}$$

where η_i and η_t are the intrinsic wave impedances of medium 1 and 2, respectively.

According to Eq. (3.91), the continuity of the tangential component of **E** at the interface requires that

$$(\mathbf{E}_{0i} + \mathbf{E}_{0r} - \mathbf{E}_{0t}) \times \hat{\mathbf{q}} = \mathbf{0} \tag{4.162}$$

Similarly, from Eq. (3.92) the continuity of the tangential component of **H** requires that

$$(\mathbf{H}_{0i} + \mathbf{H}_{0r} - \mathbf{H}_{0t}) \times \hat{\mathbf{q}} = \mathbf{0} \tag{4.163}$$

It should be noted that each of these vector equations yields only two independent relations.

When media 1 and 2 have the same magnetic properties (i.e., if $\mu_1 = \mu_2 = \mu$), the boundary condition (3.93) for the normal component of $\mathbf{B}$ becomes

$$(\mathbf{H}_{0i} + \mathbf{H}_{0r} - \mathbf{H}_{0t}) \cdot \hat{\mathbf{q}} = 0 \tag{4.164}$$

Eq. (4.163) together with Eq. (4.164) shows that the vector $(\mathbf{H}_{0i} + \mathbf{H}_{0r} - \mathbf{H}_{0t})$ is simultaneously parallel and perpendicular to the unit vector $\hat{\mathbf{q}}$. This is possible only if the vector is a zero vector, i.e.,

$$\mathbf{H}_{0i} + \mathbf{H}_{0r} - \mathbf{H}_{0t} = \mathbf{0} \tag{4.165}$$

We recall from Sec. 4.1 that the amplitude vector $\mathbf{E}_0$ of a uniform plane wave lies on a plane always perpendicular to the direction of wave propagation $\hat{\mathbf{k}}$. We may thus decompose it into a component normal to the plane of incidence called *perpendicular polarization* and a second component lying in the plane of incidence called *parallel polarization* (see Fig. 4.5). Therefore

$$\mathbf{E}_{0i} = A_\perp \mathbf{a} + A_\parallel (\hat{\mathbf{k}}_i \times \mathbf{a}) \tag{4.166}$$

$$\mathbf{E}_{0r} = B_\perp \mathbf{a} + B_\parallel (\hat{\mathbf{k}}_r \times \mathbf{a}) \tag{4.167}$$

$$\mathbf{E}_{0t} = C_\perp \mathbf{a} + C_\parallel (\hat{\mathbf{k}}_t \times \mathbf{a}) \tag{4.168}$$

where the complex constants $A_\perp$, $B_\perp$, and $C_\perp$ denote respectively the components of vector amplitudes $\mathbf{E}_{0i}$, $\mathbf{E}_{0r}$, and $\mathbf{E}_{0t}$ in the direction perpendicular to the plane of incidence and the complex constants $A_\parallel$, $B_\parallel$, and $C_\parallel$ denote the components of the same vector amplitudes in the direction parallel to the plane of incidence.

After matching the boundary conditions at the interface, we obtain a linear relationship between the components $B_\perp$, $B_\parallel$ and $A_\perp$, $A_\parallel$. In matrix form, it

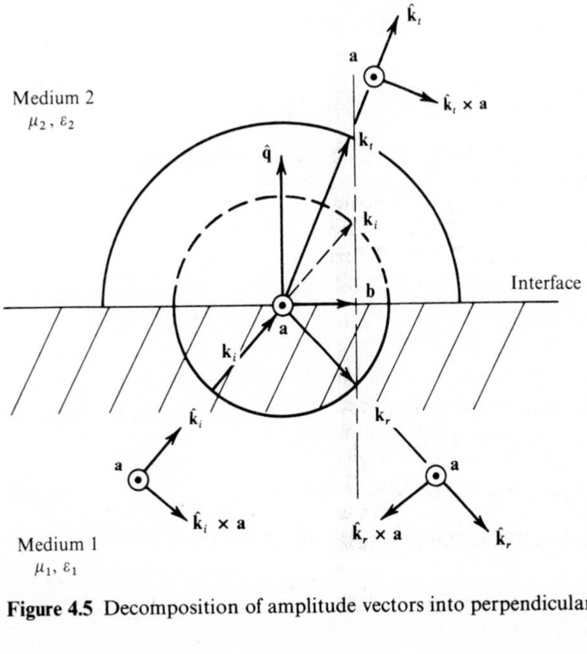

Figure 4.5 Decomposition of amplitude vectors into perpendicular and parallel polarizations.

takes

$$\begin{bmatrix} B_\perp \\ B_\parallel \end{bmatrix} = \bar{\mathbf{\Gamma}} \cdot \begin{bmatrix} A_\perp \\ A_\parallel \end{bmatrix} \tag{4.169}$$

where the 2 × 2 matrix

$$\bar{\mathbf{\Gamma}} = \begin{bmatrix} \Gamma_{11} & \Gamma_{12} \\ \Gamma_{21} & \Gamma_{22} \end{bmatrix} \tag{4.170}$$

is called the *reflection coefficient matrix;* the entries are the amplitude reflection coefficients defined by

$$\begin{aligned} \Gamma_{11} &= \frac{B_\perp}{A_\perp}\bigg|_{A_\parallel = 0} & \Gamma_{12} &= \frac{B_\perp}{A_\parallel}\bigg|_{A_\perp = 0} \\ \Gamma_{21} &= \frac{B_\parallel}{A_\perp}\bigg|_{A_\parallel = 0} & \Gamma_{22} &= \frac{B_\parallel}{A_\parallel}\bigg|_{A_\perp = 0} \end{aligned} \tag{4.171}$$

Similarly, the components $C_\perp$, $C_\parallel$ and $A_\perp$, $A_\parallel$ are related by

$$\begin{bmatrix} C_\perp \\ C_\parallel \end{bmatrix} = \bar{\mathbf{T}} \cdot \begin{bmatrix} A_\perp \\ A_\parallel \end{bmatrix} \tag{4.172}$$

where the 2 × 2 matrix

$$\bar{\mathbf{T}} = \begin{bmatrix} T_{11} & T_{12} \\ T_{21} & T_{22} \end{bmatrix} \tag{4.173}$$

is called the *transmission coefficient matrix;* the entries are the amplitude transmission coefficients defined by

$$\begin{aligned} T_{11} &= \frac{C_\perp}{A_\perp}\bigg|_{A_\parallel = 0} & T_{12} &= \frac{C_\perp}{A_\parallel}\bigg|_{A_\perp = 0} \\ T_{21} &= \frac{C_\parallel}{A_\perp}\bigg|_{A_\parallel = 0} & T_{22} &= \frac{C_\parallel}{A_\parallel}\bigg|_{A_\perp = 0} \end{aligned} \tag{4.174}$$

The explicit expressions for the amplitude reflection and transmission coefficients depend upon the electric and magnetic properties of the media under consideration and will be discussed in detail in the following sections.

4.5 FRESNEL'S EQUATIONS WHEN $\mu_1 = \mu_2$

To calculate the reflection and transmission coefficients of two isotropic media of the same permeability, we start with the boundary condition (4.165). With the aid of Maxwell's equation (4.4), we may write Eq. (4.165) as

$$\mathbf{k}_i \times \mathbf{E}_{0i} + \mathbf{k}_r \times \mathbf{E}_{0r} - \mathbf{k}_t \times \mathbf{E}_{0t} = \mathbf{0} \tag{4.175}$$

By taking the dot product of Eq. (4.175) with $\mathbf{k}_r$, $\mathbf{k}_t$, and $\mathbf{a}$ successively, we obtain

$$(\mathbf{k}_r \times \mathbf{k}_i) \cdot \mathbf{E}_{0i} - (\mathbf{k}_r \times \mathbf{k}_t) \cdot \mathbf{E}_{0t} = 0 \tag{4.176}$$

$$(\mathbf{k}_t \times \mathbf{k}_i) \cdot \mathbf{E}_{0i} + (\mathbf{k}_t \times \mathbf{k}_r) \cdot \mathbf{E}_{0r} = 0 \tag{4.177}$$

and

$$(\mathbf{a} \times \mathbf{k}_i) \cdot \mathbf{E}_{0i} + (\mathbf{a} \times \mathbf{k}_r) \cdot \mathbf{E}_{0r} - (\mathbf{a} \times \mathbf{k}_t) \cdot \mathbf{E}_{0t} = 0 \tag{4.178}$$

Next, we substitute Eqs. (4.166), (4.167), and (4.168) into Eqs. (4.176), (4.177), and (4.178). Since vectors $\mathbf{k}_\alpha \times \mathbf{k}_\beta$ and $\mathbf{k}_\alpha \times \mathbf{a}$ are perpendicular and parallel to the plane of incidence, respectively, and $(\mathbf{a} \times \mathbf{k}_\alpha)^2 = \mathbf{a}^2$ where α (orβ) denotes i, r, or t, we obtain

$$(\mathbf{a} \cdot \mathbf{k}_r \times \mathbf{k}_i)A_\perp - (\mathbf{a} \cdot \mathbf{k}_r \times \mathbf{k}_t)C_\perp = 0 \tag{4.179}$$

$$(\mathbf{a} \cdot \mathbf{k}_t \times \mathbf{k}_i)A_\perp + (\mathbf{a} \cdot \mathbf{k}_t \times \mathbf{k}_r)B_\perp = 0 \tag{4.180}$$

and

$$k_i A_{||} + k_r B_{||} - k_t C_{||} = 0 \tag{4.181}$$

Equations (4.179) and (4.180) may be combined to give

$$\frac{A_\perp}{\mathbf{a} \cdot (\mathbf{k}_r \times \mathbf{k}_t)} = \frac{B_\perp}{\mathbf{a} \cdot (\mathbf{k}_t \times \mathbf{k}_i)} = \frac{C_\perp}{\mathbf{a} \cdot (\mathbf{k}_r \times \mathbf{k}_i)} \tag{4.182}$$

which, with the aid of Eqs. (3.107) and (3.110), implies that

$$A_\perp + B_\perp - C_\perp = 0 \tag{4.183}$$

$$A_\perp \mathbf{k}_i + B_\perp \mathbf{k}_r - C_\perp \mathbf{k}_t = \mathbf{0} \tag{4.184}$$

Also, dot-multiplications of Eq. (4.184) by $\hat{\mathbf{k}}_t$ and $(\mathbf{a} \times \mathbf{k}_t)$ successively, yield

$$k_i(\hat{\mathbf{k}}_i \cdot \hat{\mathbf{k}}_t)A_\perp + k_r(\hat{\mathbf{k}}_r \cdot \hat{\mathbf{k}}_t)B_\perp - k_t C_\perp = 0 \tag{4.185}$$

and

$$(\hat{\mathbf{k}}_i \cdot \hat{\mathbf{k}}_t)(\hat{\mathbf{k}}_i \cdot \hat{\mathbf{q}})A_\perp + (\hat{\mathbf{k}}_r \cdot \hat{\mathbf{k}}_t)(\hat{\mathbf{k}}_r \cdot \hat{\mathbf{q}})B_\perp - (\hat{\mathbf{k}}_t \cdot \hat{\mathbf{q}})C_\perp = 0 \tag{4.186}$$

In obtaining Eq. (4.186), we have used the results of Eqs. (3.100), (3.101), and (4.183).

On the other hand, substituting Eqs. (4.166), (4.167), (4.168) into Eq. (4.162) and taking into account Eq. (4.183) and the fact $(\hat{\mathbf{k}}_\alpha \times \mathbf{a}) \times \hat{\mathbf{q}} = (\hat{\mathbf{k}}_\alpha \cdot \hat{\mathbf{q}})\mathbf{a}$, where $\alpha = i$, r, or t, we obtain

$$(\hat{\mathbf{k}}_i \cdot \hat{\mathbf{q}})A_{||} + (\hat{\mathbf{k}}_r \cdot \hat{\mathbf{q}})B_{||} - (\hat{\mathbf{k}}_t \cdot \hat{\mathbf{q}})C_{||} = 0 \tag{4.187}$$

Equations (4.181) and (4.187) may be solved for $A_{||}/C_{||}$ and $B_{||}/C_{||}$. The solutions are

$$\frac{A_{||}}{C_{||}} = \frac{\begin{vmatrix} k_t & k_r \\ \hat{\mathbf{k}}_t \cdot \hat{\mathbf{q}} & \hat{\mathbf{k}}_r \cdot \hat{\mathbf{q}} \end{vmatrix}}{\Delta} \tag{4.188}$$

and

$$\frac{B_{||}}{C_{||}} = \frac{\begin{vmatrix} k_i & k_t \\ \hat{\mathbf{k}}_i \cdot \hat{\mathbf{q}} & \hat{\mathbf{k}}_t \cdot \hat{\mathbf{q}} \end{vmatrix}}{\Delta} \tag{4.189}$$

where

$$\Delta = \begin{vmatrix} k_i & k_r \\ \hat{\mathbf{k}}_i \cdot \hat{\mathbf{q}} & \hat{\mathbf{k}}_r \cdot \hat{\mathbf{q}} \end{vmatrix}$$

Similarly, solving Eqs. (4.185) and (4.186) for $A_\perp/C_\perp$ and $B_\perp/C_\perp$ and then comparing the results with Eqs. (4.188) and (4.189), we find

$$\frac{A_{\parallel}/A_\perp}{\hat{\mathbf{k}}_i \cdot \hat{\mathbf{k}}_t} = \frac{B_{\parallel}/B_\perp}{\hat{\mathbf{k}}_r \cdot \hat{\mathbf{k}}_t} = \frac{C_{\parallel}/C_\perp}{\hat{\mathbf{k}}_t \cdot \hat{\mathbf{k}}_t} \tag{4.190}$$

It is noted that the denominator of the last term is one, however, we replaced it with $\hat{\mathbf{k}}_t \cdot \hat{\mathbf{k}}_t$ for the purpose of symmetry. Equation (4.190) states that the ratios of the parallel to the perpendicular components of the amplitude vectors $\mathbf{E}_0$ of the incident, the reflected, and the transmitted waves are proportional to the cosine of the angles formed respectively by the corresponding wave vectors with the wave vector of the transmitted wave.

Multiplying Eq. (4.190) by $1/k_i k_t$ and noting that $q_r = -q_i$, and

$$\begin{aligned}\mathbf{k}_\alpha \cdot \mathbf{k}_\beta &= (\mathbf{b} + q_\alpha \hat{\mathbf{q}}) \cdot (\mathbf{b} + q_\beta \hat{\mathbf{q}}) \\ &= \mathbf{b}^2 + q_\alpha q_\beta \\ &= \mathbf{a}^2 + q_\alpha q_\beta\end{aligned}$$

we may rewrite Eq. (4.190) as

$$\frac{A_{\parallel}/A_\perp}{\mathbf{a}^2 + q_i q_t} = \frac{B_{\parallel}/B_\perp}{\mathbf{a}^2 - q_i q_t} = \frac{C_{\parallel}/C_\perp}{k_i k_t}$$

In summary, for a given uniform plane wave incident on an interface of two lossless isotropic media, Eqs. (4.182) and (4.190) completely determine the amplitudes of the reflected and the transmitted waves. The reflection and transmission coefficients defined in Eqs. (4.171) and (4.174) become

$$\Gamma_{11} = \left.\frac{B_\perp}{A_\perp}\right|_{A_{\parallel}=0} = \frac{\mathbf{a} \cdot (\mathbf{k}_t \times \mathbf{k}_i)}{\mathbf{a} \cdot (\mathbf{k}_r \times \mathbf{k}_t)} \tag{4.191}$$

$$\Gamma_{12} = \Gamma_{21} = 0 \tag{4.192}$$

$$\Gamma_{22} = \left.\frac{B_{\parallel}}{A_{\parallel}}\right|_{A_\perp=0} = \frac{\hat{\mathbf{k}}_r \cdot \hat{\mathbf{k}}_t}{\hat{\mathbf{k}}_i \cdot \hat{\mathbf{k}}_t}\Gamma_{11} \tag{4.193}$$

and

$$T_{11} = \left.\frac{C_\perp}{A_\perp}\right|_{A_{\parallel}=0} = \frac{\mathbf{a} \cdot (\mathbf{k}_r \times \mathbf{k}_i)}{\mathbf{a} \cdot (\mathbf{k}_r \times \mathbf{k}_t)} \tag{4.194}$$

$$T_{12} = T_{21} = 0 \tag{4.195}$$

$$T_{22} = \left.\frac{C_{\parallel}}{A_{\parallel}}\right|_{A_\perp=0} = \frac{1}{\hat{\mathbf{k}}_i \cdot \hat{\mathbf{k}}_t} T_{11} \tag{4.196}$$

respectively. Equations (4.192) and (4.195) clearly show that the two waves—one with $\mathbf{E}_0$ perpendicular to the plane of incidence and the other with $\mathbf{E}_0$ parallel to it—are uncoupled.

Alternatively, we may express the reflection and transmission coefficients in terms of the angle of incidence and the angle of transmission. Keeping in mind that the vector $\mathbf{k}_\alpha \times \mathbf{k}_\beta$ is parallel to $\mathbf{a}$ (see Fig. 4.3), we easily find

$$\begin{aligned} \mathbf{a} \cdot (\mathbf{k}_t \times \mathbf{k}_i) &= |\mathbf{a}||\mathbf{k}_t \times \mathbf{k}_i| = |\mathbf{a}|\, k_t k_i \sin(\theta_t - \theta_i) \\ \mathbf{a} \cdot (\mathbf{k}_r \times \mathbf{k}_t) &= |\mathbf{a}||\mathbf{k}_r \times \mathbf{k}_t| = |\mathbf{a}|\, k_t k_i \sin(\theta_t + \theta_i) \\ \mathbf{a} \cdot (\mathbf{k}_r \times \mathbf{k}_i) &= |\mathbf{a}||\mathbf{k}_r \times \mathbf{k}_i| = |\mathbf{a}|\, k_i^2 \sin 2\theta_i \end{aligned} \tag{4.197}$$

and

$$\begin{aligned} \hat{\mathbf{k}}_r \cdot \hat{\mathbf{k}}_t &= \cos(\theta_r' - \theta_t) = -\cos(\theta_i + \theta_t) \\ \hat{\mathbf{k}}_i \cdot \hat{\mathbf{k}}_t &= \cos(\theta_i - \theta_t) \end{aligned} \tag{4.198}$$

Substituting Eqs. (4.197) and (4.198) into Eqs. (4.191), (4.193), (4.194), and (4.196), and using the Snell law of refraction, we obtain

$$\Gamma_{11} = \frac{\sin(\theta_t - \theta_i)}{\sin(\theta_t + \theta_i)} \tag{4.199}$$

$$\Gamma_{22} = \frac{\tan(\theta_i - \theta_t)}{\tan(\theta_i + \theta_t)} \tag{4.200}$$

$$T_{11} = \frac{2 \sin\theta_t \cos\theta_i}{\sin(\theta_i + \theta_t)} \tag{4.201}$$

$$T_{22} = \frac{2 \sin\theta_t \cos\theta_i}{\sin(\theta_i + \theta_t)\cos(\theta_i - \theta_t)} \tag{4.202}$$

Equations (4.199) to (4.202), which express the amplitudes of the reflected and transmitted waves in terms of the amplitude of the incident wave, are called *Fresnel's equations.*

The reflection and transmission coefficients may still take other forms. For instance, from

$$\begin{aligned} \mathbf{k}_\alpha \times \mathbf{k}_\beta &= (\mathbf{b} + q_\alpha \hat{\mathbf{q}}) \times (\mathbf{b} + q_\beta \hat{\mathbf{q}}) \\ &= (q_\beta - q_\alpha)(\mathbf{b} \times \hat{\mathbf{q}}) \\ &= (q_\beta - q_\alpha)\mathbf{a} \end{aligned} \tag{4.203}$$

and $q_r = -q_i$, we find

$$\begin{aligned} \mathbf{a} \cdot (\mathbf{k}_t \times \mathbf{k}_i) &= (q_i - q_t)\mathbf{a}^2 \\ \mathbf{a} \cdot (\mathbf{k}_r \times \mathbf{k}_t) &= (q_i + q_t)\mathbf{a}^2 \\ \mathbf{a} \cdot (\mathbf{k}_r \times \mathbf{k}_i) &= 2q_i \mathbf{a}^2 \end{aligned} \tag{4.204}$$

Hence, Eqs. (4.191) and (4.194) become

$$\Gamma_{11} = \frac{q_i - q_t}{q_i + q_t} = \frac{k_i \cos\theta_i - k_t \cos\theta_t}{k_i \cos\theta_i + k_t \cos\theta_t} \tag{4.205}$$

and

$$T_{11} = \frac{2q_i}{q_i + q_t} = \frac{2k_i \cos\theta_i}{k_i \cos\theta_i + k_t \cos\theta_t} \tag{4.206}$$

respectively. Also the ratio of Eqs. (4.189) and (4.188) yields

$$\Gamma_{22} = \frac{k_t(\hat{\mathbf{k}}_i \cdot \hat{\mathbf{q}}) - k_i(\hat{\mathbf{k}}_t \cdot \hat{\mathbf{q}})}{k_t(\hat{\mathbf{k}}_i \cdot \hat{\mathbf{q}}) + k_i(\hat{\mathbf{k}}_t \cdot \hat{\mathbf{q}})} = \frac{k_t \cos \theta_i - k_i \cos \theta_t}{k_t \cos \theta_i + k_i \cos \theta_t} \tag{4.207}$$

and finally, Eq. (4.188) gives

$$T_{22} = \frac{2k_i(\hat{\mathbf{k}}_i \cdot \hat{\mathbf{q}})}{k_i(\hat{\mathbf{k}}_t \cdot \hat{\mathbf{q}}) + k_t(\hat{\mathbf{k}}_t \cdot \hat{\mathbf{q}})} = \frac{2k_i \cos \theta_i}{k_i \cos \theta_t + k_t \cos \theta_i} \tag{4.208}$$

In the above, we find the amplitude vectors $\mathbf{E}_0$ of the reflected and transmitted waves in terms of the incident wave by decomposing them into components perpendicular and parallel to the plane of incidence. Alternatively, we shall now show that the results may also be expressed in terms of amplitude vectors and can be derived directly from the boundary conditions (4.162) and (4.165). To this end, let us use Maxwell's equation

$$\mathbf{E}_0 = -\frac{1}{\omega \varepsilon_0 \varepsilon}(\mathbf{k} \times \mathbf{H}_0) = -\eta(\hat{\mathbf{k}} \times \mathbf{H}_0)$$

to rewrite Eq. (4.162) as

$$\left[\mathbf{k}_i \times \mathbf{H}_{0i} + \mathbf{k}_r \times \mathbf{H}_{0r} - \frac{\varepsilon_1}{\varepsilon_2}(\mathbf{k}_t \times \mathbf{H}_{0t})\right] \times \hat{\mathbf{q}} = \mathbf{0} \tag{4.209}$$

Next, substituting $\mathbf{H}_{0r} = \mathbf{H}_{0t} - \mathbf{H}_{0i}$ into Eq. (4.209) and noting that $\mathbf{k}_i - \mathbf{k}_r = 2q_i \hat{\mathbf{q}}$, we obtain

$$2q_i[\hat{\mathbf{q}} \times (\hat{\mathbf{q}} \times \mathbf{H}_{0i})] = \hat{\mathbf{q}} \times \left[\left(\frac{\varepsilon_1}{\varepsilon_2}\mathbf{k}_t - \mathbf{k}_r\right) \times \mathbf{H}_{0t}\right] \tag{4.210}$$

To solve for $\mathbf{H}_{0t}$, we take the cross product of Eq. (4.210) first with $(\varepsilon_1 \mathbf{k}_t/\varepsilon_2 - \mathbf{k}_r)$ and then with $\mathbf{k}_t$. It yields

$$2q_i \mathbf{k}_t \times \left\{\left(\frac{\varepsilon_1}{\varepsilon_2}\mathbf{k}_t - \mathbf{k}_r\right) \times [\hat{\mathbf{q}} \times (\hat{\mathbf{q}} \times \mathbf{H}_{0i})]\right\}$$
$$= \left(\frac{\varepsilon_1}{\varepsilon_2} q_t - q_r\right)\left(\frac{\varepsilon_1}{\varepsilon_2} k_t^2 - \mathbf{k}_t \cdot \mathbf{k}_r\right)\mathbf{H}_{0t} \tag{4.211}$$

By using Eq. (3.105) and $\mathbf{k}_\alpha \cdot \mathbf{k}_\beta = \mathbf{a}^2 + q_\alpha q_\beta$, we easily establish that

$$\frac{\varepsilon_1}{\varepsilon_2} k_t^2 - \mathbf{k}_t \cdot \mathbf{k}_r = q_i(q_i + q_t)$$

$$\frac{\varepsilon_1}{\varepsilon_2} q_t - q_r = \frac{1}{k_t^2}(q_i + q_t)(\mathbf{k}_i \cdot \mathbf{k}_t)$$

$$\frac{\varepsilon_1}{\varepsilon_2}\mathbf{k}_t - \mathbf{k}_r = \frac{1}{k_t^2}(q_i + q_t)[(q_i - q_t)\mathbf{b} + (\mathbf{k}_i \cdot \mathbf{k}_t)\hat{\mathbf{q}}]$$

Substituting the above into Eq. (4.211) and expanding the vector triple product inside the square bracket, we finally obtain

$$(q_i + q_t)(\mathbf{k}_i \cdot \mathbf{k}_t)\mathbf{H}_{0t} = 2(q_i - q_t)(\mathbf{H}_{0i} \cdot \mathbf{a})\mathbf{a} - 2(\mathbf{k}_i \cdot \mathbf{k}_t)[\mathbf{k}_t \times (\hat{\mathbf{q}} \times \mathbf{H}_{0i})]$$

or

$$\mathbf{H}_{0t} = \frac{2}{q_i + q_t}\left[q_t \bar{\mathbf{I}} + (q_i - q_t)\left(\hat{\mathbf{q}}\hat{\mathbf{q}} + \frac{1}{\mathbf{k}_i \cdot \mathbf{k}_t}\,\mathbf{aa}\right)\right] \cdot \mathbf{H}_{0i} \tag{4.212}$$

Subtraction $\mathbf{H}_{0i}$ from both sides of Eq. (4.212) yields

$$\mathbf{H}_{0r} = \frac{q_t - q_i}{q_t + q_i}\left[\bar{\mathbf{I}} - 2\left(\hat{\mathbf{q}}\hat{\mathbf{q}} + \frac{1}{\mathbf{k}_i \cdot \mathbf{k}_t}\,\mathbf{aa}\right)\right] \cdot \mathbf{H}_{0i} \tag{4.213}$$

Equations (4.212) and (4.213) are the desired results. They express the amplitude vectors of the transmitted and reflected waves directly in terms of the amplitude vector of the incident wave. To show that Eqs. (4.212) and (4.213) agree with Eqs. (4.205) to (4.208), we decompose $\mathbf{H}_{0i}$ into components parallel and perpendicular to the plane of incidence [see Eq. (4.166)]:

$$\mathbf{H}_{0i} = \frac{1}{\eta_i}(\hat{\mathbf{k}}_i \times \mathbf{E}_{0i}) = \frac{1}{\eta_i}[A_\perp(\hat{\mathbf{k}}_i \times \mathbf{a}) - A_\parallel \mathbf{a}]$$

and substitute it into Eqs. (4.212) and (4.213). Since

$$\frac{2\mathbf{a}^2}{\mathbf{k}_i \cdot \mathbf{k}_t} - 1 = \frac{\hat{\mathbf{k}}_r \cdot \hat{\mathbf{k}}_t}{\hat{\mathbf{k}}_i \cdot \hat{\mathbf{k}}_t}$$

$$q_t + \frac{1}{\mathbf{k}_i \cdot \mathbf{k}_t}(q_i - q_t)\mathbf{a}^2 = \frac{q_i k_t}{k_i(\hat{\mathbf{k}}_i \cdot \hat{\mathbf{k}}_t)}$$

we obtain

$$\mathbf{H}_{0r} = \frac{q_t - q_i}{\eta_i(q_t + q_i)}\left[A_\perp(\hat{\mathbf{k}}_i \times \mathbf{a}) + \frac{\hat{\mathbf{k}}_r \cdot \hat{\mathbf{k}}_t}{\hat{\mathbf{k}}_i \cdot \hat{\mathbf{k}}_t} A_\parallel \mathbf{a} + \frac{2\mathbf{a}^2}{k_i} A_\perp \hat{\mathbf{q}}\right] \tag{4.214}$$

and

$$\mathbf{H}_{0t} = \frac{2}{\eta_i(q_i + q_t)}\left[q_t A_\perp(\hat{\mathbf{k}}_i \times \mathbf{a}) - \frac{q_i k_t}{k_i(\hat{\mathbf{k}}_i \cdot \hat{\mathbf{k}}_t)} A_\parallel \mathbf{a} - \frac{(q_i - q_t)\mathbf{a}^2}{k_i} A_\perp \hat{\mathbf{q}}\right] \tag{4.215}$$

The corresponding amplitude vectors of the electric field follow from the Maxwell equations

$$\mathbf{E}_{0r} = \frac{q_i - q_t}{q_i + q_t}\left[A_\perp \mathbf{a} + \frac{\hat{\mathbf{k}}_r \cdot \hat{\mathbf{k}}_t}{\hat{\mathbf{k}}_i \cdot \hat{\mathbf{k}}_t} A_\parallel(\hat{\mathbf{k}}_r \times \mathbf{a})\right] \tag{4.216}$$

$$\mathbf{E}_{0t} = \frac{2q_i}{q_i + q_t}\left[A_\perp \mathbf{a} + \frac{1}{\hat{\mathbf{k}}_i \cdot \hat{\mathbf{k}}_t} A_\parallel(\hat{\mathbf{k}}_t \times \mathbf{a})\right] \tag{4.217}$$

In deriving Eqs. (4.216) and (4.217), we have used the following relations

$$\frac{2\mathbf{a}^2}{k_i^2} - \hat{\mathbf{k}}_t \cdot \hat{\mathbf{k}}_r = 1$$

$$q_t(\hat{\mathbf{k}}_t \cdot \hat{\mathbf{k}}_t) + \frac{(q_i - q_t)\mathbf{a}^2}{k_i k_t} = \frac{q_i k_t}{k_i}$$

$$\frac{\eta_t k_t}{\eta_i k_i} = 1$$

Equating in Eqs. (4.216) and (4.217) the components parallel and perpendicular to the plane of incidence separately, we obtain the results given in Eqs. (4.205) to (4.208).

4.6 FRESNEL'S EQUATIONS WHEN $\mu_1 \neq \mu_2$

In the last section we have considered the Fresnel equations for media having the same magnetic permeability. However, if the magnetic permeabilities of two isotropic media are different, the Fresnel equations can no longer be expressed in the simple forms (4.182) and (4.190), since, in this case, Eq. (4.165) is not valid. We must start from the general boundary conditions (4.162) and (4.163) which can be written as

$$\mathbf{E}_{0i} + \mathbf{E}_{0r} - \mathbf{E}_{0t} = K_1 \hat{\mathbf{q}} \tag{4.218}$$

and

$$\mathbf{H}_{0i} + \mathbf{H}_{0r} - \mathbf{H}_{0t} = K_2 \hat{\mathbf{q}} \tag{4.219}$$

respectively. Where K_1 and K_2 are two arbitrary constants, the amplitude vectors of the incident, the reflected and the transmitted uniform plane waves are related by the Maxwell equations

$$\mathbf{E}_{0\alpha} = -\eta_\alpha(\hat{\mathbf{k}}_\alpha \times \mathbf{H}_{0\alpha}) \tag{4.220}$$

$$\mathbf{H}_{0\alpha} = \frac{1}{\eta_\alpha}(\hat{\mathbf{k}}_\alpha \times \mathbf{E}_{0\alpha}) \tag{4.221}$$

where $\alpha = i$, r, or t. Decomposing $\mathbf{E}_{0\alpha}$ into components perpendicular and parallel to the plane of incidence as in Eqs. (4.166) to (4.168), then using Eq. (4.221), we find the corresponding magnetic fields

$$\mathbf{H}_{0i} = \frac{1}{\eta_i}[A_\perp(\hat{\mathbf{k}}_i \times \mathbf{a}) - A_\parallel \mathbf{a}] \tag{4.222}$$

$$\mathbf{H}_{0r} = \frac{1}{\eta_i}[B_\perp(\hat{\mathbf{k}}_r \times \mathbf{a}) - B_\parallel \mathbf{a}] \tag{4.223}$$

$$\mathbf{H}_{0t} = \frac{1}{\eta_t}[C_\perp(\hat{\mathbf{k}}_t \times \mathbf{a}) - C_\parallel \mathbf{a}] \tag{4.224}$$

where

$$\eta_i = \sqrt{\frac{\mu_0 \mu_1}{\varepsilon_0 \varepsilon_1}} = \eta_r \tag{4.225}$$

and
$$\eta_t = \sqrt{\frac{\mu_0 \mu_2}{\varepsilon_0 \varepsilon_2}} \tag{4.226}$$

are the intrinsic wave impedances of medium 1 and 2 respectively. Substituting Eqs. (4.166) to (4.168) into Eq. (4.218), and Eqs. (4.222) and (4.224) into Eq. (4.219), and then taking the dot product of each of the resulting equations with $\mathbf{a}$ and $\mathbf{b} = \hat{\mathbf{q}} \times \mathbf{a}$ separately, and noting that $\hat{\mathbf{k}}_r \cdot \hat{\mathbf{q}} = -\hat{\mathbf{k}}_i \cdot \hat{\mathbf{q}}$ and

$$\mathbf{b} \cdot (\hat{\mathbf{k}}_\alpha \times \mathbf{a}) = \mathbf{a}^2(\hat{\mathbf{k}}_\alpha \cdot \hat{\mathbf{q}}) \tag{4.227}$$

we find

$$A_\perp + B_\perp - C_\perp = 0 \tag{4.228}$$

$$(\hat{\mathbf{k}}_i \cdot \hat{\mathbf{q}})A_\parallel - (\hat{\mathbf{k}}_i \cdot \hat{\mathbf{q}})B_\parallel - (\hat{\mathbf{k}}_t \cdot \hat{\mathbf{q}})C_\parallel = 0 \tag{4.229}$$

$$\eta_t A_\parallel + \eta_t B_\parallel - \eta_i C_\parallel = 0 \tag{4.230}$$

$$\eta_t(\hat{\mathbf{k}}_i \cdot \hat{\mathbf{q}})A_\perp - \eta_t(\hat{\mathbf{k}}_i \cdot \hat{\mathbf{q}})B_\perp - \eta_i(\hat{\mathbf{k}}_t \cdot \hat{\mathbf{q}})C_\perp = 0 \tag{4.231}$$

Equations (4.228) to (4.231) fall into two groups, one of which contains only the components perpendicular to the plane of incidence [Eqs. (4.228) and (4.231)], while the other contains only those parallel to the plane of incidence [Eqs. (4.229) and (4.230)]. These two groups of waves are independent of one another and can thus be treated separately.

Solving Eqs. (4.228) and (4.231) for $B_\perp/A_\perp$ and $C_\perp/A_\perp$, and Eqs. (4.229) and (4.230) for $B_\parallel/A_\parallel$ and $C_\parallel/A_\parallel$, we obtain the reflection and transmission coefficients:

$$\Gamma_{11} = \frac{\eta_t(\hat{\mathbf{k}}_i \cdot \hat{\mathbf{q}}) - \eta_i(\hat{\mathbf{k}}_t \cdot \hat{\mathbf{q}})}{\eta_t(\hat{\mathbf{k}}_i \cdot \hat{\mathbf{q}}) + \eta_i(\hat{\mathbf{k}}_t \cdot \hat{\mathbf{q}})} = \frac{\eta_t \cos\theta_i - \eta_i \cos\theta_t}{\eta_t \cos\theta_i + \eta_i \cos\theta_t}$$
$$= \frac{\cos\theta_i - \sqrt{\mu_1\varepsilon_2/\mu_2\varepsilon_1}\cos\theta_t}{\cos\theta_i + \sqrt{\mu_1\varepsilon_2/\mu_2\varepsilon_1}\cos\theta_t} \tag{4.232}$$

$$\Gamma_{22} = \frac{\eta_i(\hat{\mathbf{k}}_i \cdot \hat{\mathbf{q}}) - \eta_t(\hat{\mathbf{k}}_t \cdot \hat{\mathbf{q}})}{\eta_i(\hat{\mathbf{k}}_i \cdot \hat{\mathbf{q}}) + \eta_t(\hat{\mathbf{k}}_t \cdot \hat{\mathbf{q}})} = \frac{\eta_i \cos\theta_i - \eta_t \cos\theta_t}{\eta_i \cos\theta_i + \eta_t \cos\theta_t}$$
$$= \frac{\sqrt{\mu_1\varepsilon_2/\mu_2\varepsilon_1}\cos\theta_i - \cos\theta_t}{\sqrt{\mu_1\varepsilon_2/\mu_2\varepsilon_1}\cos\theta_i + \cos\theta_t} \tag{4.233}$$

$$T_{11} = \frac{2\eta_t(\hat{\mathbf{k}}_i \cdot \hat{\mathbf{q}})}{\eta_i(\hat{\mathbf{k}}_t \cdot \hat{\mathbf{q}}) + \eta_t(\hat{\mathbf{k}}_i \cdot \hat{\mathbf{q}})} = \frac{2\eta_t \cos\theta_i}{\eta_i \cos\theta_t + \eta_t \cos\theta_i}$$
$$= \frac{2\cos\theta_i}{\cos\theta_i + \sqrt{\mu_1\varepsilon_2/\mu_2\varepsilon_1}\cos\theta_t} \tag{4.234}$$

$$T_{22} = \frac{2\eta_t(\hat{\mathbf{k}}_i \cdot \hat{\mathbf{q}})}{\eta_i(\hat{\mathbf{k}}_i \cdot \hat{\mathbf{q}}) + \eta_t(\hat{\mathbf{k}}_t \cdot \hat{\mathbf{q}})} = \frac{2\eta_t \cos\theta_i}{\eta_i \cos\theta_i + \eta_t \cos\theta_t}$$
$$= \frac{2\cos\theta_i}{\cos\theta_t + \sqrt{\mu_1\varepsilon_2/\mu_2\varepsilon_1}\cos\theta_i} \tag{4.235}$$

and $\Gamma_{12} = \Gamma_{21} = T_{12} = T_{21} = 0$ as before. When $\mu_1 = \mu_2$, Eqs. (4.232) to (4.235) reduce to Eqs. (4.205) to (4.208) as it should be. However, in this general case we can no longer use the Snell law of refraction to express the transmission and reflection coefficients solely in terms of the angle of incidence and the angle of transmission. An exception is the case when the intrinsic wave impedances of the two media are equal, i.e., when $\eta_i = \eta_t$ or $\mu_1/\varepsilon_1 = \mu_2/\varepsilon_2$. In that case, Eqs. (4.232) to (4.235) become

$$\Gamma_{11} = \Gamma_{22} = \frac{\cos\theta_i - \cos\theta_t}{\cos\theta_i + \cos\theta_t} \tag{4.236}$$

$$T_{11} = T_{22} = \frac{2\cos\theta_i}{\cos\theta_i + \cos\theta_t} \tag{4.237}$$

which imply that the parallel and the perpendicular components of the amplitude vectors are proportional to each other, i.e.,

$$\frac{A_{\|}}{A_{\perp}} = \frac{B_{\|}}{B_{\perp}} = \frac{C_{\|}}{C_{\perp}} = K \tag{4.238}$$

In other words, all three waves—the incident, the reflected, and the transmitted—have identical polarization.

Another interesting special case is when the wave numbers (or indices of refraction) of the two media coincide: $k_i = k_t$. In this case, the ratio of the dielectric constants is inversely proportional to the ratio of the permeabilities of the two media: $\varepsilon_1/\varepsilon_2 = \mu_2/\mu_1$. The Fresnel equations (4.232) to (4.235) are reduced to

$$\Gamma_{11} = \frac{\cos\theta_i - (\mu_1/\mu_2)\cos\theta_t}{\cos\theta_i + (\mu_1/\mu_2)\cos\theta_t} \tag{4.239}$$

$$\Gamma_{22} = \frac{(\mu_1/\mu_2)\cos\theta_i - \cos\theta_t}{(\mu_1/\mu_2)\cos\theta_i + \cos\theta_t} \tag{4.240}$$

$$T_{11} = \frac{2\cos\theta_i}{\cos\theta_i + (\mu_1/\mu_2)\cos\theta_t} \tag{4.241}$$

$$T_{22} = \frac{2\cos\theta_i}{\cos\theta_t + (\mu_1/\mu_2)\cos\theta_i} \tag{4.242}$$

Here we see that the reflected and the transmitted waves depend on the ratio of μ_1/μ_2. Hence in spite of the equality of wave numbers in the two media, the presence of an interface changes the propagation characteristics of the wave.

4.7 FRESNEL'S EQUATIONS: NORMAL INCIDENCE

When a uniform plane wave is incident normally on an interface of two media, the method of decomposing the amplitude vectors into components perpendicular and parallel to the plane of incidence fails. Since $\mathbf{k}_i \| \hat{\mathbf{q}}$ means

$$\mathbf{a} = \mathbf{k}_i \times \hat{\mathbf{q}} = \mathbf{k}_r \times \hat{\mathbf{q}} = \mathbf{k}_t \times \hat{\mathbf{q}} = \mathbf{0} \tag{4.243}$$

the concept of the plane of incidence loses its meaning. Relations such as Eqs. (4.182) and (4.190) become infinite because all the denominators take the form $[\mathbf{a} \cdot (\mathbf{k}_\alpha \times \mathbf{k}_\beta)]$ and are zero. It is true that we may calculate the limits and obtain the correct results. However, it is much simpler in this case to solve the boundary conditions at the interface directly. Indeed, according to Eq. (4.243), we may write $\hat{\mathbf{k}}_i = \hat{\mathbf{k}}_t = -\hat{\mathbf{k}}_r = \hat{\mathbf{q}}$ and (see Fig. 4.6)

$$\begin{aligned} \mathbf{k}_i &= k_i \hat{\mathbf{k}}_i = k_i \hat{\mathbf{q}} \\ \mathbf{k}_r &= k_r \hat{\mathbf{k}}_r = -k_i \hat{\mathbf{q}} \\ \mathbf{k}_t &= k_t \hat{\mathbf{k}}_t = k_t \hat{\mathbf{q}} \end{aligned} \tag{4.244}$$

and, hence, from Maxwell's equation $\mathbf{k} \cdot \mathbf{E}_0 = 0$,

$$\hat{\mathbf{q}} \cdot \mathbf{E}_{0i} = \hat{\mathbf{q}} \cdot \mathbf{E}_{0r} = \hat{\mathbf{q}} \cdot \mathbf{E}_{0t} = 0 \tag{4.245}$$

at every point in space and on the interface. From Eq. (4.245), we write

$$(\mathbf{E}_{0i} + \mathbf{E}_{0r} - \mathbf{E}_{0t}) \cdot \hat{\mathbf{q}} = 0 \tag{4.246}$$

which together with the boundary condition (4.162) implies that the vector $\mathbf{E}_{0i} + \mathbf{E}_{0r} - \mathbf{E}_{0t}$ is simultaneously parallel and perpendicular to the vector $\hat{\mathbf{q}}$. This is possible only if the vector itself is a zero vector, i.e.,

$$\mathbf{E}_{0i} + \mathbf{E}_{0r} - \mathbf{E}_{0t} = \mathbf{0} \tag{4.247}$$

On the other hand, substituting $\mathbf{H}_0 = (1/\eta)\hat{\mathbf{k}} \times \mathbf{E}_0$ into Eq. (4.163) and expanding the vector triple product, after using the condition (4.245), we obtain

$$\eta_t \mathbf{E}_{0i} - \eta_t \mathbf{E}_{0r} - \eta_i \mathbf{E}_{0t} = \mathbf{0} \tag{4.248}$$

Solving Eqs. (4.247) and (4.248) simultaneously, we find

$$\mathbf{E}_{0r} = \frac{\eta_t - \eta_i}{\eta_t + \eta_i} \mathbf{E}_{0i} \tag{4.249}$$

$$\mathbf{E}_{0t} = \frac{2\eta_t}{\eta_t + \eta_i} \mathbf{E}_{0i} \tag{4.250}$$

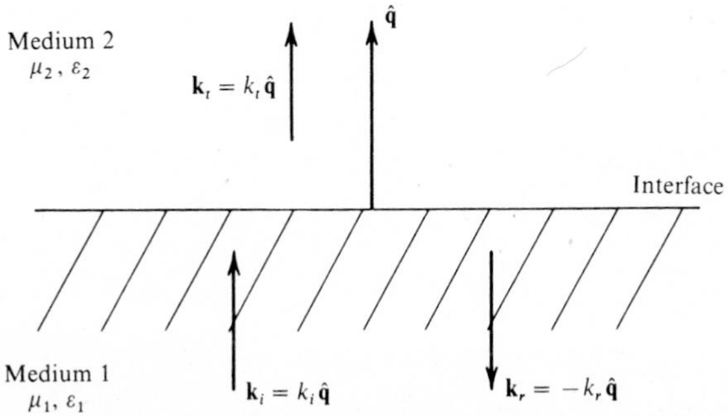

Figure 4.6 Normal incidence of a uniform plane wave on a lossless medium.

and according to the Maxwell equations, the corresponding magnetic fields for the reflected and transmitted waves:

$$\mathbf{H}_{0r} = \frac{\eta_i - \eta_t}{\eta_i + \eta_t} \mathbf{H}_{0i} \tag{4.251}$$

$$\mathbf{H}_{0t} = \frac{2\eta_i}{\eta_i + \eta_t} \mathbf{H}_{0i} \tag{4.252}$$

In the above derivation, we have used the general boundary conditions (4.162), (4.163), and the condition (4.244) for normal incidence. Therefore, Eqs. (4.249), (4.250), (4.251), and (4.252) are valid for media with different permeabilities as well as losses. For media having the same permeability, Eqs. (4.251) and (4.252) reduce to

$$\mathbf{H}_{0r} = \frac{q_t - q_i}{q_t + q_i} \mathbf{H}_{0i} \tag{4.253}$$

$$\mathbf{H}_{0t} = \frac{2q_t}{q_t + q_i} \mathbf{H}_{0i} \tag{4.254}$$

which agree with Eqs. (4.213) and (4.212) for normal incidence.

An interesting special case is when the second medium is a perfect conductor (see Fig. 4.7). Here we can obtain results from those of a lossless medium by replacing ε_2 by $\varepsilon_2(1 + i\tau)$ where $\tau = \sigma/\omega\varepsilon_0\varepsilon_2$. Thus by letting $\sigma \to \infty$, we find $\eta_t = \sqrt{\mu_0\mu_2/\varepsilon_0\varepsilon_2(1 + i\tau)} = 0$ for a perfect conductor. Substitution of $\eta_t = 0$ into Eqs. (4.250) and (4.249) yields

$$\begin{aligned} \mathbf{E}_{0t} &= \mathbf{0} \\ \mathbf{E}_{0r} &= -\mathbf{E}_{0i} \end{aligned} \tag{4.255}$$

which means that at the interface, the amplitude vector of the reflected wave is

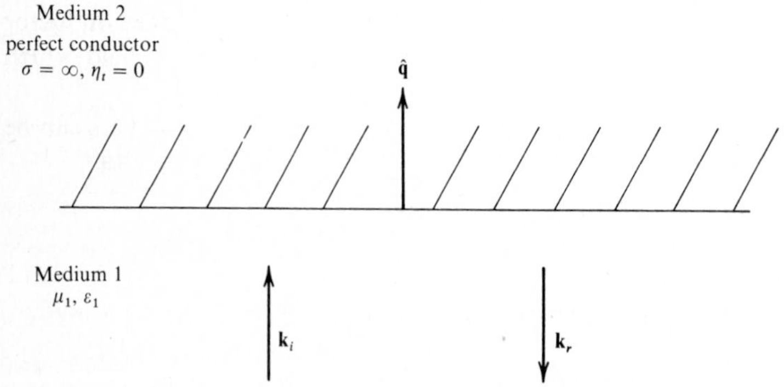

Figure 4.7 Normal incidence of a uniform plane wave on a perfect conductor.

equal and opposite to that of the incident wave. The total field in medium 1 is

$$\begin{aligned}\mathbf{E}_1^{(t)} &= \mathbf{E}_{0i}e^{i\mathbf{k}_i\cdot\mathbf{r}} + \mathbf{E}_{0r}e^{i\mathbf{k}_r\cdot\mathbf{r}} \\ &= \mathbf{E}_{0i}(e^{ik_i(\mathbf{r}\cdot\hat{\mathbf{q}})} - e^{-ik_i(\mathbf{r}\cdot\hat{\mathbf{q}})}) \\ &= 2i\mathbf{E}_{0i}\sin k_i(\mathbf{r}\cdot\hat{\mathbf{q}}) \end{aligned} \tag{4.256}$$

If $\mathbf{E}_{0i}$ is a real constant vector (i.e., a linearly polarized incident wave), the instantaneous total electric field follows from Eq. (3.1)

$$\begin{aligned}\mathscr{E}_1^{(t)} &= \operatorname{Re}(\mathbf{E}_1^{(t)}e^{-i\omega t}) \\ &= 2\mathbf{E}_{0i}\sin k_i(\mathbf{r}\cdot\hat{\mathbf{q}})\sin\omega t\end{aligned} \tag{4.257}$$

This shows that the variation of $\mathscr{E}_1^{(t)}$ in space is independent of the variation in time; in other words, in medium 1, the incident and the reflected waves combine to produce a pure standing wave. It is not a traveling wave since neither Eq. (4.256) nor Eq. (4.257) can be written as a function of the argument $(\mathbf{k}\cdot\mathbf{r} - \omega t)$, a characteristic of a traveling wave. The instantaneous electric field varies sinusoidally in the direction normal to the conducting surface and oscillates up and down according to $\sin\omega t$ at a fixed point in space.

Similarly, the total magnetic field in medium 1 is

$$\begin{aligned}\mathbf{H}_1^{(t)} &= \mathbf{H}_{0i}e^{i\mathbf{k}_i\cdot\mathbf{r}} + \mathbf{H}_{0r}e^{i\mathbf{k}_r\cdot\mathbf{r}} \\ &= \mathbf{H}_{0i}(e^{ik_i(\mathbf{r}\cdot\hat{\mathbf{q}})} + e^{-ik_i(\mathbf{r}\cdot\hat{\mathbf{q}})}) \\ &= \frac{2}{\eta_i}(\hat{\mathbf{q}}\times\mathbf{E}_{0i})\cos k_i(\mathbf{r}\cdot\hat{\mathbf{q}})\end{aligned} \tag{4.258}$$

hence the instantaneous magnetic field

$$\begin{aligned}\mathscr{H}_1^{(t)} &= \operatorname{Re}(\mathbf{H}_1^{(t)}e^{-i\omega t}) \\ &= \frac{2}{\eta_i}(\hat{\mathbf{q}}\times\mathbf{E}_{0i})\cos k_i(\mathbf{r}\cdot\hat{\mathbf{q}})\cos\omega t\end{aligned} \tag{4.259}$$

which again shows the nature of a standing wave. At the surface of the conductor the reflected magnetic field is in the same direction as the incident field which makes the total magnetic field at the surface $2\mathbf{H}_{0i}$.

It is obvious that a pure standing wave cannot transport energy. This can be seen from Eqs. (4.256) and (4.258) that $\mathbf{E}_1^{(t)}\times\mathbf{H}_1^{(t)*}$ is purely imaginary, thus

$$\langle\mathscr{P}_1^{(t)}\rangle = \tfrac{1}{2}\operatorname{Re}(\mathbf{E}_1^{(t)}\times\mathbf{H}_1^{(t)*}) = \mathbf{0} \tag{4.260}$$

4.8 POWER RELATION AT THE INTERFACE

We shall now examine the energy balance when waves are reflected and transmitted at the interface of two isotropic media. The instantaneous Poynting vectors for the incident, the reflected, and the transmitted waves are

$$\mathscr{P}_i = \mathscr{E}_i \times \mathscr{H}_i$$

$$\mathscr{P}_r = \mathscr{E}_r \times \mathscr{H}_r \tag{4.261}$$

and

$$\mathscr{P}_t = \mathscr{E}_t \times \mathscr{H}_t$$

respectively. In isotropic media, $\mathscr{P}$ points in the direction of wave vector $\mathbf{k}$. Thus in the case of oblique incidence, the Poynting vectors of all the waves have components parallel and perpendicular to the interface. There is no reason for parallel components of the Poynting vector to obey the conservation law for they are all in the same direction. On the other hand, the perpendicular components represent the power flow per unit area of the interface. It is clear that none of the power flowing normal to the interface in medium 2 can have originated anywhere but from the normal component of power in medium 1. Hence, for these normal components we would expect that the law of conservation of energy applies. That is, the power flux incident on any part of the interface must be equal to the sum of power fluxes leaving that surface:

$$(\mathscr{P}_i + \mathscr{P}_r - \mathscr{P}_t) \cdot \hat{\mathbf{q}} = 0 \tag{4.262}$$

Here $\mathscr{P}_r \cdot \hat{\mathbf{q}}$ is negative, since $\mathscr{P}_r$ and $\hat{\mathbf{q}}$ point in opposite directions. Equation (4.262) is similar to the boundary conditions for the normal components of the flux density vectors $\mathscr{B}$ and $\mathscr{D}$, and states that at the boundary between two media the normal component of the Poynting vector must be continuous.

Without solving the boundary-value problem for field vectors, we shall now establish Eq. (4.262) directly from the Maxwell equations, the boundary conditions and the definitions (4.261) of the Poynting vectors. In this approach we shall see that Eq. (4.262) is valid as long as medium 1 is lossless and isotropic and medium 2 is isotropic but may either be lossless or lossy. The plane wave in medium 2 may either be uniform or nonuniform (including the case of total reflection). At the interface of two isotropic media, the instantaneous electric and magnetic fields of the incident, the reflected, and the transmitted waves satisfy the boundary conditions:

$$(\mathscr{E}_i + \mathscr{E}_r) \times \hat{\mathbf{q}} = \mathscr{E}_t \times \hat{\mathbf{q}} \tag{4.263}$$

$$(\mathscr{H}_i + \mathscr{H}_r) \times \hat{\mathbf{q}} = \mathscr{H}_t \times \hat{\mathbf{q}} \tag{4.264}$$

Taking the cross product of Eq. (4.263) with Eq. (4.264) and expanding the vector triple products, we obtain

$$[(\mathscr{E}_i + \mathscr{E}_r) \times (\mathscr{H}_i + \mathscr{H}_r)] \cdot \hat{\mathbf{q}} = (\mathscr{E}_t \times \mathscr{H}_t) \cdot \hat{\mathbf{q}} \tag{4.265}$$

or

$$(\mathscr{P}_i + \mathscr{P}_r - \mathscr{P}_t) \cdot \hat{\mathbf{q}} = -(\mathscr{E}_i \times \mathscr{H}_r + \mathscr{E}_r \times \mathscr{H}_i) \cdot \hat{\mathbf{q}} \tag{4.266}$$

To show that the right-hand side of Eq. (4.266) vanishes, we note

$$\mathscr{E}_i = \text{Re}\,(\mathbf{E}_{0i} e^{i\phi}) = \tfrac{1}{2}(\mathbf{E}_{0i} e^{i\phi} + \mathbf{E}_{0i}^* e^{-i\phi})$$

$$\mathscr{H}_i = \text{Re}\,(\mathbf{H}_{0i} e^{i\phi}) = \tfrac{1}{2}(\mathbf{H}_{0i} e^{i\phi} + \mathbf{H}_{0i}^* e^{-i\phi})$$

and similar expressions for $\mathscr{E}_r$ and $\mathscr{H}_r$ and write

$$(\mathscr{E}_i \times \mathscr{H}_r + \mathscr{E}_r \times \mathscr{H}_i) \cdot \hat{\mathbf{q}}$$

$$= \tfrac{1}{2} \operatorname{Re} [(\mathbf{E}_{0i} \times \mathbf{H}_{0r} + \mathbf{E}_{0r} \times \mathbf{H}_{0i}) \cdot \hat{\mathbf{q}} e^{i2\phi} + (\mathbf{E}^*_{0i} \times \mathbf{H}_{0r} + \mathbf{E}_{0r} \times \mathbf{H}^*_{0i}) \cdot \hat{\mathbf{q}}] \quad (4.267)$$

But the amplitude vectors of the incident and the reflected waves satisfy the Maxwell equations:

$$\omega\varepsilon_0 \varepsilon_1 \mathbf{E}_{0i} = -(\mathbf{k}_i \times \mathbf{H}_{0i}) \quad (4.268)$$

$$-(\mathbf{k}_r \times \mathbf{H}_{0r}) = \omega\varepsilon_0 \varepsilon_1 \mathbf{E}_{0r} \quad (4.269)$$

$$\omega\mu_0 \mu_1 \mathbf{H}_{0i} = \mathbf{k}_i \times \mathbf{E}_{0i} \quad (4.270)$$

$$\mathbf{k}_r \times \mathbf{E}_{0r} = \omega\mu_0 \mu_1 \mathbf{H}_{0r} \quad (4.271)$$

Here we assume that medium 1 is lossless, thus μ_1, ε_1, and $\mathbf{k}_i$ are real. Taking the dot product of Eq. (4.268) with Eq. (4.269), and the dot product of Eq. (4.270) with Eq. (4.271) and then adding the results, we get

$$(\mathbf{E}_{0i} \times \mathbf{H}_{0r} + \mathbf{E}_{0r} \times \mathbf{H}_{0i}) \cdot (\mathbf{k}_r - \mathbf{k}_i) = 0 \quad (4.272)$$

Similarly, taking the complex conjugate of Eqs. (4.268) and (4.270) and then repeating the above procedures, we obtain

$$(\mathbf{E}^*_{0i} \times \mathbf{H}_{0r} + \mathbf{E}_{0r} \times \mathbf{H}^*_{0i}) \cdot (\mathbf{k}_r - \mathbf{k}_i) = 0 \quad (4.273)$$

Since $\mathbf{k}_\alpha = \mathbf{b} + q_\alpha \hat{\mathbf{q}}$, where $\alpha = i$ or r, thus $\mathbf{k}_r - \mathbf{k}_i = (q_r - q_i)\hat{\mathbf{q}}$. Equations (4.272) and (4.273) become

$$(\mathbf{E}_{0i} \times \mathbf{H}_{0r} + \mathbf{E}_{0r} \times \mathbf{H}_{0i}) \cdot \hat{\mathbf{q}} = 0 \quad (4.274)$$

$$(\mathbf{E}^*_{0i} \times \mathbf{H}_{0r} + \mathbf{E}_{0r} \times \mathbf{H}^*_{0i}) \cdot \hat{\mathbf{q}} = 0 \quad (4.275)$$

Substitution of the above into Eq. (4.267) shows that the right-hand side of Eq. (4.266) vanishes identically. Thus we proved Eq. (4.262). It is noted that in justifying Eq. (4.267) being zero, the nature of the transmitted wave and the property of medium 2 did not figure. Thus the validity of Eq. (4.262) does not depend on the transmitted wave being uniform or nonuniform (total reflection) or on medium 2 being lossless or lossy.

By taking the time average on both sides of Eq. (4.262), we obtain

$$\langle \mathscr{P}_i \rangle \cdot \hat{\mathbf{q}} + \langle \mathscr{P}_r \rangle \cdot \hat{\mathbf{q}} = \langle \mathscr{P}_t \rangle \cdot \hat{\mathbf{q}} \quad (4.276)$$

which indicates that the normal component of the time-averaged energy flow across the interface must also be continuous.

The ratios

$$r = -\frac{\langle \mathscr{P}_r \rangle \cdot \hat{\mathbf{q}}}{\langle \mathscr{P}_i \rangle \cdot \hat{\mathbf{q}}} \quad (4.277)$$

and

$$t = \frac{\langle \mathscr{P}_t \rangle \cdot \hat{\mathbf{q}}}{\langle \mathscr{P}_i \rangle \cdot \hat{\mathbf{q}}} \quad (4.278)$$

are called the *reflectivity* (energy reflection coefficient) and *transmissivity* (energy transmission coefficient) respectively. It follows from Eq. (4.276) that

$$r + t = 1 \tag{4.279}$$

For uniform plane waves in lossless media, the time-averaged Poynting vector is directly proportional to the square of the amplitude of the electric field:

$$\begin{aligned}\langle \mathscr{P} \rangle &= \tfrac{1}{2}\,\text{Re}\,(\mathbf{E}_0 \times \mathbf{H}_0^*) \\ &= \frac{1}{2\omega\mu_0\mu}\,\text{Re}\,[\mathbf{E}_0 \times (\mathbf{k} \times \mathbf{E}_0^*)] \\ &= \frac{|\mathbf{E}_0|^2}{2\omega\mu_0\mu}\,\mathbf{k}\end{aligned}$$

Therefore, for $\mu_1 = \mu_2 = \mu$

$$\langle \mathscr{P}_\alpha \rangle \cdot \hat{\mathbf{q}} = \frac{|\mathbf{E}_{0\alpha}|^2}{2\omega\mu_0\mu_1}\,q_\alpha \tag{4.280}$$

where $\alpha = i, r,$ or t. According to Eqs. (4.277) and (4.278), we have

$$r = \frac{|\mathbf{E}_{0r}|^2}{|\mathbf{E}_{0i}|^2} = \frac{|B_\perp|^2 + |B_\parallel|^2}{|A_\perp|^2 + |A_\parallel|^2}$$

and

$$t = \frac{q_t|\mathbf{E}_{0t}|^2}{q_i|\mathbf{E}_{0i}|^2} = \frac{q_t(|C_\perp|^2 + |C_\parallel|^2)}{q_i(|A_\perp|^2 + |A_\parallel|^2)}$$

In the special case of perpendicular polarization: $\mathbf{E}_{0i} = A_\perp \mathbf{a}$, the reflectivity denoted by $r_\perp$ becomes

$$\begin{aligned}r_\perp &= -\frac{\langle \mathscr{P}_r^\perp \rangle \cdot \hat{\mathbf{q}}}{\langle \mathscr{P}_i^\perp \rangle \cdot \hat{\mathbf{q}}} = \frac{|B_\perp|^2}{|A_\perp|^2} \\ &= |\Gamma_{11}|^2 = \frac{\sin^2(\theta_t - \theta_i)}{\sin^2(\theta_t + \theta_i)}\end{aligned} \tag{4.281}$$

and the transmissivity denoted by $t_\perp$ is

$$\begin{aligned}t_\perp &= \frac{\langle \mathscr{P}_t^\perp \rangle \cdot \hat{\mathbf{q}}}{\langle \mathscr{P}_i^\perp \rangle \cdot \hat{\mathbf{q}}} = \frac{q_t|C_\perp|^2}{q_i|A_\perp|^2} \\ &= \frac{q_t}{q_i}|T_{11}|^2 = \frac{\sin 2\theta_i \sin 2\theta_t}{\sin^2(\theta_i + \theta_t)}\end{aligned} \tag{4.282}$$

We may easily verify that

$$r_\perp + t_\perp = |\Gamma_{11}|^2 + \frac{q_t}{q_i}|T_{11}|^2 = 1 \tag{4.283}$$

Similarly, in the case of parallel polarization: $\mathbf{E}_{0i} = A_\parallel(\hat{\mathbf{k}}_i \times \mathbf{a})$, the reflectivity and the transmissivity denoted respectively by $r_{||}$ and $t_{||}$ are

$$r_{||} = -\frac{\langle \mathscr{P}_r^{||} \rangle \cdot \hat{\mathbf{q}}}{\langle \mathscr{P}_i^{||} \rangle \cdot \hat{\mathbf{q}}} = \frac{|B_{||}|^2}{|A_{||}|^2}$$

$$= |\Gamma_{22}|^2 = \frac{\tan^2(\theta_i - \theta_t)}{\tan^2(\theta_i + \theta_t)} \tag{4.284}$$

and
$$t_{||} = \frac{\langle \mathscr{P}_t^{||} \rangle \cdot \hat{\mathbf{q}}}{\langle \mathscr{P}_i^{||} \rangle \cdot \hat{\mathbf{q}}} = \frac{q_t |C_{||}|^2}{q_i |A_{||}|^2}$$

$$= \frac{q_t}{q_i} |T_{22}|^2 = \frac{\sin 2\theta_i \sin 2\theta_t}{\sin^2(\theta_i + \theta_t) \cos^2(\theta_i - \theta_t)} \tag{4.285}$$

Again we may show that

$$r_{||} + t_{||} = |\Gamma_{22}|^2 + \frac{q_t}{q_i} |T_{22}|^2 = 1 \tag{4.286}$$

In passing, note that we may use Eqs. (4.283) and (4.286) at the interface and the boundary condition (4.162) on the electric field as an alternative approach to derive the amplitude reflection and transmission coefficients Γ_{11}, Γ_{22}, T_{11}, and T_{22}.

For normal incidence, from Eqs. (4.277), (4.278), (4.249), and (4.250), we have

$$r = \frac{(\sqrt{\varepsilon_1} - \sqrt{\varepsilon_2})^2}{(\sqrt{\varepsilon_1} + \sqrt{\varepsilon_2})^2} \tag{4.287}$$

and
$$t = \frac{4\sqrt{\varepsilon_1 \varepsilon_2}}{(\sqrt{\varepsilon_1} + \sqrt{\varepsilon_2})^2} \tag{4.288}$$

An examination of Eqs. (4.281) to (4.285) shows that the denominators are finite, except when $\theta_i + \theta_t = \pi/2$. Then $\tan(\theta_i + \theta_t) \to \infty$ and hence $r_{||} = 0$ or $\Gamma_{22} = 0$ thus $B_{||} = 0$. In this case, according to Eq. (4.193), $\hat{\mathbf{k}}_r \cdot \hat{\mathbf{k}}_t = 0$, that is, the wave vectors of the reflected and the transmitted waves are perpendicular to each other as illustrated in Figs. 4.8 and 4.9. Denoting this angle of incidence by θ_B and noting that $\sin \theta_t = \sin(\pi/2 - \theta_B) = \cos \theta_B$, we obtain, from Snell's law (4.155)

$$\tan \theta_B = \frac{k_t}{k_i} = \sqrt{\frac{\varepsilon_2}{\varepsilon_1}} \tag{4.289}$$

The angle θ_B that satisfies Eq. (4.289) is known as the *Brewster angle*. It is also called the *polarizing angle*, since an unpolarized wave is incident on an interface at this angle, the reflected wave is linearly polarized with the electric field vector $\mathbf{E}_{0r} = B_\perp \mathbf{a}$ normal to the plane of incidence. This principle is sometimes used in optics to polarize natural light, although it is not very efficient in practice. On the other hand, by measuring the Brewster angle experimentally, we may determine the ratio of the dielectric constants of the two media.

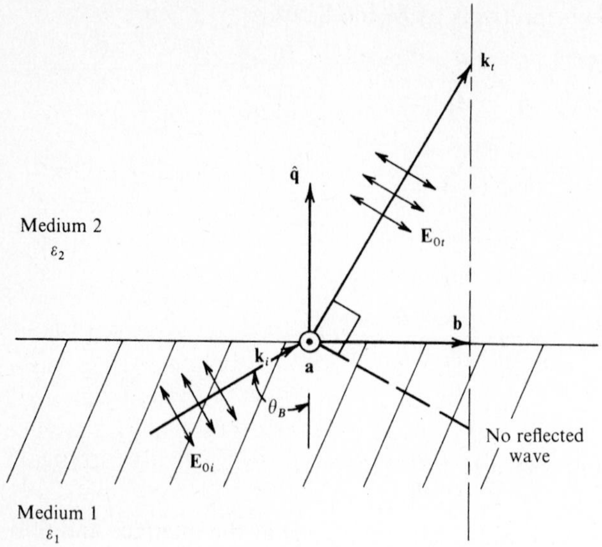

Figure 4.8 A linearly polarized wave with $\mathbf{E}_{0i} = A_{\|}(\hat{\mathbf{k}}_i \times \mathbf{a})$, incident at θ_B, has no reflected wave.

From Eq. (4.190), we see that a real ratio $A_{\|}/A_{\perp}$ gives rise to the real ratios $B_{\|}/B_{\perp}$ and $C_{\|}/C_{\perp}$. That is, if the incident wave is linearly polarized, the reflected and transmitted waves will also be linearly polarized. Also from Eq. (4.182), we see that $A_{\perp} = 0$ implies $B_{\perp} = C_{\perp} = 0$ and from Eq. (4.190) that $A_{\|} = 0$ implies

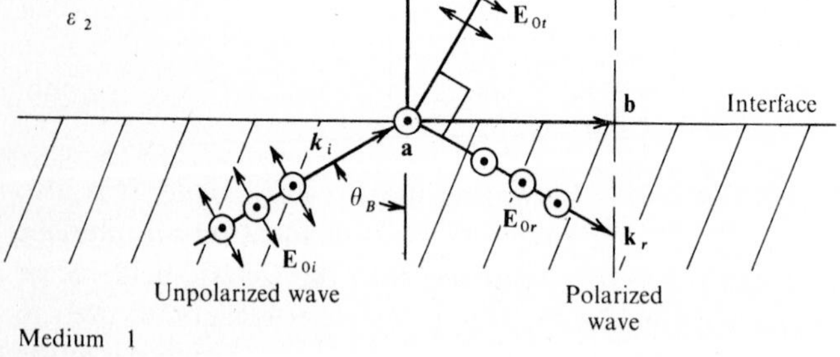

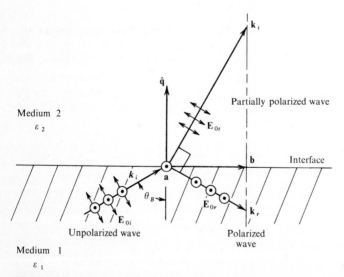

Figure 4.9 An unpolarized wave incident at θ_B, gives a polarized reflected wave and a partially polarized transmitted wave.

$B_{\|} = C_{\|} = 0$. In other words, if $\mathbf{E}_0$ of the incident wave is linearly polarized and parallel (or perpendicular) to the plane of incidence, so are the reflected and transmitted waves. The directions of oscillation in the reflected and transmitted waves as compared to the incident wave are, however, rotated in opposite directions. This can be seen from the following.

The plane containing the electric field vector and the direction of wave propagation is called the *plane of polarization*. Generally speaking, the orientation of the plane of polarization of the incident wave with respect to the plane of incidence is different from those of the reflected and transmitted waves. For a linearly polarized wave, the angle between the plane of polarization and the plane of incidence is called the *azimuthal angle*. This angle may lie in the range from $-\pi/2$ to $\pi/2$, and is defined as positive, if we rotate the plane of polarization toward the plane of incidence leading to the direction of wave propagation according to the right-hand rule. Let α_i, α_r, and α_t be the azimuthal angles of the incident, the reflected, and the transmitted waves respectively. Evidently [see Eqs. (4.166) to (4.168)],

$$\tan \alpha_i = \frac{A_\perp}{A_\|}$$

$$\tan \alpha_r = \frac{B_\perp}{B_\|} \tag{4.290}$$

and

$$\tan \alpha_t = \frac{C_\perp}{C_\|}$$

Using the Fresnel equations (4.190), we obtain

$$\tan \alpha_r = \frac{\hat{\mathbf{k}}_i \cdot \hat{\mathbf{k}}_t}{\hat{\mathbf{k}}_r \cdot \hat{\mathbf{k}}_t} \tan \alpha_i$$

$$\tan \alpha_t = (\hat{\mathbf{k}}_i \cdot \hat{\mathbf{k}}_t) \tan \alpha_i \tag{4.291}$$

or, from Eq. (4.198)

$$\tan \alpha_r = -\frac{\cos(\theta_i - \theta_t)}{\cos(\theta_i + \theta_t)} \tan \alpha_i \tag{4.292}$$

$$\tan \alpha_t = \cos(\theta_i - \theta_t) \tan \alpha_i \tag{4.293}$$

Since $0 \leq \theta_i \leq \pi/2$ and $0 < \theta_t < \pi/2$,

$$|\tan \alpha_r| \geq |\tan \alpha_i| \tag{4.294}$$

$$|\tan \alpha_t| \leq |\tan \alpha_i| \tag{4.295}$$

In Eq. (4.294) the equality sign holds only for normal ($\theta_i = \theta_t = 0$) or grazing incidence ($\theta_i = \pi/2$); in Eq. (4.295) it holds only for normal incidence. The two inequalities imply that the plane of polarization of the reflected wave rotates away from the plane of incidence whereas the plane of polarization of the transmitted wave rotates toward the plane of incidence.

As an application, the measurement of azimuthal angle of the reflected wave provides a way to determine the relative index of refraction. In fact, from Eq. (4.292), we have

$$\frac{\tan \alpha_r}{\tan \alpha_i} = \frac{\cos \theta_i \cos \theta_t + \sin \theta_i \sin \theta_t}{\sin \theta_i \sin \theta_t - \cos \theta_i \cos \theta_t} \tag{4.296}$$

Combination of Eq. (4.296) with Snell's law yields

$$\left(\frac{k_t}{k_i}\right)^2 = \sin^2 \theta_i \left[1 + \tan^2 \theta_i \left(\frac{\tan \alpha_r - \tan \alpha_i}{\tan \alpha_r + \tan \alpha_i}\right)^2\right] \tag{4.297}$$

This can be simplified greatly if we choose $\theta_i = \alpha_i = \pi/4$. In this case Eq. (4.297) becomes

$$\left(\frac{k_t}{k_i}\right)^2 = \frac{1}{2 \sin^2\left(\alpha_r + \dfrac{\pi}{4}\right)}$$

or

$$n_{21} = \frac{1}{\sqrt{2}} \csc\left(\alpha_r + \frac{\pi}{4}\right) \tag{4.298}$$

Thus, by measuring the azimuthal angle of the reflected wave we may determine the relative index of refraction of the second medium with respect to the first.

4.9. TOTAL REFLECTION

For two lossless isotropic media having the same permeability, we may write the Snell law (4.155) as

$$\sin \theta_t = \sqrt{\frac{\varepsilon_1}{\varepsilon_2}} \sin \theta_i$$

If $\varepsilon_2 > \varepsilon_1$, it follows that there corresponds to every angle of incidence θ_i a real angle of transmission θ_t. If however, $\varepsilon_1 > \varepsilon_2$, as is the case when a wave emerges from a liquid or solid dielectric into air, then θ_t is real only when

$$\sin \theta_i < \sqrt{\frac{\varepsilon_2}{\varepsilon_1}} \tag{4.299}$$

In both cases we have the ordinary reflection and transmission problem which was discussed in the previous section. On the other hand, when

$$\sin \theta_i \geq \sqrt{\frac{\varepsilon_2}{\varepsilon_1}} \tag{4.300}$$

the Snell law can be satisfied only when the complex values of θ_t and the phenomena of total reflection, which will be our next topic, occur.

The instantaneous electric field intensity of the transmitted wave in medium 2 is

$$\mathscr{E}_t = \text{Re}\,(\mathbf{E}_{0t}\, e^{i(\mathbf{k}_t \cdot \mathbf{r} - \omega t)}) \tag{4.301}$$

where the wave vector $\mathbf{k}_t$ [see Eqs. (4.151) and (4.152)] may generally be expressed as

$$\mathbf{k}_t = \mathbf{b} + q_t \hat{\mathbf{q}} \tag{4.302}$$

where
$$\mathbf{b} = \hat{\mathbf{q}} \times \mathbf{a}$$
$$q_t = \mathbf{k}_t \cdot \hat{\mathbf{q}} = \pm k_t \sqrt{1 - \frac{\varepsilon_1}{\varepsilon_2} \sin^2 \theta_i} \tag{4.303}$$

Hence when $\varepsilon_2 > \varepsilon_1$ or when the angle of incidence θ_i satisfies the inequality (4.299) in the case of $\varepsilon_1 > \varepsilon_2$, we see that q_t is real and thus $\mathbf{k}_t$ is a real vector. The transmitted wave, expressed by Eq. (4.301), propagates along the direction $\hat{\mathbf{k}}_t$ with the phase velocity ω/k_t and there is nothing new (see Fig. 4.10).

The total reflection occurs when the angle of incidence satisfies the inequality (4.300). This is possible for a wave that is passing from a medium with a larger permittivity to a medium with smaller permittivity when the angle of incidence exceeds the critical angle of incidence θ_c defined by

$$\theta_c = \sin^{-1} \sqrt{\frac{\varepsilon_2}{\varepsilon_1}} \tag{4.304}$$

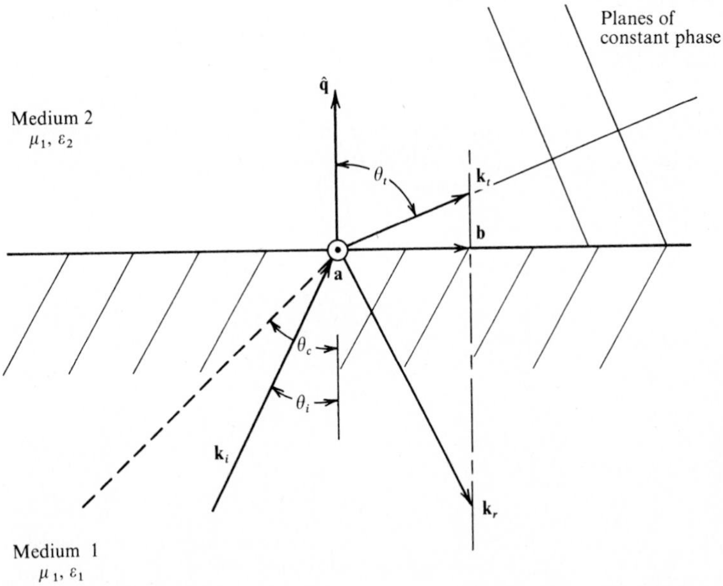

Figure 4.10 Oblique incidence inside the critical angle on an interface between two lossless media. $\varepsilon_1 > \varepsilon_2$ and $\theta_i < \theta_c$ imply $\theta_t > \theta_i$ and $\mathbf{k}_t = \mathbf{b} + q_t\hat{\mathbf{q}}$ = a real vector.

where $\varepsilon_1 > \varepsilon_2$. It is obvious that when $\theta_i = \theta_c$, $q_t = 0$, and $\mathbf{k}_t = \mathbf{b}$ and thus implies that the transmitted wave is propagating parallel to the interface. This conclusion is also confirmed by Eq. (4.301) which for $\mathbf{k}_t = \mathbf{b}$ becomes

$$\mathscr{E}_t = \text{Re}\,(\mathbf{E}_{0t}\, e^{i(\mathbf{r}\cdot\mathbf{b} - \omega t)}) \tag{4.305}$$

and states that the transmitted wave is a uniform plane wave with the planes of constant phase perpendicular to $\mathbf{b}$ or the interface (see Fig. 4.11). The time-averaged Poynting vector is then

$$\langle \mathscr{P}_t \rangle = \frac{|\mathbf{E}_{0t}|^2}{2\eta_t}\,\hat{\mathbf{b}} \tag{4.306}$$

which shows that there is no time-averaged power flowing into medium 2. When the angle of incidence exceeds the critical angle θ_c, q_t becomes purely imaginary. Since $|\mathbf{a}| = k_i \sin\theta_i$, condition (4.300) for total reflection may also be expressed as

$$|\mathbf{a}| \geq k_t \tag{4.307}$$

The wave vector of the transmitted wave, in this case, may be written as

$$\mathbf{k}_t = \mathbf{k}_{1t} + i\mathbf{k}_{2t} \tag{4.308}$$

where

$$\mathbf{k}_{1t} = \mathbf{b} = \frac{\omega}{c}\sqrt{\mu_1\varepsilon_1}\,\sin\theta_i\,\hat{\mathbf{b}} \tag{4.309}$$

and

$$\mathbf{k}_{2t} = +\sqrt{\mathbf{a}^2 - \mathbf{k}_t^2}\;\hat{\mathbf{q}} = \frac{\omega}{c}\sqrt{\mu_1\varepsilon_2}\sqrt{\frac{\varepsilon_1}{\varepsilon_2}\sin^2\theta_i - 1}\;\hat{\mathbf{q}}$$

are two real vectors. The choice of positive sign in front of the square root is dictated by the condition that the field (4.301) shall remain finite as $\mathbf{r}\cdot\hat{\mathbf{q}} \to \infty$. We

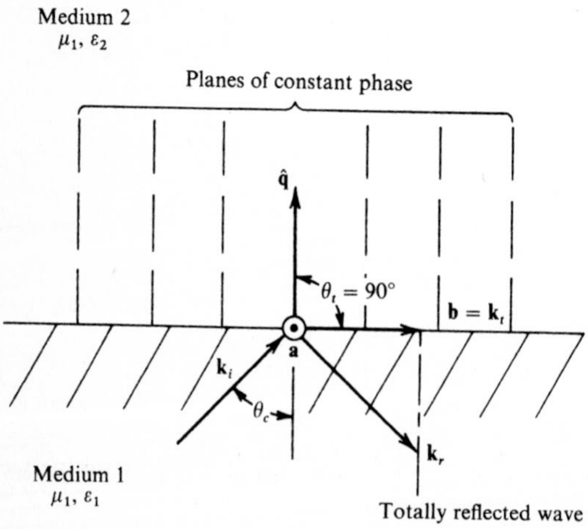

Figure 4.11 Oblique incidence at the critical angle on an interface between two lossless media. $\varepsilon_1 > \varepsilon_2$ and $\theta_i = \theta_c$ imply $\theta_t = 90°$ and $\mathbf{k}_t = \mathbf{b} =$ a real vector.

note that the wave vector $\mathbf{k}_t$ satisfies the dispersion equation $\mathbf{k}_t^2 = k_0^2 \mu_1 \varepsilon_2$ or

$$\begin{aligned} k_{1t}^2 - k_{2t}^2 &= k_0^2 \mu_1 \varepsilon_2 \\ \mathbf{k}_{1t} \cdot \mathbf{k}_{2t} &= 0 \end{aligned} \tag{4.310}$$

The field vector (4.301) now takes the form

$$\mathscr{E}_t = \text{Re}\left[\left(\mathbf{E}_{0t}\, e^{-k_{2t}(\mathbf{r}\cdot\hat{\mathbf{q}})}\right) e^{i[k_{1t}(\mathbf{r}\cdot\hat{\mathbf{b}}) - \omega t]}\right] \tag{4.311}$$

From this expression, it is apparent that the transmitted wave in the lossless medium 2 is a nonuniform plane wave. The planes of constant phase ($\mathbf{r} \cdot \hat{\mathbf{b}} =$ const) are perpendicular to the planes of constant amplitude ($\mathbf{r} \cdot \hat{\mathbf{q}} =$ const) as illustrated in Fig. 4.12. The wave is attenuated in the direction normal to the interface at a rate determined not only by the electric and magnetic properties of the medium, but also by the frequency and by the angle of incidence. It propagates parallel to the interface with a phase velocity

$$v_p = \frac{\omega}{k_{1t}} = \frac{c}{\sqrt{\mu_1 \varepsilon_1}\, \sin \theta_i} \tag{4.312}$$

explicitly dependent upon the angle of incidence. Waves of this type which propagate at a reduced velocity parallel to a surface and are attenuated normally to the surface are called *surface waves.*

The amplitudes of the reflected and transmitted field vectors are next determined from the Fresnel equations. We note that the Fresnel equations derived in

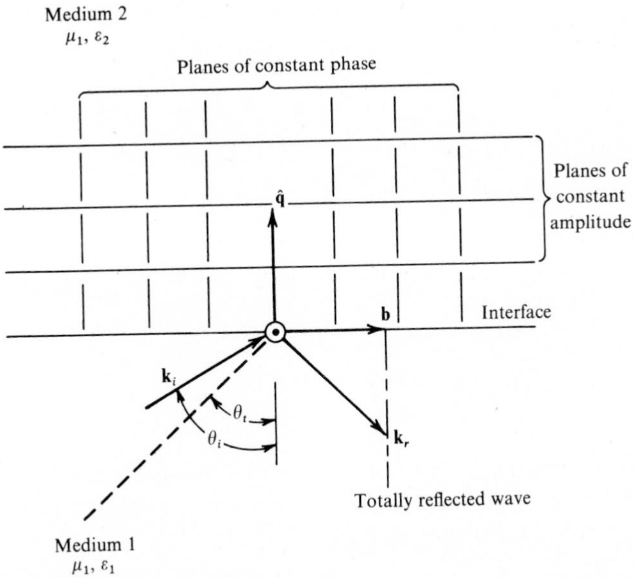

Figure 4.12 Oblique incidence beyond the critical angle on an interface between two lossless media. $\varepsilon_1 > \varepsilon_2$ and $\theta_i > \theta_c$ imply $\mathbf{k}_t = \mathbf{b} + ik_{2t}\hat{\mathbf{q}}$ = a complex vector.

the preceding sections are still valid as long as their derivations do not assume the wave vectors to be real. We should also note that in contrast to the generally accepted approach, complex angles will not be used since we have adopted a more convenient approach of complex wave vectors. Substituting Eq. (4.204) into Eq. (4.182) and noting that $q_t = ik_{2t}$, we obtain

$$\frac{A_\perp}{q_i + ik_{2t}} = \frac{B_\perp}{q_i - ik_{2t}} = \frac{C_\perp}{2q_i} \tag{4.313}$$

Similarly, from Eq. (4.190)

$$\frac{A_\parallel/A_\perp}{\mathbf{a}^2 + ik_{2t}q_i} = \frac{B_\parallel/B_\perp}{\mathbf{a}^2 - ik_{2t}q_i} = \frac{C_\parallel/C_\perp}{k_i k_t} \tag{4.314}$$

From Eq. (4.314), we see that for the real ratio $A_\parallel/A_\perp$, the ratio $B_\parallel/B_\perp$ is complex; that is, a linearly polarized incident wave, in general, gives rise to an elliptically polarized reflected wave on total reflection. Also $A_\perp = 0$ (or $A_\parallel = 0$) implies $B_\perp = 0$ (or $B_\parallel = 0$), i.e., if $\mathbf{E}_0$ of the incident wave lies on the plane of incidence (or perpendicular to the plane of incidence), so does the reflected wave. Moreover, from Eqs. (4.313) and (4.314), we obtain

$$\begin{aligned} |B_\perp| &= |A_\perp| \\ |B_\parallel| &= |A_\parallel| \end{aligned} \tag{4.315}$$

that is, for perpendicular as well as for parallel polarizations, the amplitude of the reflected wave is equal to the amplitude of the incident wave. Since

$$\begin{aligned} |\mathbf{E}_{0i}|^2 &= \mathbf{E}_{0i} \cdot \mathbf{E}_{0i}^* \\ &= [A_\perp \mathbf{a} + A_\parallel(\hat{\mathbf{k}}_i \times \mathbf{a})] \cdot [A_\perp^* \mathbf{a} + A_\parallel^*(\hat{\mathbf{k}}_i \times \mathbf{a})] \\ &= (|A_\perp|^2 + |A_\parallel|^2)\mathbf{a}^2 \end{aligned} \tag{4.316}$$

and similarly

$$|\mathbf{E}_{0r}|^2 = (|B_\perp|^2 + |B_\parallel|^2)\mathbf{a}^2$$

Thus Eq. (4.315) implies that

$$|\mathbf{E}_{0r}|^2 = |\mathbf{E}_{0i}|^2 \tag{4.317}$$

The reflectivity defined in Eq. (4.277) becomes

$$r = -\frac{\langle \mathscr{P}_r \rangle \cdot \hat{\mathbf{q}}}{\langle \mathscr{P}_i \rangle \cdot \hat{\mathbf{q}}} = \frac{|\mathbf{E}_{0r}|^2}{|\mathbf{E}_{0i}|^2} = 1 \tag{4.318}$$

that is, the time-averaged power carried by the incident wave is totally reflected back into medium 1. For this reason, the phenomenon is known as *total reflection* or *total internal reflection*. From Eq. (4.318), we have

$$(\langle \mathscr{P}_i \rangle + \langle \mathscr{P}_r \rangle) \cdot \hat{\mathbf{q}} = 0 \tag{4.319}$$

Comparing this with Eq. (4.276), we conclude that

$$\langle \mathscr{P}_t \rangle \cdot \hat{\mathbf{q}} = 0 \tag{4.320}$$

Therefore, no transfer of energy to medium 2 of lesser permittivity takes place. The result may also be proved by direct computation. In fact, according to Eq. (4.63), the time-averaged Poynting vector of the transmitted nonuniform plane wave is

$$\langle \mathscr{P}_t \rangle = \frac{e^{-2\mathbf{k}_{2t}\cdot\mathbf{r}}}{4\omega\mu_0\mu_1}\left[|\mathbf{E}_{0t}|^2(\mathbf{k}_t + \mathbf{k}_t^*) - (\mathbf{k}_t - \mathbf{k}_t^*) \times (\mathbf{E}_{0t} \times \mathbf{E}_{0t}^*)\right] \tag{4.321}$$

But from Eq. (4.308)

$$\begin{aligned} \mathbf{k}_t + \mathbf{k}_t^* &= 2\mathbf{b} \\ \mathbf{k}_t - \mathbf{k}_t^* &= i2k_{2t}\hat{\mathbf{q}} \end{aligned} \tag{4.322}$$

thus, substitution of Eq. (4.322) into Eq. (4.321) proves that $\langle \mathscr{P}_t \rangle \cdot \hat{\mathbf{q}} = 0$.

In summary, total reflection can take place at the interface between two lossless isotropic media. When a uniform plane wave is incident from a medium with a larger permittivity to a medium with a smaller permittivity at an angle equal to or exceeding the critical angle, the wave will be totally reflected and will also be accompanied by a surface wave in the medium with smaller permittivity. In practice, this phenomenon is used in the design of optical communication devices, such as optical fibers.

4.10 TRANSMISSION AND REFLECTION AT THE SURFACE OF A LOSSY MEDIUM

The Snell law and the Fresnel equations derived in previous sections for two lossless isotropic media are still valid when one or both media become lossy (conducting). However, some modifications are needed and we shall consider them here.

As shown in Fig. 4.13, we assume that the incident wave is in medium 1 which is still lossless and isotropic, and that medium 2 is isotropic but lossy with finite conductivity σ. The wave vector $\mathbf{k}_t$ of the transmitted wave is now complex (see Sec. 4.3) and can be written as

$$\begin{aligned} \mathbf{k}_t &= \mathbf{b} + q_t\hat{\mathbf{q}} \\ &= \mathbf{k}_{1t} + i\mathbf{k}_{2t} \end{aligned} \tag{4.323}$$

where $q_t = \mathbf{k}_t \cdot \hat{\mathbf{q}}$, $\mathbf{k}_{1t}$ and $\mathbf{k}_{2t}$ are two real vectors. From Eqs. (3.100) and (3.101)

$$\begin{aligned} \mathbf{k}_i \times \hat{\mathbf{q}} = \mathbf{a} &= \mathbf{k}_t \times \hat{\mathbf{q}} \\ &= (\mathbf{k}_{1t} + i\mathbf{k}_{2t}) \times \hat{\mathbf{q}} \end{aligned} \tag{4.324}$$

and the fact that **a** is a real vector, it follows that

$$\mathbf{k}_{2t} \times \hat{\mathbf{q}} = \mathbf{0}$$

or

$$\mathbf{k}_{2t} = q_2 \hat{\mathbf{q}} \tag{4.325}$$

which shows that the attenuation vector $\mathbf{k}_{2t}$ is always normal to the interface.

Using Eq. (4.323), we may write

$$q_t = \mathbf{k}_t \cdot \hat{\mathbf{q}} = q_1 + iq_2 \tag{4.326}$$

where

$$\begin{aligned} q_1 &= \mathbf{k}_{1t} \cdot \hat{\mathbf{q}} \\ q_2 &= \mathbf{k}_{2t} \cdot \hat{\mathbf{q}} \end{aligned} \tag{4.327}$$

Hence q_t is a complex number whereas it is real or imaginary as in the case of ordinary or total reflection from two lossless isotropic media. The real and imaginary parts of q_t can be determined from the dispersion equation (4.89):

$$\mathbf{k}_t^2 = k_0^2 \mu_1 \varepsilon_2 (1 + i\tau) \tag{4.328}$$

Indeed, substituting Eq. (4.323) into Eq. (4.328), we get

$$q_t = q_1 + iq_2 = \sqrt{(k_0^2 \mu_1 \varepsilon_2 - \mathbf{a}^2) + ik_0^2 \mu_1 \varepsilon_2 \tau} \tag{4.329}$$

Equating the real and imaginary parts of the square of Eq. (4.329) separately and solving the resulted equations, we find

$$\begin{aligned} q_1 &= \frac{1}{\sqrt{2}} [\sqrt{k_0^4 \mu_1^2 \varepsilon_2^2 \tau^2 + (k_0^2 \mu_1 \varepsilon_2 - \mathbf{a}^2)^2} + (k_0^2 \mu_1 \varepsilon_2 - \mathbf{a}^2)]^{1/2} \\ q_2 &= \frac{1}{\sqrt{2}} [\sqrt{k_0^4 \mu_1^2 \varepsilon_2^2 \tau^2 + (k_0^2 \mu_1 \varepsilon_2 - \mathbf{a}^2)^2} - (k_0^2 \mu_1 \varepsilon_2 - \mathbf{a}^2)]^{1/2} \end{aligned} \tag{4.330}$$

Thus the boundary conditions together with the dispersion equation uniquely determine the wave vector $\mathbf{k}_t$ of the transmitted wave. Rewriting Eq. (4.323) as

$$\mathbf{k}_t = (\mathbf{b} + q_1 \hat{\mathbf{q}}) + iq_2 \hat{\mathbf{q}} \tag{4.331}$$

we obtain the electric field intensity of the transmitted wave in lossy medium 2

$$\mathscr{E}_t = \mathrm{Re}\{(\mathbf{E}_{0t} e^{-q_2(\mathbf{r}\cdot\hat{\mathbf{q}})}) \exp i[(\mathbf{b} + q_1 \hat{\mathbf{q}}) \cdot \mathbf{r} - \omega t]\} \tag{4.332}$$

As in the case of total reflection, the transmitted wave in a lossy medium is nonuniform. The planes of constant amplitude ($\mathbf{r} \cdot \hat{\mathbf{q}} =$ const) are parallel to the interface, while the planes of constant phase defined by

$$\mathbf{r} \cdot (\mathbf{b} + q_1 \hat{\mathbf{q}}) = \text{const} \tag{4.333}$$

are normal to the direction $\mathbf{b} + q_1 \hat{\mathbf{q}}$ as shown in Fig. 4.13. The only case of a uniform plane wave in a lossy medium occurs when the wave is incident normal to the interface. In fact, a plane wave is uniform if $\mathbf{k}_t \times \mathbf{k}_t^* = 0$ (see Prob. 4.2). Combination of this condition with Eq. (4.325) shows that vectors $\mathbf{k}_{1t}$ and $\mathbf{k}_{2t}$ and thus $\mathbf{k}_t$ are parallel to $\hat{\mathbf{q}}$ which in turn implies that $\mathbf{k}_i$ and $\mathbf{k}_r$ are also parallel to $\hat{\mathbf{q}}$.

In other words, in the case of normal incidence, all three waves—the incident, the reflected, and the transmitted—are uniform plane waves.

The amplitudes of the waves are related by the Fresnel equations (4.182) and (4.190). Noting the relation (4.204) and complex nature of $q_t = q_1 + iq_2$, we repeat them below

$$\frac{A_\perp}{q_i + q_t} = \frac{B_\perp}{q_i - q_t} = \frac{C_\perp}{2q_i} \tag{4.334}$$

and

$$\frac{A_\parallel/A_\perp}{\mathbf{a}^2 + q_i q_t} = \frac{B_\parallel/B_\perp}{\mathbf{a}^2 - q_i q_t} = \frac{C_\parallel/C_\perp}{k_i k_t} \tag{4.335}$$

Here we note again as in the case of total reflection that $A_\perp = 0$ (or $A_\parallel = 0$) implies $B_\perp = C_\perp = 0$ (or $B_\parallel = C_\parallel = 0$), i.e., if $\mathbf{E}_0$ of the incident wave lies on the plane of incidence (or perpendicular to the plane of incidence), so do the reflected and the transmitted waves. Moreover, for a real ratio of $A_\parallel/A_\perp$, the ratios of $B_\parallel/B_\perp$ and $C_\parallel/C_\perp$ are complex. In other words, if a linearly polarized wave whose direction of polarization is neither normal nor parallel to the plane of incidence, is incident on a conducting surface, the reflected as well as transmitted waves will be elliptically polarized.

Unlike the case of total reflection, the normal component of the time-averaged Poynting vector in the lossy medium 2 is now different from zero. According to Eq. (4.137), we easily find

$$\langle \mathscr{P}_t \rangle \cdot \hat{\mathbf{q}} = \frac{e^{-2\mathbf{k}_{2t}\cdot\mathbf{r}}}{2\omega\mu_0\mu_1} |\mathbf{E}_{0t}|^2 q_1 \tag{4.336}$$

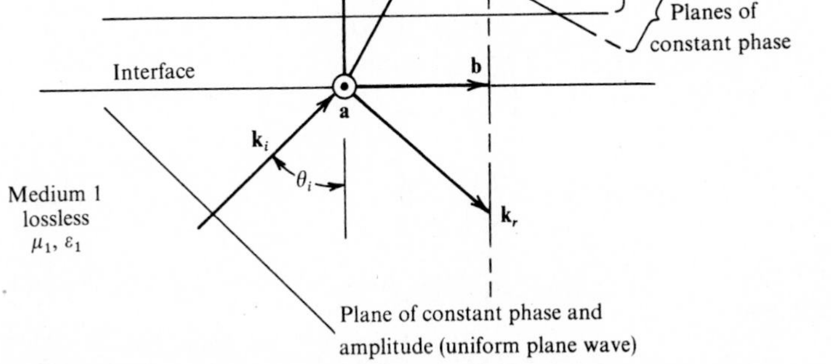

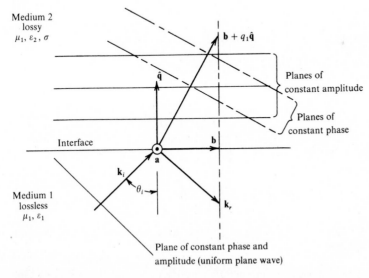

Figure 4.13 Reflection and transmission at a lossless-lossy plane interface.

From Eqs. (4.334) and (4.335), the reflectivities $r_\perp$ and $r_\parallel$ for perpendicular and parallel polarizations may be written as

$$r_\perp = -\frac{\langle \mathscr{P}_r^\perp \rangle \cdot \hat{\mathbf{q}}}{\langle \mathscr{P}_i^\perp \rangle \cdot \hat{\mathbf{q}}} = \frac{|B_\perp|^2}{|A_\perp|^2} = \frac{|q_i - q_t|^2}{|q_i + q_t|^2} \tag{4.337}$$

and

$$r_\parallel = -\frac{\langle \mathscr{P}_r^\parallel \rangle \cdot \hat{\mathbf{q}}}{\langle \mathscr{P}_i^\parallel \rangle \cdot \hat{\mathbf{q}}} = \frac{|B_\parallel|^2}{|A_\parallel|^2} = \frac{|\mathbf{a}^2 - q_i q_t|^2 |q_i - q_t|^2}{|\mathbf{a}^2 + q_i q_t|^2 |q_i + q_t|^2} \tag{4.338}$$

respectively. It is interesting to note that when $\theta_i = \pi/4$, thus $\mathbf{a}^2 = q_i^2$, we have

$$r_\parallel = r_\perp^2 \tag{4.339}$$

The fulfillment of this condition for all azimuthal angles of incidence can serve to guarantee that the reflecting medium is isotropic.

Measurements of the reflectivities for perpendicular and parallel polarizations provide a convenient and accurate way of determining the parameters ε_2 and τ for a conducting medium. To see this, from Eqs. (4.337) and (4.338), we may write

$$\frac{1 - r_\perp}{1 + r_\perp} = \frac{q_i(q_t + q_t^*)}{q_i^2 + |q_t|^2} = \frac{2q_i q_1}{q_i^2 + |q_t|^2} \tag{4.340}$$

$$\frac{r_\parallel}{r_\perp} = \frac{|\mathbf{a}^2 - q_i q_t|^2}{|\mathbf{a}^2 + q_i q_t|^2} \tag{4.341}$$

and

$$\frac{r_\perp - r_\parallel}{r_\perp + r_\parallel} = \frac{\mathbf{a}^2 q_i(q_t + q_t^*)}{\mathbf{a}^4 + q_i^2|q_t|^2} = \frac{2\mathbf{a}^2 q_i q_1}{\mathbf{a}^4 + q_i^2|q_t|^2} \tag{4.342}$$

Dividing Eq. (4.340) by Eq. (4.342), we obtain

$$\frac{\mathbf{a}^4 + q_i^2|q_t|^2}{q_i^2 + |q_t|^2} = \frac{(1 - r_\perp)(r_\perp + r_\parallel)}{(1 + r_\perp)(r_\perp - r_\parallel)}\,\mathbf{a}^2 = s \tag{4.343}$$

From this and Eq. (4.340), we find

$$|q_t|^2 = \frac{sq_i^2 - \mathbf{a}^4}{q_i^2 - s} \tag{4.344}$$

and

$$q_1 = \frac{(q_i^4 - \mathbf{a}^4)(1 - r_\perp)}{2q_i(q_i^2 - s)(1 + r_\perp)} \tag{4.345}$$

Now if the conducting medium is nonmagnetic: $\mu_1 = 1$, from Eq. (4.329) and $|q_t|^2 = q_1^2 + q_2^2$, we have

$$\varepsilon_2 = \frac{1}{k_0^2}(\mathbf{a}^2 + 2q_1^2 - |q_t|^2) \tag{4.346}$$

and

$$\varepsilon_2 \tau = \frac{2q_1}{k_0^2} \sqrt{|q_t|^2 - q_1^2} \tag{4.347}$$

Thus, to determine the complex dielectric constant $\varepsilon_c = \varepsilon_2 (1 + i\tau)$, we need only to measure the reflectivities $r_\perp$ and $r_\parallel$ for any angle of incidence different from $\pi/4$. The latter restriction is related to the fact that, when $\theta_i = \pi/4$, $r_\parallel = r_\perp^2$ and Eq. (4.342) coincides with Eq. (4.340).

PROBLEMS

4.1 A linearly polarized wave has a complex wave vector $\mathbf{k} = \mathbf{k}_1 + i\mathbf{k}_2$. Determine the relative orientation of the field vectors $\mathbf{E}_0$, $\mathbf{H}_1$, $\mathbf{H}_2$, $\mathbf{k}_1$, and $\mathbf{k}_2$, where $\mathbf{H}_1$ and $\mathbf{H}_2$ are respectively the real and imaginary parts of the complex amplitude vector $\mathbf{H}_0$ of the magnetic field. Find the locus of the tip of the vector $\mathscr{H}$ at a given point in space.

4.2 By definition, a plane wave is uniform if its amplitude vector is constant over the plane of constant phase. Show that for a complex wave vector $\mathbf{k}$, if $\mathbf{k} \times \mathbf{k}^* = \mathbf{0}$, the plane wave is uniform, otherwise, it is nonuniform.

4.3 A uniform plane wave propagating in free space is characterized by the complex electric vector

$$\mathbf{E} = (i3\hat{\mathbf{x}} + 5\hat{\mathbf{y}} - i4\hat{\mathbf{z}})e^{i0.2\pi(4x+3z)}$$

(*a*) Find the equation of the planes of constant phase.
(*b*) What is the polarization of the wave in the planes of constant phase?
(c) Find the magnetic field vector associated with the given **E**.
(*d*) Determine the electric field, the magnetic field, and the Poynting vector in the time domain.

4.4 A plane wave propagating in free space is characterized by the complex electric vector

$$\mathbf{E} = [\hat{\mathbf{x}} + E_{0y}\hat{\mathbf{y}} + (2 + i5)\hat{\mathbf{z}}]e^{i2.3(-0.6x+0.8y-i0.6z)}$$

(*a*) Is the plane wave uniform or nonuniform?
(*b*) Determine the frequency and wavelength.
(*c*) Find the component E_{0y} and the complex magnetic amplitude vector $\mathbf{H}_0$.
(*d*) Determine the polarization of the wave and its sense of rotation.

4.5 Verify Eqs. (4.21) and (4.24).

4.6 Show that if a wave of any polarization $\mathbf{E}_0$ is decomposed into two circular polarizations $\mathbf{E}_0^{(1)}$ and $\mathbf{E}_0^{(2)}$ of the same senses of rotations, then the total instantaneous energy density is

$$W = \frac{\varepsilon_0 \varepsilon}{2} |\mathbf{E}_0^{(1)} + \mathbf{E}_0^{(2)}|^2$$

4.7 Show that if a wave of elliptical polarization is decomposed into two linear polarizations oscillating along two mutually perpendicular directions, then (*a*) the instantaneous (or time-averaged) energy density of the elliptical polarization is equal to the sum of the instantaneous (or time-averaged) energy densities of the two linear polarizations, (b) the instantaneous (or time-averaged) Poynting vector of the elliptical polarization is equal to the sum of the instantaneous (or time-averaged) Poynting vectors of the two linear polarizations.

4.8 Show that the instantaneous Poynting vector of a circularly polarized wave is a constant vector (i.e., both the magnitude and direction of the instantaneous Poynting vector are constant) and is equal to the time-averaged Poynting vector.

4.9 The constitutive parameters of seawater at frequency $f = 10^6$ Hz are given by $\mu_0 = 4\pi \times 10^{-7}$ H/m, $81\varepsilon_0 = 9 \times 10^{-9}/4\pi$ F/m, and $\sigma = 4$ S/m. Calculate the attenuation constant, the phase constant, the phase velocity, the wavelength, and the intrinsic impedance. Can radar be used for underwater communication? Why? Repeat the calculations for copper whose constitutive parameters at frequency $f = 10^8$ Hz may be assumed as $\mu_0 = 4\pi \times 10^{-7}$ H/m, $\varepsilon_0 = 10^{-9}/36\pi$ F/m, and $\sigma = 5.8 \times 10^7$ S/m.

4.10 A 300-MHz uniform plane wave is traveling in a lossless medium characterized by the constitutive parameters μ_0 and $9\varepsilon_0$. If the electric field of the wave has an amplitude vector $\mathbf{E}_0 = 10\hat{\mathbf{x}}$ V/m and travels in the $+z$ direction, write the complete expressions for the field vectors in time domain and obtain the time-averaged Poynting vector.

4.11 Repeat Prob. 4.10 when the medium is lossy with conductivity $\sigma = 4$ S/m.

4.12 A uniform plane wave is propagating in a lossy medium characterized by the constitutive parameters: $4\,\mu_0$, $36\varepsilon_0$, and $\sigma = 2$ S/m. If the electric field is given by

$$\mathscr{E} = 10e^{-k_2 x} \cos(8\pi \times 10^8 t - k_1 x)\hat{\mathbf{z}} \qquad \text{V/m}$$

find k_1, k_2, and the magnetic field intensity.

4.13 A uniform plane wave is incident obliquely on the plane surface of a perfect conductor. Show that the amplitude vectors of the reflected wave are related to that of the incident wave by

$$\mathbf{E}_{0r} = (\hat{\mathbf{q}} \cdot \mathbf{E}_{0i})\hat{\mathbf{q}} - (\hat{\mathbf{q}} \times \mathbf{E}_{0i}) \times \hat{\mathbf{q}}$$

$$\hat{\mathbf{q}} \times \mathbf{H}_{0r} = \hat{\mathbf{q}} \times H_{0i}$$

where $\hat{\mathbf{q}}$ is a unit vector normal to the conducting surface. Hence the surface current density is $\mathbf{J}_s = 2(\hat{\mathbf{q}} \times \mathbf{H}_{0i})$.

4.14 A uniform plane wave is incident on the plane surface of a conductor. The wave vector $\mathbf{k}_i$ of the incident wave makes an angle 45° with the normal $\hat{\mathbf{q}}$ to the conductor (Fig. 4.14). The electric field is polarized in the direction parallel to the conducting surface. (*a*) Find the complex field vector $\mathbf{E}_r$ and $\mathbf{H}_r$ of the reflected wave. (*b*) What are the total instantaneous electric and magnetic fields? (*c*) Find the time-averaged Poynting vector.

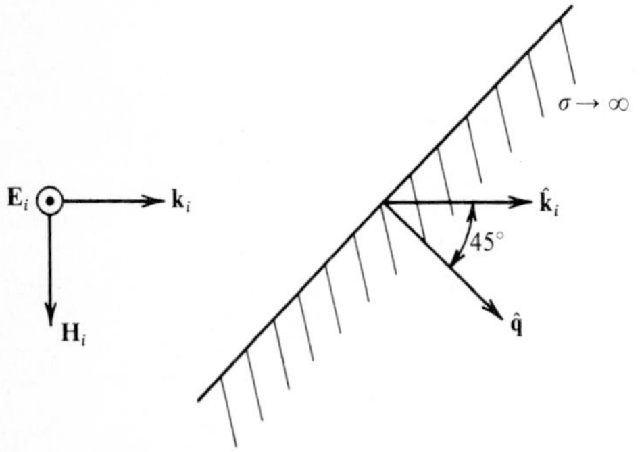

Figure 4.14 Plane wave incident on the plane surface of a conductor.

4.15 In traveling from free space into a lossless material medium at normal incidence, a uniform plane wave encounters a reflection coefficient of $-\frac{1}{4}$ and a velocity reduction of $\frac{1}{3}$. Find the permittivity and permeability of the medium.

4.16 A plane wave propagates from one dielectric into another at normal incidence. Show that the reflection and transmission coefficients are both equal to 0.5, if the ratio of the indices of refraction of the two dielectrics is 5.83.

4.17 A monochromatic plane wave with magnetic field amplitude H_0 is normally incident from free space onto the plane surface of a semi-infinite conducting medium. (*a*) Find the field vectors inside the conducting medium. (*b*) Calculate each component of the stress tensor $\bar{\bar{\mathscr{T}}}$ defined in Prob. 2.11. (*c*) If the medium is a perfect conductor, find the average pressure (or radiation pressure) exerted on the conducting surface.

4.18 (*a*) A uniform plane wave is traveling in an unbounded, isotropic medium. Show that the radiation pressure (force per unit area) exerted by the wave on the medium is

$$\frac{1}{v_p}|\tfrac{1}{2}\mathbf{E}\times\mathbf{H}^*| = \frac{\sqrt{\mu\varepsilon}}{c}|\langle\mathscr{P}\rangle|$$

where v_p is the phase velocity of the wave and $\langle\mathscr{P}\rangle$ is the time-averaged Poynting vector. (*b*) Find the radiation pressure of sunlight on an object on the earth's surface if the magnitude of the time-averaged Poynting vector of the sunlight near the earth's surface is $1.4\,\text{kW/m}^2$. Assume that the sun's radiation is all absorbed.

4.19 The low-attenuation, high-information bandwidth and immunity to electromagnetic interference make optical fiber an attractive alternative to coaxial cables for communications. An optical fiber consists of a hair-thin glass core surrounded by a dielectric, called a *cladding* (Fig. 4.15). The basis for glass fibers to guide light is the total internal reflection. The slight difference in indices of refraction between the core and the cladding makes possible the transmission of light waves through the core with minimum attenuation. (*a*) If the indices of refraction of the core and the cladding are 1.5 and 1.485 respectively, find the angle of reflection for waves in the center core. (*b*) If the cladding is free space and the index of refraction of the core is *n*, show that the condition for light to be confined in the core is $n \geq \sqrt{2}$.

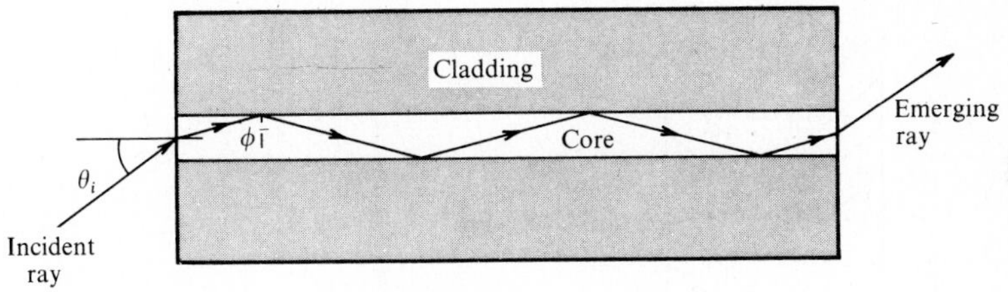

Figure 4.15 Illustration of an optical fiber.

4.20 (*a*) For a plane wave incident from free space on a lossless magnetic medium, show that Γ_{11} is zero for

$$\tan^2\theta_i = \frac{\mu_2(\mu_2-\varepsilon_2)}{\mu_2\varepsilon_2 - 1}$$

and hence that the Brewster angle exists only if $\mu_2 > \varepsilon_2$.

(*b*) Show that, similarly, Γ_{22} is zero if

$$\tan^2\theta_i = \frac{\varepsilon_2(\varepsilon_2-\mu_2)}{\mu_2\varepsilon_2 - 1}$$

In this case the Brewster angle exists only if $\varepsilon_2 > \mu_2$.

4.21 A uniform plane wave with the magnetic field normal to the plane of incidence, is incident obliquely at an angle $\theta_i = 30°$ from the free space (μ_0, ε_0) onto a lossless dielectric half-space $(\mu_0, 3.73\varepsilon_0)$. (*a*) Find the angle of reflection θ_r and the angle of transmission θ_t. (*b*) Find the values of the reflection and transmission coefficients. (*c*) Determine the fraction of the incident power that is transmitted into the dielectric half-space.

4.22 A linearly polarized wave is incident at an angle of 45° from the free space (μ_0, ε_0) onto the plane surface of polystyrene $(\mu_0, 2.7\varepsilon_0)$. The direction of the electric field vector makes an angle 45° with the plane of incidence. What is the polarization of the reflected wave? If the angle of incidence is 58.7° (Brewster angle), what is the polarization of the reflected wave?

4.23 Show that a linearly polarized wave will in general become elliptically polarized after total reflection from the surface of a dielectric medium. Under what conditions will the polarization become circular?

4.24 Calculate the critical angle of incidence for a plane wave propagating from the following dielectric media into free space (μ_0, ε_0): (*a*) mica $(\mu_0, 6\varepsilon_0)$; (*b*) distilled water $(\mu_0, 81\varepsilon_0)$.

4.25 A uniform plane wave in medium 1 is incident normally upon the plane boundary of medium 2. Let media 1 and 2 be characterized by the constitutive parameters $(\mu_0 \mu_1, \varepsilon_0 \varepsilon_1, \sigma_1)$ and $(\mu_0 \mu_2, \varepsilon_0 \varepsilon_2, \sigma_2)$ respectively. (*a*) Assuming that the incident wave is given, determine the total electric and magnetic fields in each medium. (*b*) Find the time-averaged Poynting vector in medium 1. (*c*) Determine the time-averaged Poynting vectors for the incident and the reflected waves. (*d*) Is the sum of the time-averaged Poynting vectors of the incident and the reflected waves equal to the result of (*b*)? Why? Under what condition are they equal? (*e*) Find the time-averaged Poynting vector of the transmitted wave and verify the law of conservation of energy at the interface.

4.26 An infinite dielectric layer $(\mu_0 \mu_2, \varepsilon_0 \varepsilon_2)$ of thickness d is coated on a perfectly conducting plane. A uniform plane wave in air (μ_0, ε_0) is incident normally onto the coating. (*a*) Assuming that the incident wave is given, determine the amplitude vector of the electric field in each region. (*b*) What is the time-averaged power flux density in each region?

4.27 A dielectric slab $(\mu_0, \varepsilon_0 \varepsilon)$ lies between medium 1 $(\mu_0, \varepsilon_0 \varepsilon_1)$ and medium 2 $(\mu_0, \varepsilon_0 \varepsilon_2)$ as shown in Fig. 4.16. A uniform plane wave is incident normally on the surface of the slab from the left. Find the thickness d and the dielectric constant ε for which there will be no reflection in medium 1. This technique is applied in optics to reduce reflections from lenses and is known as *lens blooming*.

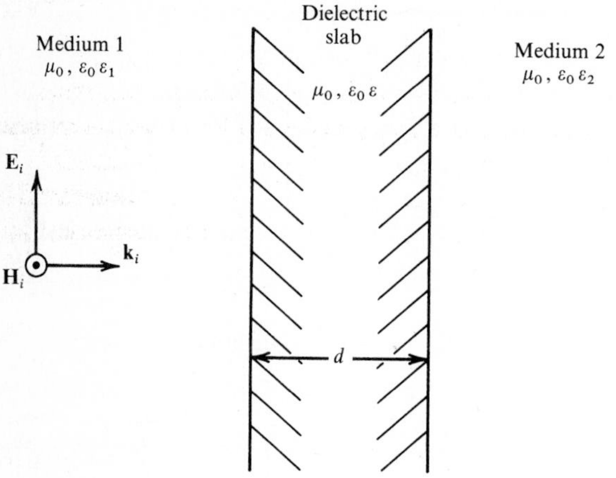

Figure 4.16 Plane wave incident normally on a dielectric slab.

4.28 A uniform plane wave in free space is incident at an angle θ_i on an infinite dielectric slab of dielectric constant ε and thickness d. The electric field vector of the incident wave is assumed to be parallel to the plane of incidence as shown in Fig. 4.17.

(*a*) Show that

$$\theta_r = \theta_t = \theta_i$$

$$\sqrt{\varepsilon} \sin \theta_1 = \sin \theta_i$$

(*b*) Find the electric and magnetic fields in each region.

(*c*) Find the range of the angle of incidence for which the plane wave in the slab is uniform.

(*d*) Find the time-averaged power flux density of the transmitted wave when the plane wave in the slab is uniform.

(*e*) Determine the phase shift of the reflected wave with respect to the incident wave.

4.29 Repeat Prob. 4.28 when the electric field vector of the incident wave is normal to the plane of incidence.

4.30 If ψ is the angle between the major semiaxis of a polarization ellipse of the reflected wave and

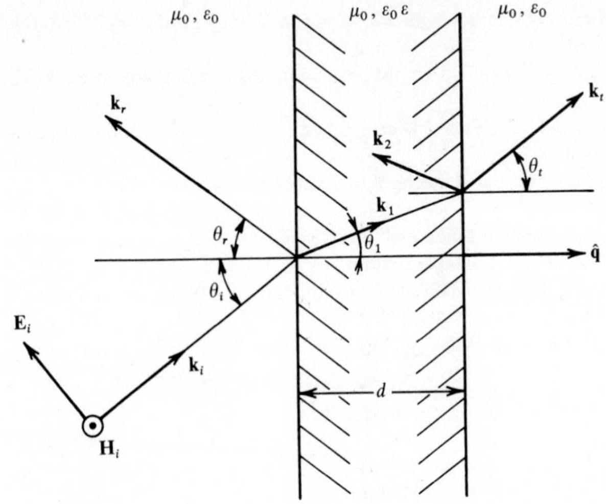

Figure 4.17 Plane wave incident obliquely on a dielectric slab.

the normal to the plane of incidence, show that

$$\tan 2\psi = \frac{g + g^*}{(1 - |g|^2)}$$

where $g = B_{\|}/B_{\perp}$, and $\mathbf{E}_{0r} = B_{\perp}\mathbf{a} + B_{\|}(\hat{\mathbf{k}}_r \times \mathbf{a})$.

4.31 An alternative method of determining the parameters ε_2 and τ for a conducting medium is to examine the polarization ellipse of the reflected wave when the incident wave is linearly polarized. If ψ_i is the angle between the direction of oscillation of the linearly polarized incident wave and the normal to the plane of incidence, and ψ is the angle defined in Prob. 4.30, show that

$$\tan 2\psi = \frac{\mathbf{a}^4 - q_i^2 |q_t|^2}{2\mathbf{a}^2 q_i q_1} \qquad \text{for } \psi_i = \pi/4$$

$$-\cos 2\psi_i = \frac{2\mathbf{a}^2 q_i q_1}{\mathbf{a}^4 + q_i^2 |q_t|^2} \qquad \text{for } \psi = \pi/4$$

or

$$|q_t|^2 = \mathbf{a}^2 \tan^2 \theta_i \frac{1 + \cos 2\psi_i \tan 2\psi}{1 - \cos 2\psi_i \tan 2\psi}$$

$$q_1 = \frac{-\mathbf{a}^2 \cos 2\psi_i}{q_i(1 - \cos 2\psi_i \tan 2\psi)}$$

thus by measuring ψ for $\psi_i = \pi/4$ and ψ_i for $\psi = \pi/4$, we may calculate $|q_t|^2$ and q_1 from the above formula and then ε_2 and τ from (4.346) and (4.347).

4.32 Consider two lossless isotropic media separated by a plane interface with the incident wave located in medium 1. If medium 1 is denser than medium 2, construct wave vector surfaces and discuss the critical angle of incidence and the phenomenon of total reflection geometrically.

REFERENCES

Adler, R. B., L. J. Chu, and R. M. Fano: *Electromagnetic Energy Transmission and Radiation*, John Wiley & Sons, Inc., New York, 1960, chaps. 7 and 8.

Born, M., and E. Wolf: *Principles of Optics*, 4th ed., Pergamon Press, New York, 1970, chap. 1.

Hauser, W.: *Introduction to the Principles of Electromagnetism*, Addison-Wesley Publishing Company, Reading, Mass., 1971, chaps. 14 and 15.

Kraus, J. D., and K. R. Carver: *Electromagnetics*, 2d ed., McGraw-Hill Book Company, New York, 1973, chaps. 10 and 12.

Lorrain, P., and D. Corson: *Electromagnetic Fields and Waves*, 2d ed., W. H. Freeman and Company, San Francisco, 1970, chaps. 11 and 12.

Mahan, A. I., and C. V. Bitterli: "Total Internal Reflection: A Deeper Look," *Appl. Optics*, vol. 17, no. 4, 1978, pp. 509–519.

Seshadri, S. R.: *Fundamentals of Transmission Lines and Electromagnetic Fields*, Addison-Wesley Publishing Company, Reading, Mass., 1971, chaps. 5 and 6.

Stratton, J. A.: *Electromagnetic Theory*, McGraw-Hill Book Company, New York, 1941, chaps. V and IX.

Wangsness, R. K.: *Electromagnetic Fields*, John Wiley & Sons, Inc., New York, 1979, chaps. 24 and 25.

CHAPTER

FIVE

PLANE WAVES IN CRYSTALS

In Chap. 4 our study of propagation and reflection of waves has been limited to isotropic media, i.e., materials whose electric and magnetic properties are the same in all directions. In many crystals, however, the electric as well as the other physical properties vary according to directions. This electric anisotropy is caused by the particular arrangement of the atoms in the crystalline lattice and affects greatly the electromagnetic wave propagation, which we shall investigate in this chapter.

We shall begin by considering some general properties of the dielectric tensor of crystals. We will then examine the uniform plane waves in unbounded anisotropic crystals which include the dispersion equation, the polarization of waves, and the determination of the directions of field vectors. We will find that the phase velocity in a crystal depends on the direction of wave propagation. We will also find that to each direction of wave propagation corresponds two linearly polarized waves traveling at different phase velocities. This phenomenon is known as *birefringence* or *double refraction*. With the aid of the wave vector surface, we can construct geometrically the direction of energy propagation which, unlike isotropic media, does not coincide with the wave normal.

Finally, we will investigate the phenomena of transmission and reflection that occur at the plane interface separating an isotropic medium and an anisotropic crystal. The wave vectors of the transmitted waves in crystals are first determined through a Booker quartic equation. The reflection and transmission coefficients which apply to all anisotropic media are then derived. The results will all be expressed in the coordinate-free forms.

5.1 PROPERTIES OF THE DIELECTRIC TENSOR OF CRYSTALS

Due to the lattice structure, crystals are electrically anisotropic. The constitutive relations for a homogeneous, lossless, and nonmagnetic crystal are

$$\mathbf{D} = \varepsilon_0 \bar{\boldsymbol{\varepsilon}} \cdot \mathbf{E} \tag{5.1}$$

and

$$\mathbf{B} = \mu_0 \mathbf{H} \tag{5.2}$$

where $\bar{\boldsymbol{\varepsilon}}$ is a real constant dyadic or tensor of rank 2† and is called the *dielectric tensor*. Equation (5.1) clearly indicates that the electric flux density $\mathbf{D}$ is not in the direction of electric field intensity $\mathbf{E}$; instead, each component of $\mathbf{D}$ is linearly related to the components of $\mathbf{E}$. Because of the tensor nature of $\bar{\boldsymbol{\varepsilon}}$, the set of nine components in the matrix

$$\bar{\boldsymbol{\varepsilon}} = [\varepsilon_{ij}] = \begin{bmatrix} \varepsilon_{11} & \varepsilon_{12} & \varepsilon_{13} \\ \varepsilon_{21} & \varepsilon_{22} & \varepsilon_{23} \\ \varepsilon_{31} & \varepsilon_{32} & \varepsilon_{33} \end{bmatrix} \tag{5.3}$$

should transform according to the transformation law of a cartesian tensor of rank 2. In other words, with respect to rotation and reflection of coordinate systems, the left- and right-hand sides of Eq. (5.1) will transform identically and, hence, the equality is always valid. The tensor nature of $\bar{\boldsymbol{\varepsilon}}$ does not impose any restriction on the values of the components ε_{ij}; it only defines the rules of transformation when the coordinate system is undergoing an orthogonal transformation. However, the components ε_{ij} are not arbitrary; they should satisfy some definite requirements, some of which are resulted from general physical laws and others are related to the symmetry properties of the crystals.

The most important physical law that the constitutive relations (5.1) and (5.2) should always obey is the law of conservation of energy. Thus according to Eq. (2.73), the dielectric tensor $\bar{\boldsymbol{\varepsilon}}$ for a lossless crystal must be symmetric, i.e.,

$$\tilde{\bar{\boldsymbol{\varepsilon}}} = \bar{\boldsymbol{\varepsilon}} \tag{5.4}$$

In this case, the time-averaged electric and magnetic energy densities in a nondispersive crystal are [cf. Eqs. (2.86) and (2.87)]

$$\langle W_e \rangle = \frac{\varepsilon_0}{4} \mathbf{E}^* \cdot \bar{\boldsymbol{\varepsilon}} \cdot \mathbf{E} \tag{5.5}$$

and

$$\langle W_m \rangle = \frac{\mu_0}{4} \mathbf{H}^* \cdot \mathbf{H} \tag{5.6}$$

respectively. Since the electric and magnetic energy densities must always be

† By a cartesian tensor of rank 2 of three-dimensional space, we mean a set of nine quantities which, under the orthogonal transformation of a coordinate system, transforms as the components of the product of two vectors (see Appendix A for detail).

positive and are zero only when **E** and **H** are zero, we conclude that $\bar{\varepsilon}$ must be a positive, definite, symmetric tensor.

According to Sec. 1.8, every real symmetric matrix can always be reduced to a diagonal form in an orthogonal coordinate system formed by the eigenvectors. Thus

$$\bar{\varepsilon} = \begin{bmatrix} \varepsilon_1 & 0 & 0 \\ 0 & \varepsilon_2 & 0 \\ 0 & 0 & \varepsilon_3 \end{bmatrix} \tag{5.7}$$

where the real constants ε_1, ε_2, and ε_3 are the eigenvalues of $\bar{\varepsilon}$ and are called the *principal dielectric constants*. The orthogonal coordinate axes in which the tensor $\bar{\varepsilon}$ takes a diagonal form are referred to as the *principal dielectric axes*. As a result of the positive definiteness of $\bar{\varepsilon}$, all the eigenvalues must be positive and nonzero. Hence the determinant of $\bar{\varepsilon}$ must always be positive and different from zero:

$$|\bar{\varepsilon}| = \varepsilon_1 \varepsilon_2 \varepsilon_3 > 0 \tag{5.8}$$

In other words, $\bar{\varepsilon}$ is nonsingular and thus the inverse of $\bar{\varepsilon}$ always exists. Because of this characteristic, the constitutive relation (5.1) may also be written as

$$\varepsilon_0 \mathbf{E} = \bar{\varepsilon}^{-1} \cdot \mathbf{D} \tag{5.9}$$

which proves to be a more convenient form to use for an electrically anisotropic medium when the derivation is based on the magnetic field instead of the electric field (see Prob. 5-10).

5.2 DISPERSION EQUATION

Here we will again seek monochromatic plane wave solution to Maxwell's equations of the form

$$\mathscr{E} = \text{Re}\,(\mathbf{E}_0 e^{i\phi}) \tag{5.10}$$

where $\phi = \mathbf{k} \cdot \mathbf{r} - \omega t$ is the phase of the wave. In a source-free region, the Maxwell equations with the above space-time variations were derived in Sec. 3.1 [cf. Eqs. (3.15), (3.16)]. For convenience, we repeat them here for nonmagnetic lossless crystals:

$$\omega \mathbf{D}_0 = \omega \varepsilon_0 \bar{\varepsilon} \cdot \mathbf{E}_0 = -\mathbf{k} \times \mathbf{H}_0 \tag{5.11}$$

$$\omega \mathbf{B}_0 = \omega \mu_0 \mathbf{H}_0 = \mathbf{k} \times \mathbf{E}_0 \tag{5.12}$$

With the aid of the antisymmetric matrix $\mathbf{k} \times \bar{\mathbf{I}}$ defined in Eq. (1.89), we may rewrite Maxwell's equations as

$$\omega \varepsilon_0 \bar{\varepsilon} \cdot \mathbf{E}_0 = -(\mathbf{k} \times \bar{\mathbf{I}}) \cdot \mathbf{H}_0 \tag{5.13}$$

$$\omega \mu_0 \mathbf{H}_0 = (\mathbf{k} \times \bar{\mathbf{I}}) \cdot \mathbf{E}_0 \tag{5.14}$$

Equations (5.13) and (5.14) are mutually coupled simultaneous equations. To decouple them, we eliminate $\mathbf{H}_0$ from Eqs. (5.13) and (5.14) and obtain an equation in $\mathbf{E}_0$ alone:

$$[k_0^2 \bar{\boldsymbol{\varepsilon}} + (\mathbf{k} \times \bar{\mathbf{I}})^2] \cdot \mathbf{E}_0 = \mathbf{0} \tag{5.15}$$

where $k_0^2 = \omega^2 \mu_0 \varepsilon_0$, or, after dot-premultiplying both sides by $\bar{\boldsymbol{\varepsilon}}^{-1}$, we have

$$[k_0^2 \bar{\mathbf{I}} + \bar{\boldsymbol{\varepsilon}}^{-1} \cdot (\mathbf{k} \times \bar{\mathbf{I}})^2] \cdot \mathbf{E}_0 = \mathbf{0} \tag{5.16}$$

Also since $\mathbf{E}_0 = \bar{\boldsymbol{\varepsilon}}^{-1} \cdot \mathbf{D}_0/\varepsilon_0$, Eq. (5.15) may be expressed as

$$[k_0^2 \bar{\mathbf{I}} + (\mathbf{k} \times \bar{\mathbf{I}})^2 \cdot \bar{\boldsymbol{\varepsilon}}^{-1}] \cdot \mathbf{D}_0 = \mathbf{0} \tag{5.17}$$

Finally, eliminating $\mathbf{E}_0$ from Eqs. (5.13) and (5.14) we find an equation in $\mathbf{H}_0$:

$$[k_0^2 \bar{\mathbf{I}} + (\mathbf{k} \times \bar{\mathbf{I}}) \cdot \bar{\boldsymbol{\varepsilon}}^{-1} \cdot (\mathbf{k} \times \bar{\mathbf{I}}] \cdot \mathbf{H}_0 = \mathbf{0} \tag{5.18}$$

For a uniform plane wave with $\mathbf{k} = k\hat{\mathbf{k}}$, we may write Eq. (5.16) in the form of an eigenvalue problem, namely,

$$[\bar{\boldsymbol{\varepsilon}}^{-1} \cdot (\hat{\mathbf{k}} \times \bar{\mathbf{I}})^2] \cdot \mathbf{E}_0 = \lambda \mathbf{E}_0 \tag{5.19}$$

where $\lambda = -k_0^2/k^2$ is an eigenvalue of the matrix $\bar{\boldsymbol{\varepsilon}}^{-1} \cdot (\hat{\mathbf{k}} \times \bar{\mathbf{I}})^2$ and $\mathbf{E}_0$ is the corresponding eigenvector. Similarly, Eqs. (5.17) and (5.18) may be written as

$$[(\hat{\mathbf{k}} \times \bar{\mathbf{I}})^2 \cdot \bar{\boldsymbol{\varepsilon}}^{-1}] \cdot \mathbf{D}_0 = \lambda \mathbf{D}_0 \tag{5.20}$$

and

$$[(\hat{\mathbf{k}} \times \bar{\mathbf{I}}) \cdot \bar{\boldsymbol{\varepsilon}}^{-1} \cdot (\hat{\mathbf{k}} \times \bar{\mathbf{I}})] \cdot \mathbf{H}_0 = \lambda \mathbf{H}_0 \tag{5.21}$$

respectively.

We note that the vector equation (5.16) [or (5.17) or (5.18)] represents a set of three linear homogeneous equations with three unknown components of the field vector $\mathbf{E}_0$ (or $\mathbf{D}_0$ or $\mathbf{H}_0$). For the homogeneous equation (5.16) to have a nonzero vector solution $\mathbf{E}_0$, it is necessary that the determinant of the coefficient matrix vanishes, i.e.,

$$|k_0^2 \bar{\mathbf{I}} + \bar{\boldsymbol{\varepsilon}}^{-1} \cdot (\mathbf{k} \times \bar{\mathbf{I}})^2| = 0 \tag{5.22}$$

This is the dispersion equation. Similarly, for nonzero $\mathbf{D}_0$ and $\mathbf{H}_0$ to exist, we must have [see Eqs. (5.17) and (5.18)]:

$$|k_0^2 \bar{\mathbf{I}} + (\mathbf{k} \times \bar{\mathbf{I}})^2 \cdot \bar{\boldsymbol{\varepsilon}}^{-1}| = 0 \tag{5.23}$$

and

$$|k_0^2 \bar{\mathbf{I}} + (\mathbf{k} \times \bar{\mathbf{I}}) \cdot \bar{\boldsymbol{\varepsilon}}^{-1} \cdot (\mathbf{k} \times \bar{\mathbf{I}})| = 0 \tag{5.24}$$

Based on the Maxwell equations (5.11) and (5.12) we can show easily that the existence of any one of the nonzero vectors $\mathbf{E}_0$, $\mathbf{D}_0$, and $\mathbf{H}_0$ implies the existence of all other nonzero vectors. That is, the dispersion equations (5.22), (5.23), and (5.24) are all equivalent. In fact, the following equalities hold:

$$\begin{aligned} |k_0^2 \bar{\mathbf{I}} + \bar{\boldsymbol{\varepsilon}}^{-1} \cdot (\mathbf{k} \times \bar{\mathbf{I}})^2| &= |k_0^2 \bar{\mathbf{I}} + (\mathbf{k} \times \bar{\mathbf{I}})^2 \cdot \bar{\boldsymbol{\varepsilon}}^{-1}| \\ &= |k_0^2 \bar{\mathbf{I}} + (\mathbf{k} \times \bar{\mathbf{I}}) \cdot \bar{\boldsymbol{\varepsilon}}^{-1} \cdot (\mathbf{k} \times \bar{\mathbf{I}})| \end{aligned} \tag{5.25}$$

To prove Eq. (5.25) we may use these facts: (*a*) The identity (1.124*b*); (*b*) tensors $\bar{\varepsilon}^{-1} \cdot (\mathbf{k} \times \bar{\mathbf{I}})^2$, $(\mathbf{k} \times \bar{\mathbf{I}})^2 \cdot \bar{\varepsilon}^{-1}$, and $(\mathbf{k} \times \bar{\mathbf{I}}) \cdot \bar{\varepsilon}^{-1} \cdot (\mathbf{k} \times \bar{\mathbf{I}})$ are the products of tensors $\bar{\varepsilon}^{-1}$, $(\mathbf{k} \times \bar{\mathbf{I}})$, and $(\mathbf{k} \times \bar{\mathbf{I}})$ in cyclic order; and (*c*) the trace and the determinant of a product of three matrices in cyclic order are invariant. In the following, we shall examine the dispersion equation expressed in various equivalent forms.

First, let us obtain the dispersion equation in explicit form. Expanding the determinant in Eq. (5.22) according to Eq. (1.124*b*), we have

$$\begin{aligned} |k_0^2 \bar{\mathbf{I}} + \bar{\varepsilon}^{-1} \cdot (\mathbf{k} \times \bar{\mathbf{I}})^2| = k_0^6 &+ [\bar{\varepsilon}^{-1} \cdot (\mathbf{k} \times \bar{\mathbf{I}})^2]_t k_0^4 \\ &+ \{\text{adj } [\bar{\varepsilon}^{-1} \cdot (\mathbf{k} \times \bar{\mathbf{I}})^2]\}_t k_0^2 \\ &+ |\bar{\varepsilon}^{-1} \cdot (\mathbf{k} \times \bar{\mathbf{I}})^2| \end{aligned} \tag{5.26}$$

From Eqs. (1.100), (1.72*j*), (1.102), (1,72*g*), and (1.95), we find

$$\begin{aligned} [\bar{\varepsilon}^{-1} \cdot (\mathbf{k} \times \bar{\mathbf{I}})^2]_t &= (\bar{\varepsilon}^{-1} \cdot \mathbf{kk} - k^2 \bar{\varepsilon}^{-1})_t \\ &= \frac{1}{|\bar{\varepsilon}|} \mathbf{k} \cdot [\text{adj } \bar{\varepsilon} - (\text{adj } \bar{\varepsilon})_t \bar{\mathbf{I}}] \cdot \mathbf{k} \\ \{\text{adj } [\bar{\varepsilon}^{-1} \cdot (\mathbf{k} \times \bar{\mathbf{I}})^2]\}_t &= [\text{adj } (\mathbf{k} \times \bar{\mathbf{I}}) \cdot \text{adj } (\mathbf{k} \times \bar{\mathbf{I}}) \cdot \text{adj } \bar{\varepsilon}^{-1}]_t \\ &= \frac{k^2}{|\bar{\varepsilon}|} (\mathbf{k} \cdot \bar{\varepsilon} \cdot \mathbf{k}) \end{aligned} \tag{5.27}$$

and $$|\bar{\varepsilon}^{-1} \cdot (\mathbf{k} \times \bar{\mathbf{I}})^2| = |\bar{\varepsilon}^{-1}| |\mathbf{k} \times \bar{\mathbf{I}}| |\mathbf{k} \times \bar{\mathbf{I}}| = 0$$

Substituting Eq. (5.27) into Eq. (5.26), we finally obtain the explicit form of the dispersion equation from Eq. (5.22), namely

$$(\mathbf{k} \cdot \bar{\varepsilon} \cdot \mathbf{k})\mathbf{k}^2 + \mathbf{k} \cdot [\text{adj } \bar{\varepsilon} - (\text{adj } \bar{\varepsilon})_t \bar{\mathbf{I}}] \cdot \mathbf{k} k_0^2 + |\bar{\varepsilon}| k_0^4 = 0 \tag{5.28}$$

or, from Eq. (1.64)

$$(\mathbf{k} \cdot \bar{\varepsilon} \cdot \mathbf{k})k^2 - \mathbf{k} \cdot \bar{\varepsilon} \cdot (\bar{\varepsilon}_t \bar{\mathbf{I}} - \bar{\varepsilon}) \cdot \mathbf{k} k_0^2 + |\bar{\varepsilon}| k_0^4 = 0$$

or, in terms of the refractive index vector $\mathbf{n} = n\hat{\mathbf{k}} = \mathbf{k}/k_0$

$$(\hat{\mathbf{k}} \cdot \bar{\varepsilon} \cdot \hat{\mathbf{k}})n^4 - \hat{\mathbf{k}} \cdot [(\text{adj } \bar{\varepsilon})_t \bar{\mathbf{I}} - \text{adj } \bar{\varepsilon}] \cdot \hat{\mathbf{k}} n^2 + |\bar{\varepsilon}| = 0 \tag{5.29}$$

where $n = k/k_0$ is the index of refraction.

We note that the same dispersion equation can be obtained by expanding the determinants in either Eq. (5.23) or (5.24). Equation (5.29) is a quadratic equation in n^2 with coefficients depending on the direction of wave normal $\hat{\mathbf{k}}$. Thus in an anisotropic medium, the index of refraction depends on the direction of wave normal, and the dispersion equation alone does not uniquely determine the vector $\mathbf{n}$.

Other equivalent forms may also be derived. To do so, let us consider for instance the homogeneous equation (5.15). The condition for the existence of nonzero $\mathbf{E}_0$ gives the dispersion equation

$$|k_0^2 \bar{\varepsilon} + (\mathbf{k} \times \bar{\mathbf{I}})^2| = 0$$

which is equivalent to Eq. (5.22). Using the identity (1.100) and then expanding the determinant according to Eq. (1.125*b*), we obtain

$$\begin{aligned}|k_0^2\bar{\varepsilon} + (\mathbf{k} \times \bar{\mathbf{I}})^2| &= |(k_0^2\bar{\varepsilon} - k^2\bar{\mathbf{I}}) + \mathbf{k}\mathbf{k}| \\ &= |k_0^2\bar{\varepsilon} - k^2\bar{\mathbf{I}}| + \mathbf{k} \cdot [\operatorname{adj}(k_0^2\bar{\varepsilon} - k^2\bar{\mathbf{I}})] \cdot \mathbf{k} = 0 \end{aligned} \tag{5.30}$$

But

$$\begin{aligned}|k_0^2\bar{\varepsilon} - k^2\bar{\mathbf{I}}| &= (\hat{\mathbf{k}} \cdot \bar{\mathbf{I}} \cdot \hat{\mathbf{k}})|k_0^2\bar{\varepsilon} - k^2\bar{\mathbf{I}}| \\ &= \hat{\mathbf{k}} \cdot (|k_0^2\bar{\varepsilon} - k^2\bar{\mathbf{I}}|\bar{\mathbf{I}}) \cdot \hat{\mathbf{k}} \\ &= \hat{\mathbf{k}} \cdot [(k_0^2\bar{\varepsilon} - k^2\bar{\mathbf{I}}) \cdot \operatorname{adj}(k_0^2\bar{\varepsilon} - k^2\bar{\mathbf{I}})] \cdot \hat{\mathbf{k}} \end{aligned} \tag{5.31}$$

Thus the dispersion equation (5.30) becomes

$$\hat{\mathbf{k}} \cdot \bar{\varepsilon} \cdot [\operatorname{adj}(k_0^2\bar{\varepsilon} - k^2\bar{\mathbf{I}})] \cdot \hat{\mathbf{k}} = 0 \tag{5.32}$$

On the other hand, if $k_0^2\bar{\varepsilon} - k^2\bar{\mathbf{I}}$ is nonsingular (that is, $|k_0^2\bar{\varepsilon} - k^2\bar{\mathbf{I}}| \neq 0$), we may divide Eq. (5.30) by $|k_0^2\bar{\varepsilon} - k^2\bar{\mathbf{I}}|$ and obtain the dispersion equation in the form of

$$1 + \mathbf{k} \cdot (k_0^2\bar{\varepsilon} - k^2\bar{\mathbf{I}})^{-1} \cdot \mathbf{k} = 0 \tag{5.33}$$

Substitution of $\bar{\varepsilon} = |\bar{\varepsilon}| \operatorname{adj} \bar{\varepsilon}^{-1}$ into Eq. (5.32) yields another form:

$$\hat{\mathbf{k}} \cdot \left[\operatorname{adj}\left(\bar{\varepsilon}^{-1} - \frac{1}{n^2}\bar{\mathbf{I}}\right)\right] \cdot \hat{\mathbf{k}} = 0 \tag{5.34}$$

Finally, if $|k_0^2\bar{\varepsilon} - k^2\bar{\mathbf{I}}| \neq 0$ thus $|\bar{\varepsilon}^{-1} - 1/n^2\bar{\mathbf{I}}| \neq 0$, we may divide Eqs. (5.32) and (5.34) by these determinants respectively and obtain still two other forms of the dispersion equation

$$\hat{\mathbf{k}} \cdot \bar{\varepsilon} \cdot (k_0^2\bar{\varepsilon} - k^2\bar{\mathbf{I}})^{-1} \cdot \hat{\mathbf{k}} = 0 \tag{5.35}$$

and

$$\hat{\mathbf{k}} \cdot \left(\bar{\varepsilon}^{-1} - \frac{1}{n^2}\bar{\mathbf{I}}\right)^{-1} \cdot \hat{\mathbf{k}} = 0 \tag{5.36}$$

With respect to the principal dielectric axes of the tensor $\bar{\varepsilon}$, we may write

$$\bar{\varepsilon}^{-1} - \frac{1}{n^2}\bar{\mathbf{I}} = \begin{bmatrix} \dfrac{1}{\varepsilon_1} - \dfrac{1}{n^2} & 0 & 0 \\ 0 & \dfrac{1}{\varepsilon_2} - \dfrac{1}{n^2} & 0 \\ 0 & 0 & \dfrac{1}{\varepsilon_3} - \dfrac{1}{n^2} \end{bmatrix} \tag{5.37}$$

Let ℓ_1, ℓ_2, and ℓ_3 be the direction cosines of the wave normal $\hat{\mathbf{k}}$, that is,

$$\hat{\mathbf{k}} = \begin{bmatrix} \ell_1 \\ \ell_2 \\ \ell_3 \end{bmatrix} \tag{5.38}$$

with $\ell_1^2 + \ell_2^2 + \ell_3^2 = 1$. Substitution of Eqs. (5.37) and (5.38) into Eq. (5.30) yields

$$\frac{\ell_1^2}{1/\varepsilon_1 - 1/n^2} + \frac{\ell_2^2}{1/\varepsilon_2 - 1/n^2} + \frac{\ell_3^2}{1/\varepsilon_3 - 1/n^2} = 0 \tag{5.39}$$

which is one of the generally used forms of the dispersion equation. With the relations established in Chap. 1 and by direct computation, we can easily verify that the various forms of the dispersion equations derived in this section are equivalent. However, Eqs. (5.28), (5.29), (5.32), and (5.34) appear to be more general in the sense that their applicability is not limited by the condition $|k_0^2\bar{\varepsilon} - k^2\bar{\mathbf{I}}| \neq 0$.

As a final note, for a given direction of wave normal $\hat{\mathbf{k}}$, the two values of n^2 obtained from the dispersion equation (5.29) are always real and positive. To see this, let us dot-premultiply Eq. (5.16) by $\mathbf{E}_0^* \cdot \bar{\varepsilon}$ and obtain

$$n^2 = \frac{\mathbf{E}_0^* \cdot \bar{\varepsilon} \cdot \mathbf{E}_0}{(\hat{\mathbf{k}} \times \mathbf{E}_0)^* \cdot (\hat{\mathbf{k}} \times \mathbf{E}_0)} \tag{5.40}$$

Since $\bar{\varepsilon}$ is a positive definite symmetric matrix, the numerator is always real and positive and hence $n^2 > 0$ for any complex vector $\mathbf{E}_0$.

5.3 POLARIZATION OF WAVES IN CRYSTALS, OPTIC AXES

Now let us examine the polarizations of field vectors in lossless crystals. For a monochromatic uniform plane wave, the Maxwell equations in lossless crystals are

$$\omega\mathbf{D}_0 = \omega\varepsilon_0\bar{\varepsilon}\cdot\mathbf{E}_0 = -\mathbf{k}\times\mathbf{H}_0 \tag{5.41}$$

$$\omega\mathbf{B}_0 = \omega\mu_0\mathbf{H}_0 = \mathbf{k}\times\mathbf{E}_0 \tag{5.42}$$

where $\mathbf{k}$ and ω are real and $\bar{\varepsilon}$ is a real positive definite symmetric matrix. The complex conjugates of Eqs. (5.41) and (5.42) give

$$\omega\mathbf{D}_0^* = \omega\varepsilon_0\bar{\varepsilon}\cdot\mathbf{E}_0^* = -\mathbf{k}\times\mathbf{H}_0^* \tag{5.43}$$

$$\omega\mathbf{B}_0^* = \omega\mu_0\mathbf{H}_0^* = \mathbf{k}\times\mathbf{E}_0^* \tag{5.44}$$

Cross-multiplying Eqs. (5.41) and (5.43) and using Eq. (1.113), we get

$$\begin{aligned}
\mathbf{D}_0 \times \mathbf{D}_0^* &= \varepsilon_0^2(\bar{\varepsilon}\cdot\mathbf{E}_0)\times(\bar{\varepsilon}\cdot\mathbf{E}_0^*) \\
&= \varepsilon_0^2(\text{adj }\bar{\varepsilon})\cdot(\mathbf{E}_0\times\mathbf{E}_0^*) \\
&= \frac{1}{\omega^2}(\mathbf{k}\times\mathbf{H}_0)\times(\mathbf{k}\times\mathbf{H}_0^*) \\
&= \frac{1}{\omega^2}\mathbf{k}\mathbf{k}\cdot(\mathbf{H}_0\times\mathbf{H}_0^*)
\end{aligned} \tag{5.45}$$

Hence

$$\mathbf{E}_0 \times \mathbf{E}_0^* = \frac{1}{\omega^2 \varepsilon_0^2 |\bar{\varepsilon}|} \bar{\varepsilon} \cdot \mathbf{kk} \cdot (\mathbf{H}_0 \times \mathbf{H}_0^*) \tag{5.46}$$

Similarly, cross-multiplying Eqs. (5.42) and (5.44) we find

$$\mathbf{H}_0 \times \mathbf{H}_0^* = \frac{1}{\omega^2 \mu_0^2} (\mathbf{k} \times \mathbf{E}_0) \times (\mathbf{k} \times \mathbf{E}_0^*)$$

$$= \frac{1}{\omega^2 \mu_0^2} \mathbf{kk} \cdot (\mathbf{E}_0 \times \mathbf{E}_0^*) \tag{5.47}$$

Eliminating either $\mathbf{H}_0 \times \mathbf{H}_0^*$ or $\mathbf{E}_0 \times \mathbf{E}_0^*$ from Eqs. (5.46) and (5.47) we obtain

$$\left(\bar{\mathbf{I}} - \frac{\mathbf{k}^2}{k_0^4 |\bar{\varepsilon}|} \bar{\varepsilon} \cdot \mathbf{kk}\right) \cdot (\mathbf{E}_0 \times \mathbf{E}_0^*) = \mathbf{0} \tag{5.48}$$

and

$$\left(\bar{\mathbf{I}} - \frac{\mathbf{k} \cdot \bar{\varepsilon} \cdot \mathbf{k}}{k_0^4 |\bar{\varepsilon}|} \mathbf{kk}\right) \cdot (\mathbf{H}_0 \times \mathbf{H}_0^*) = \mathbf{0} \tag{5.49}$$

Equations (5.48) and (5.49) are two homogeneous equations. Thus for nonzero vector $\mathbf{E}_0 \times \mathbf{E}_0^*$ or $\mathbf{H}_0 \times \mathbf{H}_0^*$ to exist, the determinant of the coefficient matrix of Eq. (5.48) or Eq. (5.49) must vanish; i.e.,

$$\left| \bar{\mathbf{I}} - \frac{\mathbf{k}^2}{k_0^4 |\bar{\varepsilon}|} \bar{\varepsilon} \cdot \mathbf{kk} \right| = 0 \tag{5.50}$$

or

$$\left| \bar{\mathbf{I}} - \frac{\mathbf{k} \cdot \bar{\varepsilon} \cdot \mathbf{k}}{k_0^4 |\bar{\varepsilon}|} \mathbf{kk} \right| = 0 \tag{5.51}$$

Calculating the determinants in Eqs. (5.50) and (5.51) according to Eq. (1.126*b*) we find that both conditions give the same result

$$(\mathbf{k} \cdot \bar{\varepsilon} \cdot \mathbf{k})\mathbf{k}^2 - k_0^4 |\bar{\varepsilon}| = 0 \tag{5.52}$$

or

$$(\hat{\mathbf{k}} \cdot \bar{\varepsilon} \cdot \hat{\mathbf{k}})n^4 - |\bar{\varepsilon}| = 0 \tag{5.53}$$

In other words, $(\mathbf{E}_0 \times \mathbf{E}_0^*)$, $(\mathbf{H}_0 \times \mathbf{H}_0^*)$, and $(\mathbf{D}_0 \times \mathbf{D}_0^*)$ are either all zero or nonzero depending on whether or not the wave vector $\mathbf{k}$ satisfies Eq. (5.52). Recalling condition (3.44) for linear polarization, we conclude that for every direction of propagation, the uniform plane waves in a lossless crystal are always linearly polarized except when the wave vector $\mathbf{k}$ satisfies the condition (5.52). In the latter case, the wave can have any polarization. But the wave vector $\mathbf{k} = k_0 \mathbf{n}$ must also satisfy the dispersion equation (5.29), the solutions of which are

$$n^2 = \frac{\hat{\mathbf{k}} \cdot [(\text{adj } \bar{\varepsilon})_t \bar{\mathbf{I}} - \text{adj } \bar{\varepsilon}] \cdot \hat{\mathbf{k}} \pm \sqrt{\Delta}}{2(\hat{\mathbf{k}} \cdot \bar{\varepsilon} \cdot \hat{\mathbf{k}})} \tag{5.54}$$

where the discriminant Δ is given by

$$\Delta = \{\hat{\mathbf{k}} \cdot [(\text{adj } \bar{\varepsilon})_t \bar{\mathbf{I}} - \text{adj } \bar{\varepsilon}] \cdot \hat{\mathbf{k}}\}^2 - 4(\hat{\mathbf{k}} \cdot \bar{\varepsilon} \cdot \hat{\mathbf{k}}) |\bar{\varepsilon}| \tag{5.55}$$

Condition (5.53) and the dispersion equation (5.29) imply that

$$\hat{\mathbf{k}} \cdot [(\text{adj } \bar{\varepsilon})_t \bar{\mathbf{I}} - \text{adj } \bar{\varepsilon}] \cdot \hat{\mathbf{k}} n^2 = 2 \,|\, \bar{\varepsilon} \,| \tag{5.56}$$

and substitution of Eqs. (5.53) and (5.56) into Eq. (5.55) yields

$$\Delta = 0 \tag{5.57}$$

That is, when the refractive index vector $\mathbf{n}$ satisfies the condition (5.53), the two solutions in n^2 of the dispersion equation become equal. Those directions of the wave normal that cause the discriminant (5.55) to vanish and yield two equal roots for n^2 are called the *optic axes* of the crystals. In other words, optic axes are the directions in the crystals for which the two values of the phase velocity $v_p = c/n$ are equal. If a crystal possesses two optic axes, it is said to be *biaxial*; if it possesses only one optic axis, it is *uniaxial.*

Aside from the optic axes, for each given direction of wave normal $\hat{\mathbf{k}}$ there are two linearly polarized waves propagating at different phase velocities in the crystals. Each wave is characterized by a set of fixed directions of amplitude vectors $\mathbf{E}_0$, $\mathbf{D}_0$, $\mathbf{H}_0$. Since the waves are linearly polarized, without loss of generality we may assume that all the amplitude vectors are real. Hence, according to Maxwell's equations (5.41) and (5.42), the vector $\mathbf{H}_0$ (and hence also $\mathbf{B}_0$) is perpendicular to $\mathbf{E}_0$, $\mathbf{D}_0$, and $\hat{\mathbf{k}}$, which must therefore be coplanar. Further, $\mathbf{D}_0$ is seen to be perpendicular to $\hat{\mathbf{k}}$. From the constitutive relation $\mathbf{D}_0 = \varepsilon_0 \bar{\varepsilon} \cdot \mathbf{E}_0$ and Eq. (5.41), we have

$$\hat{\mathbf{k}} \cdot \mathbf{D}_0 = \varepsilon_0 \hat{\mathbf{k}} \cdot \bar{\varepsilon} \cdot \mathbf{E}_0 = \varepsilon_0 \mathbf{E}_0 \cdot \bar{\varepsilon} \cdot \hat{\mathbf{k}} = 0 \tag{5.58}$$

that is, unlike isotropic media, the electric field intensity is not perpendicular to $\hat{\mathbf{k}}$, but to the vector $(\bar{\varepsilon} \cdot \hat{\mathbf{k}})$. The time-averaged Poynting vector, as before, is given by

$$\langle \mathscr{P} \rangle = \tfrac{1}{2}(\mathbf{E}_0 \times \mathbf{H}_0) \tag{5.59}$$

which is perpendicular to both $\mathbf{E}_0$ and $\mathbf{H}_0$. Figures 5.1 and 5.2 show the relative

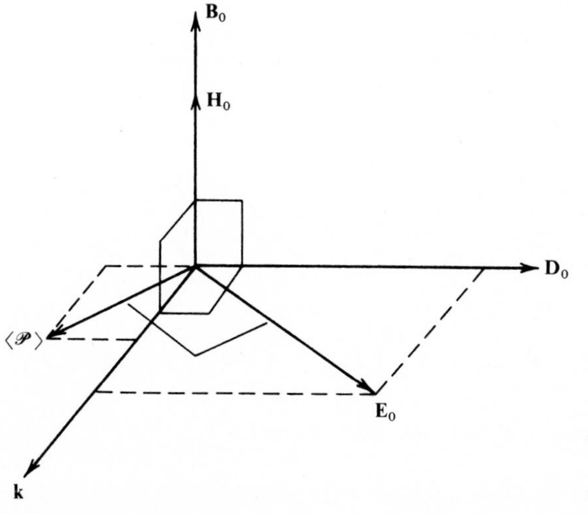

Figure 5.1 Orientations of field vectors, wave vector, and the time-averaged Poynting vector.

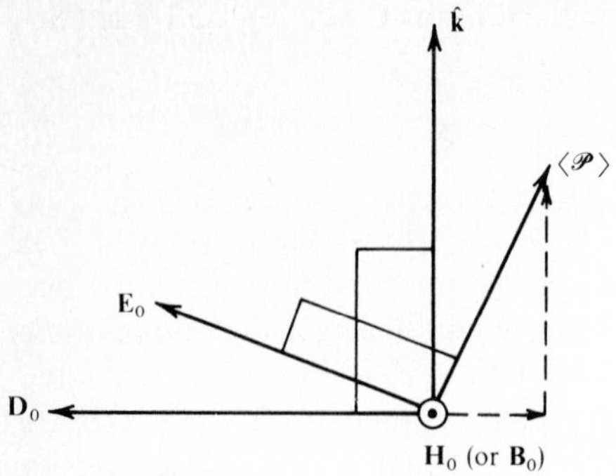

Figure 5.2 Relations among the field vectors: the wave normal $\hat{\mathbf{k}}$ is perpendicular to $\mathbf{D}_0$, and $\langle \mathscr{P} \rangle$ to $\mathbf{E}_0$, while $\langle \mathscr{P} \rangle$, $\hat{\mathbf{k}}$, $\mathbf{E}_0$, and $\mathbf{D}_0$ are coplanar. $\mathbf{H}_0$ (or $\mathbf{B}_0$) is normal to the plane of the paper.

orientation of these vectors. The angle between $\mathbf{E}_0$ and $\mathbf{D}_0$ is the same as that between $\langle \mathscr{P} \rangle$ and $\hat{\mathbf{k}}$. We see that $\mathbf{D}_0$, $\mathbf{H}_0$, and $\mathbf{k}$, on the one hand, and $\mathbf{E}_0$, $\mathbf{H}_0$, and $\langle \mathscr{P} \rangle$ on the other, form orthogonal vector triplets with the common vector $\mathbf{H}_0$ (or $\mathbf{B}_0$). An important conclusion is that in a crystal, the direction of energy propagation $\langle \mathscr{P} \rangle$ does not generally coincide with the wave normal $\hat{\mathbf{k}}$.

5.4 DETERMINATION OF WAVE AND FIELD VECTORS

We will now show that, in the general case, when any one of the vectors $\mathbf{E}_0$, $\mathbf{D}_0$, $\mathbf{H}_0$ (or $\mathbf{B}_0 = \mu_0 \mathbf{H}_0$), and $\mathbf{k}$ is given, the remaining vectors of the uniform plane wave in crystals are completely determined.

First, let us consider the case when either $\mathbf{E}_0$ or $\mathbf{D}_0$ is given. Since they are related by the constitutive relation $\mathbf{D}_0 = \varepsilon_0 \bar{\boldsymbol{\varepsilon}} \cdot \mathbf{E}_0$ once we know $\mathbf{E}_0$ we know $\mathbf{D}_0$, or vice versa. Taking the dot product of Eq. (5.41) with $\mathbf{E}_0$ and using Eq. (5.42) we find

$$\begin{aligned} \mathbf{E}_0 \cdot \mathbf{D}_0 &= -\frac{1}{\omega} \mathbf{E}_0 \cdot (\mathbf{k} \times \mathbf{H}_0) \\ &= \frac{1}{\omega} \mathbf{H}_0 \cdot (\mathbf{k} \times \mathbf{E}_0) \\ &= \mu_0 \mathbf{H}_0^2 \end{aligned}$$

or

$$\mathbf{H}_0^2 = \frac{1}{\mu_0} (\mathbf{E}_0 \cdot \mathbf{D}_0) \tag{5.60}$$

Since $\mathbf{E}_0$ is perpendicular to $\mathbf{H}_0$ ($\mathbf{E}_0 \perp \mathbf{H}_0$) and $\mathbf{H}_0$ is perpendicular to $\mathbf{D}_0$ ($\mathbf{H}_0 \perp \mathbf{D}_0$), thus $\mathbf{H}_0$ is parallel to $\mathbf{E}_0 \times \mathbf{D}_0$ [cf. Fig. 5.1], i.e.,

$$\mathbf{H}_0 \parallel \mathbf{E}_0 \times \mathbf{D}_0 \tag{5.61}$$

On the other hand, taking the dot product of Eq. (5.41) with itself and using Eq. (5.60), we obtain

$$\mathbf{k}^2 = \frac{\omega^2 \mathbf{D}_0^2}{\mathbf{H}_0^2} = \frac{\omega^2 \mu_0 \mathbf{D}_0^2}{\mathbf{E}_0 \cdot \mathbf{D}_0} \tag{5.62}$$

To find the direction of $\mathbf{k}$, we note that $\mathbf{H}_0 \perp \hat{\mathbf{k}}$ and $\hat{\mathbf{k}} \perp \mathbf{D}_0$ [cf. Figs. 5.1 and 5.2]; thus

$$\hat{\mathbf{k}} \parallel \mathbf{D}_0 \times \mathbf{H}_0 \parallel \mathbf{D}_0 \times (\mathbf{E}_0 \times \mathbf{D}_0) \tag{5.63}$$

In summary, for a given $\mathbf{E}_0$ or $\mathbf{D}_0$, we may determine the magnitude of $\mathbf{H}_0$ from Eq. (5.60) and its direction from Eq. (5.61), the magnitude of $\mathbf{k}$ from Eq. (5.62), and its direction from Eq. (5.63).

Next, suppose that $\mathbf{H}_0$ is given. As a consequence from Maxwell's equations, we see that $\mathbf{E}_0 \cdot \mathbf{H}_0 = 0$ and

$$\mathbf{H}_0 \cdot \mathbf{D}_0 = \varepsilon_0 \mathbf{H}_0 \cdot \bar{\varepsilon} \cdot \mathbf{E}_0 = \varepsilon_0 \mathbf{E}_0 \cdot \bar{\varepsilon} \cdot \mathbf{H}_0 = 0 \tag{5.64}$$

[cf. Fig. 5.1]. Consequently vector $\mathbf{E}_0$ is perpendicular to both vectors $\mathbf{H}_0$ and $\bar{\varepsilon} \cdot \mathbf{H}_0$, and hence parallel to their cross product:

$$\mathbf{E}_0 = C_1[\mathbf{H}_0 \times (\bar{\varepsilon} \cdot \mathbf{H}_0)]$$

where C_1 is an arbitrary constant. Substitution of the above equation into Eq. (5.62) yields

$$\begin{aligned}\mathbf{k}^2 &= \frac{\omega^2 \mu_0 \mathbf{D}_0^2}{\mathbf{E}_0 \cdot \mathbf{D}_0} = \frac{k_0^2(\bar{\varepsilon} \cdot \mathbf{E}_0) \cdot (\bar{\varepsilon} \cdot \mathbf{E}_0)}{\mathbf{E}_0 \cdot \bar{\varepsilon} \cdot \mathbf{E}_0} \\ &= \frac{k_0^2\{\bar{\varepsilon} \cdot [\mathbf{H}_0 \times (\bar{\varepsilon} \cdot \mathbf{H}_0)]\}^2}{[\mathbf{H}_0 \times (\bar{\varepsilon} \cdot \mathbf{H}_0)] \cdot \bar{\varepsilon} \cdot [\mathbf{H}_0 \times (\bar{\varepsilon} \cdot \mathbf{H}_0)]}\end{aligned} \tag{5.65}$$

Using the identity (1.117)

$$\bar{\mathbf{A}} \cdot (\mathbf{u} \times \bar{\mathbf{I}}) = \{[(\text{adj}\ \tilde{\bar{\mathbf{A}}}) \cdot \mathbf{u}] \times \bar{\mathbf{I}}\} \cdot \bar{\mathbf{A}}^{-1}$$

we may write

$$\bar{\varepsilon} \cdot [\mathbf{H}_0 \times (\bar{\varepsilon} \cdot \mathbf{H}_0)] = [(\text{adj}\ \bar{\varepsilon}) \cdot \mathbf{H}_0] \times \mathbf{H}_0$$

Thus Eq. (5.65) may also be expressed as:

$$\mathbf{k}^2 = \frac{k_0^2\{[(\text{adj}\ \bar{\varepsilon}) \cdot \mathbf{H}_0] \times \mathbf{H}_0\}^2}{[\mathbf{H}_0 \times (\bar{\varepsilon} \cdot \mathbf{H}_0)] \cdot \{[(\text{adj}\ \bar{\varepsilon}) \cdot \mathbf{H}_0] \times \mathbf{H}_0\}} \tag{5.66}$$

Also

$$\hat{\mathbf{k}} \parallel \mathbf{D}_0 \times \mathbf{H}_0 \parallel \{\mathbf{H}_0 \times [\mathbf{H}_0 \times (\text{adj}\ \bar{\varepsilon}) \cdot \mathbf{H}_0]\} \tag{5.67}$$

Thus when $\mathbf{H}_0$ is given, we may find the magnitude of $\mathbf{k}$ from Eq. (5.65) or Eq. (5.66) and its direction from Eq. (5.67). The vector $\mathbf{D}_0$ can then be determined from $\mathbf{D}_0 = -(\mathbf{k} \times \mathbf{H}_0)/\omega$; thus $\mathbf{E}_0 = \bar{\varepsilon}^{-1} \cdot \mathbf{D}_0/\varepsilon_0$.

Finally, let us turn to the case when the wave vector $\mathbf{k}$ is given. We will determine the directions of the field vectors $\mathbf{E}_0$, $\mathbf{D}_0$, and $\mathbf{H}_0$. According to Eq. (5.15), we have

$$(k_0^2 \bar{\varepsilon} - k^2\bar{\mathbf{I}}) \cdot \mathbf{E}_0 = -(\mathbf{k} \cdot \mathbf{E}_0)\mathbf{k} \tag{5.68}$$

or

$$(k_0^2 \bar{\varepsilon} - k^2 \bar{\mathbf{I}}) \cdot \mathbf{E}_0 \parallel \mathbf{k} \tag{5.69}$$

Thus the direction of $\mathbf{E}_0$ depends on whether $k_0^2 \bar{\varepsilon} - k^2 \bar{\mathbf{I}}$ is singular or nonsingular.

1. *Matrix* $k_0^2 \bar{\varepsilon} - k^2 \bar{\mathbf{I}}$ *is nonsingular.* In other words, we have

$$|k_0^2 \bar{\varepsilon} - k^2 \bar{\mathbf{I}}| \neq 0 \tag{5.70}$$

and the inverse exists. Dot premultiplication of Eq. (5.68) by the inverse of $k_0^2 \bar{\varepsilon} - k^2 \bar{\mathbf{I}}$ yields the direction of $\mathbf{E}_0$:

$$\mathbf{e} = [\text{adj}\,(k_0^2 \bar{\varepsilon} - k^2 \bar{\mathbf{I}})] \cdot \mathbf{k} \tag{5.71}$$

and the direction of $\mathbf{D}_0$ follows from the constitutive relation

$$\mathbf{d} = \varepsilon_0 \bar{\varepsilon} \cdot \mathbf{e} = \varepsilon_0 \bar{\varepsilon} \cdot [\text{adj}\,(k_0^2 \bar{\varepsilon} - k^2 \bar{\mathbf{I}})] \cdot \mathbf{k} \tag{5.72}$$

If we take the dot product of Eq. (5.72) with $\mathbf{k}$ and note that $\mathbf{k} \cdot \mathbf{D}_0 = 0$, we obtain

$$\hat{\mathbf{k}} \cdot \bar{\varepsilon} \cdot [\text{adj}\,(k_0^2 \bar{\varepsilon} - k^2 \bar{\mathbf{I}})] \cdot \hat{\mathbf{k}} = 0$$

which is the dispersion equation (5.32), but here it is obtained by a simple alternative method. The direction of $\mathbf{H}_0$ may be obtained from the Maxwell equations

$$\mathbf{h} = \frac{1}{\omega\mu_0} (\mathbf{k} \times \mathbf{e}) = \frac{1}{\omega\mu_0} \mathbf{k} \times \{[\text{adj}\,(k_0^2 \bar{\varepsilon} - k^2 \bar{\mathbf{I}})] \cdot \mathbf{k}\} \tag{5.73}$$

or alternatively

$$\mathbf{h} = \frac{\omega}{k^2} (\mathbf{k} \times \mathbf{d}) = \frac{\omega\varepsilon_0}{k^2} \mathbf{k} \times \{\bar{\varepsilon} \cdot [\text{adj}\,(k_0^2 \bar{\varepsilon} - k^2 \bar{\mathbf{I}})] \cdot \mathbf{k}\} \tag{5.74}$$

To summarize, when the wave vector $\mathbf{k} = k\hat{\mathbf{k}}$ is given such that $|k_0^2 \bar{\varepsilon} - k^2 \bar{\mathbf{I}}| \neq 0$, the directions of the field vectors $\mathbf{E}_0$, $\mathbf{D}_0$, and $\mathbf{H}_0$ of the wave in crystals are uniquely determined in terms of $\mathbf{k}$ by Eqs. (5.71), (5.72), and (5.73), or (5.74) respectively.

As an alternative approach, we may begin with Eq. (5.18). Expanding the product $(\mathbf{k} \times \bar{\mathbf{I}}) \cdot \bar{\varepsilon}^{-1} \cdot (\mathbf{k} \times \bar{\mathbf{I}})$ in Eq. (5.18) according to Eq. (1.120), and then dot-premultiplying the resulted equation by adj $\bar{\varepsilon}$, we obtain

$$[k_0^2(\text{adj}\,\bar{\varepsilon}) - (\mathbf{k} \cdot \bar{\varepsilon} \cdot \mathbf{k})\bar{\mathbf{I}}] \cdot \mathbf{H}_0 = -(\mathbf{k} \cdot \bar{\varepsilon} \cdot \mathbf{H}_0)\mathbf{k} \tag{5.75}$$

Now if the matrix $[k_0^2(\text{adj}\,\bar{\varepsilon}) - (\mathbf{k} \cdot \bar{\varepsilon} \cdot \mathbf{k})\bar{\mathbf{I}}]$ is nonsingular, i.e., if

$$|k_0^2(\text{adj}\,\bar{\varepsilon}) - (\mathbf{k} \cdot \bar{\varepsilon} \cdot \mathbf{k})\bar{\mathbf{I}}| \neq 0 \tag{5.76}$$

the direction of $\mathbf{H}_0$ follows from Eq. (5.75)

$$\mathbf{h}' = \{\text{adj}\,[k_0^2(\text{adj}\,\bar{\varepsilon}) - (\mathbf{k} \cdot \bar{\varepsilon} \cdot \mathbf{k})\bar{\mathbf{I}}\} \cdot \mathbf{k} \tag{5.77}$$

Thus

$$\mathbf{d}' = -\frac{1}{\omega}(\mathbf{k} \times \mathbf{h}')$$

$$= \frac{1}{\omega}\{[\text{adj}\,(k_0^2\,\text{adj}\,\bar{\boldsymbol{\varepsilon}} - \mathbf{k}\cdot\bar{\boldsymbol{\varepsilon}}\cdot\mathbf{k}\bar{\mathbf{I}})]\cdot\mathbf{k}\} \times \mathbf{k} \tag{5.78}$$

The direction of $\mathbf{E}_0$ may be found from the constitutive relation

$$\mathbf{e}' = \frac{\bar{\boldsymbol{\varepsilon}}^{-1}\cdot\mathbf{d}'}{\varepsilon_0}$$

$$= \frac{1}{\omega\varepsilon_0}\,\bar{\boldsymbol{\varepsilon}}^{-1}\cdot\{[\text{adj}\,(k_0^2\,\text{adj}\,\bar{\boldsymbol{\varepsilon}} - \mathbf{k}\cdot\bar{\boldsymbol{\varepsilon}}\cdot\mathbf{k}\bar{\mathbf{I}})]\cdot\mathbf{k}\} \times \mathbf{k} \tag{5.79}$$

By direct calculation we can easily show that for a given wave vector $\mathbf{k}$ the directions of vectors $\mathbf{E}_0$, $\mathbf{D}_0$, and $\mathbf{H}_0$ as specified by Eqs. (5.71), (5.72), (5.73), or (5.74) and Eqs. (5.79), (5.78), and (5.77) are equivalent. Furthermore, from the fact that $\mathbf{k}\cdot\mathbf{H}_0 = 0$ and Eq. (5.77) we obtain

$$\hat{\mathbf{k}}\cdot\{\text{adj}\,[k_0^2(\text{adj}\,\bar{\boldsymbol{\varepsilon}}) - (\mathbf{k}\cdot\bar{\boldsymbol{\varepsilon}}\cdot\mathbf{k})\bar{\mathbf{I}}]\}\cdot\hat{\mathbf{k}} = 0 \tag{5.80}$$

This, again, is the dispersion equation in still another form. We note that formulas (5.71) to (5.74) and (5.77) to (5.79) remain valid if we replace adj $(k_0^2\,\bar{\boldsymbol{\varepsilon}} - k^2\bar{\mathbf{I}})$ and adj $[k_0^2(\text{adj}\,\bar{\boldsymbol{\varepsilon}}) - (\mathbf{k}\cdot\bar{\boldsymbol{\varepsilon}}\cdot\mathbf{k})\bar{\mathbf{I}}]$ by the inverses $(k_0^2\,\bar{\boldsymbol{\varepsilon}} - k^2\bar{\mathbf{I}})^{-1}$ and $[k_0^2(\text{adj}\,\bar{\boldsymbol{\varepsilon}}) - (\mathbf{k}\cdot\bar{\boldsymbol{\varepsilon}}\cdot\mathbf{k})\,\bar{\mathbf{I}}]^{-1}$ respectively.

2. *Matrix $k_0^2\,\bar{\boldsymbol{\varepsilon}} - k^2\,\bar{\mathbf{I}}$ is singular.* That is, we have

$$|k_0^2\,\bar{\boldsymbol{\varepsilon}} - k^2\,\bar{\mathbf{I}}| = 0 \tag{5.81}$$

which means that k^2 is an eigenvalue of $k_0^2\,\bar{\boldsymbol{\varepsilon}}$. Because the inverse of the matrix $k_0^2\,\bar{\boldsymbol{\varepsilon}} - k^2\,\bar{\mathbf{I}}$ does not exist, the method used to determine the direction of $\mathbf{E}_0$ fails. Let $\mathbf{l}_i$ be the ith eigenvector of the matrix $k_0^2\,\bar{\boldsymbol{\varepsilon}}$ with the corresponding eigenvalue k_i^2, that is,

$$k_0^2\,\bar{\boldsymbol{\varepsilon}}\cdot\mathbf{l}_i = k_i^2\,\mathbf{l}_i \qquad \text{(no sum)} \tag{5.82}$$

where k_i^2 is a solution of Eq. (5.81). Since $(k_0^2\,\bar{\boldsymbol{\varepsilon}} - k_i^2\,\bar{\mathbf{I}})$ is a symmetric matrix, according to Sec. 1.7, we may write

$$\text{adj}\,(k_0^2\,\bar{\boldsymbol{\varepsilon}} - k_i^2\,\bar{\mathbf{I}}) = C\mathbf{l}_i\mathbf{l}_i \qquad \text{(no sum)} \tag{5.83}$$

where C is an arbitrary constant. Substitution of Eqs. (5.81) and (5.83) into the dispersion equation (5.30) yields the condition to be imposed on the direction of the wave vector $\mathbf{k}_i$:

$$\mathbf{k}_i\cdot\mathbf{l}_i = 0 \qquad \text{(no sum)} \tag{5.84}$$

i.e., the wave vector $\mathbf{k}_i$ must lie on a plane perpendicular to the ith eigenvector of $k_0^2\,\bar{\boldsymbol{\varepsilon}}$.

To determine the direction of the magnetic field intensity $\mathbf{H}_0$ we dot-premultiply Eq. (5.75) by $\mathbf{l}_i$ and then use the condition (5.84) to obtain

$$k_0^2 \mathbf{l}_i \cdot (\text{adj } \bar{\varepsilon}) \cdot \mathbf{H}_0 - (\mathbf{k}_i \cdot \bar{\varepsilon} \cdot \mathbf{k}_i)(\mathbf{l}_i \cdot \mathbf{H}_0) = 0 \tag{5.85}$$

Dot premultiplication of Eq. (5.82) by $\bar{\varepsilon}^{-1}$ gives

$$\mathbf{l}_i \cdot (\text{adj } \bar{\varepsilon}) = \frac{k_0^2 |\bar{\varepsilon}| \mathbf{l}_i}{\mathbf{k}_i^2} \tag{5.86}$$

Substituting Eq. (5.86) into Eq. (5.85), we find

$$[k_0^4 |\bar{\varepsilon}| - (\mathbf{k}_i \cdot \bar{\varepsilon} \cdot \mathbf{k}_i)\mathbf{k}_i^2](\mathbf{l}_i \cdot \mathbf{H}_0) = 0 \tag{5.87}$$

To satisfy condition (5.87), two possibilities arise: $\mathbf{H}_0 \cdot \mathbf{l}_i$ is zero or nonzero.

(*a*) $\mathbf{H}_0 \cdot \mathbf{l}_i = 0$, that is, $\mathbf{H}_0$ *is perpendicular to* $\mathbf{l}_i$. But according to Maxwell's equation $\mathbf{H}_0$ must also be perpendicular to $\mathbf{k}$. Thus the direction of $\mathbf{H}_0$ is parallel to their cross product, i.e.,

$$\mathbf{h}_i = \mathbf{k}_i \times \mathbf{l}_i \qquad \text{(no sum)} \tag{5.88}$$

Substituting Eq. (5.88) into the Maxwell equations, we find the directions of $\mathbf{D}_0$ and $\mathbf{E}_0$:

$$\mathbf{d}_i = \frac{-\mathbf{k}_i \times \mathbf{h}_i}{\omega} = \frac{k_i^2 \mathbf{l}_i}{\omega} \qquad \text{(no sum)} \tag{5.89}$$

and

$$\mathbf{e}_i = \frac{\bar{\varepsilon}^{-1} \cdot \mathbf{d}_i}{\varepsilon_0} = \omega\mu_0 \mathbf{l}_i \tag{5.90}$$

Figure 5.3 shows the relative directions of these vectors.

(*b*) $\mathbf{H}_0 \cdot \mathbf{l}_i \neq 0$. From Eq. (5.87) we must have

$$k_0^4 |\bar{\varepsilon}| - (\hat{\mathbf{k}} \cdot \bar{\varepsilon} \cdot \hat{\mathbf{k}}) k_i^4 = 0$$

which is the condition (5.53) denoting the coincidence of the given wave normal

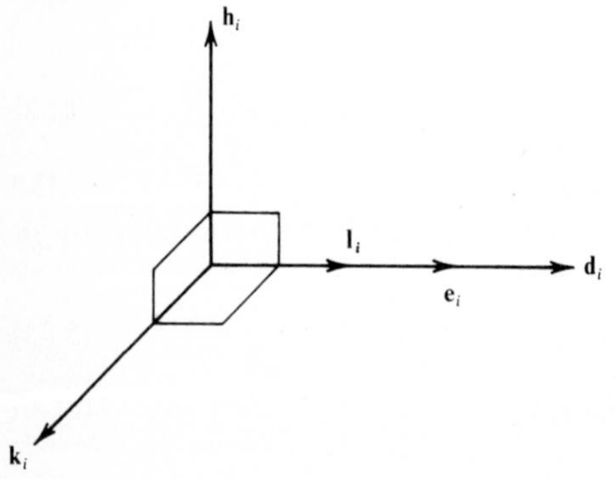

Figure 5.3 Orientations of field vectors when $k_0^2 \bar{\varepsilon} - k^2 \bar{\mathbf{I}}$ is singular.

with the optic axis. In this case, the field vectors are limited only by the conditions implied by Maxwell's equations (5.41) and (5.42):

$$\mathbf{E}_0 \cdot \bar{\varepsilon} \cdot \hat{\mathbf{k}} = 0 \qquad \mathbf{D}_0 \cdot \mathbf{H}_0 = 0 \qquad \hat{\mathbf{k}} \cdot \mathbf{H}_0 = 0 \qquad \mathbf{H}_0 \cdot \mathbf{E}_0 = 0$$

Finally, we may similarly analyze the case when

$$|k_0^2(\text{adj}\,\bar{\varepsilon}) - (\mathbf{k} \cdot \bar{\varepsilon} \cdot \mathbf{k})\bar{\mathbf{I}}| = 0 \tag{5.91}$$

or

$$\left| \frac{\text{adj}\ \bar{\varepsilon}}{\hat{\mathbf{k}} \cdot \bar{\varepsilon} \cdot \hat{\mathbf{k}}} - n^2\bar{\mathbf{I}} \right| = 0$$

Here the matrix $(\text{adj}\ \bar{\varepsilon})/(\hat{\mathbf{k}} \cdot \bar{\varepsilon} \cdot \hat{\mathbf{k}})$ and its eigenvalue play a basic role (see Prob. 5.4).

In summary, we may state that under all circumstances, a given wave vector $\mathbf{k}$ uniquely determines the directions of the field vectors of the wave in crystals except when $\hat{\mathbf{k}}$ is parallel to an optic axis. In that case, the field vectors are limited only by the conditions implied in the Maxwell equations.

5.5 ISONORMAL WAVES

According to Eq. (5.54), for every direction of wave normal $\hat{\mathbf{k}}$ there correspond two values of indices of refraction, n_+ and n_-, in crystals. Here the two values $\pm n_+$ (or $\pm n_-$) corresponding to any value of n_+^2 (or n_-^2) are counted as one since the negative value evidently belongs to the opposite direction of wave normal $\hat{\mathbf{k}}$. It follows that the structure of a crystal permits two monochromatic plane waves with two different linear polarizations and two different phase velocities c/n_+ and c/n_- to propagate in any given direction of wave normal. Such waves having the same $\hat{\mathbf{k}}$ but different phase velocities are called *isonormal waves*. We shall now show that the two directions of oscillation of $\mathbf{H}$ (and of $\mathbf{D}$) corresponding to a given direction of $\hat{\mathbf{k}}$ are perpendicular to each other. In fact, according to Eq. (5.21) the two isonormal waves $\mathbf{H}_0^+$ and $\mathbf{H}_0^-$ satisfy

$$[(\hat{\mathbf{k}} \times \bar{\mathbf{I}}) \cdot \bar{\varepsilon}^{-1} \cdot (\hat{\mathbf{k}} \times \bar{\mathbf{I}})] \cdot \mathbf{H}_0^+ = -\frac{1}{n_+^2}\mathbf{H}_0^+ \tag{5.92}$$

and

$$[(\hat{\mathbf{k}} \times \bar{\mathbf{I}}) \cdot \bar{\varepsilon}^{-1} \cdot (\mathbf{k} \times \bar{\mathbf{I}})] \cdot \mathbf{H}_0^- = -\frac{1}{n_-^2}\mathbf{H}_0^- \tag{5.93}$$

respectively. Dot-premultiply Eq. (5.92) by $\mathbf{H}_0^-$ and Eq. (5.93) by $\mathbf{H}_0^+$, and then subtract. Noting that the matrices $\bar{\varepsilon}$ and $(\hat{\mathbf{k}} \times \bar{\mathbf{I}}) \cdot \bar{\varepsilon}^{-1} \cdot (\hat{\mathbf{k}} \times \bar{\mathbf{I}})$ are symmetric and $n_+^2 \neq n_-^2$, we obtain

$$\mathbf{H}_0^+ \cdot \mathbf{H}_0^- = 0 \tag{5.94}$$

i.e., the $\mathbf{H}$ fields of the two isonormal waves are orthogonal. Also from Maxwell's equations

$$\begin{aligned} \omega\mu_0 \mathbf{H}_0^+ &= k_+ (\hat{\mathbf{k}} \times \mathbf{E}_0^+) \\ \omega\mu_0 \mathbf{H}_0^- &= k_-(\hat{\mathbf{k}} \times \mathbf{E}_0^-) \end{aligned} \tag{5.95}$$

and

$$\begin{aligned} \omega\mathbf{D}_0^+ &= -k_+ (\hat{\mathbf{k}} \times \mathbf{H}_0^+) \\ \omega\mathbf{D}_0^- &= -k_- (\hat{\mathbf{k}} \times \mathbf{H}_0^-) \end{aligned} \tag{5.96}$$

we conclude that

$$\hat{\mathbf{k}} \perp \mathbf{H}_0^+ \qquad \mathbf{H}_0^+ \perp \mathbf{H}_0^- \qquad \mathbf{H}_0^- \perp \hat{\mathbf{k}} \tag{5.97}$$

and

$$\begin{aligned} &\hat{\mathbf{k}} \perp \mathbf{D}_0^+ \qquad \mathbf{D}_0^+ \perp \mathbf{H}_0^+ \qquad \mathbf{H}_0^+ \perp \hat{\mathbf{k}} \\ &\hat{\mathbf{k}} \perp \mathbf{D}_0^- \qquad \mathbf{D}_0^- \perp \mathbf{H}_0^- \qquad \mathbf{H}_0^- \perp \hat{\mathbf{k}} \end{aligned} \tag{5.98}$$

Comparing Eqs. (5.97) and (5.98) we have

$$\mathbf{D}_0^+ \parallel \mathbf{H}_0^- \qquad \mathbf{D}_0^- \parallel \mathbf{H}_0^+ \qquad \mathbf{D}_0^+ \perp \mathbf{D}_0^- \tag{5.99}$$

Also, from Eqs. (5.95) and (5.97)

$$\begin{aligned} &\mathbf{E}_0^+ \perp \mathbf{H}_0^+ \qquad \mathbf{E}_0^+ \perp \mathbf{D}_0^- \\ &\mathbf{E}_0^- \perp \mathbf{H}_0^- \qquad \mathbf{E}_0^- \perp \mathbf{D}_0^+ \end{aligned} \tag{5.100}$$

Figure 5.4 shows the orientations of field vectors of the isonormal waves. It should be noted that the above relations are always valid except when the wave normal $\hat{\mathbf{k}}$ coincides with an optic axis. In the latter case $n_+ = n_-$, and the two isonormal waves become one.

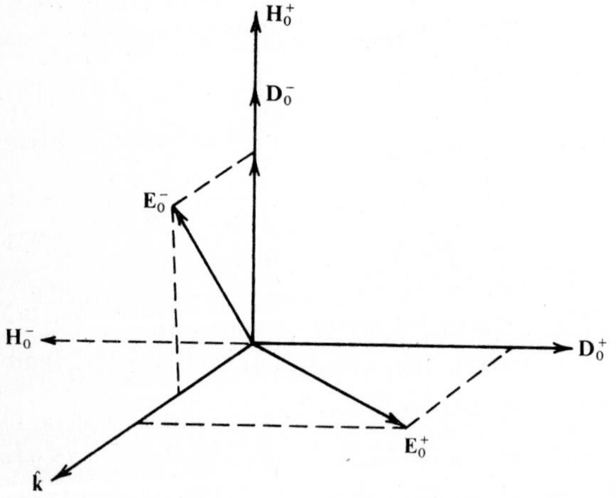

Figure 5.4 Orientations of the field vectors of the two isonormal waves.

5.6 VELOCITY OF ENERGY TRANSPORT

As we have noted before, aside from optic axes, the field vectors $\mathbf{E}_0$, $\mathbf{H}_0$, and $\mathbf{D}_0$ of the uniform plane waves in lossless crystals are always linearly polarized. Without loss of generality, we may assume that they are real vectors. As in the case of isotropic media, the theorem of equal electric and magnetic energy densities remains valid. This results from Maxwell's equations (5.11) and (5.12). Since

$$\mathbf{E}_0 \cdot \mathbf{D}_0 = -\frac{1}{\omega}\mathbf{E}_0 \cdot (\mathbf{k} \times \mathbf{H}_0)$$

$$= \frac{1}{\omega}\mathbf{H}_0 \cdot (\mathbf{k} \times \mathbf{E}_0) = \mathbf{H}_0 \cdot \mathbf{B}_0 \tag{5.101}$$

thus, from Eqs. (5.5) and (5.6)

$$\langle W_e \rangle = \tfrac{1}{4}(\mathbf{E}_0 \cdot \mathbf{D}_0) = \tfrac{1}{4}(\mathbf{H}_0 \cdot \mathbf{B}_0) = \langle W_m \rangle \tag{5.102}$$

The total time-averaged energy density is

$$\langle W \rangle = \langle W_e \rangle + \langle W_m \rangle$$

$$= \tfrac{1}{2}(\mathbf{E}_0 \cdot \mathbf{D}_0) = \tfrac{1}{2}(\mathbf{H}_0 \cdot \mathbf{B}_0) = \frac{1}{2\omega}(\mathbf{E}_0 \times \mathbf{H}_0) \cdot \mathbf{k} \tag{5.103}$$

The field vectors in Eq. (5.103), according to Eq. (5.68), are

$$\mathbf{E}_0 = -(\mathbf{k} \cdot \mathbf{E}_0)(k_0^2\bar{\varepsilon} - k^2\bar{\mathbf{I}})^{-1} \cdot \mathbf{k} \tag{5.104}$$

and

$$\mathbf{D}_0 = \varepsilon_0 \bar{\varepsilon} \cdot \mathbf{E}_0 = -\varepsilon_0(\mathbf{k} \cdot \mathbf{E}_0)\,\bar{\varepsilon} \cdot (k_0^2\bar{\varepsilon} - k^2\bar{\mathbf{I}})^{-1} \cdot \mathbf{k} \tag{5.105}$$

Substitution of Eqs. (5.104) and (5.105) into Eq. (5.103) yields

$$\langle W \rangle = \frac{\varepsilon_0(\mathbf{k} \cdot \mathbf{E}_0)^2}{2\,|k_0^2\bar{\varepsilon} - k^2\bar{\mathbf{I}}|}[\mathbf{k} \cdot \operatorname{adj}(k_0^2\bar{\varepsilon} - k^2\bar{\mathbf{I}}) \cdot \bar{\varepsilon} \cdot (k_0^2\bar{\varepsilon} - k^2\bar{\mathbf{I}})^{-1} \cdot \mathbf{k}] \tag{5.106}$$

Expanding adj $(k_0^2\bar{\varepsilon} - k^2\bar{\mathbf{I}})$ according to Eq. (1.124*c*) and using the dispersion equation (5.35) we may write

$$\mathbf{k} \cdot \operatorname{adj}(k_0^2\bar{\varepsilon} - k^2\bar{\mathbf{I}}) \cdot \bar{\varepsilon} \cdot (k_0^2\bar{\varepsilon} - k^2\bar{\mathbf{I}})^{-1} \cdot \mathbf{k} = k_0^4|\bar{\varepsilon}|[\mathbf{k} \cdot (k_0^2\bar{\varepsilon} - k^2\bar{\mathbf{I}})^{-1} \cdot \mathbf{k}]$$
$$+ k_0^2 k^2[\mathbf{k} \cdot \bar{\varepsilon}^2 \cdot (k_0^2\bar{\varepsilon} - k^2\bar{\mathbf{I}})^{-1} \cdot \mathbf{k}]$$

But from Eq. (5.35) we have

$$k_0^2\mathbf{k} \cdot \bar{\varepsilon}^2 \cdot (k_0^2\bar{\varepsilon} - k^2\bar{\mathbf{I}})^{-1} \cdot \mathbf{k} = \mathbf{k} \cdot \bar{\varepsilon} \cdot [(k_0^2\bar{\varepsilon} - k^2\bar{\mathbf{I}}) + k^2\bar{\mathbf{I}}] \cdot (k_0^2\bar{\varepsilon} - k^2\bar{\mathbf{I}})^{-1} \cdot \mathbf{k}$$
$$= \mathbf{k} \cdot \bar{\varepsilon} \cdot \mathbf{k}$$

Thus from Eq. (5.33) we finally obtain

$$\mathbf{k} \cdot \operatorname{adj}(k_0^2\bar{\varepsilon} - k^2\bar{\mathbf{I}}) \cdot \bar{\varepsilon} \cdot (k_0^2\bar{\varepsilon} - k^2\bar{\mathbf{I}})^{-1} \cdot \mathbf{k} = (\mathbf{k} \cdot \bar{\varepsilon} \cdot \mathbf{k})k^2 - k_0^4|\bar{\varepsilon}| \tag{5.107}$$

and the total time-averaged energy density

$$\langle W \rangle = \frac{\varepsilon_0(\mathbf{k}\cdot\mathbf{E}_0)^2}{2\,|k_0^2\bar{\varepsilon} - k^2\bar{\mathbf{I}}|}\,[(\mathbf{k}\cdot\bar{\varepsilon}\cdot\mathbf{k})k^2 - k_0^4\,|\bar{\varepsilon}\,|] \tag{5.108}$$

On the other hand, by taking the cross product of Eq. (5.11) with $\mathbf{k}$ and noting that $\mathbf{k}\cdot\mathbf{H} = 0$, we get

$$\mathbf{H}_0 = \frac{\omega}{k^2}\,(\mathbf{k}\times\mathbf{D}_0) \tag{5.109}$$

Substituting Eq. (5.109) into Eq. (5.59) and expanding the vector triple product, we obtain the time-averaged Poynting vector in the form [cf. Fig. 5.2]

$$\langle \mathscr{P} \rangle = \frac{\omega}{2k^2}\,[(\mathbf{E}_0\cdot\mathbf{D}_0)\mathbf{k} - (\mathbf{k}\cdot\mathbf{E}_0)\mathbf{D}_0] \tag{5.110}$$

Using the results of Eqs. (5.105) and (5.106) and noting that $\mathbf{E}_0\cdot\mathbf{D}_0 = 2\langle W\rangle$, we may rewrite Eq. (5.110) as

$$\langle \mathscr{P} \rangle = \frac{\omega\varepsilon_0(\mathbf{k}\cdot\mathbf{E}_0)^2}{2k^2\,|k_0^2\bar{\varepsilon} - k^2\bar{\mathbf{I}}|}\,\{[(\mathbf{k}\cdot\bar{\varepsilon}\cdot\mathbf{k})k^2 - k_0^4\,|\bar{\varepsilon}\,|]\mathbf{k} + \bar{\varepsilon}\cdot[\mathrm{adj}\,(k_0^2\bar{\varepsilon} - k^2\bar{\mathbf{I}})]\cdot\mathbf{k}\} \tag{5.111}$$

Again, expanding adj $(k_0^2\bar{\varepsilon} - k^2\bar{\mathbf{I}})$ according to Eq. (1.124*c*), we obtain

$$\langle \mathscr{P} \rangle = \frac{\omega\varepsilon_0(\mathbf{k}\cdot\mathbf{E}_0)^2}{2\,|k_0^2\bar{\varepsilon} - k^2\bar{\mathbf{I}}|}\,[(\mathbf{k}\cdot\bar{\varepsilon}\cdot\mathbf{k})\mathbf{k} + k_0^2\bar{\varepsilon}\cdot(\bar{\varepsilon} - \bar{\varepsilon}_t\bar{\mathbf{I}})\cdot\mathbf{k} + k^2(\bar{\varepsilon}\cdot\mathbf{k})] \tag{5.112}$$

Thus, as defined in Eq. (2.88), the velocity of energy transport for a lossless nonmagnetic crystal becomes

$$\mathbf{v}_E = \frac{\langle \mathscr{P} \rangle}{\langle W \rangle} = \omega\,\frac{(\mathbf{k}\cdot\bar{\varepsilon}\cdot\mathbf{k})\mathbf{k} + k_0^2\bar{\varepsilon}\cdot(\bar{\varepsilon} - \bar{\varepsilon}_t\bar{\mathbf{I}})\cdot\mathbf{k} + k^2(\bar{\varepsilon}\cdot\mathbf{k})}{(\mathbf{k}\cdot\bar{\varepsilon}\cdot\mathbf{k})k^2 - k_0^4\,|\bar{\varepsilon}\,|} \tag{5.113}$$

Unlike isotropic media, Eq. (5.113) clearly shows that in a crystal the energy does not generally propagate in the direction of wave normal $\hat{\mathbf{k}}$.

As a final note, we must distinguish between the phase velocity $\mathbf{v}_p$ and the velocity of energy transport $\mathbf{v}_E$. The former is in the direction of wave normal and is defined as the velocity with which the planes of constant phase travel:

$$\mathbf{v}_p = v_p\,\hat{\mathbf{k}} = \frac{\omega}{k}\,\hat{\mathbf{k}} \tag{5.114}$$

The latter is in the same direction as Poynting's vector

$$\mathbf{v}_E = \frac{\langle \mathscr{P} \rangle}{\langle W \rangle} \tag{5.115}$$

From Eqs. (5.59), (5.103), (5.114), and (5.115) we find the relation

$$v_p = \mathbf{v}_\mathbf{E}\cdot\hat{\mathbf{k}} \tag{5.116}$$

i.e., the phase velocity is the projection of the velocity of energy transport in the direction of wave normal.

5.7 DUALITY PRINCIPLE; CONICAL REFRACTION

We now define a *ray vector* $\mathbf{s}$ by

$$\mathbf{s} = \frac{\langle \mathscr{P} \rangle}{\omega \langle W \rangle} = \frac{\mathbf{v}_E}{\omega} \tag{5.117}$$

or, from Eqs. (5.59) and (5.103)

$$\mathbf{s} = \frac{\mathbf{E}_0 \times \mathbf{H}_0}{\omega(\mathbf{E}_0 \cdot \mathbf{D}_0)} \tag{5.118}$$

and derive some properties for $\mathbf{s}$. We shall first show that

$$\mathbf{k} \cdot \mathbf{s} = 1 \tag{5.119}$$

To prove this result, we take the cross product of Eq. (5.12) with $\mathbf{D}_0$ and then expand the vector triple product. Since $\mathbf{k} \cdot \mathbf{D}_0 = 0$ we have

$$\mathbf{k} = \frac{\omega(\mathbf{D}_0 \times \mathbf{B}_0)}{\mathbf{E}_0 \cdot \mathbf{D}_0} \tag{5.120}$$

Introducing Eqs. (5.120) and (5.118) into the left-hand side of Eq. (5.119) and using Eqs. (1.122) and (5.101), we readily establish (5.119).

We next show that it is possible to express Maxwell's equations (5.11) and (5.12) in terms of the ray vector $\mathbf{s}$. Indeed, by taking the cross products of Eqs. (5.11) and (5.12) respectively with $\mathbf{s}$ as defined in Eq. (5.118), we easily find

$$\mathbf{s} \times (\omega \mathbf{D}_0) = \mathbf{H}_0$$

$$\mathbf{s} \times (\omega \mathbf{B}_0) = -\mathbf{E}_0$$

For easy comparison, we will rewrite both sets below.

Set 1	*Set 2*
$\mathbf{D}_0 = -\frac{\mathbf{k}}{\omega} \times \mathbf{H}_0$	$\mathbf{E}_0 = -\omega \mathbf{s} \times \mathbf{B}_0$
$\mathbf{B}_0 = \frac{\mathbf{k}}{\omega} \times \mathbf{E}_0$	$\mathbf{H}_0 = \omega \mathbf{s} \times \mathbf{D}_0$
$\mathbf{D}_0 = \varepsilon_0 \bar{\boldsymbol{\varepsilon}} \cdot \mathbf{E}_0$	$\mathbf{E}_0 = (\varepsilon_0 \bar{\boldsymbol{\varepsilon}})^{-1} \cdot \mathbf{D}_0$
$\frac{\mathbf{k}}{\omega} = \frac{\mathbf{D}_0 \times \mathbf{B}_0}{\mathbf{E}_0 \cdot \mathbf{D}_0}$	$\omega \mathbf{s} = \frac{\mathbf{E}_0 \times \mathbf{H}_0}{\mathbf{E}_0 \cdot \mathbf{D}_0}$

(5.121)

It is clear that either set may be obtained from the other by the following systematic interchange of symbols

$$\begin{aligned} \mathbf{D}_0 &\rightleftarrows \mathbf{E}_0 \\ \mathbf{B}_0 &\rightleftarrows \mathbf{H}_0 \\ \frac{\mathbf{k}}{\omega} &\rightleftarrows \omega \mathbf{s} \\ \bar{\boldsymbol{\varepsilon}} &\rightleftarrows \bar{\boldsymbol{\varepsilon}}^{-1} \end{aligned} \tag{5.122}$$

Although the mathematical equations in sets 1 and 2 are equivalent and are mutually derivable from one another, they contain different physical quantities and have different physical constants. The formal recognition of this relation is called the *duality principle*, which may be stated generally as: if two sets of equations governing two groups of physical quantities are of the same mathematical form, and if we know a valid relation exists among one group of physical quantities, we may find an equally valid relation for the other by merely interchanging corresponding symbols. The duality principle is based solely on the mathematical symmetry of equations.

Applying this principle to the dispersion equation (5.28) we immediately obtain the required ray equation:

$$K(\mathbf{s}, \omega) = (\mathbf{s} \cdot \operatorname{adj} \bar{\boldsymbol{\varepsilon}} \cdot \mathbf{s}) k_0^4 s^2 + \mathbf{s} \cdot (\bar{\boldsymbol{\varepsilon}} - \bar{\varepsilon}_t \bar{\mathbf{I}}) \cdot \mathbf{s} k_0^2 + 1 = 0 \tag{5.123}$$

Of course, the same principle enables us to obtain other forms of the ray equation from Eqs. (5.32), (5.33), (5.34), etc. As an example, from Eq. (5.113), we may write the ray vector as

$$\mathbf{s} = \frac{(\mathbf{k} \cdot \bar{\boldsymbol{\varepsilon}} \cdot \mathbf{k})\mathbf{k} + k_0^2 \bar{\boldsymbol{\varepsilon}} \cdot (\bar{\boldsymbol{\varepsilon}} - \bar{\varepsilon}_t \bar{\mathbf{I}}) \cdot \mathbf{k} + k^2 (\bar{\boldsymbol{\varepsilon}} \cdot \mathbf{k})}{(\mathbf{k} \cdot \bar{\boldsymbol{\varepsilon}} \cdot \mathbf{k}) k^2 - k_0^4 |\bar{\boldsymbol{\varepsilon}}|} \tag{5.124}$$

Following the duality principle and interchanging the quantities in Eq. (5.124) according to Eq. (5.122), we easily obtain the wave vector expressed in terms of the ray vector:

$$\mathbf{k} = \frac{k_0^2 [k_0^2 (\mathbf{s} \cdot \operatorname{adj} \bar{\boldsymbol{\varepsilon}} \cdot \mathbf{s}) \mathbf{s} + (\bar{\boldsymbol{\varepsilon}} - \bar{\varepsilon}_t \bar{\mathbf{I}}) \cdot \mathbf{s} + k_0^2 s^2 (\operatorname{adj} \bar{\boldsymbol{\varepsilon}}) \cdot \mathbf{s}]}{k_0^4 (\mathbf{s} \cdot \operatorname{adj} \bar{\boldsymbol{\varepsilon}} \cdot \mathbf{s}) s^2 - 1} \tag{5.125}$$

In conclusion, with the aid of the duality principle, we may easily derive all the dual relations of the formulas established in this chapter.

Conical Refraction

For every given wave vector **k**, Eq. (5.124) uniquely determines the ray vector **s** except when the wave normal is parallel to an optic axis. In that case, the denominator of Eq. (5.124) vanishes and the ray vector is no longer unique. We shall now show that all the possible directions of the ray vector **s** lie on the surface of a cone.

Thus, according to Eqs. (3.117) and (5.134) we may write the group velocity as

$$\mathbf{v}_g = \frac{\partial \omega}{\partial \mathbf{k}} = -\frac{\partial F/\partial \mathbf{k}}{\partial F/\partial \omega} \tag{5.135}$$

To find the explicit forms of $\partial F/\partial \mathbf{k}$ and $\partial F/\partial \omega$ from Eq. (5.133), we note from Prob. 1-9 that

$$\frac{\partial(\mathbf{k}\cdot\bar{\varepsilon}\cdot\mathbf{k})k^2}{\partial \mathbf{k}} = 2(\mathbf{k}\cdot\bar{\varepsilon}\cdot\mathbf{k})\mathbf{k} + 2k^2\bar{\varepsilon}\cdot\mathbf{k} \tag{5.136}$$

and

$$\frac{\partial[\mathbf{k}\cdot\bar{\varepsilon}\cdot(\bar{\varepsilon}-\bar{\varepsilon}_t\bar{\mathbf{I}})\cdot\mathbf{k}]}{\partial \mathbf{k}} = 2[\bar{\varepsilon}\cdot(\bar{\varepsilon}-\bar{\varepsilon}_t\bar{\mathbf{I}})]\cdot\mathbf{k} \tag{5.137}$$

Thus

$$\frac{\partial F}{\partial \mathbf{k}} = 2\{(\mathbf{k}\cdot\bar{\varepsilon}\cdot\mathbf{k})\mathbf{k} + k^2(\bar{\varepsilon}\cdot\mathbf{k}) + k_0^2[\bar{\varepsilon}\cdot(\bar{\varepsilon}-\bar{\varepsilon}_t\bar{\mathbf{I}})]\cdot\mathbf{k}\} \tag{5.138}$$

Also, from Eq. (5.133)

$$\frac{\partial F}{\partial \omega} = \mathbf{k}\cdot[\bar{\varepsilon}\cdot(\bar{\varepsilon}-\bar{\varepsilon}_t\bar{\mathbf{I}})]\cdot\mathbf{k}\,\frac{2\omega}{c^2} + |\bar{\varepsilon}|\,\frac{4\omega^3}{c^4} \tag{5.139}$$

But according to the dispersion equation (5.133) we may express

$$\mathbf{k}\cdot[\bar{\varepsilon}\cdot(\bar{\varepsilon}-\bar{\varepsilon}_t\bar{\mathbf{I}})]\cdot\mathbf{k} = -|\bar{\varepsilon}|\,\frac{\omega^2}{c^2} - (\mathbf{k}\cdot\bar{\varepsilon}\cdot\mathbf{k})\,\frac{c^2k^2}{\omega^2} \tag{5.140}$$

Thus Eq. (5.139) becomes

$$\frac{\partial F}{\partial \omega} = -\frac{2}{\omega}\,[(\mathbf{k}\cdot\bar{\varepsilon}\cdot\mathbf{k})k^2 - k_0^4\,|\bar{\varepsilon}|] \tag{5.141}$$

Substituting Eqs. (5.138) and (5.141) into Eq. (5.135) we obtain

$$\mathbf{v}_g = \omega\,\frac{(\mathbf{k}\cdot\bar{\varepsilon}\cdot\mathbf{k})\mathbf{k} + k_0^2\bar{\varepsilon}\cdot(\bar{\varepsilon}-\bar{\varepsilon}_t\bar{\mathbf{I}})\cdot\mathbf{k} + k^2(\bar{\varepsilon}\cdot\mathbf{k})}{(\mathbf{k}\cdot\bar{\varepsilon}\cdot\mathbf{k})k^2 - k_0^4|\bar{\varepsilon}|} \tag{5.142}$$

Comparing Eq. (5.142) with Eq. (5.113), we see again that $\mathbf{v}_g = \mathbf{v}_E$, that is, the group velocity is equal to the velocity of energy transport as proved in Eq. (3.128) for the general case of lossless media. With this in mind, we may calculate the direction of ray vector $\mathbf{s}$ in a lossless medium from either $\langle\mathscr{P}\rangle$ or $\mathbf{v}_g$. The former is often simpler to use once the field vectors have been found. Since the gradient of a scalar function F is a vector whose direction is normal to F = constant surface, from Eq. (5.135) we conclude that the electromagnetic energy in crystals is propagated along the direction normal to the wave vector surface.

In summary, we have shown the following. With a known wave vector surface $k(\hat{\mathbf{k}})$, a given $\hat{\mathbf{k}}$ yields k and hence $\mathbf{k}$. The normal to the surface at $\mathbf{k}$ then gives the direction of $\mathbf{s}$ (or $\langle\mathscr{P}\rangle$, or $\mathbf{v}_E$). $\mathbf{H}_0$ is in the direction parallel to $\hat{\mathbf{k}}\times\mathbf{s}$, the plane formed by $\hat{\mathbf{k}}$ and $\mathbf{s}$ is also the plane of $\mathbf{D}_0$ and $\mathbf{E}_0$. As shown in Fig. 5.6, we can thus locate in this plane the directions of $\mathbf{D}_0$ (perpendicular to $\hat{\mathbf{k}}$) and $\mathbf{E}_0$ (perpendicular to $\mathbf{s}$).

Similar to the wave vector surface, we now define a ray vector surface as the surface traced out by the tips of the radius vector $\mathbf{s} = s\hat{\mathbf{s}}$ plotted from a fixed origin in all directions $\hat{\mathbf{s}}$ and of lengths that are proportional to the two solutions of the ray equation (5.123). Again, the ray surface is a two-sheeted surface.

Since $\mathbf{v}_E = \mathbf{v}_g = \partial\omega/\partial\mathbf{k}$, Eq. (5.117) becomes

$$\mathbf{s} = \frac{1}{\omega}\frac{\partial\omega}{\partial\mathbf{k}} \tag{5.143}$$

Using the relation (5.119), we can easily establish

$$\mathbf{k} = -\frac{1}{\omega}\frac{\partial\omega}{\partial\mathbf{s}} \tag{5.144}$$

Indeed, differentiating the implicit functional identity $\omega(\mathbf{s}) = \omega[\mathbf{k}(\mathbf{s})]$ with respect to s_i and using Eq. (5.143), we have

$$\frac{\partial\omega}{\partial s_i} = \frac{\partial\omega}{\partial\mathbf{k}} \cdot \frac{\partial\mathbf{k}}{\partial s_i} = \omega\mathbf{s} \cdot \frac{\partial\mathbf{k}}{\partial s_i} \tag{5.145}$$

But $\mathbf{k} \cdot \mathbf{s} = 1$, thus

$$\frac{\partial(\mathbf{k} \cdot \mathbf{s})}{\partial s_i} = 0 = \mathbf{s} \cdot \frac{\partial\mathbf{k}}{\partial s_i} + k_i \tag{5.146}$$

Combination of Eqs. (5.145) and (5.146) yields

$$k_i = -\frac{1}{\omega}\frac{\partial\omega}{\partial s_i}$$

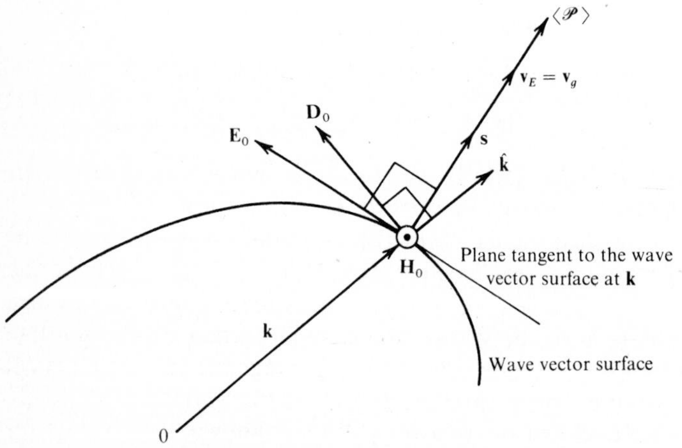

Figure 5.6 Section of a wave vector surface and orientations of vectors $\hat{\mathbf{k}}$, $\mathbf{s}$ (or $\mathbf{v}_E$ or $\langle\mathscr{P}\rangle$), $\mathbf{D}_0$, and $\mathbf{E}_0$. $\mathbf{H}_0$ (or $\mathbf{B}_0$) is out of the paper as indicated by the dot.

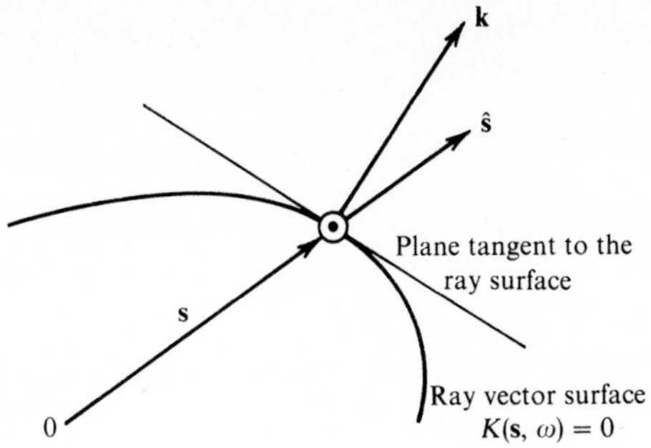

Figure 5.7 Section of a ray vector surface and the direction of the wave vector.

in index form which after converting into direct form, gives Eq. (5.144). Now taking the gradient of the ray equation $K(\mathbf{s}, \omega) = 0$ with respect to $\mathbf{s}$ we get

$$\frac{\partial K}{\partial \mathbf{s}} + \frac{\partial K}{\partial \omega}\frac{\partial \omega}{\partial \mathbf{s}} = \mathbf{0} \tag{5.147}$$

Substitution of Eq. (5.147) into Eq. (5.144) yields an alternative form

$$\mathbf{k} = \frac{1}{\omega}\frac{\partial K/\partial \mathbf{s}}{\partial K/\partial \omega} \tag{5.148}$$

Here we see that the wave vector $\mathbf{k}$ is normal to the ray vector surface as illustrated in Fig. 5.7.

As an alternative approach, from the ray equation (5.123) we easily find

$$\frac{\partial K}{\partial \mathbf{s}} = 2k_0^4\,[(\mathbf{s} \cdot \text{adj}\ \bar{\varepsilon} \cdot \mathbf{s})\mathbf{s} + \mathbf{s}^2(\text{adj}\ \bar{\varepsilon}) \cdot \mathbf{s}] + 2k_0^2(\bar{\varepsilon} - \bar{\varepsilon}_t \bar{\mathbf{I}}) \cdot \mathbf{s} \tag{5.149}$$

and

$$\frac{\partial K}{\partial \omega} = \frac{2}{\omega}\,[k_0^4(\mathbf{s} \cdot \text{adj}\ \bar{\varepsilon} \cdot \mathbf{s})\mathbf{s}^2 - 1] \tag{5.150}$$

Substitution of Eqs. (5.149) and (5.150) into Eq. (5.148) yields again the result of Eq. (5.125).

5.9 DETERMINATION OF WAVE VECTORS; THE BOOKER QUARTIC

As we have noted before, in an unbound crystal, the dispersion equation alone does not uniquely determine the wave vector. We shall now examine the problem of wave reflection from the interface of an isotropic and an anisotropic media. Assuming that the incident wave is in the isotropic medium 1, we will show how the wave vectors in the anisotropic medium 2 may be determined.

According to the laws of reflection and refraction derived in Sec. 3-5, we may represent any one of the wave vectors $\mathbf{k}_i$, $\mathbf{k}_r$, $\mathbf{k}_+$, $\mathbf{k}_-$ of the incident, the reflected, and the two transmitted waves in the general form

$$\mathbf{k}_\alpha = \mathbf{b} + q_\alpha \hat{\mathbf{q}} \tag{5.151}$$

where the subscript α denotes i, r, $+$, or $-$, and

$$\begin{aligned} \mathbf{b} &= \hat{\mathbf{q}} \times \mathbf{a} \\ \mathbf{a} &= \mathbf{k}_i \times \hat{\mathbf{q}} = \mathbf{b} \times \hat{\mathbf{q}} \end{aligned} \tag{5.152}$$

$\hat{\mathbf{q}}$ is a unit vector normal to the interface and points toward medium 2; q_α are parameters to be determined. Equation (5.151) clearly shows that from a fixed origin 0 on the interface, the tips of all the wave vectors must lie on a straight line that passes through the tip of vector $\mathbf{b}$ and is parallel to $\hat{\mathbf{q}}$ (see Fig. 3.15).

For a given incident wave, $\mathbf{b}$ is a fixed vector. Thus q_α in Eq. (5.151) may be determined from the dispersion equation of a given medium. Geometrically, this means that the tips of wave vectors must be the intersections of the straight line (5.151) and the given wave vector surface. This fact is of particular importance since it enables us to locate the wave vectors of the reflected and transmitted waves in the most direct and simple way.

To determine analytically the wave vectors in an electrically anisotropic medium 2, we substitute Eq. (5.151) into the dispersion equation (5.28) and obtain

$$Aq_\alpha^4 + Bq_\alpha^3 + Cq_\alpha^2 + Dq_\alpha + F = 0 \tag{5.153}$$

where

$$\begin{aligned} A &= \hat{\mathbf{q}} \cdot \bar{\varepsilon} \cdot \hat{\mathbf{q}} \\ B &= \mathbf{b} \cdot (\bar{\varepsilon} + \tilde{\bar{\varepsilon}}) \cdot \hat{\mathbf{q}} \\ C &= k_0^2 \hat{\mathbf{q}} \cdot [\operatorname{adj} \bar{\varepsilon} - (\operatorname{adj} \bar{\varepsilon})_t \bar{\mathbf{I}}] \cdot \hat{\mathbf{q}} + (\mathbf{b} \cdot \bar{\varepsilon} \cdot \mathbf{b}) + \mathbf{a}^2 (\hat{\mathbf{q}} \cdot \bar{\varepsilon} \cdot \hat{\mathbf{q}}) \\ D &= k_0^2 \mathbf{b} \cdot (\operatorname{adj} \bar{\varepsilon} + \operatorname{adj} \tilde{\bar{\varepsilon}}) \cdot \hat{\mathbf{q}} + \mathbf{a}^2 \mathbf{b} \cdot (\bar{\varepsilon} + \tilde{\bar{\varepsilon}}) \cdot \hat{\mathbf{q}} \\ F &= k_0^2 \mathbf{b} \cdot [\operatorname{adj} \bar{\varepsilon} - (\operatorname{adj} \bar{\varepsilon})_t \bar{\mathbf{I}}] \cdot \mathbf{b} + \mathbf{a}^2 (\mathbf{b} \cdot \bar{\varepsilon} \cdot \mathbf{b}) + k_0^4 |\bar{\varepsilon}| \end{aligned} \tag{5.154}$$

Evidently, the change of sign in $\mathbf{b}$ changes the sign in q_α in Eq. (5.153). In the case of a lossless crystal, $\bar{\varepsilon}$ is symmetric. Thus Eq. (5.154) becomes

$$\begin{aligned} A &= \hat{\mathbf{q}} \cdot \bar{\varepsilon} \cdot \hat{\mathbf{q}} \\ B &= 2(\mathbf{b} \cdot \bar{\varepsilon} \cdot \hat{\mathbf{q}}) \\ C &= k_0^2 \hat{\mathbf{q}} \cdot [\operatorname{adj} \bar{\varepsilon} - (\operatorname{adj} \bar{\varepsilon})_t \bar{\mathbf{I}}] \cdot \hat{\mathbf{q}} + (\mathbf{b} \cdot \bar{\varepsilon} \cdot \mathbf{b}) + \mathbf{a}^2 (\hat{\mathbf{q}} \cdot \bar{\varepsilon} \cdot \hat{\mathbf{q}}) \\ D &= 2k_0^2 \mathbf{b} \cdot (\operatorname{adj} \bar{\varepsilon}) \cdot \hat{\mathbf{q}} + 2\mathbf{a}^2 (\mathbf{b} \cdot \bar{\varepsilon} \cdot \hat{\mathbf{q}}) \\ F &= k_0^2 \mathbf{b} \cdot [\operatorname{adj} \bar{\varepsilon} - (\operatorname{adj} \bar{\varepsilon})_t \bar{\mathbf{I}}] \cdot \mathbf{b} + \mathbf{a}^2 (\mathbf{b} \cdot \bar{\varepsilon} \cdot \mathbf{b}) + k_0^4 |\bar{\varepsilon}| \end{aligned} \tag{5.155}$$

Equation (5.153), known as *Booker quartic*, is a complete equation of fourth degree. It reduces to a biquadratic equation in the following special cases.

Normal Incidence

In this case $\mathbf{k}_\alpha = q_\alpha \hat{\mathbf{q}}$ and Eq. (5.153) becomes

$$(\hat{\mathbf{q}} \cdot \bar{\boldsymbol{\varepsilon}} \cdot \hat{\mathbf{q}})q_\alpha^4 + k_0^2 \hat{\mathbf{q}} \cdot [\text{adj } \bar{\boldsymbol{\varepsilon}} - (\text{adj } \bar{\varepsilon})_t \bar{\mathbf{I}}] \cdot \hat{\mathbf{q}} q_\alpha^2 + k_0^4 |\bar{\boldsymbol{\varepsilon}}| = 0 \qquad (5.156)$$

which is of the same form as the dispersion equation (5.28).

Other Cases Occur When $B = D = 0$

From Eqs. (5.155) and (1.64) we see that this is possible only if

$$\mathbf{b} \cdot \hat{\mathbf{q}} = \mathbf{b} \cdot \bar{\boldsymbol{\varepsilon}} \cdot \hat{\mathbf{q}} = \mathbf{b} \cdot \bar{\boldsymbol{\varepsilon}}^2 \cdot \hat{\mathbf{q}} = 0 \qquad (5.157)$$

which means that vectors $\hat{\mathbf{q}}$, $\bar{\boldsymbol{\varepsilon}} \cdot \hat{\mathbf{q}}$, and $\bar{\boldsymbol{\varepsilon}}^2 \cdot \hat{\mathbf{q}}$ are all perpendicular to the same vector $\mathbf{b}$ and hence they must be coplanar. In other words, both vectors $\hat{\mathbf{q}} \times (\bar{\boldsymbol{\varepsilon}} \cdot \hat{\mathbf{q}})$ and $(\bar{\boldsymbol{\varepsilon}} \cdot \hat{\mathbf{q}}) \times (\bar{\boldsymbol{\varepsilon}}^2 \cdot \hat{\mathbf{q}})$ are parallel to $\mathbf{b}$. But, using Eq. (1.113) and noting that $\bar{\boldsymbol{\varepsilon}}$ is symmetric

$$(\bar{\boldsymbol{\varepsilon}} \cdot \hat{\mathbf{q}}) \times (\bar{\boldsymbol{\varepsilon}}^2 \cdot \hat{\mathbf{q}}) = (\text{adj } \bar{\boldsymbol{\varepsilon}}) \cdot [\hat{\mathbf{q}} \times (\bar{\boldsymbol{\varepsilon}} \cdot \hat{\mathbf{q}})]$$

we have

$$(\text{adj } \bar{\boldsymbol{\varepsilon}}) \cdot [\hat{\mathbf{q}} \times (\bar{\boldsymbol{\varepsilon}} \cdot \hat{\mathbf{q}})] = \lambda[\hat{\mathbf{q}} \times (\bar{\boldsymbol{\varepsilon}} \cdot \hat{\mathbf{q}})] \qquad (5.158)$$

where λ is a constant. This condition implies two possibilities: $[\hat{\mathbf{q}} \times (\bar{\boldsymbol{\varepsilon}} \cdot \hat{\mathbf{q}})]$ is zero or nonzero.

$\hat{\mathbf{q}} \times (\bar{\boldsymbol{\varepsilon}} \cdot \hat{\mathbf{q}}) = \mathbf{0}$ In this case, $\bar{\boldsymbol{\varepsilon}} \cdot \hat{\mathbf{q}}$ is parallel to $\hat{\mathbf{q}}$ and we may write

$$\bar{\boldsymbol{\varepsilon}} \cdot \hat{\mathbf{q}} = \lambda_1 \mathbf{q}$$

then $\hat{\mathbf{q}}$ is an eigenvector of $\bar{\boldsymbol{\varepsilon}}$.

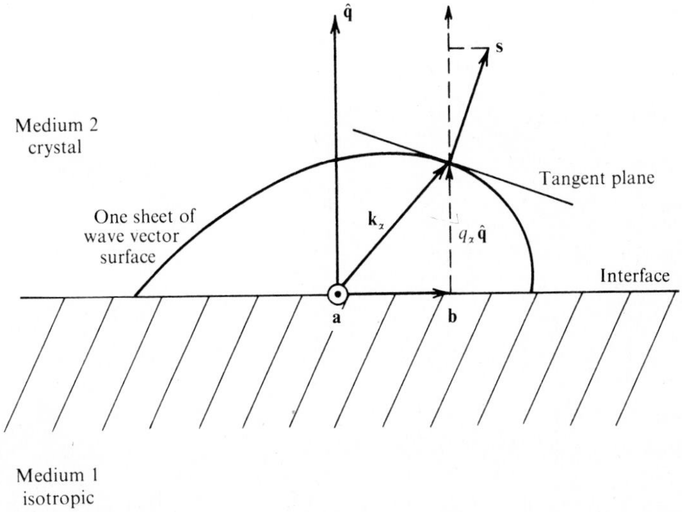

Figure 5.8 Geometrical construction of a wave vector from a given wave vector surface.

$\hat{\mathbf{q}} \times (\bar{\varepsilon} \cdot \hat{\mathbf{q}}) \neq \mathbf{0}$ Thus according to Eq. (5.158), vector $\hat{\mathbf{q}} \times (\bar{\varepsilon} \cdot \hat{\mathbf{q}})$ which is parallel to $\mathbf{b}$ is an eigenvector of adj $\bar{\varepsilon}$. In short, the Booker quartic equation reduces to biquadratic, when either $\hat{\mathbf{q}}$ is an eigenvector of $\bar{\varepsilon}$ or $\mathbf{b}$ is an eigenvector of adj $\bar{\varepsilon}$.

The quartic equation (5.153) yields four roots. But we select only those two roots which give the directions of energy propagation that are directed toward medium 2. That is, we choose those two q_α such that $\mathbf{s} \cdot \hat{\mathbf{q}} > 0$. The geometrical construction of one such point is illustrated in Fig. 5.8. Once the values of q_α are found from either Eq. (5.153) or geometrical constructions, we can determine the corresponding wave vectors from Eq. (5.151), and then the directions of field vectors in crystals according to Sec. 5.4.

5.10 TRANSMISSION AND REFLECTION COEFFICIENT MATRICES

We now proceed to determine the reflection and transmission coefficients at the interface of an isotropic-anisotropic medium. We assume that a uniform plane wave of arbitrary polarization is incident from an isotropic medium 1 on an anisotropic medium 2. The incident wave gives rise to a reflected wave in medium 1, characterized by the dielectric constant ε_1, and two transmitted waves in the anisotropic medium 2, characterized by the dielectric tensor $\bar{\varepsilon}$. Both media are assumed to have the same permeability μ_0 (see Fig. 5.9).

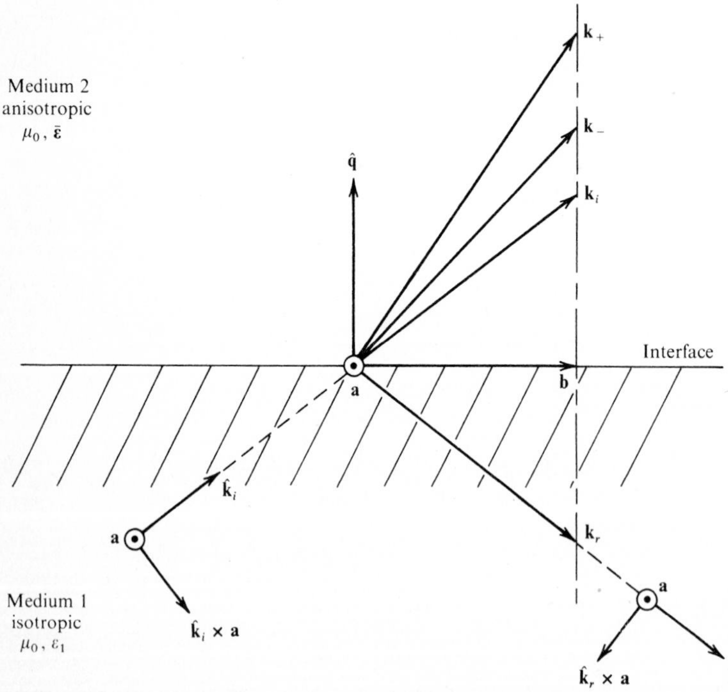

Figure 5.9 Reflection and transmission of waves at the interface of an isotropic-anisotropic medium. The dot indicates that the vector **a** is directed out of the paper. The plane of incidence lies on the paper.

Again, we decompose the amplitude vectors of the incident and the reflected waves in medium 1 into components perpendicular and parallel to the plane of incidence, namely, the incident wave

$$\mathbf{E}_{0i} = A_\perp \mathbf{a} + A_\parallel (\hat{\mathbf{k}}_i \times \mathbf{a})$$
$$\mathbf{H}_{0i} = \frac{1}{\omega\mu_0}(\mathbf{k}_i \times \mathbf{E}_{0i}) = \sqrt{\frac{\varepsilon_0 \varepsilon_1}{\mu_0}}\,[A_\perp(\mathbf{k}_i \times \mathbf{a}) - A_\parallel \mathbf{a}] \tag{5.159}$$

and the reflected wave

$$\mathbf{E}_{0r} = B_\perp \mathbf{a} + B_\parallel (\hat{\mathbf{k}}_r \times \mathbf{a})$$
$$\mathbf{H}_{0r} = \frac{1}{\omega\mu_0}(\mathbf{k}_r \times \mathbf{E}_{0r}) = \sqrt{\frac{\varepsilon_0 \varepsilon_1}{\mu_0}}\,[B_\perp(\hat{\mathbf{k}}_r \times \mathbf{a}) - B_\parallel \mathbf{a}] \tag{5.160}$$

Now, let the subscripts + and − denote the two transmitted waves in medium 2. The two solutions q_+ and q_- of the Booker quartic (5.153) are selected such that they satisfy the condition $\mathbf{s} \cdot \hat{\mathbf{q}} > 0$. The corresponding wave vectors $\mathbf{k}_+$ and $\mathbf{k}_-$ are then found from Eq. (5.151). Substituting these wave vectors separately into Eq. (5.71) or Eq. (5.90), we can finally determine the directions of the electric field intensities $\mathbf{e}_+$ and $\mathbf{e}_-$, and write the two transmitted waves as

$$\mathbf{E}_{0+} = C_+ \mathbf{e}_+$$
$$\mathbf{H}_{0+} = \frac{1}{\omega\mu_0}(\mathbf{k}_+ \times \mathbf{E}_{0+}) = C_+ \mathbf{h}_+ \tag{5.161}$$

where $$\mathbf{h}_+ = \frac{1}{\omega\mu_0}(\mathbf{k}_+ \times \mathbf{e}_+)$$

and $$\mathbf{E}_{0-} = C_- \mathbf{e}_-$$
$$\mathbf{H}_{0-} = \frac{1}{\omega\mu_0}(\mathbf{k}_- \times \mathbf{E}_{0-}) = C_- \mathbf{h}_- \tag{5.162}$$

where $$\mathbf{h}_- = \frac{1}{\omega\mu_0}(\mathbf{k}_- \times \mathbf{e}_-)$$

C_+ and C_- are two arbitrary constants.

At the interface $\mathbf{r} \cdot \hat{\mathbf{q}} = 0$ the field vectors satisfy the boundary conditions [cf. Eqs. (3.91) and (3.92)]:

$$(\mathbf{E}_{0i} + \mathbf{E}_{0r} - \mathbf{E}_{0+} - \mathbf{E}_{0-}) \times \hat{\mathbf{q}} = \mathbf{0}$$

or $$\mathbf{E}_{0i} + \mathbf{E}_{0r} - \mathbf{E}_{0+} - \mathbf{E}_{0-} = \alpha\hat{\mathbf{q}} \tag{5.163}$$

and $$(\mathbf{H}_{0i} + \mathbf{H}_{0r} - \mathbf{H}_{0+} - \mathbf{H}_{0-}) \times \hat{\mathbf{q}} = \mathbf{0}$$

or $$\mathbf{H}_{0i} + \mathbf{H}_{0r} - \mathbf{H}_{0+} - \mathbf{H}_{0-} = \beta\hat{\mathbf{q}} \tag{5.164}$$

where α and β are two constants. Substituting Eqs. (5.159), (5.162) into Eqs.

(5.163) and (5.164), we obtain

$$(A_\perp + B_\perp)\mathbf{a} + A_\parallel(\hat{\mathbf{k}}_i \times \mathbf{a}) + B_\parallel(\hat{\mathbf{k}}_r \times \mathbf{a}) - C_+\mathbf{e}_+ - C_-\mathbf{e}_- = \alpha\hat{\mathbf{q}} \tag{5.165}$$

and

$$(A_\parallel + B_\parallel)\mathbf{a} - A_\perp(\hat{\mathbf{k}}_i \times \mathbf{a}) - B_\perp(\hat{\mathbf{k}}_r \times \mathbf{a}) + \sqrt{\frac{\mu_0}{\varepsilon_0 \varepsilon_1}}\, C_+\mathbf{h}_+ + \sqrt{\frac{\mu_0}{\varepsilon_0 \varepsilon_1}}\, C_-\mathbf{h}_- = -\sqrt{\frac{\mu_0}{\varepsilon_0 \varepsilon_1}}\, \beta\hat{\mathbf{q}} \tag{5.166}$$

We will determine the amplitudes $B_\perp$, $B_\parallel$, C_+, and C_- of the reflected and transmitted waves in terms of the known amplitudes $A_\perp$, $A_\parallel$ of the incident wave. Taking the dot product of Eqs. (5.165) and (5.166) respectively, with vectors **a** and $\mathbf{b} = \hat{\mathbf{q}} \times \mathbf{a}$, we get

$$A_\perp + B_\perp - \frac{\mathbf{a}\cdot\mathbf{e}_+}{\mathbf{a}^2} C_+ - \frac{\mathbf{a}\cdot\mathbf{e}_-}{\mathbf{a}^2} C_- = 0 \tag{5.167}$$

$$A_\parallel - B_\parallel - \frac{k_i(\mathbf{b}\cdot\mathbf{e}_+)}{q_i\,\mathbf{a}^2} C_+ - \frac{k_i(\mathbf{b}\cdot\mathbf{e}_-)}{q_i\,\mathbf{a}^2} C_- = 0 \tag{5.168}$$

$$A_\parallel + B_\parallel + \frac{\omega\mu_0(\mathbf{a}\cdot\mathbf{h}_+)}{k_i\,\mathbf{a}^2} C_+ + \frac{\omega\mu_0(\mathbf{a}\cdot\mathbf{h}_-)}{k_i\,\mathbf{a}^2} C_- = 0 \tag{5.169}$$

$$A_\perp - B_\perp - \frac{q_+(\mathbf{a}\cdot\mathbf{e}_+)}{q_i\,\mathbf{a}^2} C_+ - \frac{q_-(\mathbf{a}\cdot\mathbf{e}_-)}{q_i\,\mathbf{a}^2} C_- = 0 \tag{5.170}$$

In deriving Eqs. (5.168) and (5.170), we have used the following results:

$$\mathbf{b}\cdot(\hat{\mathbf{k}}_i \times \mathbf{a}) = \frac{q_i\,\mathbf{a}^2}{k_i} = -\mathbf{b}\cdot(\hat{\mathbf{k}}_r \times \mathbf{a})$$

$$\mathbf{b}\cdot\mathbf{h}_\pm = \frac{q_\pm}{\omega\mu_0}(\mathbf{a}\cdot\mathbf{e}_\pm) \tag{5.171}$$

Eliminating $B_\perp$ or $A_\perp$ from Eqs. (5.167) and (5.170), and $B_\parallel$ or $A_\parallel$ from Eqs. (5.168) and (5.169), we have

$$\frac{(q_i + q_+)(\mathbf{a}\cdot\mathbf{e}_+)}{2q_i\,\mathbf{a}^2} C_+ + \frac{(q_i + q_-)(\mathbf{a}\cdot\mathbf{e}_-)}{2q_i\,\mathbf{a}^2} C_- = A_\perp \tag{5.172}$$

$$\frac{(q_i - q_+)(\mathbf{a}\cdot\mathbf{e}_+)}{2q_i\,\mathbf{a}^2} C_+ + \frac{(q_i - q_-)(\mathbf{a}\cdot\mathbf{e}_-)}{2q_i\,\mathbf{a}^2} C_- = B_\perp \tag{5.173}$$

$$\frac{k_i^2(\mathbf{b}\cdot\mathbf{e}_+) - \omega\mu_0\, q_i(\mathbf{a}\cdot\mathbf{h}_+)}{2k_i\, q_i\,\mathbf{a}^2} C_+ + \frac{k_i^2(\mathbf{b}\cdot\mathbf{e}_-) - \omega\mu_0\, q_i(\mathbf{a}\cdot\mathbf{h}_-)}{2k_i\, q_i\,\mathbf{a}^2} C_- = A_\parallel \tag{5.174}$$

$$-\frac{k_i^2(\mathbf{b}\cdot\mathbf{e}_+) + \omega\mu_0\, q_i(\mathbf{a}\cdot\mathbf{h}_+)}{2k_i\, q_i\,\mathbf{a}^2} C_+ - \frac{k_i^2(\mathbf{b}\cdot\mathbf{e}_-) + \omega\mu_0\, q_i(\mathbf{a}\cdot\mathbf{h}_-)}{2k_i\, q_i\,\mathbf{a}^2} C_- = B_\parallel \tag{5.175}$$

To put Eqs. (5.172) to (5.175) in a more symmetrical and compact form, we introduce vectors $\mathbf{N}_+$, $\mathbf{N}_-$, $\mathbf{F}_+$, and $\mathbf{F}_-$ defined by

$$\mathbf{N}_\pm = \mathbf{k}_i^2 \mathbf{e}_\pm + \omega\mu_0(\mathbf{k}_r \times \mathbf{h}_\pm) \tag{5.176}$$

$$\mathbf{F}_\pm = \mathbf{k}_i^2 \mathbf{e}_\pm + \omega\mu_0(\mathbf{k}_i \times \mathbf{h}_\pm) \tag{5.177}$$

Or, using Maxwell's equations and after some simplification, we arrive at

$$\begin{aligned} \mathbf{N}_\pm &= (\mathbf{k}_i^2 - \mathbf{k}_r \cdot \mathbf{k}_\pm)\mathbf{e}_\pm + (\mathbf{k}_r \cdot \mathbf{e}_\pm)\mathbf{k}_\pm \\ &= q_i(q_i + q_\pm)\mathbf{e}_\pm + (\mathbf{k}_r \cdot \mathbf{e}_\pm)\mathbf{k}_\pm \end{aligned} \tag{5.178}$$

$$\begin{aligned} \mathbf{F}_\pm &= (\mathbf{k}_i^2 - \mathbf{k}_i \cdot \mathbf{k}_\pm)\mathbf{e}_\pm + (\mathbf{k}_i \cdot \mathbf{e}_\pm)\mathbf{k}_\pm \\ &= q_i(q_i - q_\pm)\mathbf{e}_\pm + (\mathbf{k}_i \cdot \mathbf{e}_\pm)\mathbf{k}_\pm \end{aligned} \tag{5.179}$$

Thus

$$\begin{aligned} \mathbf{a} \cdot \mathbf{N}_\pm &= q_i(q_i + q_\pm)(\mathbf{a} \cdot \mathbf{e}_\pm) \\ \mathbf{a} \cdot \mathbf{F}_\pm &= q_i(q_i - q_\pm)(\mathbf{a} \cdot \mathbf{e}_\pm) \\ \mathbf{b} \cdot \mathbf{N}_\pm &= \mathbf{k}_i^2(\mathbf{b} \cdot \mathbf{e}_\pm) - \omega\mu_0 q_i(\mathbf{a} \cdot \mathbf{h}_\pm) \\ &= (\mathbf{k}_i^2 + q_i q_\pm)(\mathbf{b} \cdot \mathbf{e}_\pm) - q_i \mathbf{a}^2(\hat{\mathbf{q}} \cdot \mathbf{e}_\pm) \\ \mathbf{b} \cdot \mathbf{F}_\pm &= \mathbf{k}_i^2(\mathbf{b} \cdot \mathbf{e}_\pm) + \omega\mu_0 q_i(\mathbf{a} \cdot \mathbf{h}_\pm) \\ &= (\mathbf{k}_i^2 - q_i q_\pm)(\mathbf{b} \cdot \mathbf{e}_\pm) + q_i \mathbf{a}^2(\hat{\mathbf{q}} \cdot \mathbf{e}_\pm) \end{aligned} \tag{5.180}$$

Substituting Eqs. (5.180) into Eqs. (5.172) and (5.174), we may put the results in matrix form

$$\frac{1}{2q_i \mathbf{a}^2} \begin{bmatrix} \dfrac{1}{q_i}(\mathbf{a} \cdot \mathbf{N}_+) & \dfrac{1}{q_i}(\mathbf{a} \cdot \mathbf{N}_-) \\ \dfrac{1}{k_i}(\mathbf{b} \cdot \mathbf{N}_+) & \dfrac{1}{k_i}(\mathbf{b} \cdot \mathbf{N}_-) \end{bmatrix} \begin{bmatrix} C_+ \\ C_- \end{bmatrix} = \begin{bmatrix} A_\perp \\ A_\parallel \end{bmatrix} \tag{5.181}$$

Multiplying both sides of Eq. (5.181) by the inverse coefficient matrix, we obtain

$$\begin{bmatrix} C_+ \\ C_- \end{bmatrix} = \begin{bmatrix} T_{11} & T_{12} \\ T_{21} & T_{22} \end{bmatrix} \begin{bmatrix} A_\perp \\ A_\parallel \end{bmatrix} \tag{5.182}$$

where the transmission coefficients are given by

$$\begin{aligned} T_{11} &= \frac{2q_i^2 \mathbf{a}^2}{\Delta}(\mathbf{b} \cdot \mathbf{N}_-) \\ T_{12} &= -\frac{2q_i k_i \mathbf{a}^2}{\Delta}(\mathbf{a} \cdot \mathbf{N}_-) \\ T_{21} &= -\frac{2q_i^2 \mathbf{a}^2}{\Delta}(\mathbf{b} \cdot \mathbf{N}_+) \\ T_{22} &= \frac{2q_i k_i \mathbf{a}^2}{\Delta}(\mathbf{a} \cdot \mathbf{N}_+) \end{aligned} \tag{5.183}$$

and
$$\Delta = (\mathbf{a}\cdot\mathbf{N}_+)(\mathbf{b}\cdot\mathbf{N}_-) - (\mathbf{a}\cdot\mathbf{N}_-)(\mathbf{b}\cdot\mathbf{N}_+) \tag{5.184}$$

Similarly, using Eqs. (5.180) we may write Eqs. (5.173) and (5.175) in the following matrix form

$$\frac{1}{2q_i\mathbf{a}^2}\begin{bmatrix} \frac{1}{q_i}(\mathbf{a}\cdot\mathbf{F}_+) & \frac{1}{q_i}(\mathbf{a}\cdot\mathbf{F}_-) \\ -\frac{1}{k_i}(\mathbf{b}\cdot\mathbf{F}_+) & -\frac{1}{k_i}(\mathbf{b}\cdot\mathbf{F}_-) \end{bmatrix}\begin{bmatrix} C_+ \\ C_- \end{bmatrix} = \begin{bmatrix} B_\perp \\ B_\parallel \end{bmatrix} \tag{5.185}$$

Substituting the result of Eq. (5.182) into Eq. (5.185) and after some simplification, we obtain

$$\begin{bmatrix} B_\perp \\ B_\parallel \end{bmatrix} = \begin{bmatrix} \Gamma_{11} & \Gamma_{12} \\ \Gamma_{21} & \Gamma_{22} \end{bmatrix}\begin{bmatrix} A_\perp \\ A_\parallel \end{bmatrix} \tag{5.186}$$

where the reflection coefficients are given by

$$\begin{aligned}
\Gamma_{11} &= \frac{1}{\Delta}[(\mathbf{a}\cdot\mathbf{F}_+)(\mathbf{b}\cdot\mathbf{N}_-) - (\mathbf{a}\cdot\mathbf{F}_-)(\mathbf{b}\cdot\mathbf{N}_+)] \\
\Gamma_{12} &= -\frac{k_i}{q_i\,\Delta}[(\mathbf{a}\cdot\mathbf{F}_+)(\mathbf{a}\cdot\mathbf{N}_-) - (\mathbf{a}\cdot\mathbf{F}_-)(\mathbf{a}\cdot\mathbf{N}_+)] \\
\Gamma_{21} &= -\frac{q_i}{k_i\,\Delta}[(\mathbf{b}\cdot\mathbf{F}_+)(\mathbf{b}\cdot\mathbf{N}_-) - (\mathbf{b}\cdot\mathbf{F}_-)(\mathbf{b}\cdot\mathbf{N}_+)] \\
\Gamma_{22} &= \frac{1}{\Delta}[(\mathbf{b}\cdot\mathbf{F}_+)(\mathbf{a}\cdot\mathbf{N}_-) - (\mathbf{b}\cdot\mathbf{F}_-)(\mathbf{a}\cdot\mathbf{N}_+)]
\end{aligned} \tag{5.187}$$

and Δ is given by Eq. (5.184). We note that in the presence of an anisotropic medium 2, the reflection coefficient matrix contains both diagonal and off-diagonal elements. Thus, for a linearly polarized incident wave with the electric field intensity perpendicular to the plane of incidence, the electric vector of the reflected wave will have components both perpendicular and parallel to the plane of incidence.

For certain states of polarization of the incident wave, one of the transmitted waves will be absent. For example, from Eq. (5.182) C_- vanishes if the incident wave is polarized in such a manner that

$$\frac{A_\parallel}{A_\perp} = -\frac{T_{21}}{T_{22}} = \frac{q_i(\mathbf{b}\cdot\mathbf{N}_+)}{k_i(\mathbf{a}\cdot\mathbf{N}_+)} \tag{5.188}$$

In this case, the amplitude of the remaining transmitted wave becomes

$$C_+ = T_{11}A_\perp + T_{12}A_\parallel = \frac{2q_i^2\mathbf{a}^2}{(\mathbf{a}\cdot\mathbf{N}_+)}A_\perp \tag{5.189}$$

and the amplitudes $B_\perp$ and $B_\parallel$ of the reflected wave are

$$B_\perp = \frac{\mathbf{a}\cdot\mathbf{F}_+}{\mathbf{a}\cdot\mathbf{N}_+} A_\perp = \frac{q_i - q_+}{q_i + q_+} A_\perp \tag{5.190}$$

$$B_\parallel = -\frac{\mathbf{b}\cdot\mathbf{F}_+}{\mathbf{b}\cdot\mathbf{N}_+} A_\parallel = -\frac{q_i(\mathbf{b}\cdot\mathbf{F}_+)}{k_i(\mathbf{a}\cdot\mathbf{N}_+)} A_\perp \tag{5.191}$$

We must note that Eqs. (5.189) and (5.191) are different from those of two isotropic media. This is due to the fact that the electric field intensity in an anisotropic medium is not transverse to the wave normal. Similar results can be obtained when $C_+ = 0$.

Finally, when the incident wave is propagating normal to the interface, formulas (5.183) and (5.187) will lose their meaning because the plane of incidence can not be defined. We then have

$$\hat{\mathbf{k}}_i = \hat{\mathbf{k}}_+ = \hat{\mathbf{k}}_- = -\hat{\mathbf{k}}_r = \hat{\mathbf{q}} \tag{5.192}$$

Thus

$$\begin{aligned} \mathbf{H}_{0i} &= \frac{k_i}{\omega\mu_0}(\hat{\mathbf{q}}\times\mathbf{E}_{0i}) \\ \mathbf{H}_{0r} &= -\frac{k_i}{\omega\mu_0}(\hat{\mathbf{q}}\times\mathbf{E}_{0r}) \\ \mathbf{H}_{0+} &= C_+\mathbf{h}_+ \\ \mathbf{H}_{0-} &= C_-\mathbf{h}_- \end{aligned} \tag{5.193}$$

where $\mathbf{h}_\pm = \dfrac{k_\pm}{\omega\mu_0}(\hat{\mathbf{q}}\times\mathbf{e}_\pm)$

Substituting Eq. (5.193) into the boundary conditions

$$\hat{\mathbf{q}}\times(\mathbf{E}_{0i} + \mathbf{E}_{0r} - \mathbf{E}_{0+} - \mathbf{E}_{0-}) = \mathbf{0} \tag{5.194}$$

and

$$\mathbf{H}_{0i} + \mathbf{H}_{0r} - \mathbf{H}_{0+} - \mathbf{H}_{0-} = \mathbf{0} \tag{5.195}$$

we obtain

$$\mathbf{H}_{0i} - \mathbf{H}_{0r} - \frac{k_i}{k_+}C_+\mathbf{h}_+ - \frac{k_i}{k_-}C_-\mathbf{h}_- = \mathbf{0} \tag{5.196}$$

and

$$\mathbf{H}_{0i} + \mathbf{H}_{0r} - C_+\mathbf{h}_+ - C_-\mathbf{h}_- = \mathbf{0} \tag{5.197}$$

respectively. The boundary condition (5.195) follows from the fact that the tangential components of **H** and the normal components of **B** must be continuous across the interface and that media 1 and 2 have the same permeability μ_0. Now adding and subtracting Eqs. (5.196) and (5.197), we get

$$\mathbf{H}_{0i} = \frac{k_+ + k_i}{2k_+}C_+\mathbf{h}_+ + \frac{k_- + k_i}{2k_-}C_-\mathbf{h}_- \tag{5.198}$$

and
$$\mathbf{H}_{0r} = \frac{k_+ - k_i}{2k_+} C_+ \mathbf{h}_+ + \frac{k_- - k_i}{2k_-} C_- \mathbf{h}_- \tag{5.199}$$

respectively. So long as the incidence is normal to the interface, the two transmitted waves are isonormal waves. Thus according to Eq. (5.94), $\mathbf{h}_+$ and $\mathbf{h}_-$ are orthogonal, that is, $\mathbf{h}_+ \cdot \mathbf{h}_- = 0$. Assuming that they have been normalized, $\mathbf{h}_+^2 = \mathbf{h}_-^2 = 1$, and after taking the dot products of Eq. (5.198) with $\mathbf{h}_+$ and $\mathbf{h}_-$ respectively, we obtain

$$\begin{aligned} C_+ &= \frac{2k_+}{k_+ + k_i} (\mathbf{H}_{0i} \cdot \mathbf{h}_+) \\ C_- &= \frac{2k_-}{k_- + k_i} (\mathbf{H}_{0i} \cdot \mathbf{h}_-) \end{aligned} \tag{5.200}$$

Substitution of Eq. (5.200) into Eq. (5.199) yields

$$\mathbf{H}_{0r} = \frac{k_+ - k_i}{k_+ + k_i} (\mathbf{H}_{0i} \cdot \mathbf{h}_+)\mathbf{h}_+ + \frac{k_- - k_i}{k_- + k_i} (\mathbf{H}_{0i} \cdot \mathbf{h}_-)\mathbf{h}_- \tag{5.201}$$

where $\mathbf{h}_\pm$ are determined from the expressions given in Sec. 5.4. The other field vectors of the transmitted and the reflected waves can then be obtained from Maxwell's equations.

PROBLEMS

5.1 Show that Eqs. (5.16), (5.17), and (5.18) are equivalent; i.e., any one of the equations may be obtained from the other.

5.2 Show in detail that the equalities in Eq. (5.25) are valid.

5.3 If ε_1, ε_2, and ε_3 are the principal dielectric constants of tensor $\bar{\varepsilon}$ such that $\varepsilon_1 < \varepsilon_2 < \varepsilon_3$, and if n_1 and n_2 are two indices of refraction of Eq. (5.34), show that

$$0 < \varepsilon_1 \le n_1^2 \le \varepsilon_2 \le n_2^2 \le \varepsilon_3$$

5.4 Determine the directions of field vectors, when the given wave vector $\mathbf{k}$ satisfies the condition [cf. Eq. (5.76)]

$$|k_0^2(\text{adj }\bar{\varepsilon}) - (\mathbf{k} \cdot \bar{\varepsilon} \cdot \mathbf{k})\,\bar{\mathbf{I}}| = 0$$

5.5 For a given direction of wave propagation $\hat{\mathbf{k}}$, we can determine geometrically the two phase velocities v_p and the two directions of the electric flux densities $\mathbf{D}_0$ of the isonormal waves in crystals. For this purpose, we define the index ellipsoid

$$\mathbf{r} \cdot \bar{\varepsilon}^{-1} \cdot \mathbf{r} = 1$$

where
$$\mathbf{r} = \frac{\mathbf{D}_0}{\sqrt{2\varepsilon_0 \langle W \rangle}}$$

and $\langle W \rangle$ is the time-averaged energy density of the wave. The intersection of the ellipsoid and the plane

$$\hat{\mathbf{k}} \cdot \mathbf{r} = 0$$

which is normal to $\hat{\mathbf{k}}$ and passes through the center of the ellipsoid, is an ellipse. Using the method of Lagrange's multipliers, show that (*a*) the lengths of the semiaxes of the ellipse are equal to the two indices of refraction $n_+ = c/v_p^+$ and $n_- = c/v_p^-$ of the isonormal waves corresponding to the given wave normal $\hat{\mathbf{k}}$; (*b*) the directions of the semiaxes of the ellipse give the directions of the electric flux density vectors $\mathbf{D}_0^+$ and $\mathbf{D}_0^-$ of the isonormal waves; (*c*) the normal to the index ellipsoid at a certain point gives the direction of $\mathbf{E}_0$ corresponding to the $\mathbf{D}_0$ drawn to the same point; and (*d*) when the wave normal coincides with an optic axis, the intersection of the ellipsoid and the plane becomes a circle.

5.6 For plane waves propagating in lossless crystals, show that the electric flux density vector $\mathbf{D}_0$ satisfies the eigenvalue problem:

$$\bar{\mathbf{A}} \cdot \mathbf{D}_0 = \lambda \mathbf{D}_0$$

where

$$\bar{\mathbf{A}} = (\hat{\mathbf{k}} \times \bar{\mathbf{I}})^2 \cdot \bar{\boldsymbol{\varepsilon}}^{-1} \cdot (\hat{\mathbf{k}} \times \bar{\mathbf{I}})^2$$

is a real, symmetric tensor and $\lambda = k_0^2/k^2$. Hence show that when $\lambda \neq 0$, (*a*) $\mathbf{D}_0$ is linearly polarized; (*b*) $\mathbf{D}_0$ lies on the plane that is perpendicular to $\hat{\mathbf{k}}$; (*c*) for a given direction of wave propagation $\hat{\mathbf{k}}$, the two electric flux density vectors of the isonormal waves corresponding to two different phase velocities are mutually perpendicular; and (*d*) the eigenvector $\mathbf{D}_0$ corresponding to the eigenvalue $\lambda = 0$ must be discarded.

5.7 Show that

$$\mathbf{s} = \frac{1}{2\omega^2} \frac{\partial \omega^2}{\partial \mathbf{k}}$$

and then use the dispersion equation (5.133) to obtain the ray vector **s** as given in Eq. (5.124).

5.8 Show that

$$\mathbf{k} = \frac{\omega^2}{2} \frac{\partial \omega^{-2}}{\partial \mathbf{s}}$$

and then use the ray equation (5.123) to obtain the wave vector **k** as given in Eq. (5.125).

5.9 In the case of normal incidence from an isotropic medium on an anisotropic crystal, let the angles formed by the vectors $\mathbf{H}_{0i}$ and $\mathbf{H}_{0r}$ with $\mathbf{h}_+$ be β_i and β_r, respectively. Show that

$$\tan(\beta_r - \beta_i) = \frac{2k_i(k_- - k_+)\tan\beta_i}{(k_+ - k_i)(k_- + k_i) + (k_- - k_i)(k_+ + k_i)\tan^2\beta_i}$$

5.10 Crystals belonging to the orthorhombic, monoclinic, and triclinic systems are biaxial because they contain two optic axes. For convenience, we express the inverse dielectric tensor of such crystals by [cf. Sec. 1.8]

$$\bar{\boldsymbol{\varepsilon}}^{-1} = \alpha_0 \bar{\mathbf{I}} + \beta_0(\hat{\mathbf{c}}_1\hat{\mathbf{c}}_2 + \hat{\mathbf{c}}_2\hat{\mathbf{c}}_1)$$

where α_0 and β_0 are two constants, and $\hat{\mathbf{c}}_1$ and $\hat{\mathbf{c}}_2$ are two unit vectors. Show that in a source-free region, $\mathbf{H}_0$ satisfies the homogeneous equation

$$\bar{\mathbf{W}}_h(\mathbf{k}) \cdot \mathbf{H}_0 = \mathbf{0}$$

where

$$\bar{\mathbf{W}}_h(\mathbf{k}) = (k_0^2 - \alpha_0 \mathbf{k}^2)\bar{\mathbf{I}} - \beta_0[(\mathbf{k} \times \hat{\mathbf{c}}_1)(\mathbf{k} \times \hat{\mathbf{c}}_2) + (\mathbf{k} \times \hat{\mathbf{c}}_2)(\mathbf{k} \times \hat{\mathbf{c}}_1)]$$

and that the determinant and adjoint of $\bar{\mathbf{W}}_h(\mathbf{k})$ are

$$|\bar{\mathbf{W}}_h(\mathbf{k})| = (k_0^2 - \alpha_0 \mathbf{k}^2)\{[k_0^2 - \alpha_0 \mathbf{k}^2 - \beta_0(\mathbf{k} \times \hat{\mathbf{c}}_1) \cdot (\mathbf{k} \times \hat{\mathbf{c}}_2)]^2 - \beta_0^2(\mathbf{k} \times \hat{\mathbf{c}}_1)^2(\mathbf{k} \times \hat{\mathbf{c}}_2)^2\}$$

and

$$\begin{aligned}\text{adj } \bar{\mathbf{W}}_h(\mathbf{k}) = (k_0^2 - \alpha_0 \mathbf{k}^2)\,\{&[k_0^2 - \alpha_0 \mathbf{k}^2 - 2\beta_0(\mathbf{k} \times \hat{\mathbf{c}}_1) \cdot (\mathbf{k} \times \hat{\mathbf{c}}_2)]\bar{\mathbf{I}} \\ &+ \beta_0[(\mathbf{k} \times \hat{\mathbf{c}}_1)(\mathbf{k} \times \hat{\mathbf{c}}_2) + (\mathbf{k} \times \hat{\mathbf{c}}_2)(\mathbf{k} \times \hat{\mathbf{c}}_1)]\} \\ &- \beta_0^2[(\mathbf{k} \times \hat{\mathbf{c}}_1) \times (\mathbf{k} \times \hat{\mathbf{c}}_2)][(\mathbf{k} \times \hat{\mathbf{c}}_1) \times (\mathbf{k} \times \hat{\mathbf{c}}_2)]\end{aligned}$$

respectively. Show that the root $\mathbf{k}^2 = k_0^2/\alpha_0$ of the dispersion equation is not compatible with Maxwell's equation and must be discarded. There are, therefore, two propagating waves whose wave vectors $\mathbf{k}_+$ and $\mathbf{k}_-$ satisfy the dispersion equations

$$k_0^2 - \alpha_0 \mathbf{k}_+^2 - \beta_0(\mathbf{k}_+ \times \hat{\mathbf{c}}_1) \cdot (\mathbf{k}_+ \times \hat{\mathbf{c}}_2) - \beta_0 \sqrt{(\mathbf{k}_+ \times \hat{\mathbf{c}}_1)^2(\mathbf{k}_+ \times \hat{\mathbf{c}}_2)^2} = 0$$

and

$$k_0^2 - \alpha_0 \mathbf{k}_-^2 - \beta_0(\mathbf{k}_- \times \hat{\mathbf{c}}_1) \cdot (\mathbf{k}_- \times \hat{\mathbf{c}}_2) + \beta_0 \sqrt{(\mathbf{k}_- \times \hat{\mathbf{c}}_1)^2(\mathbf{k}_- \times \hat{\mathbf{c}}_2)^2} = 0$$

respectively. Show that $\hat{\mathbf{c}}_1$ and $\hat{\mathbf{c}}_2$ are the two optic axes of the biaxial crystal; show also that corresponding to $\mathbf{k}_+$ and $\mathbf{k}_-$, the two solutions of the homogeneous equation for $\mathbf{H}_0$ are

$$\mathbf{h}_+ = \alpha_+ \left[\frac{\mathbf{k}_+ \times \hat{\mathbf{c}}_1}{\sqrt{(\mathbf{k}_+ \times \hat{\mathbf{c}}_1)^2}} + \frac{\mathbf{k}_+ \times \hat{\mathbf{c}}_2}{\sqrt{(\mathbf{k}_+ \times \hat{\mathbf{c}}_2)^2}} \right]$$

and

$$\mathbf{h}_- = \alpha_- \left[\frac{\mathbf{k}_- \times \hat{\mathbf{c}}_1}{\sqrt{(\mathbf{k}_- \times \hat{\mathbf{c}}_1)^2}} - \frac{\mathbf{k}_- \times \hat{\mathbf{c}}_2}{\sqrt{(\mathbf{k}_- \times \hat{\mathbf{c}}_2)^2}} \right]$$

where

$$\alpha_+ = \left\{ 2 \left[1 + \frac{(\mathbf{k}_+ \times \hat{\mathbf{c}}_1) \cdot (\mathbf{k}_+ \times \hat{\mathbf{c}}_2)}{\sqrt{(\mathbf{k}_+ \times \hat{\mathbf{c}}_1)^2(\mathbf{k}_+ \times \hat{\mathbf{c}}_2)^2}} \right] \right\}^{-1/2}$$

$$\alpha_- = \left\{ 2 \left[1 - \frac{(\mathbf{k}_- \times \hat{\mathbf{c}}_1) \cdot (\mathbf{k}_- \times \hat{\mathbf{c}}_2)}{\sqrt{(\mathbf{k}_- \times \hat{\mathbf{c}}_1)^2(\mathbf{k}_- \times \hat{\mathbf{c}}_2)^2}} \right] \right\}^{-1/2}$$

Finally, show that for the two isonormal waves, $\mathbf{h}_+$ and $\mathbf{h}_-$ satisfy the orthonormal condition

$$\mathbf{h}_i \cdot \mathbf{h}_j = \delta_{ij}$$

where i and j denote either $+$ or $-$.

5.11 In Prob. 5.10, let δ_1 and δ_2 be the angles formed by the optic axes $\hat{\mathbf{c}}_1$ and $\hat{\mathbf{c}}_2$ with the wave normal $\hat{\mathbf{k}}$ and δ_3 be the angle between vectors $\hat{\mathbf{k}} \times \hat{\mathbf{c}}_1$ and $\hat{\mathbf{k}} \times \hat{\mathbf{c}}_2$, show that

$$\tan^2 \frac{\delta_3}{2} = \frac{\alpha_0 - 1/n_-^2}{1/n_+^2 - \alpha_0}$$

$$\frac{1}{n_+^2} - \frac{1}{n_-^2} = 2\beta_0 \sin \delta_1 \sin \delta_2$$

$$\left(\frac{1}{n_+^2} - \alpha_0 \right) \left(\alpha_0 - \frac{1}{n_-^2} \right) = \beta_0^2 \sin^2 \delta_1 \sin^2 \delta_2 \sin^2 \delta_3$$

where n_+ and n_- are the two indices of refraction of the two isonormal waves.

5.12 Show that in a biaxial crystal, the magnetic field intensities $\mathbf{H}_0^+$ and $\mathbf{H}_0^-$ of the two isonormal waves lie in the planes which bisect the angles between the planes formed by vectors $\hat{\mathbf{k}} \times \hat{\mathbf{c}}_1$ and $\hat{\mathbf{k}} \times \hat{\mathbf{c}}_2$.

5.13 The theory of uniform plane waves in crystals can be extended to magnetic crystals. In this case the constitutive relations are

$$\mathbf{D}_0 = \varepsilon_0 \bar{\varepsilon} \cdot \mathbf{E}_0$$

and

$$\mathbf{B}_0 = \mu_0 \bar{\mu} \cdot \mathbf{H}_0$$

where the dielectric tensor $\bar{\varepsilon}$ and the relative permeability tensor $\bar{\mu}$ are real, symmetric, and positive-definite.

(*a*) Show that the amplitude vectors $\mathbf{E}_0$, $\mathbf{H}_0$, $\mathbf{D}_0$, and $\mathbf{B}_0$ are respectively the eigenvectors of matrices $\bar{\mathbf{M}} \cdot \bar{\mathbf{N}}$, $\bar{\mathbf{N}} \cdot \bar{\mathbf{M}}$, $(\bar{\mathbf{M}} \cdot \bar{\mathbf{N}})^T$, and $(\bar{\mathbf{N}} \cdot \bar{\mathbf{M}})^T$ corresponding to the eigenvalue $\lambda = -1/n^2$ where

$$\bar{\mathbf{M}} = \bar{\varepsilon}^{-1} \cdot (\hat{\mathbf{k}} \times \bar{\mathbf{I}}) \qquad \bar{\mathbf{N}} = \bar{\mu}^{-1} \cdot \hat{\mathbf{k}} \times \bar{\mathbf{I}})$$

and $n = ck/\omega$ is the index of refraction.

(*b*) Show that the dispersion equation takes the form

$$An^4 - Bn^2 + C = 0$$

where

$$A = (\hat{\mathbf{k}} \cdot \bar{\varepsilon} \cdot \hat{\mathbf{k}})(\hat{\mathbf{k}} \cdot \bar{\mu} \cdot \hat{\mathbf{k}})$$

$$B = \hat{\mathbf{k}} \cdot \bar{\mu} \cdot \{[(\mathrm{adj}\ \bar{\varepsilon}) \cdot \bar{\mu}]_t \bar{\mathbf{I}} - (\mathrm{adj}\ \bar{\varepsilon}) \cdot \bar{\mu}\} \cdot \hat{\mathbf{k}}$$

$$C = |\bar{\mu} \cdot \bar{\varepsilon}|$$

and $\mathbf{n} = \mathbf{k}/k_0 = n\hat{\mathbf{k}}$ is the refractive index vector.

(*c*) Derive the ray vector surface.

(*d*) Show that when $(\mathrm{adj}\ \bar{\mu}) \cdot \bar{\varepsilon} - (\mathbf{n} \cdot \bar{\mu} \cdot \mathbf{n})\bar{\mathbf{I}}$ is nonsingular, the direction of $\mathbf{E}_0$ is

$$\mathbf{e} = \{\mathrm{adj}\ [(\mathrm{adj}\ \bar{\mu}) \cdot \bar{\varepsilon} - (\mathbf{n} \cdot \bar{\mu} \cdot \mathbf{n})\bar{\mathbf{I}}]\} \cdot \mathbf{n}$$

Hence the directions of the remaining field vectors are

$$\mathbf{d} = \varepsilon_0 \bar{\varepsilon} \cdot \mathbf{e}$$

$$\mathbf{b} = \frac{1}{c} (\mathbf{n} \times \mathbf{e})$$

$$\mathbf{h} = \frac{1}{\mu_0} \bar{\mu}^{-1} \cdot \mathbf{b}$$

(*e*) Determine the directions of the field vectors when $(\mathrm{adj}\ \bar{\mu}) \cdot \bar{\varepsilon} - (\mathbf{n} \cdot \bar{\mu} \cdot \mathbf{n})\bar{\mathbf{I}}$ is singular.

5.14 In a magnetic crystal, show that

(*a*)
$$\mathbf{D}_0 \times \mathbf{D}_0^* = \varepsilon_0^2 (\mathrm{adj}\ \bar{\varepsilon}) \cdot (\mathbf{E}_0 \times \mathbf{E}_0^*)$$

$$\mathbf{B}_0 \times \mathbf{B}_0^* = \frac{1}{\omega^2} \mathbf{kk} \cdot (\mathbf{E}_0 \times \mathbf{E}_0^*)$$

$$\mathbf{H}_0 \times \mathbf{H}_0^* = \frac{1}{\omega^2 \mu_0^2 |\bar{\mu}|} \bar{\mu} \cdot \mathbf{kk} \cdot (\mathbf{E}_0 \times \mathbf{E}_0^*)$$

Hence if the vector $\mathbf{E}_0$ is linearly polarized so are the field vectors $\mathbf{D}_0$, $\mathbf{B}_0$, and $\mathbf{H}_0$; (*b*) the plane waves are always linearly polarized except when the wave is propagating along the directions of optic axes. In that case the wave can have any polarization.

5.15 In a doubly refracting magnetic crystal if the principal coordinates of the tensors $\bar{\varepsilon}$ and $\bar{\mu}$ coincide, and if the principal permittivities ε_1, ε_2, and ε_3 and the principal permeabilities μ_1, μ_2, and μ_3 satisfy the condition:

$$\frac{\varepsilon_1}{\mu_1} = \frac{\varepsilon_2}{\mu_2} = \frac{\varepsilon_3}{\mu_3}$$

show that the dispersion equation becomes

$$\sqrt{\mu_1 \varepsilon_1}\, n_1^2 + \sqrt{\mu_2 \varepsilon_2}\, n_2^2 + \sqrt{\mu_3 \varepsilon_3}\, n_3^2 = \sqrt{\mu_1 \mu_2 \mu_3 \varepsilon_1 \varepsilon_2 \varepsilon_3}$$

Thus for each direction of wave normal $\hat{\mathbf{k}}$, there is only one index of refraction (a singly refracting medium). In other words, the two linearly polarized waves travel with the same velocity.

5.16 In a magnetic crystal, show that the field vectors of the two isonormal waves are related by

$$\mathbf{E}_0^+ \cdot \mathbf{D}_0^- = \mathbf{E}_0^- \cdot \mathbf{D}_0^+ = \mathbf{H}_0^+ \cdot \mathbf{B}_0^- = \mathbf{H}_0^- \cdot \mathbf{B}_0^+ = 0$$

Hence

$$\mathbf{D}_0^+ \parallel \mathbf{B}_0^- \quad \text{and} \quad \mathbf{D}_0^- \parallel \mathbf{B}_0^+$$

5.17 In Prob. 5.13, assume that the dielectric tensor $\bar{\varepsilon}$ and the permeability tensor $\bar{\mu}$ are positive definite hermitian matrices. Show that the square of the index of refraction is always real and positive.

REFERENCES

Agranovich, V. M., and V. L. Ginzburg: *Spatial Dispersion in Crystal Optics and the Theory of Excitons*, Interscience Publishers, John Wiley & Sons, Inc., New York, 1966.

Born, M., and E. Wolf: *Principles of Optics*, 4th ed., Pergamon Press, New York, 1970, chap. 14.

Chen, H. C.: "A Coordinate-Free Approach to Wave Reflection from an Anisotropic Medium," *Radio Science*, vol. 16, no. 6, 1981, pp. 1213–1215.

Curry, C. E.: *Electromagnetic Theory of Light*, Macmillan and Co., New York, 1905, chap. 8.

Ditchburn, R. W.: *Light*, 2d ed., Interscience Publishers, John Wiley & Sons, Inc., New York, 1963, chap. 16.

Juretschke, H. J.: *Crystal Physics*, W. A. Benjamin, Inc., Reading, Mass., 1974, chap. 9.

Klein, M. V.: *Optics*, John Wiley & Sons, Inc., New York, 1970, pp. 596–616.

Kunz, K. S.: "Treatment of Optical Propagation in Crystals Using Projection Dyadics," *Am. J. Phys.*, vol. 45, no. 3, 1977, pp. 267–269.

Landau, L. D., and E. M. Lifshitz: *Electrodynamics of Continuous Media*, Pergamon Press, New York, 1960, chap. 11.

Mason, M. P.: *Crystal Physics of Interaction Processes*, Academic Press, New York, 1966, chap. 7 and appendix A.

Nelson, D. F.: *Electric, Optic, and Acoustic Interactions in Dielectrics*, John Wiley & Sons, Inc., New York, 1979, chap. 9.

Nye, J. F.: *Physical Properties of Crystals*, Oxford University Press, London, 1957.

Ramachandran, G. N., and S. Ramaseshan: "Crystal Optics," in S. Flügge (ed.), *Handbuch der Physik*, vol. XXV/1, Springer, Berlin, 1961.

Ramo, S., J. R. Whinnery, and T. Van Duzer: *Fields and Waves in Communication Electronics*, John Wiley & Sons, Inc., New York, 1965, chap. 9.

Sommerfeld, A.: *Optics*, Academic Press, New York, 1954, chap. 4.

Stone, J. M.: *Radiation and Optics*, McGraw-Hill Book Company, New York, 1963, chap. 17.

CHAPTER

SIX

PLANE WAVES IN UNIAXIAL MEDIA

In the previous chapter, some general results concerning the propagation of waves in lossless nonmagnetic crystals were obtained. We found that the electric property of a crystal is characterized by a real, positive definite, symmetric dielectric tensor $\bar{\varepsilon}$. Thus the tensor can always be transformed into a diagonal form, and the three diagonal elements corresponding to the eigenvalues of $\bar{\varepsilon}$ are in general distinct (biaxial). Because of symmetry, crystals such as trigonal, tetragonal, and hexagonal structures are characterized either by a diagonal dielectric tensor with two equal diagonal elements or by a single axis of symmetry called an *optic axis*. Media with such a property are called *uniaxial*.

In this chapter we shall study wave propagation in a uniaxial medium. We shall see that in such a medium, the dispersion equation can be factored into a product of two terms which in turn indicates the existence of two waves: ordinary and extraordinary. In this case the results obtained in the preceding chapter can be expressed in explicit and considerably simplified forms.

6.1 WAVE MATRIX OF A UNIAXIAL MEDIUM

With respect to an orthogonal coordinate system, the dielectric tensor of a uniaxial medium takes the matrix form

$$\bar{\varepsilon} = \begin{bmatrix} \varepsilon_\perp & 0 & 0 \\ 0 & \varepsilon_\perp & 0 \\ 0 & 0 & \varepsilon_\parallel \end{bmatrix} \tag{6.1}$$

where two of the three diagonal elements are equal. According to Sec. 1.8, we may express Eq. (6.1) in dyadic form, namely,

$$\bar{\boldsymbol{\varepsilon}} = \varepsilon_\perp \bar{\mathbf{I}} + (\varepsilon_{||} - \varepsilon_\perp)\hat{\mathbf{c}}\hat{\mathbf{c}} \tag{6.2}$$

where $\hat{\mathbf{c}}$ is a unit eigenvector of $\bar{\boldsymbol{\varepsilon}}$ corresponding to the nonrepeated eigenvalue $\varepsilon_{||}$, and $\varepsilon_\perp$ (repeated) is the other eigenvalue of $\bar{\boldsymbol{\varepsilon}}$. From Eqs. (6.2) and (1.126), we easily find

$$\begin{aligned} |\,\bar{\boldsymbol{\varepsilon}}\,| &= \varepsilon_\perp^2 \varepsilon_{||} \\ \text{adj } \bar{\boldsymbol{\varepsilon}} &= \varepsilon_\perp[\varepsilon_{||}\bar{\mathbf{I}} + (\varepsilon_\perp - \varepsilon_{||})\hat{\mathbf{c}}\hat{\mathbf{c}}] \\ (\text{adj } \bar{\boldsymbol{\varepsilon}})_t \bar{\mathbf{I}} - \text{adj } \bar{\boldsymbol{\varepsilon}} &= \varepsilon_\perp(\varepsilon_{||}\bar{\mathbf{I}} + \bar{\boldsymbol{\varepsilon}}) \end{aligned} \tag{6.3}$$

Substitution of Eq. (6.2) into Eq. (5.15) gives

$$\bar{\mathbf{W}}_u(\mathbf{k}) \cdot \mathbf{E}_0 = \mathbf{0} \tag{6.4}$$

where $\bar{\mathbf{W}}_u(\mathbf{k})$, the wave matrix of a uniaxial medium, is a function of $\mathbf{k}$ and is given by

$$\bar{\mathbf{W}}_u(\mathbf{k}) = (k_0^2 \varepsilon_\perp - \mathbf{k}^2)\bar{\mathbf{I}} + \mathbf{k}\mathbf{k} + k_0^2(\varepsilon_{||} - \varepsilon_\perp)\hat{\mathbf{c}}\hat{\mathbf{c}} \tag{6.5}$$

Since multiplication of any solution of the homogeneous equation (6.4) by a constant yields another solution, Eq. (6.4) uniquely determines only the direction of $\mathbf{E}_0$ not its magnitude. Using the results of Example 1.4, we find the determinant and adjoint of the wave matrix $\bar{\mathbf{W}}_u(\mathbf{k})$:

$$|\,\bar{\mathbf{W}}_u(\mathbf{k})\,| = k_0^2(\mathbf{k}^2 - k_0^2\varepsilon_\perp)[(\mathbf{k}\cdot\bar{\boldsymbol{\varepsilon}}\cdot\mathbf{k}) - k_0^2\varepsilon_\perp\varepsilon_{||}] \tag{6.6}$$

and

$$\begin{aligned} \text{adj } \bar{\mathbf{W}}_u(\mathbf{k}) = {} & (k_0^2\varepsilon_\perp - \mathbf{k}^2)[k_0^2\varepsilon_{||}\bar{\mathbf{I}} - \mathbf{k}\mathbf{k} - k_0^2(\varepsilon_{||} - \varepsilon_\perp)\hat{\mathbf{c}}\hat{\mathbf{c}}] \\ & + k_0^2(\varepsilon_{||} - \varepsilon_\perp)(\mathbf{k}\times\hat{\mathbf{c}})(\mathbf{k}\times\hat{\mathbf{c}}) \end{aligned} \tag{6.7}$$

A nonzero solution $\mathbf{E}_0$ of Eq. (6.4) exists provided that $|\,\bar{\mathbf{W}}_u(\mathbf{k})\,| = 0$. This is the dispersion equation which relates the wave vector to the source frequency ω and the properties of the medium. For those values of $\mathbf{k}$ we obtain nonzero solutions which are proportional to the columns of adj $\bar{\mathbf{W}}(\mathbf{k})$, i.e.,

$$\mathbf{E}_0 = [\text{adj } \bar{\mathbf{W}}_u(\mathbf{k})] \cdot \mathbf{u} \tag{6.8}$$

where $\mathbf{u}$ is an arbitrary vector. Once the direction of $\mathbf{E}_0$ is determined, the remaining field vectors may be found from Maxwell's equations and the constitutive relations:

$$\begin{aligned} \mathbf{H}_0 &= \frac{1}{\omega\mu_0}(\mathbf{k}\times\mathbf{E}_0) \\ \mathbf{B}_0 &= \mu_0\mathbf{H}_0 \\ \mathbf{D}_0 &= \varepsilon_0\bar{\boldsymbol{\varepsilon}}\cdot\mathbf{E}_0 = -\frac{1}{\omega}(\mathbf{k}\times\mathbf{H}_0) \end{aligned} \tag{6.9}$$

6.2 DETERMINATION OF THE DIRECTIONS OF THE FIELD VECTORS

In the case of uniaxial media, we see that the determinant of the wave matrix (6.6) consists of a product of two terms. By setting them equal to zero, we obtain the dispersion equations

$$\mathbf{k}^2 = k_0^2 \varepsilon_\perp \tag{6.10}$$

and

$$\mathbf{k} \cdot \bar{\boldsymbol{\varepsilon}} \cdot \mathbf{k} = k_0^2 \varepsilon_\perp \varepsilon_{||} \tag{6.11}$$

For a given direction of wave normal $\hat{\mathbf{k}}$, these equations determine two values of wave numbers k; thus, also, the two phase velocities. Since the wave number defined by Eq. (6.10) does not depend on the direction of wave normal $\hat{\mathbf{k}}$ as in the case of isotropic media, the corresponding wave is called the *ordinary wave.* Denoting this wave number by k_+, we have

$$k_+ = k_0 \sqrt{\varepsilon_\perp} \tag{6.12}$$

On the other hand, the wave number defined by Eq. (6.11) does depend on the direction of wave normal $\hat{\mathbf{k}}$ and thus the corresponding wave is called the extraordinary wave. According to Eq. (6.11), the wave number of the extraordinary wave denoted by k_- is

$$k_- = k_0 \sqrt{\frac{\varepsilon_\perp \varepsilon_{||}}{\hat{\mathbf{k}}_- \cdot \bar{\boldsymbol{\varepsilon}} \cdot \hat{\mathbf{k}}_-}} \tag{6.13}$$

Substituting Eq. (6.2) into Eq. (6.13), we obtain

$$k_- = k_0 \sqrt{\frac{\varepsilon_\perp \varepsilon_{||}}{\varepsilon_\perp (\hat{\mathbf{k}}_- \times \hat{\mathbf{c}})^2 + \varepsilon_{||} (\hat{\mathbf{k}}_- \cdot \hat{\mathbf{c}})^2}} \tag{6.14}$$

If the wave normal $\hat{\mathbf{k}}_-$ coincides with the vector $\hat{\mathbf{c}}$, then $(\hat{\mathbf{k}}_- \cdot \hat{\mathbf{c}})^2 = 1$, $(\hat{\mathbf{k}}_- \times \hat{\mathbf{c}})^2 = 0$, and

$$k_- = k_0 \sqrt{\varepsilon_\perp} = k_+$$

On the other hand, if we let $k_+ = k_-$, a comparison of Eq. (6.12) and Eq. (6.14) shows that $(\hat{\mathbf{k}}_- \times \hat{\mathbf{c}})^2 = 0$ and $\hat{\mathbf{k}}_- \cdot \hat{\mathbf{c}} = 1$, that is, $\hat{\mathbf{k}}_- = \hat{\mathbf{c}}$. In other words, the wave numbers, and hence the phase velocities, of the two isonormal waves coincide only when wave normal $\hat{\mathbf{k}}_-$ is along $\hat{\mathbf{c}}$, an eigenvector of the dielectric tensor $\bar{\boldsymbol{\varepsilon}}$. According to the definition given in Sec. 5.3, $\hat{\mathbf{c}}$ is therefore an optic axis. As it turns out in this case that $\hat{\mathbf{c}}$ is the only optic axis of the medium, we thus call such a medium *uniaxial.*

For the ordinary wave $\mathbf{k}_+ = k_+ \hat{\mathbf{k}}_+$, the adjoint of the wave matrix (6.7) becomes

$$\operatorname{adj} \bar{\mathbf{W}}_u(\mathbf{k}_+) = k_0^2(\varepsilon_{||} - \varepsilon_\perp)(\mathbf{k}_+ \times \hat{\mathbf{c}})(\mathbf{k}_+ \times \hat{\mathbf{c}}) \tag{6.15}$$

According to Eq. (6.8), we may then choose the direction of $\mathbf{E}_0$ as

$$\mathbf{e}_+ = \mathbf{k}_+ \times \hat{\mathbf{c}} \tag{6.16}$$

The directions of the remaining field vectors are determined from Eq. (6.9):

$$\mathbf{h}_+ = \frac{1}{\omega\mu_0}[\mathbf{k}_+ \times (\mathbf{k}_+ \times \hat{\mathbf{c}})]$$

$$\mathbf{b}_+ = \frac{1}{\omega}[\mathbf{k}_+ \times (\mathbf{k}_+ \times \hat{\mathbf{c}})] \tag{6.17}$$

$$\mathbf{d}_+ = \varepsilon_0 \varepsilon_\perp (\mathbf{k}_+ \times \hat{\mathbf{c}})$$

In summary, for the ordinary wave in uniaxial media, $\mathbf{e}_+$ and $\mathbf{d}_+$ are perpendicular to the plane formed by vectors $\mathbf{k}_+$ and $\hat{\mathbf{c}}$ while $\mathbf{h}_+$ and $\mathbf{b}_+$ lie on the plane as illustrated in Fig. 6.1.

For the extraordinary wave $\mathbf{k}_- = k_- \hat{\mathbf{k}}_-$ the adjoint of the wave matrix is

$$\begin{aligned}\operatorname{adj} \bar{\mathbf{W}}_u(\mathbf{k}_-) = (k_0^2 \varepsilon_\perp - k_-^2)[k_0^2 \varepsilon_{\|} \bar{\mathbf{I}} - \mathbf{k}_-\mathbf{k}_- - k_0^2(\varepsilon_{\|} - \varepsilon_\perp)\hat{\mathbf{c}}\hat{\mathbf{c}}] \\ + k_0^2(\varepsilon_{\|} - \varepsilon_\perp)(\mathbf{k}_- \times \hat{\mathbf{c}})(\mathbf{k}_- \times \hat{\mathbf{c}})\end{aligned}$$

For simplicity we choose the arbitrary vector $\mathbf{u} = \hat{\mathbf{c}}$ in Eq. (6.8). The direction of $\mathbf{E}_0$ for the extraordinary wave then becomes

$$\mathbf{e}_- = k_0^2 \varepsilon_\perp \hat{\mathbf{c}} - (\mathbf{k}_- \cdot \hat{\mathbf{c}})\mathbf{k}_- \tag{6.18}$$

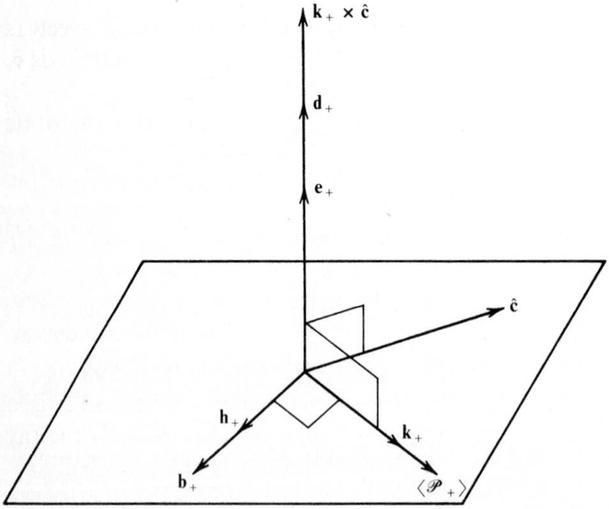

Figure 6.1 Orientations of field vectors of the ordinary wave in a uniaxial medium.

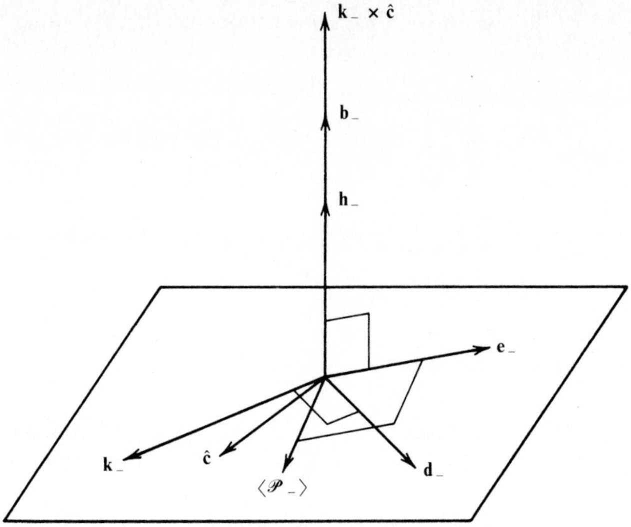

Figure 6.2 Orientations of field vectors of the extraordinary wave in a uniaxial medium.

Substituting Eq. (6.18) into Eq. (6.9), we obtain the directions of the remaining field vectors

$$\begin{aligned} \mathbf{h}_- &= \omega\varepsilon_0\,\varepsilon_\perp(\mathbf{k}_- \times \hat{\mathbf{c}}) \\ \mathbf{b}_- &= \omega\mu_0\,\varepsilon_0\,\varepsilon_\perp(\mathbf{k}_- \times \hat{\mathbf{c}}) \\ \mathbf{d}_- &= -\varepsilon_0\,\varepsilon_\perp[\mathbf{k}_- \times (\mathbf{k}_- \times \hat{\mathbf{c}})] \end{aligned} \tag{6.19}$$

In summary, for the extraordinary wave propagating in uniaxial media, vectors $\mathbf{e}_-$ and $\mathbf{d}_-$ lie on the plane formed by the wave vector $\mathbf{k}_-$ and the optic axis $\hat{\mathbf{c}}$, but $\mathbf{h}_-$ and $\mathbf{b}_-$ are perpendicular to it as shown in Fig. 6.2.

Finally, we note that in uniaxial media, the vectors $\mathbf{E}_0$, $\mathbf{D}_0$, and the optic axis $\hat{\mathbf{c}}$ of either the ordinary or extraordinary waves are coplanar.

6.3 WAVE AND RAY VECTOR SURFACES IN UNIAXIAL CRYSTALS

In Chap. 5, we have defined the wave and ray vector surfaces for anisotropic crystals. In the special case of uniaxial crystals they are considerably simpler.

For the ordinary wave, the wave vector surface defined by Eq. (6.12) is a surface of a sphere of radius $k_0\sqrt{\varepsilon_\perp}$. By duality principle we obtain easily the ray vector surface

$$s_+^2 = \frac{1}{k_0^2\,\varepsilon_\perp} \tag{6.20}$$

which again is a sphere, but with radius $1/k_0\sqrt{\varepsilon_\perp}$. Furthermore, the Poynting vector from Eqs. (6.16) and (6.17) is

$$\begin{aligned}\langle \mathscr{P}_+ \rangle &= \tfrac{1}{2}\,(\mathbf{e}_+ \times \mathbf{h}_+) \\ &= \frac{(\mathbf{k}_+ \times \hat{\mathbf{c}})^2}{2\omega\mu_0}\,\mathbf{k}_+ \end{aligned} \tag{6.21}$$

The equality of electric and magnetic energy densities now becomes

$$\langle W_{e+} \rangle = \langle W_{m+} \rangle = \frac{\varepsilon_0\,\varepsilon_\perp}{4}\,(\mathbf{k}_+ \times \hat{\mathbf{c}})^2 \tag{6.22}$$

Thus the total time-averaged energy density is

$$\langle W_+ \rangle = \frac{\varepsilon_0\,\varepsilon_\perp}{2}\,(\mathbf{k}_+ \times \hat{\mathbf{c}})^2 \tag{6.23}$$

The velocity of energy transport

$$\mathbf{v}_{E+} = \frac{\langle \mathscr{P}_+ \rangle}{\langle W_+ \rangle} = \frac{c}{\sqrt{\varepsilon_\perp}}\,\hat{\mathbf{k}}_+ \tag{6.24}$$

is again equal to the group velocity

$$\mathbf{v}_{g+} = \frac{\partial\omega}{\partial\mathbf{k}} = \frac{c}{\sqrt{\varepsilon_\perp}}\,\hat{\mathbf{k}}_+ \tag{6.25}$$

Moreover, the dot product of the phase velocity

$$\mathbf{v}_{p+} = \frac{\omega}{k_+}\,\hat{\mathbf{k}}_+ = \frac{c}{\sqrt{\varepsilon_\perp}}\,\hat{\mathbf{k}}_+ \tag{6.26}$$

and the group velocity gives

$$\mathbf{v}_{g+} \cdot \mathbf{v}_{p+} = \left(\frac{\omega}{k_+}\right)^2 = v_{p+}^2 \tag{6.27}$$

Yet the dot product of the ray vector obtained from Eq. (6.24)

$$\mathbf{s}_+ = \frac{c}{\omega\sqrt{\varepsilon_\perp}}\,\hat{\mathbf{k}}_+ \tag{6.28}$$

and the wave vector **k** satisfies

$$\mathbf{s}_+ \cdot \mathbf{k}_+ = 1$$

We may thus conclude that the ordinary wave in uniaxial crystals behaves as waves in isotropic media.

On the other hand, for the extraordinary wave, the wave vector surface defined by Eq. (6.14) may be written as

$$\frac{(\mathbf{k}_- \times \hat{\mathbf{c}})^2}{k_0^2\,\varepsilon_{||}} + \frac{(\mathbf{k}_- \cdot \hat{\mathbf{c}})^2}{k_0^2\,\varepsilon_\perp} = 1 \tag{6.29}$$

which represents a surface of revolution about the optic axis $\hat{\mathbf{c}}$ and is an ellipsoid. By duality principle we obtain the ray vector surface

$$\frac{(\mathbf{s}_- \times \hat{\mathbf{c}})^2}{(1/k_0\sqrt{\varepsilon_{\|}})^2} + \frac{(\mathbf{s}_- \cdot \hat{\mathbf{c}})^2}{(1/k_0\sqrt{\varepsilon_{\perp}})^2} = 1 \tag{6.30}$$

which again is an ellipsoid. The Poynting vector, from Eqs. (6.18) and (6.19), is

$$\begin{aligned} \langle \mathscr{P}_- \rangle &= \tfrac{1}{2}(\mathbf{e}_- \times \mathbf{h}_-) \\ &= \frac{\omega\varepsilon_0\,\varepsilon_{\perp}(\mathbf{k}_- \times \hat{\mathbf{c}})^2}{2\varepsilon_{\|}}\,(\bar{\varepsilon}\cdot\mathbf{k}_-) \end{aligned} \tag{6.31}$$

and the equality of electric and magnetic energy densities becomes

$$\langle W_{e-} \rangle = \langle W_{m-} \rangle = \frac{k_0^2\,\varepsilon_{\perp}^2\,\varepsilon_0}{4}\,(\mathbf{k}_- \times \hat{\mathbf{c}})^2 \tag{6.32}$$

Thus the total time-averaged energy density is

$$\langle W_- \rangle = \frac{k_0^2\,\varepsilon_{\perp}^2\,\varepsilon_0}{2}\,(\mathbf{k}_- \times \hat{\mathbf{c}})^2 \tag{6.33}$$

In this case the velocity of energy transport

$$\mathbf{v}_{E-} = \frac{\langle \mathscr{P}_- \rangle}{\langle W_- \rangle} = \frac{c^2}{\omega\varepsilon_{\perp}\,\varepsilon_{\|}}\,(\bar{\varepsilon}\cdot\mathbf{k}_-) \tag{6.34}$$

is again equal to the group velocity

$$\mathbf{v}_{g-} = \frac{\partial\omega}{\partial\mathbf{k}} = \frac{c^2}{\omega\varepsilon_{\perp}\,\varepsilon_{\|}}\,(\bar{\varepsilon}\cdot\mathbf{k}_-) \tag{6.35}$$

but is not equal to the phase velocity

$$\begin{aligned} \mathbf{v}_{p-} &= \frac{\omega}{k_-}\,\hat{\mathbf{k}}_- \\ &= c\left[\frac{(\hat{\mathbf{k}}_- \times \hat{\mathbf{c}})^2}{\varepsilon_{\|}} + \frac{(\hat{\mathbf{k}}_- \cdot \hat{\mathbf{c}})^2}{\varepsilon_{\perp}}\right]^{1/2}\hat{\mathbf{k}}_- \end{aligned} \tag{6.36}$$

However, the dot product of group and phase velocities still yields

$$\mathbf{v}_{g-}\cdot\mathbf{v}_{p-} = v_{p-}^2 = \left(\frac{\omega}{k_-}\right)^2 \tag{6.37}$$

whereas the dot product of the ray vector

$$\mathbf{s}_- = \frac{1}{k_0^2\,\varepsilon_{\perp}\,\varepsilon_{\|}}\,(\bar{\varepsilon}\cdot\mathbf{k}_-) \tag{6.38}$$

and the wave vector $\mathbf{k}_-$ satisfies

$$\mathbf{s}_- \cdot \mathbf{k}_- = 1$$

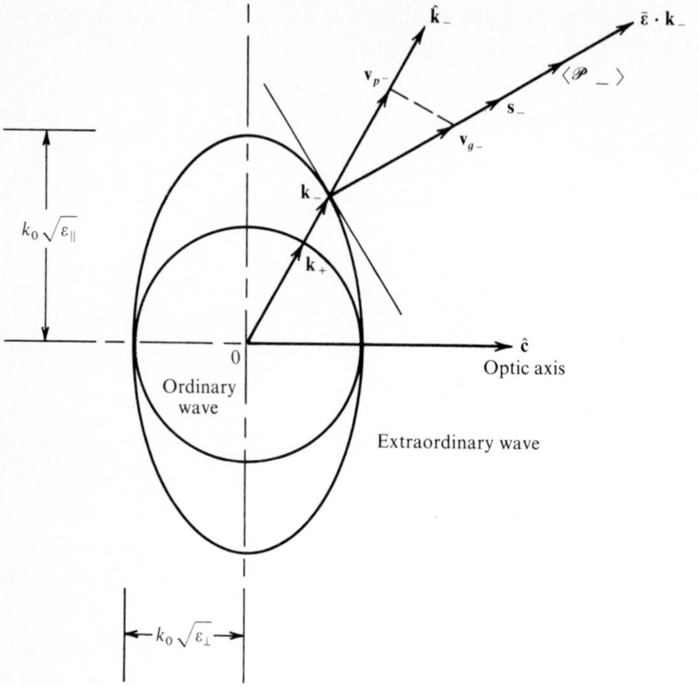

Figure 6.3 Wave vector surfaces for a positive uniaxial crystal: $\varepsilon_{\|} > \varepsilon_{\perp}$ (e.g., quartz).

We may now illustrate graphically the results. Figure 6.3 shows a cross-sectional view of a wave vector surface when $\varepsilon_{\|} > \varepsilon_{\perp}$. We see that the ordinary wave travels faster than the extraordinary wave, thus justifying the term "positive uniaxial crystal" (e.g., quartz). Conversely, when $\varepsilon_{\|} < \varepsilon_{\perp}$, the ordinary wave in crystal travels slower than the extraordinary wave, as is shown in Fig. 6.4, and hence the term "negative uniaxial crystal" (e.g., calcite). In Figs. 6.3 and 6.4 we also show directions of some related vectors. We note from Eq. (6.37) that the projection of the group velocity of the extraordinary wave in the direction of wave normal gives the phase velocity of the wave (see Fig. 6.3).

6.4 DETERMINATION OF WAVE VECTORS AT ISOTROPIC-UNIAXIAL INTERFACE

We shall now examine the problem of reflection and transmission of waves at the interface of an isotropic medium and a uniaxial crystal. Following the same procedures as in Secs. 5.9 and 5.10, we shall first find the wave vectors of the transmitted and the reflected waves in this section and leave the determination of the transmission and reflection coefficients to the following sections. We shall

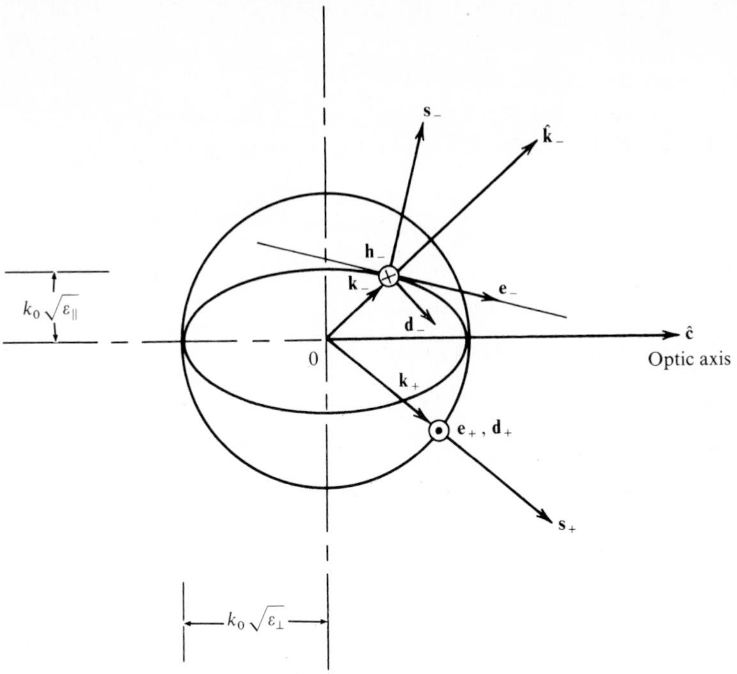

Figure 6.4 Wave vector surfaces for a negative uniaxial crystal: $\varepsilon_\parallel < \varepsilon_\perp$ (e.g., calcite). Vectors $\mathbf{h}_-$ and $\mathbf{b}_-$ are directed into the paper. $\mathbf{s}_-$ is normal to the wave vector surface, $\mathbf{e}_-$ is perpendicular to $\mathbf{s}_-$, and $\mathbf{d}_-$ is perpendicular to $\mathbf{k}_-$. Vectors $\mathbf{e}_+$ and $\mathbf{d}_+$ are directed out of the paper.

assume that both media 1 and 2 are nonmagnetic and that the given incident wave is in the isotropic medium 1. Medium 2 is uniaxial crystal characterized by the dielectric tensor (6.2).

To determine the wave vector

$$\mathbf{k}_+ = \mathbf{b} + q_+\hat{\mathbf{q}} \tag{6.39}$$

of the transmitted ordinary wave, we substitute Eq. (6.39) into the dispersion equation (6.10) and find

$$q_+ = +\sqrt{k_0^2\,\varepsilon_\perp - \mathbf{a}^2} \tag{6.40}$$

The choice of the positive sign in front of the square root is dictated by the fact that the energy carried by the transmitted wave should flow toward medium 2, that is $\mathbf{s}_+ \cdot \hat{\mathbf{q}} > 0$. Since the ray vector (6.28) of the ordinary wave is parallel to the wave normal $\hat{\mathbf{k}}_+$, condition $\mathbf{s}_+ \cdot \hat{\mathbf{q}} > 0$ implies $\mathbf{k}_+ \cdot \hat{\mathbf{q}} = q_+ > 0$.

Similarly, substituting the wave vector

$$\mathbf{k}_- = \mathbf{b} + q_-\hat{\mathbf{q}} \tag{6.41}$$

of the transmitted extraordinary wave into the dispersion equation (6.11), we obtain

$$(\hat{\mathbf{q}} \cdot \bar{\varepsilon} \cdot \hat{\mathbf{q}})q_-^2 + 2(\mathbf{b} \cdot \bar{\varepsilon} \cdot \hat{\mathbf{q}})q_- + (\mathbf{b} \cdot \bar{\varepsilon} \cdot \mathbf{b}) - k_0^2 \varepsilon_\perp \varepsilon_\parallel = 0 \tag{6.42}$$

which is a quadratic equation in q_-. Again we must choose solution q_- so that $\mathbf{s}_- \cdot \hat{\mathbf{q}} > 0$. From Eq. (6.38) we see that this condition corresponds to $\mathbf{k}_- \cdot \bar{\varepsilon} \cdot \hat{\mathbf{q}} > 0$, or

$$q_- > -\frac{\mathbf{b} \cdot \bar{\varepsilon} \cdot \hat{\mathbf{q}}}{\hat{\mathbf{q}} \cdot \bar{\varepsilon} \cdot \hat{\mathbf{q}}} \tag{6.43}$$

From Eqs. (6.42) and (6.43) we may now obtain

$$q_- = \frac{-(\mathbf{b} \cdot \bar{\varepsilon} \cdot \hat{\mathbf{q}}) + \zeta}{\hat{\mathbf{q}} \cdot \bar{\varepsilon} \cdot \hat{\mathbf{q}}} \tag{6.44}$$

where

$$\zeta = [(\mathbf{b} \cdot \bar{\varepsilon} \cdot \hat{\mathbf{q}})^2 - (\hat{\mathbf{q}} \cdot \bar{\varepsilon} \cdot \hat{\mathbf{q}})(\mathbf{b} \cdot \bar{\varepsilon} \cdot \mathbf{b} - k_0^2 \varepsilon_\perp \varepsilon_\parallel)]^{1/2}$$

The expression inside the square root may be simplified further since

$$(\mathbf{b} \cdot \bar{\varepsilon} \cdot \hat{\mathbf{q}})^2 - (\hat{\mathbf{q}} \cdot \bar{\varepsilon} \cdot \hat{\mathbf{q}})(\mathbf{b} \cdot \bar{\varepsilon} \cdot \mathbf{b}) = (\hat{\mathbf{q}} \times \mathbf{b}) \cdot [(\bar{\varepsilon} \cdot \mathbf{b}) \times (\bar{\varepsilon} \cdot \hat{\mathbf{q}})]$$

and, from Eq. (1.104),

$$(\bar{\varepsilon} \cdot \mathbf{b}) \times (\bar{\varepsilon} \cdot \hat{\mathbf{q}}) = (\text{adj } \bar{\varepsilon}) \cdot (\mathbf{b} \times \hat{\mathbf{q}})$$

Thus

$$\zeta = [k_0^2 \varepsilon_\perp \varepsilon_\parallel (\hat{\mathbf{q}} \cdot \bar{\varepsilon} \cdot \hat{\mathbf{q}}) - \mathbf{a} \cdot (\text{adj } \bar{\varepsilon}) \cdot \mathbf{a}]^{1/2}$$

where $\mathbf{b} \times \hat{\mathbf{q}} = \mathbf{a}$ is used. Substituting Eq. (6.44) into Eq. (6.41) and noting that

$$\begin{aligned}\mathbf{b}(\hat{\mathbf{q}} \cdot \bar{\varepsilon} \cdot \hat{\mathbf{q}}) - \hat{\mathbf{q}}(\mathbf{b} \cdot \bar{\varepsilon} \cdot \hat{\mathbf{q}}) &= (\mathbf{b}\hat{\mathbf{q}} - \hat{\mathbf{q}}\mathbf{b}) \cdot (\bar{\varepsilon} \cdot \hat{\mathbf{q}}) \\ &= [(\hat{\mathbf{q}} \times \mathbf{b}) \times \bar{\mathbf{I}}] \cdot (\bar{\varepsilon} \cdot \hat{\mathbf{q}}) \\ &= (\bar{\varepsilon} \cdot \hat{\mathbf{q}}) \times \mathbf{a}\end{aligned}$$

we finally obtain

$$\mathbf{k}_- = \frac{(\bar{\varepsilon} \cdot \hat{\mathbf{q}}) \times \mathbf{a} + \zeta\hat{\mathbf{q}}}{\hat{\mathbf{q}} \cdot \bar{\varepsilon} \cdot \hat{\mathbf{q}}} \tag{6.45}$$

Since medium 1 is isotropic, the results obtained in Sec. 4.4 for $\mathbf{k}_i$ and $\mathbf{k}_r$ remain valid, that is,

$$\begin{aligned}\mathbf{k}_i &= \mathbf{b} + q_i \hat{\mathbf{q}} \\ \mathbf{k}_r &= \mathbf{b} - q_i \hat{\mathbf{q}}\end{aligned} \tag{6.46}$$

where

$$q_i = \sqrt{k_0^2 \varepsilon_1 - \mathbf{a}^2} \tag{6.47}$$

and ε_1 is the dielectric constant of the isotropic medium 1. The geometrical determination of wave vectors at the interface of an isotropic medium and uniaxial crystal is shown in Fig. 6.5.

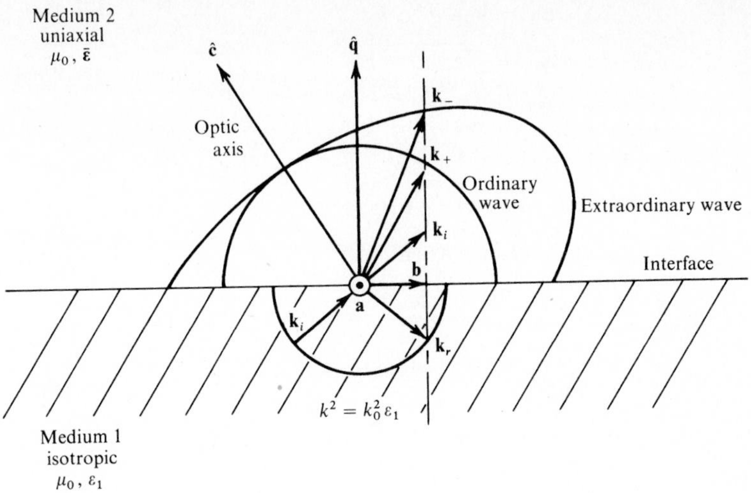

Figure 6.5 Geometrical determination of wave vectors at the interface of an isotropic and a uniaxial medium. Vector **a** is directed out of the paper.

6.5 REFLECTION AND TRANSMISSION COEFFICIENTS

Knowing the wave vectors of all the waves, we may now proceed to use the results of Sec. 5.10 to calculate the reflection and transmission coefficients. For easy reference, we summarize the waves in media 1 and 2 as follows:

Incident wave in isotropic medium 1

$$\mathbf{E}_{0i} = A_\perp \mathbf{a} + A_\parallel (\hat{\mathbf{k}}_i \times \mathbf{a})$$
$$\mathbf{H}_{0i} = \sqrt{\frac{\varepsilon_0 \varepsilon_1}{\mu_0}} [A_\perp (\hat{\mathbf{k}}_i \times \mathbf{a}) - A_\parallel \mathbf{a}] \tag{6.48}$$

Reflected wave in isotropic medium 1

$$\mathbf{E}_{0r} = B_\perp \mathbf{a} + B_\parallel (\hat{\mathbf{k}}_r \times \mathbf{a})$$
$$\mathbf{H}_{0r} = \sqrt{\frac{\varepsilon_0 \varepsilon_1}{\mu_0}} [B_\perp (\hat{\mathbf{k}}_r \times \mathbf{a}) - B_\parallel \mathbf{a}] \tag{6.49}$$

Transmitted waves in uniaxial medium 2

Ordinary wave:

$$\mathbf{E}_{0+} = C_+ \mathbf{e}_+$$
$$\mathbf{H}_{0+} = C_+ \mathbf{h}_+ \tag{6.50}$$

where

$$\mathbf{e}_+ = \mathbf{k}_+ \times \hat{\mathbf{c}}$$
$$\mathbf{h}_+ = \frac{1}{\omega\mu_0} (\mathbf{k}_+ \times \mathbf{e}_+) = \frac{1}{\omega\mu_0} [\mathbf{k}_+ \times (\mathbf{k}_+ \times \hat{\mathbf{c}})] \tag{6.51}$$

and

$$\begin{aligned} \mathbf{k}_+ &= \mathbf{b} + q_+\hat{\mathbf{q}} \\ q_+ &= \sqrt{k_0^2 \varepsilon_\perp - \mathbf{a}^2} \\ k_+ &= k_0\sqrt{\varepsilon_\perp} \end{aligned} \tag{6.52}$$

Extraordinary wave:

$$\begin{aligned} \mathbf{E}_{0-} &= C_-\mathbf{e}_- \\ \mathbf{H}_{0-} &= C_-\mathbf{h}_- \end{aligned} \tag{6.53}$$

where

$$\begin{aligned} \mathbf{e}_- &= k_0^2 \varepsilon_\perp \hat{\mathbf{c}} - (\mathbf{k}_- \cdot \hat{\mathbf{c}})\mathbf{k}_- \\ \mathbf{h}_- &= \frac{1}{\omega\mu_0}(\mathbf{k}_- \times \mathbf{e}_-) = \omega\varepsilon_0 \varepsilon_\perp(\mathbf{k}_- \times \hat{\mathbf{c}}) \end{aligned} \tag{6.54}$$

and

$$\begin{aligned} \mathbf{k}_- &= \frac{(\boldsymbol{\varepsilon} \cdot \hat{\mathbf{q}}) \times \mathbf{a} + \zeta\hat{\mathbf{q}}}{\hat{\mathbf{q}} \cdot \bar{\boldsymbol{\varepsilon}} \cdot \hat{\mathbf{q}}} \\ \zeta &= [k_0^2 \varepsilon_\perp \varepsilon_\parallel (\hat{\mathbf{q}} \cdot \bar{\boldsymbol{\varepsilon}} \cdot \hat{\mathbf{q}}) - \mathbf{a} \cdot (\operatorname{adj} \bar{\boldsymbol{\varepsilon}}) \cdot \mathbf{a}]^{1/2} \\ k_- &= k_0 \sqrt{\frac{\varepsilon_\perp \varepsilon_\parallel}{\hat{\mathbf{k}}_- \cdot \bar{\boldsymbol{\varepsilon}} \cdot \hat{\mathbf{k}}_-}} \end{aligned} \tag{6.55}$$

where the constants $A_\perp$ and $A_\parallel$ are assumed to be given, and the constants $B_\perp$, $B_\parallel$, C_+, and C_- are to be determined. Figure 6.6 shows the orientation of the

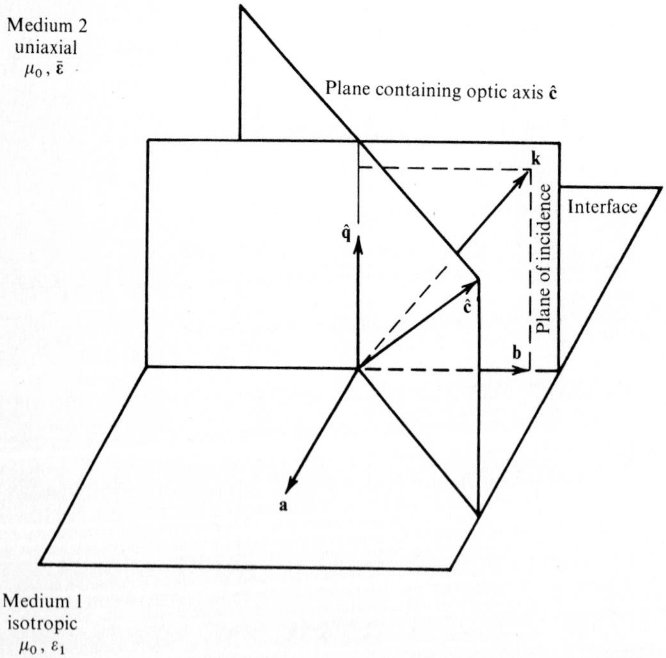

Figure 6.6 Orientations of the plane of incidence, the interface and the plane containing the optic axis $\hat{\mathbf{c}}$.

optic axis with respect to the plane of incidence and the interface, and Fig. 6.7 shows the wave vectors in both media. From Eqs. (6.51) and (6.54), we easily find

$$\begin{aligned}
\mathbf{a}\cdot\mathbf{e}_+ &= -\mathbf{k}_+\cdot(\mathbf{a}\times\hat{\mathbf{c}})\\
\mathbf{a}\cdot\mathbf{e}_- &= k_0^2\varepsilon_\perp(\mathbf{a}\cdot\hat{\mathbf{c}})\\
\mathbf{a}\cdot\mathbf{h}_+ &= -\omega\varepsilon_0\varepsilon_\perp(\mathbf{a}\cdot\hat{\mathbf{c}})\\
\mathbf{a}\cdot\mathbf{h}_- &= -\omega\varepsilon_0\varepsilon_\perp[\mathbf{k}_-\cdot(\mathbf{a}\times\hat{\mathbf{c}})]\\
\mathbf{b}\cdot\mathbf{e}_+ &= q_+(\mathbf{a}\cdot\hat{\mathbf{c}})\\
\mathbf{b}\cdot\mathbf{e}_- &= q_+^2(\mathbf{b}\cdot\hat{\mathbf{c}}) - q_-\mathbf{a}^2(\hat{\mathbf{q}}\cdot\hat{\mathbf{c}})
\end{aligned} \tag{6.56}$$

Substitution of Eq. (6.56) into Eq. (5.180) yields

$$\begin{aligned}
\mathbf{a}\cdot\mathbf{N}_+ &= -q_i(q_i+q_+)[\mathbf{k}_+\cdot(\mathbf{a}\times\hat{\mathbf{c}})]\\
\mathbf{a}\cdot\mathbf{N}_- &= k_0^2\varepsilon_\perp q_i(q_i+q_-)(\mathbf{a}\cdot\hat{\mathbf{c}})\\
\mathbf{b}\cdot\mathbf{N}_+ &= (\varepsilon_1 q_+ + \varepsilon_\perp q_i)k_0^2(\mathbf{a}\cdot\hat{\mathbf{c}})\\
\mathbf{b}\cdot\mathbf{N}_- &= k_i^2[q_+^2(\mathbf{b}\cdot\hat{\mathbf{c}}) - q_-\mathbf{a}^2(\hat{\mathbf{q}}\cdot\hat{\mathbf{c}})] + k_0^2 q_i\varepsilon_\perp[\mathbf{k}_-\cdot(\mathbf{a}\times\hat{\mathbf{c}})]\\
\mathbf{a}\cdot\mathbf{F}_+ &= q_i(q_+-q_i)[\mathbf{k}_+\cdot(\mathbf{a}\times\hat{\mathbf{c}})]\\
\mathbf{a}\cdot\mathbf{F}_- &= k_0^2\varepsilon_\perp q_i(q_i-q_-)(\mathbf{a}\cdot\hat{\mathbf{c}})\\
\mathbf{b}\cdot\mathbf{F}_+ &= (\varepsilon_1 q_+ - \varepsilon_\perp q_i)k_0^2(\mathbf{a}\cdot\hat{\mathbf{c}})\\
\mathbf{b}\cdot\mathbf{F}_- &= k_i^2[q_+^2(\mathbf{b}\cdot\hat{\mathbf{c}}) - q_-\mathbf{a}^2(\hat{\mathbf{q}}\cdot\hat{\mathbf{c}})] - k_0^2 q_i\varepsilon_\perp[\mathbf{k}_-\cdot(\mathbf{a}\times\hat{\mathbf{c}})]
\end{aligned} \tag{6.57}$$

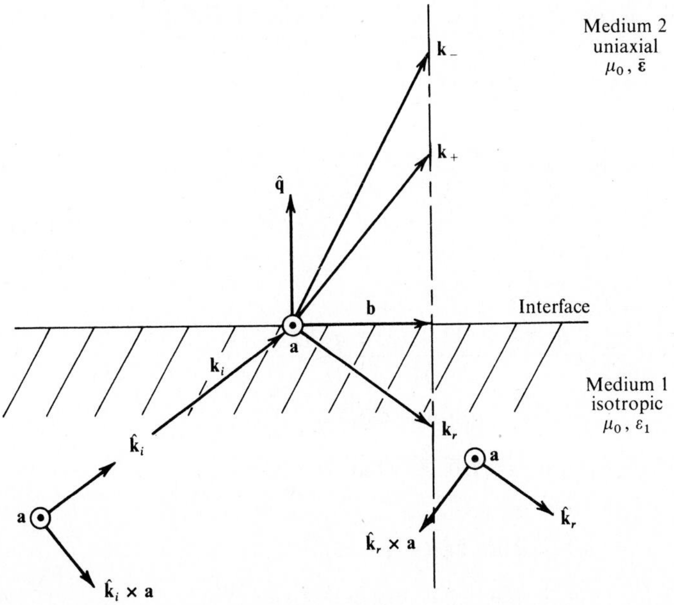

Figure 6.7 The wave vectors of the incident, the reflected, and the transmitted waves. The dot indicates that vector **a** is directed out of the paper.

Finally, substituting Eq. (6.57) into Eq. (5.183), we find the transmission coefficients when medium 2 is a uniaxial crystal:

$$T_{11} = \frac{M_{11}}{\Delta_1} \qquad T_{12} = \frac{M_{12}}{\Delta_1}$$
$$T_{21} = \frac{M_{21}}{\Delta_1} \qquad T_{22} = \frac{M_{22}}{\Delta_1} \tag{6.58}$$

where

$$\begin{aligned} M_{11} &= \frac{-2q_i\mathbf{a}^2(X+Y)}{\mathbf{k}_+ \cdot (\mathbf{a} \times \hat{\mathbf{c}})} \\ M_{21} &= \frac{2q_i\mathbf{a}^2(U+Z)}{k_0^2\varepsilon_\perp(\mathbf{a}\cdot\hat{\mathbf{c}})} \\ &= 2q_i\mathbf{a}^2(k_i^2 q_+ + k_0^2\varepsilon_\perp q_i)(\mathbf{a}\cdot\hat{\mathbf{c}}) \\ M_{12} &= 2k_0^2 k_i \mathbf{a}^2\varepsilon_\perp q_i(q_i + q_-)(\mathbf{a}\cdot\hat{\mathbf{c}}) \\ M_{22} &= 2k_i\mathbf{a}^2 q_i(q_i+q_+)[\mathbf{k}_+\cdot(\mathbf{a}\times\hat{\mathbf{c}})] \end{aligned} \tag{6.59}$$

and

$$\begin{aligned} \Delta_1 &= (q_i + q_+)(X+Y) + (q_i + q_-)(U+Z) \\ X &= k_0^2 q_i \varepsilon_\perp[\mathbf{k}_+\cdot(\mathbf{a}\times\hat{\mathbf{c}})][\mathbf{k}_-\cdot(\mathbf{a}\times\hat{\mathbf{c}})] \\ Y &= k_i^2[\mathbf{k}_+\cdot(\mathbf{a}\times\hat{\mathbf{c}})][q_+^2(\mathbf{b}\cdot\hat{\mathbf{c}}) - q_-\mathbf{a}^2(\hat{\mathbf{q}}\cdot\hat{\mathbf{c}})] \\ U &= k_0^2 k_i^2 \varepsilon_\perp q_+(\mathbf{a}\cdot\hat{\mathbf{c}})^2 \\ Z &= k_0^4\varepsilon_\perp^2 q_i(\mathbf{a}\cdot\hat{\mathbf{c}})^2 \end{aligned} \tag{6.60}$$

Similarly, substituting Eq. (6.57) into Eq. (5.187), we obtain the reflection coefficients

$$\begin{aligned} \Gamma_{11} &= \frac{(q_i - q_+)(X+Y) + (q_i - q_-)(U+Z)}{\Delta_1} \\ \Gamma_{12} &= \frac{2(q_+ - q_-)(V-L)}{\Delta_1} \\ \Gamma_{21} &= \frac{2(q_- - q_+)(V+L)}{\Delta_1} \\ \Gamma_{22} &= \frac{(q_i + q_+)(X-Y) + (q_i + q_-)(Z-U)}{\Delta_1} \end{aligned} \tag{6.61}$$

where

$$\begin{aligned} V &= k_0^2\varepsilon_\perp k_i q_i q_+(\mathbf{a}\cdot\hat{\mathbf{c}})(\mathbf{b}\cdot\hat{\mathbf{c}}) \\ L &= k_0^2\varepsilon_\perp k_i q_i \mathbf{a}^2(\mathbf{a}\cdot\hat{\mathbf{c}})(\hat{\mathbf{q}}\cdot\hat{\mathbf{c}}) \end{aligned} \tag{6.62}$$

and

$$\begin{aligned} V - L &= k_0^2\varepsilon_\perp k_i q_i(\mathbf{a}\cdot\hat{\mathbf{c}})[\mathbf{k}_+\cdot(\mathbf{a}\times\hat{\mathbf{c}})] \\ V + L &= k_0^2\varepsilon_\perp k_i q_i(\mathbf{a}\cdot\hat{\mathbf{c}})[q_+(\mathbf{b}\cdot\hat{\mathbf{c}}) + \mathbf{a}^2(\hat{\mathbf{q}}\cdot\hat{\mathbf{c}})] \end{aligned} \tag{6.63}$$

Equations (6.58) and (6.61) give, respectively, the transmission and reflection coefficients of the wave that is incident from an isotropic to a uniaxial medium. These formulas correspond to Fresnel's equations of two isotropic media discussed in Sec. 4.5. The results are obtained in the most general vector form with an incident wave of any polarization and optic axis $\hat{\mathbf{c}}$ arbitrarily oriented with respect to the interface and the plane of incidence.

Let us now examine in more detail the effect of the direction of the optic axis on the field vectors. From Eq. (6.60) it is clear that, when the direction of $\hat{\mathbf{c}}$ is reversed, the reflection coefficients given in Eq. (6.61) and the reflected field vectors in Eq. (6.49) remain unchanged. On the other hand, from Eq. (6.59) we see that the transmission coefficients given in Eq. (6.58) change signs when we reverse the direction of $\hat{\mathbf{c}}$. However, according to Eqs. (6.51) and (6.54) the field vectors $\mathbf{E}_{0+}$, $\mathbf{H}_{0+}$, $\mathbf{E}_{0-}$, and $\mathbf{H}_{0-}$ of the transmitted wave again remain unchanged. In other words, the sign of $\hat{\mathbf{c}}$ may be chosen arbitrarily. To be specific, we shall assume that $\hat{\mathbf{c}}$ is directed from the interface toward the anisotropic crystal as shown in Fig. 6.6.

Next, we will show that the transmission and reflection coefficients (6.58) and (6.61) reduce to the usual Fresnel formulas (4.205) to (4.208) when medium 2 becomes isotropic. To obtain this special case, we note that $\varepsilon_\perp = \varepsilon_\parallel = \varepsilon_2$; hence the dielectric tensor of medium 2 takes the form of a scalar tensor $\bar{\varepsilon} = \varepsilon_2 \bar{\mathbf{I}}$. Therefore,

$$\begin{aligned} q_+ &= q_- = \sqrt{k_0^2 \varepsilon_2 - \mathbf{a}^2} = q_t \\ k_+^2 &= k_-^2 = k_0^2 \varepsilon_2 = k_t^2 \end{aligned} \tag{6.64}$$

and

$$\mathbf{k}_+ = \mathbf{k}_- = \mathbf{k}_t = \mathbf{b} + q_t \hat{\mathbf{q}}$$

Furthermore,

$$\begin{aligned} \mathbf{E}_{0-} &= C_-[k_0^2 \varepsilon_\perp \hat{\mathbf{c}} - (\mathbf{k}_- \cdot \hat{\mathbf{c}})\mathbf{k}_-] \\ &= -C_-[\mathbf{k}_t \times (\mathbf{k}_t \times \hat{\mathbf{c}})] \end{aligned} \tag{6.65}$$

Since $\hat{\mathbf{c}}$ may now be in any direction and in order to identify $\mathbf{E}_{0+}$ and $\mathbf{E}_{0-}$ with the components perpendicular and parallel to the plane of incidence, we choose $\hat{\mathbf{c}} = \hat{\mathbf{q}}$; thus

$$\begin{aligned} \mathbf{E}_{0+} &= C_+ \mathbf{a} \\ \mathbf{E}_{0-} &= -C_-(\mathbf{k}_t \times \mathbf{a}) \end{aligned} \tag{6.66}$$

Substituting the above into Eqs. (6.58) to (6.63), we obtain

$$\frac{C_+}{A_\perp} = \frac{2q_i}{q_i + q_t} \qquad \frac{C_-}{A_\parallel} = \frac{-2k_i q_i}{k_t^2 q_i + k_i^2 q_t} \qquad T_{12} = T_{21} = 0 \tag{6.67}$$

and

$$\Gamma_{11} = \frac{q_i - q_t}{q_i + q_t} \qquad \Gamma_{22} = \frac{k_t^2 q_i - k_i^2 q_t}{k_t^2 q_i + k_i^2 q_t} \qquad \Gamma_{12} = \Gamma_{21} = 0 \tag{6.68}$$

Comparing Eq. (6.66) with Eq. (4.168), we see that $C_+ = C_\perp$ and $C_- = -C_\parallel/k_t$. Thus Eqs. (6.67) and (6.68) reduce to Eqs. (4.205) to (4.208).

In the case of normal incidence, formulas (6.58) to (6.63) are no longer valid because the concept of the plane of incidence loses its meaning. Equally meaningless is decomposing the field vectors into components perpendicular and parallel to the plane of incidence. In this case, the wave vectors take the form

$$\mathbf{k}_i = k_i \hat{\mathbf{q}} = -\mathbf{k}_r \qquad \mathbf{k}_+ = k_+ \hat{\mathbf{q}} \qquad \mathbf{k}_- = k_- \hat{\mathbf{q}} \tag{6.69}$$

where

$$k_i = k_0\sqrt{\varepsilon_1} \qquad k_+ = k_0\sqrt{\varepsilon_\perp} \qquad k_- = k_0\sqrt{\frac{\varepsilon_\perp \varepsilon_\parallel}{\hat{\mathbf{q}} \cdot \bar{\boldsymbol{\varepsilon}} \cdot \hat{\mathbf{q}}}} \tag{6.70}$$

As an alternative approach to Sec. 5.9, we may treat the plane formed by vectors $\hat{\mathbf{c}}$ and $\hat{\mathbf{q}}$ as though it were the plane of incidence. Henceforth the subscripts $\perp$ and $\parallel$ will be used in this sense. We now decompose the field vectors of the incident and the reflected waves into components perpendicular and parallel to the plane formed by vectors $\hat{\mathbf{c}}$ and $\hat{\mathbf{q}}$ as shown in Figs. 6.8 and 6.9. Thus, the incident wave becomes

$$\begin{aligned} \mathbf{E}_{0i} &= A_\perp(\hat{\mathbf{q}} \times \hat{\mathbf{c}}) + A_\parallel[\hat{\mathbf{q}} \times (\hat{\mathbf{q}} \times \hat{\mathbf{c}})] \\ \mathbf{H}_{0i} &= \frac{1}{\omega\mu_0}(\mathbf{k}_i \times \mathbf{E}_{0i}) = \sqrt{\frac{\varepsilon_0\varepsilon_1}{\mu_0}}\{A_\perp[\hat{\mathbf{q}} \times (\hat{\mathbf{q}} \times \hat{\mathbf{c}})] - A_\parallel(\hat{\mathbf{q}} \times \hat{\mathbf{c}})\} \end{aligned} \tag{6.71}$$

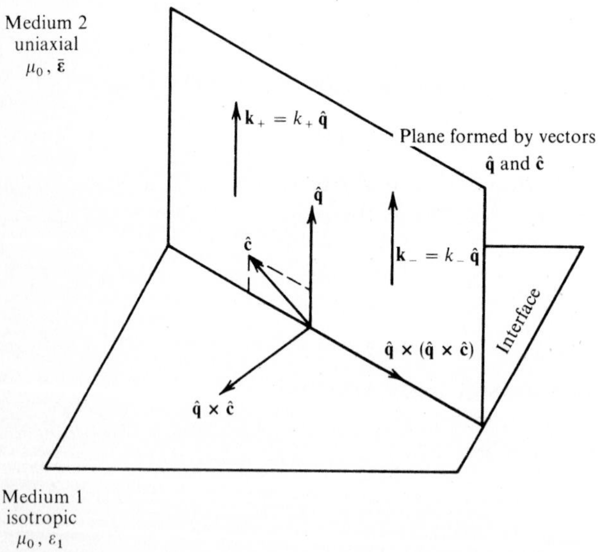

Figure 6.8 Orientations of the interface and the plane formed by vectors $\hat{\mathbf{q}}$ and $\hat{\mathbf{c}}$ in the case of normal incidence.

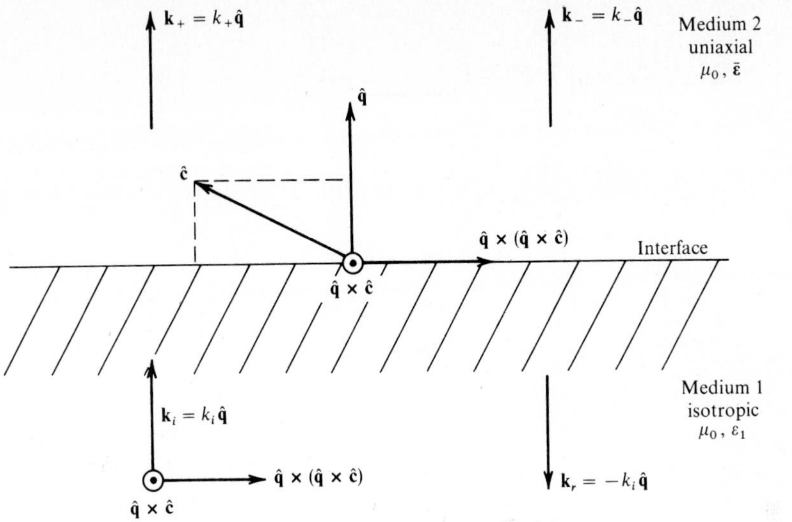

Figure 6.9 The wave vectors of the incident, the reflected, and the transmitted waves in the case of normal incidence.

The reflected wave takes the form

$$\begin{aligned} \mathbf{E}_{0r} &= B_\perp(\hat{\mathbf{q}} \times \hat{\mathbf{c}}) + B_\parallel[\hat{\mathbf{q}} \times (\hat{\mathbf{q}} \times \hat{\mathbf{c}})] \\ \mathbf{H}_{0r} &= \frac{1}{\omega\mu_0}(\mathbf{k}_r \times \mathbf{E}_{0r}) = \sqrt{\frac{\varepsilon_0\varepsilon_1}{\mu_0}}\{-B_\perp[\hat{\mathbf{q}} \times (\hat{\mathbf{q}} \times \hat{\mathbf{c}})] + B_\parallel(\hat{\mathbf{q}} \times \hat{\mathbf{c}})\} \end{aligned} \tag{6.72}$$

and the transmitted waves, according to Eqs. (6.51) and (6.54), are

Ordinary wave:

$$\begin{aligned} \mathbf{E}_{0+} &= C_+(\hat{\mathbf{q}} \times \hat{\mathbf{c}}) \\ \mathbf{H}_{0+} &= \frac{1}{\omega\mu_0}(\mathbf{k}_+ \times \mathbf{E}_{0+}) = \sqrt{\frac{\varepsilon_0\varepsilon_1}{\mu_0}}\, C_+[\hat{\mathbf{q}} \times (\hat{\mathbf{q}} \times \hat{\mathbf{c}})] \end{aligned} \tag{6.73}$$

Extraordinary wave:

$$\begin{aligned} \mathbf{E}_{0-} &= C_-[k_0^2\varepsilon_\perp\hat{\mathbf{c}} - k_-^2(\hat{\mathbf{q}} \cdot \hat{\mathbf{c}})\hat{\mathbf{q}}] \\ \mathbf{H}_{0-} &= \frac{1}{\omega\mu_0}(\mathbf{k}_- \times \mathbf{E}_{0-}) = \omega\varepsilon_0\varepsilon_\perp k_- C_-(\hat{\mathbf{q}} \times \hat{\mathbf{c}}) \end{aligned} \tag{6.74}$$

Substituting Eqs. (6.71) to (6.74) into the boundary conditions (5.163) and (5.164), we obtain

$$(A_\perp + B_\perp - C_+)[\hat{\mathbf{q}} \times (\hat{\mathbf{q}} \times \hat{\mathbf{c}})] - (A_\parallel + B_\parallel + k_0^2\varepsilon_\perp C_-)(\hat{\mathbf{q}} \times \hat{\mathbf{c}}) = \mathbf{0} \tag{6.75}$$

and

$$\left(A_\perp - B_\perp - \sqrt{\frac{\varepsilon_\perp}{\varepsilon_1}}\,C_+\right)[\hat{\mathbf{q}} \times (\hat{\mathbf{q}} \times \hat{\mathbf{c}})] + \left(B_\parallel - A_\parallel - \frac{k_0\,\varepsilon_\perp k_-}{\sqrt{\varepsilon_1}}\,C_-\right)(\hat{\mathbf{q}} \times \hat{\mathbf{c}}) = \mathbf{0} \tag{6.76}$$

respectively. Since $\hat{\mathbf{q}} \times \hat{\mathbf{c}}$ and $\hat{\mathbf{q}} \times (\hat{\mathbf{q}} \times \hat{\mathbf{c}})$ are two linearly independent vectors, it follows that

$$\begin{aligned} A_\perp + B_\perp - C_+ &= 0 \\ A_\perp - B_\perp - \sqrt{\frac{\varepsilon_\perp}{\varepsilon_1}}\,C_+ &= 0 \end{aligned} \tag{6.77}$$

and

$$\begin{aligned} A_\parallel + B_\parallel + k_0^2\,\varepsilon_\perp C_- &= 0 \\ A_\parallel - B_\parallel + \frac{k_0\,\varepsilon_\perp k_-}{\sqrt{\varepsilon_1}}\,C_- &= 0 \end{aligned} \tag{6.78}$$

which can easily be solved, yielding

$$\frac{C_+}{A_\perp} = \frac{2k_i}{k_i + k_+} \qquad \frac{C_-}{A_\parallel} = \frac{-2k_i}{k_0^2\,\varepsilon_\perp(k_i + k_-)} \tag{6.79}$$

$$\frac{B_\perp}{A_\perp} = \frac{k_i - k_+}{k_i + k_+} \qquad \frac{B_\parallel}{A_\parallel} = \frac{k_i - k_-}{k_i + k_-} \tag{6.80}$$

Relations (6.79) and (6.80) together with Eqs. (6.58) and (6.61) completely solve the problem of wave reflection from a uniaxial crystal. However, we should note that in the special case when the wave normals $\hat{\mathbf{k}}_+$ and $\hat{\mathbf{k}}_-$ of the two transmitted waves coincide with the optic axis $\hat{\mathbf{c}}$, formulas (6.58) and (6.61) again lose their validity. Since in this case $k_+^2 = k_-^2 = k_0^2\,\varepsilon_\perp = k_t^2$ and $\mathbf{k}_+ = \mathbf{k}_- = k_t\hat{\mathbf{c}} = \mathbf{k}_t$, it follows that $\mathbf{e}_+$ and $\mathbf{e}_-$ in Eqs. (6.51) and (6.54) become zero. But the condition $\mathbf{k}_t \cdot \mathbf{D}_0 = 0$ implies $\mathbf{c} \cdot \mathbf{E}_0 = 0$. Thus the electric and the magnetic field vectors of the transmitted wave are perpendicular to the wave normal $\hat{\mathbf{k}}_t = \hat{\mathbf{c}}$ and may have any directions in the plane perpendicular to $\hat{\mathbf{c}}$. Analogous to the case of isotropic media, we may now decompose the transmitted wave into components perpendicular and parallel to the plane of incidence. Without repeating the derivation of Sec. 4.5, we use the results of Eqs. (4.191) to (4.196) directly and note that $\mathbf{k}_t = k_0\sqrt{\varepsilon_\perp}\,\hat{\mathbf{c}}$. Thus

$$\begin{aligned} \frac{B_\perp}{A_\perp} &= \frac{\mathbf{a} \cdot (\hat{\mathbf{c}} \times \mathbf{k}_i)}{\mathbf{a} \cdot (\mathbf{k}_r \times \hat{\mathbf{c}})} \\ \frac{B_\parallel}{A_\parallel} &= \frac{(\hat{\mathbf{k}}_r \cdot \hat{\mathbf{c}})[\mathbf{a} \cdot (\hat{\mathbf{c}} \times \mathbf{k}_i)]}{(\hat{\mathbf{k}}_i \cdot \hat{\mathbf{c}})[\mathbf{a} \cdot (\mathbf{k}_r \times \hat{\mathbf{c}})]} \\ \frac{C_\perp}{A_\perp} &= \frac{\mathbf{a} \cdot (\mathbf{k}_r \times \mathbf{k}_i)}{k_0\sqrt{\varepsilon_\perp}[\mathbf{a} \cdot (\mathbf{k}_r \times \hat{\mathbf{c}})]} \\ \frac{C_\parallel}{A_\parallel} &= \frac{\mathbf{a} \cdot (\mathbf{k}_r \times k_i)}{k_0\sqrt{\varepsilon_\perp}(\mathbf{k}_i \cdot \hat{\mathbf{c}})[\mathbf{a} \cdot (\mathbf{k}_r \times \hat{\mathbf{c}})]} \end{aligned} \tag{6.81}$$

We have thus far considered the incident wave propagating either obliquely or normally to the interface. In the following section, we shall examine some interesting special orientations of the optic axis with respect to the interface and the plane of incidence.

6.6 SOME SPECIAL CASES

Optic Axis Parallel to the Interface

When the optic axis of a uniaxial crystal is parallel to the interface but is arbitrarily oriented with respect to the plane of incidence as shown in Fig. 6.10, we have the condition $\hat{\mathbf{q}} \cdot \hat{\mathbf{c}} = 0$. Then

$$\begin{aligned} \mathbf{k}_{\pm} \cdot (\mathbf{a} \times \hat{\mathbf{c}}) &= q_{\pm}(\mathbf{b} \cdot \hat{\mathbf{c}}) \\ \mathbf{b} \cdot \bar{\boldsymbol{\varepsilon}} \cdot \hat{\mathbf{q}} &= 0 \\ \hat{\mathbf{q}} \cdot \bar{\boldsymbol{\varepsilon}} \cdot \hat{\mathbf{q}} &= \varepsilon_{\perp} \end{aligned} \tag{6.82}$$

Both Eqs. (6.60) and (6.62) now reduce to

$$\begin{aligned} X \pm Y &= q_{+}(\mathbf{b} \cdot \hat{\mathbf{c}})^2 (k_0^2 \varepsilon_{\perp} q_i q_{-} \pm k_i^2 q_{+}^2) \\ Z \pm U &= k_0^2 \varepsilon_{\perp} (\mathbf{a} \cdot \hat{\mathbf{c}})^2 (k_0^2 \varepsilon_{\perp} q_i \pm k_i^2 q_{+}) \\ L &= 0 \end{aligned} \tag{6.83}$$

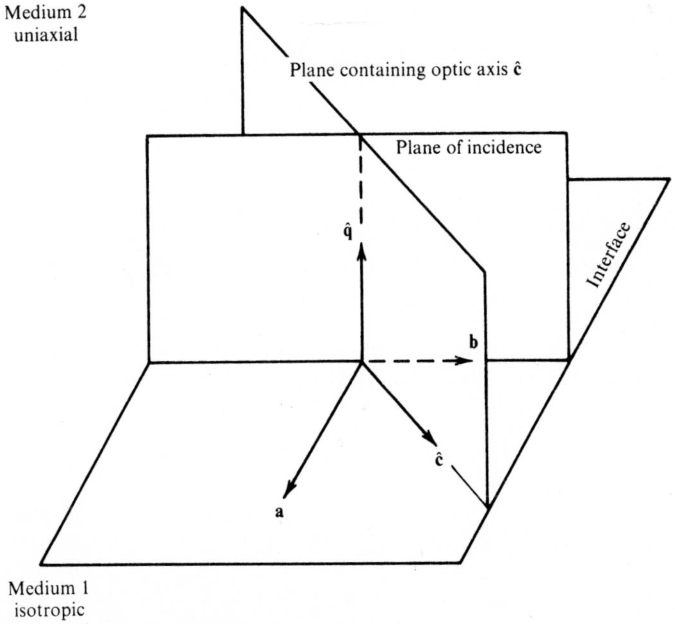

Figure 6.10 Optic axis $\hat{\mathbf{c}}$ of the uniaxial crystal lies on the interface but is arbitrarily oriented with respect to the plane of incidence.

and $$(\mathbf{b} \cdot \hat{\mathbf{c}})^2 = (\mathbf{a} \times \hat{\mathbf{c}})^2 = \mathbf{a}^2 - (\mathbf{a} \cdot \hat{\mathbf{c}})^2$$

We may write Eq. (6.44) as

$$q_- = \frac{1}{\varepsilon_\perp} \sqrt{\varepsilon_\perp \varepsilon_\parallel (k_0^2 \varepsilon_\perp - \mathbf{a}^2) - \varepsilon_\perp (\varepsilon_\perp - \varepsilon_\parallel)(\mathbf{a} \cdot \hat{\mathbf{c}})^2} \tag{6.84}$$

Substituting Eqs. (6.82) and (6.83) into Eqs. (6.58) and (6.61), we obtain, respectively, the transmission coefficients

$$\begin{aligned} T_{11} &= \frac{-2q_i \mathbf{a}^2 (\mathbf{b} \cdot \hat{\mathbf{c}})(k_0^2 \varepsilon_\perp q_i q_- + k_i^2 q_+^2)}{\Delta_1} \\ T_{12} &= \frac{2k_0^2 \varepsilon_\perp k_i q_i \mathbf{a}^2 (\mathbf{a} \cdot \hat{\mathbf{c}})(q_i + q_-)}{\Delta_1} \\ T_{21} &= \frac{2q_i \mathbf{a}^2 (\mathbf{a} \cdot \hat{\mathbf{c}})(k_i^2 q_+ + k_0^2 \varepsilon_\perp q_i)}{\Delta_1} \\ T_{22} &= \frac{2k_i q_i q_+ \mathbf{a}^2 (\mathbf{b} \cdot \hat{\mathbf{c}})(q_i + q_+)}{\Delta_1} \end{aligned} \tag{6.85}$$

and the reflection coefficients

$$\begin{aligned} \Gamma_{11} &= \frac{(q_i - q_+)(X + Y) + (q_i - q_-)(Z + U)}{\Delta_1} \\ \Gamma_{12} &= \frac{2(q_+ - q_-) k_0^2 \varepsilon_\perp k_i q_i q_+ (\mathbf{a} \cdot \hat{\mathbf{c}})(\mathbf{b} \cdot \hat{\mathbf{c}})}{\Delta_1} \\ \Gamma_{21} &= \frac{2(q_- - q_+) k_0^2 \varepsilon_\perp k_i q_i q_+ (\mathbf{a} \cdot \hat{\mathbf{c}})(\mathbf{b} \cdot \hat{\mathbf{c}})}{\Delta_1} \\ \Gamma_{22} &= \frac{(q_i + q_+)(X - Y) + (q_i + q_-)(Z - U)}{\Delta_1} \end{aligned} \tag{6.86}$$

where $$\Delta_1 = (q_i + q_+)(X + Y) + (q_i + q_-)(U + Z)$$

Expressions (6.85) and (6.86) may further be simplified in the following two cases.

Optic axis parallel to the plane of incidence Figure 6.11 shows the orientation of the optic axis. In this case $\hat{\mathbf{c}}$ is parallel to $\mathbf{b}$; thus $\mathbf{b} \cdot \hat{\mathbf{c}} = |\mathbf{a}|$ and $\mathbf{a} \cdot \hat{\mathbf{c}} = 0$. We now have

$$\begin{aligned} q_- &= \sqrt{\frac{\varepsilon_\parallel}{\varepsilon_\perp}} \sqrt{k_0^2 \varepsilon_\perp - \mathbf{a}^2} = \sqrt{\frac{\varepsilon_\parallel}{\varepsilon_\perp}}\, q_+ \\ U &= Z = V = L = 0 \\ \Delta_1 &= q_+(q_i + q_+)\mathbf{a}^2 (k_0^2 \varepsilon_\perp q_i q_- + k_i^2 q_+^2) \end{aligned} \tag{6.87}$$

and the transmission and reflection coefficients become

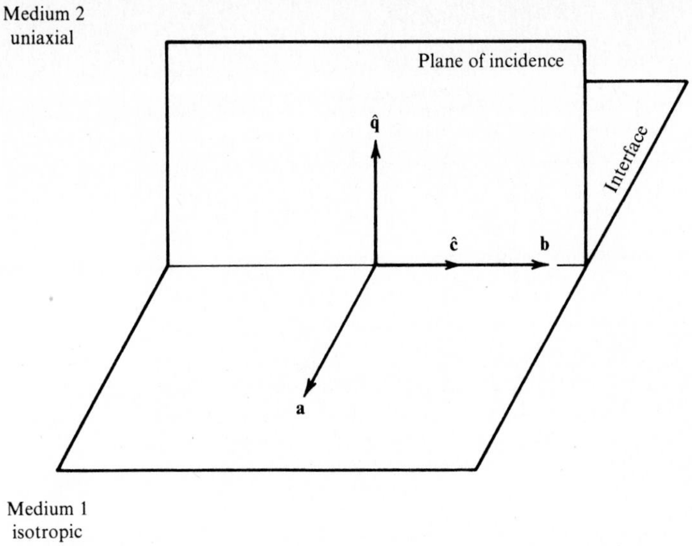

Figure 6.11 Optic axis $\hat{\mathbf{c}}$ is parallel to the interface and the plane of incidence.

$$
\begin{aligned}
T_{11} &= -\frac{2q_i|\mathbf{a}|}{q_+(q_i + q_+)} \\
T_{22} &= \frac{2k_i q_i|\mathbf{a}|}{q_+(k_0^2\sqrt{\varepsilon_\perp \varepsilon_\parallel}\, q_i + k_i^2 q_+)} \\
T_{12} &= T_{21} = 0
\end{aligned}
\tag{6.88}
$$

and

$$
\begin{aligned}
\Gamma_{11} &= \frac{q_i - q_+}{q_i + q_+} \\
\Gamma_{22} &= \frac{\sqrt{\varepsilon_\perp \varepsilon_\parallel}\, q_i - \varepsilon_1 q_+}{\sqrt{\varepsilon_\perp \varepsilon_\parallel}\, q_i + \varepsilon_1 q_+} \\
\Gamma_{12} &= \Gamma_{21} = 0
\end{aligned}
\tag{6.89}
$$

respectively.

Optic axis perpendicular to the plane of incidence Figure 6.12 shows the orientation of the optic axis. In this case, $\hat{\mathbf{c}}$ is parallel to $\mathbf{a}$. Thus $\mathbf{a} \cdot \hat{\mathbf{c}} = |\mathbf{a}|$, $\mathbf{b} \cdot \hat{\mathbf{c}} = \hat{\mathbf{q}} \cdot \hat{\mathbf{c}} = 0$, and we have

$$
\begin{aligned}
q_+ &= \sqrt{k_0^2 \varepsilon_\perp - \mathbf{a}^2} \qquad k_+^2 = k_0^2 \varepsilon_\perp \\
q_- &= \sqrt{k_0^2 \varepsilon_\parallel - \mathbf{a}^2} \qquad k_-^2 = k_0^2 \varepsilon_\parallel \\
X &= Y = V = L = 0 \\
\Delta_1 &= (q_i + q_-)k_0^2 \varepsilon_\perp \mathbf{a}^2(k_0^2 \varepsilon_\perp q_i + k_i^2 q_+)
\end{aligned}
\tag{6.90}
$$

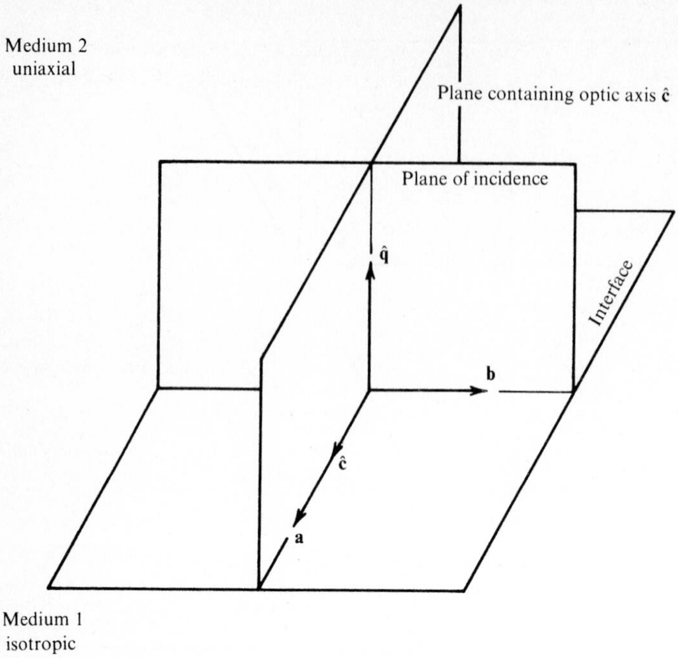

Figure 6.12 Optic axis ĉ is parallel to the interface but perpendicular to the plane of incidence.

Then the transmission and reflection coefficients reduce to

$$T_{12} = \frac{2k_i q_i |\mathbf{a}|}{k_0^2 \varepsilon_\perp q_i + k_i^2 q_+}$$

$$T_{21} = \frac{2q_i |\mathbf{a}|}{k_0^2 \varepsilon_\perp (q_i + q_-)} \tag{6.91}$$

$$T_{11} = T_{22} = 0$$

and

$$\Gamma_{11} = \frac{q_i - q_-}{q_i + q_-}$$

$$\Gamma_{22} = \frac{\varepsilon_\perp q_i - \varepsilon_1 q_+}{\varepsilon_\perp q_i + \varepsilon_1 q_+} \tag{6.92}$$

$$\Gamma_{12} = \Gamma_{21} = 0$$

respectively. Here we note that

$$k_0^2(\varepsilon_\perp q_i \pm \varepsilon_1 q_+) = (q_i \pm q_+)(\mathbf{a}^2 \pm q_i q_+) \tag{6.93}$$

Optic Axis Parallel to the Plane of Incidence

When the optic axis of the uniaxial crystal is parallel to the plane of incidence but is arbitrarily oriented with respect to the interface as illustrated in Fig. 6.13, we have $\mathbf{a} \cdot \hat{\mathbf{c}} = 0$. From Eqs. (6.60) and (6.62), we obtain

$$U = Z = V = L = 0 \tag{6.94}$$

$$\Delta_1 = (q_i + q_+)(X + Y) \tag{6.95}$$

Also, from Eq. (6.44)

$$q_- = \frac{-(\mathbf{b} \cdot \bar{\varepsilon} \cdot \hat{\mathbf{q}}) + \zeta}{\hat{\mathbf{q}} \cdot \bar{\varepsilon} \cdot \hat{\mathbf{q}}} \tag{6.96}$$

where $\zeta = [k_0^2 \varepsilon_\perp \varepsilon_\parallel (\hat{\mathbf{q}} \cdot \bar{\varepsilon} \cdot \hat{\mathbf{q}}) - \mathbf{a} \cdot (\text{adj } \bar{\varepsilon}) \cdot \mathbf{a}]^{1/2}$

In this case, we also have

$$\mathbf{a} \cdot (\text{adj } \bar{\varepsilon}) \cdot \mathbf{a} = \varepsilon_\perp \varepsilon_\parallel \mathbf{a}^2$$

Thus ζ becomes

$$\zeta = \{\varepsilon_\perp \varepsilon_\parallel [k_0^2(\hat{\mathbf{q}} \cdot \bar{\varepsilon} \cdot \hat{\mathbf{q}}) - \mathbf{a}^2]\}^{1/2} \tag{6.97}$$

Since, according to Eq. (6.45),

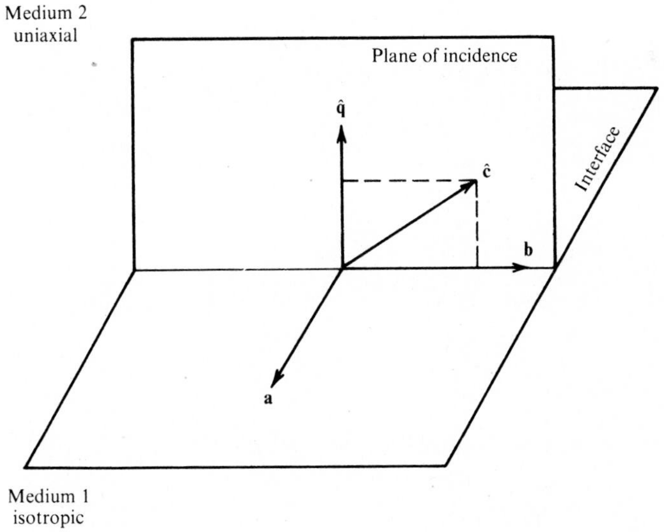

Figure 6.13 Optic axis $\hat{\mathbf{c}}$ is parallel to the plane of incidence but is arbitrarily oriented with respect to the interface.

$$\mathbf{k}_- = \frac{(\bar{\varepsilon} \cdot \hat{\mathbf{q}}) \times \mathbf{a} + \zeta \hat{\mathbf{q}}}{\hat{\mathbf{q}} \cdot \bar{\varepsilon} \cdot \hat{\mathbf{q}}} \tag{6.98}$$

it follows that

$$\begin{aligned}
\mathbf{k}_- \cdot (\mathbf{a} \times \hat{\mathbf{c}}) &= \frac{1}{\hat{\mathbf{q}} \cdot \bar{\varepsilon} \cdot \hat{\mathbf{q}}} \{[(\bar{\varepsilon} \cdot \hat{\mathbf{q}}) \times \mathbf{a}] \cdot (\mathbf{a} \times \hat{\mathbf{c}}) + \zeta[(\hat{\mathbf{q}} \times \mathbf{a}) \cdot \hat{\mathbf{c}}]\} \\
&= \frac{1}{\hat{\mathbf{q}} \cdot \bar{\varepsilon} \cdot \hat{\mathbf{q}}} [\zeta(\mathbf{b} \cdot \hat{\mathbf{c}}) - \mathbf{a}^2(\hat{\mathbf{c}} \cdot \bar{\varepsilon} \cdot \hat{\mathbf{q}})] \\
&= \frac{1}{\hat{\mathbf{q}} \cdot \bar{\varepsilon} \cdot \hat{\mathbf{q}}} [\zeta(\mathbf{b} \cdot \hat{\mathbf{c}}) - \varepsilon_{\|} \mathbf{a}^2(\hat{\mathbf{q}} \cdot \hat{\mathbf{c}})]
\end{aligned} \tag{6.99}$$

Substitution of q_+ as given in Eq. (6.52) and q_- as given in Eq. (6.96) into the following expression contained in Y yields

$$\begin{aligned}
q_+^2(\mathbf{b} \cdot \hat{\mathbf{c}}) - q_-\mathbf{a}^2(\hat{\mathbf{q}} \cdot \hat{\mathbf{c}}) = \frac{1}{\hat{\mathbf{q}} \cdot \bar{\varepsilon} \cdot \hat{\mathbf{q}}} \{&k_0^2 \varepsilon_\perp (\mathbf{b} \cdot \hat{\mathbf{c}})(\hat{\mathbf{q}} \cdot \bar{\varepsilon} \cdot \hat{\mathbf{q}}) \\
&+ [(\hat{\mathbf{q}} \cdot \hat{\mathbf{c}})(\mathbf{b} \cdot \bar{\varepsilon} \cdot \hat{\mathbf{q}}) - (\mathbf{b} \cdot \hat{\mathbf{c}})(\hat{\mathbf{q}} \cdot \bar{\varepsilon} \cdot \hat{\mathbf{q}})]\mathbf{a}^2 - \zeta \mathbf{a}^2(\hat{\mathbf{q}} \cdot \hat{\mathbf{c}})\}
\end{aligned}$$

But

$$\begin{aligned}
(\hat{\mathbf{q}} \cdot \hat{\mathbf{c}})(\mathbf{b} \cdot \bar{\varepsilon} \cdot \hat{\mathbf{q}}) - (\mathbf{b} \cdot \hat{\mathbf{c}})(\hat{\mathbf{q}} \cdot \bar{\varepsilon} \cdot \hat{\mathbf{q}}) &= [(\hat{\mathbf{q}} \cdot \hat{\mathbf{c}})\mathbf{b} - (\mathbf{b} \cdot \hat{\mathbf{c}})\hat{\mathbf{q}}] \cdot (\bar{\varepsilon} \cdot \hat{\mathbf{q}}) \\
&= (\hat{\mathbf{q}} \times \mathbf{b}) \cdot [\hat{\mathbf{c}} \times (\bar{\varepsilon} \cdot \hat{\mathbf{q}})] \\
&= -\varepsilon_\perp (\mathbf{b} \times \hat{\mathbf{q}}) \cdot (\hat{\mathbf{c}} \times \hat{\mathbf{q}}) \\
&= -\varepsilon_\perp (\hat{\mathbf{c}} \cdot \mathbf{b})
\end{aligned}$$

Thus

$$\begin{aligned}
q_+^2(\mathbf{b} \cdot \hat{\mathbf{c}}) - q_- \mathbf{a}^2(\hat{\mathbf{q}} \cdot \hat{\mathbf{c}}) &= \frac{1}{\hat{\mathbf{q}} \cdot \bar{\varepsilon} \cdot \hat{\mathbf{q}}} \{\varepsilon_\perp (\mathbf{b} \cdot \hat{\mathbf{c}})[k_0^2(\hat{\mathbf{q}} \cdot \bar{\varepsilon} \cdot \hat{\mathbf{q}}) - \mathbf{a}^2] - \zeta \mathbf{a}^2(\hat{\mathbf{q}} \cdot \hat{\mathbf{c}})\} \\
&= \frac{\zeta}{\varepsilon_{\|}(\hat{\mathbf{q}} \cdot \bar{\varepsilon} \cdot \hat{\mathbf{q}})} [\zeta(\mathbf{b} \cdot \hat{\mathbf{c}}) - \varepsilon_{\|} \mathbf{a}^2(\hat{\mathbf{q}} \cdot \hat{\mathbf{c}})] \\
&= \frac{\zeta}{\varepsilon_{\|}} (\mathbf{k}_- \cdot \mathbf{a} \times \hat{\mathbf{c}})
\end{aligned} \tag{6.100}$$

In obtaining Eq. (6.100), we have used the results of Eq. (6.99) and

$$k_0^2(\hat{\mathbf{q}} \cdot \bar{\varepsilon} \cdot \hat{\mathbf{q}}) - \mathbf{a}^2 = \frac{\zeta^2}{\varepsilon_\perp \varepsilon_{\|}}$$

Consequently, Y in Eq. (6.60) becomes

$$Y = \frac{k_i^2 \zeta}{\varepsilon_\parallel} [\mathbf{k}_+ \cdot (\mathbf{a} \times \hat{\mathbf{c}})][\mathbf{k}_- \cdot (\mathbf{a} \times \hat{\mathbf{c}})]$$

and

$$X \pm Y = \frac{1}{\varepsilon_\parallel} [\mathbf{k}_+ \cdot (\mathbf{a} \times \hat{\mathbf{c}})][\mathbf{k}_- \cdot (\mathbf{a} \times \hat{\mathbf{c}})](k_0^2 q_i \varepsilon_\perp \varepsilon_\parallel \pm k_i^2 \zeta) \tag{6.101}$$

Substituting Eqs. (6.94), (6.95), and (6.101) into Eqs. (6.58) and (6.61), we obtain the transmission coefficients

$$\begin{aligned} T_{11} &= -\frac{2q_i \mathbf{a}^2}{(q_i + q_+)[\mathbf{k}_+ \cdot (\mathbf{a} \times \hat{\mathbf{c}})]} \\ T_{22} &= \frac{2k_i q_i \varepsilon_\parallel \mathbf{a}^2}{[\mathbf{k}_- \cdot (\mathbf{a} \times \hat{\mathbf{c}})](k_0^2 \varepsilon_\perp \varepsilon_\parallel q_i + k_i^2 \zeta)} \\ T_{12} &= T_{21} = 0 \end{aligned} \tag{6.102}$$

and the reflection coefficients

$$\begin{aligned} \Gamma_{11} &= \frac{q_i - q_+}{q_i + q_+} \\ \Gamma_{22} &= \frac{X - Y}{X + Y} = \frac{k_0^2 \varepsilon_\perp \varepsilon_\parallel q_i - k_i^2 \zeta}{k_0^2 \varepsilon_\perp \varepsilon_\parallel q_i + k_i^2 \zeta} \\ &= \frac{\sqrt{\varepsilon_\perp \varepsilon_\parallel}\, q_i - \varepsilon_1 \sqrt{k_0^2(\hat{\mathbf{q}} \cdot \bar{\varepsilon} \cdot \hat{\mathbf{q}}) - \mathbf{a}^2}}{\sqrt{\varepsilon_\perp \varepsilon_\parallel}\, q_i + \varepsilon_1 \sqrt{k_0^2(\hat{\mathbf{q}} \cdot \bar{\varepsilon} \cdot \hat{\mathbf{q}}) - \mathbf{a}^2}} \\ \Gamma_{12} &= \Gamma_{21} = 0 \end{aligned} \tag{6.103}$$

When the optic axis is normal to the interface as shown in Fig. 6.14, we have two possible situations: oblique incidence and normal incidence.

Oblique incidence (see Fig. 6.15) Since $\hat{\mathbf{c}} = \hat{\mathbf{q}}$, $\mathbf{a} \cdot \hat{\mathbf{c}} = \mathbf{b} \cdot \hat{\mathbf{c}} = 0$, $\mathbf{a} \times \hat{\mathbf{c}} = -\mathbf{b}$, and

$$\begin{aligned} \mathbf{k}_+ \cdot (\mathbf{a} \times \hat{\mathbf{c}}) &= \mathbf{k}_- \cdot (\mathbf{a} \times \hat{\mathbf{c}}) = -\mathbf{a}^2 \\ \hat{\mathbf{q}} \cdot \bar{\varepsilon} \cdot \hat{\mathbf{q}} &= \varepsilon_\parallel \\ \mathbf{b} \cdot \bar{\varepsilon} \cdot \hat{\mathbf{q}} &= 0 \end{aligned} \tag{6.104}$$

Substitution of Eq. (6.104) into Eqs. (6.97), (6.96), and (6.101) yields

$$\begin{aligned} \zeta &= \sqrt{\varepsilon_\perp \varepsilon_\parallel}\, \sqrt{k_0^2 \varepsilon_\parallel - \mathbf{a}^2} \\ q_- &= \frac{\zeta}{\varepsilon_\parallel} \\ X \pm Y &= k_0^2 \mathbf{a}^4 (\varepsilon_\perp q_i \pm \varepsilon_1 q_-) \end{aligned} \tag{6.105}$$

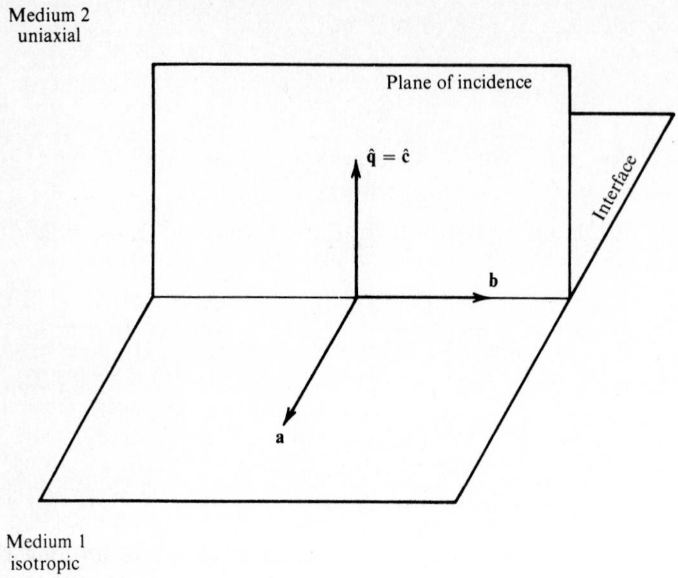

Figure 6.14 Optic axis $\hat{\mathbf{c}}$ is normal to the interface.

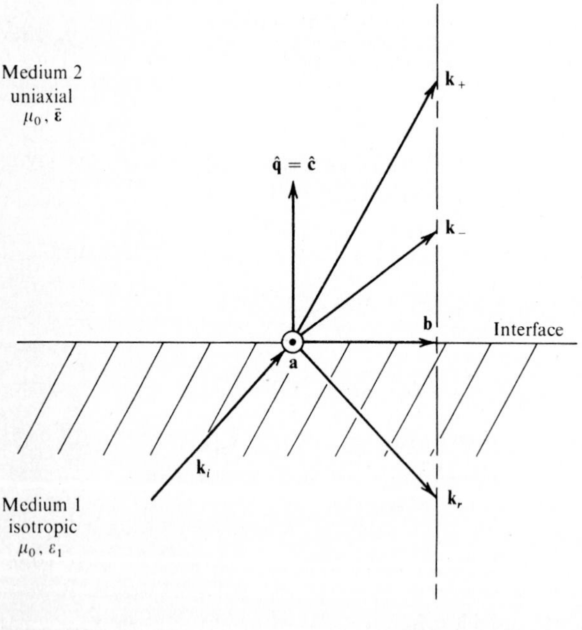

Figure 6.15 Orientations of the wave vectors of the incident, the reflected, and the transmitted waves when $\hat{\mathbf{c}} = \hat{\mathbf{q}}$ and at an oblique incidence.

The transmission coefficients given in Eq. (6.102) now become

$$T_{11} = \frac{2q_i}{q_i + q_+}$$
$$T_{22} = -\frac{2k_i q_i}{k_0^2 \varepsilon_\perp q_i + k_i^2 q_-} \tag{6.106}$$

and the reflection coefficients in Eq. (6.103) reduce to

$$\Gamma_{11} = \frac{q_i - q_+}{q_i + q_+}$$
$$\Gamma_{22} = \frac{\varepsilon_\perp q_i - \varepsilon_1 q_-}{\varepsilon_\perp q_i + \varepsilon_1 q_-} = \frac{\sqrt{\varepsilon_\perp \varepsilon_\parallel}\, q_i - \varepsilon_1 \sqrt{k_0^2 \varepsilon_\parallel - \mathbf{a}^2}}{\sqrt{\varepsilon_\perp \varepsilon_\parallel}\, q_i + \varepsilon_1 \sqrt{k_0^2 \varepsilon_\parallel - \mathbf{a}^2}} \tag{6.107}$$

Normal incidence For $\hat{\mathbf{c}} = \hat{\mathbf{q}}$ we see that formulas (6.71) to (6.74) are no longer useful. In this case (see Fig. 6.16), we have $\mathbf{k}_+ = \mathbf{k}_- = k_0\sqrt{\varepsilon_\perp}\ \hat{\mathbf{c}} = \mathbf{k}_t$. Hence $\mathbf{k}_t \cdot \mathbf{D}_{0t} = 0$ implies $\hat{\mathbf{c}} \cdot \mathbf{E}_{0t} = 0$. Following the derivation of normal incidence on the interface of two isotropic media [cf. Eqs. (4.249) to (4.252)], we obtain

$$\mathbf{E}_{0r} = \frac{k_i - k_t}{k_i + k_t}\mathbf{E}_{0i} \qquad \mathbf{H}_{0r} = \frac{k_t - k_i}{k_t + k_i}\mathbf{H}_{0i}$$
$$\mathbf{E}_{0t} = \frac{2k_i}{k_i + k_t}\mathbf{E}_{0i} \qquad \mathbf{H}_{0t} = \frac{2k_t}{k_i + k_t}\mathbf{H}_{0i} \tag{6.108}$$

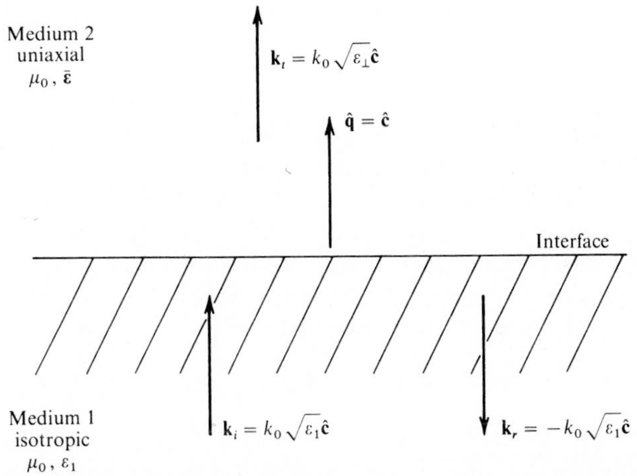

Figure 6.16 Orientations of the wave vectors of the incident, the reflected, and the transmitted waves when $\hat{\mathbf{c}} = \hat{\mathbf{q}}$ and at the normal incidence.

6.7 ROTATION OF THE PLANE OF POLARIZATION UPON REFLECTION, BREWSTER'S ANGLE

With the aid of the results obtained in previous sections, we shall now consider the problem of rotation of the plane of polarization, defined by vectors $\mathbf{E}_0$ and $\mathbf{k}$, of a linearly polarized wave upon reflection from a uniaxial medium. According to Sec. 4.8, the azimuthal angles α_i and α_r of the incident and the reflected waves with the amplitude vectors

$$\mathbf{E}_{0i} = A_\perp \mathbf{a} + A_{||}(\hat{\mathbf{k}}_i \times \mathbf{a})$$

and

$$\mathbf{E}_{0r} = B_\perp \mathbf{a} + B_{||}(\hat{\mathbf{k}}_r \times \mathbf{a}) \tag{6.109}$$

are defined by

$$\tan \alpha_i = \frac{A_\perp}{A_{||}}$$

and

$$\tan \alpha_r = \frac{B_\perp}{B_{||}} \tag{6.110}$$

respectively. The amplitudes of the perpendicular and the parallel components of the incident and the reflected waves are related by the reflection coefficient matrix:

$$\begin{bmatrix} B_\perp \\ B_{||} \end{bmatrix} = \begin{bmatrix} \Gamma_{11} & \Gamma_{12} \\ \Gamma_{21} & \Gamma_{22} \end{bmatrix} \begin{bmatrix} A_\perp \\ A_{||} \end{bmatrix} \tag{6.111}$$

where Γ_{11}, Γ_{12}, Γ_{21}, and Γ_{22} are given by Eq. (6.61).

Using the results of Eqs. (6.110) and (6.111), we write

$$\tan \alpha_r = \frac{\Gamma_{12} + \Gamma_{11} \tan \alpha_i}{\Gamma_{22} + \Gamma_{21} \tan \alpha_i} \tag{6.112}$$

which clearly shows that, for a linearly polarized incident wave, the plane of polarization of the reflected wave is different from that of the incident wave.

We shall next examine the condition under which an unpolarized wave upon reflection from a uniaxial medium becomes a linearly polarized wave. The angle of incidence which satisfies this condition is called the Brewster angle as defined earlier in Sec. 4.8. Evidently, the reflected wave will be linearly polarized if α_r is a constant. That is, we are looking for a condition such that α_r is constant for all α_i. An examination of Eq. (6.112) shows that this is possible if

$$\Gamma_{11}\Gamma_{22} = \Gamma_{12}\Gamma_{21} \tag{6.113}$$

Under this condition, Eq. (6.112) becomes

$$\tan \alpha_r = \frac{\Gamma_{12}}{\Gamma_{22}} = \frac{\Gamma_{11}}{\Gamma_{21}} \tag{6.114}$$

that is, α_r is a constant, independent of α_i. Substituting Eq. (6.61) into Eq. (6.113) and noting that

$$XU - YZ = \frac{(q_- - q_+)(V^2 - L^2)}{q_i}$$

and

$$2(q_i^2 - q_+q_-)(XZ - YU) + 2q_i(q_+ - q_-)(YZ - XU)$$
$$= (q_i + q_+)(q_i - q_-)(X + Y)(Z - U) + (q_i - q_+)(q_i + q_-)(X - Y)(Z + U)$$

after some simplification, we obtain

$$[(q_i - q_+)(X - Y) + (q_i - q_-)(Z - U)]\Delta_1 = 0$$

Since Δ_1, the denominator of the reflection coefficients, cannot be zero, the condition (6.113) for a linearly polarized, reflected wave becomes

$$(q_i - q_+)(X - Y) + (q_i - q_-)(Z - U) = 0 \tag{6.115}$$

or

$$(q_i - q_+)X + (q_i - q_-)Z = (q_i - q_+)Y + (q_i - q_-)U \tag{6.116}$$

Under this condition, we find from Eq. (6.114) that

$$\tan \alpha_r = \frac{(q_+ - q_-)(V - L)}{q_+(X - Y) + q_-(Z - U)}$$

or

$$\tan \alpha_r = \frac{(q_i - q_+)X + (q_i - q_-)Z}{(q_- - q_+)(V + L)} \tag{6.117}$$

$$= \frac{(q_i - q_+)Y + (q_i - q_-)U}{(q_- - q_+)(V + L)}$$

In other words, the reflected wave is linearly polarized with the electric field intensity making an angle α_r with the plane of incidence.
Next, let us examine the following important special cases.

Optic Axis Parallel to the Plane of Incidence

When the optic axis of the uniaxial crystal is parallel to the plane of incidence but arbitrarily oriented with respect to the interface as shown in Fig. 6.13, we have, from Eqs. (6.103) and (6.112),

$$\tan \alpha_r = \frac{(q_i - q_+)[q_i\sqrt{\varepsilon_\perp \varepsilon_\parallel} + \varepsilon_1\sqrt{k_0^2(\hat{\mathbf{q}} \cdot \bar{\boldsymbol{\varepsilon}} \cdot \hat{\mathbf{q}}) - \mathbf{a}^2}]}{(q_i + q_+)[q_i\sqrt{\varepsilon_\perp \varepsilon_\parallel} - \varepsilon_1\sqrt{k_0^2(\hat{\mathbf{q}} \cdot \bar{\boldsymbol{\varepsilon}} \cdot \hat{\mathbf{q}}) - \mathbf{a}^2}]} \tan \alpha_i \tag{6.118}$$

In this case, condition (6.115) for a linearly polarized reflected wave becomes

$$q_i\sqrt{\varepsilon_\perp\varepsilon_\parallel} - \varepsilon_1\sqrt{k_0^2(\hat{\mathbf{q}} \cdot \bar{\boldsymbol{\varepsilon}} \cdot \hat{\mathbf{q}}) - \mathbf{a}^2} = 0 \tag{6.119}$$

Under this condition, we see from either Eq. (6.117) or Eq. (6.118) that $\alpha_r = \pi/2$.

That is, the reflected wave is linearly polarized with the electric field intensity normal to the plane of incidence.

To determine the Brewster angle θ_B, we eliminate q_i from Eq. (6.119) and $q_i^2 = k_i^2 - \mathbf{a}^2$, and obtain

$$\mathbf{a}^2 = \frac{k_0^2 \varepsilon_1[\varepsilon_\perp \varepsilon_\| - \varepsilon_1(\hat{\mathbf{q}} \cdot \bar{\varepsilon} \cdot \hat{\mathbf{q}})]}{\varepsilon_\perp \varepsilon_\| - \varepsilon_1^2} \tag{6.120}$$

Since $\mathbf{a}^2 = k_i^2 \sin^2 \theta_i$,

$$\sin^2 \theta_B = \frac{\mathbf{a}^2}{k_i^2} = \frac{\varepsilon_\perp \varepsilon_\| - \varepsilon_1(\hat{\mathbf{q}} \cdot \bar{\varepsilon} \cdot \hat{\mathbf{q}})}{\varepsilon_\perp \varepsilon_\| - \varepsilon_1^2}$$

or, Brewster's law,

$$\tan^2 \theta_B = \frac{\varepsilon_\perp \varepsilon_\| - \varepsilon_1(\hat{\mathbf{q}} \cdot \bar{\varepsilon} \cdot \hat{\mathbf{q}})}{\varepsilon_1(\hat{\mathbf{q}} \cdot \bar{\varepsilon} \cdot \hat{\mathbf{q}} - \varepsilon_1)} \tag{6.121}$$

Evidently, the Brewster angle exists only when $\tan^2 \theta_B > 0$, which, according to Eq. (6.121), is equivalent to either

$$\varepsilon_1 < \hat{\mathbf{q}} \cdot \bar{\varepsilon} \cdot \hat{\mathbf{q}} < \frac{\varepsilon_\perp \varepsilon_\|}{\varepsilon_1} \tag{6.122}$$

or

$$\frac{\varepsilon_\perp \varepsilon_\|}{\varepsilon_1} < \hat{\mathbf{q}} \cdot \bar{\varepsilon} \cdot \hat{\mathbf{q}} < \varepsilon_1 \tag{6.123}$$

With the aid of Eq. (6.120) we also find

$$q_i^2 = k_i^2 - \mathbf{a}^2 = \frac{k_0^2 \varepsilon_1^2(\hat{\mathbf{q}} \cdot \bar{\varepsilon} \cdot \hat{\mathbf{q}} - \varepsilon_1)}{\varepsilon_\perp \varepsilon_\| - \varepsilon_1^2} \tag{6.124}$$

and

$$q_+^2 = k_0^2 \varepsilon_\perp - \mathbf{a}^2 = \frac{k_0^2 \varepsilon_\perp \varepsilon_\|(\varepsilon_\perp - \varepsilon_1) + k_0^2 \varepsilon_1^2(\hat{\mathbf{q}} \cdot \bar{\varepsilon} \cdot \hat{\mathbf{q}} - \varepsilon_\perp)}{\varepsilon_\perp \varepsilon_\| - \varepsilon_1^2} \tag{6.125}$$

with q_- given by Eq. (6.96).

Furthermore, we have two subcases: optic axis is parallel or perpendicular to the interface.

Optic axis parallel to the interface In this case, $\hat{\mathbf{c}}$ is parallel to $\mathbf{b}$ as shown in Fig. 6.11; thus $\mathbf{b} \cdot \hat{\mathbf{c}} = |\mathbf{a}|$, $\mathbf{a} \cdot \hat{\mathbf{c}} = \hat{\mathbf{q}} \cdot \hat{\mathbf{c}} = 0$ and

$$\begin{aligned} \hat{\mathbf{q}} \cdot \bar{\varepsilon} \cdot \hat{\mathbf{q}} &= \varepsilon_\perp \\ \mathbf{b} \cdot \bar{\varepsilon} \cdot \hat{\mathbf{q}} &= 0 \end{aligned} \tag{6.126}$$

Consequently, Eq. (6.118) becomes

$$\tan \alpha_r = \frac{(q_i - q_+)(\sqrt{\varepsilon_\perp \varepsilon_\|}\, q_i + \varepsilon_1 q_+)}{(q_i + q_+)(\sqrt{\varepsilon_\perp \varepsilon_\|}\, q_i - \varepsilon_1 q_+)} \tan \alpha_i \tag{6.127}$$

and condition (6.119) for a linearly polarized reflected wave is reduced to

$$q_i = \frac{\varepsilon_1 q_+}{\sqrt{\varepsilon_\perp \varepsilon_\|}} \tag{6.128}$$

Here, we see again, under condition (6.128), Eq. (6.127) gives $\alpha_r = \pi/2$. Now substituting Eq. (6.126) into Eqs. (6.121), (6.122), and (6.123), we obtain, respectively, Brewster's law:

$$\tan^2 \theta_B = \frac{\varepsilon_\perp(\varepsilon_\| - \varepsilon_1)}{\varepsilon_1(\varepsilon_\perp - \varepsilon_1)} \tag{6.129}$$

and the condition for the existence of the Brewster angle θ_B:

$$\begin{aligned} &\varepsilon_1 < \varepsilon_\| \quad \text{and} \quad \varepsilon_1 < \varepsilon_\perp \\ \text{or} \quad &\varepsilon_1 > \varepsilon_\| \quad \text{and} \quad \varepsilon_1 > \varepsilon_\perp \end{aligned} \tag{6.130}$$

In other words, when the dielectric constant ε_1 of the isotropic medium lies between $\varepsilon_\|$ and $\varepsilon_\perp$, the Brewster angle does not exist. Thus, unlike the case of two isotropic media, the Brewster angle defined by $\tan \theta_B = \sqrt{\varepsilon_2/\varepsilon_1}$ always exists.

Substitution of Eq. (6.126) into Eqs. (6.120), (6.124), (6.125), and (6.96) or (6.87) yields

$$\begin{aligned} \mathbf{a}^2 &= \frac{k_0^2 \varepsilon_\perp \varepsilon_1(\varepsilon_\| - \varepsilon_1)}{\varepsilon_\perp \varepsilon_\| - \varepsilon_1^2} \\ q_i^2 &= \frac{k_0^2 \varepsilon_1^2(\varepsilon_\perp - \varepsilon_1)}{\varepsilon_\perp \varepsilon_\| - \varepsilon_1^2} \\ q_+^2 &= \frac{k_0^2 \varepsilon_\perp \varepsilon_\|(\varepsilon_\perp - \varepsilon_1)}{\varepsilon_\perp \varepsilon_\| - \varepsilon_1^2} \\ \text{and} \quad q_-^2 &= \frac{k_0^2 \varepsilon_\|^2(\varepsilon_\perp - \varepsilon_1)}{\varepsilon_\perp \varepsilon_\| - \varepsilon_1^2} \end{aligned} \tag{6.131}$$

In Sec. 4-8, we have shown that for two isotropic media, the Brewster angle is characterized by the fact that the wave vectors of the reflected and the transmitted waves are perpendicular to each other, i.e.,

$$\mathbf{k}_r \cdot \mathbf{k}_t = \mathbf{a}^2 - q_i q_t = 0$$

We will now show that in the case of wave reflection from a uniaxial medium, the above statement is not valid. Indeed, using Eq. (6.131), we easily find

$$\begin{aligned} &\mathbf{k}_r \cdot \mathbf{k}_+ = \mathbf{a}^2 - q_i q_+ \neq 0 \\ \text{and} \quad &\mathbf{k}_r \cdot \mathbf{k}_- = \mathbf{a}^2 - q_i q_- \neq 0 \end{aligned}$$

Therefore, unlike two isotropic media, in this case the wave vector of the reflected wave is not perpendicular to that of the ordinary wave or the extraordinary wave.

In addition to the foregoing, we have the interesting result that when the dielectric constant ε_1 of the isotropic medium 1 is the geometric mean of $\varepsilon_{\parallel}$ and $\varepsilon_{\perp}$, i.e.,

$$\varepsilon_1 = \sqrt{\varepsilon_{\perp}\,\varepsilon_{\parallel}} \tag{6.132}$$

then it follows from Eq. (6.127) that

$$\tan \alpha_r = \tan \alpha_i \tag{6.133}$$

As a result, in this special case (that is, $\hat{\mathbf{c}} = \hat{\mathbf{b}}$, $\varepsilon_1 = \sqrt{\varepsilon_{\perp}\,\varepsilon_{\parallel}}$), there is no rotation of the plane of polarization upon reflection for any angle of incidence. This property proves to be useful in practical application when the direction of propagation of a linearly polarized wave must be changed without simultaneously changing the azimuthal angle.

Optic axis perpendicular to the interface In this case $\hat{\mathbf{c}} = \hat{\mathbf{q}}$ as shown in Fig. 6.14; thus,

$$\begin{aligned} \hat{\mathbf{q}} \cdot \bar{\boldsymbol{\varepsilon}} \cdot \hat{\mathbf{q}} &= \varepsilon_{\parallel} \\ \mathbf{b} \cdot \bar{\boldsymbol{\varepsilon}} \cdot \hat{\mathbf{q}} &= 0 \end{aligned} \tag{6.134}$$

From Eq. (6.105), Eq. (6.118) becomes

$$\tan \alpha_r = \frac{(q_i - q_+)(\varepsilon_{\perp} q_i + \varepsilon_1 q_-)}{(q_i + q_+)(\varepsilon_{\perp} q_i - \varepsilon_1 q_-)} \tan \alpha_i \tag{6.135}$$

and condition (6.119) for a linearly polarized reflected wave is reduced to

$$\varepsilon_{\perp} q_i - \varepsilon_1 q_- = 0 \tag{6.136}$$

Again, under condition (6.136), we have $\alpha_r = \pi/2$.

Substituting Eq. (6.134) into Eqs. (6.120), (6.124), (6.125), (6.105), and (6.121), we obtain

$$\mathbf{a}^2 = \frac{k_0^2 \varepsilon_{\parallel} \varepsilon_1 (\varepsilon_{\perp} - \varepsilon_1)}{\varepsilon_{\perp} \varepsilon_{\parallel} - \varepsilon_1^2} \tag{6.137}$$

$$q_i^2 = \frac{k_0^2 \varepsilon_1^2 (\varepsilon_{\parallel} - \varepsilon_1)}{\varepsilon_{\perp} \varepsilon_{\parallel} - \varepsilon_1^2} \tag{6.138}$$

$$q_+^2 = \frac{k_0^2 \varepsilon_{\perp} \varepsilon_{\parallel} (\varepsilon_{\perp} - \varepsilon_1) + k_0^2 \varepsilon_1^2 (\varepsilon_{\parallel} - \varepsilon_{\perp})}{\varepsilon_{\perp} \varepsilon_{\parallel} - \varepsilon_1^2} \tag{6.139}$$

$$q_-^2 = \frac{k_0^2 \varepsilon_{\perp}^2 (\varepsilon_{\parallel} - \varepsilon_1)}{\varepsilon_{\perp} \varepsilon_{\parallel} - \varepsilon_1^2} \tag{6.140}$$

and Brewster's law,

$$\tan^2 \theta_B = \frac{\varepsilon_{\parallel} (\varepsilon_{\perp} - \varepsilon_1)}{\varepsilon_1 (\varepsilon_{\parallel} - \varepsilon_1)} \tag{6.141}$$

It is interesting to note that by interchanging $\varepsilon_\perp$ and $\varepsilon_{||}$ in Eq. (6.141), we obtain expression (6.129) for the case where the optic axis is parallel to the interface. Thus condition (6.130) for the existance of the Brewster angle is also valid in this case. Using Eqs. (6.137) to (6.140) we again easily show that $\mathbf{k}_r \cdot \mathbf{k}_+ \neq 0$ and $\mathbf{k}_r \cdot \mathbf{k}_- \neq 0$.

Optic Axis Perpendicular to the Plane of Incidence

In this case, $\hat{\mathbf{c}}$ is parallel to $\mathbf{a}$ as shown in Fig. 6.12; thus, $\mathbf{a} \cdot \hat{\mathbf{c}} = |\mathbf{a}|$, $\mathbf{b} \cdot \hat{\mathbf{c}} = \hat{\mathbf{q}} \cdot \hat{\mathbf{c}} = 0$. Substituting Eq. (6.92) into Eq. (6.112) and using Eq. (6.93), we obtain

$$\tan \alpha_r = \frac{(q_i - q_-)(q_i + q_+)(\mathbf{a}^2 + q_i q_+)}{(q_i + q_-)(q_i - q_+)(\mathbf{a}^2 - q_i q_+)} \tan \alpha_i \tag{6.142}$$

and condition (6.115) now becomes

$$(q_i - q_-)(Z - U) = k_0^2 \varepsilon_\perp \mathbf{a}^2 (q_i - q_-)(q_i - q_+)(\mathbf{a}^2 - q_i q_+) = 0 \tag{6.143}$$

Assuming $q_i \neq q_+$ and $q_i \neq q_-$, we obtain, from Eq. (6.143),

$$\mathbf{a}^2 = q_i q_+ \tag{6.144}$$

which shows that $\alpha_r = \pi/2$ and that the wave vector of the reflected wave is perpendicular to the wave vector of the transmitted ordinary wave, i.e.,

$$\mathbf{k}_r \cdot \mathbf{k}_+ = \mathbf{a}^2 - q_i q_+ = 0 \tag{6.145}$$

From Eqs. (6.90), (6.144), and $q_i = \sqrt{k_0^2 \varepsilon_1 - \mathbf{a}^2}$, we find

$$\mathbf{a}^2 = \frac{k_0^2 \varepsilon_\perp \varepsilon_1}{\varepsilon_\perp + \varepsilon_1} \tag{6.146}$$

$$q_i^2 = \frac{k_0^2 \varepsilon_1^2}{\varepsilon_\perp + \varepsilon_1} \tag{6.147}$$

$$q_-^2 = \frac{k_0^2(\varepsilon_{||} \varepsilon_1 + \varepsilon_\perp \varepsilon_{||} - \varepsilon_\perp \varepsilon_1)}{\varepsilon_\perp + \varepsilon_1}$$

and

$$q_+^2 = \frac{k_0^2 \varepsilon_\perp^2}{\varepsilon_+ + \varepsilon_1} \tag{6.148}$$

The Brewster angle θ_B, determined from Eq. (6.146), satisfies

$$\tan \theta_B = \sqrt{\frac{\varepsilon_\perp}{\varepsilon_1}} \tag{6.149}$$

and, as in the case of two isotropic media, always exists.

In all the special cases considered so far, we find that the azimuthal angle α_r of the reflected wave is $\pi/2$. That is, the reflected wave is linearly polarized with its electric field intensity normal to the plane of incidence. This, however, is not true in general, as for example, when the plane containing the optic axis is

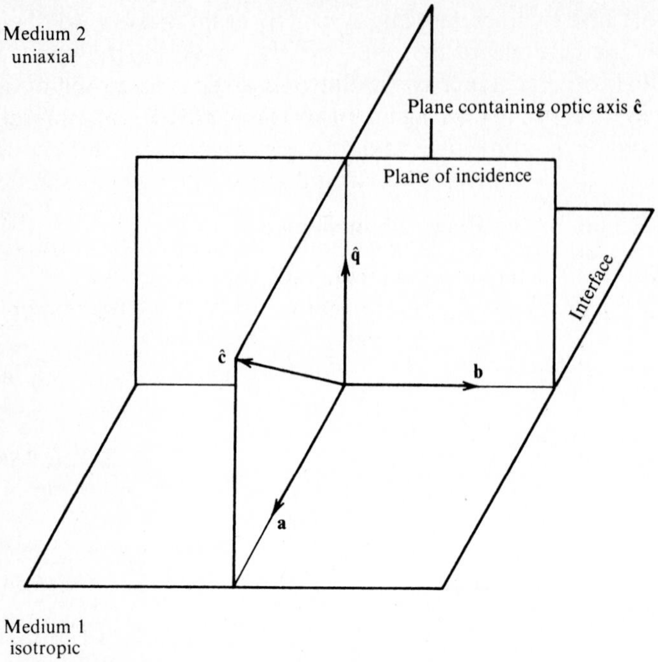

Figure 6.17 Optic axis lies in the plane which is perpendicular to the plane of incidence.

perpendicular to the plane of incidence as illustrated in Fig. 6.17. In this case, $\mathbf{b} \cdot \hat{\mathbf{c}} = 0$; thus,

$$\begin{aligned} \mathbf{k}_+ \cdot (\mathbf{a} \times \hat{\mathbf{c}}) = \mathbf{k}_- \cdot (\mathbf{a} \times \hat{\mathbf{c}}) &= -\mathbf{a}^2(\hat{\mathbf{q}} \cdot \hat{\mathbf{c}}) \\ (\mathbf{a} \cdot \hat{\mathbf{c}})^2 &= \mathbf{a}^2(\hat{\mathbf{q}} \times \hat{\mathbf{c}})^2 \end{aligned} \tag{6.150}$$

Substitution of the above into Eqs. (6.60) and (6.62) yields

$$\begin{aligned} X &= k_0^2 \varepsilon_\perp q_i \mathbf{a}^4(\hat{\mathbf{q}} \cdot \hat{\mathbf{c}})^2 \qquad Y = k_0^2 \varepsilon_1 q_- \mathbf{a}^4(\hat{\mathbf{q}} \cdot \hat{\mathbf{c}})^2 \\ Z &= k_0^4 \varepsilon_\perp^2 q_i \mathbf{a}^2(\hat{\mathbf{q}} \times \hat{\mathbf{c}})^2 \qquad U = k_0^4 \varepsilon_1 \varepsilon_\perp q_+ \mathbf{a}^2(\hat{\mathbf{q}} \times \hat{\mathbf{c}})^2 \qquad V = 0 \end{aligned} \tag{6.151}$$

Condition (6.115) for a linearly polarized reflected wave becomes

$$(q_i - q_+)\mathbf{a}^2(\hat{\mathbf{q}} \cdot \hat{\mathbf{c}})^2(\varepsilon_\perp q_i - \varepsilon_1 q_-) + (q_i - q_-)k_0^2 \varepsilon_\perp(\hat{\mathbf{q}} \times \hat{\mathbf{c}})^2(\varepsilon_\perp q_i - \varepsilon_1 q_+) = 0 \tag{6.152}$$

Under this condition, we obtain the azimuthal angle of the reflected wave according to Eq. (6.117):

$$\tan \alpha_r = \frac{(q_i - q_+)\mathbf{a}^2(\hat{\mathbf{q}} \cdot \hat{\mathbf{c}})^2 + (q_i - q_-)k_0^2 \varepsilon_\perp(\hat{\mathbf{q}} \times \hat{\mathbf{c}})^2}{(q_- - q_+)k_i(\mathbf{a} \cdot \hat{\mathbf{c}})(\hat{\mathbf{q}} \cdot \hat{\mathbf{c}})} \tag{6.153}$$

which clearly shows that the denominator does not generally vanish and hence $\alpha_r \neq \pi/2$.

6.8 SOME SPECIAL CASES

In the previous section we have examined various orientations of the optic axis with respect to the interface and the plane of incidence. We shall now consider cases where some special relationships exist between the dielectric constant of the isotropic medium and that of the uniaxial crystal. One such case was given in Eq. (6.132). Another can easily be seen from Eqs. (6.147), (6.143), and (6.142): When $\varepsilon_1 = \varepsilon_{||}$, and hence $q_i = q_-$, we obtain a linearly polarized reflected wave with the electric field intensity parallel to the plane of incidence.

Still another interesting case is when the dielectric constant of the isotropic medium is equal to $\varepsilon_\perp$ of the uniaxial crystal. Here $\varepsilon_1 = \varepsilon_\perp$ and

$$\begin{aligned} k_i^2 &= k_r^2 = k_0^2\,\varepsilon_\perp = k_+^2 \\ q_i &= \sqrt{k_0^2\,\varepsilon_\perp - \mathbf{a}^2} = q_+ \\ \mathbf{k}_i &= \mathbf{b} + q_i\hat{\mathbf{q}} = \mathbf{k}_+ \end{aligned} \tag{6.154}$$

$$[\mathbf{k}_i \cdot (\mathbf{a} \times \hat{\mathbf{c}})]^2 = k_i^2(\mathbf{a} \times \hat{\mathbf{c}})^2 - \mathbf{a}^2(\mathbf{k}_i \cdot \hat{\mathbf{c}})^2 = [\mathbf{k}_+ \cdot (\mathbf{a} \times \hat{\mathbf{c}})]^2$$

Thus, from Eqs. (6.60) and (6.62)

$$\begin{aligned} Z = U &= k_0^4\,\varepsilon_\perp^2\,q_i(\mathbf{a} \cdot \hat{\mathbf{c}})^2 \\ X + Y &= k_i^2(q_i + q_-)[\mathbf{k}_i \cdot (\mathbf{a} \times \hat{\mathbf{c}})]^2 \\ X - Y &= k_i^2(q_i - q_-)[\mathbf{k}_i \cdot (\mathbf{a} \times \hat{\mathbf{c}})][\mathbf{k}_r \cdot (\mathbf{a} \times \hat{\mathbf{c}})] \\ V + L &= -k_i^3\,q_i(\mathbf{a} \cdot \hat{\mathbf{c}})[\mathbf{k}_r \cdot (\mathbf{a} \times \hat{\mathbf{c}})] \\ V - L &= k_i^3\,q_i(\mathbf{a} \cdot \hat{\mathbf{c}})[\mathbf{k}_i \cdot (\mathbf{a} \times \hat{\mathbf{c}})] \\ \Delta_1 &= 2k_i^2\,\mathbf{a}^2 q_i(q_i + q_-)(\mathbf{k}_i \times \hat{\mathbf{c}})^2 \end{aligned} \tag{6.155}$$

Substituting the above into the general formulas (6.61) and (6.58), we obtain, respectively, the reflection coefficients

$$\begin{aligned} \Gamma_{11} &= \frac{k_i^2(q_i - q_-)(\mathbf{a} \cdot \hat{\mathbf{c}})^2}{\mathbf{a}^2(q_i + q_-)(\mathbf{k}_i \times \hat{\mathbf{c}})^2} \\ \Gamma_{12} &= \frac{k_i(q_i - q_-)(\mathbf{a} \cdot \hat{\mathbf{c}})[\mathbf{k}_i \cdot (\mathbf{a} \times \hat{\mathbf{c}})]}{\mathbf{a}^2(q_i + q_-)(\mathbf{k}_i \times \hat{\mathbf{c}})^2} \\ \Gamma_{21} &= \frac{k_i(q_i - q_-)(\mathbf{a} \cdot \hat{\mathbf{c}})[\mathbf{k}_r \cdot (\mathbf{a} \times \hat{\mathbf{c}})]}{\mathbf{a}^2(q_i + q_-)(\mathbf{k}_i \times \hat{\mathbf{c}})^2} \\ \Gamma_{22} &= \frac{(q_i - q_-)[\mathbf{k}_i \cdot (\mathbf{a} \times \hat{\mathbf{c}})][\mathbf{k}_r \cdot (\mathbf{a} \times \hat{\mathbf{c}})]}{\mathbf{a}^2(q_i + q_-)(\mathbf{k}_i \times \hat{\mathbf{c}})^2} \end{aligned} \tag{6.156}$$

and the transmission coefficients

$$
\begin{aligned}
T_{11} &= -\frac{\mathbf{k}_i \cdot (\mathbf{a} \times \hat{\mathbf{c}})}{(\mathbf{k}_i \times \hat{\mathbf{c}})^2} \\
T_{12} &= \frac{k_i(\mathbf{a} \cdot \hat{\mathbf{c}})}{(\mathbf{k}_i \times \hat{\mathbf{c}})^2} \\
T_{21} &= \frac{2q_i(\mathbf{a} \cdot \hat{\mathbf{c}})}{(q_i + q_-)(\mathbf{k}_i \times \hat{\mathbf{c}})^2} \\
T_{22} &= \frac{2q_i[\mathbf{k}_i \cdot (\mathbf{a} \times \hat{\mathbf{c}})]}{k_i(q_i + q_-)(\mathbf{k}_i \times \hat{\mathbf{c}})^2}
\end{aligned}
\tag{6.157}
$$

Since $q_i = q_+$ and $Z = U$, condition (6.115) can always be satisfied. Furthermore, substitution of Eq. (6.156) into Eq. (6.112) gives

$$
\tan \alpha_r = \frac{k_i(\mathbf{a} \cdot \hat{\mathbf{c}})}{\mathbf{k}_r \cdot (\mathbf{a} \times \hat{\mathbf{c}})} = -\frac{k_i(\mathbf{a} \cdot \hat{\mathbf{c}})}{\mathbf{a}^2(\hat{\mathbf{q}} \cdot \hat{\mathbf{c}}) + q_i(\mathbf{b} \cdot \hat{\mathbf{c}})} \tag{6.158}
$$

Now we let Ψ_1 be the angle between the vector $\mathbf{b}$ and the projection of $\hat{\mathbf{c}}$ on the interface and let Ψ_2 be the angle between vectors $\hat{\mathbf{q}}$ and $\hat{\mathbf{c}}$. Since $q_i = \mathbf{k}_i \cdot \hat{\mathbf{q}} = k_i \cos \theta_i$, we may rewrite Eq. (6.158) as

$$
\tan \alpha_r = -\frac{\sin \Psi_1 \sin \Psi_2}{\cos \theta_i \cos \Psi_1 \sin \Psi_2 + \sin \theta_i \cos \Psi_2} \tag{6.159}
$$

which shows the remarkable property that α_r does not depend on α_i nor $\varepsilon_\perp$ nor $\varepsilon_\parallel$. In other words, Eq. (6.158) is a universal relation as long as $\varepsilon_1 = \varepsilon_\perp$.

From Eq. (6.158) we may also conclude that $\tan \alpha_r$ is real, implying that the reflected wave is always linearly polarized. Furthermore, the azimuthal angle α_r of the reflected wave may assume any value from 0 to $\pi/2$; α_r is zero when the optic axis lies on the plane of incidence (that is, $\hat{\mathbf{c}} \cdot \mathbf{a} = 0$), and is $\pi/2$ when the optic axis is parallel to $\mathbf{a}$.

From Eq. (6.156) we see that the reflected wave will be totally absent (that is, $\mathbf{E}_{0r} = 0$) if the wave vector $\mathbf{k}_r$ of the reflected wave is parallel to the optic axis. On the other hand, if the incident wave is linearly polarized in the direction perpendicular to the plane of incidence (that is, $A_\parallel = 0$), and if the optic axis of the uniaxial crystal lies on the plane of incidence (that is, $\hat{\mathbf{c}} \cdot \mathbf{a} = 0$), then according to Eqs. (6.156) and (6.157), we have

$$
\begin{aligned}
\Gamma_{11} = \Gamma_{12} = \Gamma_{21} &= 0 \\
T_{12} = T_{21} &= 0
\end{aligned}
$$

which implies that $\mathbf{E}_{0r} = \mathbf{0}$, $C_- = 0$, and

$$
C_+ = \frac{\mathbf{a} \cdot (\mathbf{k}_i \times \hat{\mathbf{c}})}{(\mathbf{k}_i \times \hat{\mathbf{c}})^2} A_\perp \tag{6.160}
$$

Thus the electric field vector of the transmitted wave is

$$\mathbf{E}_{0t} = C_+(\mathbf{k}_i \times \hat{\mathbf{c}}) = A_\perp \mathbf{a} = \mathbf{E}_{0i} \tag{6.161}$$

That is, under the stated conditions the wave passes through the interface of an isotropic and a uniaxial medium without change.

6.9 ENERGY RELATIONS

Similar to the treatment of two isotropic media, we shall now establish energy balance across the interface of an isotropic and a uniaxial medium, namely,

$$(\mathscr{P}_i + \mathscr{P}_r - \mathscr{P}_+ - \mathscr{P}_-) \cdot \hat{\mathbf{q}} = 0 \tag{6.162}$$

$\mathscr{P}_i$, $\mathscr{P}_r$, $\mathscr{P}_+$, and $\mathscr{P}_-$ are respectively the instantaneous Poynting vectors of the incident, the reflected, and the transmitted ordinary and extraordinary waves. Physically, Eq. (6.162) states that the power flux incident on any part of the interface must be equal to the sum of the power fluxes leaving that surface. Thus it represents the law of conservation of energy. To prove Eq. (6.162), we again use Maxwell's equations and the boundary conditions

$$(\mathscr{E}_i + \mathscr{E}_r) \times \hat{\mathbf{q}} = (\mathscr{E}_+ + \mathscr{E}_-) \times \hat{\mathbf{q}} \tag{6.163}$$

and

$$\mathscr{H}_i + \mathscr{H}_r = \mathscr{H}_+ + \mathscr{H}_- \tag{6.164}$$

Now dot-multiplying Eqs. (6.163) and (6.164) and noting that

$$\mathscr{E}_i = \operatorname{Re}\,(\mathbf{E}_{0i}e^{i\phi}) = \tfrac{1}{2}(\mathbf{E}_{0i}e^{i\phi} + \mathbf{E}_{0i}^* e^{-i\phi})$$

$$\mathscr{H}_i = \operatorname{Re}\,(\mathbf{H}_{0i}e^{i\phi}) = \tfrac{1}{2}(\mathbf{H}_{0i}e^{i\phi} + \mathbf{H}_{0i}^* e^{-i\phi})$$

and with similar expressions for $\mathscr{E}_r$, $\mathscr{H}_r$, $\mathscr{E}_+$, $\mathscr{H}_+$, $\mathscr{E}_-$, and $\mathscr{H}_-$, we obtain

$$\begin{aligned}
(\mathscr{P}_i + \mathscr{P}_r &- \mathscr{P}_+ - \mathscr{P}_-) \cdot \hat{\mathbf{q}} \\
&= -(\mathscr{E}_i \times \mathscr{H}_r + \mathscr{E}_r \times \mathscr{H}_i) \cdot \hat{\mathbf{q}} + (\mathscr{E}_+ \times \mathscr{H}_- + \mathscr{E}_- \times \mathscr{H}_+) \cdot \hat{\mathbf{q}} \\
&= \tfrac{1}{2} \operatorname{Re} \{[(\mathbf{E}_{0+} \times \mathbf{H}_{0-} + \mathbf{E}_{0-} \times \mathbf{H}_{0+}) \cdot \hat{\mathbf{q}} - (\mathbf{E}_{0i} \times \mathbf{H}_{0r} + \mathbf{E}_{0r} \times \mathbf{H}_{0i}) \cdot \hat{\mathbf{q}}]e^{i2\phi} \\
&\quad + [(\mathbf{E}_{0+} \times \mathbf{H}_{0-}^* + \mathbf{E}_{0-}^* \times \mathbf{H}_{0+}) \cdot \hat{\mathbf{q}} - (\mathbf{E}_{0i} \times \mathbf{H}_{0r}^* + \mathbf{E}_{0r}^* \times \mathbf{H}_{0i}) \cdot \hat{\mathbf{q}}]\}
\end{aligned} \tag{6.165}$$

Writing Maxwell's equations for the incident and the reflected waves in isotropic medium 1 as

$$\begin{aligned}
k_i^2 \mathbf{E}_{0i} &= -\omega\mu_0(\mathbf{k}_i \times \mathbf{H}_{0i}) \\
-\mathbf{k}_r \times \mathbf{H}_{0r} &= \frac{k_i^2}{\omega\mu_0} \mathbf{E}_{0r}
\end{aligned} \tag{6.166}$$

and

$$\begin{aligned}
\omega\mu_0 \mathbf{H}_{0i} &= \mathbf{k}_i \times \mathbf{E}_{0i} \\
\frac{1}{\omega\mu_0}(\mathbf{k}_r \times \mathbf{E}_{0r}) &= \mathbf{H}_{0r}
\end{aligned} \tag{6.167}$$

and the transmitted ordinary and extraordinary waves in the uniaxial crystal in the form

$$\omega\varepsilon_0 \bar{\varepsilon} \cdot \mathbf{E}_{0+} = -\mathbf{k}_+ \times \mathbf{H}_{0+}$$
$$\frac{1}{\omega\varepsilon_0} \bar{\varepsilon}^{-1} \cdot (\mathbf{k}_- \times \mathbf{H}_{0-}) = -\mathbf{E}_{0-} \tag{6.168}$$

and

$$\omega\mu_0 \mathbf{H}_{0+} = \mathbf{k}_+ \times \mathbf{E}_{0+}$$
$$\frac{1}{\omega\mu_0} (\mathbf{k}_- \times \mathbf{E}_{0-}) = \mathbf{H}_{0-} \tag{6.169}$$

dot-multiplying two equations in (6.166) and two equations in (6.167) respectively and then adding the results, noting that $\mathbf{k}_r - \mathbf{k}_i = -2q_i\hat{\mathbf{q}}$, we obtain

$$(\mathbf{E}_{0i} \times \mathbf{H}_{0r} + \mathbf{E}_{0r} \times \mathbf{H}_{0i}) \cdot \hat{\mathbf{q}} = 0 \tag{6.170}$$

Next, we dot-multiply the complex conjugate of one equation in Eq. (6.166) with the other, and repeat the process for Eq. (6.167). Summing the results, we get

$$(\mathbf{E}_{0i} \times \mathbf{H}_{0r}^* + \mathbf{E}_{0r}^* \times \mathbf{H}_{0i}) \cdot \hat{\mathbf{q}} = 0 \tag{6.171}$$

Likewise, from Eqs. (6.168) and (6.169), we find

$$(\mathbf{E}_{0+} \times \mathbf{H}_{0-} + \mathbf{E}_{0-} \times \mathbf{H}_{0+}) \cdot \hat{\mathbf{q}} = 0 \tag{6.172}$$

and

$$(\mathbf{E}_{0+} \times \mathbf{H}_{0-}^* + \mathbf{E}_{0-}^* \times \mathbf{H}_{0+}) \cdot \hat{\mathbf{q}} = 0 \tag{6.173}$$

Substituting Eqs. (6.170) to (6.173) into the right-hand side of Eq. (6.165), we have thus proved the law of conservation of energy, Eq. (6.162). By taking the time average of Eq. (6.162), we obtain

$$\langle \mathscr{P}_i \rangle \cdot \hat{\mathbf{q}} + \langle \mathscr{P}_r \rangle \cdot \hat{\mathbf{q}} - \langle \mathscr{P}_+ \rangle \cdot \hat{\mathbf{q}} - \langle \mathscr{P}_- \rangle \cdot \hat{\mathbf{q}} = 0 \tag{6.174}$$

which indicates that the normal component of the time-averaged energy flowing across the interface must also be continuous.

For convenience, we repeat in the following the reflectivities defined in Secs. 4.8:

$$r_\perp = -\frac{\langle \mathscr{P}_r^\perp \rangle \cdot \hat{\mathbf{q}}}{\langle \mathscr{P}_i^\perp \rangle \cdot \hat{\mathbf{q}}} = \frac{|B_\perp|^2}{|A_\perp|^2} \tag{6.175}$$

$$r_\parallel = -\frac{\langle \mathscr{P}_r^\parallel \rangle \cdot \hat{\mathbf{q}}}{\langle \mathscr{P}_i^\parallel \rangle \cdot \hat{\mathbf{q}}} = \frac{|B_\parallel|^2}{|A_\parallel|^2} \tag{6.176}$$

and

$$r = -\frac{\langle \mathscr{P}_r \rangle \cdot \hat{\mathbf{q}}}{\langle \mathscr{P}_i \rangle \cdot \hat{\mathbf{q}}} = \frac{|B_\perp|^2 + |B_\parallel|^2}{|A_\perp|^2 + |A_\parallel|^2} \tag{6.177}$$

where $r_\perp$ and $r_\parallel$ are the reflectivities corresponding to, respectively, the perpendicular and the parallel polarizations, and r is the total reflectivity. The transmissivities of the ordinary and extraordinary waves, denoted respectively by t_+ and t_-, are similarly defined:

$$t_+ = \frac{\langle \mathscr{P}_+ \rangle \cdot \hat{\mathbf{q}}}{\langle \mathscr{P}_i \rangle \cdot \hat{\mathbf{q}}} \tag{6.178}$$

$$t_- = \frac{\langle \mathscr{P}_- \rangle \cdot \hat{\mathbf{q}}}{\langle \mathscr{P}_i \rangle \cdot \hat{\mathbf{q}}} \tag{6.179}$$

and the total transmissivity is

$$t = \frac{(\langle \mathscr{P}_+ \rangle + \langle \mathscr{P}_- \rangle) \cdot \hat{\mathbf{q}}}{\langle \mathscr{P}_i \rangle \cdot \hat{\mathbf{q}}} \tag{6.180}$$

Analogous to the case of two isotropic media, it follows from Eq. (6.174) that

$$r + t = 1 \tag{6.181}$$

A comparison of the reflectivities for the cases of (*a*) isotropic medium–isotropic medium and (*b*) isotropic medium–anisotropic medium shows the following basic differences:

1. In case *a*, the reflectivities

$$\begin{aligned} r_\perp &= |\Gamma_{11}|^2 \\ r_\parallel &= |\Gamma_{22}|^2 \end{aligned} \tag{6.182}$$

do not depend on the azimuthal angle of the incident wave. However, in case *b*, because

$$\begin{aligned} r_\perp &= \frac{(\Gamma_{12} + \Gamma_{11} \tan \alpha_i)^2}{\tan^2 \alpha_i} \\ r_\parallel &= (\Gamma_{22} + \Gamma_{21} \tan \alpha_i)^2 \end{aligned} \tag{6.183}$$

they do depend on α_i.

2. The total reflectivity in case *a* is given by

$$r = \frac{\Gamma_{22}^2 + \Gamma_{11}^2 \tan^2 \alpha_i}{1 + \tan^2 \alpha_i} \tag{6.184}$$

which depends on α_i and varies from $r_\parallel$ (for $\alpha_i = 0$) to $r_\perp$ (for $\alpha_i = \pi/2$). Thus

$$r_\parallel \le r \le r_\perp \tag{6.185}$$

On the other hand, in case *b* the extremum (maximum or minimum) of r again depends on α_i but is not equal to $r_\parallel$ or $r_\perp$. Using Eq. (6.111), we may write Eq. (6.177) as

$$r = \frac{a_0 \tan^2 \alpha_i + 2b_0 \tan \alpha_i + c_0}{1 + \tan^2 \alpha_i} \tag{6.186}$$

where

$$\begin{aligned} a_0 &= \Gamma_{11}^2 + \Gamma_{21}^2 \\ b_0 &= \Gamma_{12}\Gamma_{11} + \Gamma_{21}\Gamma_{22} \\ c_0 &= \Gamma_{12}^2 + \Gamma_{22}^2 \end{aligned} \tag{6.187}$$

and the extremum of r occurs when the azimuthal angle of the incident wave satisfies

$$\tan^2 \alpha_i - \frac{a_0 - c_0}{b_0} \tan \alpha_i - 1 = 0 \tag{6.188}$$

that is, when

$$\tan \alpha_i^{(1)} = \frac{a_0 - c_0}{2b_0} + \sqrt{\left(\frac{a_0 - c_0}{2b_0}\right)^2 + 1}$$

or

$$\tan \alpha_i^{(2)} = \frac{a_0 - c_0}{2b_0} - \sqrt{\left(\frac{a_0 - c_0}{2b_0}\right)^2 + 1} \tag{6.189}$$

In this case, if the minimum of r occurs at $\alpha_i^{(1)}$, then the maximum must occur at $\alpha_i^{(2)}$. Also, since

$$\tan \alpha_i^{(1)} \tan \alpha_i^{(2)} = -1 \tag{6.190}$$

which implies that $\alpha_i^{(2)} = \alpha_i^{(1)} + \pi/2$, a similar relation exists for the azimuthal angles of the reflected wave. To see this, we note from Eq. (6.112) that

$$\tan \alpha_r^{(1)} = \frac{\Gamma_{12} + \Gamma_{11} \tan \alpha_i^{(1)}}{\Gamma_{22} + \Gamma_{21} \tan \alpha_i^{(1)}}$$

and

$$\tan \alpha_r^{(2)} = \frac{\Gamma_{12} \tan \alpha_i^{(1)} - \Gamma_{11}}{\Gamma_{22} \tan \alpha_i^{(1)} - \Gamma_{21}}$$

Thus

$$\tan \alpha_r^{(2)} \tan \alpha_r^{(1)} = -1 \tag{6.191}$$

or $\alpha_r^{(2)} = \alpha_r^{(1)} + \pi/2$, where $\alpha_i^{(1)}$ satisfies Eq. (6.188).

3. When the dielectric constant ε_1 of the isotropic medium is equal to $\varepsilon_\perp$ of the uniaxial crystal, it is possible that the reflectivity r vanishes for all azimuthal angles of the incident wave. To see this, let us substitute Eq. (6.156) into Eq. (6.177) and note that

$$k_i^2(\mathbf{a} \cdot \hat{\mathbf{c}})^2 + [\mathbf{k}_r \cdot (\mathbf{a} \times \hat{\mathbf{c}})]^2 = \mathbf{a}^2(\mathbf{k}_r \times \hat{\mathbf{c}})^2$$

We obtain

$$r = \frac{(q_i - q_-)^2(\mathbf{k}_r \times \hat{\mathbf{c}})^2[k_i(\mathbf{a} \cdot \hat{\mathbf{c}})A_\perp + (\mathbf{k}_i \cdot \mathbf{a} \times \hat{\mathbf{c}})A_\parallel]^2}{\mathbf{a}^2(\mathbf{k}_i \times \hat{\mathbf{c}})^4(q_i + q_-)^2(A_\perp^2 + A_\parallel^2)} \tag{6.192}$$

Evidently, $r = 0$ if $\mathbf{k}_r \times \hat{\mathbf{c}} = \mathbf{0}$, i.e., the reflectivity is zero (minimum) if the wave vector of the reflected wave is parallel to the optic axis. From Eq. (6.192), we see that r is also zero if

$$k_i(\mathbf{a}\cdot\hat{\mathbf{c}})A_\perp + (\mathbf{k}_i\cdot\mathbf{a}\times\hat{\mathbf{c}})A_\parallel = 0$$

or

$$\tan\alpha_i = \frac{A_\perp}{A_\parallel} = -\frac{\mathbf{k}_i\cdot(\mathbf{a}\times\hat{\mathbf{c}})}{k_i(\mathbf{a}\cdot\hat{\mathbf{c}})} \tag{6.193}$$

According to Eq. (6.190), the maximum value of r occurs when

$$\tan\alpha_i = \frac{k_i(\mathbf{a}\cdot\hat{\mathbf{c}})}{\mathbf{k}_i\cdot(\mathbf{a}\times\hat{\mathbf{c}})} \tag{6.194}$$

Substituting Eq. (6.194) into Eq. (6.192), we find the maximum reflectivity:

$$r_{\max} = \frac{(\mathbf{k}_r\times\hat{\mathbf{c}})^2(q_i - q_-)^2}{(\mathbf{k}_i\times\hat{\mathbf{c}})^2(q_i + q_-)^2} \tag{6.195}$$

6.10 TOTAL REFLECTION

To conclude this chapter we shall examine briefly conditions under which total reflection from a uniaxial crystal may occur. Evidently, we may have either one or both wave vectors of the transmitted waves become complex. From Eq. (6.52), the total reflection for the ordinary wave occurs if

$$k_0^2\varepsilon_\perp - \mathbf{a}^2 \le 0 \tag{6.196}$$

This condition is no different from the similar condition for isotropic media (see Sec. 4.9). The critical angle of incidence for the total reflection of the ordinary wave is thus found from the condition $k_0^2\varepsilon_\perp - \mathbf{a}^2 = 0$ and is given by

$$\theta_{c+} = \sin^{-1}\sqrt{\frac{\varepsilon_\perp}{\varepsilon_1}} \tag{6.197}$$

Similarly, according to Eq. (6.55), total reflection for the extraordinary wave may occur at the interface between an isotropic medium and a uniaxial crystal, provided that

$$\zeta^2 = k_0^2\varepsilon_\perp\varepsilon_\parallel(\hat{\mathbf{q}}\cdot\bar{\varepsilon}\cdot\hat{\mathbf{q}}) - \mathbf{a}\cdot(\operatorname{adj}\bar{\varepsilon})\cdot\mathbf{a} \le 0 \tag{6.198}$$

Thus the critical angle of incidence for the total reflection of the extraordinary wave is defined by the equation

$$k_0^2\varepsilon_\perp\varepsilon_\parallel(\hat{\mathbf{q}}\cdot\bar{\varepsilon}\cdot\hat{\mathbf{q}}) - \mathbf{a}\cdot(\operatorname{adj}\bar{\varepsilon})\cdot\mathbf{a} = 0 \tag{6.199}$$

or

$$\sin^2\theta_{c-} = \frac{\varepsilon_\perp\varepsilon_\parallel(\hat{\mathbf{q}}\cdot\bar{\varepsilon}\cdot\hat{\mathbf{q}})}{\varepsilon_1\hat{\mathbf{a}}\cdot(\operatorname{adj}\bar{\varepsilon})\cdot\hat{\mathbf{a}}} \tag{6.200}$$

where $\hat{\mathbf{a}}$ is a unit vector in the direction of $\mathbf{a}$.

If the optic axis lies on the plane of incidence, that is, $\hat{\mathbf{c}}\cdot\hat{\mathbf{a}} = 0$, Eq. (6.200) is reduced to

$$\sin^2 \theta_{c-} = \frac{\hat{\mathbf{q}} \cdot \bar{\varepsilon} \cdot \hat{\mathbf{q}}}{\varepsilon_1} \tag{6.201}$$

On the other hand, if the optic axis lies on a plane perpendicular to the plane of incidence, that is, $\hat{\mathbf{c}} \cdot \mathbf{b} = 0$ (see Fig. 6.17), then $(\hat{\mathbf{a}} \cdot \hat{\mathbf{c}})^2 = 1 - (\hat{\mathbf{q}} \cdot \hat{\mathbf{c}})^2$, and Eq. (6.200) becomes

$$\sin^2 \theta_{c-} = \frac{\varepsilon_{\|}}{\varepsilon_1} \tag{6.202}$$

Finally, it is interesting to note that from Eq. (6.55) we have

$$\mathbf{k}_- \cdot \bar{\varepsilon} \cdot \hat{\mathbf{q}} = \zeta$$

We may therefore express condition (6.199) for the total reflection of the extraordinary wave in the following compact form:

$$\hat{\mathbf{q}} \cdot \bar{\varepsilon} \cdot \mathbf{k}_- = 0 \tag{6.203}$$

In short, total reflection occurs if the ray vector [cf. Eq. (6.38)] of the extraordinary wave is parallel to the interface.

PROBLEMS

6.1 In a uniaxial crystal, (*a*) show that $\mathbf{H}_0$ satisfies

$$\left[(k_0^2 \varepsilon_\perp - k^2)\bar{\mathbf{I}} - \frac{\varepsilon_\perp - \varepsilon_\|}{\varepsilon_\|}(\mathbf{k} \times \hat{\mathbf{c}})(\mathbf{k} \times \hat{\mathbf{c}})\right] \cdot \mathbf{H}_0 = 0$$

Thus for an ordinary wave: $k_+ = k_0\sqrt{\varepsilon_\perp}$, the equation becomes

$$(\mathbf{k}_+ \times \hat{\mathbf{c}}) \cdot \mathbf{H}_0 = 0$$

and the solution is

$$\mathbf{h}_+ = \frac{1}{\omega\mu_0}[\mathbf{k}_+ \times (\mathbf{k}_+ \times \hat{\mathbf{c}})]$$

On the other hand, for an extraordinary wave:

$$k_- = k_0\sqrt{\frac{\varepsilon_\perp \varepsilon_\|}{\hat{\mathbf{k}}_- \cdot \bar{\varepsilon} \cdot \hat{\mathbf{k}}_-}}$$

the equation for $\mathbf{H}_0$ is reduced to

$$[(\hat{\mathbf{k}}_- \times \hat{\mathbf{c}})^2\bar{\mathbf{I}} - (\hat{\mathbf{k}}_- \times \hat{\mathbf{c}})(\hat{\mathbf{k}}_- \times \hat{\mathbf{c}})] \cdot \mathbf{H}_0 = \mathbf{0}$$

and the solution is

$$\mathbf{h}_- = \omega\varepsilon_0 \varepsilon_\perp(\mathbf{k}_- \times \hat{\mathbf{c}})$$

(*b*) Show that the corresponding electric field vectors are

$$\mathbf{e}_+ = \mathbf{k}_+ \times \hat{\mathbf{c}}$$

and $$\mathbf{e}_- = k_0^2 \varepsilon_\perp \hat{\mathbf{c}} - (\mathbf{k}_- \cdot \hat{\mathbf{c}})\mathbf{k}_-$$

respectively. (*c*) Find the electric flux density vectors.

6.2 If the direction of propagation of the extraordinary wave in a uniaxial crystal makes an angle α with the optic axis, find (*a*) the angle between the wave vector and the electric field intensity, and (*b*) the angle between the ray vector (or Poynting's vector) and the optic axis of the crystal.

6.3 Show that in a uniaxial medium, the directions of the electric and magnetic field vectors of the extraordinary wave may be expressed as

$$\mathbf{e}_- = -k_0^4 \varepsilon_\perp \varepsilon_\| [\mathbf{s}_- \times (\mathbf{s}_- \times \hat{\mathbf{c}})]$$

and $$\mathbf{h}_- = \omega\varepsilon_0 k_0^2 \varepsilon_\perp \varepsilon_\| (\mathbf{s}_- \times \hat{\mathbf{c}})$$

respectively, where $\mathbf{s}_-$ is the ray vector of the extraordinary wave.

6.4 A monochromatic plane wave in free space is incident at an angle θ_i on the plane surface of a uniaxial crystal. The optic axis of the crystal is perpendicular to its surface. Find the directions of the ray vectors of the ordinary and extraordinary waves in the crystal.

6.5 Repeat Prob. 6.4 when the optic axis of the crystal is parallel to its surface, and makes an angle α_0 with the plane of incidence.

6.6 Verify Eq. (6.108).

6.7 Using the results of Sec. 6.5, show that in the special case of two isotropic media

$$\mathbf{a}^2 = \frac{k_0^2 \varepsilon_1 \varepsilon_2}{\varepsilon_1 + \varepsilon_2} \qquad q_i^2 = \frac{k_0^2 \varepsilon_1^2}{\varepsilon_1 + \varepsilon_2} \qquad q_t^2 = \frac{k_0^2 \varepsilon_2^2}{\varepsilon_1 + \varepsilon_2}$$

where ε_1 and ε_2 are the dielectric constants of medium 1 and 2 respectively. Show also that Brewster's law becomes

$$\tan \theta_B = \sqrt{\frac{\varepsilon_2}{\varepsilon_1}}$$

and $\mathbf{k}_r \cdot \mathbf{k}_t = 0$.

6.8 In Sec. 6.5 we obtained the reflection and transmission coefficients for the problem of wave reflection from the plane boundary of a uniaxial medium. We assumed that the incident wave was located in the isotropic medium. In application such as in designing crystal lasers, the reversed incidence is also important.

Consider an ordinary wave in a uniaxial medium 1 being incident obliquely on the plane interface of an isotropic medium 2. The uniaxial medium is characterized by the dielectric tensor (6.2), and the isotropic medium is characterized by the dielectric constant ε_2. Both media have the same permeability μ_0. The two solutions of the dispersion equation of the uniaxial medium indicate the existence of two reflected waves: ordinary and extraordinary. The transmitted wave in isotropic medium 2 can be decomposed into components parallel and perpendicular to the plane of incidence.

(*a*) Assuming the incident ordinary wave is given, find the amplitude vectors of the reflected and transmitted waves.

(*b*) If the wave normals of the reflected waves coincide with the optic axis $\hat{\mathbf{c}}$ of the uniaxial medium, find the amplitude vectors of the reflected and transmitted waves.

(*c*) If the incident wave is propagating normal to the interface, find the amplitude vectors of the reflected and transmitted waves.

6.9 Repeat Prob. 6.8 when the incident wave in the uniaxial medium 1 is an extraordinary wave.

6.10 If the incident wave in Prob. 6.8 is propagating along the direction of optic axis, find the amplitude vectors of the reflected and transmitted waves.

6.11 Verify Eqs. (6.170), (6.171), (6.172), and (6.173).

6.12 Verify Eq. (6.188).

6.13 Verify Eqs. (6.192) and (6.195).

6.14 The theory of uniform plane waves in uniaxial media developed in this chapter can be extended to nonuniform plane waves. As in Chap. 4, we write the complex wave vector of the nonuniform plane wave as

$$\mathbf{k} = \mathbf{k}_1 + i\mathbf{k}_2$$

(*a*) Show that for the ordinary wave,

$$\mathbf{k}_+ = \mathbf{b} + i\sqrt{\mathbf{a}^2 - k_0^2 \varepsilon_\perp}\,\hat{\mathbf{q}}$$

the phase constant k_{1+} and the attenuation constant k_{2+} are given by

$$k_{1+} = k_0\sqrt{\varepsilon_1}\,\sin\theta_i$$

and

$$k_{2+} = k_0\sqrt{\varepsilon_1 \sin^2\theta_i - \varepsilon_\perp}$$

respectively.

(*b*) Show that for the extraordinary wave

$$\mathbf{k}_- = \frac{(\bar{\boldsymbol{\varepsilon}} \cdot \hat{\mathbf{q}}) \times \mathbf{a} + i\sqrt{\mathbf{a} \cdot (\operatorname{adj} \bar{\boldsymbol{\varepsilon}}) \cdot \mathbf{a} - k_0^2 \varepsilon_\perp \varepsilon_\| (\hat{\mathbf{q}} \cdot \bar{\boldsymbol{\varepsilon}} \cdot \hat{\mathbf{q}})}\,\hat{\mathbf{q}}}{\hat{\mathbf{q}} \cdot \bar{\varepsilon} \cdot \hat{\mathbf{q}}}$$

the phase constant k_{1-} and the attenuation constant k_{2-} are given by

$$k_{1-} = \frac{k_0\sqrt{\varepsilon_1}\,\sin\theta_i}{\hat{\mathbf{q}} \cdot \bar{\boldsymbol{\varepsilon}} \cdot \hat{\mathbf{q}}} \sqrt{\varepsilon_\perp (\hat{\mathbf{q}} \cdot \bar{\boldsymbol{\varepsilon}} \cdot \hat{\mathbf{q}}) + (\varepsilon_\| - \varepsilon_\perp)[\varepsilon_\| - (\varepsilon_\| - \varepsilon_\perp)(\hat{\mathbf{q}} \times \hat{\mathbf{c}})^2 \sin^2\alpha](\hat{\mathbf{q}} \cdot \hat{\mathbf{c}})^2}$$

and

$$k_{2-} = \frac{\sqrt{\varepsilon_\perp}\sqrt{k_0^2 \varepsilon_1 \sin^2\theta_i [\varepsilon_\| - (\varepsilon_\| - \varepsilon_\perp)(\hat{\mathbf{q}} \times \hat{\mathbf{c}})^2 \sin^2\alpha] - k_0^2 \varepsilon_\| (\hat{\mathbf{q}} \cdot \bar{\boldsymbol{\varepsilon}} \cdot \hat{\mathbf{q}})}}{\hat{\mathbf{q}} \cdot \bar{\boldsymbol{\varepsilon}} \cdot \hat{\mathbf{q}}}$$

respectively, where α is the angle between the plane of incidence and the plane containing vectors $\hat{\mathbf{q}}$ and $\hat{\mathbf{c}}$, and

$$\hat{\mathbf{q}} \cdot \bar{\boldsymbol{\varepsilon}} \cdot \hat{\mathbf{q}} = \varepsilon_\perp (\hat{\mathbf{q}} \times \hat{\mathbf{c}})^2 + \varepsilon_\| (\hat{\mathbf{q}} \cdot \hat{\mathbf{c}})^2$$

From the expressions for k_{1-} and k_{2-}, we may thus conclude that for a given angle of incidence, the phase and attenuation constants depend on α and the angle between vectors $\hat{\mathbf{q}}$ and $\hat{\mathbf{c}}$. That is, they depend on the orientation of the optic axis with respect to the plane of incidence and the interface.

REFERENCES

See references cited in Chapter 5.

Mosteller, L. P., Jr., and F. Wooten: "Optical Properties and Reflectance of Uniaxial Absorbing Crystals," *J. Opt. Soc. Am.*, vol. 58, no. 4, 1968, pp. 511–518.

CHAPTER

SEVEN

PLANE WAVES IN PLASMAS AND FERRITES

It is known that a constant magnetic field can render electric and magnetic properties of a medium anisotropic. In this chapter, we shall consider two such media—plasma and ferrite—with particular emphasis on plasma. Ferrites are not analyzed since their analysis is closely parallel to that of plasmas. In considering wave propagation in plasmas and ferrites, it is convenient to treat them as material media. We will first derive the dielectric tensor of a plasma and the permeability tensor of a ferrite. These tensors describe the response of charge systems to electric and magnetic fields. When inserted into the Maxwell equations, these tensors allow us to determine the electromagnetic fields which the plasma or ferrite will support and propagate. The problem of transmission and reflection of waves from a plasma half-space is then carried out by the methods similar to those discussed in the previous chapters.

7.1 DIELECTRIC TENSOR OF A MAGNETOPLASMA

Macroscopically, a plasma has been defined as a neutral ionized gas consisting principally of free electrons and positive ions. When a high-frequency electromagnetic wave passes through a plasma, since the ions are much heavier than the electrons, only the interaction between the wave and the free electrons need be considered.

We will therefore consider a medium which consists of N_0 free electrons per unit volume, each with charge e (a negative number) and mass m, immersed in an externally applied constant magnetic field $\mathbf{B}_c$. Each electron is acted upon by a force $e\mathscr{E}$ arising from the electric field of the wave and a force $e(\mathscr{V} \times \mathbf{B}_c)$ arising from the motion of the electron with the average velocity $\mathscr{V}$ through the constant magnetic field $\mathbf{B}_c$. For simplicity, we shall ignore collisions among particles by assuming that the medium is lossless. In this case, the equation of motion of an electron is

$$m \frac{d\mathscr{V}}{dt} = e(\mathscr{E} + \mathscr{V} \times \mathbf{B}_c) \tag{7.1}$$

For a monochromatic field with time factor $e^{-i\omega t}$, Eq. (7.1) reduces to

$$-i\omega m\mathbf{v} = e(\mathbf{E} + \mathbf{v} \times \mathbf{B}_c) \tag{7.2}$$

and yields the following expression for the velocity:

$$\mathbf{v} = \frac{ie}{m\omega}\left[\bar{\mathbf{I}} - i\frac{\omega_b}{\omega}(\hat{\mathbf{b}}_c \times \bar{\mathbf{I}})\right]^{-1} \cdot \mathbf{E} \tag{7.3}$$

where $\hat{\mathbf{b}}_c$ is a unit vector in the direction of $\mathbf{B}_c = B_c\,\hat{\mathbf{b}}_c$ and where

$$\omega_b = -\frac{eB_c}{m} \tag{7.4}$$

is the gyrofrequency of the electrons. Since the electronic convection current density is

$$\mathbf{J}_c = N_0 e\mathbf{v} = \bar{\boldsymbol{\sigma}} \cdot \mathbf{E} \tag{7.5}$$

it follows from Eqs. (7.3) and (7.5) that the conductivity tensor of the plasma is given by

$$\bar{\boldsymbol{\sigma}} = i\omega\varepsilon_0 X_0[\bar{\mathbf{I}} - iY_0(\hat{\mathbf{b}}_c \times \bar{\mathbf{I}})]^{-1} \tag{7.6}$$

where

$$X_0 = \frac{\omega_p^2}{\omega^2} \qquad Y_0 = \frac{\omega_b}{\omega} \tag{7.7}$$

and

$$\omega_p = \left(\frac{N_0 e^2}{m\varepsilon_0}\right)^{1/2}$$

is the plasma frequency. Carrying out the inverse in Eq. (7.6) according to Example 1.5 and substituting the result into Eq. (2.52), we finally obtain the explicit form of the dielectric tensor of the plasma:

$$\bar{\boldsymbol{\varepsilon}} = \varepsilon_\perp(\bar{\mathbf{I}} - \hat{\mathbf{b}}_c\hat{\mathbf{b}}_c) + i\varepsilon_\times(\hat{\mathbf{b}}_c \times \bar{\mathbf{I}}) + \varepsilon_\parallel \hat{\mathbf{b}}_c\hat{\mathbf{b}}_c \tag{7.8}$$

where

$$\varepsilon_\perp = 1 - \frac{X_0}{1 - Y_0^2} = 1 - \frac{\omega_p^2}{\omega^2 - \omega_b^2}$$

$$\varepsilon_\times = -\frac{X_0 Y_0}{1 - Y_0^2} = -\frac{\omega_b \omega_p^2}{\omega(\omega^2 - \omega_b^2)} \tag{7.9}$$

$$\varepsilon_\parallel = 1 - X_0 = 1 - \frac{\omega_p^2}{\omega^2}$$

We note that the dielectric tensor $\bar{\boldsymbol{\varepsilon}}$ in Eq. (7.8) is expressed in dyadic form, i.e., coordinate-independent form. However, if we choose a cartesian coordinate system whose z axis is parallel to $\mathbf{B}_c$, that is $\hat{\mathbf{b}}_c = (0, 0, 1)$, then the matrix representation of Eq. (7.8) takes the form

$$\bar{\boldsymbol{\varepsilon}} = \begin{bmatrix} \varepsilon_\perp & -i\varepsilon_\times & 0 \\ i\varepsilon_\times & \varepsilon_\perp & 0 \\ 0 & 0 & \varepsilon_\parallel \end{bmatrix} \tag{7.10}$$

From Eqs. (7.8) and (7.9), we observe the following important properties of the dielectric tensor: (1) If we reverse the direction of the constant magnetic field $\mathbf{B}_c$, the gyrofrequency ω_b, and hence $\varepsilon_\times$, change signs and thus $\bar{\boldsymbol{\varepsilon}}$ becomes

$$\bar{\boldsymbol{\varepsilon}}(-\mathbf{B}_c) = \varepsilon_\perp(\bar{\mathbf{I}} - \hat{\mathbf{b}}_c\hat{\mathbf{b}}_c) - i\varepsilon_\times(\hat{\mathbf{b}}_c \times \bar{\mathbf{I}}) + \varepsilon_\parallel \hat{\mathbf{b}}_c\hat{\mathbf{b}}_c$$

But tensor $\hat{\mathbf{b}}_c \times \bar{\mathbf{I}}$ is antisymmetric, i.e.,

$$(\hat{\mathbf{b}}_c \times \bar{\mathbf{I}})^T = -(\hat{\mathbf{b}}_c \times \bar{\mathbf{I}})$$

Therefore, the dielectric tensor $\bar{\boldsymbol{\varepsilon}}$ satisfies the symmetry relation

$$\bar{\boldsymbol{\varepsilon}}(-\mathbf{B}_c) = \tilde{\bar{\boldsymbol{\varepsilon}}}(\mathbf{B}_c) \tag{7.11}$$

which is generally valid for any medium whose anisotropy is due to an externally applied constant magnetic field. (2) For a lossless plasma, we see that $\varepsilon_\perp$, $\varepsilon_\times$, and $\varepsilon_\parallel$ are real quantities. Thus the transpose and complex conjugate of tensor $\bar{\boldsymbol{\varepsilon}}$ in Eq. (7.8) is equal to itself:

$$\tilde{\bar{\boldsymbol{\varepsilon}}}^* = \bar{\boldsymbol{\varepsilon}} \tag{7.12}$$

In other words, the dielectric tensor of a lossless plasma is hermitian. (3) Since $\bar{\boldsymbol{\varepsilon}}$ is a function of frequency, a plasma is a frequency-dispersive medium. The real part of $\bar{\boldsymbol{\varepsilon}}$ is an even function of ω:

$$\operatorname{Re}\,\bar{\boldsymbol{\varepsilon}}(\omega) = \varepsilon_\perp\bar{\mathbf{I}} + (\varepsilon_\parallel - \varepsilon_\perp)\hat{\mathbf{b}}_c\hat{\mathbf{b}}_c = \operatorname{Re}\,\bar{\boldsymbol{\varepsilon}}(-\omega) \tag{7.13}$$

and the imaginary part of $\bar{\boldsymbol{\varepsilon}}$ is an odd function of ω:

$$\operatorname{Im}\,\bar{\boldsymbol{\varepsilon}}(\omega) = \varepsilon_\times(\hat{\mathbf{b}}_c \times \bar{\mathbf{I}}) = -\operatorname{Im}\,\bar{\boldsymbol{\varepsilon}}(-\omega) \tag{7.14}$$

Also we observe the following two limiting cases. When the constant magnetic field $\mathbf{B}_c$ vanishes, $\varepsilon_\times$ vanishes and $\varepsilon_\perp = \varepsilon_\parallel$, and thus the dielectric tensor (7.8) reduces to

$$\bar{\boldsymbol{\varepsilon}} = \varepsilon_\parallel\bar{\mathbf{I}} \tag{7.15}$$

That is, when $\mathbf{B}_c = \mathbf{0}$, the plasma becomes isotropic and is characterized by the dielectric constant

$$\varepsilon_\parallel = 1 - \frac{\omega_p^2}{\omega^2} \tag{7.16}$$

On the other hand, if the applied constant magnetic field $\mathbf{B}_c$ is very strong, that is, $\mathbf{B}_c \to \infty$ or $\omega_b \gg \omega, \omega_p$, then $\varepsilon_\times = 0$, $\varepsilon_\perp = 1$, and

$$\bar{\boldsymbol{\varepsilon}} = \bar{\mathbf{I}} + (\varepsilon_\parallel - 1)\hat{\mathbf{b}}_c\hat{\mathbf{b}}_c \tag{7.17}$$

That is, in a strong constant magnetic field the plasma becomes a uniaxial medium.

Returning to the dielectric tensor (7.8) and using the results of Example 1.6, we easily find

$$\begin{aligned} |\bar{\boldsymbol{\varepsilon}}| &= \varepsilon_{\|}(\varepsilon_{\perp}^2 - \varepsilon_{\times}^2) \\ \operatorname{adj} \bar{\boldsymbol{\varepsilon}} &= \varepsilon_{\perp}\varepsilon_{\|}\bar{\mathbf{I}} - i\varepsilon_{\times}\varepsilon_{\|}(\hat{\mathbf{b}}_c \times \bar{\mathbf{I}}) + (\varepsilon_{\perp} - \varepsilon_{\|})\hat{\mathbf{b}}_c\hat{\mathbf{b}}_c \\ (\operatorname{adj} \bar{\boldsymbol{\varepsilon}})_t &= \varepsilon_{\perp}^2 - \varepsilon_{\times}^2 + 2\varepsilon_{\perp}\varepsilon_{\|} \end{aligned} \tag{7.18}$$

In summary, when a plasma is immersed in a constant magnetic field, *magnetoplasma*, it may be described as a continuous dielectric medium characterized in the frequency domain by the constitutive relations

$$\begin{aligned} \mathbf{B} &= \mu_0 \mathbf{H} \\ \mathbf{D} &= \varepsilon_0 \bar{\boldsymbol{\varepsilon}} \cdot \mathbf{E} \end{aligned} \tag{7.19}$$

where the magnetic property of the plasma remains the same as free space, but the electric property of plasma is determined by the dielectric tensor given by Eq. (7.8).

7.2 PERMEABILITY TENSOR OF A FERRITE

According to the macroscopic theory, the magnetic property of a ferrite may be attributed to its bound spinning electrons. A spinning electron is characterized by a magnetic dipole moment **m** and angular momentum **L**. Because of the negative charge on the electron, **m** and **L** are in opposite directions and are related by

$$\mathbf{m} = -\gamma' \mathbf{L} \tag{7.20}$$

where $\gamma' = g|e|/2m$ is the gyromagnetic ratio, $|e|$ is the absolute value of the electronic charge, m is the mass of the electron, and g is the spectroscopic splitting factor approximately equal to 2. If the total magnetic field intensity in the vicinity of the spinning electron is $\mathbf{H}_T$, then the torque **T** acting on it is

$$\mathbf{T} = \mu_0 \mathbf{m} \times \mathbf{H}_T \tag{7.21}$$

But from Newton's equation of motion for a rotating body, we also have

$$\frac{d\mathbf{L}}{dt} = \mathbf{T} \tag{7.22}$$

Combination of Eqs. (7.20), (7.21), and (7.22) yields the equation of motion for a single magnetic dipole:

$$\frac{d\mathbf{m}}{dt} = -\gamma(\mathbf{m} \times \mathbf{H}_T) \tag{7.23}$$

where $\gamma = \mu_0 \gamma' = 2.21 \times 10^5$ (rad/s)/(A · turns/m). If there are N_0 magnetic dipoles per unit volume, the total magnetization or total magnetic moment per unit volume is $\mathbf{M}_T = N_0 \mathbf{m}$. Thus Eq. (7.23) becomes

$$\frac{d\mathbf{M}_T}{dt} = -\gamma(\mathbf{M}_T \times \mathbf{H}_T) \tag{7.24}$$

which is the equation of motion of the magnetization vector and is the basis for studying the magnetic property of a ferrite medium. Here again, we have ignored the magnetic losses in Eq. (7.24) [cf. Prob. 7.4].

When both constant and time-varying magnetic fields are present in a ferrite, the total $\mathbf{H}_T$ and $\mathbf{M}_T$ take the form

$$\begin{aligned} \mathbf{H}_T &= \mathbf{H}_c + \mathbf{H}e^{-i\omega t} \\ \mathbf{M}_T &= \mathbf{M}_c + \mathbf{M}e^{-i\omega t} \end{aligned} \tag{7.25}$$

where the saturated magnetization vector $\mathbf{M}_c$ is in the same direction as the applied constant magnetic field $\mathbf{H}_c$, that is, $\mathbf{M}_c = M_c \hat{\mathbf{b}}_c$ and $\mathbf{H}_c = H_c \hat{\mathbf{b}}_c$. $\mathbf{H}e^{-i\omega t}$ and $\mathbf{M}e^{-i\omega t}$ are, respectively, the time-varying magnetic field and magnetization, and are such that $|\mathbf{M}| \ll |\mathbf{M}_c|$ and $|\mathbf{H}| \ll |\mathbf{H}_c|$. Substituting Eq. (7.25) into Eq. (7.24) and neglecting the second-order term in the product of small quantities $\mathbf{H}$ and $\mathbf{M}$, we then obtain the linearized equation of motion:

$$i\omega\mathbf{M} = \gamma(\mathbf{M} \times \mathbf{H}_c) + \gamma(\mathbf{M}_c \times \mathbf{H})$$

or

$$[i\omega\bar{\mathbf{I}} + \omega_0(\hat{\mathbf{b}}_c \times \bar{\mathbf{I}})] \cdot \mathbf{M} = \omega_m(\hat{\mathbf{b}}_c \times \bar{\mathbf{I}}) \cdot \mathbf{H} \tag{7.26}$$

where

$$\omega_m = \gamma M_c$$

and

$$\omega_0 = \gamma H_c \tag{7.27}$$

is the Larmor precessional frequency of the electron in the magnetic field $\mathbf{H}_c$. Solving Eq. (7.26) for $\mathbf{M}$,

$$\mathbf{M} = \omega_m[i\omega\bar{\mathbf{I}} + \omega_0(\hat{\mathbf{b}}_c \times \bar{\mathbf{I}})]^{-1} \cdot (\hat{\mathbf{b}}_c \times \bar{\mathbf{I}}) \cdot \mathbf{H} \tag{7.28}$$

and carrying out the inverse in Eq. (7.28) according to the results of Example 1.5, and keeping in mind that

$$\mathbf{B} = \mu_0(\mathbf{H} + \mathbf{M}) = \mu_0 \bar{\boldsymbol{\mu}} \cdot \mathbf{H} \tag{7.29}$$

we finally obtain the permeability tensor for the ferrite:

$$\bar{\boldsymbol{\mu}} = \mu_\perp(\bar{\mathbf{I}} - \hat{\mathbf{b}}_c \hat{\mathbf{b}}_c) + i\mu_\times(\hat{\mathbf{b}}_c \times \bar{\mathbf{I}}) + \mu_\parallel \hat{\mathbf{b}}_c \hat{\mathbf{b}}_c \tag{7.30}$$

where

$$\begin{aligned} \mu_\perp &= 1 + \frac{\omega_0 \omega_m}{\omega_0^2 - \omega^2} \\ \mu_\times &= \frac{\omega\omega_m}{\omega_0^2 - \omega^2} \\ \mu_\parallel &= 1 \end{aligned} \tag{7.31}$$

We note that the permeability tensor (7.30) of a ferrite is identical in form to the dielectric tensor (7.8) of a magnetoplasma. The components of $\bar{\mu}$ are related to ω, ω_c, and ω_m and thus are frequency-dispersive. The properties stated for the plasma also apply here. The constitutive relations for a ferrite medium are

$$\mathbf{D} = \varepsilon_0 \varepsilon \mathbf{E}$$
$$\mathbf{B} = \mu_0 \bar{\mu} \cdot \mathbf{H} \tag{7.32}$$

where the dielectric constant ε ranges from 5 to 20. The anisotropy is maintained by the term ω_m, which depends upon the saturation magnetization of the material and is therefore a property of a particular ferrite used.

7.3 DISPERSION EQUATION OF A MAGNETOPLASMA

We will now examine the propagation and polarization properties of a monochromatic uniform plane wave in a homogeneous, lossless magnetoplasma. We consider the plasma as a continuous medium characterized by the constitutive relations

$$\mathbf{D}_0 = \varepsilon_0 \bar{\varepsilon} \cdot \mathbf{E}_0$$
$$\mathbf{B}_0 = \mu_0 \mathbf{H}_0 \tag{7.33}$$

where the dielectric tensor $\bar{\varepsilon}$ is given by Eq. (7.8) and is hermitian. The Maxwell equations for a monochromatic plane wave and the constitutive relations (7.33) give

$$\omega \mathbf{D}_0 = \omega \varepsilon_0 \bar{\varepsilon} \cdot \mathbf{E}_0 = -\mathbf{k} \times \mathbf{H}_0$$
$$\omega \mu_0 \mathbf{H}_0 = \mathbf{k} \times \mathbf{E}_0 \tag{7.34}$$

Here we assume that both $\mathbf{k}$ and ω are real. The problem is to determine the wave vector $\mathbf{k}$ which describes the propagation properties of the wave, and the direction of vector $\mathbf{E}_0$ which describes the polarization of the wave. To this end, we eliminate $\mathbf{H}_0$ from Eq. (7.34) and obtain

$$[k_0^2 \bar{\varepsilon} + (\mathbf{k} \times \bar{\mathbf{I}})^2] \cdot \mathbf{E}_0 = (k_0^2 \bar{\varepsilon} - k^2 \bar{\mathbf{I}} + \mathbf{k}\mathbf{k}) \cdot \mathbf{E}_0 = \mathbf{0} \tag{7.35}$$

Equation (7.35) is a homogeneous equation. A nonzero solution $\mathbf{E}_0$ exists only if the determinant of the coefficient matrix vanishes, i.e.,

$$| k_0^2 \bar{\varepsilon} - k^2 \bar{\mathbf{I}} + \mathbf{k}\mathbf{k} | = 0 \tag{7.36}$$

But according to Eqs. (5.30) and (5.31), we have

$$| k_0^2 \bar{\varepsilon} - k^2 \bar{\mathbf{I}} + \mathbf{k}\mathbf{k} | = \hat{\mathbf{k}} \cdot (k_0^2 \bar{\varepsilon}) \cdot \operatorname{adj} (k_0^2 \bar{\varepsilon} - k^2 \bar{\mathbf{I}}) \cdot \hat{\mathbf{k}} \tag{7.37}$$

Thus, the dispersion equation (7.36) may also be written as

$$\hat{\mathbf{k}} \cdot \bar{\varepsilon} \cdot \operatorname{adj} (k_0^2 \bar{\varepsilon} - k^2 \bar{\mathbf{I}}) \cdot \hat{\mathbf{k}} = 0 \tag{7.38}$$

From Eq. (1.124*c*),

$$\operatorname{adj} (k_0^2 \bar{\varepsilon} - k^2 \bar{\mathbf{I}}) = k^4 \bar{\mathbf{I}} + k_0^2 k^2 (\bar{\varepsilon} - \bar{\varepsilon}_t \bar{\mathbf{I}}) + k_0^4 \operatorname{adj} \bar{\varepsilon} \tag{7.39}$$

Hence Eq. (7.38) becomes

$$(\hat{\mathbf{k}} \cdot \bar{\varepsilon} \cdot \hat{\mathbf{k}})n^4 + \hat{\mathbf{k}} \cdot [\text{adj } \bar{\varepsilon} - (\text{adj } \bar{\varepsilon})_t \bar{\mathbf{I}}] \cdot \hat{\mathbf{k}} n^2 + |\bar{\varepsilon}| = 0 \tag{7.40}$$

where $n = k/k_0$ is the index of refraction. Here we give an alternative derivation of the dispersion equation (5.29). As in the case of a crystal, n^2 of a hermitian dielectric tensor $\bar{\varepsilon}$ is always real. Hence n is either real (propagation) or imaginary (nonpropagation). To see this, we dot-multiply Eq. (7.35) by $\mathbf{E}_0^*$ and obtain

$$n^2 = \frac{\mathbf{E}_0^* \cdot \bar{\varepsilon} \cdot \mathbf{E}_0}{(\hat{\mathbf{k}} \times \mathbf{E}_0) \cdot (\hat{\mathbf{k}} \times \mathbf{E}_0)^*} \tag{7.41}$$

Since $\bar{\varepsilon}$ is hermitian, the hermitian form $(\mathbf{E}_0^* \cdot \bar{\varepsilon} \cdot \mathbf{E}_0)$ is always real, which in turn shows that n^2 is real.

For a lossless magnetoplasma, from Eqs. (7.8) and (7.18), we find

$$\begin{aligned} \hat{\mathbf{k}} \cdot \bar{\varepsilon} \cdot \hat{\mathbf{k}} &= \varepsilon_\perp (\hat{\mathbf{k}} \times \hat{\mathbf{b}}_c)^2 + \varepsilon_\| (\hat{\mathbf{k}} \cdot \hat{\mathbf{b}}_c)^2 \\ \hat{\mathbf{k}} \cdot [\text{adj } \bar{\varepsilon} - (\text{adj } \bar{\varepsilon})_t \bar{\mathbf{I}}] \cdot \hat{\mathbf{k}} &= (\varepsilon_\| - \varepsilon_\perp)(\hat{\mathbf{k}} \times \hat{\mathbf{b}}_c)^2 - 2\varepsilon_\perp \varepsilon_\| \\ |\bar{\varepsilon}| &= \varepsilon_\| (\varepsilon_\perp^2 - \varepsilon_\times^2) \end{aligned} \tag{7.42}$$

Substitution of Eq. (7.42) into Eq. (7.40) yields the required dispersion equation relating the wave number to the source frequency in a magnetoplasma. We note that the solution in n^2 depends on the direction of wave propagation relative to the applied constant magnetic field. If θ is the angle between the vectors $\hat{\mathbf{k}}$ and $\hat{\mathbf{b}}_c$, then Eq. (7.42) becomes

$$\begin{aligned} \hat{\mathbf{k}} \cdot \bar{\varepsilon} \cdot \hat{\mathbf{k}} &= \varepsilon_\perp \sin^2 \theta + \varepsilon_\| \cos^2 \theta \\ \hat{\mathbf{k}} \cdot [\text{adj } \bar{\varepsilon} - (\text{adj } \bar{\varepsilon})_t \bar{\mathbf{I}}] \cdot \hat{\mathbf{k}} &= (\varepsilon_\times^2 - \varepsilon_\perp^2 - \varepsilon_\perp \varepsilon_\|) \sin^2 \theta - 2\varepsilon_\perp \varepsilon_\| \cos^2 \theta \\ |\bar{\varepsilon}| &= \varepsilon_\| (\varepsilon_\perp^2 - \varepsilon_\times^2) \sin^2 \theta + \varepsilon_\| (\varepsilon_\perp^2 - \varepsilon_\times^2) \cos^2 \theta \end{aligned} \tag{7.43}$$

Substituting Eq. (7.43) into Eq. (7.40) and solving for the direction of propagation θ in terms of the index of refraction, we obtain an alternative form:

$$\tan^2 \theta = -\frac{\varepsilon_\| (n^2 - \varepsilon_\perp - \varepsilon_\times)(n^2 - \varepsilon_\perp + \varepsilon_\times)}{(n^2 - \varepsilon_\|)(\varepsilon_\perp n^2 - \varepsilon_\perp^2 + \varepsilon_\times^2)} \tag{7.44}$$

This equation determines two values of n^2 for each value of θ.

In the case where the propagation is parallel to the constant magnetic field ($\hat{\mathbf{k}} = \hat{\mathbf{b}}_c$), we have $\theta = 0$; accordingly, Eq. (7.44) yields two solutions:

$$k_L = k_0 \sqrt{\varepsilon_\perp - \varepsilon_\times} = k_0 \sqrt{1 - \frac{\omega_p^2}{\omega(\omega + \omega_b)}} \tag{7.45}$$

and

$$k_R = k_0 \sqrt{\varepsilon_\perp + \varepsilon_\times} = k_0 \sqrt{1 - \frac{\omega_p^2}{\omega(\omega - \omega_b)}} \tag{7.46}$$

We will see later that the wave numbers k_L and k_R correspond respectively to a left- and a right-handed circularly polarized wave. Moreover, when the propagation is perpendicular to the constant magnetic field ($\hat{\mathbf{k}} \cdot \hat{\mathbf{b}}_c = 0$), θ is equal to $\pi/2$

and in this case the two solutions of Eq. (7.44) are

$$k_{0r} \doteq k_0\sqrt{\varepsilon_{\parallel}} = k_0\sqrt{1 - \frac{\omega_p^2}{\omega^2}} \tag{7.47}$$

and

$$k_e = k_0\sqrt{\frac{\varepsilon_{\perp}^2 - \varepsilon_{\times}^2}{\varepsilon_{\perp}}} = k_0\sqrt{1 - \frac{\omega_p^2(\omega^2 - \omega_p^2)}{\omega^2(\omega^2 - \omega_p^2 - \omega_b^2)}} \tag{7.48}$$

We note that the wave number k_{0r} does not depend on the constant magnetic field $\mathbf{B}_c$ and is called an ordinary wave, whereas k_e depends on the constant magnetic field and is thus called an extraordinary wave. In each case ($\theta = 0$ or $\theta = \pi/2$) there are two uncoupled waves, all of which are termed *principal waves.*

In general, when θ is arbitrary, we have the two solutions of Eq. (7.44) denoted by the subscripts + and −:

$$k_+ = k_0\left[1 - \frac{X_0}{1 - \dfrac{Y_0^2 \sin^2\theta}{2(1 - X_0)} + \sqrt{\dfrac{Y_0^4 \sin^4\theta}{4(1 - X_0)^2} + Y_0^2 \cos^2\theta}}\right]^{1/2} \tag{7.49}$$

and

$$k_- = k_0\left[1 - \frac{X_0}{1 - \dfrac{Y_0^2 \sin^2\theta}{2(1 - X_0)} - \sqrt{\dfrac{Y_0^4 \sin^4\theta}{4(1 - X_0)^2} + Y_0^2 \cos^2\theta}}\right]^{1/2} \tag{7.50}$$

Thus, there are two waves traveling with different velocities in an arbitrary direction θ. One of the waves has wave number k_+, and hence the phase velocity $v_{p+} = \omega/k_+$, while the other has a wave number k_-, and hence the phase velocity $v_{p-} = \omega/k_-$.

7.4 DETERMINATION OF THE DIRECTIONS OF FIELD VECTORS

To each solution of the dispersion equation (7.40), we may determine the corresponding direction of the electric field intensity from Eq. (7.35) and thus the state of polarization of the wave. Following Sec. 5.4, we rewrite Eq. (7.35) as

$$(k_0^2\bar{\varepsilon} - k^2\bar{\mathbf{I}}) \cdot \mathbf{E}_0 = -(\mathbf{k} \cdot \mathbf{E}_0)\mathbf{k} \tag{7.51}$$

and consider the following two cases:

Case 1 If the matrix $k_0^2\bar{\varepsilon} - k^2\bar{\mathbf{I}}$ is nonsingular, i.e., if k^2 is not an eigenvalue of $k_0^2\bar{\varepsilon}$, we find from Eq. (7.51) the direction of $\mathbf{E}_0$:

$$\mathbf{e} = [\text{adj}\,(k_0^2\bar{\varepsilon} - k^2\bar{\mathbf{I}})] \cdot \mathbf{k} \tag{7.52}$$

Using the explicit form (7.8) of the dielectric tensor of the magnetoplasma and noting from Eq. (7.9) that

$$\varepsilon_\perp^2 - \varepsilon_\times^2 - \varepsilon_\perp \varepsilon_\parallel = \varepsilon_\perp - \varepsilon_\parallel \tag{7.53}$$

we obtain, according to Eq. (1.133*c*),

$$\begin{aligned}\text{adj}\,(k_0^2 \bar{\boldsymbol{\varepsilon}} - k^2 \bar{\mathbf{I}}) = (k^2 - k_0^2 \varepsilon_\parallel)[(k^2 - k_0^2 \varepsilon_\perp)\bar{\mathbf{I}} + i\varepsilon_\times k_0^2 (\hat{\mathbf{b}}_c \times \bar{\mathbf{I}})] \\ + (\varepsilon_\parallel - \varepsilon_\perp)(k^2 - k_0^2)k_0^2 \hat{\mathbf{b}}_c \hat{\mathbf{b}}_c\end{aligned} \tag{7.54}$$

Thus the direction of the electric field intensity corresponding to the wave number given by Eq. (7.49) is

$$\begin{aligned}\mathbf{e}_+ = (k_+^2 - k_0^2 \varepsilon_\parallel)[(k_+^2 - k_0^2 \varepsilon_\perp)\mathbf{k}_+ - i\varepsilon_\times k_0^2(\mathbf{k}_+ \times \hat{\mathbf{b}}_c)] \\ + (\varepsilon_\parallel - \varepsilon_\perp)(k_+^2 - k_0^2)k_0^2(\mathbf{k}_+ \cdot \hat{\mathbf{b}}_c)\hat{\mathbf{b}}_c\end{aligned} \tag{7.55}$$

We see from the above that, in a magnetoplasma, vector $\mathbf{e}_+$ is completely determined by the wave vector $\mathbf{k}_+$ and the direction of the constant magnetic field $\hat{\mathbf{b}}_c$. In general, $\mathbf{e}_+$ is elliptically polarized and has a component parallel to $\mathbf{k}_+$. The boundary conditions can only affect the amplitude vector $\mathbf{E}_{0+}$ but not the direction $\mathbf{e}_+$. Substituting Eq. (7.55) into Maxwell's equations, we easily find

$$\begin{aligned}\mathbf{h}_+ &= \frac{1}{\omega\mu_0}(\mathbf{k}_+ \times \mathbf{e}_+) \\ &= \omega\varepsilon_0 \{(\varepsilon_\parallel - \varepsilon_\perp)(k_+^2 - k_0^2)(\mathbf{k}_+ \cdot \hat{\mathbf{b}}_c)(\mathbf{k}_+ \times \hat{\mathbf{b}}_c) \\ &\quad - i\varepsilon_\times(k_+^2 - k_0^2 \varepsilon_\parallel)[\mathbf{k}_+ \times (\mathbf{k}_+ \times \hat{\mathbf{b}}_c)]\}\end{aligned} \tag{7.56}$$

and

$$\begin{aligned}\mathbf{d}_+ &= -\frac{1}{\omega}(\mathbf{k}_+ \times \mathbf{h}_+) \\ &= \varepsilon_0 \{(\varepsilon_\perp - \varepsilon_\parallel)(k_+^2 - k_0^2)(\mathbf{k}_+ \cdot \hat{\mathbf{b}})[\mathbf{k}_+ \times (\mathbf{k}_+ \times \hat{\mathbf{b}}_c)] \\ &\quad - i\varepsilon_\times(k_+^2 - k_0^2 \varepsilon_\parallel)k_+^2(\mathbf{k}_+ \times \hat{\mathbf{b}}_c)\}\end{aligned} \tag{7.57}$$

Similarly, corresponding to the wave number k_- given in Eq. (7.50), we obtain expressions for $\mathbf{e}_-$, $\mathbf{h}_-$, and $\mathbf{d}_-$ by replacing the subscript + by − in Eqs. (7.55), (7.56), and (7.57).

When the propagation is perpendicular to the constant magnetic field, that is, $\mathbf{k} \cdot \hat{\mathbf{b}}_c = 0$, expressions (7.55), (7.56), and (7.57) fail to give the correct field vectors for the ordinary wave characterized by the wave number $k_{0r} = k_0 \sqrt{\varepsilon_\parallel}$, since in this case k_{0r}^2 is an eigenvalue of $k_0^2 \bar{\boldsymbol{\varepsilon}}$. On the other hand, for the extraordinary wave with wave number $k_e = k_0 \sqrt{(\varepsilon_\perp^2 - \varepsilon_\times^2)/\varepsilon_\perp}$, we obtain, after dropping the factor, $-i\varepsilon_\times k_0^2(k_e^2 - k_0^2 \varepsilon_\parallel)/\varepsilon_\perp$,

$$\begin{aligned}\mathbf{e}_e &= \varepsilon_\perp(\mathbf{k}_e \times \hat{\mathbf{b}}_c) - i\varepsilon_\times \mathbf{k}_e \\ \mathbf{h}_e &= -\frac{\varepsilon_\perp k_e^2}{\omega\mu_0}\hat{\mathbf{b}}_c\end{aligned} \tag{7.58}$$

and
$$\mathbf{d}_e = \frac{\varepsilon_\perp k_e^2}{\omega^2 \mu_0}(\mathbf{k}_e \times \hat{\mathbf{b}}_c)$$

We see that $\mathbf{e}_e$ is elliptically polarized in the plane perpendicular to the constant magnetic field $\mathbf{B}_c$ and has a component in the direction of $\mathbf{k}_e$ which will lead to space charge effects (see Fig. 7.1). We note that the polarization for this wave does depend on the values of $\varepsilon_\perp$ and $\varepsilon_\times$, and thus on the properties of the plasma.

Case 2 When the matrix $k_0^2\bar{\boldsymbol{\varepsilon}} - k^2\bar{\mathbf{I}}$ is singular, i.e., when k^2 is an eigenvalue of $k_0^2\bar{\boldsymbol{\varepsilon}}$. According to Eq. (1.133*b*, *c*), the determinant and the adjoint of the matrix

$$\bar{\boldsymbol{\varepsilon}} - \lambda\bar{\mathbf{I}} = (\varepsilon_\perp - \lambda)\bar{\mathbf{I}} + i\varepsilon_\times(\hat{\mathbf{b}}_c \times \bar{\mathbf{I}}) + (\varepsilon_\parallel - \varepsilon_\perp)\hat{\mathbf{b}}_c\hat{\mathbf{b}}_c \tag{7.59}$$

are

$$|\bar{\boldsymbol{\varepsilon}} - \lambda\bar{\mathbf{I}}| = (\varepsilon_\parallel - \lambda)(\varepsilon_\perp - \varepsilon_\times - \lambda)(\varepsilon_\perp + \varepsilon_\times - \lambda) \tag{7.60}$$

and

$$\text{adj}\,(\bar{\boldsymbol{\varepsilon}} - \lambda\bar{\mathbf{I}}) = (\lambda - \varepsilon_\parallel)[(\lambda - \varepsilon_\perp)\bar{\mathbf{I}} + i\varepsilon_\times(\hat{\mathbf{b}}_c \times \bar{\mathbf{I}})] + (\varepsilon_\parallel - \varepsilon_\perp)(\lambda - 1)\hat{\mathbf{b}}_c\hat{\mathbf{b}}_c \tag{7.61}$$

respectively. By setting Eq. (7.60) equal to zero, we find the three eigenvalues of $\bar{\boldsymbol{\varepsilon}}$:

$$\lambda_1 = \varepsilon_\perp + \varepsilon_\times \tag{7.62}$$

$$\lambda_2 = \varepsilon_\perp - \varepsilon_\times \tag{7.63}$$

and

$$\lambda_3 = \varepsilon_\parallel \tag{7.64}$$

The eigenvector corresponding to the eigenvalue λ_1 is in the direction of vector $[\text{adj}\,(\bar{\boldsymbol{\varepsilon}} - \lambda_1\bar{\mathbf{I}})] \cdot \mathbf{a}_1$, where $\mathbf{a}_1$ is an arbitrary vector. Thus, from Eq. (7.61), we find

$$\mathbf{l}_1 = [(\hat{\mathbf{b}}_c \times \bar{\mathbf{I}})^2 - i(\hat{\mathbf{b}}_c \times \bar{\mathbf{I}})] \cdot \mathbf{a}_1 = i(\hat{\mathbf{b}}_c \times \mathbf{l}_1) \tag{7.65}$$

In a similar manner, the eigenvectors corresponding to the eigenvalues λ_2 and λ_3

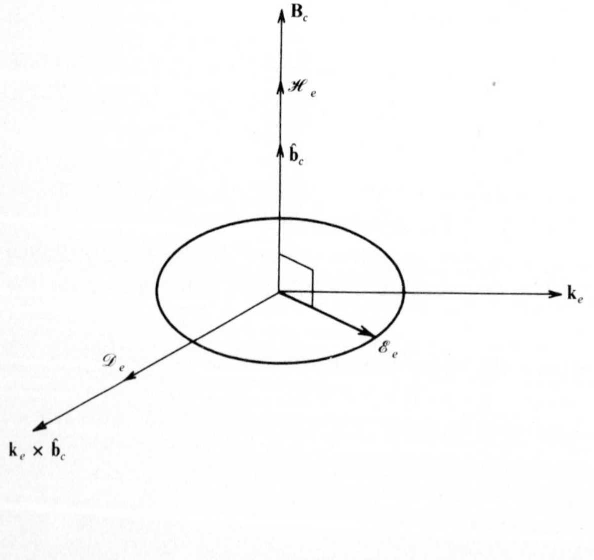

Figure 7.1 Extraordinary wave, elliptically polarized.

are found to be

$$\mathbf{l}_2 = [(\hat{\mathbf{b}}_c \times \bar{\mathbf{I}})^2 + i(\hat{\mathbf{b}}_c \times \bar{\mathbf{I}})] \cdot \mathbf{a}_2 = -i(\hat{\mathbf{b}}_c \times \mathbf{l}_2) \tag{7.66}$$

and

$$\mathbf{l}_3 = \hat{\mathbf{b}}_c \tag{7.67}$$

respectively, where $\mathbf{a}_2$ is again an arbitrary vector. Since $\bar{\varepsilon}$ is a hermitian matrix, the eigenvalues given by Eqs. (7.62) to (7.64) are all real and the eigenvectors given by Eqs. (7.65) to (7.67) are orthogonal in the hermitian sense, that is, $\mathbf{l}_i \cdot \mathbf{l}_j^* = 0$ for $i \neq j$.

We will now find the directions of the field vectors when $k_0^2 \bar{\varepsilon} - k^2 \bar{\mathbf{I}}$ is singular. Following Sec. 5.4, for $\lambda_3 = \varepsilon_{\parallel}$ or $k_{0r} = k_0 \sqrt{\varepsilon_{\parallel}}$ [the subscript $0r$ is used here to be consistent with Eq. (7.47)], the condition (5.84) imposed on the direction of the wave vector $\mathbf{k}_{0r}$ becomes

$$\mathbf{k}_{0r} \cdot \mathbf{l}_3 = \mathbf{k}_{0r} \cdot \hat{\mathbf{b}}_c = 0 \tag{7.68}$$

That is, the propagation must be perpendicular to the constant magnetic field. In this case, since

$$k_0^4 |\bar{\varepsilon}| - (\mathbf{k}_{0r} \cdot \bar{\varepsilon} \cdot \mathbf{k}_{0r}) k_{0r}^2 = k_0^4 \varepsilon_{\parallel} (\varepsilon_{\perp} - \varepsilon_{\parallel}) \neq 0$$

from Eqs. (5.88), (5.89), and (5.90), we have

$$\mathbf{h}_{0r} = \mathbf{k}_{0r} \times \mathbf{l}_3 = \mathbf{k}_{0r} \times \hat{\mathbf{b}}_c = k_0 \sqrt{\varepsilon_{\parallel}}\, (\hat{\mathbf{k}} \times \hat{\mathbf{b}}_c)$$

$$\mathbf{d}_{0r} = \frac{k_{0r}^2}{\omega} \mathbf{l}_3 = \omega \mu_0 \varepsilon_0 \varepsilon_{\parallel} \hat{\mathbf{b}}_c \tag{7.69}$$

and

$$\mathbf{e}_{0r} = \omega \mu_0 \mathbf{l}_3 = \omega \mu_0 \hat{\mathbf{b}}_c$$

which shows that the ordinary wave is linearly polarized with the electric field intensity parallel to $\mathbf{B}_c$. We note that the propagation constant $k_{0r} = k_0 \sqrt{\varepsilon_{\parallel}}$ is independent of B_c and all the field vectors are perpendicular to the direction of wave propagation (a TEM wave) as in an isotropic medium (see Fig. 7.2).

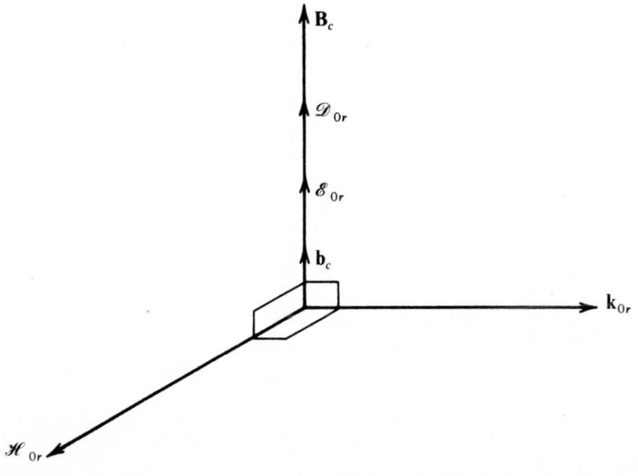

Figure 7.2 Ordinary wave, linearly polarized.

For $\lambda_2 = \varepsilon_\perp - \varepsilon_\times$ or $k_L = k_0\sqrt{\varepsilon_\perp - \varepsilon_\times}$ [again the subscript L is used to be consistent with Eq. (7.45)], the condition (5.84) becomes

$$\mathbf{k}_L \cdot \mathbf{l}_2 = -i(\mathbf{k}_L \times \hat{\mathbf{b}}_c) \cdot \mathbf{l}_2 = 0$$

or

$$\hat{\mathbf{k}} = \hat{\mathbf{b}}_c \tag{7.70}$$

In other words, the wave must propagate parallel to the constant magnetic field. In this case, from Eqs. (5.88), (5.89), and (5.90), we find the directions of the field vectors

$$\mathbf{h}_L = \mathbf{k}_L \times \mathbf{l}_2 = -i(\hat{\mathbf{b}}_c \times \mathbf{h}_L)$$

$$\mathbf{d}_L = \frac{k_L^2}{\omega}\,\mathbf{l}_2 = -i(\hat{\mathbf{b}}_c \times \mathbf{d}_L) \tag{7.71}$$

and

$$\mathbf{e}_L = \omega\mu_0\,\mathbf{l}_2 = -i(\hat{\mathbf{b}}_c \times \mathbf{e}_L)$$

which, according to Eq. (3.54), clearly shows that the wave is a left-handed, circularly polarized TEM wave (see Fig. 7.3), hence justifying the use of subscript L.

Finally, for $\lambda_1 = \varepsilon_\perp + \varepsilon_\times$ or $k_R = k_0\sqrt{\varepsilon_\perp + \varepsilon_\times}$ [cf. (7.46)], condition (5.84) implies again that

$$\hat{\mathbf{k}} = \hat{\mathbf{b}}_c \tag{7.72}$$

However, the directions of the field vectors are now given by

$$\mathbf{h}_R = i(\hat{\mathbf{b}}_c \times \mathbf{h}_R)$$

$$\mathbf{d}_R = i(\hat{\mathbf{b}}_c \times \mathbf{d}_R) \tag{7.73}$$

and

$$\mathbf{e}_R = i(\hat{\mathbf{b}}_c \times \mathbf{e}_R)$$

Consequently, the wave that travels parallel to $\mathbf{B}_c$ is a right-handed, circularly polarized TEM wave as illustrated in Fig. 7.4.

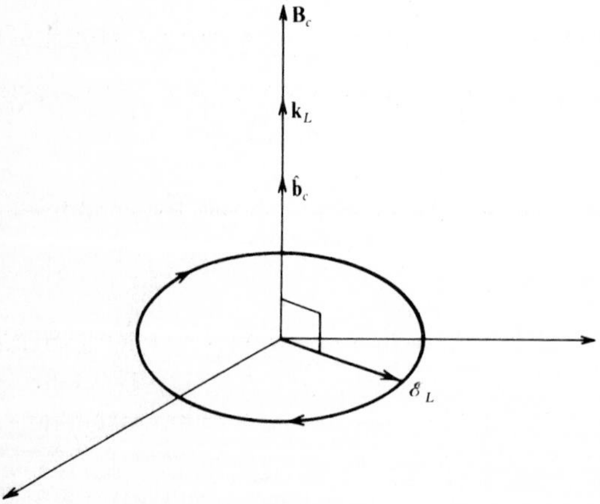

Figure 7.3 Left-handed circularly polarized wave.

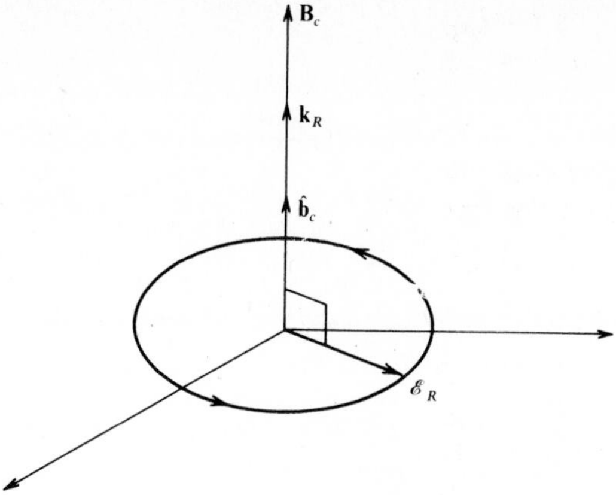

Figure 7.4 Right-handed circularly polarized wave.

7.5 FARADAY ROTATION

According to Sec. 3.4, a linearly polarized wave may be regarded as the sum of two circularly polarized waves of equal amplitude and opposite sense of rotation. If the two circularly polarized waves propagate with different phase velocities in a given medium, the plane of polarization (the plane containing **E** and **k**) of the resultant linearly polarized wave rotates as the wave propagates. This phenomenon is known as the *Faraday rotation.*

To study the Faraday rotation, we consider the electric vectors of two circularly polarized waves traveling in the same direction $\hat{\mathbf{k}}$:

$$\begin{aligned} \mathbf{E}_R &= (\mathbf{a} + i\hat{\mathbf{k}} \times \mathbf{a})\, e^{i(\mathbf{k}_R \cdot \mathbf{r} - \omega t)} \\ \mathbf{E}_L &= (\mathbf{a} - i\hat{\mathbf{k}} \times \mathbf{a})\, e^{i(\mathbf{k}_L \cdot \mathbf{r} - \omega t)} \end{aligned} \tag{7.74}$$

where vector **a** is perpendicular to $\hat{\mathbf{k}}$, that is, $\hat{\mathbf{k}} \cdot \mathbf{a} = 0$, $\mathbf{k}_R = k_R \hat{\mathbf{k}}$, and $\mathbf{k}_L = k_L \hat{\mathbf{k}}$. We can easily check with Eq. (3.54) that $\mathbf{E}_R$ is a right-handed, circularly polarized wave traveling with the phase velocity ω/k_R, whereas $\mathbf{E}_L$ is a left-handed, circularly polarized wave traveling with phase velocity ω/k_L. The instantaneous electric field intensity of the sum of these two equal amplitude waves is

$$\begin{aligned} \mathscr{E} &= \text{Re}\,(\mathbf{E}_R + \mathbf{E}_L) \\ &= 2\{\mathbf{a} \cos\left[\tfrac{1}{2}(k_R - k_L)(\mathbf{r} \cdot \hat{\mathbf{k}})\right] - (\hat{\mathbf{k}} \times \mathbf{a}) \sin\left[\tfrac{1}{2}(k_R - k_L)(\mathbf{r} \cdot \hat{\mathbf{k}})\right]\} \\ &\quad \times \cos\left[\tfrac{1}{2}(k_R + k_L)(\mathbf{r} \cdot \hat{\mathbf{k}}) - \omega t\right] \end{aligned} \tag{7.75}$$

which shows that the two terms are in time phase and thus that $\mathscr{E}$ is linearly polarized. The resultant $\mathscr{E}$ makes an angle θ'_F with respect to the vector **a** (see Fig. 7.5) and

$$\tan \theta'_F = -\tan \frac{(k_R - k_L)(\mathbf{r} \cdot \hat{\mathbf{k}})}{2}$$

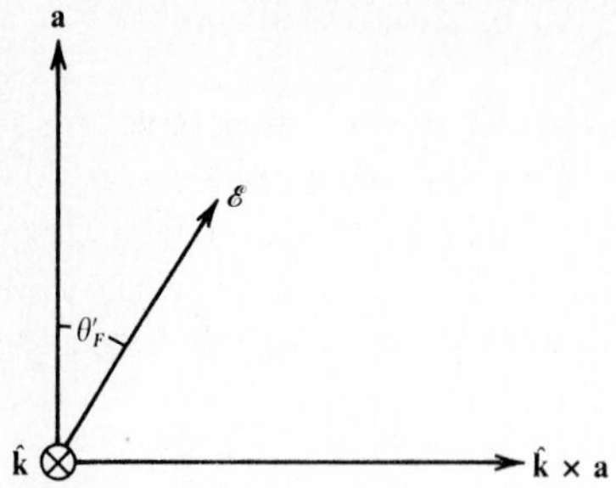

Figure 7.5 Faraday rotation.

Thus the angle θ_F, through which the resultant vector $\mathscr{E}$ rotates as the wave travels a unit distance in the direction of $\hat{\mathbf{k}}$, is given by

$$\theta_F = \tfrac{1}{2}(k_L - k_R) \tag{7.76}$$

Returning now to the case of wave propagation in a magnetoplasma along the direction of the constant magnetic field, and using Eqs. (7.45) and (7.46), we may write θ_F in the form

$$\theta_F = \frac{k_0}{2}\left[\sqrt{1 - \frac{\omega_p^2}{\omega(\omega + \omega_b)}} - \sqrt{1 - \frac{\omega_p^2}{\omega(\omega - \omega_b)}}\right] \tag{7.77}$$

It shows the dependence of the Faraday rotation θ_F on frequencies ω, ω_p, and ω_b. We note that when a wave is propagated parallel to $\mathbf{B}_c$, the electric field vector undergoes a clockwise Faraday rotation. In reversing the direction of wave propagation, the clockwise rotation becomes counterclockwise, and vice versa. Thus if the plane of polarization of a wave traveling parallel to $\mathbf{B}_c$ is rotated through a certain angle, then upon reflection it will be rotated still further and the rotation for the round trip will be twice that of a single trip.

In the case of the source frequency ω being much higher than the plasma frequency ω_p and gyrofrequency ω_b, we may approximate Eqs. (7.45) and (7.46) by the binomial expansions

$$k_L \cong k_0\left[1 - \frac{1}{2}\frac{\omega_p^2}{\omega(\omega + \omega_b)}\right]$$

and

$$k_R \cong k_0\left[1 - \frac{1}{2}\frac{\omega_p^2}{\omega(\omega - \omega_b)}\right] \tag{7.78}$$

respectively. Substituting Eq. (7.78) into Eq. (7.76), we obtain

$$\theta_F \cong \frac{1}{2c}\left(\frac{\omega_p}{\omega}\right)^2 \omega_b \tag{7.79}$$

which shows that the Faraday rotation θ_F for $\omega_b \ll \omega$ and $\omega_p \ll \omega$ is linearly proportional to ω_b and hence linearly proportional to the constant magnetic field B_c.

7.6 ENERGY DENSITY AND POYNTING'S VECTOR

Different from a crystal, a lossless magnetoplasma is a frequency-dispersive anisotropic medium characterized by a complex hermitian dielectric tensor given by Eq. (7.8). In general, waves in a magnetoplasma are elliptically polarized; hence the field vectors are complex. Methods used in Secs. 5.6 and 5.8 for crystals may be extended to the case of a magnetoplasma. Based on the Maxwell equations (7.34), we easily establish that

$$\mathbf{E}_0^* \cdot \mathbf{D}_0 = \frac{1}{\omega}\,\mathbf{k} \cdot (\mathbf{E}_0 \times \mathbf{H}_0^*) = \mathbf{H}_0^* \cdot \mathbf{B}_0 \tag{7.80}$$

According to Eq. (2.85), the time-averaged energy density for a magnetoplasma, a sum of electric and magnetic energy densities, is given by

$$\langle W \rangle = \frac{\varepsilon_0}{4}\,\mathbf{E}_0^* \cdot \frac{\partial(\omega\bar{\boldsymbol{\varepsilon}})}{\partial\omega} \cdot \mathbf{E}_0 + \frac{\mu_0}{4}\,\mathbf{H}_0 \cdot \mathbf{H}_0^* \tag{7.81}$$

Substituting Eq. (7.80) into the last term of Eq. (7.81), we may write

$$\langle W \rangle = \frac{\varepsilon_0}{4}\,\mathbf{E}_0^* \cdot \frac{\partial(\omega^2\bar{\boldsymbol{\varepsilon}})}{\omega\,\partial\omega} \cdot \mathbf{E}_0 \tag{7.82}$$

On the other hand, by taking the cross product of the complex conjugate of Eq. (7.34) with $\mathbf{k}$ and noting that $\mathbf{k} \cdot \mathbf{H}_0^* = 0$, we have

$$\mathbf{H}_0^* = \frac{\omega}{k^2}\,(\mathbf{k} \times \mathbf{D}_0^*) \tag{7.83}$$

Substituting Eq. (7.83) into the expression for the time-averaged Poynting vector and then expanding the vector triple product, we obtain

$$\langle \mathscr{P} \rangle = \tfrac{1}{2}\,\mathrm{Re}\,(\mathbf{E}_0 \times \mathbf{H}_0^*)$$

$$= \frac{\omega}{2k^2}\,\mathrm{Re}\,[(\mathbf{E}_0 \cdot \mathbf{D}_0^*)\mathbf{k} - (\mathbf{k} \cdot \mathbf{E}_0)\mathbf{D}_0^*] \tag{7.84}$$

or, after using Eqs. (7.80) and (7.34),

$$\langle \mathscr{P} \rangle = \frac{2\omega\langle W_m \rangle}{k^2}\,\mathbf{k} + \mathrm{Re}\left[\frac{\mathbf{k} \cdot \mathbf{E}_0}{2k^2}\,(\mathbf{k} \times \mathbf{H}_0^*)\right] \tag{7.85}$$

where

$$\langle W_m \rangle = \frac{\mu_0}{4}\,\mathbf{H}_0 \cdot \mathbf{H}_0^* \tag{7.86}$$

is the time-averaged magnetic energy density. In isotropic media, we have $\mathbf{k} \cdot \mathbf{E}_0 = 0$ and thus $\langle \mathscr{P} \rangle$ is in the direction of $\mathbf{k}$. However, in the magnetoplasma, Eq. (7.85) shows that the two directions are, in general, different.

Let us consider the principal waves in the magnetoplasma. The right- and left-handed circularly polarized waves (7.73) and (7.71) propagating along $\mathbf{B}_c$ and the ordinary wave (7.69) propagating across $\mathbf{B}_c$ have their electric fields perpendicular to $\mathbf{k}$, that is, $\mathbf{k} \cdot \mathbf{E}_0 = 0$. However, for the extraordinary wave propagating across $\mathbf{B}_c$, $(\mathbf{k} \cdot \mathbf{E}_0)$ is not zero. Substituting Eq. (7.58) into the last term of Eq. (7.85), we see that the quantity inside the square bracket is purely imaginary. Thus, for all the principal waves, we have

$$\langle \mathscr{P} \rangle = \frac{2\omega \langle W_m \rangle}{k^2} \mathbf{k} \tag{7.87}$$

Similar to the isotropic media, the time-averaged Poynting vector is in the direction of $\mathbf{k}$, or

$$\langle \mathscr{P} \rangle = 2\langle W_m \rangle \mathbf{v}_p \tag{7.88}$$

where $\mathbf{v}_p = \omega\mathbf{k}/k^2$ is the phase velocity vector of the wave.

However, according to Eq. (3.128), the velocity of energy transport in a magnetoplasma is

$$\mathbf{v}_E = \frac{\langle \mathscr{P} \rangle}{\langle W \rangle} = \mathbf{v}_g \tag{7.89}$$

where $\mathbf{v}_g$ is the group velocity vector defined in Eq. (3.117). Thus, in the general case, the time-averaged Poynting vector, a product of the time-averaged energy density and the group velocity, is not in the direction of wave vector $\mathbf{k}$ but in the direction of the group velocity. Since the group velocity vector $\mathbf{v}_g = \partial\omega/\partial\mathbf{k}$ is perpendicular to the wave vector surface, it follows that $\mathbf{k}$ and $\mathbf{v}_g$ are parallel only when the magnitude k is independent of direction. In other words, this holds when the wave vector surface is a sphere as in the cases of principal waves and the waves in isotropic media.

7.7 WAVES IN AN ISOTROPIC PLASMA

In the absence of an externally applied constant magnetic field $\mathbf{B}_c$, the dielectric tensor (7.8) of a plasma reduces to a scalar as given by Eq. (7.15):

$$\bar{\boldsymbol{\varepsilon}} = \varepsilon_{\parallel} \bar{\mathbf{I}} \tag{7.90}$$

where

$$\varepsilon_{\parallel} = 1 - \frac{\omega_p^2}{\omega^2} \tag{7.91}$$

In this case the field vectors are all perpendicular to the wave vector $\mathbf{k}$, a TEM wave, and the homogeneous equation (7.35) for $\mathbf{E}_0$ becomes

$$(k_0^2 \varepsilon_{\parallel} - k^2)\mathbf{E}_0 = \mathbf{0} \tag{7.92}$$

By setting the coefficient of $\mathbf{E}_0$ to zero, we obtain the condition for the existence of nonzero $\mathbf{E}_0$, i.e., the dispersion equation

$$\omega^2 = \omega_p^2 + c^2k^2 \tag{7.93}$$

and thus the wave number

$$k = \begin{cases} \dfrac{\omega}{c}\left(1 - \dfrac{\omega_p^2}{\omega^2}\right)^{1/2} & \text{for } \omega > \omega_p \\ 0 & \text{for } \omega = \omega_p \\ i\dfrac{\omega}{c}\left(\dfrac{\omega_p^2}{\omega^2} - 1\right)^{1/2} & \text{for } \omega < \omega_p \end{cases} \tag{7.94}$$

The above equation shows that the propagation of a wave in a plasma depends on the source frequency—a marked difference between a dielectric and a plasma. When $\omega > \omega_p$, k is real, and thus the wave travels without attenuation in the plasma. On the other hand, when $\omega < \omega_p$, k is purely imaginary and the wave is evanescent, carrying no power. At $\omega = \omega_p$ the wave is cut off, the magnetic field is zero, and the electric field is parallel to $\mathbf{k}$. Hence, at cutoff a TEM wave can not exist. However, a longitudinal wave, sometimes called *plasma oscillation,* can exist. Figure 7.6 shows a plot of ω vs. k.

In the propagation region, $\omega > \omega_p$, the phase velocity of a wave in a plasma, is greater than that of light in free space

$$\mathbf{v}_p = \frac{\omega}{k}\hat{\mathbf{k}} = \frac{c}{\sqrt{1 - \omega_p^2/\omega^2}}\hat{\mathbf{k}} \tag{7.95}$$

However, the group velocity of the wave cannot exceed the velocity of light.

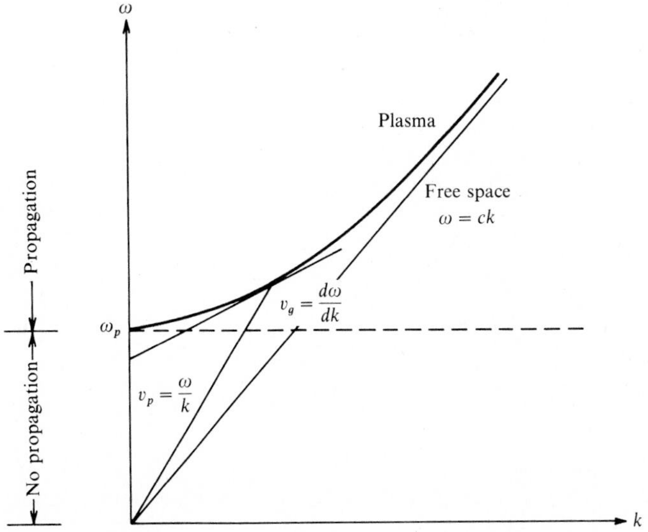

Figure 7.6 Dispersion curve for TEM wave in an isotropic plasma.

From Eq. (7.93), we find

$$\mathbf{v}_g = \frac{\partial \omega}{\partial \mathbf{k}} = \frac{c^2 k}{\omega} \hat{\mathbf{k}} = c \left(1 - \frac{\omega_p^2}{\omega^2}\right)^{1/2} \hat{\mathbf{k}} \tag{7.96}$$

and the magnitudes of the two velocities are related by

$$v_g v_p = c^2 \tag{7.97}$$

so that v_g is less than c whenever v_p is greater than c as shown in Fig. 7.7. The time-averaged Poynting vector is

$$\langle \mathscr{P} \rangle = \frac{1}{2} \sqrt{\frac{\varepsilon_0 \varepsilon_\parallel}{\mu_0}} (\mathbf{E}_0 \cdot \mathbf{E}_0^*) \hat{\mathbf{k}} \tag{7.98}$$

and the time-averaged energy density (7.82) in this case reduces to

$$\langle W \rangle = \frac{\varepsilon_0}{4} \frac{\partial(\omega^2 \varepsilon_\parallel)}{\omega \, \partial \omega} (\mathbf{E}_0 \cdot \mathbf{E}_0^*) \tag{7.99}$$

Therefore, the velocity of energy transport assumes the form

$$\mathbf{v}_E = \frac{\langle \mathscr{P} \rangle}{\langle W \rangle} = \frac{2c\sqrt{\varepsilon_\parallel}}{\partial(\omega^2 \varepsilon_\parallel)/\omega \, \partial \omega} \hat{\mathbf{k}} \tag{7.100}$$

Substituting Eq. (7.91) into Eq. (7.100), we find that

$$\mathbf{v}_E = c\sqrt{\varepsilon_\parallel} \, \hat{\mathbf{k}} \tag{7.101}$$

which, as expected, is identical to the group velocity given in Eq. (7.96).

It is interesting to note that the time-averaged electric energy density given

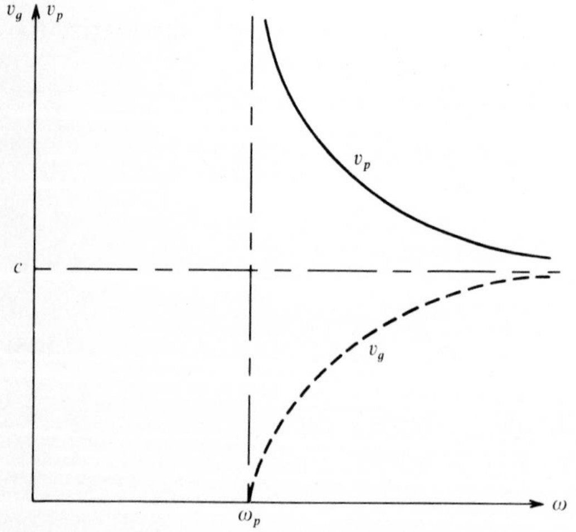

Figure 7.7 Phase and group velocities for TEM wave in an isotropic plasma.

by the first term on the right-hand side of Eq. (7.81) may be written as

$$\langle W_e \rangle = \frac{\varepsilon_0}{4} \frac{\partial(\omega\varepsilon_{\parallel})}{\partial\omega} \mathbf{E}_0 \cdot \mathbf{E}_0^*$$

$$= \frac{\varepsilon_0}{4} \mathbf{E}_0 \cdot \mathbf{E}_0^* + \frac{\varepsilon_0}{4} \frac{\omega_p^2}{\omega^2} \mathbf{E}_0 \cdot \mathbf{E}_0^* \tag{7.102}$$

Evidently, the first term corresponds to the electric energy density in free space. To interpret the second term, we note from the equation of motion (7.2) of an electron in the absence of a constant magnetic field that

$$-i\omega m\mathbf{v} = e\mathbf{E}$$

The time-averaged kinetic energy density is thus given by

$$\langle W_k \rangle = \tfrac{1}{4} N_0 m \mathbf{v} \cdot \mathbf{v}^* = \frac{\varepsilon_0}{4} \frac{\omega_p^2}{\omega^2} \mathbf{E}_0 \cdot \mathbf{E}_0^* \tag{7.103}$$

In other words, the time-averaged electric energy density is the sum of the electric energy in free space and the kinetic energy of the electrons.

7.8 WAVES IN A UNIAXIAL PLASMA

In a very strong constant magnetic field, plasma may be considered as a medium characterized by the uniaxial dielectric tensor

$$\bar{\boldsymbol{\varepsilon}} = \varepsilon_\perp \bar{\mathbf{I}} + (\varepsilon_\parallel - \varepsilon_\perp)\hat{\mathbf{b}}_c \hat{\mathbf{b}}_c \tag{7.104}$$

where

$$\varepsilon_\perp = 1$$

$$\varepsilon_\parallel = 1 - \frac{\omega_p^2}{\omega^2}$$

Following the derivations in Secs. 6.1, 6.2, and 6.3, we can now summarize the results for the ordinary and extraordinary waves as follows:

Ordinary wave:

$$k_+ = \frac{\omega}{c}\sqrt{\varepsilon_\perp}$$

$$\mathbf{e}_+ = \mathbf{k}_+ \times \hat{\mathbf{b}}_c$$

$$\mathbf{h}_+ = \frac{1}{\omega\mu_0}[\mathbf{k}_+ \times (\mathbf{k}_+ \times \hat{\mathbf{b}}_c)] \tag{7.105}$$

$$\mathbf{d}_+ = \varepsilon_0 \varepsilon_\perp (\mathbf{k}_+ \times \hat{\mathbf{b}}_c)$$

$$\langle \mathscr{P}_+ \rangle = \frac{(\mathbf{k}_+ \times \hat{\mathbf{b}}_c)^2}{2\omega\mu_0} \mathbf{k}_+$$

Since the medium now is dispersive, the time-averaged energy density is given by

$$\langle W_+ \rangle = \frac{\varepsilon_0}{4}\, \mathbf{e}_+^* \cdot \frac{\partial(\omega^2 \bar{\boldsymbol{\varepsilon}})}{\omega\, \partial\omega} \cdot \mathbf{e}_+$$

$$= \frac{\varepsilon_0}{4} \frac{\partial(\omega^2 \varepsilon_\perp)}{\omega\, \partial\omega} (\mathbf{k}_+ \times \hat{\mathbf{b}}_c)^2 \tag{7.106}$$

and the velocity of energy transport is

$$\mathbf{v}_{E+} = \frac{\langle \mathscr{P}_+ \rangle}{\langle W_+ \rangle} = \frac{2c\sqrt{\varepsilon_\perp}}{\partial(\omega^2 \varepsilon_\perp)/\omega\, \partial\omega}\, \hat{\mathbf{k}} \tag{7.107}$$

From the dispersion equation $c^2k^2 = \omega^2 \varepsilon_\perp$, we easily obtain the group velocity

$$\mathbf{v}_{g+} = \frac{\partial\omega}{\partial\mathbf{k}} = \frac{2c\sqrt{\varepsilon_\perp}}{\partial(\omega^2\varepsilon_\perp)/\omega\, \partial\omega}\, \hat{\mathbf{k}} \tag{7.108}$$

which shows that $\mathbf{v}_{g+} = \mathbf{v}_{E+}$, and that $\mathbf{v}_{g+}$ is also equal to the phase velocity of the wave.

Extraordinary wave:

$$\begin{aligned}
k_- &= k_0 \sqrt{\frac{\varepsilon_\perp \varepsilon_\parallel}{\hat{\mathbf{k}} \cdot \bar{\boldsymbol{\varepsilon}} \cdot \hat{\mathbf{k}}}} \\
\mathbf{e}_- &= k_0^2 \varepsilon_\perp \hat{\mathbf{b}}_c - (\mathbf{k}_- \cdot \hat{\mathbf{b}}_c)\mathbf{k}_- \\
\mathbf{h}_- &= \omega\varepsilon_0 \varepsilon_\perp (\mathbf{k}_- \times \hat{\mathbf{b}}_c) \\
\mathbf{d}_- &= -\varepsilon_0 \varepsilon_\perp [\mathbf{k}_- \times (\mathbf{k}_- \times \hat{\mathbf{b}}_c)] \\
\langle \mathscr{P}_- \rangle &= \frac{\omega\varepsilon_0 \varepsilon_\perp (\mathbf{k}_- \times \hat{\mathbf{b}}_c)^2}{2\varepsilon_\parallel} (\bar{\boldsymbol{\varepsilon}} \cdot \mathbf{k}_-)
\end{aligned} \tag{7.109}$$

The time-averaged energy density in this case is

$$\begin{aligned}
\langle W_- \rangle &= \frac{\varepsilon_0}{4}\, \mathbf{e}_-^* \cdot \frac{\partial(\omega^2 \bar{\boldsymbol{\varepsilon}})}{\omega\, \partial\omega} \cdot \mathbf{e}_- \\
&= \frac{\varepsilon_0 \varepsilon_\perp (\mathbf{k}_- \times \hat{\mathbf{b}}_c)^2}{4\varepsilon_\parallel} \left[\frac{\varepsilon_\parallel}{\varepsilon_\perp} \frac{\partial(\omega^2\varepsilon_\perp)}{\omega\, \partial\omega} (\mathbf{k}_- \cdot \hat{\mathbf{b}}_c)^2 \right. \\
&\quad \left. + \frac{\varepsilon_\perp}{\varepsilon_\parallel} \frac{\partial(\omega^2 \varepsilon_\parallel)}{\omega\, \partial\omega} (\mathbf{k}_- \times \hat{\mathbf{b}}_c)^2 \right]
\end{aligned} \tag{7.110}$$

Thus the velocity of energy transport for the extraordinary wave is

$$\mathbf{v}_{E-} = \frac{\langle \mathscr{P}_- \rangle}{\langle W_- \rangle} = \frac{2(\bar{\boldsymbol{\varepsilon}} \cdot \mathbf{k}_-)}{\left[\dfrac{\varepsilon_\perp}{\varepsilon_\parallel} \dfrac{\partial(\omega^2 \varepsilon_\parallel)}{\omega^2\, \partial\omega} (\mathbf{k}_- \times \hat{\mathbf{b}}_c)^2 + \dfrac{\varepsilon_\parallel}{\varepsilon_\perp} \dfrac{\partial(\omega^2\varepsilon_\perp)}{\omega^2\, \partial\omega} (\mathbf{k}_- \times \hat{\mathbf{b}}_c)^2 \right]} \tag{7.111}$$

From the dispersion equation for the extraordinary wave

$$F(\mathbf{k}, \omega) = \mathbf{k}_- \cdot \bar{\boldsymbol{\varepsilon}} \cdot \mathbf{k}_- - \frac{\omega^2}{c^2} \varepsilon_\perp \varepsilon_\parallel = 0$$

we find

$$\frac{\partial F}{\partial \mathbf{k}} = 2(\bar{\boldsymbol{\varepsilon}} \cdot \mathbf{k}_-)$$

and

$$\frac{\partial F}{\partial \omega} = -\frac{1}{\omega^2}\left[\frac{\varepsilon_\perp}{\varepsilon_\parallel}\frac{\partial(\omega^2\varepsilon_\parallel)}{\partial\omega}(\mathbf{k}_- \times \hat{\mathbf{b}}_c)^2 + \frac{\varepsilon_\parallel}{\varepsilon_\perp}\frac{\partial(\omega^2\varepsilon_\perp)}{\partial\omega}(\mathbf{k}_- \cdot \hat{\mathbf{b}}_c)^2\right]$$

Thus we see again that the group velocity vector

$$\mathbf{v}_{g-} = \frac{\partial \omega}{\partial \mathbf{k}} = -\frac{\partial F/\partial \mathbf{k}}{\partial F/\partial \omega}$$

is equal to the velocity of energy transport.

7.9 BOOKER QUARTIC FOR A MAGNETOPLASMA

We shall now consider the problem of plane wave transmission and reflection from a sharply bounded, homogeneous plasma half-space. The geometry of the problem is shown in Fig. 7.8. In this section, we shall determine the wave vectors

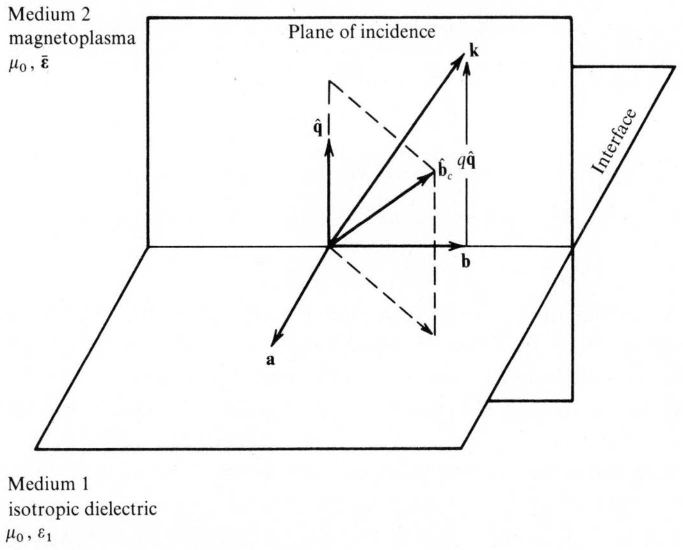

Figure 7.8 The geometry used for the study of transmission and reflection of waves from a plasma half-space.

in the magnetoplasma and leave the amplitude relations to the next section. According to Sec. 3.4, the boundary conditions at the interface of two media require that the wave vectors have the form

$$\mathbf{k} = \mathbf{b} + q\hat{\mathbf{q}} \tag{7.112}$$

On the other hand, for waves to exist in an anisotropic medium characterized by the dielectric tensor $\bar{\varepsilon}$, the wave vector must also satisfy the dispersion equation (5.28) or (7.40). Combination of Eq. (7.112) and the dispersion equation yields the Booker quartic equation (5.153) for the determination of q and thus the wave vector $\mathbf{k}$. If the dielectric tensor is hermitian, that is, $\tilde{\bar{\varepsilon}} = \bar{\varepsilon}^*$, the Booker quartic (5.153) becomes

$$Aq^4 + Bq^3 + Cq^2 + Dq + F = 0 \tag{7.113}$$

where

$$\begin{aligned}
A &= (\hat{\mathbf{q}} \cdot \bar{\varepsilon} \cdot \hat{\mathbf{q}}) \\
B &= 2\mathbf{b} \cdot (\text{Re}\ \bar{\varepsilon}) \cdot \hat{\mathbf{q}} \\
C &= \mathbf{b} \cdot \bar{\varepsilon} \cdot \mathbf{b} + \mathbf{a}^2(\hat{\mathbf{q}} \cdot \bar{\varepsilon} \cdot \hat{\mathbf{q}}) + k_0^2 \hat{\mathbf{q}} \cdot [\text{adj}\ \bar{\varepsilon} - (\text{adj}\ \bar{\varepsilon})_t \bar{\mathbf{I}}] \cdot \hat{\mathbf{q}} \\
D &= 2\mathbf{a}^2[\mathbf{b} \cdot (\text{Re}\ \bar{\varepsilon}) \cdot \hat{\mathbf{q}}] + 2k_0^2 \mathbf{b} \cdot [\text{Re}\ (\text{adj}\ \bar{\varepsilon})] \cdot \hat{\mathbf{q}} \\
F &= \mathbf{a}^2(\mathbf{b} \cdot \bar{\varepsilon} \cdot \mathbf{b}) + k_0^2 \mathbf{b} \cdot [\text{adj}\ \bar{\varepsilon} - (\text{adj}\ \bar{\varepsilon})_t \bar{\mathbf{I}}] \cdot \mathbf{b} + k_0^4 |\bar{\varepsilon}|
\end{aligned} \tag{7.114}$$

In the case of magnetoplasma, we substitute Eqs. (7.8) and (7.18) into Eq. (7.114) and obtain

$$\begin{aligned}
A &= \varepsilon_\perp + (\varepsilon_\parallel - \varepsilon_\perp)(\hat{\mathbf{q}} \cdot \hat{\mathbf{b}}_c)^2 \\
B &= 2(\varepsilon_\parallel - \varepsilon_\perp)(\mathbf{b} \cdot \hat{\mathbf{b}}_c)(\hat{\mathbf{q}} \cdot \hat{\mathbf{b}}_c) \\
C &= 2\varepsilon_\perp \mathbf{a}^2 + (\varepsilon_\times^2 - \varepsilon_\perp^2 - \varepsilon_\perp \varepsilon_\parallel)k_0^2 \\
&\quad + (\varepsilon_\parallel - \varepsilon_\perp)[(\mathbf{b} \cdot \hat{\mathbf{b}}_c)^2 + (\mathbf{a}^2 - k_0^2)(\hat{\mathbf{q}} \cdot \hat{\mathbf{b}}_c)^2] \\
D &= 2(\varepsilon_\parallel - \varepsilon_\perp)(\mathbf{a}^2 - k_0^2)(\mathbf{b} \cdot \hat{\mathbf{b}}_c)(\hat{\mathbf{q}} \cdot \hat{\mathbf{b}}_c) \\
F &= (\mathbf{a}^2 - k_0^2 \varepsilon_\parallel)[\varepsilon_\perp \mathbf{a}^2 - k_0^2(\varepsilon_\perp^2 - \varepsilon_\times^2)] \\
&\quad + (\varepsilon_\parallel - \varepsilon_\perp)(\mathbf{a}^2 - k_0^2)(\mathbf{b} \cdot \hat{\mathbf{b}}_c)^2
\end{aligned} \tag{7.115}$$

which shows that the coefficients of the Booker quartic depend on the orientation of the constant magnetic field with respect to the three mutually perpendicular vectors $\hat{\mathbf{q}}$, $\mathbf{a}$, and $\mathbf{b}$. The quartic equation (7.113) gives four values of q. For transmission and reflection from a sharply bounded plasma half-space, the two values of q_+ and q_- which correspond to two waves propagating into plasma must be selected. From them the wave vectors $\mathbf{k}_+$ and $\mathbf{k}_-$ may in turn be found from Eq. (7.112).

The quartic equation (7.113) reduces to a quadratic equation in q^2 when the coefficients B and D vanish. An examination of B and D in Eq. (7.115) gives the following four cases. The first is when the constant magnetic field is absent so that $\varepsilon_\times = 0$ and $\varepsilon_\perp = \varepsilon_\parallel$. In this case, the solution of Eq. (7.113) becomes

$$q^2 = k_0^2 \varepsilon_\parallel - \mathbf{a}^2 \tag{7.116}$$

which is just the result for an isotropic plasma. The second case is for normal incidence; thus $\mathbf{a} = \mathbf{b} = \mathbf{0}$ and $\mathbf{k} = q\hat{\mathbf{q}}$. The solutions of Eq. (7.113) are then simply the square of Eqs. (7.49) and (7.50).

The third case is when $\hat{\mathbf{b}}_c \cdot \mathbf{b} = 0$, which means that the constant magnetic field lies on a plane perpendicular to $\mathbf{b}$ (see Fig. 7.9). The solutions of the quadratic equation may, in general, be written in the form

$$q^2 = k_0^2 - \mathbf{a}^2 - \frac{2[(k_0^2 - \mathbf{a}^2)^2 A + (k_0^2 - \mathbf{a}^2)C + F]}{C + 2(k_0^2 - \mathbf{a}^2)A \pm \sqrt{C^2 - 4AF}} \tag{7.117}$$

where, in this case,

$$(k_0^2 - \mathbf{a}^2)^2 A + (k_0^2 - \mathbf{a}^2)C + F = k_0^4(1 - \varepsilon_\parallel)(\varepsilon_\times^2 - \varepsilon_\perp^2 + \varepsilon_\perp)$$

$$C + 2(k_0^2 - \mathbf{a}^2)A = 2k_0^2(\varepsilon_\times^2 - \varepsilon_\perp^2 + \varepsilon_\perp) - (\varepsilon_\parallel - \varepsilon_\perp)[k_0^2 + (\mathbf{a}^2 - k_0^2)(\hat{\mathbf{q}} \cdot \hat{\mathbf{b}}_c)^2]$$

and

$$C^2 - 4AF = (\varepsilon_\parallel - \varepsilon_\perp)^2[k_0^2 + (\mathbf{a}^2 - k_0^2)(\hat{\mathbf{q}} \cdot \hat{\mathbf{b}}_c)^2]^2 + 4k_0^2(\varepsilon_\parallel - \varepsilon_\perp)(k_0^2\varepsilon_\parallel - \mathbf{a}^2)(\varepsilon_\times^2 - \varepsilon_\perp^2 + \varepsilon_\perp)(\hat{\mathbf{q}} \cdot \hat{\mathbf{b}}_c)^2 \tag{7.118}$$

Using Eq. (7.9), we find

$$1 - \varepsilon_\parallel = X_0$$

$$\frac{\varepsilon_\parallel - \varepsilon_\perp}{\varepsilon_\times^2 - \varepsilon_\perp^2 + \varepsilon_\perp} = \frac{Y_0^2}{1 - X_0} \tag{7.119}$$

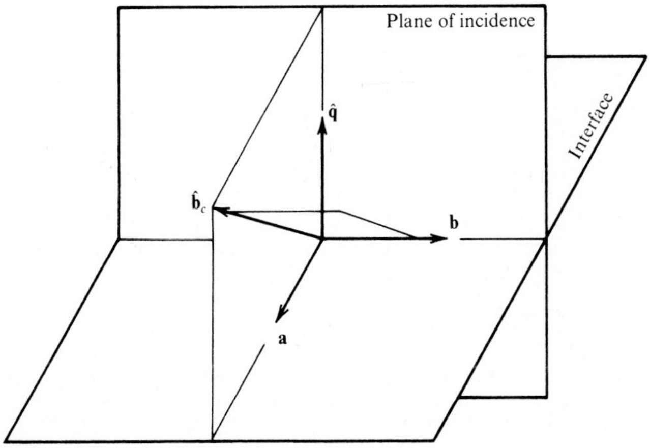

Figure 7.9 The constant magnetic field lies on a plane which is perpendicular to the plane of incidence and the interface.

and
$$\frac{\varepsilon_{\parallel}(\varepsilon_{\parallel}-\varepsilon_{\perp})}{\varepsilon_{\times}^2-\varepsilon_{\perp}^2+\varepsilon_{\perp}} = Y_0^2$$

Hence Eq. (7.117) may be expressed as

$$q^2 = k_0^2 - \mathbf{a}^2 - \left[k_0^2 X_0 \Big/ \left(1 - \frac{Y_0^2[k_0^2 + (\mathbf{a}^2 - k_0^2)(\hat{\mathbf{q}}\cdot\hat{\mathbf{b}}_c)^2]}{2k_0^2(1-X_0)} \pm \sqrt{\frac{Y_0^4[k_0^2 + (\mathbf{a}^2 - k_0^2)(\hat{\mathbf{q}}\cdot\hat{\mathbf{b}}_c)^2]^2}{4k_0^4(1-X_0)^2} + \frac{Y_0^2(k_0^2\varepsilon_{\parallel} - \mathbf{a}^2)(\hat{\mathbf{q}}\cdot\hat{\mathbf{b}}_c)^2}{k_0^2(1-X_0)}}\right)\right] \tag{7.120}$$

Finally, the fourth case is when $\hat{\mathbf{b}}_c \cdot \hat{\mathbf{q}} = 0$, which means that the constant magnetic field is parallel to the interface (see Fig. 7.10). The solutions of the quadratic equation still take the form of Eq. (7.117), but in this case,

$$(k_0^2 - \mathbf{a}^2)^2 A + (k_0^2 - \mathbf{a}^2)C + F = k_0^4(1-\varepsilon_{\parallel})(\varepsilon_{\times}^2 - \varepsilon_{\perp}^2 + \varepsilon_{\perp})$$
$$C + 2(k_0^2 - \mathbf{a}^2)A = 2k_0^2(\varepsilon_{\times}^2 - \varepsilon_{\perp}^2 + \varepsilon_{\perp}) - (\varepsilon_{\parallel} - \varepsilon_{\perp})[k_0^2 - (\mathbf{b}\cdot\hat{\mathbf{b}}_c)^2]$$

and

$$C^2 - 4AF = (\varepsilon_{\parallel} - \varepsilon_{\perp})^2[k_0^2 - (\mathbf{b}\cdot\hat{\mathbf{b}}_c)^2]^2 + 4k_0^2(\varepsilon_{\parallel} - \varepsilon_{\perp})(\varepsilon_{\times}^2 - \varepsilon_{\perp}^2 + \varepsilon_{\perp})(\mathbf{b}\cdot\hat{\mathbf{b}}_c)^2 \tag{7.121}$$

Using Eq. (7.119), we may write the solutions as

$$q^2 = k_0^2 - \mathbf{a}^2 - \left[k_0^2 X_0 \Big/ \left(1 - \frac{Y_0^2[k_0^2 - (\mathbf{b}\cdot\hat{\mathbf{b}}_c)^2]}{2k_0^2(1-X_0)} \pm \sqrt{\frac{Y_0^4[k_0^2 - (\mathbf{b}\cdot\hat{\mathbf{b}}_c)^2]^2}{4k_0^4(1-X_0)^2} + \frac{Y_0^2(\mathbf{b}\cdot\hat{\mathbf{b}}_c)^2}{k_0^2(1-X_0)}}\right)\right] \tag{7.122}$$

If in addition $\hat{\mathbf{b}}_c \cdot \mathbf{b} = 0$, that is, if the constant magnetic field is parallel to the

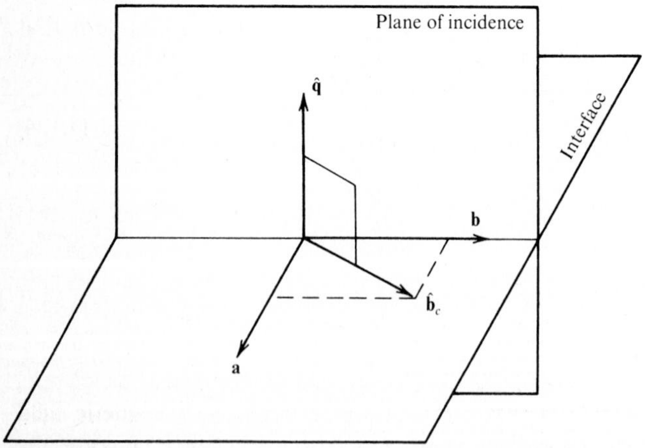

Figure 7.10 The constant magnetic field is parallel to the interface.

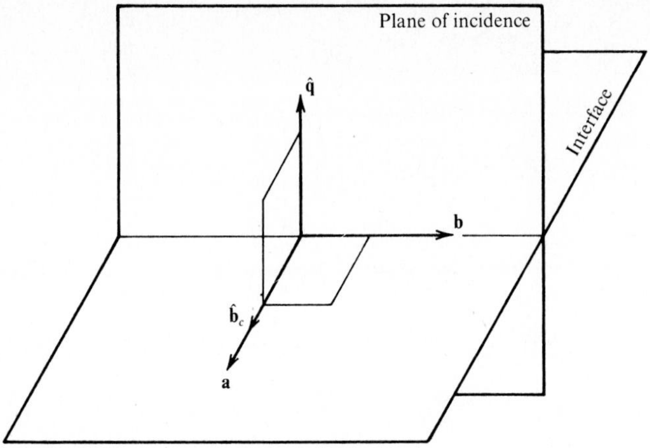

Figure 7.11 The constant magnetic field is parallel to the interface but perpendicular to the plane of incidence.

interface but perpendicular to the plane of incidence (see Fig. 7.11), the solutions (7.122) reduce to

$$q_+^2 = k_0^2(1 - X_0) - \mathbf{a}^2$$

and

$$q_-^2 = k_0^2 - \mathbf{a}^2 - \frac{k_0^2 X_0}{1 - Y_0^2/(1 - X_0)} \tag{7.123}$$

A comparison of Eq. (7.123) with Eqs. (7.47) and (7.48) shows that the first corresponds to the ordinary wave and the second to the extraordinary wave. In this special case, the wave is propagating normal to the constant magnetic field.

The above four cases are the only ones for which the coefficients B and D in Eq. (7.115) are zero, so that the quartic equation (7.113) is a quadratic in q^2, and the roots of the quartic occur in pairs with equal values but opposite signs.

The condition that one root of the quartic equation (7.113) is infinite is that the coefficient $A = 0$. This requires that

$$X_0 = \frac{1 - Y_0^2}{1 - Y_0^2(\hat{\mathbf{q}} \cdot \hat{\mathbf{b}}_c)^2} \tag{7.124}$$

This condition is independent of the angle of incidence.

7.10 REFLECTION AND TRANSMISSION COEFFICIENT MATRICES

Having determined the wave vectors of the transmitted waves from Eqs. (7.112), (7.113), and (7.115) and thus the directions of the electric and magnetic field vectors from Eqs. (7.55) and (7.56) in a magnetoplasma, we can now solve the boundary-value problem of an isotropic medium–magnetoplasma interface as

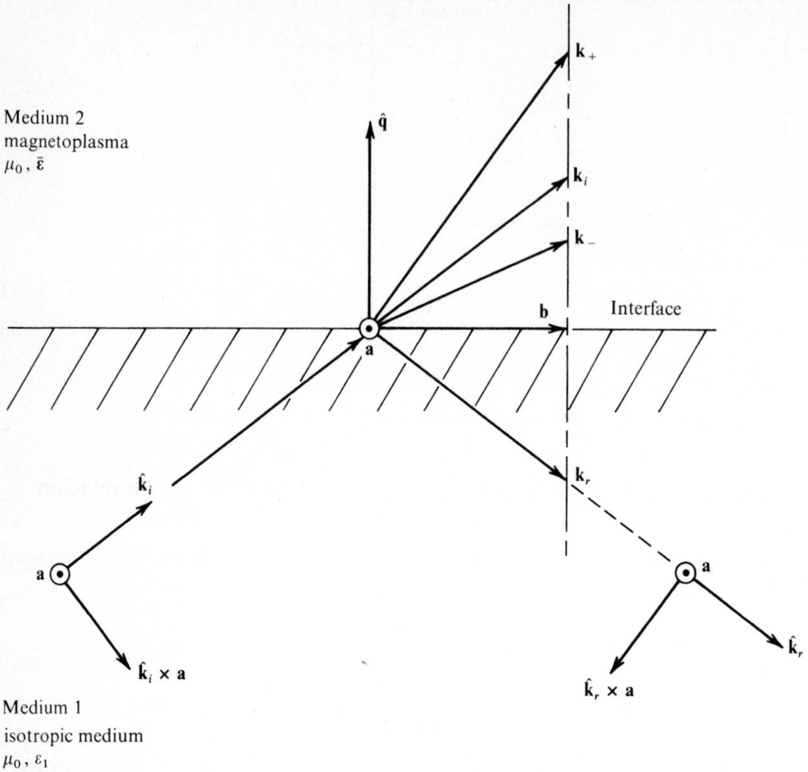

Figure 7.12 Decomposition of field vectors in an isotropic medium into parallel and perpendicular polarizations.

shown in Fig. 7.12. We recall that in deriving the reflection and transmission coefficient matrices in Sec. 5.10, no assumption was made with regard to the realness of the dielectric tensor $\bar{\varepsilon}$. Therefore, the results obtained in that section are valid also for magnetoplasma. For easy reference, we summarize them below:

Incident wave in isotropic medium 1:

$$\mathbf{E}_{0i} = A_\perp \mathbf{a} + A_\parallel (\hat{\mathbf{k}}_i \times \mathbf{a}) \tag{7.125}$$

Reflected wave in isotropic medium 1:

$$\mathbf{E}_{0r} = B_\perp \mathbf{a} + B_\parallel (\hat{\mathbf{k}}_r \times \mathbf{a}) \tag{7.126}$$

Transmitted waves in magnetoplasma 2:

$$\begin{aligned} \mathbf{E}_{0+} &= C_+ \mathbf{e}_+ \qquad \mathbf{h}_+ = \frac{\mathbf{k}_+ \times \mathbf{e}_+}{\omega\mu_0} \\ \mathbf{E}_{0-} &= C_- \mathbf{e}_- \qquad \mathbf{h}_- = \frac{\mathbf{k}_- \times \mathbf{e}_-}{\omega\mu_0} \end{aligned} \tag{7.127}$$

The 2 × 2 reflection coefficient matrix $\bar{\Gamma}$ is defined by the relation

$$\begin{bmatrix} B_\perp \\ B_\| \end{bmatrix} = \begin{bmatrix} \Gamma_{11} & \Gamma_{12} \\ \Gamma_{21} & \Gamma_{22} \end{bmatrix} \begin{bmatrix} A_\perp \\ A_\| \end{bmatrix} \tag{7.128}$$

where

$$\begin{aligned} \Gamma_{11} &= \frac{(\mathbf{a}\cdot\mathbf{F}_+)(\mathbf{b}\cdot\mathbf{N}_-) - (\mathbf{a}\cdot\mathbf{F}_-)(\mathbf{b}\cdot\mathbf{N}_+)}{\Delta} \\ \Gamma_{12} &= -\frac{k_i[(\mathbf{a}\cdot\mathbf{F}_+)(\mathbf{a}\cdot\mathbf{N}_-) - (\mathbf{a}\cdot\mathbf{F}_-)(\mathbf{a}\cdot\mathbf{N}_+)]}{q_i\,\Delta} \\ \Gamma_{21} &= -\frac{q_i[(\mathbf{b}\cdot\mathbf{F}_+)(\mathbf{b}\cdot\mathbf{N}_-) - (\mathbf{b}\cdot\mathbf{F}_-)(\mathbf{b}\cdot\mathbf{N}_+)]}{k_i\,\Delta} \\ \Gamma_{22} &= \frac{(\mathbf{b}\cdot\mathbf{F}_+)(\mathbf{a}\cdot\mathbf{N}_-) - (\mathbf{b}\cdot\mathbf{F}_-)(\mathbf{a}\cdot\mathbf{N}_+)}{\Delta} \end{aligned} \tag{7.129}$$

Similarly, the 2 × 2 transmission coefficient matrix $\bar{\mathbf{T}}$ is defined by the relation

$$\begin{bmatrix} C_+ \\ C_- \end{bmatrix} = \begin{bmatrix} T_{11} & T_{12} \\ T_{21} & T_{22} \end{bmatrix} \begin{bmatrix} A_\perp \\ A_\| \end{bmatrix} \tag{7.130}$$

where

$$\begin{aligned} T_{11} &= \frac{2q_i^2\,\mathbf{a}^2(\mathbf{b}\cdot\mathbf{N}_-)}{\Delta} \\ T_{12} &= -\frac{2q_i\,k_i\,\mathbf{a}^2(\mathbf{a}\cdot\mathbf{N}_-)}{\Delta} \\ T_{21} &= -\frac{2q_i^2\,\mathbf{a}^2(\mathbf{b}\cdot\mathbf{N}_+)}{\Delta} \\ T_{22} &= \frac{2q_i\,k_i\,\mathbf{a}^2(\mathbf{a}\cdot\mathbf{N}_+)}{\Delta} \end{aligned} \tag{7.131}$$

and

$$\Delta = (\mathbf{a}\cdot\mathbf{N}_+)(\mathbf{b}\cdot\mathbf{N}_-) - (\mathbf{a}\cdot\mathbf{N}_-)(\mathbf{b}\cdot\mathbf{N}_+) \tag{7.132}$$

In Eqs. (7.129), (7.131), and (7.132) we have defined

$$\begin{aligned} \mathbf{a}\cdot\mathbf{N}_\pm &= q_i(q_i + q_\pm)(\mathbf{a}\cdot\mathbf{e}_\pm) \\ \mathbf{a}\cdot\mathbf{F}_\pm &= q_i(q_i - q_\pm)(\mathbf{a}\cdot\mathbf{e}_\pm) \\ \mathbf{b}\cdot\mathbf{N}_\pm &= k_i^2(\mathbf{b}\cdot\mathbf{e}_\pm) - \omega\mu_0\,q_i(\mathbf{a}\cdot\mathbf{h}_\pm) \\ \mathbf{b}\cdot\mathbf{F}_\pm &= k_i^2(\mathbf{b}\cdot\mathbf{e}_\pm) + \omega\mu_0\,q_i(\mathbf{a}\cdot\mathbf{h}_\pm) \end{aligned} \tag{7.133}$$

For a magnetoplasma $\mathbf{e}_+$ and $\mathbf{h}_+$ are given by Eqs. (7.55) and (7.56) respectively. Similarly, expressions for $\mathbf{e}_-$ and $\mathbf{h}_-$ are obtained by replacing the subscript + by − in Eqs. (7.55) and (7.56).

As an example, let us consider the case where the constant magnetic field is parallel to the interface but perpendicular to the plane of incidence (see Fig. 7.11).

For the ordinary wave, according to Eqs. (7.123) and (7.69), we have

$$
\begin{aligned}
q_+ &= \sqrt{k_0^2\,\varepsilon_\parallel - \mathbf{a}^2} \\
k_+ &= k_0\,\sqrt{\varepsilon_\parallel} \\
\mathbf{k}_+ &= \mathbf{b} + q_+\,\hat{\mathbf{q}}
\end{aligned} \tag{7.134}
$$

and

$$
\begin{aligned}
\mathbf{e}_+ &= \omega\mu_0\,\hat{\mathbf{b}}_c \\
\mathbf{h}_+ &= \mathbf{k}_+ \times \hat{\mathbf{b}}_c
\end{aligned} \tag{7.135}
$$

and for the extraordinary wave, according to Eqs. (7.123) and (7.58), we have

$$
\begin{aligned}
q_- &= \sqrt{\frac{k_0^2(\varepsilon_\perp^2 - \varepsilon_\times^2)}{\varepsilon_\perp} - \mathbf{a}^2} \\
k_- &= k_0\,\sqrt{\frac{\varepsilon_\perp^2 - \varepsilon_\times^2}{\varepsilon_\perp}} \\
\mathbf{k}_- &= \mathbf{b} + q_-\,\hat{\mathbf{q}}
\end{aligned} \tag{7.136}
$$

and

$$
\begin{aligned}
\mathbf{e}_- &= \varepsilon_\perp(\mathbf{k}_- \times \hat{\mathbf{b}}_c) - i\varepsilon_\times\,\mathbf{k}_- \\
\mathbf{h}_- &= -\frac{\varepsilon_\perp\,k_-^2\,\hat{\mathbf{b}}_c}{\omega\mu_0}
\end{aligned} \tag{7.137}
$$

Substitution of Eqs. (7.135) and (7.137) into Eqs. (7.133) and (7.132) yields

$$
\begin{aligned}
\mathbf{a}\cdot\mathbf{N}_+ &= \omega\mu_0\,q_i(q_i + q_+)\,|\mathbf{a}| \\
\mathbf{b}\cdot\mathbf{N}_- &= (k_i^2\,q_- + k_-^2\,q_i)\varepsilon_\perp\,|\mathbf{a}| - i\varepsilon_\times\,k_i^2\,|\mathbf{a}|^2 \\
\mathbf{a}\cdot\mathbf{F}_+ &= \omega\mu_0\,q_i(q_i - q_+)\,|\mathbf{a}| \\
\mathbf{b}\cdot\mathbf{F}_- &= (k_i^2\,q_- - k_-^2\,q_i)\varepsilon_\perp\,|\mathbf{a}| - i\varepsilon_\times\,k_i^2\,|\mathbf{a}|^2 \\
\mathbf{a}\cdot\mathbf{N}_- &= \mathbf{b}\cdot\mathbf{N}_+ = \mathbf{a}\cdot\mathbf{F}_- = \mathbf{b}\cdot\mathbf{F}_+ = 0
\end{aligned} \tag{7.138}
$$

and

$$
\Delta = (\mathbf{a}\cdot\mathbf{N}_+)(\mathbf{b}\cdot\mathbf{N}_-)
$$

Finally, substituting Eq. (7.138) into Eqs. (7.129) and (7.131), we obtain the reflection coefficients

$$
\begin{aligned}
\Gamma_{11} &= \frac{q_i - q_+}{q_i + q_+} \\
\Gamma_{22} &= \frac{(k_-^2\,q_i - k_i^2\,q_-)\varepsilon_\perp + i\varepsilon_\times\,k_i^2\,|\mathbf{a}|}{(k_-^2\,q_i + k_i^2\,q_-)\varepsilon_\perp - i\varepsilon_\times\,k_i^2\,|\mathbf{a}|} \\
\Gamma_{12} &= \Gamma_{21} = 0
\end{aligned} \tag{7.139}
$$

and the transmission coefficients

$$\begin{aligned} T_{11} &= \frac{2q_i |\mathbf{a}|}{\omega\mu_0(q_i + q_+)} \\ T_{22} &= \frac{2q_i k_i |\mathbf{a}|}{(k_i^2 q_- + k_-^2 q_i)\varepsilon_\perp - i\varepsilon_\times k_i^2 |\mathbf{a}|} \\ T_{12} &= T_{21} = 0 \end{aligned} \tag{7.140}$$

where $k_i = k_0 \sqrt{\varepsilon_1}$, $q_i = \sqrt{k_0^2 \varepsilon_1 - \mathbf{a}^2}$, and $\mathbf{a} = \mathbf{k}_i \times \hat{\mathbf{q}}$. We note that in this special case, there is no coupling between the parallel and perpendicular polarizations as in the case of two isotropic media.

PROBLEMS

7.1 To take into account the collision losses in a plasma we add a damping term to the right-hand side of Eq. (7.1). Thus the equation of motion of an electron becomes

$$m \frac{d\mathscr{V}}{dt} = e(\mathscr{E} + \mathscr{V} \times \mathbf{B}_c) - m\nu_e \mathscr{V}$$

where ν_e is the effective collision frequency. Show that, in this case, the dielectric tensor takes the same form as Eq. (7.8), i.e.,

$$\bar{\boldsymbol{\varepsilon}}' = \varepsilon'_\perp(\bar{\mathbf{I}} - \hat{\mathbf{b}}_c \hat{\mathbf{b}}_c) + i\varepsilon'_\times(\hat{\mathbf{b}}_c \times \bar{\mathbf{I}}) + \varepsilon'_\| \hat{\mathbf{b}}_c \hat{\mathbf{b}}_c$$

but the components are given by

$$\begin{aligned} \varepsilon'_\perp &= 1 - \frac{\omega_p^2(\omega + i\nu_e)}{\omega[(\omega + i\nu_e)^2 - \omega_b^2]} \\ \varepsilon'_\times &= -\frac{\omega_b \omega_p^2}{\omega[(\omega + i\nu_e)^2 - \omega_b^2]} \\ \varepsilon'_\| &= 1 - \frac{\omega_p^2}{\omega(\omega + i\nu_e)} \end{aligned}$$

where the prime is used here to distinguish the lossy components from the lossless ones given in Eq. (7.9). Show also that

$$\tilde{\bar{\boldsymbol{\varepsilon}}}'(\mathbf{B}_c) = \bar{\boldsymbol{\varepsilon}}'(-\mathbf{B}_c)$$

as in the lossless case, but $\bar{\boldsymbol{\varepsilon}}'$ is no longer a hermitian matrix.

7.2 In the case of low-frequency electromagnetic waves that are passing through a magnetoplasma, the motion of the plasma ions must also be included in the analysis. Assuming that collisions may be neglected, show that the dielectric tensor is given by

$$\bar{\boldsymbol{\varepsilon}} = \varepsilon_\perp(\bar{\mathbf{I}} - \hat{\mathbf{b}}_c \hat{\mathbf{b}}_c) + i\varepsilon_\times(\hat{\mathbf{b}}_c \times \bar{\mathbf{I}}) + \varepsilon_\| \hat{\mathbf{b}}_c \hat{\mathbf{b}}_c$$

where

$$\varepsilon_\perp = 1 - \frac{\omega_p^2}{\omega^2 - \omega_b^2} - \frac{\omega_{pi}^2}{\omega^2 - \omega_{bi}^2}$$

$$\varepsilon_\times = -\frac{\omega_p^2 \omega_b}{\omega(\omega^2 - \omega_b^2)} - \frac{\omega_{pi}^2 \omega_{bi}}{\omega(\omega^2 - \omega_{bi}^2)}$$

$$\varepsilon_\| = 1 - \frac{\omega_p^2}{\omega^2} - \frac{\omega_{pi}^2}{\omega^2}$$

and ω_{pi} and ω_{bi} are, respectively, the plasma ionic and gyrofrequencies, and are defined by

$$\omega_{pi} = \left(\frac{N_{0i} e_i}{m_i \varepsilon_0}\right)^{1/2}$$

$$\omega_{bi} = \frac{e_i B_c}{m_i}$$

Here m_i denotes the ionic mass, e_i the ionic charge, and N_{0i} the ionic density. Show also that $\bar{\varepsilon}$ is hermitian and satisfies the symmetry relation

$$\bar{\varepsilon}(-\mathbf{B}_c) = \tilde{\bar{\varepsilon}}(\mathbf{B}_c)$$

7.3 When the pressure in an electron gas is included, the linearized equation of motion of an electron takes the form

$$m \frac{\partial \mathscr{V}}{\partial t} = -\frac{1}{N_0} \nabla p + e(\mathscr{E} + \mathscr{V} \times \mathbf{B}_c) - m v_e \mathscr{V}$$

In the above equation, the pressure p satisfies the ideal gas law

$$p = a^2 m n$$

where a is the acoustic velocity in electron gas, n is the excess electron number density over the equilibrium number density N_0 and is related to $\mathscr{V}$ by the equation of continuity

$$\frac{\partial n}{\partial t} + N_0 \nabla \cdot \mathscr{V} = 0$$

Since the pressure is due to the existence of thermal velocities in an electron gas, it is termed "warm" plasma. Show that the dielectric tensor of a warm plasma is given by

$$\bar{\varepsilon} = \varepsilon_\perp(\bar{\mathbf{I}} - \hat{\mathbf{b}}_c \hat{\mathbf{b}}_c) + i\varepsilon_\times(\hat{\mathbf{b}}_c \times \bar{\mathbf{I}}) + \varepsilon_\| \hat{\mathbf{b}}_c \hat{\mathbf{b}}_c$$
$$+ \alpha_0[(1 - \varepsilon_\perp)(\bar{\mathbf{I}} - \hat{\mathbf{k}}\hat{\mathbf{k}}) - i\varepsilon_\times(\hat{\mathbf{b}}_c \cdot \hat{\mathbf{k}})(\hat{\mathbf{k}} \times \bar{\mathbf{I}})]$$

where

$$\varepsilon_\perp = 1 - \frac{X_0}{U_0 \Delta_0} \qquad \varepsilon_\times = -\frac{X_0 Y_0}{U_0^2 \Delta_0}$$

$$\varepsilon_\| = 1 - \frac{X_0(U_0^2 - Y_0^2)}{U_0^3 \Delta_0} \qquad \alpha_0 = \frac{a^2 k^2}{\omega^2 U_0}$$

$$\Delta_0 = 1 - \frac{Y_0^2}{U_0^2} - \alpha_0\left[1 - \frac{Y_0^2}{U_0^2}(\hat{\mathbf{b}}_c \cdot \hat{\mathbf{k}})^2\right]$$

and

$$U_0 = 1 + \frac{i v_e}{\omega}$$

7.4 In the derivation of the permeability tensor (7.30), we have neglected losses associated with the motion of the dipoles in the medium. Macroscopically, losses in a ferrite may be accounted for as in the case of a plasma, by introducing into the equation of motion a damping term. The following modified equation of motion, due to Landau and Lifshitz, has often been used in practice:

$$\frac{d\mathbf{M}_t}{dt} = -\gamma(\mathbf{M}_T \times \mathbf{H}_T) - \frac{\alpha}{|\mathbf{M}_T|}\mathbf{M}_T \times \frac{d\mathbf{M}_T}{dt}$$

where α is a dimensionless damping constant. Show that the permeability tensor for a lossy ferrite is

$$\bar{\boldsymbol{\mu}}' = \mu'_\perp(\bar{\mathbf{I}} - \hat{\mathbf{b}}_c\hat{\mathbf{b}}_c) + i\mu'_\times(\hat{\mathbf{b}}_c \times \bar{\mathbf{I}}) + \mu'_{||}\hat{\mathbf{b}}_c\hat{\mathbf{b}}_c$$

where

$$\mu'_\perp = 1 + \frac{\omega_m(\omega_0 + i\alpha\omega)}{(\omega_0 + i\alpha\omega)^2 - \omega^2}$$

$$\mu'_\times = \frac{\omega_m\omega}{(\omega_0 + i\alpha\omega)^2 - \omega^2}$$

$$\mu'_{||} = 1$$

7.5 (*a*) Show that the solution to the dispersion equation (7.40) may be written in the form

$$n^2 = 1 - \frac{2\{\hat{\mathbf{k}}\cdot[\text{adj }\bar{\varepsilon} - (\text{adj }\bar{\varepsilon})_t\bar{\mathbf{I}}]\cdot\hat{\mathbf{k}} + \hat{\mathbf{k}}\cdot\bar{\varepsilon}\cdot\hat{\mathbf{k}} + |\bar{\varepsilon}|\}}{\hat{\mathbf{k}}\cdot[\text{adj }\bar{\varepsilon} - (\text{adj }\bar{\varepsilon})_t\bar{\mathbf{I}}]\cdot\hat{\mathbf{k}} + 2(\hat{\mathbf{k}}\cdot\bar{\varepsilon}\cdot\hat{\mathbf{k}}) \pm \sqrt{\Delta}}$$

where

$$\Delta = \{\hat{\mathbf{k}}\cdot[\text{adj }\bar{\varepsilon} - (\text{adj }\bar{\varepsilon})_t\bar{\mathbf{I}}]\cdot\hat{\mathbf{k}}\}^2 - 4(\hat{\mathbf{k}}\cdot\bar{\varepsilon}\cdot\hat{\mathbf{k}})\,|\bar{\varepsilon}|$$

(*b*) In the case of a magnetoplasma, show that the discriminant Δ is reduced to

$$\Delta = (\varepsilon_\perp^2 - \varepsilon_\times^2 - \varepsilon_\perp\varepsilon_{||})^2(\hat{\mathbf{k}} \times \hat{\mathbf{b}}_c)^4 + 4\varepsilon_{||}^2\varepsilon_\times^2(\hat{\mathbf{k}}\cdot\hat{\mathbf{b}}_c)^2$$

and is positive. Hence

$$n^2 = 1 - \frac{X_0}{1 - \dfrac{Y_0^2}{2(1 - X_0)}(\hat{\mathbf{k}} \times \hat{\mathbf{b}}_c)^2 \pm \sqrt{\dfrac{Y_0^4(\hat{\mathbf{k}} \times \hat{\mathbf{b}}_c)^4}{4(1 - X_0)^2} + Y_0^2(\hat{\mathbf{k}}\cdot\hat{\mathbf{b}}_c)^2}}$$

This is known in magnetoionic theory as the *Appleton-Hartree equation*, and from which we easily obtain Eqs. (7.49) and (7.50).

(*c*) Show that the Appleton-Hartree equation may be expressed as

$$c^2k^2 = \omega^2 - \frac{\omega\omega_p^2(\omega^2 - \omega_p^2)}{\omega(\omega^2 - \omega_p^2) - \frac{1}{2}\omega_b^2\omega\sin\theta \pm \sqrt{\frac{1}{4}\omega_b^4\omega^2\sin^4\theta + (\omega^2 - \omega_p^2)\omega_b^2\cos^2\theta}}$$

where θ is the angle which the direction of wave propagation makes with the constant magnetic field.

7.6 Using the Appleton-Hartree equation, find the value of n^2 for a wave of 1 MHz that is propagating at an angle 45° with the constant magnetic field. The magnitude of the constant magnetic field is $B_0 = 50 \times 10^{-6}$ Wb/m^2 and the electron density is $N_0 = 1.5 \times 10^{12}$ m^{-3}.

7.7 Discuss and plot the variation of the dispersion equation in Prob. 7.5: (*a*) n^2 vs. ω; (*b*) ck vs. ω; (*c*) n^2 vs. X_0, Y_0 is regarded as fixed.

7.8 For fixed values of the wave, plasma, and gyrofrequencies, discuss and plot the variation of the index of refraction against angle θ.

7.9 A monochromatic plane wave propagates in a ferrite medium. The direction of wave propagation makes an angle θ with the constant magnetic field. The constitutive relations for the ferrite medium are given by Eq. (7.32) with the permeability tensor given by Eq. (7.30). Find the phase velocities of the wave.

7.10 In a magnetoplasma, (*a*) show that $\mathbf{H}_0$ satisfies the equation

$$\bar{\mathbf{W}}(\mathbf{k}) \cdot \mathbf{H}_0 = \mathbf{0}$$

where the wave matrix is given by

$$\bar{\mathbf{W}}(\mathbf{k}) = (k_0^2 \,|\bar{\varepsilon}| - k^2 \varepsilon_\perp \varepsilon_{||})\bar{\mathbf{I}} + (\varepsilon_{||} - \varepsilon_\perp)(\mathbf{k} \times \hat{\mathbf{b}}_c)(\mathbf{k} \times \hat{\mathbf{b}}_c) + i\varepsilon_\times \varepsilon_{||}(\mathbf{k} \cdot \hat{\mathbf{b}}_c)(\mathbf{k} \times \bar{\mathbf{I}})$$

(*b*) Show that the determinant and adjoint of the wave matrix $\bar{\mathbf{W}}(\mathbf{k})$ are given by

$$|\bar{\mathbf{W}}(\mathbf{k})| = |\bar{\varepsilon}|\,(k_0^2\,|\bar{\varepsilon}| - k^2\varepsilon_\perp \varepsilon_{||})\{[\varepsilon_\perp(\hat{\mathbf{k}} \times \hat{\mathbf{b}}_c)^2 + \varepsilon_{||}(\hat{\mathbf{k}} \cdot \hat{\mathbf{b}}_c)^2]k^4$$
$$+ [(\varepsilon_{||} - \varepsilon_\perp)(\hat{\mathbf{k}} \times \hat{\mathbf{b}}_c)^2 - 2\varepsilon_\perp \varepsilon_{||}]k_0^2 k^2 + k_0^4\,|\bar{\varepsilon}|\}$$

and

$$\operatorname{adj} \bar{\mathbf{W}}(\mathbf{k}) = (k_0^2\,|\bar{\varepsilon}| - k^2\varepsilon_\perp\varepsilon_{||})^2\bar{\mathbf{I}} - \varepsilon_\times^2 \varepsilon_{||}^2(\hat{\mathbf{k}} \cdot \hat{\mathbf{b}}_c)^2\mathbf{k}\mathbf{k}$$
$$+ (k_0^2\,|\bar{\varepsilon}| - k^2\varepsilon_\perp\varepsilon_{||})[(\varepsilon_{||} - \varepsilon_\perp)(\mathbf{k} \times \hat{\mathbf{b}}_c)^2\bar{\mathbf{I}}$$
$$- i\varepsilon_\times \varepsilon_{||}(\mathbf{k} \cdot \hat{\mathbf{b}}_c)(\mathbf{k} \times \bar{\mathbf{I}}) - (\varepsilon_{||} - \varepsilon_\perp)(\mathbf{k} \times \hat{\mathbf{b}}_c)(\mathbf{k} \times \hat{\mathbf{b}}_c)]$$

respectively.

(*c*) Using the results of (*b*), obtain the dispersion equation and the direction of $\mathbf{H}_0$.

(*d*) Find the directions of $\mathbf{E}_0$ and $\mathbf{D}_0$.

7.11 An example of plasma which occurs in nature is the earth's ionosphere. This is a region of the upper atmosphere extending from about 50 to 1000 km above the earth's surface. In this region the constitute gases are ionized as a result of ultraviolet radiation from the sun. The electron density in the ionosphere exists in several layers known as D, E, F_1, and F_2 in order of height. The ionization in each layer varies with the hour of the day, the season, the geographical region, and the 11-year sunspot cycle. The ionosphere tends to bend back waves incident from the earth and hence has been used for point-to-point communication on the earth's surface.

(*a*) Using Snell's law of refraction, show that a wave of frequency f which is incident obliquely at an angle θ_0 (see Fig. 7.13) with the normal to the boundary is reflected when

$$f_p = f \cos \theta_0$$

where

$$f_p = \frac{1}{2\pi}\left(\frac{N_0 q^2}{m\varepsilon_0}\right)^{1/2} \cong 9\sqrt{N_0}$$

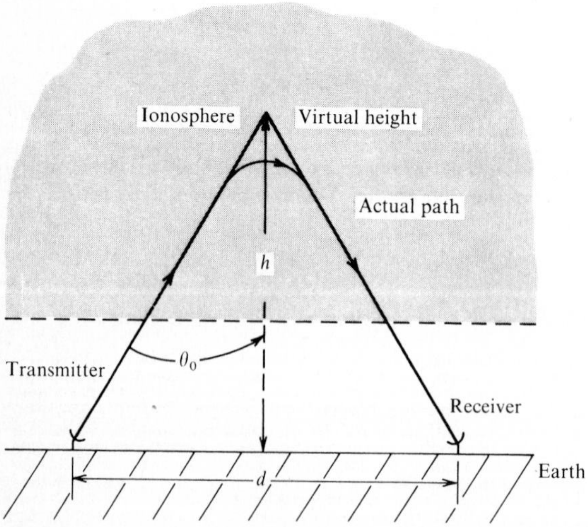

Figure 7.13 Communication between two points via reflection from ionosphere.

is the plasma frequency in hertz and N_0 is the electron density in electrons per cubic meter. The frequency f which satisfies the above relation is called the *maximum usable frequency* (MUF). For the special case of normal incidence on the ionosphere (that is, $\theta_0 = 0$) the wave is reflected from a level at which the wave frequency is equal to the maximum plasma frequency. Hence vertically incident waves of frequencies less than MUF are reflected and those greater than MUF pass through.

(*b*) Find the MUF for a distance $d = 1.5 \times 10^6$ m by the reflection of F_2 layer ($h = 300$ km) of electron density $N_0 = 6 \times 10^{11}\ \text{m}^{-3}$.

7.12 A stationary communication satellite is placed at a height of 1000 km above the earth's surface (Fig. 7.14). A 20-MHz signal from the satellite passes through the ionospheric layer of uniform plasma frequency $f_p = 11$ MHz. Find the true elevation angle of the satellite as seen by the receiver.

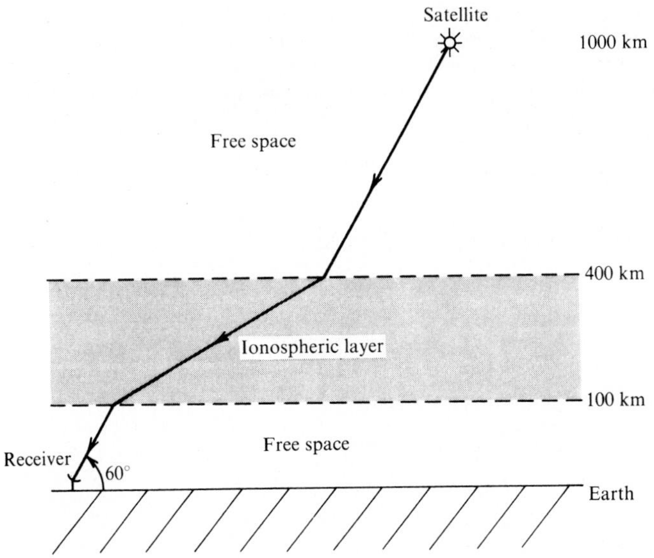

Figure 7.14 A satellite located above the ionospheric layer.

7.13 An ionospheric layer 200 km thick is located at heights between 100 and 300 km. The layer has a uniform electron density $N_0 = 1.12 \times 10^{11}\ \text{m}^{-3}$. The earth's magnetic field is vertical and has the magnitude of $B_0 = 10^{-4}\ \text{Wb/m}^2$. (*a*) Find the electrical-path length (in terms of free-space wavelengths) for a 14-MHz wave traveling through the layer at vertical incidence. (*b*) Under what condition can a path length (in wavelengths) be smaller in the denser medium? (*c*) Find the Faraday rotation through the ionospheric layer.

7.14 Repeat Prob. 7.13 when the earth's magnetic field is horizontal.

7.15 Show that the time-averaged Poynting vector in a magnetoplasma can be expressed as

$$\langle \mathscr{P} \rangle = \frac{\omega \varepsilon_0}{2} \,|\, k_0^2 \bar{\varepsilon} - k^2 \bar{\mathbf{I}} \,|\, \{(\mathbf{k} \cdot \bar{\varepsilon} \cdot \mathbf{k})\bar{\mathbf{I}} + k^2(\text{Re}\ \bar{\varepsilon}) + k_0^2\ \text{Re}\ [\bar{\varepsilon} \cdot (\bar{\varepsilon} - \bar{\varepsilon}_t \bar{\mathbf{I}})]\} \cdot \mathbf{k}$$

7.16 From the dispersion equation

$$F(\mathbf{k}, \omega) = (\mathbf{k} \cdot \bar{\varepsilon} \cdot \mathbf{k})k^2 + \mathbf{k} \cdot \bar{\varepsilon} \cdot (\bar{\varepsilon} - \bar{\varepsilon}_t \bar{\mathbf{I}}) \cdot \mathbf{k}\, \frac{\omega^2}{c^2} + |\,\bar{\varepsilon}\,|\, \frac{\omega^4}{c^4} = 0$$

where $\bar{\varepsilon}$ is the hermitian matrix given by Eq. (7.8), show that the group velocity $\mathbf{v}_g$ is parallel to the vector

$$(\mathbf{k} \cdot \bar{\varepsilon} \cdot \mathbf{k})\mathbf{k} + k^2 \operatorname{Re}(\bar{\varepsilon} \cdot \mathbf{k}) + k_0^2 \operatorname{Re}[\bar{\varepsilon} \cdot (\bar{\varepsilon} - \bar{\varepsilon}_t \bar{\mathbf{I}}) \cdot \mathbf{k}]$$

A comparison with Prob. 7.15 shows that $\mathbf{v}_g$ has the same direction as $\langle \mathscr{P} \rangle$. Show also that for the principal waves, the group velocities are in the direction of wave vector $\mathbf{k}$.

7.17 Show that the magnitude of the group velocity for the extraordinary wave in the uniaxial plasma is less than the velocity of light in free space: $v_{g-} < c$.

7.18 Derive the coefficients in Eq. (7.115).

7.19 Derive Eqs. (7.118) and (7.120).

7.20 Derive Eqs. (7.121) and (7.122).

7.21 In the case of normal incidence, there is no unique plane of incidence. Therefore, the method used in Sec. 7.10 by decomposing the incident and the reflected waves into perpendicular and parallel polarizations fails. In this case, however, the plane formed by vectors $\hat{\mathbf{q}}$ and $\hat{\mathbf{b}}_c$ may be treated as though it were the plane of incidence. Decompose the incident and reflected waves into components perpendicular and parallel to this plane:

$$\mathbf{E}_{0i} = A'_\perp(\hat{\mathbf{q}} \times \hat{\mathbf{b}}_c) + A'_\parallel[\hat{\mathbf{q}} \times (\hat{\mathbf{q}} \times \hat{\mathbf{b}}_c)]$$

and

$$\mathbf{E}_{0r} = B'_\perp(\hat{\mathbf{q}} \times \hat{\mathbf{b}}_c) + B'_\parallel[\hat{\mathbf{q}} \times (\hat{\mathbf{q}} \times \hat{\mathbf{b}}_c)]$$

Show that the transmission coefficient matrix $\bar{\mathbf{T}}'$ is defined by the relation

$$\begin{bmatrix} C_+ \\ C_- \end{bmatrix} = \begin{bmatrix} T'_{11} & T'_{12} \\ T'_{21} & T'_{22} \end{bmatrix} \begin{bmatrix} A'_\perp \\ A'_\parallel \end{bmatrix}$$

and the reflection coefficient matrix $\bar{\mathbf{\Gamma}}'$ is defined by the relation

$$\begin{bmatrix} B'_\perp \\ B'_\parallel \end{bmatrix} = \begin{bmatrix} \Gamma'_{11} & \Gamma'_{12} \\ \Gamma'_{21} & \Gamma'_{22} \end{bmatrix} \begin{bmatrix} A'_\perp \\ A'_\parallel \end{bmatrix}$$

where

$$T'_{11} = \frac{2g_{22}^+}{\Delta'} \qquad T'_{12} = -\frac{2g_{12}^+}{\Delta'}$$

$$T'_{21} = -\frac{2g_{21}^+}{\Delta'} \qquad T'_{22} = \frac{2g_{11}^+}{\Delta'}$$

$$\Gamma'_{11} = \frac{g_{11}^- g_{22}^+ - g_{12}^- g_{21}^+}{\Delta'}$$

$$\Gamma'_{12} = \frac{g_{12}^- g_{11}^+ - g_{11}^- g_{12}^+}{\Delta'}$$

$$\Gamma'_{21} = \frac{g_{21}^- g_{22}^+ - g_{22}^- g_{21}^+}{\Delta'}$$

$$\Gamma'_{22} = \frac{g_{22}^- g_{11}^+ - g_{21}^- g_{12}^+}{\Delta'}$$

$$\Delta' = g_{11}^+ g_{22}^+ - g_{12}^+ g_{21}^+$$

and

$$g_{11}^\pm = \frac{(k_i \pm k_+)[(\hat{\mathbf{q}} \times \hat{\mathbf{b}}_c) \cdot \mathbf{e}_+]}{k_i(\hat{\mathbf{q}} \times \hat{\mathbf{b}}_c)^2}$$

$$g_{12}^{\pm} = \frac{(k_i \pm k_-)[(\hat{\mathbf{q}} \times \hat{\mathbf{b}}_c) \cdot \mathbf{e}_-]}{k_i(\hat{\mathbf{q}} \times \hat{\mathbf{b}}_c)^2}$$

$$g_{21}^{\pm} = \frac{(k_i \pm k_+)\{[\hat{\mathbf{q}} \times (\hat{\mathbf{q}} \times \hat{\mathbf{b}}_c)] \cdot \mathbf{e}_+\}}{k_i(\hat{\mathbf{q}} \times \hat{\mathbf{b}}_c)^2}$$

$$g_{22}^{\pm} = \frac{(k_i \pm k_-)\{[\hat{\mathbf{q}} \times (\hat{\mathbf{q}} \times \hat{\mathbf{b}}_c)] \cdot \mathbf{e}_-\}}{k_i(\hat{\mathbf{q}} \times \hat{\mathbf{b}}_c)^2}$$

7.22 In Prob. 7.21, if the constant magnetic field $\mathbf{B}_c$ is perpendicular to $\hat{\mathbf{q}}$, show that the reflection coefficients become

$$\Gamma'_{11} = \frac{k_i - k_-}{k_i + k_-} \qquad \Gamma'_{22} = \frac{k_i - k_+}{k_i + k_+} \qquad \Gamma'_{12} = \Gamma'_{21} = 0$$

where

$$k_i = k_0\sqrt{\varepsilon_1} \qquad k_+ = k_0\sqrt{1 - X_0}$$

$$k_- = k_0\sqrt{1 - \frac{X_0(1 - X_0)}{1 - X_0 - Y_0^2}}$$

7.23 A plane wave in free space is incident normally on a semi-infinite isotropic plasma (that is, $B_0 = 0$). If the electron density of the plasma is $N_0 = 10^{17}\ \text{m}^{-3}$, find the reflection and transmission coefficients. Determine the cutoff frequency f_c at which there is no transmitted wave in plasma. What are the electric and magnetic fields at this frequency?

7.24 A circularly polarized plane wave propagating in free space is incident normally on the plane boundary of a ferrite medium. The ferrite is magnetized in the direction of propagation of the incident wave. Determine the polarizations and the amplitudes of the reflected and transmitted waves.

7.25 Repeat Prob. 7.24 when the incident wave is linearly polarized.

REFERENCES

Allis, W. P., S. J. Buchsbaum, and A. Bers: *Waves in Anisotropic Plasmas*, The MIT Press, Cambridge, Mass., 1963.

Bohn, E. V.: *Introduction to Electromagnetic Fields and Waves*, Addision-Wesley Publishing Company, Reading, Mass., 1968, chap. 12.

Budden, K. G.: *Radio Waves in the Ionosphere*, Cambridge University Press, New York, 1961.

Chen, F. F.: *Introduction to Plasma Physics*, Plenum Press, New York, 1974.

Chen, H. C.: "Compressivity Tensor and Dispersion Relation for a Plasma," *Am. J. Phys.*, vol. 37, no. 10, 1969, pp. 1022–1028.

Chen, H. C.: "An Invariant Theory of the Constitutive Parameters of a Plasma and a Ferrite," *Appl. Phys.*, vol. 4, 1974, pp. 175–179.

Chen, H. C.: "Transmission and Reflection of Waves from a Magnetized Ferrite Surface by a Coordinate-Free Method," in *Research Topics in Electromagnetic Wave Theory*, edited by J. A. Kong, Interscience-Wiley, New York, 1981, chap. 4, pp. 61–80.

Clemmow, P. C., and J. P. Dougherty: *Electrodynamics of Particles and Plasmas*, Addison-Wesley Publishing Company, Reading, Mass., 1969.

Ginzburg, V. L.: *The Propagation of Electromagnetic Waves in Plasmas*, 2d ed., Pergamon Press, New York, 1970.

Graf, K. A., and M. P. Bachynski: "Transmission and Reflection of Electromagnetic Waves at a Plasma Boundary for Arbitrary Angles of Incidence," *Can. J. Phys.*, vol. 39, 1961, pp. 1544–1562.

Graf, K. A. and M. P. Bachynski: "Electromagnetic Waves in a Bounded Anisotropic Plasma," *Can. J. Phys.*, vol. 40, 1962, pp. 887–905.

Gurevich, A. G.: *Ferrites at Microwave Frequencies*, Boston Technical Publishers, Inc., Cambridge, Mass., 1965.

Helszajn, J.: *Principles of Microwave Ferrite Engineering*, Interscience-Wiley, New York, 1969.

Hlawiczka, P.: *Gyrotropic Waveguides*, Academic Press, New York, 1981.

Holt, E. H., and R. E. Haskell: *Foundations of Plasma Dynamics*, The Macmillan Co., New York, 1965.

Kelso, J. M.: *Radio Ray Propagation in the Ionosphere*, McGraw-Hill Book Company, New York, 1964.

Lax, B., and K. J. Button: *Microwave Ferrites and Ferrimagnetics*, McGraw-Hill Book Company, New York, 1962.

Mueller, R. S.: "Reflection and Refraction of Plane Waves from Plane Ferrite Surfaces," *J. Appl. Phys.*, vol. 42, no. 6, 1971, pp. 2264–2273.

Papas, C. H.: *Theory of Electromagnetic Wave Propagation*, McGraw-Hill Book Company, New York, 1965, chap. 6.

Ratcliffe, J. A.: *The Magneto-Ionic Theory and Its Application to the Ionosphere*, Cambridge University Press, London, 1959.

Sodha, M. S., and N. C. Srivastava: *Microwave Propagation in Ferrimagnetics*, Plenum Press, New York, 1981.

Soohoo, R. F.: *Theory and Application of Ferrites*, Prentice-Hall, Inc., Englewood Cliffs, N. J., 1960.

Stix, T. H.: *The Theory of Plasma Waves*, McGraw-Hill Book Company, New York, 1962.

Tanenbaum, B. S.: *Plasma Physics*, McGraw-Hill Book Company, New York, 1967.

Turner, C. H. M.: "Birefringence in Crystals and in the Ionosphere," *Can. J. Phys.*, vol. 32, 1954, pp. 16–34.

Tyras, G.: *Radiation and Propagation of Electromagnetic Waves*, Academic Press, New York, 1969, chap. 3.

Wait, J. R.: *Electromagnetics and Plasmas*, Holt, Rinehart and Winston, Inc. New York, 1968.

CHAPTER

EIGHT

PLANE WAVES IN MOVING MEDIA

So far we have considered the problem of wave propagation in media which are stationary with respect to an observer. In this chapter, we shall examine the effects of the motion of media on wave propagation. Based on the transformation laws derived in Sec. 2-8, we shall first present the basic equations of the electrodynamics of moving media formulated by Minkowski. We will then apply the results to wave propagation in unbounded moving media and to transmission and reflection of waves from a moving half-space.

8.1 MINKOWSKI CONSTITUTIVE RELATIONS OF A MOVING ISOTROPIC MEDIUM

Consider two inertial systems Σ and Σ' in which the two origins coincide at the instant $t = t' = 0$ and the system Σ' is moving with uniform velocity $\mathbf{v}$ with respect to Σ. According to the principle of relativity, the Maxwell equations are covariant under the Lorentz transformation. In other words, when the coordinate $\mathbf{r}$ and time t undergo the Lorentz transformation expressed by Eq. (2.107):

$$\begin{aligned} \mathbf{r}' &= \bar{\mathbf{L}} \cdot \mathbf{r} - \gamma \mathbf{v} t \\ t' &= \gamma\left(t - \frac{\mathbf{r} \cdot \mathbf{v}}{c^2}\right) \end{aligned} \tag{8.1}$$

where

$$\gamma = \frac{1}{\sqrt{1 - \beta^2}} \qquad \beta = \frac{v}{c} \tag{8.2}$$

and

$$\bar{\mathbf{L}} = \bar{\mathbf{I}} + (\gamma - 1)\hat{\mathbf{v}}\hat{\mathbf{v}} \tag{8.3}$$

the Maxwell equations with respect to Σ

$$\begin{aligned} \nabla \times \mathscr{E} &= -\frac{\partial \mathscr{B}}{\partial t} \\ \nabla \times \mathscr{H} &= \frac{\partial \mathscr{D}}{\partial t} \\ \nabla \cdot \mathscr{B} &= 0 \\ \nabla \cdot \mathscr{D} &= 0 \end{aligned} \tag{8.4}$$

transform into the Maxwell equations with respect to Σ′

$$\begin{aligned} \nabla' \times \mathscr{E}' &= -\frac{\partial \mathscr{B}'}{\partial t'} \\ \nabla' \times \mathscr{H}' &= \frac{\partial \mathscr{D}'}{\partial t'} \\ \nabla' \cdot \mathscr{B}' &= 0 \\ \nabla' \cdot \mathscr{D}' &= 0 \end{aligned} \tag{8.5}$$

provided that the primed quantities are related to the unprimed quantities by

$$\begin{aligned} \mathscr{E}' &= (\mathscr{E} \cdot \hat{\mathbf{v}})\hat{\mathbf{v}} + \gamma[(\bar{\mathbf{I}} - \hat{\mathbf{v}}\hat{\mathbf{v}}) \cdot \mathscr{E} + \mathbf{v} \times \mathscr{B}] \\ \mathscr{B}' &= (\mathscr{B} \cdot \hat{\mathbf{v}})\hat{\mathbf{v}} + \gamma\left[(\bar{\mathbf{I}} - \hat{\mathbf{v}}\hat{\mathbf{v}}) \cdot \mathscr{B} - \frac{\mathbf{v} \times \mathscr{E}}{c^2}\right] \end{aligned} \tag{8.6}$$

and

$$\begin{aligned} \mathscr{H}' &= (\mathscr{H} \cdot \hat{\mathbf{v}})\hat{\mathbf{v}} + \gamma[(\bar{\mathbf{I}} - \hat{\mathbf{v}}\hat{\mathbf{v}}) \cdot \mathscr{H} - \mathbf{v} \times \mathscr{D}] \\ \mathscr{D}' &= (\mathscr{D} \cdot \hat{\mathbf{v}})\hat{\mathbf{v}} + \gamma\left[(\bar{\mathbf{I}} - \hat{\mathbf{v}}\hat{\mathbf{v}}) \cdot \mathscr{D} + \frac{\mathbf{v} \times \mathscr{H}}{c^2}\right] \end{aligned} \tag{8.7}$$

Equations (8.6) and (8.7) summarize the transformation laws given in Eqs. (2.138) and (2.135) respectively.

Now, if an isotropic lossless medium is stationary relative to the primed system Σ′ (called the rest frame of the medium), we write the constitutive relations

$$\begin{aligned} \mathscr{D}' &= \varepsilon_0\, \varepsilon\, \mathscr{E}' \\ \mathscr{B}' &= \mu_0\, \mu\, \mathscr{H}' \end{aligned} \tag{8.8}$$

where ε and μ are respectively the dielectric constant and the relative permeability of the medium in the primed system. In this case, the medium is moving at a uniform velocity **v** with respect to the system Σ (called laboratory frame). To obtain the constitutive relations in Σ, we substitute Eqs. (8.6) and (8.7) into Eq. (8.8) and obtain

$$\mathscr{D} + \frac{\mathbf{v} \times \mathscr{H}}{c^2} = \varepsilon_0\, \varepsilon\, (\mathscr{E} + \mathbf{v} \times \mathscr{B}) \tag{8.9}$$

$$\mathscr{B} - \frac{\mathbf{v} \times \mathscr{E}}{c^2} = \mu_0 \mu (\mathscr{H} - \mathbf{v} \times \mathscr{D}) \tag{8.10}$$

To put the above constitutive relations in a more convenient form, we take the cross products of Eqs. (8.9) and (8.10) respectively with $\mathbf{v}$ and obtain

$$\begin{aligned} \mathbf{v} \times \mathscr{D} + \beta^2 \hat{\mathbf{v}} \times (\hat{\mathbf{v}} \times \mathscr{H}) &= \varepsilon_0 \varepsilon \, [\mathbf{v} \times \mathscr{E} + \mathbf{v} \times (\mathbf{v} \times \mathscr{B})] \\ \mathbf{v} \times \mathscr{B} - \beta^2 \hat{\mathbf{v}} \times (\hat{\mathbf{v}} \times \mathscr{E}) &= \mu_0 \mu [\mathbf{v} \times \mathscr{H} - \mathbf{v} \times (\mathbf{v} \times \mathscr{D})] \end{aligned} \tag{8.11}$$

Elimination of $(\mathbf{v} \times \mathscr{B})$ or $(\mathbf{v} \times \mathscr{D})$ yields

$$\mathbf{v} \times \mathscr{D} = \beta \mathbf{m} \times (\hat{\mathbf{v}} \times \mathscr{H}) + \varepsilon_0 \varepsilon \, \alpha (\mathbf{v} \times \mathscr{E})$$

or

$$\mathbf{v} \times \mathscr{B} = -\beta \mathbf{m} \times (\hat{\mathbf{v}} \times \mathscr{E}) + \mu_0 \mu \, \alpha (\mathbf{v} \times \mathscr{H}) \tag{8.12}$$

where

$$\alpha = \frac{1 - \beta^2}{1 - \mu\varepsilon\beta^2}$$

$$\mathbf{m} = m \hat{\mathbf{v}} \tag{8.13}$$

$$m = \frac{\beta(\mu\varepsilon - 1)}{1 - \mu\varepsilon\beta^2}$$

Substituting Eq. (8.12) into Eqs. (8.9) and (8.10), we finally obtain the Minkowski constitutive relations for a moving medium:

$$\mathscr{D} = \varepsilon_0 \varepsilon \, \bar{\boldsymbol{\alpha}} \cdot \mathscr{E} + \frac{1}{c} (\mathbf{m} \times \bar{\mathbf{I}}) \cdot \mathscr{H} \tag{8.14}$$

$$\mathscr{B} = -\frac{1}{c} (\mathbf{m} \times \bar{\mathbf{I}}) \cdot \mathscr{E} + \mu_0 \mu \, \bar{\boldsymbol{\alpha}} \cdot \mathscr{H} \tag{8.15}$$

where

$$\bar{\boldsymbol{\alpha}} = \alpha \bar{\mathbf{I}} + (1 - \alpha) \hat{\mathbf{v}} \hat{\mathbf{v}} \tag{8.16}$$

From the above constitutive relations we note that the flux density $\mathscr{D}$ (or $\mathscr{B}$) depends simultaneously on the field intensities $\mathscr{E}$ and $\mathscr{H}$; hence a moving dielectric is a magnetoelectric medium. We also note that $\mathscr{D}$ (or $\mathscr{B}$) is not in the direction of $\mathscr{E}$ or $\mathscr{H}$. In this sense, a moving medium is anisotropic even if it is isotropic in the rest frame of the medium. And lastly, we observe that the tensor $\bar{\boldsymbol{\alpha}}$ is symmetric and the tensor connecting $\mathscr{D}$ and $\mathscr{H}$ in Eq. (8.14) is the transpose of the tensor connecting $\mathscr{B}$ and $\mathscr{E}$ in Eq. (8.15). In other words, the constitutive tensors satisfy the conditions stated in Eq. (2.73).

When $\mathbf{v}$ is zero or $\mu = 1$, and $\varepsilon = 1$, Eqs. (8.14) and (8.15) reduce to the constitutive relations of a stationary isotropic medium or to those of the free space as they should.

8.2 BOUNDARY CONDITIONS

The Maxwell equations (8.4) together with the constitutive relations (8.14) and (8.15), form the basic set of equations for the electrodynamics of a moving medium and are sufficient for the determination of wave propagation problems in an unbounded moving isotropic medium. However, if there are boundaries in the moving medium, then the system of equations (8.14), (8.15), and (8.4) must be supplemented by the boundary conditions which we shall discuss in this section.

As with the case of stationary media, from equations $\nabla \cdot \mathscr{D} = 0$ and $\nabla \cdot \mathscr{B} = 0$ the continuity of the normal components of $\mathscr{D}$ and $\mathscr{B}$ across a moving interface follows as in Sec. 2.3:

$$\begin{aligned} \mathscr{D}_1 \cdot \hat{\mathbf{q}} &= \mathscr{D}_2 \cdot \hat{\mathbf{q}} \\ \mathscr{B}_1 \cdot \hat{\mathbf{q}} &= \mathscr{B}_2 \cdot \hat{\mathbf{q}} \end{aligned} \tag{8.17}$$

where the subscripts 1 and 2 denote the fields in media 1 and 2 respectively. To derive the boundary conditions on the tangential components, we need the following subsidiary theorem:

$$\frac{d}{dt} \int_{s_1(t)} \mathbf{A} \cdot d\mathbf{s} = \int_{s_1(t)} \left[\frac{\partial \mathbf{A}}{\partial t} - \nabla \times (\mathbf{v} \times \mathbf{A}) + (\nabla \cdot \mathbf{A})\mathbf{v} \right] \cdot d\mathbf{s} \tag{8.18}$$

This expresses the total time rate of change of the flux of a vector field **A** through a moving surface $s_1(t)$ in terms of a surface integral of a function of **A**. To show Eq. (8.18), let us consider the surface s_1 to move to s_2 in the time Δt, as shown in Fig. 8.1. By the rules for differentiation, we have

$$\frac{\Delta}{\Delta t} \int \mathbf{A} \cdot d\mathbf{s} = \frac{1}{\Delta t} \left[\int_{s_2} \mathbf{A}(t + \Delta t) \cdot d\mathbf{s} - \int_{s_1} \mathbf{A}(t) \cdot d\mathbf{s} \right] \tag{8.19}$$

In order to evaluate this, we apply the divergence theorem to the volume V

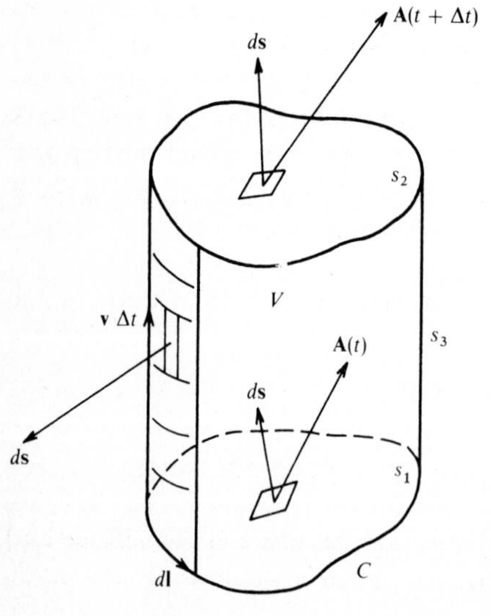

Figure 8.1 Change in flux which crosses a moving surface.

enclosed by the surfaces s_1, s_2, and s_3 (see Fig. 8.1), and obtain

$$\int_V \nabla \cdot \mathbf{A}(t)\, d^3r = \int_{s2} \mathbf{A}(t) \cdot d\mathbf{s} - \int_{s1} \mathbf{A}(t) \cdot d\mathbf{s} + \int_{s3} \mathbf{A}(t) \cdot d\mathbf{s}$$

Now from Fig. 8.1, the surface element on the side surface s_3 generated by the motion of the boundary curve C is

$$d\mathbf{s} = d\mathbf{l} \times (\mathbf{v}\, \Delta t) = -(\mathbf{v} \times d\mathbf{l})\, \Delta t$$

and the volume element is

$$d^3r = \mathbf{v} \cdot d\mathbf{s}\, \Delta t$$

Hence, we have

$$\int_{s2} \mathbf{A}(t) \cdot d\mathbf{s} - \int_{s1} \mathbf{A}(t) \cdot d\mathbf{s} = \left\{ \int_{s1} [\nabla \cdot \mathbf{A}(t)]\mathbf{v} \cdot d\mathbf{s} + \oint_c [\mathbf{A}(t) \times \mathbf{v}] \cdot d\mathbf{l} \right\} \Delta t \tag{8.20}$$

Expanding $A(t + \Delta t)$ in Eq. (8.19) by Taylor's theorem,

$$\mathbf{A}(t + \Delta t) = \mathbf{A}(t) + \frac{\partial \mathbf{A}}{\partial t} \Delta t + \cdots \tag{8.21}$$

then substituting Eqs. (8.20) and (8.21) into Eq. (8.19), we obtain, in the limit,

$$\frac{d}{dt} \int_{s1(t)} \mathbf{A} \cdot d\mathbf{s} = \int_{s1(t)} \left[\frac{\partial \mathbf{A}}{\partial t} + (\nabla \cdot \mathbf{A})\mathbf{v} \right] \cdot d\mathbf{s} - \oint_{c(t)} (\mathbf{v} \times \mathbf{A}) \cdot d\mathbf{l} \tag{8.22}$$

Finally, by applying Stokes' theorem to the last term on the right-hand side of Eq. (8.22), we prove the desired relation (8.18).

The first term on the right-hand side of Eq. (8.18) represents the change in flux due to the time variation of the vector field, while the second term represents the flux loss across the boundary of the moving surface. The third term arises from the passage of surface S through an inhomogeneous vector field.

We now apply theorem (8.18) to the magnetic flux density $\mathscr{B}$:

$$\frac{d}{dt} \int_{s1(t)} \mathscr{B} \cdot d\mathbf{s} = \int_{s1(t)} \left[\frac{\partial \mathscr{B}}{\partial t} - \nabla \times (\mathbf{v} \times \mathscr{B}) + (\nabla \cdot \mathscr{B})\mathbf{v} \right] \cdot d\mathbf{s} \tag{8.23}$$

Making use of the Maxwell equations (8.4), and Stokes' theorem, we obtain from Eq. (8.23) Faraday's law for a moving surface, namely

$$\oint_{c(t)} (\mathscr{E} + \mathbf{v} \times \mathscr{B}) \cdot d\mathbf{l} = -\frac{d}{dt} \int_{s1(t)} \mathscr{B} \cdot d\mathbf{s} \tag{8.24}$$

To determine the boundary condition on the tangential components of the field vectors, we consider a moving surface $S(t)$ across which the field vectors may possess discontinuities and draw an arbitrary infinitesimal plane rectangular loop $C(t)$ around a point p on the moving surface $S(t)$ (see Fig. 2.4). We then apply Eq. (8.24) to the rectangular loop $C(t)$ that we have just defined and the surface $S_1(t)$ is now the plane area contained by $C(t)$. We shall assume that $C(t)$ always moves with the surface $S(t)$ in such a manner that it preserves its orientation. Vector $\mathbf{v}$ becomes the velocity of the discontinuity surface $S(t)$. Following the argument used in the derivation of Eq. (2.25), we conclude that the tangential component of the vector $\mathscr{E} + \mathbf{v} \times \mathscr{B}$ must be continuous across the moving surface $S(t)$, that is,

$$\hat{\mathbf{q}} \times (\mathscr{E}_1 + \mathbf{v} \times \mathscr{B}_1) = \hat{\mathbf{q}} \times (\mathscr{E}_2 + \mathbf{v} \times \mathscr{B}_2) \tag{8.25}$$

Expanding the vector triple product in this equation and using the condition (8.17), we finally have

$$\hat{\mathbf{q}} \times (\mathscr{E}_1 - \mathscr{E}_2) - (\mathbf{v} \cdot \hat{\mathbf{q}})(\mathscr{B}_1 - \mathscr{B}_2) = \mathbf{0} \tag{8.26}$$

Similarly, applying theorem (8.18) to the electric flux density $\mathscr{D}$ and using the Maxwell equations (8.4), we obtain

$$\oint_{c(t)} (\mathscr{H} - \mathbf{v} \times \mathscr{D}) \cdot d\mathbf{l} = \frac{d}{dt} \int_{s_1(t)} \mathscr{D} \cdot d\mathbf{s} \tag{8.27}$$

By essentially the same process as the one used in deriving the boundary condition (8.25), we then readily arrive at the result that the tangential component of the vector $\mathscr{H} - \mathbf{v} \times \mathscr{D}$ must be continuous across the moving surface $S(t)$, namely

$$\hat{\mathbf{q}} \times (\mathscr{H}_1 - \mathbf{v} \times \mathscr{D}_1) = \hat{\mathbf{q}} \times (\mathscr{H}_2 - \mathbf{v} \times \mathscr{D}_2) \tag{8.28}$$

or, after using Eq. (8.17),

$$\hat{\mathbf{q}} \times (\mathscr{H}_1 - \mathscr{H}_2) + (\mathbf{v} \cdot \hat{\mathbf{q}})(\mathscr{D}_1 - \mathscr{D}_2) = \mathbf{0} \tag{8.29}$$

In the case when the interface is moving parallel to itself, $\mathbf{v} \cdot \hat{\mathbf{q}} = 0$, the boundary conditions (8.26) and (8.29) become

$$\begin{aligned} \hat{\mathbf{q}} \times (\mathscr{E}_1 - \mathscr{E}_2) &= \mathbf{0} \\ \hat{\mathbf{q}} \times (\mathscr{H}_1 - \mathscr{H}_2) &= \mathbf{0} \end{aligned} \tag{8.30}$$

which are the same as for a stationary interface.

In the derivation of Eqs. (8.26) and (8.29), we have assumed that there are no surface currents or charges on the moving interface.

8.3 PLANE WAVE SOLUTIONS IN A MOVING MEDIUM

In a source-free region, the field vectors in a moving isotropic medium satisfy the Maxwell equations (8.4) and are related by the constitutive relations (8.14) and (8.15). To study the problem of plane wave propagation in such a medium, we shall again seek a solution of the form

$$\mathscr{E} = \text{Re}\,(\mathbf{E}_0\, e^{i(\mathbf{k} \cdot \mathbf{r} - \omega t)}) \tag{8.31}$$

On substituting Eq. (8.31) and similar expressions for $\mathscr{H}$, $\mathscr{B}$, and $\mathscr{D}$ into Eqs. (8.4), (8.14), and (8.15), and then eliminating $\mathbf{B}_0$ and $\mathbf{D}_0$ from the resulted equations, we obtain

$$\mathbf{p} \times \mathbf{E}_0 = \omega\mu_0 \mu \bar{\boldsymbol{\alpha}} \cdot \mathbf{H}_0 \tag{8.32}$$

$$\mathbf{p} \times \mathbf{H}_0 = -\omega\varepsilon_0 \varepsilon \bar{\boldsymbol{\alpha}} \cdot \mathbf{E}_0 \tag{8.33}$$

where

$$\mathbf{p} = \mathbf{k} + k_0 \mathbf{m} \tag{8.34}$$

From Eq. (8.32), we find

$$\mathbf{H}_0 = \frac{1}{\omega\mu_0 \mu} \bar{\boldsymbol{\alpha}}^{-1} \cdot (\mathbf{p} \times \mathbf{E}_0) \tag{8.35}$$

Substitution of Eq. (8.35) into Eq. (8.33) yields the equation containing $\mathbf{E}_0$ alone:

$$[k_0^2 \mu\varepsilon |\bar{\boldsymbol{\alpha}}| \bar{\boldsymbol{\alpha}} + (\mathbf{p} \times \bar{\mathbf{I}}) \cdot (\operatorname{adj} \bar{\boldsymbol{\alpha}}) \cdot (\mathbf{p} \times \bar{\mathbf{I}})] \cdot \mathbf{E}_0 = \mathbf{0} \tag{8.36}$$

or, after expanding the second term according to Eq. (1.120),

$$\bar{\mathbf{W}}_m(\mathbf{p}) \cdot \mathbf{E}_0 = \mathbf{0} \tag{8.37}$$

where

$$\bar{\mathbf{W}}_m(\mathbf{p}) = (k_0^2 \mu\varepsilon |\bar{\boldsymbol{\alpha}}| - \mathbf{p} \cdot \bar{\boldsymbol{\alpha}} \cdot \mathbf{p})\bar{\mathbf{I}} + \mathbf{pp} \cdot \bar{\boldsymbol{\alpha}} \tag{8.38}$$

is the wave matrix of a moving medium.

Equation (8.37) has nontrivial solutions $\mathbf{E}_0$ if the determinant of the wave matrix vanishes. From Eq. (1.127*b*), we may write the determinant of $\bar{\mathbf{W}}_m(\mathbf{p})$ as

$$|\bar{\mathbf{W}}_m(\mathbf{p})| = k_0^2 \mu\varepsilon |\bar{\boldsymbol{\alpha}}| (k_0^2 \mu\varepsilon |\bar{\boldsymbol{\alpha}}| - \mathbf{p} \cdot \bar{\boldsymbol{\alpha}} \cdot \mathbf{p})^2 \tag{8.39}$$

Thus by setting Eq. (8.39) to zero, we obtain the dispersion equation for the monochromatic plane waves in a moving medium:

$$\mathbf{p} \cdot \bar{\boldsymbol{\alpha}} \cdot \mathbf{p} - k_0^2 \mu\varepsilon |\bar{\boldsymbol{\alpha}}| = 0 \tag{8.40}$$

Substituting the condition (8.40) into Eq. (8.37), we find

$$\mathbf{pp} \cdot \bar{\boldsymbol{\alpha}} \cdot \mathbf{E}_0 = \mathbf{0} \tag{8.41}$$

Thus any nonzero vector $\mathbf{E}_0$ is a solution of the homogeneous equation (8.37) if it is perpendicular to the vector $(\mathbf{p} \cdot \bar{\boldsymbol{\alpha}})$, that is,

$$\mathbf{p} \cdot \bar{\boldsymbol{\alpha}} \cdot \mathbf{E}_0 = 0 \tag{8.42}$$

But

$$\mathbf{p} \cdot \bar{\boldsymbol{\alpha}} = \alpha[\hat{\mathbf{v}} \times (\mathbf{p} \times \hat{\mathbf{v}})] + (\mathbf{p} \cdot \hat{\mathbf{v}})\hat{\mathbf{v}}$$

We thus choose the direction of $\mathbf{E}_0$ as

$$\mathbf{e}_1 = \mathbf{p} \times \hat{\mathbf{v}} = \mathbf{k} \times \hat{\mathbf{v}} \tag{8.43}$$

which evidently satisfies the orthogonality condition (8.42) and thus is a solution of Eq. (8.37). The magnetic field intensity $\mathbf{h}_1$ corresponding to $\mathbf{e}_1$ can be found from the Maxwell equation (8.35):

$$\mathbf{h}_1 = \frac{1}{\omega\mu_0 \mu} \bar{\boldsymbol{\alpha}}^{-1} \cdot (\mathbf{p} \times \mathbf{e}_1) \tag{8.44}$$

Since Eq. (8.40) is a repeated root of Eq. (8.39), we may choose another solution of Eq. (8.37), namely,

$$\mathbf{e}_2 = \bar{\alpha}^{-1} \cdot [\mathbf{p} \times (\mathbf{p} \times \hat{\mathbf{v}})] = \bar{\alpha}^{-1} \cdot (\mathbf{p} \times \mathbf{e}_1) \tag{8.45}$$

which again satisfies the orthogonality condition (8.42) and is linearly independent of $\mathbf{e}_1$. To express $\mathbf{e}_2$ in an alternative form, we note from Eqs. (8.16) and (1.126) that

$$\begin{aligned} |\bar{\alpha}| &= \alpha^2 \\ \operatorname{adj} \bar{\alpha} &= \alpha\bar{\mathbf{I}} - \alpha(1 - \alpha)\hat{\mathbf{v}}\hat{\mathbf{v}} \end{aligned} \tag{8.46}$$

Thus

$$\bar{\alpha}^{-1} = \frac{1}{\alpha}\bar{\mathbf{I}} + \left(1 - \frac{1}{\alpha}\right)\hat{\mathbf{v}}\hat{\mathbf{v}} \tag{8.47}$$

Substituting Eq. (8.47) into Eq. (8.45) and making use of the dispersion equation, we obtain

$$\mathbf{e}_2 = \frac{1}{\alpha}[(\mathbf{p} \cdot \hat{\mathbf{v}})\mathbf{p} - k_0^2\alpha^2\mu\varepsilon\hat{\mathbf{v}}] \tag{8.48}$$

The magnetic field intensity again follows from Eq. (8.35):

$$\begin{aligned} \mathbf{h}_2 &= \frac{1}{\omega\mu_0\mu}\bar{\alpha}^{-1} \cdot (\mathbf{p} \times \mathbf{e}_2) \\ &= \frac{1}{\omega\mu_0\mu}\bar{\alpha}^{-1} \cdot (\mathbf{p} \times \bar{\mathbf{I}}) \cdot \bar{\alpha}^{-1} \cdot (\mathbf{p} \times \bar{\mathbf{I}}) \cdot \mathbf{e}_1 \end{aligned} \tag{8.49}$$

or, after making use of Eq. (1.120) and the dispersion equation (8.40),

$$\mathbf{h}_2 = -\omega\varepsilon_0\varepsilon\mathbf{e}_1 \tag{8.50}$$

Comparing Eqs. (8.44) and (8.45), we see that $\mathbf{h}_1$ is simply related to $\mathbf{e}_2$ by

$$\mathbf{h}_1 = \frac{1}{\omega\mu_0\mu}\mathbf{e}_2 \tag{8.51}$$

8.4 WAVE VECTOR AND NORMAL SURFACES

We shall first examine the shape of the wave vector surface for various velocities of the moving medium. On substituting Eq. (8.34) into the dispersion equation (8.40) and noting that

$$\begin{aligned} \hat{\mathbf{v}} \cdot \bar{\alpha} \cdot \hat{\mathbf{v}} &= 1 \\ \hat{\mathbf{v}} \cdot \bar{\alpha} \cdot \hat{\mathbf{k}} &= \hat{\mathbf{k}} \cdot \hat{\mathbf{v}} \end{aligned} \tag{8.52}$$

we obtain the equation that determines the wave number:

$$(\hat{\mathbf{k}} \cdot \bar{\bar{\alpha}} \cdot \hat{\mathbf{k}})k^2 + 2k_0 m(\hat{\mathbf{k}} \cdot \hat{\mathbf{v}})k + k_0^2(m^2 - \alpha^2 \mu\varepsilon) = 0 \tag{8.53}$$

Now, letting θ be the angle between the wave vector $\mathbf{k}$ and the velocity $\mathbf{v}$ of the moving medium and noting that

$$\begin{aligned}\hat{\mathbf{k}} \cdot \bar{\bar{\alpha}} \cdot \hat{\mathbf{k}} &= \alpha(\hat{\mathbf{k}} \times \hat{\mathbf{v}})^2 + (\hat{\mathbf{k}} \cdot \hat{\mathbf{v}})^2 \\ &= \alpha + (1 - \alpha)(\hat{\mathbf{k}} \cdot \hat{\mathbf{v}})^2\end{aligned} \tag{8.54}$$

we find the solutions of Eq. (8.53):

$$\frac{k_\pm}{k_0} = \frac{-\beta\xi \cos\theta \pm \sqrt{1 + \xi(1 - \beta^2 \cos^2\theta)}}{1 - \xi\beta^2 \cos^2\theta} \tag{8.55}$$

where

$$\xi = (\mu\varepsilon - 1)\,\gamma^2 \tag{8.56}$$

Equation (8.55) shows that the wave number, and thus in turn the phase velocity of a wave in a moving medium, depends on the angle θ between the directions of the wave vector $\mathbf{k}$ and the velocity $\mathbf{v}$ of the moving medium. In this sense, a moving medium is also anisotropic even when it is isotropic in the rest frame of the medium. At first glance, we might conclude from Eq. (8.55) that the plus and the minus signs correspond to two different solutions. This, however, is not so. Actually, the two solutions k_+ and k_- are connected with each other by the relation

$$k_+(\pi - \theta) = -k_-(\theta)$$

Thus, we choose the solution that yields $k/k_0 = +\sqrt{\mu\varepsilon}$ for $v = 0$, namely

$$\frac{k(\theta)}{k_0} = \frac{-\beta\xi \cos\theta + \sqrt{1 + \xi(1 - \beta^2 \cos^2\theta)}}{1 - \xi\beta^2 \cos^2\theta} \tag{8.57}$$

For each direction $\hat{\mathbf{k}}$ which makes an angle θ with the velocity of the moving medium, we plot the value of $k(\theta)$. The surface thus obtained is the wave vector surface. Since $k(\theta)$ depends only on θ, and $k(\theta) = k(-\theta)$, the wave vector surface is a surface of revolution about the velocity vector $\mathbf{v}$ of the moving medium, and the plane $\theta = 0$ is a plane of symmetry.

The anisotropy of a moving medium is different from that of a crystal. In both cases, the wave numbers depend on the direction of wave propagation. But in crystals, to a given direction $\hat{\mathbf{k}}$, we can have two plane waves propagating with two different phase velocities and each wave has its own polarization. On the other hand, in a moving medium, to every direction $\hat{\mathbf{k}}$ corresponds one phase velocity and thus one wave which can have an arbitrary polarization provided that it satisfies the orthogonality condition (8.42).

To plot the wave vector surfaces for various velocities, we note that

$$\begin{aligned}(\hat{\mathbf{k}} \cdot \hat{\mathbf{v}})^2 &= k^2 \cos^2\theta \\ (\hat{\mathbf{k}} \times \hat{\mathbf{v}})^2 &= k^2 \sin^2\theta\end{aligned} \tag{8.58}$$

and substitute Eq. (8.54) into Eq. (8.53) and obtain

$$\frac{(\hat{\mathbf{k}} \cdot \hat{\mathbf{v}} + k_0 m)^2}{\alpha^2 k_0^2 \mu\varepsilon} + \frac{(\mathbf{k} \times \hat{\mathbf{v}})^2}{\alpha k_0^2 \mu\varepsilon} = 1 \tag{8.59}$$

or

$$(1 - \mu\varepsilon\beta^2)(\mathbf{k} \cdot \hat{\mathbf{v}})^2 + 2k_0 \beta(\mu\varepsilon - 1)(\mathbf{k} \cdot \hat{\mathbf{v}}) + (1 - \beta^2)(\mathbf{k} \times \hat{\mathbf{v}})^2 + k_0^2(\beta^2 - \mu\varepsilon) = 0 \tag{8.60}$$

In the case of a stationary medium ($\beta = 0$), we see from Eq. (8.60) that the wave vector surface is a sphere centered at the origin with the radius $k_0\sqrt{\mu\varepsilon}$ as it should be (see Fig. 8.2*a*). As the velocity of the moving medium increases from zero to less than that of the phase velocity of the wave in the stationary medium ($0 < \beta < 1/\sqrt{\mu\varepsilon}$), we have from Eq. (8.13) that α is positive; thus the wave vector surface (8.59) is an ellipsoid of revolution about $\hat{\mathbf{v}}$ axis. This ellipsoid is displaced in the opposite direction of motion of the medium as β increases (see Figs. 8.2*b* and 8.2*c*).

When the velocity of the moving medium is equal to the phase velocity of the wave in the stationary medium ($\beta = 1/\sqrt{\mu\varepsilon}$), we see from Eq. (8.60) that the coefficient of $(\mathbf{k} \cdot \hat{\mathbf{v}})^2$ vanishes; thus the wave vector surface becomes a paraboloid (see Fig. 8.2*d*). As the velocity of the moving medium increases from $c/\sqrt{\mu\varepsilon}$ to that less than c ($1/\sqrt{\mu\varepsilon} < \beta < 1$), we have from Eq. (8.13) that α is negative; thus

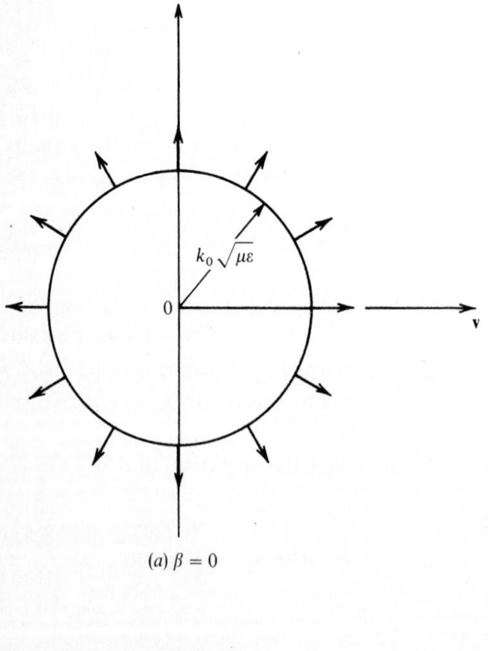

Figure 8.2 Wave vector surfaces of a moving medium. The index of refraction in the rest frame of a nondispersive medium is chosen as 4 (that is, $\sqrt{\mu\varepsilon} = 4$). The arrows normal to the surfaces indicate the direction of the group velocity of the wave.

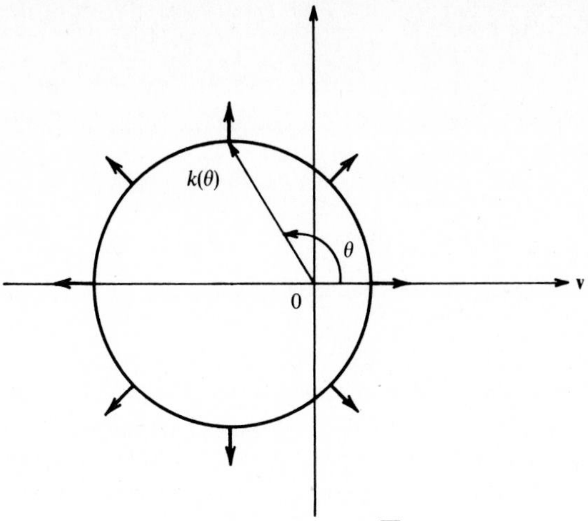

(b) $\beta = 0.1 < 1/\sqrt{\mu\varepsilon}$

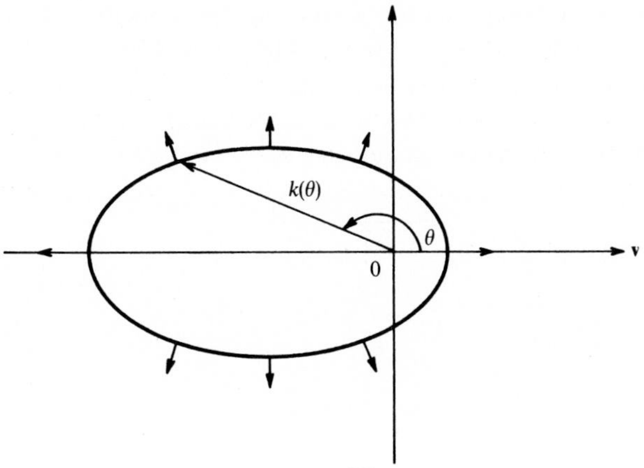

(c) $\beta = 0.2 < 1/\sqrt{\mu\varepsilon}$

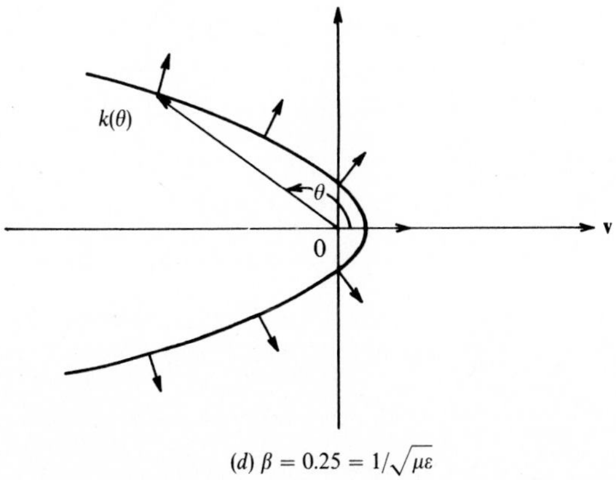

(d) $\beta = 0.25 = 1/\sqrt{\mu\varepsilon}$

Figure 8.2 (Continued)

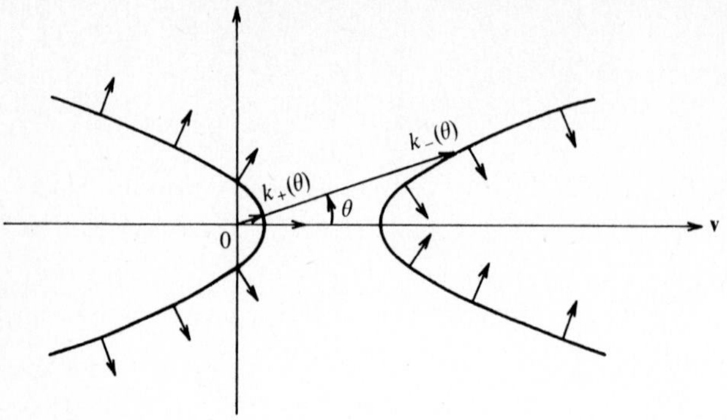

(e) $\beta = 0.3 > 1/\sqrt{\mu\varepsilon}$

Figure 8.2 (Continued)

the wave vector surface (8.59) is a hyperboloid of revolution about $\hat{\mathbf{v}}$ axis (see Figs. 8.2*e* to *g*).

Finally, if the velocity of the moving medium is equal to the velocity of light ($\beta = 1$), the wave vector surface (8.60) becomes a plane. This plane is perpendicular to the direction of motion of the medium at a distance k_0 from the origin (see Fig. 8.2*h*). In Fig. 8.2, the arrows indicate the directions of the group velocity.

From Eq. (8.57) we easily obtain the phase velocity of the wave in a moving medium:

$$v_p = \frac{ck_0}{k} = c\,\frac{\beta\xi\cos\theta + \sqrt{1 + \xi(1 - \beta^2\cos^2\theta)}}{1 + \xi} \tag{8.61}$$

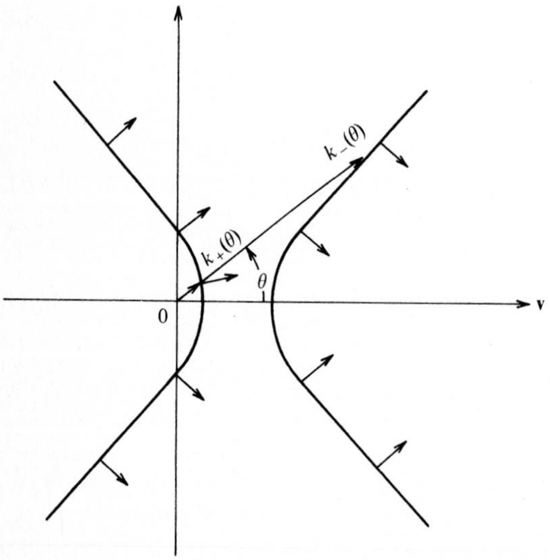

(f) $\beta = 0.6 > 1/\sqrt{\mu\varepsilon}$

Figure 8.2 (Continued)

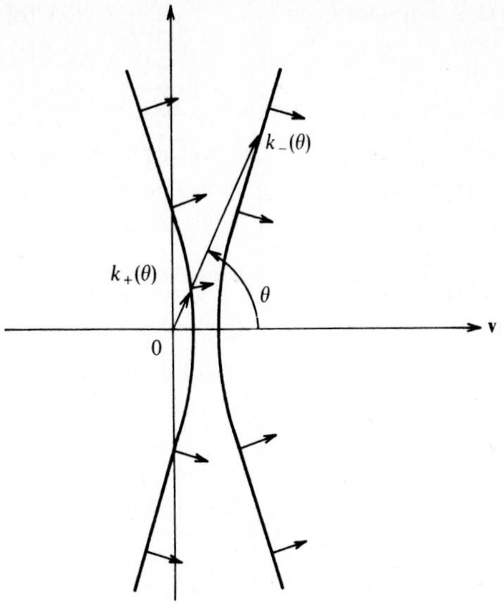

(*g*) $\beta = 0.9 > 1/\sqrt{\mu\varepsilon}$

Figure 8.2 (Continued)

When $\beta^2 \ll 1$, Eq. (8.61) reduces to

$$v_p = \frac{c}{\sqrt{\mu\varepsilon}} + \left(1 - \frac{1}{\mu\varepsilon}\right) v \cos\theta \tag{8.62}$$

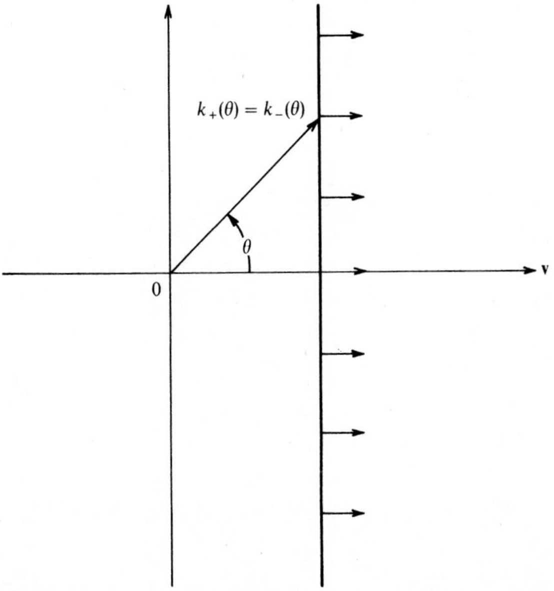

(*h*) $\beta = 1$

Figure 8.2 (Continued)

This is the well-known formula of Fresnel. The coefficient $1 - 1/\mu\varepsilon$ is called the *Fresnel drag coefficient.*

For each given direction $\hat{\mathbf{k}}$ which makes an angle θ with the velocity of the moving medium, we mark off length corresponding to the phase velocity $v_p(\theta)$ in that direction. The surface thus obtained is called a *normal surface* or a *phase velocity surface.* Similar to the wave vector surface, we note from Eq. (8.61) that the normal surface is also a surface of revolution about the velocity vector $\mathbf{v}$ of the moving medium and the plane $\theta = 0$ is a plane of symmetry.

We shall now examine the shape of the normal surface for various velocities of the moving medium. When the medium is stationary ($\beta = 0$), the normal surface is of course a sphere showing a phase velocity $c/\sqrt{\mu\varepsilon}$ in all directions [see Fig. 8.3*a*]. As the velocity of the moving medium increases from zero to less than $c/\sqrt{\mu\varepsilon}$, this sphere which contains the origin inside moves in the direction of motion, at first with little distortion and then becomes dimpled on one side, that is, the phase velocity of a wave whose direction of propagation makes an acute angle with the velocity of the medium is larger than the phase velocity of a wave propagating at an obtuse angle to the velocity of the moving medium (see Figs. 8.3*b* and 8.3*c*).

When the velocity of the moving medium is equal to the phase velocity of the wave in the stationary medium ($\beta = 1/\sqrt{\mu\varepsilon}$), the normal surface has the shape shown in Fig. 8.3*d*. On this surface there is no point of intersection, that is, in any given direction, $v_p(\theta)$ has one positive and one negative value. When the velocity of the moving medium exceeds $c/\sqrt{\mu\varepsilon}$, the normal surface becomes double and has a self-intersection point at the origin as shown in Figs. 8.3*e* to *g*. At the origin, the normal surface is tangent to a cone whose half-angle θ_0 (see Fig. 8.3*f*)

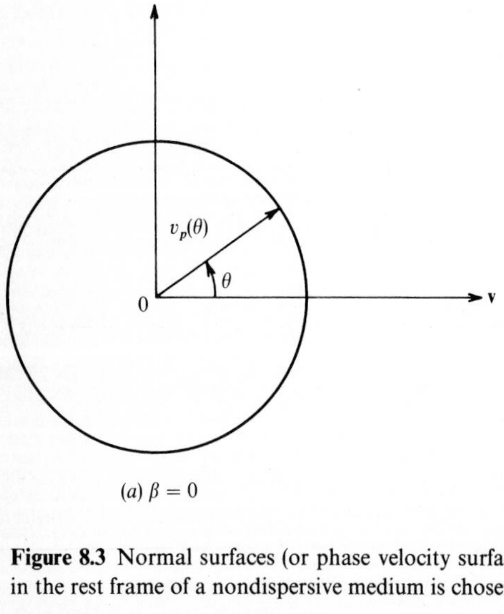

(*a*) $\beta = 0$

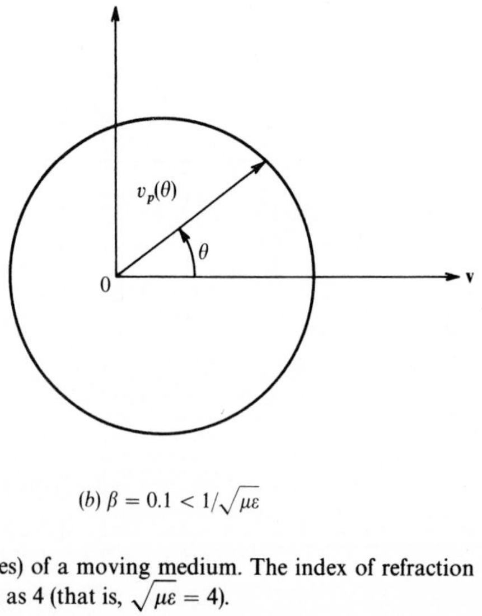

(*b*) $\beta = 0.1 < 1/\sqrt{\mu\varepsilon}$

Figure 8.3 Normal surfaces (or phase velocity surfaces) of a moving medium. The index of refraction in the rest frame of a nondispersive medium is chosen as 4 (that is, $\sqrt{\mu\varepsilon} = 4$).

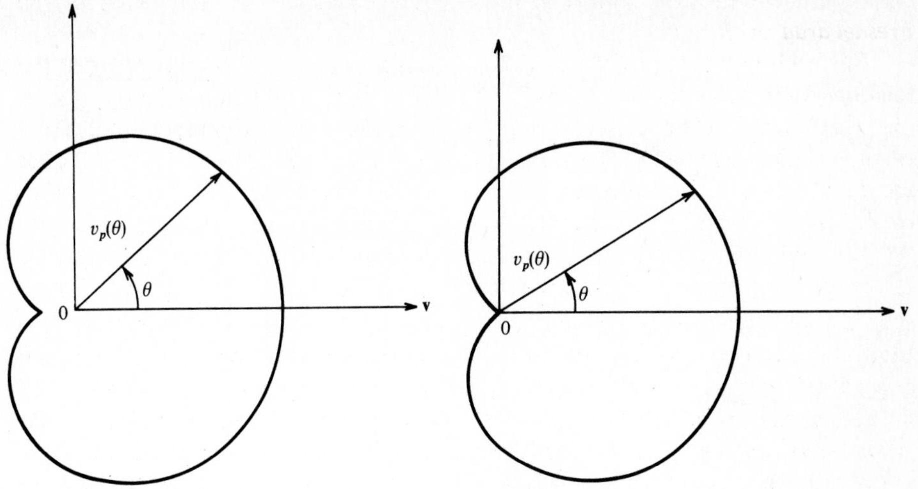

(c) $\beta = 0.2 < 1/\sqrt{\mu\varepsilon}$ (d) $\beta = 0.25 = 1/\sqrt{\mu\varepsilon}$

Figure 8.3 (Continued)

is determined by the equation

$$\tan \theta_0 = \pm \sqrt{\frac{\beta^2 \mu\varepsilon - 1}{1 - \beta^2}} \tag{8.63}$$

For directions lying inside the cone with half-angle θ_0, there will correspond two values of the phase velocity $v_p(\theta)$ of the same sign. But to directions lying outside the cone, the two values of the phase velocity have opposite signs.

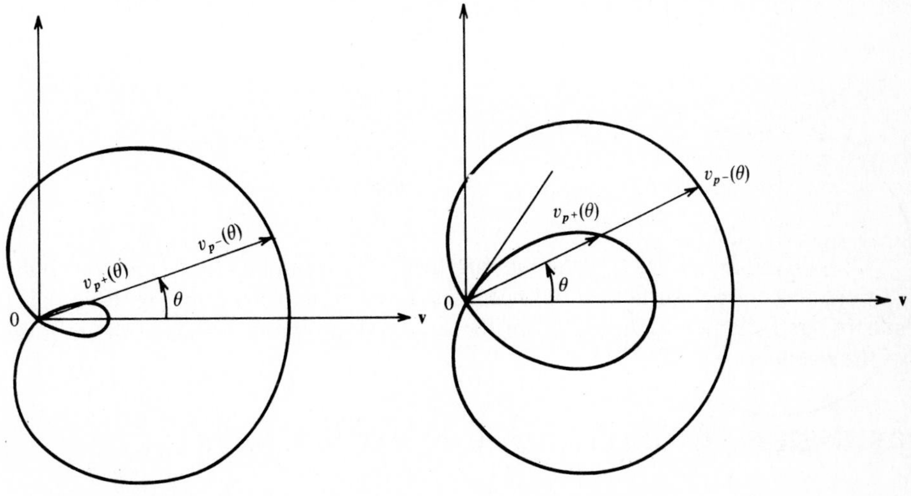

(e) $\beta = 0.30 > 1/\sqrt{\mu\varepsilon}$ (f) $\beta = 0.6 > 1/\sqrt{\mu\varepsilon}$

Figure 8.3 (Continued)

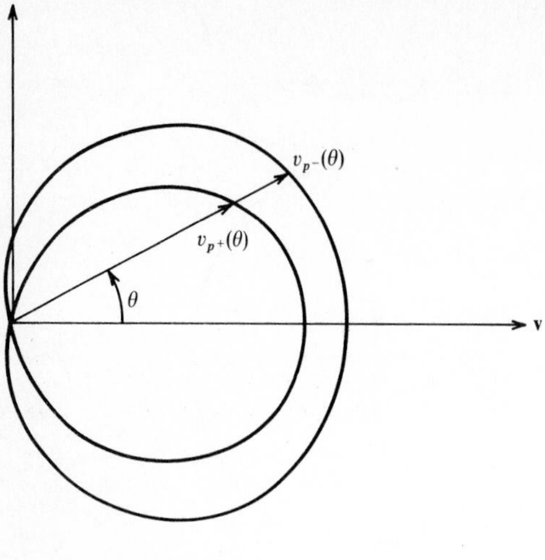

(g) $\beta = 0.9 > 1/\sqrt{\mu\varepsilon}$

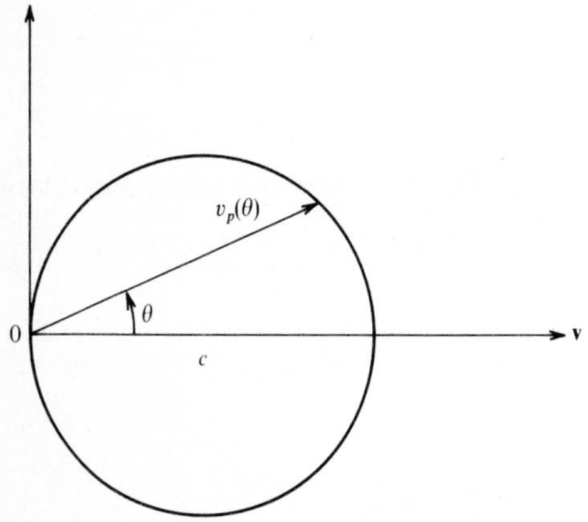

(h) $\beta = 1$

Figure 8.3 (Continued)

Finally, if the velocity of the moving medium is equal to the velocity of light ($\beta = 1$), the normal surface (8.61) becomes a sphere as shown in Fig. 8-3*h*. This sphere has a diameter c drawn from the origin along the direction of the motion of the medium.

8.5 GROUP VELOCITY AND RAY VECTOR SURFACE

Similar to the case of stationary media, by combining Minkowski's constitutive relations (8.14) and (8.15) with the Maxwell equations (8.4), we easily establish the

Poynting theorem for a moving medium:

$$-\nabla \cdot \mathscr{P} = \frac{\partial}{\partial t}\left(\tfrac{1}{2}\,\mathscr{E} \cdot \mathscr{D} + \tfrac{1}{2}\,\mathscr{H} \cdot \mathscr{B}\right) \tag{8.64}$$

where

$$\mathscr{P} = \mathscr{E} \times \mathscr{H} \tag{8.65}$$

is the Poynting vector. Thus

$$W = \tfrac{1}{2}\,(\mathscr{E} \cdot \mathscr{D} + \mathscr{H} \cdot \mathscr{B}) \tag{8.66}$$

may again be interpreted as the stored energy density. For a monochromatic plane wave of the form (8.31), we obtain the time-averaged values of Eqs. (8.65) and (8.66), namely,

$$\langle \mathscr{P} \rangle = \tfrac{1}{2}\,\mathrm{Re}\,(\mathbf{E}_0 \times \mathbf{H}_0^*) \tag{8.67}$$

and

$$\langle W \rangle = \tfrac{1}{4}\,\mathrm{Re}\,(\mathbf{E}_0^* \cdot \mathbf{D}_0 + \mathbf{H}_0^* \cdot \mathbf{B}_0) \tag{8.68}$$

To express the time-averaged Poynting vector and energy density in terms of the wave vector and the velocity of the moving medium, we note that

$$\mathbf{E}_0 = C_1 \mathbf{e}_1 + C_2 \mathbf{e}_2 \tag{8.69}$$

Thus

$$\begin{aligned} \mathbf{H}_0 &= \frac{1}{\omega \mu_0 \mu}\,\bar{\boldsymbol{\alpha}}^{-1} \cdot (\mathbf{p} \times \mathbf{E}_0) \\ &= \frac{C_1}{\omega \mu_0 \mu}\,\mathbf{e}_2 - \omega \varepsilon_0 \varepsilon\, C_2\, \mathbf{e}_1 \end{aligned} \tag{8.70}$$

and

$$\begin{aligned} \mathbf{D}_0 &= -\frac{1}{\omega}\,(\mathbf{k} \times \mathbf{H}_0) \\ &= \frac{-C_1}{\omega^2 \mu_0 \mu}\,(\mathbf{k} \times \mathbf{e}_2) + \varepsilon_0 \varepsilon C_2 (\mathbf{k} \times \mathbf{e}_1) \end{aligned} \tag{8.71}$$

where C_1 and C_2 are two arbitrary constants, and $\mathbf{e}_1$ and $\mathbf{e}_2$ are given respectively by Eqs. (8.43) and (8.45). Substituting Eqs. (8.69) and (8.70) into Eq. (8.67) we have, after some simplification,

$$\langle \mathscr{P} \rangle = \frac{(|C_1|^2 + k_0^2 \mu\varepsilon |C_2|^2)(\mathbf{k} \times \hat{\mathbf{v}})^2}{2\omega \mu_0 \mu}\,[\mathbf{k} + \xi c^{-2}(\omega - \mathbf{k} \cdot \mathbf{v})\mathbf{v}] \tag{8.72}$$

On the other hand, from the Maxwell equations (8.4) for a plane wave of the form (8.31), we obtain

$$\mathbf{E}_0^* \cdot \mathbf{D}_0 = \mathbf{H}_0 \cdot \mathbf{B}_0^* \tag{8.73}$$

Thus the time-averaged energy density (8.68) becomes

$$\langle W \rangle = \tfrac{1}{2}\,\mathrm{Re}\,(\mathbf{E}_0^* \cdot \mathbf{D}_0) \tag{8.74}$$

Again, substituting Eqs. (8.69) and (8.71) into Eq. (8.74) and using the result of the dispersion equation, we may write Eq. (8.74) as

$$\langle W \rangle = \frac{(|C_1|^2 + k_0^2 \mu\varepsilon |C_2|^2)(\mathbf{k} \times \hat{\mathbf{v}})^2}{2\omega c \mu_0 \mu} \left[\frac{\omega}{c} + \xi c^{-1}(\omega - \mathbf{k} \cdot \mathbf{v}) \right] \tag{8.75}$$

Therefore, the velocity of energy transport is

$$\mathbf{v}_E = \frac{\langle \mathscr{P} \rangle}{\langle W \rangle} = c\,\frac{\mathbf{k} + \xi c^{-2}(\omega - \mathbf{k} \cdot \mathbf{v})\mathbf{v}}{\omega/c + \xi c^{-1}(\omega - \mathbf{k} \cdot \mathbf{v})} \tag{8.76}$$

To obtain the group velocity of a wave in a moving nondispersive medium, we rewrite the wave vector surface given by Eq. (8.60), in the form

$$F(\mathbf{k}, \omega) = \mathbf{k}^2 - \frac{\xi}{c^2}(\omega - \mathbf{k} \cdot \mathbf{v})^2 - \frac{\omega^2}{c^2} = 0 \tag{8.77}$$

Thus

$$\frac{\partial F}{\partial \mathbf{k}} = 2[\mathbf{k} + \xi c^{-2}(\omega - \mathbf{k} \cdot \mathbf{v})\mathbf{v}] \tag{8.78}$$

and

$$\frac{\partial F}{\partial \omega} = -\frac{2}{c}\left[\frac{\omega}{c} + \xi c^{-1}(\omega - \mathbf{k} \cdot \mathbf{v}) \right] \tag{8.79}$$

Substituting Eqs. (8.78) and (8.79) into the definition of the group velocity,

$$\mathbf{v}_g = \frac{\partial \omega}{\partial \mathbf{k}} = -\frac{\partial F/\partial \mathbf{k}}{\partial F/\partial \omega} \tag{8.80}$$

and comparing the result with Eq. (8.76), we see again that the group velocity is equal to the velocity of energy transport, and coincides in direction with the normal to the wave vector surface as indicated by the arrows in Fig. 8.2.

We recall that in the rest frame Σ', the medium is isotropic; thus the group velocity coincides with the phase velocity both in magnitude and direction. The motion of the medium produces a peculiar anisotropy and leads to a difference between the two velocities. A comparison of Eq. (8.76) with the phase velocity,

$$\mathbf{v}_p = \frac{\omega}{k}\hat{\mathbf{k}} \tag{8.81}$$

of the wave shows that the phase velocity is directed along the wave vector **k**, but the group velocity has components both along the wave vector **k** and along the velocity **v** of the moving medium. The latter fact indicates drag of the electromagnetic energy by the moving medium. With the aid of the dispersion equation (8.77), we easily establish from Eqs. (8.76) and (8.81) the following relation between the phase and group velocities in a moving medium:

$$\mathbf{v}_g \cdot \mathbf{v}_p = v_p^2 \tag{8.82}$$

In other words, the projection of the group velocity on the direction of wave normal gives the phase velocity.

From the group velocity, we find the ray vector $\mathbf{s} = s\hat{\mathbf{s}}$ in a moving medium according to Eq. (5.117):

$$\mathbf{s} = \frac{\mathbf{v}_g}{\omega} = \frac{c^2[\mathbf{k} + \xi c^{-2}(\omega - \mathbf{k} \cdot \mathbf{v})\mathbf{v}]}{\omega[\omega + \xi(\omega - \mathbf{k} \cdot \mathbf{v})]} \tag{8.83}$$

To derive the equation satisfied by the ray vector **s**, we dot-multiply Eq. (8.83) by **k** and get

$$\mathbf{k} \cdot \mathbf{v} = \frac{\omega[(1 + \xi)(\mathbf{s} \cdot \mathbf{v})\omega - \xi \mathbf{v}^2]}{c^2 + \xi\omega(\mathbf{s} \cdot \mathbf{v}) - \xi \mathbf{v}^2} \tag{8.84}$$

Eliminating **k** and $\mathbf{k} \cdot \mathbf{v}$ from Eqs. (8.83), (8.84), and the dispersion equation (8.77), we obtain after some simplification, the equation for the ray vector surface:

$$(1 - \beta^2)\mu\varepsilon\omega^2(\mathbf{s} \times \hat{\mathbf{v}})^2 + (\mu\varepsilon - \beta^2)\omega^2(\mathbf{s} \cdot \hat{\mathbf{v}})^2 - 2(\mu\varepsilon - 1)\omega\beta c(\mathbf{s} \cdot \hat{\mathbf{v}}) - c^2(1 - \mu\varepsilon\beta^2) = 0 \tag{8.85}$$

or

$$\frac{(\mathbf{s} \times \hat{\mathbf{v}})^2}{\left(\dfrac{c}{\omega\gamma\sqrt{\mu\varepsilon - \beta^2}}\right)^2} + \frac{\left[(\mathbf{s} \cdot \hat{\mathbf{v}}) - \dfrac{\beta c(\mu\varepsilon - 1)}{\omega(\mu\varepsilon - \beta^2)}\right]^2}{\left[\dfrac{c\sqrt{\mu\varepsilon}}{\omega\gamma^2(\mu\varepsilon - \beta^2)}\right]^2} = 1 \tag{8.86}$$

This clearly shows that the ray surface is an ellipsoid of revolution about the velocity vector **v**. The ellipsoid is centered at $\beta c(\mu\varepsilon - 1)/\omega(\mu\varepsilon - \beta^2)$ and has the semiaxes $c/[\omega\gamma\sqrt{\mu\varepsilon - \beta^2}]$ and $c\sqrt{\mu\varepsilon}/[\omega\gamma^2(\mu\varepsilon - \beta^2)]$. If Ψ is the angle between the ray vector **s** and the velocity **v** of the moving medium, we may solve Eq. (8.85) for s to yield another form of the equation for the ray surface:

$$s = \frac{c}{\omega} \frac{(\mu\varepsilon - 1)\beta \cos \Psi \pm \sqrt{\mu\varepsilon(1 - \beta^2)[1 - \beta^2 - \beta^2(\mu\varepsilon - 1) \sin^2 \Psi]}}{\mu\varepsilon(1 - \beta^2) \sin^2 \Psi + (\mu\varepsilon - \beta^2) \cos^2 \Psi} \tag{8.87}$$

The ray vector surface for various velocities of the moving medium is plotted in Fig. 8.4. A comparison of Fig. 8.3 and Fig. 8.4 shows that at low velocities ($\beta \cong 0$), the ray vector surface and the normal surface are much the same, i.e., the group velocity and phase velocity are nearly identical. As the velocity of the moving medium increases toward $c/\sqrt{\mu\varepsilon}$, the ray surface flattened along the direction of the velocity vector **v** remains more or less spherical and contains the origin, which indicates that the group velocity is possible in all directions. When the velocity of the moving medium is equal to $c/\sqrt{\mu\varepsilon}$, there is a definite difference between the two surfaces (cf. Figs. 8.3*d* and 8.4*d*). The normal surface still lies partly to the left of the origin showing that it is possible for the phase velocity vector $\mathbf{v}_p$ to have a component in the opposite direction to the motion of the medium. However, this is not the case with the ray surface which lies entirely to the right of the origin as we should expect. Thus the group velocity can make

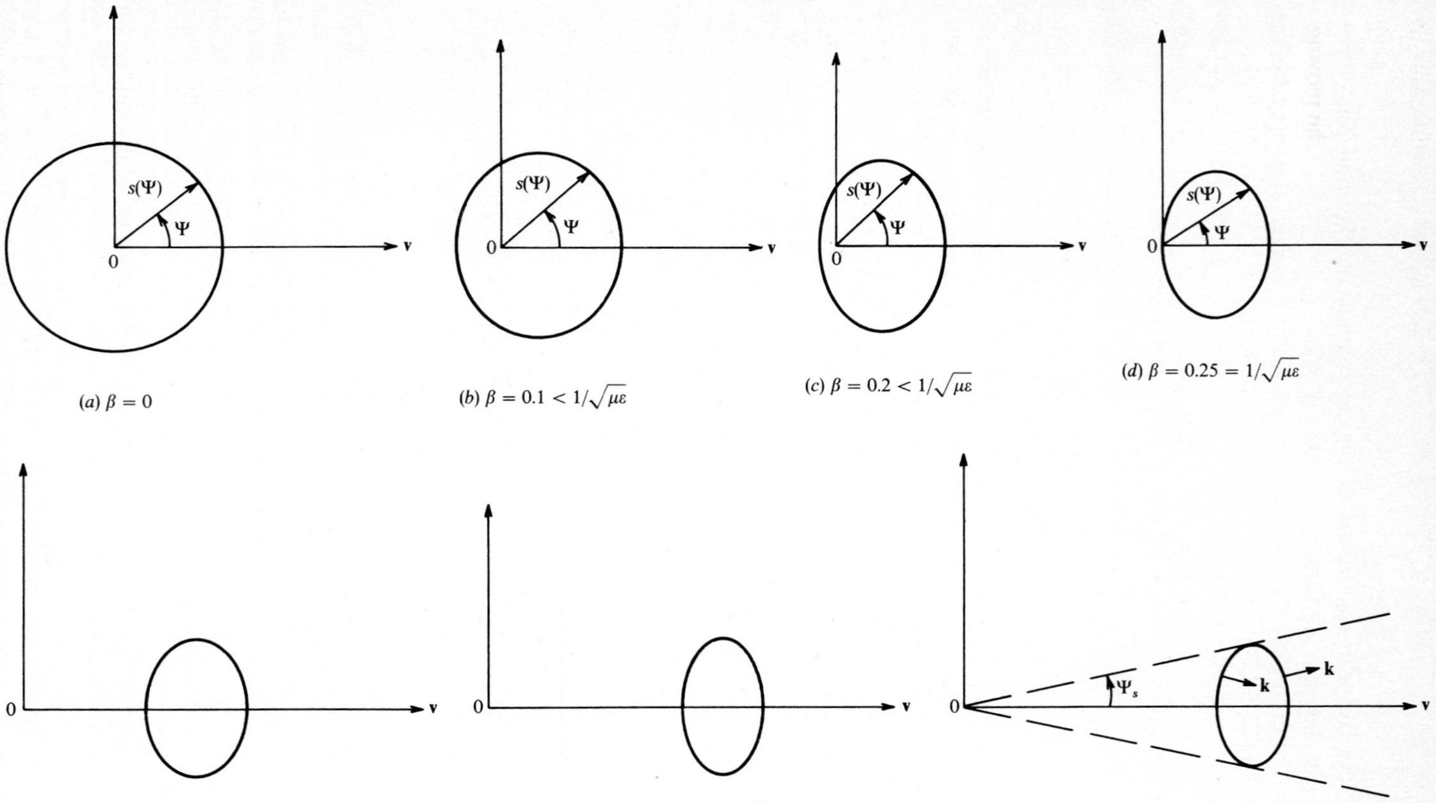

(a) $\beta = 0$

(b) $\beta = 0.1 < 1/\sqrt{\mu\varepsilon}$

(c) $\beta = 0.2 < 1/\sqrt{\mu\varepsilon}$

(d) $\beta = 0.25 = 1/\sqrt{\mu\varepsilon}$

(e) $\beta = 0.30 > 1/\sqrt{\mu\varepsilon}$

(f) $\beta = 0.60 > 1/\sqrt{\mu\varepsilon}$

(g) $\beta = 0.9 > 1/\sqrt{\mu\varepsilon}$

Figure 8.4 Ray vector surfaces of a moving medium. The index of refraction in the rest frame of a nondispersive medium is chosen to be 4 (that is, $\sqrt{\mu\varepsilon} = 4$).

only an acute angle with the velocity of the moving medium. In other words, all the energy propagates in the direction of motion once the velocity of the moving medium reaches $c/\sqrt{\mu\varepsilon}$.

When the velocity of the moving medium exceeds $c/\sqrt{\mu\varepsilon}$, the ray vector surface no longer contains the origin (see Figs. 8.4*e* to *g*). In this case, there are two possible group velocities for each forward direction of propagation and, furthermore, the group velocity vectors lie inside a cone with half-angle Ψ_s defined by

$$\tan \Psi_s = \sqrt{\frac{1-\beta^2}{\beta^2\mu\varepsilon - 1}} \tag{8.88}$$

Outside this cone, Eq. (8.87) shows that s becomes complex. In other words, there is no group velocity corresponding to directions making an angle greater than Ψ_s, and there is only one group velocity for this direction. This, together with the fact that the wave vector is always normal to the ray surface (cf. Sec. 5.8), is illustrated in Fig. 8.4*g*. In summary, for $\beta = 1/\sqrt{\mu\varepsilon}$, the cone is actually the entire right half-space. For $\beta > 1/\sqrt{\mu\varepsilon}$, the half-angle of the cone decreases as the velocity of the medium increases. Of course, for $\beta < 1/\sqrt{\mu\varepsilon}$, there is no cone; the energy propagates in all directions.

8.6 LAWS OF REFLECTION AND REFRACTION

We shall next consider the problem of reflection and transmission of a plane wave by a uniformly moving medium. The geometry of the problem is shown in Fig. 8.5. Medium 1 is free space and medium 2, characterized by the dielectric constant ε and permeability μ, is moving uniformly parallel to the interface with velocity $\mathbf{v}$.

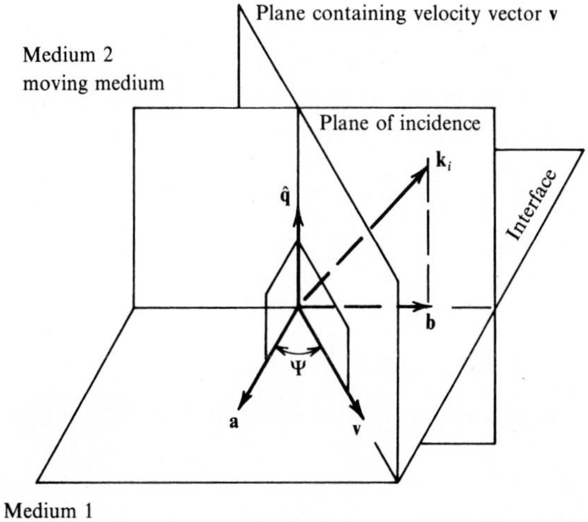

Figure 8.5 Geometry of the problem: orientation of the velocity vector $\mathbf{v}$ of the moving medium 2.

A monochromatic plane wave traveling in free space in an arbitrary direction is incident upon the interface; as a result there will be a reflected wave and a transmitted wave (cf. Secs. 8.3 and 8.4). According to Eq. (8.30), when the medium is moving parallel to the interface, the boundary conditions on the field vectors are the same as for stationary media. Thus, by matching the phase of the waves at the interface as in Sec. 3.5, we conclude that the wave vectors of the incident, the reflected, and the transmitted waves must satisfy

$$\mathbf{k}_r \times \hat{\mathbf{q}} = \mathbf{k}_t \times \hat{\mathbf{q}} = \mathbf{k}_i \times \hat{\mathbf{q}} = \mathbf{a} \tag{8.89}$$

or

$$\begin{aligned} \mathbf{k}_i &= \mathbf{b} + q_i \hat{\mathbf{q}} \\ \mathbf{k}_r &= \mathbf{b} + q_r \hat{\mathbf{q}} \\ \mathbf{k}_t &= \mathbf{b} + q_t \hat{\mathbf{q}} \end{aligned} \tag{8.90}$$

where q_i, q_r, q_t, and $\mathbf{b}$ are defined by Eqs. (3.105) and (3.106) respectively. Combination of Eq. (8.90) with the dispersion equation in medium 1 (free space) yields

$$\begin{aligned} k_i &= k_r = k_0 = \omega\sqrt{\mu_0 \varepsilon_0} \\ q_i &= -q_r = \sqrt{k_0^2 - \mathbf{a}^2} \end{aligned} \tag{8.91}$$

Again, Eqs. (8.89) to (8.91) show that the three wave vectors $\mathbf{k}_i$, $\mathbf{k}_r$, and $\mathbf{k}_t$ lie on the same plane, the plane of incidence, and the angle of reflection is equal to the angle of incidence. These two results are no different from the case in which medium 2 is stationary. However, Snell's law of refraction, which relates the angle of transmission to the angle of incidence, is modified. By taking the magnitude on both sides of the second equal sign in Eq. (8.89), we get

$$k_0 \sin\theta_i = k_t \sin\theta_t \tag{8.92}$$

where θ_i and θ_t are respectively the angle of incidence and the angle of transmission. Since medium 2 is moving parallel to the interface, $\hat{\mathbf{v}} \cdot \hat{\mathbf{q}} = 0$ and

$$\hat{\mathbf{k}}_t \cdot \hat{\mathbf{v}} = \frac{k_0}{k_t}(\hat{\mathbf{k}}_i \cdot \hat{\mathbf{v}}) \tag{8.93}$$

Substituting Eq. (8.93) into the dispersion equation (8.60) of the moving medium we obtain, after some simplifications, the wave number of the transmitted wave:

$$k_t = k_0\sqrt{1 + \xi[1 - \beta(\hat{\mathbf{k}}_i \cdot \hat{\mathbf{v}})]^2} \tag{8.94}$$

or, after making use of Eq. (8.92),

$$\sin\theta_t = \frac{\sin\theta_i}{\sqrt{1 + \xi[1 - \beta(\hat{\mathbf{k}}_i \cdot \hat{\mathbf{v}})]^2}} \tag{8.95}$$

This is the desired Snell's law of refraction for a moving half-space. From Eqs. (8.90) and (8.94), we easily obtain

$$q_t = \sqrt{q_i^2 + \xi[k_0 - \beta(\mathbf{k}_i \cdot \hat{\mathbf{v}})]^2} \tag{8.96}$$

In other words, for a given incident wave, the wave vector $\mathbf{k}_t$ of the transmitted wave in the moving medium is uniquely determined by Eqs. (8.96) and (8.90).

8.7 REFLECTION AND TRANSMISSION COEFFICIENTS

Having determined the wave vectors in both media, we may now proceed to find the reflection and transmission coefficients by a moving half-space. Since medium 1 is isotropic, we may again decompose the incident and the reflected waves into perpendicular and parallel polarizations (see Fig. 8.6), namely,

Incident wave

$$\begin{aligned}\mathbf{E}_{0i} &= A_\perp \mathbf{a} + A_\parallel(\hat{\mathbf{k}}_i \times \mathbf{a})\\ \mathbf{H}_{0i} &= \frac{1}{\omega\mu_0}(\mathbf{k}_i \times \mathbf{E}_{0i})\\ &= \frac{1}{\eta_0}[A_\perp(\hat{\mathbf{k}}_i \times \mathbf{a}) - A_\parallel \mathbf{a}]\end{aligned} \tag{8.97}$$

Reflected wave

$$\begin{aligned}\mathbf{E}_{0r} &= B_\perp \mathbf{a} + B_\parallel(\hat{\mathbf{k}}_r \times \mathbf{a})\\ \mathbf{H}_{0r} &= \frac{1}{\omega\mu_0}(\mathbf{k}_r \times \mathbf{E}_{0r})\\ &= \frac{1}{\eta_0}[B_\perp(\hat{\mathbf{k}}_r \times \mathbf{a}) - B_\parallel \mathbf{a}]\end{aligned} \tag{8.98}$$

where

$$\eta_0 = \sqrt{\frac{\mu_0}{\varepsilon_0}}$$

Since the electric field intensity in a moving medium is no longer transverse to the direction of wave propagation [cf. Eq. (8.48)], little is gained by decomposing it into parallel and perpendicular polarizations. Thus, for a transmitted wave, we take

$$\mathbf{E}_{0t} = C_1\mathbf{e}_1 + C_2\mathbf{e}_2 \tag{8.99}$$

where

$$\begin{aligned}\mathbf{e}_1 &= \mathbf{k}_t \times \hat{\mathbf{v}}\\ \mathbf{e}_2 &= \frac{1}{\alpha}[(\mathbf{p}_t \cdot \hat{\mathbf{v}})\mathbf{p}_t - k_0^2\alpha^2\mu\varepsilon\hat{\mathbf{v}}]\end{aligned} \tag{8.100}$$

and

$$\begin{aligned}\mathbf{p}_t &= \mathbf{k}_t + k_0 m\hat{\mathbf{v}}\\ \mathbf{k}_t &= \mathbf{b} + q_t\hat{\mathbf{q}}\\ q_t &= \sqrt{q_i^2 + \xi[k_0 - \beta(\mathbf{k}_i \cdot \hat{\mathbf{v}})]^2}\end{aligned} \tag{8.101}$$

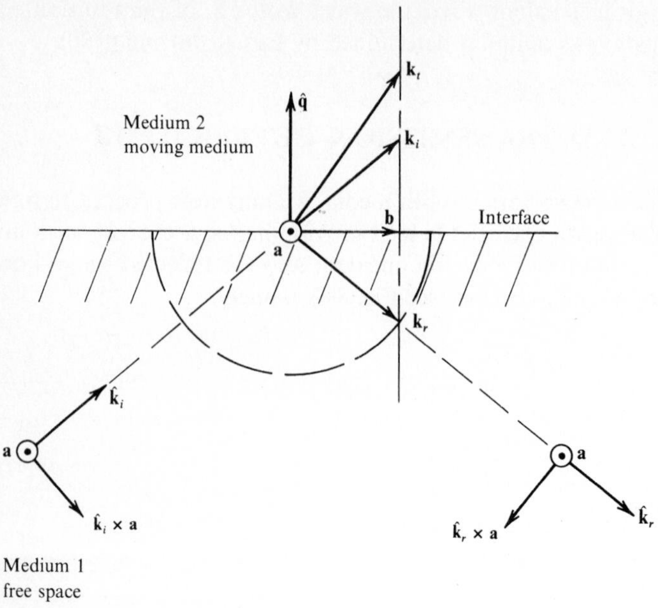

Figure 8.6 Oblique incidence: decomposition of field vectors into perpendicular and parallel polarizations.

The magnetic field corresponding to Eq. (8.99) is

$$\mathbf{H}_{0t} = C_1\mathbf{h}_1 + C_2\mathbf{h}_2 \tag{8.102}$$

where

$$\mathbf{h}_1 = \frac{1}{\omega\mu_0\mu}\mathbf{e}_2 \qquad \mathbf{h}_2 = -\omega\varepsilon_0\varepsilon\mathbf{e}_1 \tag{8.103}$$

and C_1 and C_2 are two arbitrary constants to be determined.

At the interface, the field vectors satisfy the boundary conditions [cf. (8.30)]:

$$\mathbf{E}_{0i} + \mathbf{E}_{0r} - \mathbf{E}_{0t} = m_1\hat{\mathbf{q}} \qquad \mathbf{H}_{0i} + \mathbf{H}_{0r} - \mathbf{H}_{0t} = m_2\hat{\mathbf{q}} \tag{8.104}$$

where m_1 and m_2 are two constants. Substituting Eqs. (8.97), (8.98), (8.99), and (8.102) into Eq. (8.104) and then taking the dot product of the resulted equations with **a** and **b** respectively, we obtain

$$(A_\perp + B_\perp)\mathbf{a}^2 - (\mathbf{a}\cdot\mathbf{e}_1)C_1 - (\mathbf{a}\cdot\mathbf{e}_2)C_2 = 0$$
$$(A_\perp - B_\perp)\mathbf{a}^2 - \frac{\mathbf{b}\cdot\mathbf{e}_2}{q_i\mu}C_1 + \frac{k_0^2\varepsilon(\mathbf{b}\cdot\mathbf{e}_1)}{q_i}C_2 = 0 \tag{8.105}$$

and

$$(A_{\parallel} + B_{\parallel})\mathbf{a}^2 + \frac{\mathbf{a}\cdot\mathbf{e}_2}{k_0\,\mu}\,C_1 - k_0\,\varepsilon(\mathbf{a}\cdot\mathbf{e}_1)C_2 = 0$$
$$(A_{\parallel} - B_{\parallel})\mathbf{a}^2 - \frac{k_0(\mathbf{b}\cdot\mathbf{e}_1)}{q_i}\,C_1 - \frac{k_0(\mathbf{b}\cdot\mathbf{e}_2)}{q_i}\,C_2 = 0 \qquad (8.106)$$

In deriving Eqs. (8.105) and (8.106), we have used the following relations:

$$\mathbf{b}\cdot(\hat{\mathbf{k}}_i \times \mathbf{a}) = -\mathbf{b}\cdot(\hat{\mathbf{k}}_r \times \mathbf{a}) = \frac{q_i\,\mathbf{a}^2}{k_0}$$
$$\eta_0(\mathbf{a}\cdot\mathbf{h}_1) = \frac{\mathbf{a}\cdot\mathbf{e}_2}{k_0\,\mu}$$
$$\eta_0(\mathbf{a}\cdot\mathbf{h}_2) = -k_0\,\varepsilon(\mathbf{a}\cdot\mathbf{e}_2) \qquad (8.107)$$
$$\eta_0(\mathbf{b}\cdot\mathbf{h}_1) = \frac{\mathbf{b}\cdot\mathbf{e}_2}{k_0\,\mu}$$
$$\eta_0(\mathbf{b}\cdot\mathbf{h}_2) = -k_0\,\varepsilon(\mathbf{b}\cdot\mathbf{e}_2)$$

But from Eq. (8.100) we also have

$$\mathbf{a}\cdot\mathbf{e}_1 = -q_t(\mathbf{b}\cdot\hat{\mathbf{v}})$$
$$\mathbf{b}\cdot\mathbf{e}_1 = q_t(\mathbf{a}\cdot\hat{\mathbf{v}})$$
$$\mathbf{a}\cdot\mathbf{e}_2 = \frac{(\mathbf{p}_t\cdot\hat{\mathbf{v}})(\mathbf{p}_t\cdot\mathbf{a}) - (\mathbf{p}_t\cdot\bar{\alpha}\cdot\mathbf{p}_t)(\mathbf{a}\cdot\hat{\mathbf{v}})}{\alpha} \qquad (8.108)$$
$$\mathbf{b}\cdot\mathbf{e}_2 = \frac{(\mathbf{p}_t\cdot\hat{\mathbf{v}})(\mathbf{p}_t\cdot\mathbf{b}) - (\mathbf{p}_t\cdot\bar{\alpha}\cdot\mathbf{p}_t)(\mathbf{b}\cdot\hat{\mathbf{v}})}{\alpha}$$

where q_t and $\mathbf{p}_t$ are given in Eq. (8.101). Now, adding and subtracting the two equations in Eq. (8.105) and in Eq. (8.106), respectively, we may write the final results in the following matrix forms:

$$\frac{1}{2\mathbf{a}^2}\begin{bmatrix} J_+ & K_+ \\ L_+ & M_+ \end{bmatrix}\begin{bmatrix} C_1 \\ C_2 \end{bmatrix} = \begin{bmatrix} A_\perp \\ A_\parallel \end{bmatrix} \qquad (8.109)$$

and

$$\frac{1}{2\mathbf{a}^2}\begin{bmatrix} J_- & K_- \\ L_- & M_- \end{bmatrix}\begin{bmatrix} C_1 \\ C_2 \end{bmatrix} = \begin{bmatrix} B_\perp \\ B_\parallel \end{bmatrix} \qquad (8.110)$$

where

$$
\begin{aligned}
J_{\pm} &= \mathbf{a} \cdot \mathbf{e}_1 \pm \frac{\mathbf{b} \cdot \mathbf{e}_2}{q_i \mu} \\
K_{\pm} &= \mathbf{a} \cdot \mathbf{e}_2 \mp \frac{k_0^2 \varepsilon(\mathbf{b} \cdot \mathbf{e}_1)}{q_i} \\
L_{\pm} &= -\frac{\mathbf{a} \cdot \mathbf{e}_2}{k_0 \mu} \pm \frac{k_0(\mathbf{b} \cdot \mathbf{e}_1)}{q_i} \\
M_{\pm} &= k_0 \varepsilon(\mathbf{a} \cdot \mathbf{e}_1) \pm \frac{k_0(\mathbf{b} \cdot \mathbf{e}_2)}{q_i}
\end{aligned}
\tag{8.111}
$$

Multiplication of Eq. (8.109) by the inverse of the 2 × 2 coefficient matrix yields

$$
\begin{bmatrix} C_1 \\ C_2 \end{bmatrix} = \begin{bmatrix} T_{11} & T_{12} \\ T_{21} & T_{22} \end{bmatrix} \begin{bmatrix} A_\perp \\ A_\parallel \end{bmatrix} \tag{8.112}
$$

where the transmission coefficients are given by

$$
\begin{aligned}
T_{11} &= \frac{2\mathbf{a}^2 M_+}{\Delta_m} \\
T_{12} &= \frac{-2\mathbf{a}^2 K_+}{\Delta_m} \\
T_{21} &= \frac{-2\mathbf{a}^2 L_+}{\Delta_m} \\
T_{22} &= \frac{2\mathbf{a}^2 J_+}{\Delta_m}
\end{aligned}
\tag{8.113}
$$

and

$$
\Delta_m = J_+ M_+ - K_+ L_+ \tag{8.114}
$$

Substitution of Eq. (8.112) into Eq. (8.110) gives

$$
\begin{bmatrix} B_\perp \\ B_\parallel \end{bmatrix} = \begin{bmatrix} \Gamma_{11} & \Gamma_{12} \\ \Gamma_{21} & \Gamma_{22} \end{bmatrix} \begin{bmatrix} A_\perp \\ A_\parallel \end{bmatrix} \tag{8.115}
$$

where the reflection coefficients are

$$
\begin{aligned}
\Gamma_{11} &= \frac{M_+ J_- - L_+ K_-}{\Delta_m} \\
\Gamma_{12} &= \frac{J_+ K_- - K_+ J_-}{\Delta_m} \\
\Gamma_{21} &= \frac{M_+ L_- - L_+ M_-}{\Delta_m} \\
\Gamma_{22} &= \frac{J_+ M_- - K_+ L_-}{\Delta_m}
\end{aligned}
\tag{8.116}
$$

We see from Eqs. (8.112) and (8.115) that the boundary conditions at the moving interface give rise to the transmission and reflection coefficient matrices which contain both the diagonal and off-diagonal terms. In other words, if the incident wave has the electric field intensity perpendicular to the plane of incidence, the transmitted and the reflected waves will have components both perpendicular and parallel to the plane of incidence.

The transmission and reflection coefficients given in Eqs. (8.113) and (8.116) are valid for the general case when the medium is moving parallel to the interface in an arbitrary direction with respect to the plane of incidence. They are considerably simplified in the following two special cases.

Case 1. The medium is moving uniformly parallel to the plane of incidence In this case $\hat{\mathbf{v}} = \hat{\mathbf{b}}$, and the wave vector in the moving medium is

$$\mathbf{k}_t = \mathbf{b} + q_t\hat{\mathbf{q}} = k_t\hat{\mathbf{k}}_t \tag{8.117}$$

where

$$\begin{aligned} k_t &= k_0\sqrt{1 + \xi(1 - \beta \sin \theta_i)^2} \\ q_t &= \sqrt{q_i^2 + k_0^2\,\xi(1 - \beta \sin \theta_i)^2} \end{aligned} \tag{8.118}$$

and Eq. (8.108) becomes

$$\begin{aligned} \mathbf{a} \cdot \mathbf{e}_1 &= -aq_t \\ \mathbf{b} \cdot \mathbf{e}_2 &= -aq_t^2 \\ \mathbf{a} \cdot \mathbf{e}_2 &= \mathbf{b} \cdot \mathbf{e}_1 = 0 \end{aligned} \tag{8.119}$$

Substituting Eq. (8.119) into Eq. (8.111), we have

$$\begin{aligned} J_\pm &= -aq_t\left(1 \pm \frac{q_t}{\mu q_i}\right) \\ M_\pm &= -ak_0\,q_t\left(\varepsilon \pm \frac{q_t}{q_i}\right) \\ K_\pm &= L_\pm = 0 \end{aligned} \tag{8.120}$$

Thus the transmission coefficients (8.113) reduce to

$$\begin{aligned} T_{11} &= \frac{-2a\mu q_i}{q_t(q_t + \mu q_i)} \\ T_{22} &= \frac{-2aq_i}{k_0\,q_t(q_t + \varepsilon q_i)} \\ T_{12} &= T_{21} = 0 \end{aligned} \tag{8.121}$$

and the reflection coefficients (8.116) become

$$\Gamma_{11} = \frac{\mu q_i - q_t}{\mu q_i + q_t}$$
$$\Gamma_{22} = \frac{\varepsilon q_i - q_t}{\varepsilon q_i + q_t} \tag{8.122}$$
$$\Gamma_{12} = \Gamma_{21} = 0$$

We note that in this case a perpendicularly (or parallel) polarized incident wave gives rise to the perpendicularly (or parallel) polarized transmitted and reflected waves.

Case 2. The medium is moving uniformly perpendicular to the plane of incidence Here we have $\hat{\mathbf{v}} = \hat{\mathbf{a}}$, and thus

$$k_t = k_0\sqrt{1+\xi} = k_0\,\gamma\sqrt{\mu\varepsilon - \beta^2}$$
$$q_t = \sqrt{q_i^2 + k_0^2\,\xi} = \sqrt{k_0^2\,\gamma^2(\mu\varepsilon - \beta^2) - \mathbf{a}^2} \tag{8.123}$$

and, from Eq. (8.108)

$$\mathbf{a}\cdot\mathbf{e}_1 = 0$$
$$\mathbf{b}\cdot\mathbf{e}_1 = aq_t$$
$$\mathbf{a}\cdot\mathbf{e}_2 = -ak_t^2 \tag{8.124}$$
$$\mathbf{b}\cdot\mathbf{e}_2 = \frac{k_0\,m\mathbf{a}^2}{\alpha}$$

Substitution of Eq. (8.124) into Eq. (8.111) yields

$$J_+ = -J_- = \frac{k_0\,m\mathbf{a}^2}{\mu\alpha q_i}$$
$$M_+ = -M_- = \frac{k_0^2\,m\mathbf{a}^2}{\alpha q_i}$$
$$K_\pm = -a\left(k_t^2 \pm \frac{k_0^2\,\varepsilon q_t}{q_i}\right) \tag{8.125}$$
$$L_\pm = a\left(\frac{k_t^2}{k_0\,\mu} \pm \frac{k_0\,q_t}{q_i}\right)$$

Hence the transmission and reflection coefficients follow from Eqs. (8.113) and (8.116) respectively.

8.8 NORMAL INCIDENCE

When the incidence is normal to the interface, the results obtained in the previous section fail due to the fact that there is no unique plane of incidence. In this case

We note that $\Gamma_{11}^{(n)}$ and $\Gamma_{22}^{(n)}$ are not equal, which is contrary to the nonmoving case.

8.9 BREWSTER'S ANGLE FOR A MOVING DIELECTRIC

When a monochromatic plane wave is incident on an interface of two stationary media, there exists an angle of incidence called the *Brewster angle* under which the reflected wave becomes completely linearly polarized. We shall now investigate this phenomenon for a dielectric ($\mu = 1$) half-space moving uniformly parallel to its interface.

As discussed in Sec. 6.7, the reflected wave will be linearly polarized if the reflection coefficients satisfy the condition

$$\Gamma_{11}\Gamma_{22} - \Gamma_{12}\Gamma_{21} = 0 \tag{8.138}$$

which in turn determines the Brewster angle. Under this condition, the azimuthal angle α_r of the reflected wave can be found from

$$\tan \alpha_r = \frac{\Gamma_{12}}{\Gamma_{22}} = \frac{\Gamma_{11}}{\Gamma_{21}} \tag{8.139}$$

Substitution of Eq. (8.116) into Eqs. (8.138) and (8.139) gives the results in terms of J_-, M_-, K_-, and L_-, namely,

$$J_- M_- - K_- L_- = 0 \tag{8.140}$$

and

$$\tan \alpha_r = \frac{J_-}{L_-} \tag{8.141}$$

In the following, we shall consider the two cases discussed in Sec. 8.7 in detail.

Case 1. The dielectric medium ($\mu = 1$) is moving uniformly parallel to the plane of incidence In this case, two possibilities arise—$\hat{\mathbf{v}}$ being either parallel or antiparallel to $\hat{\mathbf{b}}$. First, if $\hat{\mathbf{v}} = \hat{\mathbf{b}}$, condition (8.140) with the aid of Eq. (8.120) becomes

$$q_t = \varepsilon q_i \tag{8.142}$$

Substituting Eq. (8.118) into Eq. (8.142) and noting that $q_i = k_0 \cos \theta_B$, we obtain, after some algebraic simplifications,

$$\tan \theta_B = \gamma(\beta\sqrt{\varepsilon + 1} \pm \sqrt{\varepsilon}) \tag{8.143}$$

where θ_B is the Brewster angle. We note here that when $\beta = 0$, Eq. (8.143) is reduced to the familiar equation $\tan \theta_B = \sqrt{\varepsilon}$ for the stationary medium [cf. (4.289)].

Since the Brewster angles are measured from the normal of the interface and are positive values (that is, $0 \leq \theta_B \leq \pi/2$), for $\beta \leq \sqrt{\varepsilon}/\sqrt{\varepsilon + 1}$, only the plus sign in Eq. (8.143) is admissible. However, when $\sqrt{\varepsilon}/\sqrt{\varepsilon + 1} < \beta \leq 1$, both the plus and minus signs are admissible.

Secondly, if $\hat{\mathbf{v}} = -\hat{\mathbf{b}}$, following the same procedure as before we obtain

$$\tan \theta_B = \gamma(-\beta\sqrt{\varepsilon + 1} + \sqrt{\varepsilon}) \tag{8.144}$$

This equation gives one solution for $0 \leq \beta \leq \sqrt{\varepsilon}/\sqrt{\varepsilon + 1}$. From Eqs. (8.120) and (8.141), we see that $\alpha_r = \pi/2$. Thus the reflected wave is linearly polarized with the electric field intensity normal to the plane of incidence.

Case 2. The medium is moving uniformly perpendicular to the plane of incidence Here, we substitute Eq. (8.125) into Eq. (8.140) and get

$$\frac{m^2 a^2}{\alpha^2 q_i^2} + \left(\frac{k_t^2}{k_0^2} - \varepsilon \frac{q_t}{q_i}\right)\left(\frac{k_t^2}{k_0^2} - \frac{q_t}{q_i}\right) = 0 \tag{8.145}$$

To obtain the Brewster angle, we note from Eq. (8.123) that

$$\frac{k_t^2}{k_0^2} = \gamma^2(\varepsilon - \beta^2)$$

and

$$\frac{q_t}{q_i} = \gamma\sqrt{\varepsilon - \beta^2 + (\varepsilon - 1)\tan^2\theta_B} \tag{8.146}$$

thus Eq. (8.145) becomes

$$\tan^4\theta_B + [1 - \varepsilon + (1 + \varepsilon)\beta^2]\tan^2\theta_B - (\varepsilon - \beta^2 - \varepsilon\beta^2) = 0 \tag{8.147}$$

Hence

$$\tan \theta_B = \sqrt{\varepsilon - (1 + \varepsilon)\beta^2} \tag{8.148}$$

This equation has an admissible solution only when $\beta \leq \sqrt{\varepsilon}/\sqrt{\varepsilon + 1}$. It should be noted that Eq. (8.148) is also valid when $\hat{\mathbf{v}} = -\hat{\mathbf{a}}$.

To find the polarization of the reflected wave, we substitute Eq. (8.125) into Eq. (8.141) and get

$$\tan \alpha_r = -\frac{1}{\beta}\tan \theta_B \tag{8.149}$$

That is, the reflected wave is linearly polarized with the electric field intensity making an angle α_r with the plane of incidence.

From these two special cases we may conclude in general that both the Brewster angle and the state of polarization of the reflected wave vary with the dielectric constant, the velocity and the relative orientation of the plane of incidence with respect to the velocity vector of the moving medium.

8.10 TOTAL REFLECTION

Finally, we shall conclude this chapter by examining how the phenomenon of total reflection is affected by the motion of medium 2. For this purpose, we assume that medium 1, characterized by the dielectric constant ε_1 and permeability μ_0, is at rest with respect to Σ, and medium 2, characterized by the dielectric constant ε_2 and permeability μ_0, is moving uniformly parallel to the interface but otherwise in an arbitrary direction as shown in Fig. 8.5. The transmitted wave in the moving medium 2 takes the form

$$\mathscr{E}_t = \text{Re}\,[\mathbf{E}_{0t}\, e^{i(\mathbf{k}_t \cdot \mathbf{r} - \omega t)}] \tag{8.150}$$

where the wave vector $\mathbf{k}_t = \mathbf{b} + q_t \hat{\mathbf{q}}$ satisfies the dispersion equation (8.60). Since medium 2 moves parallel to the interface, $\mathbf{v} \cdot \hat{\mathbf{q}} = 0$ and hence

$$\mathbf{k}_t \cdot \hat{\mathbf{v}} = \mathbf{k}_i \cdot \hat{\mathbf{v}} = k_0 \sqrt{\varepsilon_1} (\hat{\mathbf{k}}_i \cdot \hat{\mathbf{v}}) \tag{8.151}$$

Substituting Eq. (8.151) into the dispersion equation (8.60) of a moving medium, we obtain

$$\begin{aligned} q_t &= \sqrt{k_t^2 - \mathbf{a}^2} \\ &= \sqrt{k_0^2 - \mathbf{a}^2 + k_0^2 \gamma^2 (\varepsilon_2 - 1)[1 - \beta\sqrt{\varepsilon_1}(\hat{\mathbf{k}}_i \cdot \hat{\mathbf{v}})]^2} \end{aligned} \tag{8.152}$$

Noting that

$$\mathbf{a}^2 = k_0^2 \varepsilon_1 \sin^2 \theta_i$$

$$\hat{\mathbf{k}}_i \cdot \hat{\mathbf{v}} = \sin \theta_i \sin \Psi$$

where θ_i is the angle of incidence and Ψ is the angle between vectors **a** and **v**, we may write Eq. (8.152) as

$$q_t = k_0 \sqrt{1 - \varepsilon_1 \sin^2 \theta_i + \gamma^2 (\varepsilon_2 - 1)(1 - \beta\sqrt{\varepsilon_1} \sin \theta_i \sin \Psi)^2} \tag{8.153}$$

As in the case of stationary media, total reflection from a moving medium occurs when q_t is imaginary. Rearranging the terms inside the square root sign, we see that total reflection is possible if the angle of incidence satisfies the following inequality:

$$N_1 \sin^2 \theta_i - 2N_2 \sin \theta_i + N_3 \le 0 \tag{8.154}$$

or

$$N_1 (\sin \theta_i - \zeta_+)(\sin \theta_i - \zeta_-) \le 0 \tag{8.155}$$

where

$$\begin{aligned} N_1 &= [1 + (\varepsilon_2 - 1)\sin^2 \Psi]\beta^2 - 1 \\ N_2 &= \frac{\beta(\varepsilon_2 - 1)\sin \Psi}{\sqrt{\varepsilon_1}} \\ N_3 &= \frac{\varepsilon_2 - \beta^2}{\varepsilon_1} \end{aligned} \tag{8.156}$$

and

$$\zeta_\pm = \frac{N_2 \pm \sqrt{N_2^2 - N_1 N_3}}{N_1} \tag{8.157}$$

Here we assume that $\varepsilon_1, \varepsilon_2 > 1$, $0 \le \theta_i \le \pi/2$, and ζ_+ and ζ_- take real values.

From Eq. (8.154) or Eq. (8.155), we may study the following cases as β varies from 0 to 1.

Case 1. $N_1 < 0$ **or** $0 \le \beta < 1/\sqrt{1 + (\varepsilon_2 - 1)\sin^2 \Psi}$ In this case, ζ_+ is negative and ζ_- is positive. Depending upon the value of ζ_-, we further find the following two possiblities: (*a*) if $\zeta_- > 1$, there is no total reflection; (*b*) if $\zeta_- \le 1$, total reflection occurs when $\zeta_- \le \sin\theta_i \le 1$.

Case 2. $N_1 = 0$ **or** $\beta = 1/\sqrt{1 + (\varepsilon_2 - 1)\sin^2 \Psi}$ When $\pi < \Psi < 2\pi$, there is no total reflection. On the other hand, when $0 < \Psi < \pi$, we have the following two possibilities: (*a*) if $N_3/2N_2 > 1$, there is no total reflection; (*b*) if $N_3/2N_2 < 1$, total reflection is possible for $N_3/2N_2 < \sin\theta_i < 1$.

Case 3. $N_1 > 0$ **or** $1/\sqrt{1 + (\varepsilon_2 - 1)\sin^2 \Psi} < \beta \le 1$ When $\pi < \Psi < 2\pi$, both ζ_+ and ζ_- are negative; thus there is no total reflection. However, when $0 < \Psi < \pi$, ζ_+ and ζ_- are positive. Depending upon the values of ζ_+ and ζ_-, we have the following three possibilities: (*a*) if $\zeta_- > 1$, there is no total reflection; (*b*) if $\zeta_- \le 1$ and $\zeta_+ > 1$, total reflection occurs when $\zeta_- \le \sin\theta_i \le 1$; (*c*) if $\zeta_+ \le 1$, total reflection occurs when $\zeta_- \le \sin\theta_i \le \zeta_+$.

Finally, in the special case of $\Psi = 0$ or π, total reflection occurs when

$$\frac{\gamma\sqrt{\varepsilon_2 - \beta^2}}{\sqrt{\varepsilon_1}} \le \sin\theta_i \le 1 \tag{8.158}$$

or

$$0 \le \beta \le \frac{\sqrt{\varepsilon_1 - \varepsilon_2}}{\sqrt{\varepsilon_1 - 1}} \tag{8.159}$$

with

$$\varepsilon_1 > \varepsilon_2 \ge 1$$

and does not occur otherwise.

PROBLEMS

8.1 Define dual quantities by comparing the constitutive relations (8.9) and (8.10). Apply duality to Eq. (8.14) to obtain Eq. (8.15) and vice versa.

8.2 Show that the transformation formulas (8.6) for the field vectors from Σ to Σ' may be written as

$$\begin{bmatrix} c\mathscr{D}' \\ \mathscr{H}' \end{bmatrix} = \begin{bmatrix} \bar{\mathbf{A}}_{11} & \bar{\mathbf{A}}_{12} \\ -\bar{\mathbf{A}}_{12} & \bar{\mathbf{A}}_{11} \end{bmatrix} \cdot \begin{bmatrix} c\mathscr{D} \\ \mathscr{H} \end{bmatrix}$$

and

$$\begin{bmatrix} \mathscr{E}' \\ c\mathscr{B}' \end{bmatrix} = \begin{bmatrix} \bar{\mathbf{A}}_{11} & \bar{\mathbf{A}}_{12} \\ -\bar{\mathbf{A}}_{12} & \bar{\mathbf{A}}_{11} \end{bmatrix} \cdot \begin{bmatrix} \mathscr{E} \\ c\mathscr{B} \end{bmatrix}$$

where

$$\bar{\mathbf{A}}_{11} = \gamma\bar{\mathbf{L}}^{-1} = \gamma\bar{\mathbf{I}} + (1 - \gamma)\hat{\mathbf{v}}\hat{\mathbf{v}}$$

$$\bar{\mathbf{A}}_{12} = \beta\gamma(\hat{\mathbf{v}} \times \bar{\mathbf{I}})$$

Then show that

(*a*) $\bar{\mathbf{A}}_{11}$ is symmetric and $\bar{\mathbf{A}}_{12}$ is antisymmetric

(*b*) $\bar{\mathbf{A}}_{11}^2 = \gamma^2(\bar{\mathbf{I}} - \beta^2\hat{\mathbf{v}}\hat{\mathbf{v}})$ and $\bar{\mathbf{A}}_{12}^2 = \beta^2\gamma^2(\hat{\mathbf{v}}\hat{\mathbf{v}} - \bar{\mathbf{I}})$

(*c*) $\bar{\mathbf{A}}_{11}$ and $\bar{\mathbf{A}}_{12}$ are commutative:

$$\bar{\mathbf{A}}_{11} \cdot \bar{\mathbf{A}}_{12} = \bar{\mathbf{A}}_{12} \cdot \bar{\mathbf{A}}_{11} = \gamma \bar{\mathbf{A}}_{12}$$

(*d*)
$$\begin{bmatrix} \bar{\mathbf{A}}_{11} & \bar{\mathbf{A}}_{12} \\ -\bar{\mathbf{A}}_{12} & \bar{\mathbf{A}}_{11} \end{bmatrix}^{-1} = \begin{bmatrix} \bar{\mathbf{A}}_{11} & -\bar{\mathbf{A}}_{12} \\ \bar{\mathbf{A}}_{12} & \bar{\mathbf{A}}_{11} \end{bmatrix}$$

8.3 An electrically anisotropic medium is at rest in the primed system Σ' and has the constitutive relations

$$\mathscr{D}' = \varepsilon_0 \bar{\varepsilon} \cdot \mathscr{E}'$$

$$\mathscr{B}' = \mu_0 \mathscr{H}'$$

Using the results of Prob. 8.2, show that when the medium moves with velocity $\mathbf{v}$, the constitutive relations in the laboratory frame take the form

$$\mathscr{D} = \varepsilon_0(\bar{\mathbf{A}}_{11} \cdot \bar{\varepsilon} \cdot \bar{\mathbf{A}}_{11} + \bar{\mathbf{A}}_{12}^2) \cdot \mathscr{E} + \frac{1}{\eta_0}(\bar{\mathbf{A}}_{11} \cdot \bar{\varepsilon} \cdot \bar{\mathbf{A}}_{12} - \gamma \bar{\mathbf{A}}_{12}) \cdot \mathscr{B}$$

$$\mathscr{H} = \frac{1}{\eta_0}(\bar{\mathbf{A}}_{12} \cdot \bar{\varepsilon} \cdot \bar{\mathbf{A}}_{11} - \gamma \bar{\mathbf{A}}_{12}) \cdot \mathscr{E} + \frac{1}{\mu_0}(\bar{\mathbf{A}}_{12} \cdot \bar{\varepsilon} \cdot \bar{\mathbf{A}}_{12} + \bar{\mathbf{A}}_{11}^2) \cdot \mathscr{B}$$

where $\eta_0 = \sqrt{\mu_0/\varepsilon_0}$.

8.4 If the velocity v is much smaller than the velocity of light $c = 1/\sqrt{\mu_0 \varepsilon_0}$, terms of second order and higher in β may be neglected when compared with unity.

(*a*) Show that the first order relativistic transformation of field vectors is given by

$$\mathbf{E}' = \mathbf{E} + \mathbf{v} \times \mathbf{B}$$

$$\mathbf{B}' = \mathbf{B} - \mu_0 \varepsilon_0 \mathbf{v} \times \mathbf{E}$$

$$\mathbf{H}' = \mathbf{H} - \mathbf{v} \times \mathbf{D}$$

$$\mathbf{D}' = \mathbf{D} + \mu_0 \varepsilon_0 \mathbf{v} \times \mathbf{H}$$

$$\mathbf{J}' = \mathbf{J} - \rho \mathbf{v}$$

$$\rho' = \rho - \mu_0 \varepsilon_0 \mathbf{v} \cdot \mathbf{J}$$

(*b*) Assuming that there is no free charge in the rest frame of medium, show that the constitutive relations of a moving, isotropic, conducting homogeneous medium are

$$\mathbf{D} = \varepsilon_0 \varepsilon \mathbf{E} + \mathbf{\Lambda} \times \mathbf{H}$$

$$\mathbf{B} = \mu_0 \mu \mathbf{H} - \mathbf{\Lambda} \times \mathbf{E}$$

$$\mathbf{J} = \sigma(\mathbf{E} + \mu_0 \mu \mathbf{v} \times \mathbf{H})$$

where
$$\mathbf{\Lambda} = \mu_0 \varepsilon_0(\mu\varepsilon - 1)\mathbf{v}$$

(*c*) Show that the Maxwell-Minkowski equations take the form

$$(\nabla + i\omega\mathbf{\Lambda}) \times \mathbf{E} = i\omega\mu_0 \mu \mathbf{H}$$

$$(\nabla + i\omega\mathbf{\Lambda} - \mu_0 \mu \sigma \mathbf{v}) \times \mathbf{H} = (\sigma - i\omega\varepsilon_0 \varepsilon)\mathbf{E}$$

$$(\nabla + i\omega\mathbf{\Lambda}) \cdot \mathbf{H} = 0$$

$$(\nabla + i\omega\mathbf{\Lambda} - \mu_0 \mu \sigma \mathbf{v}) \cdot \mathbf{E} = 0$$

Hence, deduce the following wave equation for $\mathbf{E}$:

$$[\nabla^2 + (i2\omega\mathbf{\Lambda} - \mu_0 \mu \sigma \mathbf{v}) \cdot \nabla + k_0^2 \mu \varepsilon_f]\mathbf{E} = 0$$

where $$k_0^2 = \omega^2 \mu_0 \varepsilon_0 \qquad \varepsilon_f = \varepsilon\left(1 + i\frac{\sigma}{\omega\,\varepsilon_0\,\varepsilon}\right)$$

and a similar wave equation for **H**.

(*d*) Show that the set of Maxwell-Minkowski equations in (*c*) can also be solved by potential functions. In this case, the magnetic and electric field vectors are given by

$$\mathbf{H} = \exp(-i\omega\boldsymbol{\Lambda}\cdot\mathbf{r})\nabla \times \mathbf{A}$$

$$\mathbf{E} = \frac{\exp(-i\omega\boldsymbol{\Lambda}\cdot\mathbf{r})}{-i\omega\varepsilon_0\,\varepsilon_f}(\nabla - \mu_0\,\mu\sigma\mathbf{v}) \times (\nabla \times \mathbf{A})$$

where the vector potential **A** satisfies the equation

$$\nabla^2\mathbf{A} - \mu_0\,\mu\sigma(\mathbf{v}\cdot\nabla)\mathbf{A} + k_0^2\,\mu\varepsilon_f\,\mathbf{A} = \mathbf{0}$$

8.5 Use the fact that a monochromatic plane wave is invariant under the Lorentz transformation, i.e., a plane wave remains a plane wave in all inertial systems.

(*a*) Derive the transformation formulas for the amplitude vector $\mathbf{E}_0$, the wave vector **k**, and the angular frequency ω between two inertial systems.

(*b*) Show that $[k_\alpha] = (k_1, k_2, k_3, i\omega/c) = (\mathbf{k}, i\omega/c)$ is a 4-vector.

(*c*) Hence show that the phase of a uniform plane wave in a homogeneous medium is invariant under the Lorentz transformation.

8.6 Using the results of Prob. 8.5, obtain the dispersion equation (8.60) of a moving medium directly from the dispersion equation $k'^2 - \omega'^2\mu\varepsilon/c^2 = 0$ for the stationary medium.

8.7 Show that the phase velocity of a plane wave in a moving medium is given by Eq. (8.61).

8.8 Consider a monochromatic plane wave traveling in a homogeneous medium. The medium has an index of refraction n and moves with a velocity **v**. Find the transformation formulas for the frequency (Doppler effect), the angle between the wave vector and the direction of motion of the medium (aberration of the wave vector).

8.9 Show that in a conducting medium, a uniform plane wave in one inertial frame will generally appear as a nonuniform plane wave in another.

8.10 Verify Eqs. (8.72) and (8.75).

8.11 Verify Eqs. (8.78) and (8.79).

8.12 In a moving isotropic medium, show that the phase velocity is equal to the projection of the group velocity on the direction of wave normal.

8.13 A stationary observer finds that a certain region of space contains a plane wave with field components

$$E_y(x, t) = E_0 \exp[i\omega(\sqrt{\mu_0\,\varepsilon_0\,\mu\varepsilon}\; x - t)]$$

$$H_z(x, t) = E_0\sqrt{\frac{\varepsilon_0\,\varepsilon}{\mu_0\,\mu}} \exp[i\omega(\sqrt{\mu_0\,\varepsilon_0\,\mu\varepsilon}\; x - t)]$$

(*a*) How do these field components appear to an observer moving through the region with a uniform velocity $\mathbf{v} = -v\hat{\mathbf{x}}$.

(*b*) Find the relationship between the phase velocities observed by both observers.

8.14 A spacecraft moves away from an observer at a velocity v emits a signal of frequency f_s and Poynting's vector $\mathscr{P}_s$, both measured in the rest frame of the craft. Show that the observed frequency f_0 and the magnitude of Poynting's vector $\mathscr{P}_0$ are given by

$$f_0 = \frac{\sqrt{1-\beta^2}}{1+\beta} f_s$$

and $$\mathscr{P}_0 = \frac{1-\beta}{1+\beta}\mathscr{P}_3$$

respectively.

8.15 Verify Eqs. (8.85) and (8.86).

8.16 Derive Eq. (8.88).

8.17 A plane mirror moves with a velocity $\mathbf{v}$ in the direction of its normal. A monochromatic plane wave of angular frequency ω_0 strikes the moving mirror at an angle of incidence θ_0 with $\mathbf{v}$ as shown in Fig. 8.8. Find the angle of reflection and the frequency of the reflected wave.

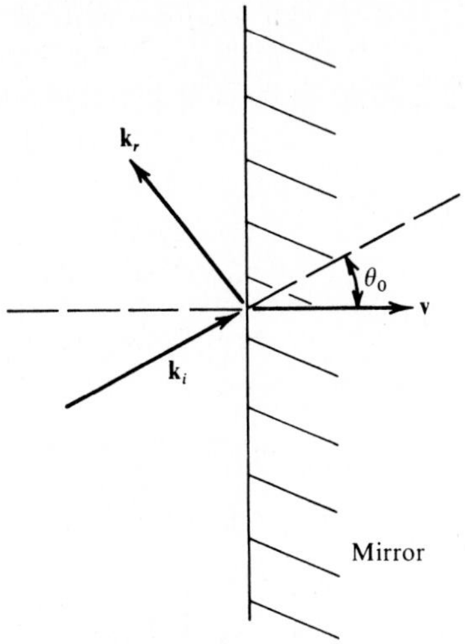

Figure 8.8 A plane mirror moves perpendicular to its plane.

8.18 Repeat Prob. 8.17 when the mirror moves in a direction parallel to its own plane as shown in Fig. 8.9.

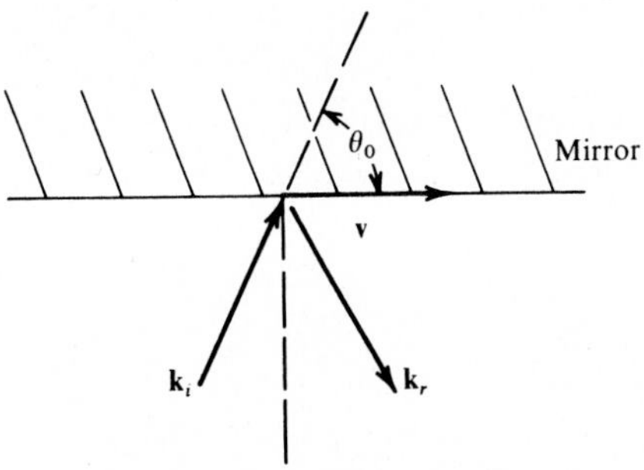

Figure 8.9 A plane mirror moves parallel to its plane.

8.19 Show that the Brewster angle is given by Eq. (8.143) when the dielectric is moving parallel to the plane of incidence, and by Eq. (8.148) when it is moving perpendicular to the plane of incidence.

8.20 The problem of wave reflection from a moving isotropic, dielectric half-space can also be solved by using the transformations derived in Prob. 8.5, that is,

$$\mathbf{E}'_0 = (\mathbf{E}_0 \cdot \hat{\mathbf{v}})\hat{\mathbf{v}} + \gamma[(\bar{\mathbf{I}} - \hat{\mathbf{v}}\hat{\mathbf{v}}) \cdot \mathbf{E}_0 + \mathbf{v} \times \mathbf{B}_0]$$

$$= \left[\gamma\left(1 - \frac{\mathbf{k} \cdot \mathbf{v}}{\omega}\right)\bar{\mathbf{I}} + (1 - \gamma)\hat{\mathbf{v}}\hat{\mathbf{v}} + \frac{\gamma\mathbf{k}\mathbf{v}}{\omega}\right] \cdot \mathbf{E}_0$$

$$\mathbf{k}' = \gamma\left(\mathbf{k} \cdot \hat{\mathbf{v}} - \frac{\omega v}{c^2}\right)\hat{\mathbf{v}} + (\bar{\mathbf{I}} - \hat{\mathbf{v}}\hat{\mathbf{v}}) \cdot \mathbf{k}$$

$$\omega' = \gamma(\omega - \mathbf{k} \cdot \mathbf{v})$$

where the primed quantities are referred to the frame Σ' and the unprimed quantities are referred to Σ. To find the reflected and transmitted waves in the laboratory frame Σ with respect to which the medium is moving, we first transform all the given quantities of the incident wave from Σ to the rest frame of the medium Σ'. In the frame Σ', the reflected and transmitted waves can be calculated by the method of Chap. 4 for stationary media. Then, by transforming the results from Σ' back to Σ, we obtain the desired solutions.

Consider the problem with geometry as shown in Fig. 8.10. The dielectric half-space with permittivity $\varepsilon_0 \varepsilon$ and permeability μ_0 which occupies the half-space $z > 0$, is moving with uniform velocity v parallel to the interface in the x direction, that is, $\mathbf{v} = v\hat{\mathbf{x}}$. In the frames of reference Σ and Σ', the electric fields of the incident, the reflected, and the transmitted waves take the forms

$$\mathbf{E}^{(i)} = \hat{\mathbf{y}} A_\perp \exp\,[i(\mathbf{k} \cdot \mathbf{r} - \omega t)]$$

$$\mathbf{E}^{(r)} = \hat{\mathbf{y}} B_\perp \exp\,[i(\mathbf{k}^{(r)} \cdot \mathbf{r} - \omega^{(r)} t)]$$

$$\mathbf{E}^{(t)} = \hat{\mathbf{y}} C_\perp \exp\,[i(\mathbf{k}^{(t)} \cdot \mathbf{r} - \omega^{(t)} t)]$$

and

$$\mathbf{E}^{(i)\prime} = \hat{\mathbf{y}} A'_\perp \exp\,[i(\mathbf{k}' \cdot \mathbf{r}' - \omega' t')]$$

$$\mathbf{E}^{(r)\prime} = \hat{\mathbf{y}} B'_\perp \exp\,[i(\mathbf{k}^{(r)\prime} \cdot \mathbf{r}' - \omega^{(r)\prime} t')]$$

$$\mathbf{E}^{(t)\prime} = \hat{\mathbf{y}} C'_\perp \exp\,[i(\mathbf{k}^{(t)\prime} \cdot \mathbf{r}' - \omega^{(t)\prime} t')]$$

respectively.

(*a*) Show that

$$\omega^{(r)\prime} = \omega^{(t)\prime} = \omega' \qquad C'_\perp = \frac{2k'_z}{k'_z - q'_t} A'_\perp$$

$$\mathbf{k}^{(r)\prime} = k'_x \hat{\mathbf{x}} - k'_z \hat{\mathbf{z}}$$

$$\mathbf{k}^{(t)\prime} = k'_x \hat{\mathbf{x}} + q'_t \hat{\mathbf{z}} \qquad B'_\perp = \frac{k'_z + q'_t}{k'_z - q'_t} A'_\perp$$

where $\mathbf{k}' = k'_x \hat{\mathbf{x}} + k'_z \hat{\mathbf{z}}$ is the wave vector of the incident wave with respect to Σ' and $q'_t = (\omega'^2 \mu_0 \varepsilon_0 \varepsilon - k'^2_x)^{1/2}$.

(*b*) Show that

$$\omega^{(r)} = \omega^{(t)} = \gamma(\omega' + k'_x v)$$

$$k^{(r)}_x = k^{(t)}_x = \gamma\left(k'_x + \frac{\omega' v}{c^2}\right)$$

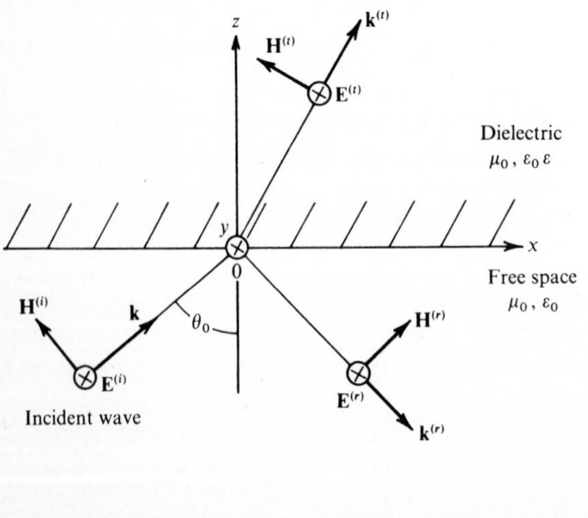

Figure 8.10 A moving dielectric half-space.

$$k_z^{(r)} = -k_z'$$

$$q_t = q_t' = (\omega'^2 \mu_0 \varepsilon_0 \varepsilon - k_x'^2)^{1/2}$$

$$B_\perp = \gamma B_\perp' \left(1 + \frac{k_x' v}{\omega'}\right)$$

$$C_\perp = \gamma C_\perp' \left(1 + \frac{k_x' v}{\omega'}\right)$$

(*c*) Hence show that

$$\omega^{(r)} = \omega^{(t)} = \omega$$

$$k_x^{(r)} = k_x^{(t)} = k_x = k_0 \sin \theta_0$$

$$k_z^{(r)} = -k_z = -k_0 \cos \theta_0$$

$$q_t = k_0 \gamma \sqrt{\varepsilon(1 - \beta \sin \theta_0)^2 - (\sin \theta_0 - \beta)^2}$$

$$B_\perp = \frac{\cos \theta_0 + q_t/k_0}{\cos \theta_0 - q_t/k_0} A_\perp$$

$$C_\perp = \frac{2 \cos \theta_0}{\cos \theta_0 - q_t/k_0} A_\perp$$

where ω, θ_0, and $\mathbf{k} = k_x \hat{\mathbf{x}} + k_z \hat{\mathbf{z}}$ are respectively the angular frequency, the angle of incidence, and the wave vector of the incident wave with respect to the frame Σ. We note that in this case, there is no Doppler shift in frequency for the reflected and the transmitted waves. Also the angle of incidence is equal to the angle of reflection. The angle of transmission defined by $\theta_t = \tan^{-1} |k_x^{(t)}/q_t|$ depends on v and ε.

8.21 Repeat Prob. 8.20 when the dielectric half-space moves parallel to the interface but perpendicular to the plane of incidence, that is, $\mathbf{v} = v\hat{\mathbf{y}}$.

8.22 In Prob. 8.20, if the dielectric half-space moves uniformly perpendicular to the interface, that is, $\mathbf{v} = v\hat{\mathbf{z}}$, (*a*) show that

$$\omega^{(r)} = \gamma(\omega' - k_z' v)$$

$$k_x^{(r)} = k_x^{(t)} = k_x'$$

$$k_z^{(r)} = \gamma\left(-k_z' + \frac{\omega' v}{c^2}\right)$$

$$\omega^{(t)} = \gamma(\omega' + q_t' v)$$

$$k_z^{(t)} = \gamma\left(q_t' + \frac{\omega' v}{c^2}\right)$$

$$B_\perp = \gamma B_\perp' \left(1 - \frac{k_z' v}{\omega'}\right)$$

$$C_\perp = \gamma C_\perp' \left(1 + \frac{q_t' v}{\omega'}\right)$$

(*b*) Hence show that

$$\omega^{(r)} = \gamma^2(1 + \beta^2 - 2\beta \cos \theta_0)\omega$$

$$k_x^{(r)} = k_x^{(t)} = k_x = k_0 \sin \theta_0$$

$$k_z^{(r)} = -k_0 \gamma^2[(1 + \beta^2) \cos \theta_0 - 2\beta]$$

$$\omega^{(t)} = \gamma^2(1 - \beta \cos \theta_0 + \beta M)\omega$$

$$k_z^{(t)} = k_0 \gamma^2[M + \beta(1 - \beta \cos \theta_0)]$$

$$B_\perp = \frac{\omega^{(r)} (\cos \theta_0 - \beta + M)}{\omega (\cos \theta_0 - \beta - M)} A_\perp$$

$$C_\perp = \frac{2\omega^{(t)} (\cos \theta_0 - \beta)}{\omega (\cos \theta_0 - \beta - M)} A_\perp$$

where

$$M = [\varepsilon (1 - \beta \cos \theta_0)^2 - \sin^2 \theta_0(1 - \beta^2)]^{1/2}$$

Unlike Prob. 8.20, there is Doppler shift in frequency for the reflected and transmitted waves. The frequency shift for the reflected wave is independent of ε, while the frequency shift for the transmitted wave depends on ε. Also, the angle of reflection defined by $\theta_r = \tan^{-1} |k_x^{(r)}/k_z^{(r)}|$, depends on v. Consequently, the angle of reflection is not equal to the angle of incidence.

REFERENCES

Chen, H. C.: "Constitutive Relations of a Moving Gyro-Electric-Magnetic Medium," *Int. J. Electronics*, vol. 36, no. 3, 1974, pp. 319–328.

Chen, H. C., and D. K. Cheng: "A Useful Matrix Inversion Formula and Its Applications," *Proc. IEEE*, vol. 55, no. 5, 1967, pp. 705–706.

Chuang, C. W., and H. C. Ko: "On the Generalized Brewster's Law for a Moving Dielectric Medium," *J. Appl. Phys.*, vol. 45, no. 3, 1974, pp. 1154–1157.

Costen, R. C., and D. Adamson: "Three-Dimensional Derivation of the Electrodynamic Jump Conditions and Momentum-Energy Laws at a Moving Boundary," *Proc. IEEE*, vol. 53, no. 9, 1965, pp. 1181–1196.

Daly, P., and H. Gruenberg: "Energy Relations for Plane Waves Reflected from Moving Media," *J. Appl. Phys.*, vol. 38, no. 11, 1967, pp. 4486–4489.

Kunz, K. S.: "Plane Electromagnetic Waves in Moving Media and Reflections from Moving Interfaces," *J. Appl. Phys.*, vol. 51, no. 2, 1980, pp. 873–884.

Landau, L. D., and E. M. Lifshitz: *Electrodynamics of Continuous Media*, Pergamon Press, New York, 1960, pp. 243–247.

Lee, S. W., and Y. T. Lo: "Reflection and Transmission of Electromagnetic Waves by a Moving Uniaxially Anisotropic Medium," *J. Appl. Phys.*, vol. 38, no. 2, 1967, pp. 870–875.

McKenzie, J. F.: "Electromagnetic Waves in Uniformly Moving Media," *Proc. Phys. Soc.*, vol. 91, 1967, pp. 532–536.

O'Dell, T. H.: *The Electrodynamics of Magneto-Electric Media*, North-Holland Publishing Co., New York, 1970, chap. 3.

Pyati, V. P.: "Reflection and Refraction of Electromagnetic Waves by a Moving Dielectric Medium," *J. Appl. Phys.*, vol. 38, no. 2, 1967, pp. 652–655.

Shiozawa, T., and K. Hazama: "General Solution to the Problem of Reflection and Transmission by a Moving Dielectric Medium," *Radio Science*, vol. 3, no. 6, 1968, pp. 569–576.

Shiozawa, T., K. Hazama, and N. Kumagai: "Reflection and Transmission of Electromagnetic Waves by a Dielectric Half-Space Moving Perpendicular to the Plane of Incidence," *J. Appl. Phys.*, vol. 38, no. 11, 1967, pp. 4459–4461.

Shiozawa, T., and N. Numagai: "Total Reflection at the Interface between Relatively Moving Media," *Proc. IEEE*, vol. 55, no. 7, 1967, pp. 1243–1244.

Tai, C. T.: "A Study of Electrodynamics of Moving Media," *Proc. IEEE*, vol. 52, no. 6, 1964, pp. 685–689.

Tai, C. T.: "Electrodynamics of Moving Anisotropic Media: The First-Order Theory," *Radio Science*, vol. 69D, no. 3, 1965, pp. 401–405.

Yeh, C.: "Reflection and Transmission of Electromagnetic Waves by a Moving Dielectric Medium," *J. Appl. Phys.*, vol. 36, no. 11, 1965, pp. 3513–3517.

Yeh, C.: "Brewster Angle for a Dielectric Medium Moving at Relativistic Speed," *J. Appl. Phys.*, vol. 38, no. 13, 1967, pp. 5194–5200.

CHAPTER

NINE

RADIATION IN AN ISOTROPIC MEDIUM

So far we have studied the propagation of electromagnetic waves in various media and also the reflection and transmission of waves from the surface of such media without discussing how they are excited. Starting with this chapter, we shall consider the basic problem of determining the electromagnetic fields generated by a given distribution of sources. Our method of approach is that of the dyadic Green function, which yields the field directly in terms of the source current. In this chapter, we shall concentrate our discussion on radiation of sources in isotropic media.

9.1 DYADIC GREEN'S FUNCTION FOR AN ISOTROPIC MEDIUM

In a region with an external source, the two curl Maxwell equations, (2.1) and (2.2), for an isotropic medium become

$$\nabla \times \mathscr{E} = -\mu_0 \frac{\partial \mathscr{H}}{\partial t} \tag{9.1}$$

and

$$\nabla \times \mathscr{H} = \varepsilon_0 \varepsilon \frac{\partial \mathscr{E}}{\partial t} + \mathscr{J} \tag{9.2}$$

where $\mathscr{J}$ is the externally applied current density. In order to solve for field vectors in terms of the given current distribution, it is convenient to use the Fourier transform. Let $\mathscr{V}(t)$ be an arbitrary function of time which is square-integrable, the Fourier transform of $\mathscr{V}(t)$ with respect to t, denoted by $V(\omega)$, is defined by

$$V(\omega) = \int_{-\infty}^{+\infty} \mathscr{V}(t)e^{i\omega t}\, dt \tag{9.3}$$

and according to the Fourier theorem, the inverse transformation is

$$\mathscr{V}(t) = \frac{1}{2\pi}\int_{-\infty}^{+\infty} V(\omega)e^{-i\omega t}\, d\omega \tag{9.4}$$

which expresses $\mathscr{V}(t)$ as a linear superposition of monochromatic waves. To verify the Fourier theorem directly, we substitute Eq. (9.3) into the right-hand side of Eq. (9.4), interchange the orders of integration and make use of the fact that

$$\frac{1}{2\pi}\int_{-\infty}^{+\infty} e^{-i(t-t')\omega}\, d\omega = \delta(t - t') \tag{9.5}$$

where $\delta(t)$ is the delta function (see Appendix B for details). Now we let $\mathbf{E}(\mathbf{r})$, $\mathbf{H}(\mathbf{r})$, and $\mathbf{J}(\mathbf{r})$ be the Fourier transforms of $\mathscr{E}$, $\mathscr{H}$, and $\mathscr{J}$ respectively. For physical reasons it is clear that each of these vectors is square-integrable, and thus the Fourier transforms exist. By taking the Fourier transform of Eqs. (9.1) and (9.2), noting that the differentiation with respect to t becomes multiplication by $-i\omega$, we obtain Maxwell's equation in the frequency domain:

$$\nabla \times \mathbf{E}(\mathbf{r}) = i\omega\mu_0 \mathbf{H}(\mathbf{r}) \tag{9.6}$$

$$\nabla \times \mathbf{H}(\mathbf{r}) = -i\omega\varepsilon_0 \varepsilon\mathbf{E}(\mathbf{r}) + \mathbf{J}(\mathbf{r}) \tag{9.7}$$

Here we note that for a monochromatic wave with a time dependence $e^{-i\omega t}$, the Maxwell equations take the same form as Eqs. (9.6) and (9.7).

Eliminating $\mathbf{H}(\mathbf{r})$ from Eqs. (9.6) and (9.7), we obtain

$$(\nabla\nabla - \nabla^2\bar{\mathbf{I}} - k_0^2\varepsilon\bar{\mathbf{I}}) \cdot \mathbf{E}(\mathbf{r}) = i\omega\mu_0 \mathbf{J}(\mathbf{r}) \tag{9.8}$$

Since Eq. (9.8) is linear, we may write

$$\mathbf{E}(\mathbf{r}) = i\omega\mu_0 \int_V \bar{\mathbf{G}}(\mathbf{r}, \mathbf{r}') \cdot \mathbf{J}(\mathbf{r}')\, d^3r' \tag{9.9}$$

where $\bar{\mathbf{G}}(\mathbf{r}, \mathbf{r}')$ is called the dyadic Green function which directly relates the field to the source current. It is a function of both the field point $\mathbf{r}$ and the source point $\mathbf{r}'$, whereas d^3r' is a shorthand of the volume element $dx'\, dy'\, dz'$, and the integration is performed over the volume V containing the source currents. To determine the unknown dyadic Green function $\bar{\mathbf{G}}(\mathbf{r}, \mathbf{r}')$ we substitute Eq. (9.9) into Eq. (9.8) and interchange the order of integration and differentiation, and obtain

$$\int_V [(\nabla\nabla - \nabla^2\bar{\mathbf{I}} - k_0^2\varepsilon\bar{\mathbf{I}}) \cdot \bar{\mathbf{G}}(\mathbf{r}, \mathbf{r}')] \cdot \mathbf{J}(\mathbf{r}')\, d^3r' = \mathbf{J}(\mathbf{r}) \tag{9.10}$$

With the aid of the three-dimensional delta function (see Appendix B) which permits $\mathbf{J}(\mathbf{r})$ to be represented as the volume integral

$$\mathbf{J}(\mathbf{r}) = \int_V \bar{\mathbf{I}} \cdot \mathbf{J}(\mathbf{r}')\, \delta(\mathbf{r} - \mathbf{r}')\, d^3r' \tag{9.11}$$

we see that Eq. (9.10) can be written as

$$\int_V [(\nabla\nabla - \nabla^2\bar{\mathbf{I}} - k_0^2\,\varepsilon\bar{\mathbf{I}}) \cdot \bar{\mathbf{G}}(\mathbf{r}, \mathbf{r}') - \bar{\mathbf{I}}\,\delta(\mathbf{r} - \mathbf{r}')] \cdot \mathbf{J}(\mathbf{r}')\, d^3r' = \mathbf{0} \tag{9.12}$$

From this it follows that $\bar{\mathbf{G}}(\mathbf{r}, \mathbf{r}')$ must satisfy the differential equation

$$(\nabla\nabla - \nabla^2\bar{\mathbf{I}} - k_0^2\,\varepsilon\bar{\mathbf{I}}) \cdot \bar{\mathbf{G}}(\mathbf{r}, \mathbf{r}') = \bar{\mathbf{I}}\,\delta(\mathbf{r} - \mathbf{r}') \tag{9.13}$$

A comparison of Eq. (9.13) with Eq. (9.8) shows that the dyadic Green function $\bar{\mathbf{G}}$ satisfies the same differential equation as $\mathbf{E}$ with the source term replaced by the delta function. Thus, if the Green function is known, we may obtain the solution to Eq. (9.8) by carrying out the integral (9.9) for an arbitrary source $\mathbf{J}(\mathbf{r})$. Alternatively, in the terminology of system theory, we may consider $\mathbf{J}$ as an "input" and $\mathbf{E}$ as the corresponding "output." The differential operator describing the propagation of electromagnetic waves in an isotropic medium is merely a "black box" transforming every permissible input into a corresponding output. The properties of the black box are embodied in the Green function.

To solve Eq. (9.13), let us denote the three-dimensional Fourier transform of $f(\mathbf{r})$ with respect to $\mathbf{r}$ by $f(\mathbf{k})$; thus

$$f(\mathbf{k}) = \int_{-\infty}^{+\infty} f(\mathbf{r}) e^{-i\mathbf{k}\cdot\mathbf{r}}\, d^3r \tag{9.14}$$

which may be inverted to yield

$$f(\mathbf{r}) = \frac{1}{(2\pi)^3} \int_{-\infty}^{+\infty} f(\mathbf{k}) e^{i\mathbf{k}\cdot\mathbf{r}}\, d^3k \tag{9.15}$$

In establishing the compatibility of Eqs. (9.14) and (9.15), we use the relation

$$\frac{1}{(2\pi)^3} \int_{-\infty}^{+\infty} e^{i(\mathbf{r}-\mathbf{r}')\cdot\mathbf{k}}\, d^3k = \delta(\mathbf{r} - \mathbf{r}') \tag{9.16}$$

where $\delta(\mathbf{r})$ is the three-dimensional delta function as defined in Appendix *B*. Equation (9.15) may be interpreted as the expansion of $f(\mathbf{r})$ in terms of plane waves. Now let us take the Fourier transform of Eq. (9.13) with respect to $\mathbf{r}$. Since the operation of taking the gradient becomes multiplication by $i\mathbf{k}$, we obtain easily

$$\bar{\mathbf{W}}_i(\mathbf{k}) \cdot \bar{\mathbf{G}}(\mathbf{k}, \mathbf{r}') = \bar{\mathbf{I}} e^{-i\mathbf{k}\cdot\mathbf{r}'} \tag{9.17}$$

where the wave matrix $\bar{\mathbf{W}}_i(\mathbf{k})$ of an isotropic medium is given by

$$\bar{\mathbf{W}}_i(\mathbf{k}) = (k^2 - k_0^2\,\varepsilon)\bar{\mathbf{I}} - \mathbf{k}\mathbf{k} \tag{9.18}$$

Premultiplication of Eq. (9.17) by the inverse matrix $\bar{\mathbf{W}}_i^{-1}$ yields

$$\bar{\mathbf{G}}(\mathbf{k}, \mathbf{r}') = \bar{\mathbf{W}}_i^{-1}(\mathbf{k})e^{-i\mathbf{k}\cdot\mathbf{r}'} \tag{9.19}$$

Taking the inverse Fourier transform of Eq. (9.19), we obtain

$$\bar{\mathbf{G}}(\mathbf{r}, \mathbf{r}') = \frac{1}{(2\pi)^3}\int_{-\infty}^{+\infty} \bar{\mathbf{W}}_i^{-1}(\mathbf{k})e^{i\mathbf{k}\cdot(\mathbf{r}-\mathbf{r}')}\, d^3k \tag{9.20}$$

Using Eq. (9.18) and the result of Eq. (1.126), we find the explicit form of the inverse matrix of $\bar{\mathbf{W}}_i(\mathbf{k})$:

$$\bar{\mathbf{W}}_i^{-1}(\mathbf{k}) = \frac{k_0^2\,\varepsilon\bar{\mathbf{I}} - \mathbf{k}\mathbf{k}}{k_0^2\,\varepsilon(k^2 - k_0^2\,\varepsilon)} \tag{9.21}$$

Thus Eq. (9.20) becomes

$$\bar{\mathbf{G}}(\mathbf{r}, \mathbf{r}') = \frac{1}{(2\pi)^3 k_0^2\,\varepsilon}\int_{-\infty}^{+\infty} \frac{(k_0^2\,\varepsilon\bar{\mathbf{I}} - \mathbf{k}\mathbf{k})e^{i\mathbf{k}\cdot(\mathbf{r}-\mathbf{r}')}}{k^2 - k_0^2\,\varepsilon}\, d^3k \tag{9.22}$$

Because of the exponential function in the integrand, the wave vector $\mathbf{k}$ in the numerator may be replaced by $-i\nabla$. Interchanging the order of differentiation and integration, we have

$$\bar{\mathbf{G}}(\mathbf{R}, \omega) = \left(\bar{\mathbf{I}} + \frac{1}{k_0^2\,\varepsilon}\nabla\nabla\right) g(\mathbf{R}, \omega) \tag{9.23}$$

where
$$g(\mathbf{R}, \omega) = \frac{1}{(2\pi)^3}\int_{-\infty}^{+\infty} \frac{e^{i\mathbf{k}\cdot\mathbf{R}}}{k^2 - k_0^2\,\varepsilon}\, d^3k \tag{9.24}$$

is called the scalar Green function and $\mathbf{R} = \mathbf{r} - \mathbf{r}'$, $k_0^2 = \omega^2/c^2$. Equation (9.23) is a solution of Eq. (9.13). To satisfy the prescribed boundary conditions, solutions to the homogeneous equation may be added to it. Since in this chapter we are concerned only with sources radiating in an unbounded isotropic medium, we must choose a solution which represents a wave traveling away from the source (or outgoing wave).

Once the electric field intensity is determined from Eq. (9.9), the magnetic field intensity follows from Maxwell's equation (9.6):

$$\begin{aligned}\mathbf{H}(\mathbf{r}) &= \frac{1}{i\omega\mu_0}\nabla \times \mathbf{E}(\mathbf{r}) \\ &= \int_{-\infty}^{+\infty} \nabla \times [\bar{\mathbf{G}}(\mathbf{r}, \mathbf{r}') \cdot \mathbf{J}(\mathbf{r}')]\, d^3r' \end{aligned} \tag{9.25}$$

But $\bar{\mathbf{G}}(\mathbf{r}, \mathbf{r}')$ is given by Eq. (9.23) and consequently

$$\mathbf{H}(\mathbf{r}) = \int_{-\infty}^{+\infty} \nabla g(\mathbf{R}, \omega) \times \mathbf{J}(\mathbf{r}')\, d^3r' \tag{9.26}$$

Thus for a given current distribution $\mathbf{J}(\mathbf{r}')$, the electric and magnetic fields depend on the integral $g(\mathbf{R}, \omega)$ defined in Eq. (9.24). A detailed evaluation of the integral will be carried out in the following section.

9.2 EVALUATION OF THE SCALAR GREEN FUNCTION

To evaluate the scalar Green function, we take the inverse Fourier transform of Eq. (9.24) according to Eq. (9.4) and consider the following integral:

$$g(\mathbf{R}, t) = \frac{1}{2\pi} \int_{-\infty}^{+\infty} g(\mathbf{R}, \omega) e^{-i\omega t}\, d\omega$$

$$= \frac{1}{(2\pi)^4} \int_{-\infty}^{+\infty} \frac{e^{i(\mathbf{k}\cdot\mathbf{R} - \omega t)}}{\Delta\,(\omega, \mathbf{k})}\, d^3k\, d\omega \tag{9.27}$$

where $\Delta\,(\omega, \mathbf{k}) = k^2 - \omega^2 \varepsilon / c^2$. First, we can readily verify by direct substitution that $g(\mathbf{R}, t)$ is a solution of the scalar wave equation with delta functions as its source:

$$\left(\nabla^2 - \frac{\varepsilon}{c^2} \frac{\partial^2}{\partial t^2}\right) g(\mathbf{R}, t) = -\delta(\mathbf{R})\, \delta(t) \tag{9.28}$$

hence justifying the name of scalar Green's function. Next, in order to carry out the integration with respect to ω in Eq. (9.27) by the method of residues, two questions must be answered: how to close the contour in the complex ω-plane, and how to avoid the poles that lie on the path of integration. Regarding the first question, we note that

$$|e^{-i\omega t}| = |e^{-it(\mathrm{Re}\,\omega + i\,\mathrm{Im}\,\omega)}| = e^{t\,\mathrm{Im}\,\omega} \tag{9.29}$$

Thus, for $t < 0$, the path of integration must be closed with a semicircle of infinite radius that lies in the upper half of the complex ω-plane. Then, $\mathrm{Im}\,\omega > 0$ on the semicircle C_1, and its contribution to the integral (9.27) vanishes exponentially. On the other hand, for $t > 0$, the contour must be closed in the lower half of the complex ω-plane as shown in Fig. 9.1 by path C_2.

The second question can be answered by imposing a causality requirement. Since the scalar Green function $g(\mathbf{R}, t)$ satisfying Eq. (9.28) represents the wave disturbance caused by a point source located at $\mathbf{R} = \mathbf{0}$ and is turned on only for an infinitesimal time interval at $t = 0$, it is then natural to require that before the source is turned on, the disturbance be set identically to zero. This requirement provides a rule to bypass poles that lie on the path of integration. In fact, in order to make $g(\mathbf{R}, t)$ vanish for $t < 0$, we must assume that the poles are displaced slightly into the lower half-plane as shown in Fig. 9.1. Then the integral over C_2 for $t > 0$ will give a nonvanishing contribution, while the integral over C_1 for $t < 0$ will vanish. The lowering of poles below the real axis is equivalent to raising the path of integration above the real axis. This is achieved mathematically by changing ω to $\omega + i\sigma$ and then considering the limit as σ approaches zero. Based on the above prescriptions for carrying out the ω-integration in Eq.

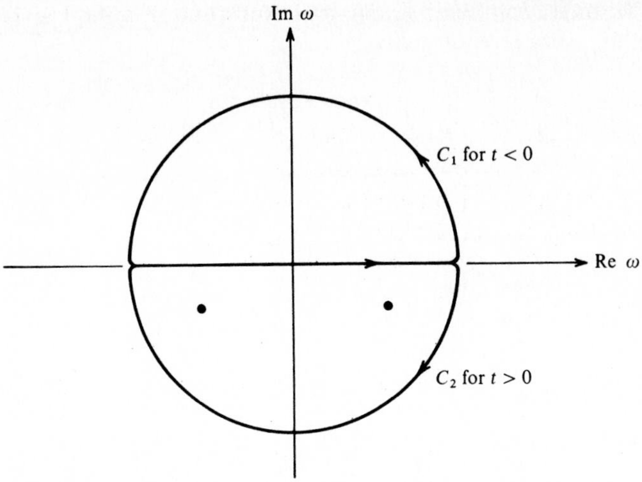

Figure 9.1 Complex ω-plane with contour C_1 for $t < 0$ and contour C_2 for $t > 0$.

(9.27), we obtain

$$g(\mathbf{R}, t) = \lim_{\sigma \to 0} \frac{U(t)}{(2\pi)^4} \int_{-\infty}^{+\infty} e^{i\mathbf{k}\cdot\mathbf{R}}\, d^3k \int_{-\infty}^{+\infty} \frac{e^{-i\omega t}}{\Delta\,(\omega + i\sigma, \mathbf{k})}\, d\omega \tag{9.30}$$

where σ is a small positive quantity, and $U(t)$ is the unit step function defined by

$$U(t) = \begin{cases} 1 & \text{for } t > 0 \\ 0 & \text{for } t < 0 \end{cases} \tag{9.31}$$

Expanding the denominator of Eq. (9.30) into powers of σ and retaining only up to the first order term, we get

$$g(\mathbf{R}, t) = \lim_{\sigma \to 0} \frac{U(t)}{(2\pi)^4} \int_{-\infty}^{+\infty} \frac{e^{i(\mathbf{k}\cdot\mathbf{R} - \omega t)}}{\Delta(\omega, \mathbf{k} + i\sigma(\partial\Delta/\partial\omega)}\, d^3k\, d\omega \tag{9.32}$$

After comparing with Eq. (9.27), we obtain

$$g(\mathbf{R}, \omega) = \lim_{\sigma \to 0} \frac{1}{(2\pi)^3} \int_{-\infty}^{+\infty} \frac{e^{i\mathbf{k}\cdot\mathbf{R}}}{\Delta(\omega, \mathbf{k}) + i\sigma(\partial\Delta/\partial\omega)}\, d^3k \tag{9.33}$$

Based on the prescription of Eq. (9.33), we may now proceed to evaluate the integral (9.24). Let us introduce spherical coordinates in **k**-space with **R** along the k_z axis. We have

$$\begin{aligned} \mathbf{k}\cdot\mathbf{R} &= kR\cos\theta \\ d^3k &= k^2 \sin\theta\, dk\, d\theta\, d\phi \end{aligned} \tag{9.34}$$

where $R = |\mathbf{R}|$ and θ is the angle between $\mathbf{k}$ and k_z axis. Substitution of Eq. (9.34) into Eq. (9.24) yields

$$g(\mathbf{R}, \omega) = \lim_{\sigma \to 0} \frac{1}{(2\pi)^3} \int_0^\infty \frac{k^2\, dk}{k^2 - k_1^2} \int_0^\pi e^{ikR\cos\theta} \sin\theta\, d\theta \int_0^{2\pi} d\phi \tag{9.35}$$

where

$$k_1 = k_0\sqrt{\varepsilon}\left(1 + i\,\frac{2\sigma}{\omega}\right)^{1/2}$$

$$\cong k_0\sqrt{\varepsilon}\left(1 + i\,\frac{\sigma}{\omega}\right) \tag{9.36}$$

Since

$$\int_0^\pi e^{ikR\cos\theta} \sin\theta\, d\theta = \frac{2\sin kR}{kR} \tag{9.37}$$

Eq. (9.35) becomes

$$g(\mathbf{R}, \omega) = \lim_{\sigma \to 0} \frac{1}{2\pi^2 R} \int_0^\infty \frac{\sin kR}{k^2 - k_1^2}\, k\, dk$$

$$= \lim_{\sigma \to 0} \frac{1}{i4\pi^2 R} \int_{-\infty}^{+\infty} \frac{k e^{ikR}}{k^2 - k_1^2}\, dk \tag{9.38}$$

The remaining integral in Eq. (9.38) may again be evaluated by the method of residues. We note that R is a distance and is always nonnegative; also

$$|e^{ikR}| = e^{-R\,\mathrm{Im}\,k} \tag{9.39}$$

Thus the path of integration must be closed with a semicircle lying in the upper half of the complex k-plane as shown in Fig. 9.2. Then $\mathrm{Im}\,k > 0$ on the semicircle,

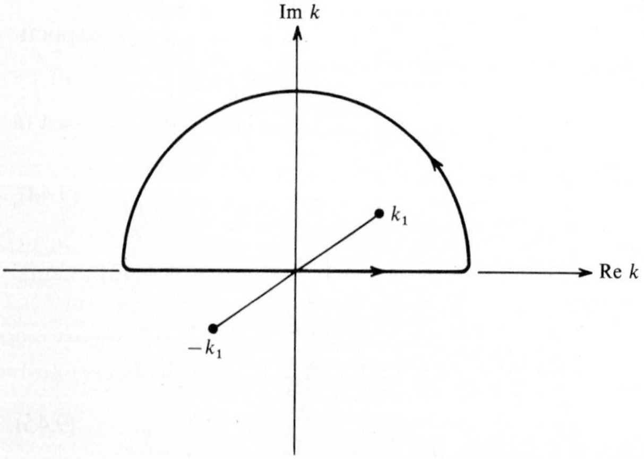

Figure 9.2 Complex k-plane with location of poles.

and its contribution to Eq. (9.38) vanishes exponentially. The pole in the upper half-plane occurs at $k_1 = k_0\sqrt{\varepsilon}(1 + i\sigma/\omega)$ and the residue theorem finally gives

$$g(\mathbf{R}, \omega) = \lim_{\sigma\to 0} \frac{1}{4\pi R} \exp\left(\frac{i\omega\sqrt{\varepsilon}R - \sigma\sqrt{\varepsilon}R}{c}\right)$$
$$= \frac{1}{4\pi R} \exp\left(\frac{i\omega\sqrt{\varepsilon}R}{c}\right) \tag{9.40}$$

To see whether Eq. (9.40) is correct physically, let us evaluate $g(\mathbf{R}, t)$ in time domain according to Eq. (9.27):

$$g(\mathbf{R}, t) = \frac{1}{2\pi} \int_{-\infty}^{+\infty} g(\mathbf{R}, \omega) e^{-i\omega t}\, d\omega$$
$$= \frac{1}{4\pi R} \frac{1}{2\pi} \int_{-\infty}^{+\infty} \exp\left[-i\omega\left(t - \frac{\sqrt{\varepsilon}R}{c}\right)\right] d\omega \tag{9.41}$$

Using the result of Eq. (9.5), we find that

$$g(\mathbf{R}, t) = \frac{1}{4\pi R}\, \delta\left(t - \frac{\sqrt{\varepsilon}R}{c}\right) \tag{9.42}$$

Hence we see that $g(\mathbf{R}, t)$ corresponds to a spherical pulse which expands away from the origin $\mathbf{R} = \mathbf{0}$ as time increases. In other words, it is an outgoing wave solution as desired.

9.3 DYADIC GREEN'S FUNCTION IN COORDINATE-FREE FORM

To find the dyadic Green function of an isotropic medium in an explicit coordinate-free form, we will need the following relations:

$$\nabla R = \hat{\mathbf{R}} \tag{9.43}$$

$$\nabla \hat{\mathbf{R}} = \frac{1}{R}(\bar{\mathbf{I}} - \hat{\mathbf{R}}\hat{\mathbf{R}}) \tag{9.44}$$

which can be demonstrated by writing $\mathbf{R}$ in terms of the cartesian components and then carrying out the differentiation. Since

$$\begin{aligned} \mathbf{R} &= \mathbf{r} - \mathbf{r}' = (x - x')\hat{\mathbf{x}} + (y - y')\hat{\mathbf{y}} + (z - z')\hat{\mathbf{z}} \\ R &= |\mathbf{R}| = \sqrt{(x - x')^2 + (y - y')^2 + (z - z')^2} \\ \hat{\mathbf{R}} &= \frac{\mathbf{R}}{R} \end{aligned} \tag{9.45}$$

it follows that

$$\frac{\partial R}{\partial x} = \frac{x - x'}{R} \qquad \frac{\partial R}{\partial y} = \frac{y - y'}{R} \qquad \frac{\partial R}{\partial z} = \frac{z - z'}{R} \tag{9.46}$$

and

$$\begin{aligned} \frac{\partial \hat{\mathbf{R}}}{\partial x} &= \frac{R\hat{\mathbf{x}} - (x - x')\hat{\mathbf{R}}}{R^2} \\ \frac{\partial \hat{\mathbf{R}}}{\partial y} &= \frac{R\hat{\mathbf{y}} - (y - y')\hat{\mathbf{R}}}{R^2} \\ \frac{\partial \hat{\mathbf{R}}}{\partial z} &= \frac{R\hat{\mathbf{z}} - (z - z')\hat{\mathbf{R}}}{R^2} \end{aligned} \tag{9.47}$$

Substituting Eqs. (9.46) and (9.47) into the definition of the gradient operator and noting that $\bar{\mathbf{I}} = \hat{\mathbf{x}}\hat{\mathbf{x}} + \hat{\mathbf{y}}\hat{\mathbf{y}} + \hat{\mathbf{z}}\hat{\mathbf{z}}$, we thus prove the desired results.

With the aid of Eqs. (9.43), (9.44), and the chain rule that if f is a function of u and u is a function of $\mathbf{r}$, we have

$$\nabla f = \frac{df}{du} \nabla u \tag{9.48}$$

Furthermore, we may establish that

$$\begin{aligned} \nabla \frac{e^{ik_0\sqrt{\varepsilon}R}}{R} &= \left(ik_0\sqrt{\varepsilon} - \frac{1}{R} \right) \frac{e^{ik_0\sqrt{\varepsilon}R}}{R} \hat{\mathbf{R}} \\ \nabla \frac{e^{ik_0\sqrt{\varepsilon}R}}{R^2} &= \left(\frac{ik_0\sqrt{\varepsilon}}{R} - \frac{2}{R^2} \right) \frac{e^{ik_0\sqrt{\varepsilon}R}}{R} \hat{\mathbf{R}} \end{aligned} \tag{9.49}$$

and

$$\nabla\nabla \frac{e^{ik_0\sqrt{\varepsilon}R}}{R} = \left[\left(\frac{ik_0\sqrt{\varepsilon}}{R} - \frac{1}{R^2} \right) (\bar{\mathbf{I}} - 3\hat{\mathbf{R}}\hat{\mathbf{R}}) - k_0^2 \varepsilon \hat{\mathbf{R}}\hat{\mathbf{R}} \right] \frac{e^{ik_0\sqrt{\varepsilon}R}}{R} \tag{9.50}$$

Now, substituting Eq. (9.40) into Eq. (9.23) and using the result of Eq. (9.50), we finally obtain the dyadic Green function in the explicit coordinate-free form:

$$\bar{\mathbf{G}}(\mathbf{R}, \omega) = (\bar{\mathbf{I}} - \hat{\mathbf{R}}\hat{\mathbf{R}})g(\mathbf{R}, \omega) + \frac{i}{k_0\sqrt{\varepsilon}R} (\bar{\mathbf{I}} - 3\hat{\mathbf{R}}\hat{\mathbf{R}})g(\mathbf{R}, \omega) - \frac{1}{k_0^2 \varepsilon R^2} (\bar{\mathbf{I}} - 3\hat{\mathbf{R}}\hat{\mathbf{R}})g(\mathbf{R}, \omega) \tag{9.51}$$

where

$$g(\mathbf{R}, \omega) = \frac{e^{ik_0\sqrt{\varepsilon}R}}{4\pi R} \tag{9.52}$$

In terms of the dyadic Green function (9.51), the electric field produced by an arbitrary current distribution $\mathbf{J}$ is given by

$$\mathbf{E}(\mathbf{r}) = i\omega\mu_0 \int_V \bar{\mathbf{G}}(\mathbf{R}, \omega) \cdot \mathbf{J}(\mathbf{r}')\, d^3r' \tag{9.53}$$

for **r** lying outside the source region.† This is the case that is of interest in this chapter.

Also from Eq. (9.51), we see that the dyadic Green function satisfies the reciprocity relationship,

$$\bar{\mathbf{G}}(\mathbf{r}\,|\,\mathbf{r}', \omega) = \tilde{\bar{\mathbf{G}}}(\mathbf{r}'\,|\,\mathbf{r}, \omega) \tag{9.54}$$

and has the following symmetry property:

$$\bar{\mathbf{G}}(\mathbf{R}, \omega) = \tilde{\bar{\mathbf{G}}}(\mathbf{R}, \omega) \tag{9.55}$$

In addition, Eq. (9.51) consists of three terms. The first term that varies as $1/R$ is called the *radiation term* since it dominates at distances far from the source. Because the dyad $\bar{\mathbf{I}} - \hat{\mathbf{R}}\hat{\mathbf{R}}$ projects any vector in the direction perpendicular to $\hat{\mathbf{R}}$, the radiation field is thus purely transverse to the direction $\hat{\mathbf{R}}$. The second term varies as $1/R^2$ and is called the *induction term* since it dominates at an intermediate distance from the source. At a distance very close to the source, the $1/R^3$ term dominates and is often called the *electrostatic term.*

9.4 RADIATION FROM AN OSCILLATING ELECTRIC DIPOLE

To illustrate the application of the dyadic Green function (9.51), we will consider the simplest radiating source of an oscillating electric dipole and evaluate the electromagnetic field generated by it. The current distribution of a point dipole with electric dipole moment $\mathbf{p}_0$ located at the origin (Fig. 9.3) may be expressed as

$$\mathbf{J}(\mathbf{r}') = -\, i\omega\mathbf{p}_0\, \delta(\mathbf{r}') \tag{9.56}$$

Substituting Eq. (9.56) into Eq. (9.9) and using the property of the delta function, we get

$$\mathbf{E}(\mathbf{r}) = \omega^2\mu_0\, \bar{\mathbf{G}}(\mathbf{r}\,|\,\mathbf{0}, \omega) \cdot \mathbf{p}_0 \tag{9.57}$$

where $\bar{\mathbf{G}}(\mathbf{r}\,|\,\mathbf{0}, \omega)$ is the dyadic Green function (9.51) evaluated at $\mathbf{r}' = \mathbf{0}$, that is,

$$\bar{\mathbf{G}}(\mathbf{r}\,|\,\mathbf{0}, \omega) = (\bar{\mathbf{I}} - \hat{\mathbf{r}}\hat{\mathbf{r}})g(\mathbf{r}, \omega) + \frac{i}{k_0\sqrt{\varepsilon}\, r}(\bar{\mathbf{I}} - 3\hat{\mathbf{r}}\hat{\mathbf{r}})g(\mathbf{r}, \omega) - \frac{1}{k_0^2\, \varepsilon r^2}(\bar{\mathbf{I}} - 3\hat{\mathbf{r}}\hat{\mathbf{r}})g(\mathbf{r}, \omega) \tag{9.58}$$

and

$$g(\mathbf{r}, \omega) = \frac{e^{ik_0\sqrt{\varepsilon}r}}{4\pi r} \tag{9.59}$$

† To obtain the field inside the source region, the principal value of the integral (9.53) is required. For detail see, for example, J. Van Bladel (1961) and A. Yaghjian (1980).

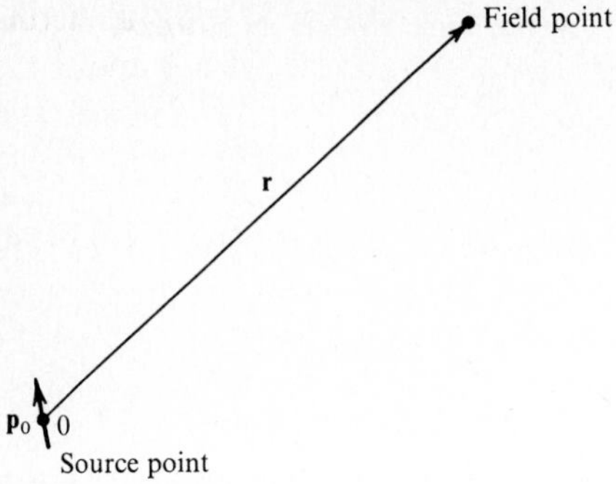

Figure 9.3 An oscillating electric dipole with dipole moment $\mathbf{p}_0$.

where $r = |\mathbf{r}|$, and $\hat{\mathbf{r}} = \mathbf{r}/r$. With the aid of Eq. (9.58), we may write Eq. (9.57) as

$$\mathbf{E}(\mathbf{r}) = \frac{1}{\varepsilon_0}\left\{\frac{1}{\varepsilon r^2}[3(\mathbf{p}_0 \cdot \hat{\mathbf{r}})\hat{\mathbf{r}} - \mathbf{p}_0] - \frac{ik_0}{\sqrt{\varepsilon}r}[3(\mathbf{p}_0 \cdot \hat{\mathbf{r}})\hat{\mathbf{r}} - \mathbf{p}_0] - k_0^2[\hat{\mathbf{r}} \times (\hat{\mathbf{r}} \times \mathbf{p}_0)]\right\} g(\mathbf{r}, \omega) \tag{9.60}$$

or

$$\mathbf{E}(\mathbf{r}) = \frac{1}{\varepsilon_0}\left[\left(k_0^2 + \frac{ik_0}{\sqrt{\varepsilon}r} - \frac{1}{\varepsilon r^2}\right)\mathbf{p}_0 - \left(k_0^2 + \frac{i3k_0}{\sqrt{\varepsilon}r} - \frac{3}{\varepsilon r^2}\right)(\mathbf{p}_0 \cdot \hat{\mathbf{r}})\hat{\mathbf{r}}\right] g(\mathbf{r}, \omega) \tag{9.61}$$

or

$$\mathbf{E}(\mathbf{r}) = \frac{1}{\varepsilon_0}\left\{2\left(\frac{1}{\varepsilon r^2} - \frac{ik_0}{\sqrt{\varepsilon}r}\right)(\mathbf{p}_0 \cdot \hat{\mathbf{r}})\hat{\mathbf{r}} + \left(\frac{1}{\varepsilon r^2} - \frac{ik_0}{\sqrt{\varepsilon}r} - k_0^2\right)[\hat{\mathbf{r}} \times (\hat{\mathbf{r}} \times \mathbf{p}_0)]\right\} g(\mathbf{r}, \omega) \tag{9.62}$$

To find the magnetic field intensity **H**, we substitute Eq. (9.56) into Eq. (9.26), make use of result (9.49), and thus obtain

$$\mathbf{H}(\mathbf{r}) = i\omega\left(\frac{1}{r} - ik_0\sqrt{\varepsilon}\right) g(\mathbf{r}, \omega)(\hat{\mathbf{r}} \times \mathbf{p}_0) \tag{9.63}$$

The terms in Eqs. (9.60) and (9.63) have been arranged in inverse powers of r and the ratio of the magnitudes of successive terms is $k_0\sqrt{\varepsilon}r$ or $2\pi\sqrt{\varepsilon}(r/\lambda)$. In the immediate neighborhood of the dipole, the electrostatic and induction fields in $1/r^3$ and $1/r^2$ dominate, while in the far-zone where $r \gg \lambda$, only the radiation field in $1/r$ is important. Several features of a dipole field are easily noted from Eqs. (9.60), (9.61), (9.62), and (9.63). They are: (1) The electric field intensity lies in the plane formed by vectors $\mathbf{p}_0$ and $\hat{\mathbf{r}}$, and the magnetic field intensity is perpendicular to this plane (Fig. 9.3); (2) at zero frequency, $\omega = 0$ (or $k_0 = 0$), all terms vanish with the exception of the terms in $1/r^3$ in Eq. (9.61). Thus, $\mathbf{H} = \mathbf{0}$ and

$$\mathbf{E} = \frac{1}{4\pi\varepsilon_0 \varepsilon r^3} [3(\mathbf{p}_0 \cdot \hat{\mathbf{r}})\hat{\mathbf{r}} - \mathbf{p}_0] \tag{9.64}$$

which is the electrostatic field produced by an electric dipole. (3) At distances very close to the dipole, the electric field intensity has the same structure, apart from the factor $e^{ik_0\sqrt{\varepsilon}r}$, as Eq. (9.64). (4) At large distances from the dipole ($k_0\sqrt{\varepsilon}\, r \gg 1$), Eqs. (9.62) and (9.63) become approximately

$$\mathbf{E}(\mathbf{r}) \cong -\frac{k_0^2\, e^{ik_0\sqrt{\varepsilon}r}}{4\pi\varepsilon_0 r} [\hat{\mathbf{r}} \times (\hat{\mathbf{r}} \times \mathbf{p}_0)]$$

and

$$\mathbf{H}(\mathbf{r}) \cong \frac{k_0\, \omega\sqrt{\varepsilon}\, e^{ik_0\sqrt{\varepsilon}r}}{4\pi r} (\hat{\mathbf{r}} \times \mathbf{p}_0) \tag{9.65}$$

Thus in the far-zone the radiation field has the structure of a TEM wave, and the field vectors are related by the simple relation

$$\hat{\mathbf{r}} \times \mathbf{E}(\mathbf{r}) = \sqrt{\frac{\mu_0}{\varepsilon_0\, \varepsilon}}\, \mathbf{H}(\mathbf{r}) \tag{9.66}$$

Furthermore, if we let $r \to \infty$ and at the same time increase $\mathbf{p}_0$ so that the magnitudes of the field vectors remain constant, we obtain in the limit the plane wave.

Let us calculate next the power radiated by an oscillating electric dipole. This can be found by integrating the Poynting vector over a spherical surface centered on the dipole. Using the electric and magnetic fields of Eqs. (9.61) and (9.63), we find the complex Poynting vector $\mathbf{P}$:

$$\begin{aligned} \mathbf{P} &= \tfrac{1}{2}(\mathbf{E} \times \mathbf{H}^*) \\ &= \frac{\omega k_0^3 \sqrt{\varepsilon}}{32\varepsilon_0 \pi^2 r^2} (|\mathbf{p}_0|^2 - |\mathbf{p}_0 \cdot \hat{\mathbf{r}}|^2)\hat{\mathbf{r}} + \frac{i\omega}{32\varepsilon_0 \varepsilon \pi^2 r^5} (|\mathbf{p}_0|^2 - |\mathbf{p}_0 \cdot \hat{\mathbf{r}}|^2)\hat{\mathbf{r}} \\ &\quad + \frac{i\omega}{16\varepsilon_0 \pi^2} \left(\frac{k_0^2}{r^3} - \frac{1}{\varepsilon r^5}\right) (\mathbf{p}_0 \cdot \hat{\mathbf{r}})[\mathbf{p}_0^* - (\mathbf{p}_0^* \cdot \hat{\mathbf{r}})\hat{\mathbf{r}}] \end{aligned} \tag{9.67}$$

Next, the radial component of $\mathbf{P}$ is integrated over the surface of a sphere of radius r:

$$\int_0^{2\pi} \int_0^{\pi} (\mathbf{P} \cdot \hat{\mathbf{r}}) r^2 \sin\theta \, d\theta \, d\phi = \frac{\omega^4 \mu_0 \sqrt{\mu_0 \varepsilon_0 \varepsilon}\, |\mathbf{p}_0|^2}{12\pi} + \frac{i\omega |\mathbf{p}_0|^2}{12\pi\varepsilon_0 \varepsilon r^3} \tag{9.68}$$

The real part of Eq. (9.68) gives the total time-averaged power $\langle \mathscr{S} \rangle$ radiated into space by the dipole:

$$\langle \mathscr{S} \rangle = \frac{\omega^4 \mu_0 \sqrt{\mu_0 \varepsilon_0 \varepsilon}\, |\mathbf{p}_0|^2}{12\pi} \tag{9.69}$$

If we wish to express $\langle \mathscr{S} \rangle$ in terms of a uniform current I that flows along an infinitesimal length $\mathbf{l}$, we must substitute $\mathbf{p}_0$ by $-I\mathbf{l}/i\omega$ (see Prob. 9.6), and then

$$\langle \mathscr{S} \rangle = \frac{\pi |I|^2}{3} \sqrt{\frac{\mu_0}{\varepsilon_0}} \sqrt{\varepsilon} \left(\frac{l}{\lambda}\right)^2 \tag{9.70}$$

where λ is the wavelength and is related to k_0 by $2\pi/\lambda$. As far as the dipole is concerned, this radiated power is lost in the same way as if it were dissipated in a resistance R_r:

$$\langle \mathscr{S} \rangle = \tfrac{1}{2} |I|^2 R_r \tag{9.71}$$

where this equivalent resistance is called the *radiation resistance* and is given by

$$R_r = \frac{2\pi}{3} \sqrt{\frac{\mu_0}{\varepsilon_0}} \sqrt{\varepsilon} \left(\frac{l}{\lambda}\right)^2 \tag{9.72}$$

These results are only true for point dipoles, where l is much less than a wavelength ($l \ll \lambda$).

It is interesting to note that only the terms that vary as $1/r$ enter into the expression (9.69) for the radiated power. In other words, the far fields alone contribute to the time-averaged power flow. The near and induction fields contribute only to the imaginary part in Eq. (9.68) representing the reactive power. This power, which oscillates back and forth between the source and the region of space surrounding the source, accounts for the energy stored in the field without ever being lost from the system.

9.5 FAR-ZONE APPROXIMATION

We now return to the general case of arbitrary current distribution and consider some general characteristics of the far-zone field. The far-zone region corresponds to the region in which the radiation field predominates and hence is the region of most interest in connection with antennas—devices whose primary purpose is to radiate or receive electromagnetic energy. The far-zone region is defined by the conditions

$$r \gg r' \qquad \text{and} \qquad k_0 \sqrt{\varepsilon}\, r \gg 1 \tag{9.73}$$

where $r = |\mathbf{r}| = \sqrt{\mathbf{r} \cdot \mathbf{r}}$ and $r' = |\mathbf{r}'| = \sqrt{\mathbf{r}' \cdot \mathbf{r}'}$. Using binomial expansion, we obtain the following approximation:

$$R = |\mathbf{r} - \mathbf{r}'| = \sqrt{r^2 + r'^2 - 2\mathbf{r} \cdot \mathbf{r}'} \cong r - \mathbf{r}' \cdot \hat{\mathbf{r}} \tag{9.74}$$

In this approximation we may replace Eq. (9.52) by

$$g_f(\mathbf{R}, \omega) = g(\mathbf{r}, \omega) e^{-ik_0\sqrt{\varepsilon}\mathbf{r}' \cdot \hat{\mathbf{r}}} \tag{9.75}$$

where $g(\mathbf{r}, \omega)$ is given by Eq. (9.59). Accordingly the dyadic Green function (9.51) in the far-zone by including terms of the order $1/r$ only becomes

$$\bar{\mathbf{G}}_f(\mathbf{R}, \omega) = (\bar{\mathbf{I}} - \hat{\mathbf{r}}\hat{\mathbf{r}})g_f(\mathbf{R}, \omega) \tag{9.76}$$

Substituting this expression into Eq. (9.53) we obtain the following representation for the far-zone electric field intensity:

$$\begin{aligned} \mathbf{E}(\mathbf{r}) &= i\omega\mu_0\, g(\mathbf{r}, \omega)(\bar{\mathbf{I}} - \hat{\mathbf{r}}\hat{\mathbf{r}}) \cdot \mathbf{N} \\ &= -i\omega\mu_0\, g(\mathbf{r}, \omega)[\hat{\mathbf{r}} \times (\hat{\mathbf{r}} \times \mathbf{N})] \end{aligned} \tag{9.77}$$

where the *radiation vector* $\mathbf{N}$ is defined by the integral

$$\mathbf{N} = \int_V \mathbf{J}(\mathbf{r}')e^{-ik_0\sqrt{\varepsilon}\,\mathbf{r}'\cdot\hat{\mathbf{r}}}\, d^3r' \tag{9.78}$$

In Eq. (9.78), the integration is performed over the volume V containing the source currents. The far-zone magnetic field intensity is found by substituting Eq. (9.75) into Eq. (9.26) and making use of result (9.49). Thus to the same order of approximation,

$$\mathbf{H}(\mathbf{r}) = ik_0\sqrt{\varepsilon}\; g(\mathbf{r}, \omega)(\hat{\mathbf{r}} \times \mathbf{N}) \tag{9.79}$$

Comparing Eq. (9.77) and (9.79) we see that in the far-zone

$$\mathbf{E}(\mathbf{r}) = \sqrt{\frac{\mu_0}{\varepsilon_0\,\varepsilon}}\, \mathbf{H}(\mathbf{r}) \times \hat{\mathbf{r}} \tag{9.80}$$

or

$$\mathbf{H}(\mathbf{r}) = \sqrt{\frac{\varepsilon_0\,\varepsilon}{\mu_0}}\, \hat{\mathbf{r}} \times \mathbf{E}(\mathbf{r}) \tag{9.81}$$

is always valid. In other words, the far-zone $\mathbf{E}(\mathbf{r})$ and $\mathbf{H}(\mathbf{r})$ are perpendicular to each other and to the direction of wave propagation $\hat{\mathbf{r}}$. Furthermore, Eqs. (9.80) and (9.81) yield the following expression for the far-zone complex Poynting vector:

$$\begin{aligned} \mathbf{P} &= \tfrac{1}{2}(\mathbf{E} \times \mathbf{H}^*) \\ &= \frac{k_0^2}{32\pi^2 r^2}\sqrt{\frac{\mu_0\,\varepsilon}{\varepsilon_0}}\, |\hat{\mathbf{r}} \times \mathbf{N}|^2\hat{\mathbf{r}} \end{aligned} \tag{9.82}$$

From Eq. (9.82) we see that for a real ε, $\mathbf{P}$ is purely real and always pointing in the direction parallel to $\hat{\mathbf{r}}$. Hence, the integration of the radial component of $\mathbf{P}$ over the surface S of a sphere of radius r gives the time-averaged power $\langle \mathscr{S} \rangle$ radiated by the source:

$$\begin{aligned} \langle \mathscr{S} \rangle &= \mathrm{Re} \oint_S \mathbf{P} \cdot d\mathbf{s} \\ &= \frac{k_0^2}{32\pi^2}\sqrt{\frac{\mu_0\,\varepsilon}{\varepsilon_0}} \int |\hat{\mathbf{r}} \times \mathbf{N}|^2\, d\Omega \end{aligned} \tag{9.83}$$

where $d\Omega$ is an element of solid angle. The above procedure of calculating the radiated power is often called the *Poynting vector method.*

To illustrate the use of the method, we calculate again the time-averaged power radiated by an electric dipole. Substituting $\mathbf{J}(\mathbf{r}') = I\mathbf{l}\ \delta(\mathbf{r}')$ into Eq. (9.78) we find the radiation vector $\mathbf{N} = I\mathbf{l}$. Accordingly, after carrying out the integration in Eq. (9.83), we obtain the result given by Eq. (9.70).

9.6 RADIATION FROM A THIN-WIRE ANTENNA

In Sec. 9.4 we considered radiation from an electric dipole. A more realistic and practical source of radiation is the thin-wire antenna. The field excited by such an antenna can be obtained by using the formulas derived in the previous section.

Figure 9.4 illustrates a straight-wire antenna of length $2L$ that is excited in the center across a very small gap. Since the wire is thin and since we wish to calculate only the far-zone field it is reasonable to assume that the antenna current density follows filamentary distribution:

$$\mathbf{J}(\mathbf{r}') = I(l')\ \delta(x')\ \delta(y')\hat{\mathbf{l}} \tag{9.84}$$

where $\hat{\mathbf{l}}$ is the unit vector pointing along the direction of the thin wire. Substitution of Eq. (9.84) into Eq. (9.78) yields the radiation vector:

$$\mathbf{N} = \hat{\mathbf{l}} \int_{-L}^{L} I(l')e^{-ik_0\sqrt{\varepsilon}\, l'\cos\theta}\, dl' \tag{9.85}$$

Since the current vanishes at the ends of the wire, we may further approximate the filamentary current $I(l')$ by a sinusoidal distribution:

$$I(l') = I_0 \sin k_0\sqrt{\varepsilon}(L - |l'|) \tag{9.86}$$

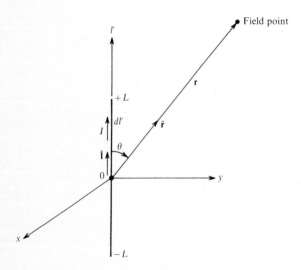

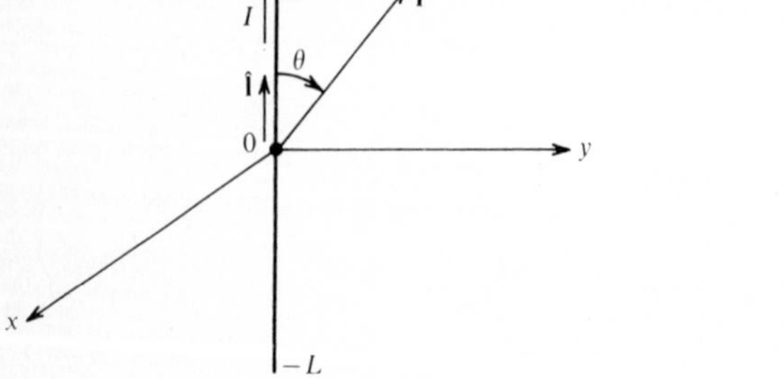

Figure 9.4 A thin-wire antenna.

With the aid of the integration formula,

$$\int \sin(ax+\alpha)\cos(bx+\beta)\,dx = -\frac{\cos[(a+b)x+\alpha+\beta]}{2(a+b)} - \frac{\cos[(a-b)x+\alpha-\beta]}{2(a-b)} \tag{9.87}$$

the radiation vector (9.85) can now be evaluated as

$$\mathbf{N} = 2I_0 \int_0^L \sin k_0\sqrt{\varepsilon}(L-l')\cos(k_0\sqrt{\varepsilon}\,l'\cos\theta)\,dl'\hat{\mathbf{l}}$$

$$= 2I_0 \frac{\cos(k_0\sqrt{\varepsilon}\,L\cos\theta) - \cos k_0\sqrt{\varepsilon}\,L}{k_0\sqrt{\varepsilon}\sin^2\theta}\hat{\mathbf{l}} \tag{9.88}$$

From Eqs. (9.79) and (9.80), the far-zone magnetic and electric fields are given by

$$\mathbf{H}(\mathbf{r}) = i2I_0\, g(\mathbf{r},\omega)\frac{\cos(k_0\sqrt{\varepsilon}\,L\cos\theta) - \cos k_0\sqrt{\varepsilon}\,L}{\sin^2\theta}(\hat{\mathbf{r}}\times\hat{\mathbf{l}}) \tag{9.89}$$

and

$$\mathbf{E}(\mathbf{r}) = i\sqrt{\frac{\mu_0}{\varepsilon_0\varepsilon}}\,2I_0\, g(\mathbf{r},\omega)\frac{\cos(k_0\sqrt{\varepsilon}\,L\cos\theta) - \cos k_0\sqrt{\varepsilon}\,L}{\sin^2\theta}[(\hat{\mathbf{r}}\times\hat{\mathbf{l}})\times\hat{\mathbf{r}}] \tag{9.90}$$

respectively, where $g(\mathbf{r},\omega)$ is given by Eq. (9.59). According to Eqs. (9.82) and (9.83), the complex Poynting vector becomes

$$\mathbf{P} = \sqrt{\frac{\mu_0}{\varepsilon_0\varepsilon}}\frac{I_0^2}{8\pi^2 r^2}\left[\frac{\cos(k_0\sqrt{\varepsilon}\,L\cos\theta) - \cos k_0\sqrt{\varepsilon}\,L}{\sin\theta}\right]^2\hat{\mathbf{r}} \tag{9.91}$$

and the time-averaged power radiated by the antenna is

$$\langle\mathscr{S}\rangle = \frac{I_0^2}{4\pi}\sqrt{\frac{\mu_0}{\varepsilon_0\varepsilon}}\int_0^\pi \frac{[\cos(k_0\sqrt{\varepsilon}\,L\cos\theta) - \cos k_0\sqrt{\varepsilon}\,L]^2}{\sin\theta}\,d\theta \tag{9.92}$$

To evaluate the integral in Eq. (9.92) we introduce the new variable $u = k_0\sqrt{\varepsilon}\,L\cos\theta$. Thus

$$K = \int_0^\pi \frac{[\cos(k_0\sqrt{\varepsilon}\,L\cos\theta) - \cos k_0\sqrt{\varepsilon}\,L]^2}{\sin\theta}\,d\theta$$

$$= \frac{1}{2}\int_{-k_0\sqrt{\varepsilon}L}^{k_0\sqrt{\varepsilon}L}(\cos u - \cos k_0\sqrt{\varepsilon}\,L)^2\left(\frac{1}{k_0\sqrt{\varepsilon}\,L+u} + \frac{1}{k_0\sqrt{\varepsilon}\,L-u}\right)du$$

$$= \int_{-k_0\sqrt{\varepsilon}L}^{k_0\sqrt{\varepsilon}L}\frac{(\cos u - \cos k_0\sqrt{\varepsilon}\,L)^2}{k_0\sqrt{\varepsilon}\,L-u}\,du \tag{9.93}$$

Now replacing u by $k_0\sqrt{\varepsilon}\,L - v$, expanding the square, and regrouping the terms, we obtain

$$K = (1 + \cos 2k_0\sqrt{\varepsilon}\,L) \int_0^{2k_0\sqrt{\varepsilon}\,L} \frac{1 - \cos v}{v}\,dv$$

$$- \sin 2k_0\sqrt{\varepsilon}\,L \int_0^{2k_0\sqrt{\varepsilon}\,L} \frac{\sin v}{v}\,dv + \tfrac{1}{2}\sin 2k_0\sqrt{\varepsilon}\,L \int_0^{2k_0\sqrt{\varepsilon}\,L} \frac{\sin 2v}{v}\,dv$$

$$- \frac{\cos 2k_0\sqrt{\varepsilon}\,L}{2} \int_0^{2k_0\sqrt{\varepsilon}\,L} \frac{1 - \cos 2v}{v}\,dv \tag{9.94}$$

and hence

$$\langle \mathscr{S} \rangle = \frac{I_0^2}{4\pi}\sqrt{\frac{\mu_0}{\varepsilon_0\,\varepsilon}} \Bigg\{ C + \ln (2k_0\sqrt{\varepsilon}\,L) - Ci(2k_0\sqrt{\varepsilon}\,L)$$

$$+ \frac{\sin 2k_0\sqrt{\varepsilon}\,L}{2}\,[Si(4k_0\sqrt{\varepsilon}\,L) - 2Si(2k_0\sqrt{\varepsilon}\,L)]$$

$$+ \frac{\cos 2k_0\sqrt{\varepsilon}\,L}{2}\,[C + \ln k_0\sqrt{\varepsilon}\,L + Ci(4k_0\sqrt{\varepsilon}\,L) - 2Ci(2k_0\sqrt{\varepsilon}\,L)]\Bigg\} \tag{9.95}$$

where

$$Si(x) = \int_0^x \frac{\sin \xi}{\xi}\,d\xi = \text{sine integral}$$

$$Ci(x) = -\int_x^\infty \frac{\cos \xi}{\xi}\,d\xi = \text{cosine integral}$$

$$= C + \ln x - \int_0^x \frac{1 - \cos \xi}{\xi}\,d\xi$$

and $C = 0.5722$ is Euler's constant. With the aid of a table of sine and cosine integrals, we can easily calculate the time-averaged power from Eq. (9.95) and thus the radiation resistance R_r from definition (9.71).

9.7 RADIATION RESISTANCE BY THE EMF METHOD

As an alternative way of calculating the time-averaged power (thus the radiation resistance) radiated by electric current sources, we will consider the complex Poynting theorem in the presence of an externally applied current source **J** (cf. Sec. 2.5):

$$\nabla \cdot \mathbf{P} = -\tfrac{1}{2}\mathbf{J}^* \cdot \mathbf{E} + i2\omega(\langle W_m \rangle - \langle W_e \rangle) \tag{9.96}$$

where

$$\langle W_m \rangle = \tfrac{1}{4}\mu_0\,\mathbf{H} \cdot \mathbf{H}^*$$

and $$\langle W_e \rangle = \tfrac{1}{4}\varepsilon_0 \,\varepsilon \mathbf{E} \cdot \mathbf{E}^*$$

are the time-averaged magnetic and electric energy densities respectively. Here we assume that ε is real. The real part of Eq. (9.96), after being integrated throughout a volume V bounded by a closed surface A which completely encloses the volume V_0 occupied by the current $\mathbf{J}$, gives

$$\operatorname{Re} \int_V \nabla \cdot \mathbf{P} \, d^3r = -\tfrac{1}{2} \operatorname{Re} \int_{V_0} \mathbf{J}^* \cdot \mathbf{E} \, d^3r \tag{9.97}$$

Converting the volume integral on the left-hand side to a surface integral by the divergence theorem, we obtain

$$\begin{aligned} \langle \mathscr{S} \rangle &= \operatorname{Re} \oint_A \mathbf{P} \cdot d\mathbf{s} \\ &= -\tfrac{1}{2} \operatorname{Re} \int_{V_0} \mathbf{J}^* \cdot \mathbf{E} \, d^3r \end{aligned} \tag{9.98}$$

Thus we see that the time-averaged power $\langle \mathscr{S} \rangle$ radiated by the source can be computed by the Poynting vector method. In this method, we integrate the real part of the normal component of the complex Poynting vector $\mathbf{P}$ over a closed surface A enclosing all the sources. Alternatively, we may use the EMF method by integrating $-\tfrac{1}{2} \operatorname{Re} (\mathbf{J}^* \cdot \mathbf{E})$ throughout the volume V_0 occupied by the current $\mathbf{J}$.

In some cases it is more convenient to express Eq. (9.98) in terms of the wave matrix and the Fourier transform $\mathbf{J}(\mathbf{k})$ of the current density:

$$\mathbf{J}(\mathbf{r}) = \frac{1}{(2\pi)^3} \int_{-\infty}^{+\infty} \mathbf{J}(\mathbf{k}) e^{i\mathbf{k}\cdot\mathbf{r}} \, d^3k \tag{9.99}$$

To this end we substitute Eqs. (9.20), (9.9), and (9.99) into Eq. (9.98) and get

$$\begin{aligned} \langle \mathscr{S} \rangle &= -\frac{1}{2} \operatorname{Re} \int_{V_0} \mathbf{J}^*(\mathbf{r}) \cdot \mathbf{E}(\mathbf{r}) \, d^3r \\ &= \operatorname{Re}\left[-\frac{1}{2} \frac{i\omega\mu_0}{(2\pi)^9} \int \mathbf{J}^*(\mathbf{k}') \cdot \bar{\mathbf{W}}_i^{-1}(\mathbf{k}) \cdot \mathbf{J}(\mathbf{k}'') e^{i\mathbf{r}\cdot(\mathbf{k}-\mathbf{k}')} \right. \\ &\qquad \left. \cdot \, e^{i\mathbf{r}'\cdot(\mathbf{k}''-\mathbf{k})} \, d^3k \, d^3k' \, d^3k'' \, d^3r \, d^3r' \right] \end{aligned} \tag{9.100}$$

Recalling the relation

$$\frac{1}{(2\pi)^3} \int_{-\infty}^{+\infty} e^{i\mathbf{r}\cdot(\mathbf{k}-\mathbf{k}')} \, d^3r = \delta(\mathbf{k} - \mathbf{k}') \tag{9.101}$$

we can reduce the fifteen integrals in Eq. (9.100) to three and obtain the formula

$$\langle \mathscr{S} \rangle = -\tfrac{1}{2} \operatorname{Re} \left[\frac{i\omega\mu_0}{(2\pi)^3} \int_{-\infty}^{+\infty} \mathbf{J}^*(\mathbf{k}) \cdot \bar{\mathbf{W}}_i^{-1}(\mathbf{k}) \cdot \mathbf{J}(\mathbf{k}) \, d^3k \right] \tag{9.102}$$

which, of course, may be written down almost immediately, if Parseval's formula in the Fourier transform theory is used.

As an application of Eq. (9.102), we compute the radiation resistance of an electric dipole by the EMF method. The Fourier transform of $\mathbf{J}(\mathbf{r}') = I\mathbf{l}\,\delta(\mathbf{r}')$ yields

$$\mathbf{J}(\mathbf{k}) = I\mathbf{l} \tag{9.103}$$

Substituting this into Eq. (9.102) and comparing the result with Eq. (9.71), we can write the radiation resistance in the following quadratic form:

$$R_r = \mathbf{l} \cdot (\text{Re}\ \bar{\mathbf{Z}}) \cdot \mathbf{l} \tag{9.104}$$

where the components of the complex matrix

$$\bar{\mathbf{Z}} = -\frac{i\omega\mu_0}{(2\pi)^3} \int_{-\infty}^{+\infty} \bar{\mathbf{W}}_i^{-1}(\mathbf{k})\, d^3k \tag{9.105}$$

have the dimension of impedance units per unit area.

To calculate the matrix $\bar{\mathbf{Z}}$ for the case of an electric dipole radiating in an isotropic medium, we substitute Eq. (9.21) into Eq. (9.105) and carry out the integration in **k**-space (see Prob. 9.22):

$$\bar{\mathbf{Z}} = \frac{2\pi}{3} \sqrt{\frac{\mu_0}{\varepsilon_0}} \frac{\sqrt{\varepsilon}}{\lambda^2} \bar{\mathbf{I}} \tag{9.106}$$

Substitution of Eq. (9.106) into Eq. (9.104) yields the same formula (9.72) for the radiation resistance as computed by the Poynting vector method. Similarly, Eq. (9.71) results in the same formula for the time-averaged power radiated by the dipole.

9.8 HUYGEN'S PRINCIPLE

As another application of the dyadic Green function, we conclude this chapter with the mathematical formulation of Huygen's principle. Consider a homogeneous, isotropic, source-free dielectric region of space V bounded by two surfaces S and S_∞ as shown in Fig. 9.5. The current source is assumed to be confined within the surface S, and S_∞ is the surface of a sphere with infinite radius.

Integrating the vector identity

$$\mathbf{F} \cdot [\nabla \times (\nabla \times \mathbf{A})] - \mathbf{A} \cdot [\nabla \times (\nabla \times \mathbf{F})] = \nabla \cdot [\mathbf{A} \times (\nabla \times \mathbf{F}) - \mathbf{F} \times (\nabla \times \mathbf{A})] \tag{9.107}$$

over the volume V and then applying the divergence theorem to the result, we

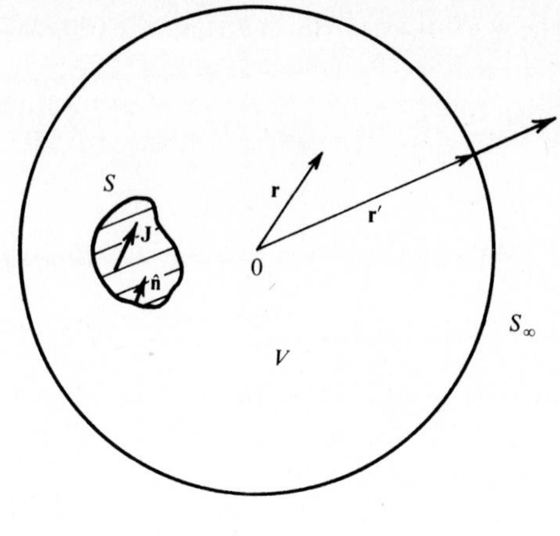

Figure 9.5 A region of space V bounded by surfaces S and S_∞ with the current sources lying inside the surface S.

obtain the vector Green theorem:

$$\int_V \{\mathbf{F} \cdot [\nabla \times (\nabla \times \mathbf{A})] - [\nabla \times (\nabla \times \mathbf{F})] \cdot \mathbf{A}\}\, d^3r$$

$$= \oint_{S+S_\infty} [\mathbf{A} \times (\nabla \times \mathbf{F}) - \mathbf{F} \times (\nabla \times \mathbf{A})] \cdot d\mathbf{s}$$

$$= -\oint_{S+S_\infty} [\hat{\mathbf{n}} \times (\nabla \times \mathbf{F}) \cdot \mathbf{A} + (\hat{\mathbf{n}} \times \mathbf{F}) \cdot (\nabla \times \mathbf{A})]\, ds \tag{9.108}$$

where $\hat{\mathbf{n}}$ is an outward unit vector normal to the surface. Now letting $\mathbf{F} = \mathbf{E}(\mathbf{r})$ and $\mathbf{A} = \bar{\mathbf{G}}(\mathbf{r}, \mathbf{r}') \cdot \mathbf{a}$, where $\mathbf{a}$ is an arbitrary constant vector, we have

$$\int_V \{\mathbf{E} \cdot [\nabla \times (\nabla \times \bar{\mathbf{G}} \cdot \mathbf{a})] - [\nabla \times (\nabla \times \mathbf{E})] \cdot \bar{\mathbf{G}} \cdot \mathbf{a}\}\, d^3r$$

$$= -\oint_{S+S_\infty} [\hat{\mathbf{n}} \times (\nabla \times \mathbf{E}) \cdot (\bar{\mathbf{G}} \cdot \mathbf{a}) + (\hat{\mathbf{n}} \times \mathbf{E}) \cdot \nabla \times (\bar{\mathbf{G}} \cdot \mathbf{a})]\, ds \tag{9.109}$$

or

$$\int_V \{\mathbf{E}(\mathbf{r}) \cdot [\nabla \times \nabla \times \bar{\mathbf{G}}(\mathbf{r}, \mathbf{r}')] - [\nabla \times \nabla \times \mathbf{E}(\mathbf{r})] \cdot \bar{\mathbf{G}}(\mathbf{r}, \mathbf{r}')\}\, d^3r$$

$$= -\oint_{S+S_\infty} \{\hat{\mathbf{n}} \times [\nabla \times \mathbf{E}(\mathbf{r})] \cdot \bar{\mathbf{G}}(\mathbf{r}, \mathbf{r}') + [\hat{\mathbf{n}} \times \mathbf{E}(\mathbf{r})] \cdot \nabla \times \bar{\mathbf{G}}(\mathbf{r}, \mathbf{r}')\}\, ds \tag{9.110}$$

Since there is no source current in V, the field vectors satisfy the homogeneous Maxwell equations

$$\begin{aligned} \nabla \times \mathbf{E} &= i\omega\mu_0 \mathbf{H} \\ \nabla \times \mathbf{H} &= -i\omega\varepsilon_0\, \varepsilon \mathbf{E} \end{aligned} \tag{9.111}$$

Elimination of $\mathbf{H}$ from the above equations yields

$$\nabla \times \nabla \times \mathbf{E}(\mathbf{r}) - k_0^2\, \varepsilon \mathbf{E}(\mathbf{r}) = \mathbf{0} \tag{9.112}$$

Substituting Eqs. (9.111) and (9.112) into Eq. (9.110), noting that the dyadic Green function satisfies Eq. (9.13):

$$\nabla \times \nabla \times \bar{\mathbf{G}}(\mathbf{r}, \mathbf{r}') - k_0^2 \varepsilon \bar{\mathbf{G}}(\mathbf{r}, \mathbf{r}') = \delta(\mathbf{r} - \mathbf{r}')\bar{\mathbf{I}} \tag{9.113}$$

and noting also the fact that

$$\int_V \mathbf{E}(\mathbf{r}) \cdot \bar{\mathbf{I}}\delta(\mathbf{r} - \mathbf{r}')\, d^3r = \mathbf{E}(\mathbf{r}') \tag{9.114}$$

we obtain

$$\mathbf{E}(\mathbf{r}') = -\oint_{S+S_\infty} \{i\omega\mu_0[\hat{\mathbf{n}} \times \mathbf{H}(\mathbf{r})] \cdot \bar{\mathbf{G}}(\mathbf{r}, \mathbf{r}') + [\hat{\mathbf{n}} \times \mathbf{E}(\mathbf{r})] \cdot \nabla \times \bar{\mathbf{G}}(\mathbf{r}, \mathbf{r}')\}\, ds \tag{9.115}$$

Interchanging the primed and the unprimed variables, we get

$$\begin{aligned}\mathbf{E}(\mathbf{r}) = &-\oint_S \{i\omega\mu_0[\hat{\mathbf{n}} \times \mathbf{H}(\mathbf{r}')] \cdot \bar{\mathbf{G}}(\mathbf{r}', \mathbf{r}) + [\hat{\mathbf{n}} \times \mathbf{E}(\mathbf{r}')] \cdot \nabla' \times \bar{\mathbf{G}}(\mathbf{r}', \mathbf{r})\}\, ds' \\ &-\oint_{S_\infty} \{i\omega\mu_0[\hat{\mathbf{n}} \times \mathbf{H}(\mathbf{r}')] \cdot \bar{\mathbf{G}}(\mathbf{r}', \mathbf{r}) + [\hat{\mathbf{n}} \times \mathbf{E}(\mathbf{r}')] \cdot \nabla' \times \bar{\mathbf{G}}(\mathbf{r}', \mathbf{r})\}\, ds'\end{aligned} \tag{9.116}$$

in which the dyadic Green function $\bar{\mathbf{G}}$ is given by Eq. (9.23), or

$$\bar{\mathbf{G}}(\mathbf{r}', \mathbf{r}) \cdot \mathbf{a} = \left(\bar{\mathbf{I}} + \frac{1}{k_0^2 \varepsilon} \nabla'\nabla'\right) g(\mathbf{r}', \mathbf{r}) \cdot \mathbf{a} \tag{9.117}$$

for an arbitrary constant vector **a**, where

$$g(\mathbf{r}', \mathbf{r}) = \frac{e^{ik_0 \sqrt{\varepsilon}\, |\mathbf{r}' - \mathbf{r}|}}{4\pi |\mathbf{r}' - \mathbf{r}|} \tag{9.118}$$

Since the operator "curl grad" is identically zero, and since

$$\nabla g(\mathbf{r}', \mathbf{r}) = -\nabla' g(\mathbf{r}', \mathbf{r}) \tag{9.119}$$

we have

$$\begin{aligned}\nabla' \times \bar{\mathbf{G}}(\mathbf{r}', \mathbf{r}) \cdot \mathbf{a} &= \nabla' \times [g(\mathbf{r}', \mathbf{r})\mathbf{a}] = -\mathbf{a} \times \nabla' g(\mathbf{r}', \mathbf{r}) \\ &= -\nabla g(\mathbf{r}', \mathbf{r}) \times \mathbf{a}\end{aligned} \tag{9.120}$$

or

$$\nabla' \times \bar{\mathbf{G}}(\mathbf{r}', \mathbf{r}) = -\nabla g(\mathbf{r}', \mathbf{r}) \times \bar{\mathbf{I}} \tag{9.121}$$

Let us now consider the second surface integral in Eq. (9.116). On the large spherical surface S_∞, $r' \gg r$, the far-zone approximation of Sec. 9.5 is valid. Thus for large r', the scalar and dyadic Green's function become

$$g_f(\mathbf{r}', \mathbf{r}) = \frac{e^{ik_0 \sqrt{\varepsilon}\, r'}}{4\pi r'} e^{-ik_0 \sqrt{\varepsilon}\, \mathbf{r} \cdot \hat{\mathbf{r}}'} \tag{9.122}$$

and

$$\bar{\mathbf{G}}_f(\mathbf{r}', \mathbf{r}) = (\bar{\mathbf{I}} - \hat{\mathbf{r}}'\hat{\mathbf{r}}')g_f(\mathbf{r}', \mathbf{r}) \tag{9.123}$$

respectively. Hence, from Eqs. (9.121), (9.122), and (9.123), it follows that

$$\begin{aligned}\nabla' \times \bar{\mathbf{G}}_f(\mathbf{r}', \mathbf{r}) &= ik_0\sqrt{\varepsilon}\, g_f(\mathbf{r}', \mathbf{r})(\hat{\mathbf{r}}' \times \bar{\mathbf{I}}) \\ &= ik_0\sqrt{\varepsilon}\, \hat{\mathbf{r}}' \times \bar{\mathbf{G}}_f(\mathbf{r}', \mathbf{r})\end{aligned} \tag{9.124}$$

The field vectors on S_∞ are related by

$$\hat{\mathbf{r}}' \times \mathbf{H}(\mathbf{r}') = -\sqrt{\frac{\varepsilon_0\,\varepsilon}{\mu_0}}\,\mathbf{E}(\mathbf{r}') \tag{9.125}$$

Substituting Eqs. (9.124) and (9.125) into the second surface integral in Eq. (9.116) and noting that $\hat{\mathbf{r}}' \cdot \mathbf{E}(\mathbf{r}') = 0$ on S_∞, we find that the integrand over the large spherical surface vanishes:

$$\begin{aligned}&i\omega\mu_0[\hat{\mathbf{n}} \times \mathbf{H}(\mathbf{r}')] \cdot \bar{\mathbf{G}}(\mathbf{r}', \mathbf{r}) + [\hat{\mathbf{n}} \times \mathbf{E}(\mathbf{r}')] \cdot \nabla' \times \bar{\mathbf{G}}(\mathbf{r}', \mathbf{r}) \\ &\qquad \cong -ik_0\sqrt{\varepsilon}\,\mathbf{E}(\mathbf{r}') \cdot \bar{\mathbf{G}}_f(\mathbf{r}', \mathbf{r}) + [\hat{\mathbf{r}}' \times \mathbf{E}(\mathbf{r}')] \cdot [ik_0\sqrt{\varepsilon}\,\hat{\mathbf{r}}' \times \bar{\mathbf{G}}_f(\mathbf{r}', \mathbf{r})] \\ &\qquad = -ik_0\sqrt{\varepsilon}\,\mathbf{E}(\mathbf{r}') \cdot \bar{\mathbf{G}}_f(\mathbf{r}', \mathbf{r}) - ik_0\sqrt{\varepsilon}\,\hat{\mathbf{r}}' \times [\hat{\mathbf{r}}' \times \mathbf{E}(\mathbf{r}')] \cdot \bar{\mathbf{G}}_f(\mathbf{r}', \mathbf{r}) \\ &\qquad = \mathbf{0}\end{aligned}$$

We finally obtain the result

$$\mathbf{E}(\mathbf{r}) = -\oint_s \{i\omega\mu_0[\hat{\mathbf{n}} \times \mathbf{H}(\mathbf{r}')] \cdot \bar{\mathbf{G}}(\mathbf{r}', \mathbf{r}) + [\hat{\mathbf{n}} \times \mathbf{E}(\mathbf{r}')] \cdot \nabla' \times \bar{\mathbf{G}}(\mathbf{r}', \mathbf{r})\}\, ds' \tag{9.126}$$

Following the same procedure by letting $\mathbf{F} = \mathbf{H}(\mathbf{r})$ in the vector Green's theorem and making use of the fact that

$$\nabla \times \nabla \times \mathbf{H}(\mathbf{r}) - k_0^2\,\varepsilon\mathbf{H}(\mathbf{r}) = \mathbf{0} \tag{9.127}$$

we have

$$\mathbf{H}(\mathbf{r}) = \oint_s \{i\omega\varepsilon_0\,\varepsilon[\hat{\mathbf{n}} \times \mathbf{E}(\mathbf{r}')] \cdot \bar{\mathbf{G}}(\mathbf{r}', \mathbf{r}) - [\hat{\mathbf{n}} \times \mathbf{H}(\mathbf{r}')] \cdot \nabla' \times \bar{\mathbf{G}}(\mathbf{r}', \mathbf{r})\}\, ds' \tag{9.128}$$

Equation (9.128) may also be obtained from Eq. (9.126) by applying the duality principle, i.e., by replacing **E** by **H**, **H** by $-\mathbf{E}$, and μ_0 by $\varepsilon_0\,\varepsilon$.

Equation (9.126) or Eq. (9.128) is a mathematical statement of *Huygen's principle* in an isotropic, stationary medium. It states that the electromagnetic field at any point in a source-free region is determined by the dyadic Green function and the tangential components of **E** and **H** on a surface which encloses the given current sources.

PROBLEMS

9.1 Show that Eqs. (9.6) and (9.7) are respectively the Fourier transforms of the Maxwell equations (9.1) and (9.2) with respect to the time t.

9.2 Verify Eqs. (9.17) and (9.21).

9.3 Verify Eq. (9.26).

9.4 Verify Eq. (9.28).

9.5 Verify Eqs. (9.49) and (9.50).

9.6 An oscillating electric dipole consists of two equal and opposite time-varying charges $q_1 = q_0 e^{-i\omega t}$ and $q_2 = -q_0 e^{-i\omega t}$ separated by an infinitesimal distance **l**. To satisfy the law of conservation of charge, we connect the two charges by a filamentary wire so that the current I will flow in the wire. Show that

$$\mathbf{p}_0 = -\frac{I\mathbf{l}}{i\omega}$$

Thus the current distribution of an oscillating dipole located at the origin may also be expressed as

$$\mathbf{J}(\mathbf{r}') = I\mathbf{l}\ \delta(\mathbf{r}')$$

9.7 Verify Eqs. (9.62) and (9.63).

9.8 Verify Eq. (9.67).

9.9 (*a*) Show that the solutions of the inhomogeneous wave equations for the vector and scalar potentials (see Prob. 2.10) can be represented respectively by the integrals

$$\mathcal{A}(\mathbf{r}, t) = \mu_0 \mu \int_{-\infty}^{+\infty} g(\mathbf{R}, \tau)\mathcal{J}(\mathbf{r}', t')\, d^3r'\, dt'$$

and

$$\phi(\mathbf{r}, t) = \frac{1}{\varepsilon_0 \varepsilon} \int_{-\infty}^{+\infty} g(\mathbf{R}, \tau)\rho(\mathbf{r}', t')\, d^3r'\, dt'$$

where $\mathbf{R} = \mathbf{r} - \mathbf{r}'$ and $\tau = t - t'$. In the above expressions, $g(\mathbf{R}, \tau)$ is a scalar Green's function of an isotropic medium satisfying the equation [cf. Eq. (9.28)]

$$\left(\nabla^2 - \frac{\mu\varepsilon}{c^2}\frac{\partial^2}{\partial t^2}\right) g(\mathbf{R}, \tau) = -\delta(\mathbf{R})\ \delta(\tau)$$

(*b*) Show that the solution $g(\mathbf{R}, \tau)$ of the differential equation subjected to the causality condition, $g(\mathbf{R}, \tau) = 0$ for $\tau < 0$, is given by

$$g(\mathbf{R}, \tau) = \frac{1}{4\pi R}\,\delta\left[t' - \left(t - \frac{\sqrt{\mu\varepsilon}R}{c}\right)\right]$$

where $R = |\mathbf{R}| = |\mathbf{r} - \mathbf{r}'|$.

(*c*) Show that for monochromatic sources, the potential integrals of (*a*) become

$$\mathbf{A}(\mathbf{r}) = \frac{\mu_0 \mu}{4\pi} \int_{-\infty}^{+\infty} \frac{\mathbf{J}(\mathbf{r}')e^{ik_0\sqrt{\mu\varepsilon}R}}{R}\, d^3r'$$

$$\phi(\mathbf{r}) = \frac{1}{4\pi\varepsilon_0 \varepsilon} \int_{-\infty}^{+\infty} \frac{\rho(\mathbf{r}')e^{ik_0\sqrt{\mu\varepsilon}R}}{R}\, d^3r'$$

where the integrations extend throughout the region occupied by the sources.

9.10 An oscillating electric dipole of dipole moment $I\mathbf{l}$ radiates in free space.

(*a*) Show that the vector potential, the magnetic and electric fields are given respectively by

$$\mathbf{A}(\mathbf{r}) = \mu_0\, g(\mathbf{r}) I\mathbf{l}$$

$$\mathbf{H}(\mathbf{r}) = \frac{1}{\mu_0} \nabla \times \mathbf{A}(\mathbf{r}) = -\left(ik_0 - \frac{1}{r}\right) g(\mathbf{r})(I\mathbf{l} \times \hat{\mathbf{r}})$$

and $$\mathbf{E}(\mathbf{r}) = i\omega\mathbf{A}(\mathbf{r}) + \frac{i}{\omega\mu_0\varepsilon_0}\nabla\nabla\cdot\mathbf{A}(\mathbf{r})$$

$$= i\eta_0\left[\left(k_0 + \frac{i}{r} - \frac{1}{k_0 r^2}\right)I\mathbf{l} - \left(k_0 + \frac{3i}{r} - \frac{3}{k_0 r^2}\right)(I\mathbf{l}\cdot\hat{\mathbf{r}})\hat{\mathbf{r}}\right]g(\mathbf{r})$$

where $$\eta_0 = \sqrt{\frac{\mu_0}{\varepsilon_0}} \qquad g(\mathbf{r}) = \frac{e^{ik_0 r}}{4\pi r}$$

(*b*) When the dipole moment is oriented along the z axis, show that the spherical components of the field vectors become

$$H_r = H_\theta = E_\phi = 0$$

$$H_\phi = -Il\left(ik_0 - \frac{1}{r}\right)g(\mathbf{r})\sin\theta$$

$$E_r = \eta_0 Il\left(\frac{2}{r} + \frac{2i}{k_0 r^2}\right)g(\mathbf{r})\cos\theta$$

$$E_\theta = \eta_0 Il\left(-ik_0 + \frac{1}{r} + \frac{i}{k_0 r^2}\right)g(\mathbf{r})\sin\theta$$

(*c*) When the dipole moment is oriented along the x axis, show that the spherical components of the field vectors reduce to

$$H_r = 0$$

$$H_\theta = Il\left(ik_0 - \frac{1}{r}\right)g(\mathbf{r})\sin\phi$$

$$H_\phi = Il\left(ik_0 - \frac{1}{r}\right)g(\mathbf{r})\cos\theta\cos\phi$$

$$E_r = i\eta_0 Il\left(-\frac{2i}{r} + \frac{2}{k_0 r^2}\right)g(\mathbf{r})\sin\theta\cos\phi$$

$$E_\theta = i\eta_0 Il\left(k_0 + \frac{i}{r} - \frac{1}{k_0 r^2}\right)g(\mathbf{r})\cos\theta\cos\phi$$

$$E_\phi = -i\eta_0 Il\left(k_0 + \frac{i}{r} - \frac{1}{k_0 r^2}\right)g(\mathbf{r})\sin\phi$$

9.11 The *gain* $g_a(\theta, \phi)$ of an antenna in a given direction θ, ϕ is defined as the ratio of the total power which an isotropic (nondirectional) antenna would radiate to the power radiated by the antenna in question so that both antennas deliver the same power densities in the given direction. Thus

$$g_a(\theta, \phi) = \frac{4\pi r^2\langle\mathscr{P}\rangle\cdot\hat{\mathbf{r}}}{\langle\mathscr{S}\rangle}$$

is a measure of the directional property of the antenna.

(*a*) Show that the gain of an oscillating electric dipole is $1.5\sin^2\theta$.

(*b*) Calculate the gains of a half-wave dipole ($k_0\sqrt{\varepsilon}L = \pi/2$) and a full-wave dipole.

9.12 Radar is an electromagnetic system used to determine range and location of objects. It operates by transmitting a particular type of signal and detects the nature of the echo. If a common transmitting and receiving radar antenna of gain G transmits power S_T, and if a target located at a distance R

from the radar has a radar cross section σ (a measure of both receiving area and of reradiating characteristics), show that the power S_R received by the receiver is

$$S_R = \frac{S_T G^2 \lambda^2 \sigma}{(4\pi)^3 R^4} \qquad \text{W}$$

where λ is the wavelength of the wave.

9.13 An airport surveillance radar has a receiver sensitivity of -115 dBm (dBm is dB with respect to 1 milliwatt), and a common transmitting and receiving antenna with a gain of 45 dB. If an airplane 100 mi away has a radar cross section of 1 m^2, what is the transmitter power required for a frequency of 8 GHz?

9.14 Consider a circular loop antenna of radius a located in the xy plane and centered at the origin as shown in Fig. 9.6.

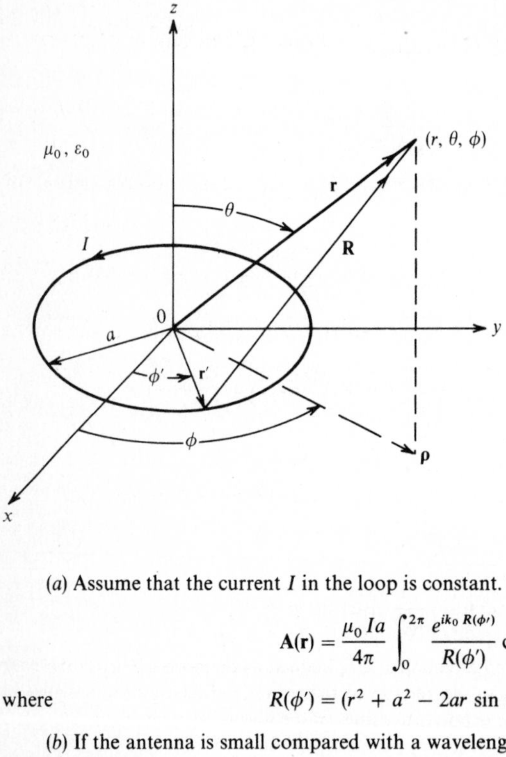

Figure 9.6 A circular loop antenna.

(*a*) Assume that the current I in the loop is constant. Show that the vector potential is

$$\mathbf{A}(\mathbf{r}) = \frac{\mu_0 I a}{4\pi} \int_0^{2\pi} \frac{e^{ik_0 R(\phi')}}{R(\phi')} \cos \phi' \, d\phi' \, \hat{\boldsymbol{\phi}}$$

where

$$R(\phi') = (r^2 + a^2 - 2ar \sin \theta \cos \phi')^{1/2}$$

(*b*) If the antenna is small compared with a wavelength ($k_0 a \ll 1$) and $a \ll r$, show that the vector potential becomes

$$\mathbf{A}(\mathbf{r}) = \frac{\mu_0 I S e^{ik_0 r}}{4\pi r} \left(-ik_0 + \frac{1}{r} \right) \sin \theta \, \hat{\boldsymbol{\phi}}$$

where $S = \pi a^2$ is the area of the loop.

(*c*) Show that the spherical components of the field vectors are

$$E_r = E_\theta = H_\phi = 0$$

$$E_\phi = IS\eta_0 k_0^2 \left(1 - \frac{1}{ik_0 r} \right) g(\mathbf{r}) \sin \theta$$

$$H_r = 2ISk_0\left(\frac{1}{ir} + \frac{1}{k_0 r^2}\right)g(\mathbf{r}) \cos\theta$$

$$H_\theta = -ISk_0^2\left(1 - \frac{1}{ik_0 r} - \frac{1}{k_0^2 r^2}\right)g(\mathbf{r}) \sin\theta$$

where $$\eta_0 = \sqrt{\frac{\mu_0}{\varepsilon_0}} \quad \text{and} \quad g(\mathbf{r}) = \frac{e^{ik_0 r}}{4\pi r}$$

(*d*) Show that the radiation resistance and the gain of the loop antenna are given by

$$R_r = \frac{k_0^4 S^2 \eta_0}{6\pi} = 320\pi^6\left(\frac{a}{\lambda}\right)^4$$

and $$g_a(\theta) = 1.5 \sin^2\theta$$

respectively.

9.15 (*a*) Comparing the field equations of an electric current source only with those of magnetic current source only:

Electric current only	*Magnetic current only*
$\nabla \times \mathbf{H} = -i\omega\varepsilon_0\varepsilon\mathbf{E} + \mathbf{J}$	$-\nabla \times \mathbf{E} = -i\omega\mu_0\mu\mathbf{H} + \mathbf{J}_m$
$-\nabla \times \mathbf{E} = -i\omega\mu_0\mu\mathbf{H}$	$\nabla \times \mathbf{H} = -i\omega\mu_0\varepsilon\mathbf{E}$
$H = \frac{1}{\mu_0\mu}\nabla \times \mathbf{A}$	$\mathbf{E} = -\frac{1}{\varepsilon_0\varepsilon}\nabla \times \mathbf{F}$

justify the following dual quantities:

Electric type	$\mathbf{H}$	$\mathbf{E}$	$\mathbf{J}$	$\mathbf{A}$	μ	ε	η_0	k_0	I
Magnetic type	$-\mathbf{E}$	$\mathbf{H}$	$\mathbf{J}_m$	$\mathbf{F}$	ε	μ	$1/\eta_0$	k_0	I_m

where $\mathbf{F}$ is an electric vector potential and I_m is the magnetic current.

(*b*) Making use of the duality, show that the fields due to an oscillating magnetic dipole in free space are

$$\mathbf{E} = \left(ik_0 - \frac{1}{r}\right)g(\mathbf{r})\,(I_m\mathbf{l} \times \hat{\mathbf{r}})$$

$$\mathbf{H} = \frac{i}{\eta_0}\left[\left(k_0 + \frac{i}{r} - \frac{1}{k_0 r^2}\right)I_m\mathbf{l} - \left(k_0 + \frac{3i}{r} - \frac{3}{k_0 r^2}\right)(I_m\mathbf{l}\cdot\hat{\mathbf{r}})\hat{\mathbf{r}}\right]g(\mathbf{r})$$

where $$g(\mathbf{r}) = \frac{e^{ik_0 r}}{4\pi r}$$

(*c*) Compare the results of part (*b*) for the magnetic dipole oriented along z axis with the fields due to a small electric current loop (Prob. 9.14). What is your conclusion?

9.16 An oscillating electric dipole of moment $I_z l$ and a small current loop of magnetic moment $I_\phi S$ exist simultaneously as shown in Fig. 9.7.

(*a*) Find the far-zone electric and magnetic fields.

(*b*) Discuss the polarization of the electric field in the far-zone for the following cases: (*i*) the currents in the loop and in the dipole are 90° out of the time phase; (*ii*) the currents in the loop and in the dipole are in time phase; (*iii*) the currents in the loop and in the dipole are in time phase, and the moments are related by

$$k_0 I_\phi S = I_z l$$

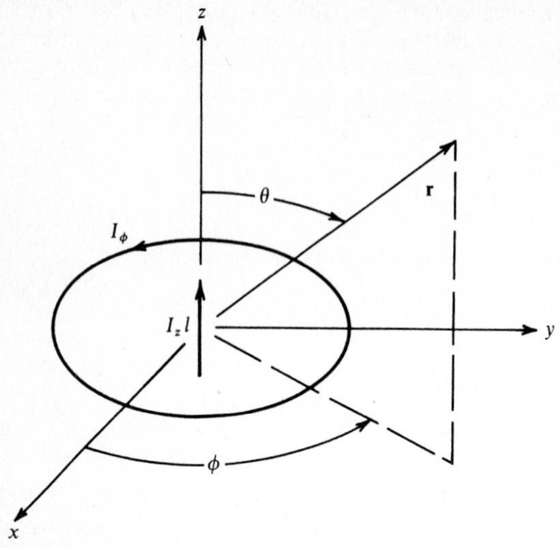

Figure 9.7 A current loop with an electric dipole oriented along z axis at its center.

(*c*) Find the time-averaged Poynting vector and the total time-averaged power radiated by the antennas. (*d*) What fraction of the total power radiated is supplied by the dipole? (*e*) Find the gain of the antennas.

9.17 Two oscillating electric dipoles carry currents of equal magnitudes but 90° out of the time phase. The dipoles form a cross at the origin as shown in Fig. 9.8. This arrangement of antennas is known as *crossed dipoles* or a *turnstile array*. (*a*) Find the far-zone electric and magnetic fields. (*b*) Discuss in detail the polarization of the electric field in the far-zone at the points on (*i*) xz plane; (*ii*) xy plane; (*iii*) yz plane. (*c*) Find the time-averaged Poynting vector and the total time-averaged power radiated by the dipoles. (*d*) What fraction of the total power radiated is supplied by each dipole? (*e*) Find the gain and the directions for maximum gain of the dipoles.

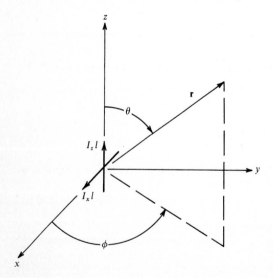

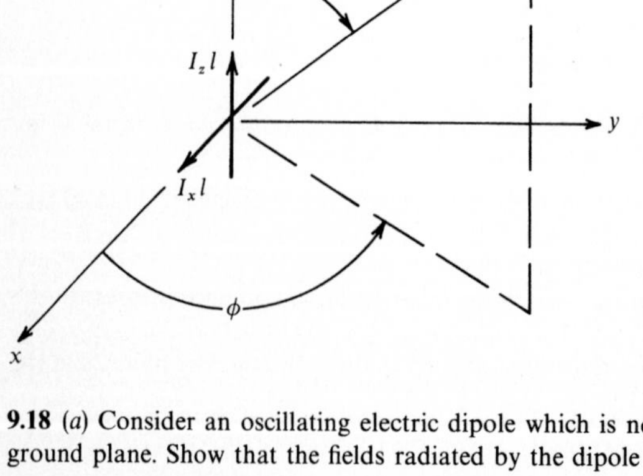

Figure 9.8 Turnstile array of two electric dipoles.

9.18 (*a*) Consider an oscillating electric dipole which is normal to and above a perfectly conducting ground plane. Show that the fields radiated by the dipole can be calculated by removing the ground and replacing its effect by an "image" dipole oriented in the same direction as shown in Fig. 9.9. (*b*) Repeat (*a*) but with a magnetic dipole. The source and image in this case are oppositely directed (see Fig. 9.9). *Hint*: Show that the total field due to the source and its image satisfies Maxwell's equations and that the tangential component of **E** is zero over the conducting plane. Hence, by the uniqueness theorem (see Prob. 2.9), the results follow.

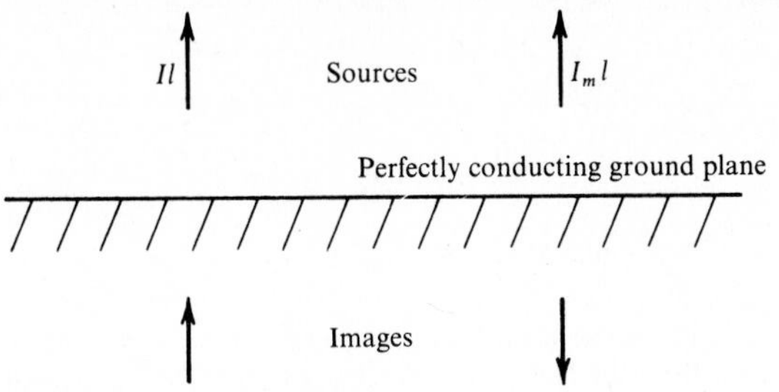

Figure 9.9 Images of vertically directed electric and magnetic dipoles.

9.19 Repeat Prob. 9.18 when the dipoles are parallel to the conducting ground plane.

9.20 An electric dipole of moment $I\mathbf{l}$ is located at a height h and normal to a conducting ground plane. (*a*) Using the result of Prob. 9.18, find the far-zone electric and magnetic fields. (*b*) Obtain the radiation resistance and the gain of the antenna.

9.21 Repeat Prob. 9.20 but with a small electric current loop of moment IS. The loop is parallel to the conducting ground plane.

9.22 For an electric dipole radiating in an isotropic medium, show that

$$(a)\quad k^2\bar{\mathbf{W}}_i^{-1}(\mathbf{k}) = -\frac{1}{k_0^2\,\varepsilon}\left[k^2\hat{\mathbf{k}}\hat{\mathbf{k}} + k_0^2\,\varepsilon(\hat{\mathbf{k}}\hat{\mathbf{k}} - \bar{\mathbf{I}}) + \frac{k_0^4\,\varepsilon^2(\hat{\mathbf{k}}\hat{\mathbf{k}} - \bar{\mathbf{I}})}{k^2 - k_0^2\,\varepsilon}\right]$$

$$(b)\quad \int \hat{\mathbf{k}}\hat{\mathbf{k}}\; d\Omega = \frac{4\pi}{3}\,\bar{\mathbf{I}}$$

and

$$(c)\quad \int \bar{\mathbf{W}}_i^{-1}(\mathbf{k})\; d^3k = \frac{i4\pi^2 k_0\sqrt{\varepsilon}}{3}\,\bar{\mathbf{I}}$$

9.23 Let R be the magnitude of the distance vector $\mathbf{R} = \mathbf{r} - \mathbf{r}'$, and let $g(R) = \exp(ikR)/4\pi R$ be a scalar Green's function. Show that

$$(a)\quad \nabla^2\left(\frac{1}{R}\right) = -4\pi\,\delta(\mathbf{R})$$

$$(b)\quad \nabla^2 g(R) = -\delta(\mathbf{R}) - k^2 g(R)$$

and

(*c*) $g(R)$ satisfies the Sommerfeld radiation condition:

$$\lim_{R\to\infty} R\left(\frac{\partial g}{\partial R} - ikg\right) = 0$$

9.24 (*a*) Make use of the divergence theorem of vector analysis to establish the following Green theorem:

$$\int_V (\phi\nabla^2\Psi - \Psi\nabla^2\phi)\; d^3r = \oint_S (\phi\nabla\Psi - \Psi\nabla\phi)\cdot d\mathbf{s}$$

where ϕ and Ψ are two scalar fields.

(*b*) Consider a region V bounded by the surface S' and a spherical surface S_∞ of infinite radius. If, in Green's theorem, we choose ϕ to be the scalar Green function g as defined in Prob. 9.23 and Ψ satisfying the homogeneous Helmholtz equation

$$\nabla^2\Psi + k^2\Psi = 0$$

in V and the Sommerfeld radiation condition

$$\lim_{R\to\infty} R\left(\frac{\partial\Psi}{\partial R} - ik\Psi\right) = 0$$

at infinity. Show that

$$\Psi(\mathbf{r}) = \oint_{S'} [g\nabla'\Psi(\mathbf{r}') - \Psi(\mathbf{r}')\nabla' g] \cdot d\mathbf{s}'$$

This is called the *Kirchhoff integral* expressing the scalar field Ψ at any point in V in terms of the values of the field and its normal derivative on the boundary surface S'. Since Ψ and $\nabla\Psi$ are related by the Helmholtz equation, so a specification of either quantity on S' is sufficient to determine Ψ in V.

9.25 (*a*) Considering Ψ in Prob. 9.24*b* as a cartesian component of the electric field intensity $\mathbf{E}$ establish

$$\mathbf{E}(\mathbf{r}) = \oint_{S'} [g(\hat{\mathbf{n}}\cdot\nabla')\mathbf{E}(\mathbf{r}') - \mathbf{E}(\mathbf{r}')(\hat{\mathbf{n}}\cdot\nabla')g]\, ds'$$

where $\hat{\mathbf{n}}$ is a unit vector normal to ds'.

(*b*) Show that $\mathbf{E}(\mathbf{r})$ in (*a*) can also be written as

$$\mathbf{E}(\mathbf{r}) = -\oint_{S'} [(\hat{\mathbf{n}}\times\mathbf{E})\times\nabla' g + (\hat{\mathbf{n}}\cdot\mathbf{E})\nabla' g + i\omega\mu_0\, g(\hat{\mathbf{n}}\times\mathbf{H})]\, ds'$$

This equation shows that the field in V can be expressed in terms of the values of the normal and tangential components of fields on S'.

(*c*) When the surface S' is a perfectly conducting plane of infinite extent with apertures on it, show that the field in the half-space that is of interest becomes

$$\mathbf{E}(\mathbf{r}) = -2\nabla\times\int_{S'_1} \hat{\mathbf{n}}\times\mathbf{E}(\mathbf{r}')g\, ds'$$

where the integration is carried out over the apertures only.

9.26 Using the results of Sec. 9.5, show that the electromagnetic field satisfies the following radiation conditions:

$$\lim_{r\to\infty} r\,(\omega\mu_0\hat{\mathbf{r}}\times\mathbf{H} + k_0\sqrt{\varepsilon}\mathbf{E}) = \mathbf{0}$$

$$\lim_{r\to\infty} r\,(k_0\sqrt{\varepsilon}\hat{\mathbf{r}}\times\mathbf{E} - \omega\mu_0\mathbf{H}) = \mathbf{0}$$

These conditions essentially state that the field solution must represent an outward traveling wave at infinity.

9.27 Verify Eq. (9.124).

9.28 Using Green's first identity,

$$\int_V [(\nabla\times\mathbf{A})\cdot(\nabla\times\mathbf{F}) - \mathbf{F}\cdot\nabla\times(\nabla\times\mathbf{A})]\, d^3r = \oint_S [\mathbf{F}\times(\nabla\times\mathbf{A})]\cdot d\mathbf{s}$$

show that

$$[\nabla'\times\bar{\mathbf{G}}(\mathbf{r}',\mathbf{r})]^T = \nabla\times\bar{\mathbf{G}}(\mathbf{r},\mathbf{r}')$$

where $\nabla' \times$ means to take curl with respect to the primed variable. Thus Eqs. (9.126) and (9.128) may be expressed as

$$\mathbf{E}(\mathbf{r}) = -\oint_s \{i\omega\mu_0 \bar{\mathbf{G}}(\mathbf{r}, \mathbf{r}') \cdot [\hat{\mathbf{n}} \times \mathbf{H}(\mathbf{r}')] + \nabla \times \bar{\mathbf{G}}(\mathbf{r}, \mathbf{r}') \cdot [\hat{\mathbf{n}} \times \mathbf{E}(\mathbf{r}')]\}\, ds'$$

and

$$\mathbf{H}(\mathbf{r}) = \oint_s \{i\omega\varepsilon_0 \varepsilon \bar{\mathbf{G}}(\mathbf{r}, \mathbf{r}') \cdot [\hat{\mathbf{n}} \times \mathbf{E}(\mathbf{r}')] - \nabla \times \bar{\mathbf{G}}(\mathbf{r}, \mathbf{r}') \cdot [\hat{\mathbf{n}} \times \mathbf{H}(\mathbf{r}')]\}\, ds'$$

respectively.

REFERENCES

Abramowitz, M., and I. A. Stegun: *Handbook of Mathematical Functions*, Dover Publications, Inc., New York, 1965.

Baker, B. B., and E. T. Copson: *The Mathematical Theory of Huggens' Principle*, 2d ed., Clarendon Press, Oxford, 1950.

Butkov, E.: *Mathematical Physics*, Addison-Wesley Publishing Company, Reading, Mass., 1968, chaps. 12 and 14.

Byron, F. W., Jr., and R. W. Fuller: *Mathematics of Classical and Quantum Physics*, vol. 2, Addison-Wesley Publishing Company, Reading, Mass., 1970, chap. 7.

Friedman, B.: *Principles and Techniques of Applied Mathematics*, John Wiley & Sons, New York, 1956, chap. 3.

Good, R. H., and T. J. Nelson: *Classical Theory of Electric and Magnetic Fields*, Academic Press, New York, 1971, chaps. 7 and 9.

Jordan, E. C., and K. G. Balmain: *Electromagnetic Waves and Radiating Systems*, 2d ed., Prentice-Hall, Inc., Englewood Cliffs, N.J., 1968, chaps. 10, 11, 12, and 13.

King, R. W. P., and C. W. Harrison: *Antennas and Waves: A Modern Approach*, The MIT Press, Cambridge, Mass., 1969, chap. 1.

Kraut, E. G.: *Fundamentals of Mathematical Physics*, McGraw-Hill Book Company, 1967, chap. 7.

Morse, P. M., and H. Feshback: *Methods of Mathematical Physics*, vol. 1, McGraw-Hill Book Company, New York, 1953, chap. 7.

Papas, C. H.: *Theory of Electromagnetic Wave Propagation*, McGraw-Hill Book Company, New York, 1965, chaps. 2 and 3.

Stakgold, I.: *Boundary Value Problems of Mathematical Physics*, vols. 1 and 2, The Macmillan Company, New York, 1967.

Stratton, J. A.: *Electromagnetic Theory*, McGraw-Hill Book Company, New York, 1941, chap. 8.

Tai, C. T.: *Dyadic Green's Functions in Electromagnetic Theory*, Intext Educational Publishers, Scranton, Pa., 1971, chap. 4.

Van Bladel, J., "Some Remarks on Green's Dyadic for Infinite Space," *IRE Trans. Antennas Propagation*, vol. AP-9, no. 6, 1961, pp. 563–566.

Van Bladel, J.: *Electromagnetic Fields*, McGraw-Hill Book Company, New York, 1964, chap. 7.

Vladimirov, V. S.: *Equations of Mathematical Physics*, Marcel Dekker, Inc., New York, 1971.

Yaghjian, A., "Electric Dyadic Green's Functions in the Source Region," *Proc. IEEE*, vol. 68, no. 2, 1980, pp. 248–263.

CHAPTER

TEN

RADIATION IN A UNIAXIAL MEDIUM

As was shown in the previous chapter, the electromagnetic fields radiated by a given current distribution in an isotropic medium can be calculated successfully by the dyadic Green function method. In this chapter, we shall extend the method to include anisotropic media.

We shall begin with the formulation of a dyadic Green function for an anisotropic medium. It is expected that the results will be more complicated than in isotropic media. In order to obtain solutions in a manageable form, we shall concentrate our discussion on uniaxially anisotropic media. In this case we find that integrals in the solution can be carried out explicitly without resorting to approximation. After having obtained the dyadic Green function in closed form, we will apply it to evaluate fields of an electric dipole.

10.1 DYADIC GREEN'S FUNCTION OF A UNIAXIAL MEDIUM

For a given current distribution immersed in an unbounded anisotropic medium characterized by a dielectric tensor $\bar{\boldsymbol{\varepsilon}}$, according to Eq. (9.3) the Fourier transforms of the Maxwell equations are

$$\nabla \times \mathbf{E}(\mathbf{r}) = i\omega\mu_0 \mathbf{H}(\mathbf{r}) \tag{10.1}$$

$$\nabla \times \mathbf{H}(\mathbf{r}) = -i\omega\varepsilon_0 \bar{\boldsymbol{\varepsilon}} \cdot \mathbf{E}(\mathbf{r}) + \mathbf{J}(\mathbf{r}) \tag{10.2}$$

Taking the curl of Eq. (10.1) and inserting the result into Eq. (10.2), we obtain

$$(\nabla\nabla - \nabla^2 \bar{\mathbf{I}} - k_0^2 \bar{\boldsymbol{\varepsilon}}) \cdot \mathbf{E}(\mathbf{r}) = i\omega\mu_0 \mathbf{J}(\mathbf{r}) \tag{10.3}$$

where $k_0^2 = \omega^2 \mu_0 \varepsilon_0$. Because of the linearity of Eq. (10.3), we may express the desired solution as

$$\mathbf{E}(\mathbf{r}) = i\omega\mu_0 \int \bar{\mathbf{G}}(\mathbf{r}, \mathbf{r}') \cdot \mathbf{J}(\mathbf{r}')\, d^3r' \tag{10.4}$$

where the integration extends throughout the region occupied by the current. Following the same procedure as in Sec. 9.1, we find that the dyadic Green function $\bar{\mathbf{G}}(\mathbf{r}, \mathbf{r}')$ of an anisotropic medium satisfies the differential equation

$$(\nabla\nabla - \nabla^2\bar{\mathbf{I}} - k_0^2\,\bar{\boldsymbol{\varepsilon}}) \cdot \bar{\mathbf{G}}(\mathbf{r}, \mathbf{r}') = \bar{\mathbf{I}}\,\delta(\mathbf{r} - \mathbf{r}') \tag{10.5}$$

To facilitate the construction of the dyadic Green function, we take the three-dimensional Fourier transform of Eq. (10.5) with respect to $\mathbf{r}$. Thus, according to Eqs. (9.14) and (9.15), we obtain

$$\bar{\mathbf{G}}(\mathbf{R}) = \frac{1}{(2\pi)^3} \int \bar{\mathbf{W}}^{-1}(\mathbf{k}) e^{i\mathbf{k}\cdot\mathbf{R}}\, d^3k \tag{10.6}$$

where $\mathbf{R} = \mathbf{r} - \mathbf{r}'$ and the wave matrix $\bar{\mathbf{W}}(\mathbf{k})$ of the anisotropic medium is given by

$$\bar{\mathbf{W}}(\mathbf{k}) = k^2\bar{\mathbf{I}} - \mathbf{k}\mathbf{k} - k_0^2\,\bar{\boldsymbol{\varepsilon}} \tag{10.7}$$

The integral representation (10.6) of the dyadic Green function satisfies the radiation condition and hence constitutes the only solution of Eq. (10.5) that leads to a physically acceptable result.

To carry out the integration in Eq. (10.6), we shall consider uniaxially anisotropic media. In this case, the dielectric tensor $\bar{\boldsymbol{\varepsilon}}$ takes the form

$$\bar{\boldsymbol{\varepsilon}} = \varepsilon_\perp \bar{\mathbf{I}} + (\varepsilon_\parallel - \varepsilon_\perp)\hat{\mathbf{c}}\hat{\mathbf{c}} \tag{10.8}$$

where $\hat{\mathbf{c}}$ is the unit vector in the direction of the optic axis. Substitution of Eq. (10.8) into Eq. (10.7) yields the wave matrix of a uniaxially anisotropic medium:

$$\bar{\mathbf{W}}_u(\mathbf{k}) = (k^2 - k_0^2\,\varepsilon_\perp)\bar{\mathbf{I}} - \mathbf{k}\mathbf{k} + (\varepsilon_\perp - \varepsilon_\parallel)k_0^2\,\hat{\mathbf{c}}\hat{\mathbf{c}} \tag{10.9}$$

Using the results of Example 1.4, we find the determinant and adjoint of the wave matrix $\bar{\mathbf{W}}_u(\mathbf{k})$:

$$|\bar{\mathbf{W}}_u(\mathbf{k})| = -k_0^2(k^2 - k_0^2\,\varepsilon_\perp)(\mathbf{k} \cdot \bar{\boldsymbol{\varepsilon}} \cdot \mathbf{k} - k_0^2\,\varepsilon_\perp\varepsilon_\parallel) \tag{10.10}$$

and

$$\operatorname{adj} \bar{\mathbf{W}}_u(\mathbf{k}) = (k^2 - k_0^2\,\varepsilon_\perp)\left(\mathbf{k}\mathbf{k} - \frac{k_0^2}{\varepsilon_\perp}\operatorname{adj}\bar{\boldsymbol{\varepsilon}}\right) - k_0^2(\varepsilon_\perp - \varepsilon_\parallel)(\mathbf{k} \times \hat{\mathbf{c}})(\mathbf{k} \times \hat{\mathbf{c}}) \tag{10.11}$$

respectively. Hence

$$\begin{aligned}
\bar{\mathbf{W}}_u^{-1}(\mathbf{k}) &= \frac{\operatorname{adj} \bar{\mathbf{W}}_u(\mathbf{k})}{|\bar{\mathbf{W}}_u(\mathbf{k})|} \\
&= \frac{\varepsilon_\perp\varepsilon_\parallel\bar{\boldsymbol{\varepsilon}}^{-1}}{\mathbf{k} \cdot \bar{\boldsymbol{\varepsilon}} \cdot \mathbf{k} - k_0^2\,\varepsilon_\perp\varepsilon_\parallel} - \frac{\mathbf{k}\mathbf{k}}{k_0^2(\mathbf{k} \cdot \bar{\boldsymbol{\varepsilon}} \cdot \mathbf{k} - k_0^2\,\varepsilon_\perp\varepsilon_\parallel)} \\
&\quad + \frac{(\mathbf{k} \times \hat{\mathbf{c}})(\mathbf{k} \times \hat{\mathbf{c}})}{(\mathbf{k} \times \hat{\mathbf{c}})^2(k^2 - k_0^2\,\varepsilon_\perp)} - \frac{\varepsilon_\parallel(\mathbf{k} \times \hat{\mathbf{c}})(\mathbf{k} \times \hat{\mathbf{c}})}{(\mathbf{k} \times \hat{\mathbf{c}})^2(\mathbf{k} \cdot \bar{\boldsymbol{\varepsilon}} \cdot \mathbf{k} - k_0^2\,\varepsilon_\perp\varepsilon_\parallel)}
\end{aligned} \tag{10.12}$$

Inserting Eq. (10.12) into Eq. (10.6), replacing the wave vector $\mathbf{k}$ in the numerator by $-i\nabla$ as in Sec. 9.1 and, finally, interchanging the order of differentiation and integration, we obtain

$$\bar{\mathbf{G}}(\mathbf{R}) = \left(\varepsilon_\perp \varepsilon_\| \bar{\boldsymbol{\varepsilon}}^{-1} + \frac{1}{k_0^2} \nabla\nabla\right) I_1(\mathbf{R}) + \bar{\mathbf{I}}_2(\mathbf{R}) - \varepsilon_\| \bar{\mathbf{I}}_3(\mathbf{R}) \tag{10.13}$$

where

$$I_1(\mathbf{R}) = \frac{1}{(2\pi)^3} \int_{-\infty}^{+\infty} \frac{e^{i\mathbf{k}\cdot\mathbf{R}}}{\mathbf{k}\cdot\bar{\boldsymbol{\varepsilon}}\cdot\mathbf{k} - k_0^2 \varepsilon_\perp \varepsilon_\|}\, d^3k \tag{10.14}$$

$$\bar{\mathbf{I}}_2(\mathbf{R}) = \frac{1}{(2\pi)^3} \int_{-\infty}^{+\infty} \frac{(\mathbf{k}\times\hat{\mathbf{c}})(\mathbf{k}\times\hat{\mathbf{c}})}{(\mathbf{k}\times\hat{\mathbf{c}})^2} \frac{e^{i\mathbf{k}\cdot\mathbf{R}}}{k^2 - k_0^2 \varepsilon_\perp}\, d^3k \tag{10.15}$$

and

$$\bar{\mathbf{I}}_3(\mathbf{R}) = \frac{1}{(2\pi)^3} \int_{-\infty}^{+\infty} \frac{(\mathbf{k}\times\hat{\mathbf{c}})(\mathbf{k}\times\hat{\mathbf{c}})}{(\mathbf{k}\times\hat{\mathbf{c}})^2} \frac{e^{i\mathbf{k}\cdot\mathbf{R}}}{\mathbf{k}\cdot\bar{\boldsymbol{\varepsilon}}\cdot\mathbf{k} - k_0^2 \varepsilon_\perp \varepsilon_\|}\, d^3k \tag{10.16}$$

Equation (10.13) gives the dyadic Green function of a uniaxial medium. The detailed evaluation of integrals $I_1(\mathbf{R})$, $\bar{\mathbf{I}}_2(\mathbf{R})$, and $\bar{\mathbf{I}}_3(\mathbf{R})$ will be carried out in the next section. It is interesting to note that in the special case of isotropic media, $\varepsilon_\perp = \varepsilon_\| = \varepsilon$, the last two terms in Eq. (10.13) cancel out and the first term reduces to the result of Eq. (9.23) as expected.

10.2 EVALUATION OF INTEGRALS

To carry out integrations in $I_1(\mathbf{R})$, $\bar{\mathbf{I}}_2(\mathbf{R})$, and $\bar{\mathbf{I}}_3(\mathbf{R})$, let us first consider the following:

$$M_1 = \frac{1}{(2\pi)^3} \int_{-\infty}^{+\infty} \frac{e^{i\mathbf{k}\cdot\mathbf{r}}}{\mathbf{k}\cdot\bar{\boldsymbol{\varepsilon}}\cdot\mathbf{k} - a^2}\, d^3k \tag{10.17}$$

This integral may be evaluated in cylindrical coordinates. After the result is obtained, we may rewrite it back in a coordinate-free form. For convenience, we choose $\hat{\mathbf{c}} = \hat{\mathbf{z}}$ as shown in Fig. 10.1. Hence

$$\mathbf{k} = k_\rho \cos\Phi\, \hat{\mathbf{x}} + k_\rho \sin\Phi\, \hat{\mathbf{y}} + k_z \hat{\mathbf{z}} \tag{10.18}$$

$$\mathbf{r} = \rho \cos\phi\, \hat{\mathbf{x}} + \rho \sin\phi\, \hat{\mathbf{y}} + z\hat{\mathbf{z}} \tag{10.19}$$

and

$$\mathbf{k}\cdot\mathbf{r} = k_\rho \rho \cos(\Phi - \phi) + k_z z \tag{10.20}$$

while the differential volume element is $d^3k = k_\rho\, dk_\rho\, dk_z\, d\Phi$. To cover the whole **k**-space, the limits for k_z, k_ρ, and Φ are from $-\infty$ to $+\infty$, from 0 to $+\infty$, and from 0 to 2π, respectively. Also,

$$\mathbf{k}\cdot\bar{\boldsymbol{\varepsilon}}\cdot\mathbf{k} = \varepsilon_\perp k_\rho^2 + \varepsilon_\| k_z^2 \tag{10.21}$$

Substituting Eq. (10.20) into Eq. (10.17), and noting that the denominator of the

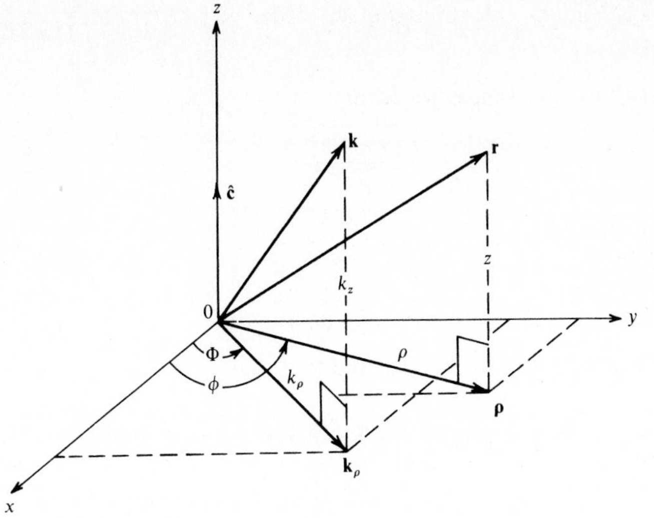

Figure 10.1 Coordinate system used in evaluating integrals.

integrand is a function of k_z and k_ρ, we can readily perform the integration over Φ according to the known formula

$$\int_0^{2\pi} e^{i\alpha\cos(\Phi-\phi)}\,d\Phi = 2\pi J_0(\alpha) \tag{10.22}$$

where $J_0(\alpha)$ is the Bessel function of the first kind of order zero (see Appendix C). Hence Eq. (10.17) becomes

$$M_1 = \frac{1}{4\pi^2\varepsilon_\perp}\int_{-\infty}^{+\infty} e^{izk_z}\,dk_z \int_0^\infty \frac{J_0(\rho k_\rho)k_\rho\,dk_\rho}{k_\rho^2 + (\varepsilon_\parallel/\varepsilon_\perp)k_z^2 - a^2/\varepsilon_\perp} \tag{10.23}$$

To carry out integrations with respect to k_ρ and k_z, we use the known integral formulas

$$\int_0^\infty \frac{J_0(\alpha\lambda)\lambda\,d\lambda}{\lambda^2+\beta^2} = K_0(\alpha\beta) \tag{10.24}$$

and

$$\int_{-\infty}^{+\infty} K_0(\rho\sqrt{\lambda^2-\beta^2})e^{iz\lambda}\,d\lambda = \frac{\pi}{\sqrt{\rho^2+z^2}}\,e^{i\beta\sqrt{\rho^2+z^2}} \tag{10.25}$$

where $K_0(\alpha\beta)$ is the modified Bessel function of the second kind of order zero (see Appendix C). Since

$$|\bar{\boldsymbol{\varepsilon}}| = \varepsilon_\perp^2\,\varepsilon_\parallel \tag{10.26}$$

and
$$\frac{1}{\varepsilon_\perp}\rho^2 + \frac{1}{\varepsilon_\parallel}z^2 = \mathbf{r}\cdot\bar{\boldsymbol{\varepsilon}}^{-1}\cdot\mathbf{r} \tag{10.27}$$

we finally obtain the result in coordinate-free form:

$$M_1 = \frac{\exp\left(ia\sqrt{\mathbf{r}\cdot\bar{\boldsymbol{\varepsilon}}^{-1}\cdot\mathbf{r}}\right)}{4\pi\sqrt{|\bar{\boldsymbol{\varepsilon}}|}\sqrt{\mathbf{r}\cdot\bar{\boldsymbol{\varepsilon}}^{-1}\cdot\mathbf{r}}} \tag{10.28}$$

Next, let us consider the integral

$$\bar{\mathbf{M}}_2 = \frac{1}{(2\pi)^3}\int_{-\infty}^{+\infty}\frac{(\mathbf{k}\times\hat{\mathbf{c}})(\mathbf{k}\times\hat{\mathbf{c}})}{(\mathbf{k}\times\hat{\mathbf{c}})^2}\,\frac{e^{i\mathbf{k}\cdot\mathbf{r}}}{\mathbf{k}\cdot\bar{\boldsymbol{\varepsilon}}\cdot\mathbf{k}-a^2}\,d^3k \tag{10.29}$$

Here, we again refer to Fig. 10.1 and Eq. (10.18) and find that

$$\frac{(\mathbf{k}\times\hat{\mathbf{c}})(\mathbf{k}\times\hat{\mathbf{c}})}{(\mathbf{k}\times\hat{\mathbf{c}})^2} = \sin^2\Phi\,\hat{\mathbf{x}}\hat{\mathbf{x}} - \sin\Phi\cos\Phi\,(\hat{\mathbf{x}}\hat{\mathbf{y}}+\hat{\mathbf{y}}\hat{\mathbf{x}}) + \cos^2\Phi\,\hat{\mathbf{y}}\hat{\mathbf{y}} \tag{10.30}$$

Substituting Eq. (10.30) into Eq. (10.29) and carrying out the integrations over Φ according to the known formulas

$$\int_0^{2\pi}\cos^2\Phi\,e^{i\beta\cos(\Phi-\phi)}\,d\Phi = 2\pi\left[\cos^2\phi\,J_0(\beta) - \frac{\cos 2\phi\,J_1(\beta)}{\beta}\right]$$

$$\int_0^{2\pi}\sin^2\Phi\,e^{i\beta\cos(\Phi-\phi)}\,d\Phi = 2\pi\left[\sin^2\phi\,J_0(\beta) + \frac{\cos 2\phi\,J_1(\beta)}{\beta}\right] \tag{10.31}$$

and

$$\int_0^{2\pi}\sin\Phi\cos\Phi\,e^{i\beta\cos(\Phi-\phi)}\,d\Phi = \pi\left[\sin 2\phi\,J_0(\beta) - \frac{2\sin 2\phi\,J_1(\beta)}{\beta}\right]$$

where $J_1(\beta)$ is the Bessel function of the first kind of order one (see Appendix C), we obtain

$$\begin{aligned}\bar{\mathbf{M}}_2 = {} & M_1[\sin^2\phi\,\hat{\mathbf{x}}\hat{\mathbf{x}} + \cos^2\phi\,\hat{\mathbf{y}}\hat{\mathbf{y}} - \sin\phi\cos\phi\,(\hat{\mathbf{x}}\hat{\mathbf{y}}+\hat{\mathbf{y}}\hat{\mathbf{x}})] \\ & + M_3[\cos 2\phi\,\hat{\mathbf{x}}\hat{\mathbf{x}} - \cos 2\phi\,\hat{\mathbf{y}}\hat{\mathbf{y}} + \sin 2\phi\,(\hat{\mathbf{x}}\hat{\mathbf{y}}+\hat{\mathbf{y}}\hat{\mathbf{x}})]\end{aligned} \tag{10.32}$$

where M_1 is given by Eq. (10.23) and

$$M_3 = \frac{1}{4\pi^2\varepsilon_\parallel\rho}\int_0^\infty J_1(\rho k_\rho)\,dk_\rho\int_{-\infty}^{\infty}\frac{e^{izk_z}\,dk_z}{k_z^2 + (\varepsilon_\perp/\varepsilon_\parallel)k_\rho^2 - a^2/\varepsilon_\parallel} \tag{10.33}$$

The integrations with respect to k_z and k_ρ in Eq. (10.33) may be performed by using the formulas

$$\int_{-\infty}^{+\infty}\frac{e^{iz\lambda}}{\lambda^2+\gamma^2}\,d\lambda = \frac{\pi e^{-\gamma|z|}}{\gamma} \qquad \operatorname{Re}\gamma > 0 \tag{10.34}$$

and

$$\int_0^\infty \frac{J_1(\rho\lambda)\exp(-|z|\sqrt{\lambda^2-b^2})}{\sqrt{\lambda^2-b^2}}\,d\lambda = \frac{1}{i\rho b}\left[\exp(ib\sqrt{\rho^2+z^2}) - \exp(ibz)\right] \tag{10.35}$$

respectively. Thus

$$M_3 = \frac{1}{i4\pi\sqrt{\varepsilon_\|}\,a(\mathbf{r}\times\hat{\mathbf{c}})^2}\left[\exp(ia\sqrt{\mathbf{r}\cdot\bar{\boldsymbol{\varepsilon}}^{-1}\cdot\mathbf{r}}) - \exp\left(\frac{ia\mathbf{r}\cdot\hat{\mathbf{c}}}{\sqrt{\varepsilon_\|}}\right)\right] \tag{10.36}$$

To express Eq. (10.32) in a coordinate-free form, we note from Eq. (10.19) that

$$\frac{(\mathbf{r}\times\hat{\mathbf{c}})(\mathbf{r}\times\hat{\mathbf{c}})}{(\mathbf{r}\times\hat{\mathbf{c}})^2} = \sin^2\phi\,\hat{\mathbf{x}}\hat{\mathbf{x}} - \sin\phi\cos\phi\,(\hat{\mathbf{x}}\hat{\mathbf{y}} + \hat{\mathbf{y}}\hat{\mathbf{x}}) + \cos^2\phi\,\hat{\mathbf{y}}\hat{\mathbf{y}} \tag{10.37}$$

But,

$$\bar{\mathbf{I}} = \hat{\mathbf{x}}\hat{\mathbf{x}} + \hat{\mathbf{y}}\hat{\mathbf{y}} + \hat{\mathbf{z}}\hat{\mathbf{z}} \tag{10.38}$$

Hence,

$$\bar{\mathbf{I}} - \hat{\mathbf{c}}\hat{\mathbf{c}} - \frac{2(\mathbf{r}\times\hat{\mathbf{c}})(\mathbf{r}\times\hat{\mathbf{c}})}{(\mathbf{r}\times\hat{\mathbf{c}})^2} = \cos 2\phi\,\hat{\mathbf{x}}\hat{\mathbf{x}} - \cos 2\phi\,\hat{\mathbf{y}}\hat{\mathbf{y}} + \sin 2\phi\,(\hat{\mathbf{x}}\hat{\mathbf{y}} + \hat{\mathbf{y}}\hat{\mathbf{x}}) \tag{10.39}$$

Replacing the terms in Eq. (10.32) by Eqs. (10.37) and (10.39), we finally obtain the result for integral $\bar{\mathbf{M}}_2$ in the coordinate-free form

$$\bar{\mathbf{M}}_2 = M_1\frac{(\mathbf{r}\times\hat{\mathbf{c}})(\mathbf{r}\times\hat{\mathbf{c}})}{(\mathbf{r}\times\hat{\mathbf{c}})^2} + M_3\left[\bar{\mathbf{I}} - \hat{\mathbf{c}}\hat{\mathbf{c}} - \frac{2(\mathbf{r}\times\hat{\mathbf{c}})(\mathbf{r}\times\hat{\mathbf{c}})}{(\mathbf{r}\times\hat{\mathbf{c}})^2}\right] \tag{10.40}$$

where M_1 and M_3 are given by Eqs. (10.28) and (10.36) respectively.

With the results of Eqs. (10.28) and (10.40), it is now a simple matter to evaluate integrals $I_1(\mathbf{R})$, $\bar{\mathbf{I}}_2(\mathbf{R})$, and $\bar{\mathbf{I}}_3(\mathbf{R})$. Replacing $\mathbf{r}$ by $\mathbf{R} = \mathbf{r} - \mathbf{r}'$ and letting $a = k_0\sqrt{\varepsilon_\perp\varepsilon_\|}$ in Eqs. (10.28) and (10.40), we obtain

$$I_1(\mathbf{R}) = \frac{e^{ik_0 R_e}}{4\pi\sqrt{\varepsilon_\perp}\,R_e} \tag{10.41}$$

and

$$\bar{\mathbf{I}}_3(\mathbf{R}) = I_1(\mathbf{R})\frac{(\mathbf{R}\times\hat{\mathbf{c}})(\mathbf{R}\times\hat{\mathbf{c}})}{(\mathbf{R}\times\hat{\mathbf{c}})^2} + I_4(\mathbf{R})\left[\bar{\mathbf{I}} - \hat{\mathbf{c}}\hat{\mathbf{c}} - \frac{2(\mathbf{R}\times\hat{\mathbf{c}})(\mathbf{R}\times\hat{\mathbf{c}})}{(\mathbf{R}\times\hat{\mathbf{c}})^2}\right] \tag{10.42}$$

respectively. In Eq. (10.42), $I_4(\mathbf{R})$ is defined by

$$I_4(\mathbf{R}) = \frac{e^{ik_0 R_e} - e^{ik_0\sqrt{\varepsilon_\perp}(\mathbf{R}\cdot\hat{\mathbf{c}})}}{i4\pi k_0\,\varepsilon_\|\sqrt{\varepsilon_\perp}(\mathbf{R}\times\hat{\mathbf{c}})^2} \tag{10.43}$$

where

$$R_e = \sqrt{\varepsilon_\perp\varepsilon_\|(\mathbf{R}\cdot\bar{\boldsymbol{\varepsilon}}^{-1}\cdot\mathbf{R})} \tag{10.44}$$

Similarly, replacing $\bar{\varepsilon}$ by $\bar{\mathbf{I}}$ and a by $k_0\sqrt{\varepsilon_\perp}$ in Eq. (10.40), we get

$$\bar{\mathbf{I}}_2(\mathbf{R}) = N_1(\mathbf{R}) \frac{(\mathbf{R}\times\hat{\mathbf{c}})(\mathbf{R}\times\hat{\mathbf{c}})}{(\mathbf{R}\times\hat{\mathbf{c}})^2} + N_2(\mathbf{R})\left[\bar{\mathbf{I}} - \hat{\mathbf{c}}\hat{\mathbf{c}} - \frac{2(\mathbf{R}\times\hat{\mathbf{c}})(\mathbf{R}\times\hat{\mathbf{c}})}{(\mathbf{R}\times\hat{\mathbf{c}})^2}\right] \tag{10.45}$$

where

$$N_1(\mathbf{R}) = \frac{\sqrt{\varepsilon_\perp}\, e^{ik_0 R_0}}{4\pi R_0}$$

$$N_2(\mathbf{R}) = \frac{e^{ik_0 R_0} - e^{ik_0\sqrt{\varepsilon_\perp}(\mathbf{R}\cdot\hat{\mathbf{c}})}}{i4\pi k_0\sqrt{\varepsilon_\perp}(\mathbf{R}\times\hat{\mathbf{c}})^2} \tag{10.46}$$

and

$$R_0 = \sqrt{\varepsilon_\perp \mathbf{R}^2} \tag{10.47}$$

Here we note that the integrals given by Eqs. (10.41), (10.42), and (10.45) are all in coordinate-free forms.

10.3 DYADIC GREEN'S FUNCTION IN EXPLICIT COORDINATE-FREE FORM

In the previous two sections we derived the dyadic Green function of a uniaxial medium and carried out some related integrations. In this section we shall summarize the result and express it in a coordinate-free form.

Substituting Eqs. (10.41), (10.42), and (10.45) into Eq. (10.13), we obtain, after some simplification,

$$\begin{aligned}\bar{\mathbf{G}}(\mathbf{R}) = \frac{1}{4\pi\sqrt{\varepsilon_\perp}}\Bigg\{&\left(\varepsilon_\perp\varepsilon_\parallel \bar{\varepsilon}^{-1} + \frac{1}{k_0^2}\nabla\nabla\right)\frac{e^{ik_0 R_e}}{R_e} \\ &+ \left(\frac{\varepsilon_\perp e^{ik_0 R_0}}{R_0} - \frac{\varepsilon_\parallel e^{ik_0 R_e}}{R_e}\right)\frac{(\mathbf{R}\times\hat{\mathbf{c}})(\mathbf{R}\times\hat{\mathbf{c}})}{(\mathbf{R}\times\hat{\mathbf{c}})^2} \\ &+ \frac{e^{ik_0 R_0} - e^{ik_0 R_e}}{ik_0(\mathbf{R}\times\hat{\mathbf{c}})^2}\left[\bar{\mathbf{I}} - \hat{\mathbf{c}}\hat{\mathbf{c}} - 2\frac{(\mathbf{R}\times\hat{\mathbf{c}})(\mathbf{R}\times\hat{\mathbf{c}})}{(\mathbf{R}\times\hat{\mathbf{c}})^2}\right]\Bigg\}\end{aligned} \tag{10.48}$$

where

$$\begin{aligned} R_0 &= \sqrt{\varepsilon_\perp \mathbf{R}^2} = \sqrt{\varepsilon_\perp(\mathbf{R}\times\hat{\mathbf{c}})^2 + \varepsilon_\perp(\mathbf{R}\cdot\hat{\mathbf{c}})^2} \\ R_e &= \sqrt{\varepsilon_\perp\varepsilon_\parallel(\mathbf{R}\cdot\bar{\varepsilon}^{-1}\cdot\mathbf{R})} = \sqrt{\varepsilon_\parallel(\mathbf{R}\times\hat{\mathbf{c}})^2 + \varepsilon_\perp(\mathbf{R}\cdot\hat{\mathbf{c}})^2}\end{aligned} \tag{10.49}$$

and

$$\mathbf{R} = \mathbf{r} - \mathbf{r}'$$

To evaluate the double gradient in Eq. (10.48), we need the identity

$$\nabla R_e = \frac{\varepsilon_\perp\varepsilon_\parallel}{R_e}(\bar{\varepsilon}^{-1}\cdot\mathbf{R}) \tag{10.50}$$

which can easily be established by using Eq. (9.48), and the result of Prob. 1.9, and also the fact that for a uniaxial medium $\bar{\varepsilon}^{-1}$ is symmetric.

With the aid of Eqs. (9.48) and (10.50), we then obtain

$$\nabla\left(\frac{e^{ik_0 R_e}}{R_e}\right) = \left(\frac{ik_0}{R_e} - \frac{1}{R_e^2}\right)\frac{e^{ik_0 R_e}}{R_e}\,(\varepsilon_\perp \varepsilon_\parallel \bar{\boldsymbol{\varepsilon}}^{-1} \cdot \mathbf{R}) \tag{10.51}$$

and

$$\nabla\left[\left(\frac{ik_0}{R_e} - \frac{1}{R_e^2}\right)\frac{e^{ik_0 R_e}}{R_e}\right] = \frac{(3 - i3k_0 R_e - k_0^2 R_e^2)e^{ik_0 R_e}}{R_e^5}\,(\varepsilon_\perp \varepsilon_\parallel \bar{\boldsymbol{\varepsilon}}^{-1} \cdot \mathbf{R}) \tag{10.52}$$

By taking the gradient on both sides of Eq. (10.51) and making use of the identities

$$\nabla(\phi \mathbf{B}) = (\nabla\phi)\,\mathbf{B} + \phi \nabla \mathbf{B} \tag{10.53}$$

where ϕ is a scalar function of **R** and **B** is a vector function of **R**, and

$$\nabla\,(\bar{\mathbf{A}} \cdot \mathbf{R}) = \bar{\mathbf{A}} \tag{10.54}$$

for a constant dyadic $\bar{\mathbf{A}}$, we finally obtain

$$\nabla\nabla\left(\frac{e^{ik_0 R_e}}{R_e}\right) = \frac{(3 - i3k_0 R_e - k_0^2 R_e^2)e^{ik_0 R_e}}{R_e^3}\,\frac{\varepsilon_\perp^2 \varepsilon_\parallel^2 (\bar{\boldsymbol{\varepsilon}}^{-1} \cdot \mathbf{R})(\bar{\boldsymbol{\varepsilon}}^{-1} \cdot \mathbf{R})}{R_e^2} + \frac{(ik_0 R_e - 1)e^{ik_0 R_e}}{R_e^3}\,\varepsilon_\perp \varepsilon_\parallel \bar{\boldsymbol{\varepsilon}}^{-1} \tag{10.55}$$

Also, from Eq. (10.49) we see that

$$(\mathbf{R} \times \hat{\mathbf{c}})^2 = \frac{R_e^2 - R_0^2}{\varepsilon_\parallel - \varepsilon_\perp} \tag{10.56}$$

Hence we may write that

$$\frac{e^{ik_0 R_0} - e^{ik_0 R_e}}{ik_0 (\mathbf{R} \times \hat{\mathbf{c}})^2} = (\varepsilon_\perp - \varepsilon_\parallel)\,\frac{\exp\,[ik_0(R_e + R_0)/2]}{R_e + R_0}\,\frac{\sin\,[k_0(R_e - R_0)/2]}{k_0(R_e - R_0)/2} \tag{10.57}$$

Substituting Eqs. (10.55) and (10.57) into Eq. (10.48), we obtain the dyadic Green function of a uniaxial medium in the following coordinate-free form:

$$\begin{aligned}
\bar{\mathbf{G}}(\mathbf{R}) = \frac{1}{4\pi\sqrt{\varepsilon_\perp}} \Bigg\{ & \Bigg[\left(1 - \frac{1}{ik_0 R_e} - \frac{1}{k_0^2 R_e^2}\right) \varepsilon_\perp \varepsilon_\parallel \bar{\boldsymbol{\varepsilon}}^{-1} \\
& - \left(1 - \frac{3}{ik_0 R_e} - \frac{3}{k_0^2 R_e^2}\right) \frac{\varepsilon_\perp^2 \varepsilon_\parallel^2 (\bar{\boldsymbol{\varepsilon}}^{-1} \cdot \mathbf{R})(\bar{\boldsymbol{\varepsilon}}^{-1} \cdot \mathbf{R})}{R_e^2}\Bigg] \frac{e^{ik_0 R_e}}{R_e} \\
& + \left(\frac{\varepsilon_\perp e^{ik_0 R_0}}{R_0} - \frac{\varepsilon_\parallel e^{ik_0 R_e}}{R_e}\right) \frac{(\mathbf{R} \times \hat{\mathbf{c}})(\mathbf{R} \times \hat{\mathbf{c}})}{(\mathbf{R} \times \hat{\mathbf{c}})^2} \\
& + (\varepsilon_\perp - \varepsilon_\parallel)\,\frac{e^{ik_0 (R_e + R_0)/2}}{R_e + R_0}\,\frac{\sin\,[k_0(R_e - R_0)/2]}{k_0(R_e - R_0)/2} \\
& \left[\bar{\mathbf{I}} - \hat{\mathbf{c}}\hat{\mathbf{c}} - 2\,\frac{(\mathbf{R} \times \hat{\mathbf{c}})(\mathbf{R} \times \hat{\mathbf{c}})}{(\mathbf{R} \times \hat{\mathbf{c}})^2}\right]\Bigg\}
\end{aligned} \tag{10.58}$$

In the special case of an isotropic medium [$\bar{\varepsilon} = \varepsilon\bar{\mathbf{I}}$, $\varepsilon_\perp = \varepsilon_\| = \varepsilon$, $\bar{\varepsilon}^{-1} = (1/\varepsilon)\bar{\mathbf{I}}$, $R_e = R_0 = \sqrt{\varepsilon}R$], dyadic Green's function (10.58) reduces to Eq. (9.51) as expected.

10.4 RADIATION FROM AN OSCILLATING ELECTRIC DIPOLE

We shall now consider the field radiated by an oscillating electric dipole located in a uniaxially anisotropic medium. As in the case of an isotropic medium, we write the current distribution of a dipole at the origin as

$$\mathbf{J}(\mathbf{r}') = -i\omega\mathbf{p}_0\,\delta(\mathbf{r}') \tag{10.59}$$

where $\mathbf{p}_0$ denotes the electric dipole moment. Substituting the current distribution (10.59) into Eq. (10.4) and making use of the property of the delta function, we obtain

$$\mathbf{E}(\mathbf{r}) = \omega^2\mu_0\,\bar{\mathbf{G}}(\mathbf{r}, \mathbf{0}) \cdot \mathbf{p}_0 \tag{10.60}$$

where the dyadic Green function is evaluated at $\mathbf{r}' = \mathbf{0}$.

To cast the electric field in a convenient form, we need the following formulas:

$$\bar{\mathbf{I}} - \hat{\mathbf{c}}\hat{\mathbf{c}} - 2\,\frac{(\mathbf{r}\times\hat{\mathbf{c}})(\mathbf{r}\times\hat{\mathbf{c}})}{(\mathbf{r}\times\hat{\mathbf{c}})^2} = \frac{[\hat{\mathbf{c}}\times(\mathbf{r}\times\hat{\mathbf{c}})][\hat{\mathbf{c}}\times(\mathbf{r}\times\hat{\mathbf{c}})] - (\mathbf{r}\times\hat{\mathbf{c}})(\mathbf{r}\times\hat{\mathbf{c}})}{(\mathbf{r}\times\hat{\mathbf{c}})^2} \tag{10.61}$$

and

$$\bar{\varepsilon}^{-1} - \frac{\varepsilon_\perp\varepsilon_\|(\bar{\varepsilon}^{-1}\cdot\mathbf{r})(\bar{\varepsilon}^{-1}\cdot\mathbf{r})}{r_e^2} - \frac{(\mathbf{r}\times\hat{\mathbf{c}})(\mathbf{r}\times\hat{\mathbf{c}})}{\varepsilon_\perp(\mathbf{r}\times\hat{\mathbf{c}})^2} = \frac{[\mathbf{r}\times(\mathbf{r}\times\hat{\mathbf{c}})][\mathbf{r}\times(\mathbf{r}\times\hat{\mathbf{c}})]}{r_e^2(\mathbf{r}\times\hat{\mathbf{c}})^2} \tag{10.62}$$

where

$$r_e = \sqrt{\varepsilon_\perp\varepsilon_\|(\mathbf{r}\cdot\bar{\varepsilon}^{-1}\cdot\mathbf{r})} \tag{10.63}$$

To verify Eqs. (10.61) and (10.62), we note from Fig. 10.1 that

$$\mathbf{r} = \boldsymbol{\rho} + z\hat{\mathbf{c}} \tag{10.64}$$

Here $z = \mathbf{r}\cdot\hat{\mathbf{c}}$, and hence,

$$\begin{aligned} \boldsymbol{\rho} &= \mathbf{r} - z\hat{\mathbf{c}} = \hat{\mathbf{c}}\times(\mathbf{r}\times\hat{\mathbf{c}}) \\ \rho^2 &= (\mathbf{r}\times\hat{\mathbf{c}})^2 \end{aligned} \tag{10.65}$$

Because any two vectors of $\hat{\mathbf{c}}$, $\hat{\boldsymbol{\rho}}$, and $(\mathbf{r}\times\hat{\mathbf{c}})$ are mutually orthogonal, it follows that

$$\bar{\mathbf{I}} = \hat{\mathbf{c}}\hat{\mathbf{c}} + \frac{\boldsymbol{\rho}\boldsymbol{\rho}}{\rho^2} + \frac{(\mathbf{r}\times\hat{\mathbf{c}})(\mathbf{r}\times\hat{\mathbf{c}})}{\rho^2} \tag{10.66}$$

Also, from Eq. (6.3), the inverse of the dielectric tensor is

$$\bar{\varepsilon}^{-1} = \frac{1}{\varepsilon_\perp}\bar{\mathbf{I}} + \left(\frac{1}{\varepsilon_\parallel} - \frac{1}{\varepsilon_\perp}\right)\hat{\mathbf{c}}\hat{\mathbf{c}} \tag{10.67}$$

Thus we have the relations

$$\varepsilon_\perp \varepsilon_\parallel \bar{\varepsilon}^{-1} \cdot \mathbf{r} = \varepsilon_\parallel \boldsymbol{\rho} + \varepsilon_\perp z\hat{\mathbf{c}} \tag{10.68}$$

and

$$\varepsilon_\perp \rho^2 \bar{\varepsilon}^{-1} - (\mathbf{r} \times \hat{\mathbf{c}})(\mathbf{r} \times \hat{\mathbf{c}}) = \boldsymbol{\rho}\boldsymbol{\rho} + \frac{\varepsilon_\perp \rho^2}{\varepsilon_\parallel}\hat{\mathbf{c}}\hat{\mathbf{c}} \tag{10.69}$$

Substituting Eqs. (10.61) and (10.62) into dyadic Green's function (10.58), evaluated at $\mathbf{r}' = \mathbf{0}$, and using Eq. (10.60) we finally obtain the electric field of a dipole immersed in a uniaxial medium:

$$\begin{aligned}\mathbf{E}(\mathbf{r}) = \frac{\omega^2 \mu_0 \sqrt{\varepsilon_\perp}}{4\pi} &\left(\frac{\varepsilon_\parallel \exp(ik_0 r_e)}{r_e}\left\{\frac{[\mathbf{r} \times (\mathbf{r} \times \hat{\mathbf{c}})][\mathbf{r} \times (\mathbf{r} \times \hat{\mathbf{c}})]}{r_e^2(\mathbf{r} \times \hat{\mathbf{c}})^2}\right.\right. \\ &- \left.\left(\frac{1}{ik_0 r_e} + \frac{1}{k_0^2 r_e^2}\right)\left[\bar{\varepsilon}^{-1} - \frac{3\varepsilon_\perp \varepsilon_\parallel(\bar{\varepsilon}^{-1} \cdot \mathbf{r})(\bar{\varepsilon}^{-1} \cdot \mathbf{r})}{r_e^2}\right]\right\} \\ &+ \frac{\exp(ik_0 r_0)}{r_0}\frac{(\mathbf{r} \times \hat{\mathbf{c}})(\mathbf{r} \times \hat{\mathbf{c}})}{(\mathbf{r} \times \hat{\mathbf{c}})^2} \\ &+ \frac{(\varepsilon_\perp - \varepsilon_\parallel)}{\varepsilon_\perp}\frac{\exp[ik_0(r_e + r_0)/2]}{r_e + r_0}\frac{\sin[k_0(r_e - r_0)/2]}{k_0(r_e - r_0)/2} \\ &\cdot \left.\frac{[\hat{\mathbf{c}} \times (\mathbf{r} \times \hat{\mathbf{c}})][\hat{\mathbf{c}} \times (\mathbf{r} \times \hat{\mathbf{c}})] - (\mathbf{r} \times \hat{\mathbf{c}})(\mathbf{r} \times \hat{\mathbf{c}})}{(\mathbf{r} \times \hat{\mathbf{c}})^2}\right) \cdot \mathbf{p}_0\end{aligned} \tag{10.70}$$

where

$$r_0 = \sqrt{\varepsilon_\perp \mathbf{r}^2} \tag{10.71}$$

To find the corresponding magnetic field, we use the Maxwell equation and note that for a constant vector **a**,

$$\nabla \times (\bar{\mathbf{G}} \cdot \mathbf{a}) = (\nabla \times \bar{\mathbf{G}}) \cdot \mathbf{a} \tag{10.72}$$

Hence,

$$\begin{aligned}\mathbf{H}(\mathbf{r}) &= \frac{1}{i\omega\mu_0}\nabla \times \mathbf{E}(\mathbf{r}) \\ &= -i\omega(\nabla \times \bar{\mathbf{G}}) \cdot \mathbf{p}_0\end{aligned} \tag{10.73}$$

The curl of $\bar{\mathbf{G}}$ may be carried out term by term. Making use of identities (9.48), (10.54),

$$\nabla \times (\phi\bar{\mathbf{U}}) = \nabla\phi \times \bar{\mathbf{U}} + \phi\nabla \times \bar{\mathbf{U}} \tag{10.74}$$

and

$$\nabla \times (\mathbf{u}\mathbf{v}) = (\nabla \times \mathbf{u})\mathbf{v} - \mathbf{u} \times \nabla\mathbf{v} \tag{10.75}$$

we obtain

$$\begin{aligned}\nabla r_0 &= \sqrt{\varepsilon_\perp}\,\hat{\mathbf{r}} \\ \nabla r_e &= \frac{\varepsilon_\perp \varepsilon_\parallel}{r_e}(\bar{\boldsymbol{\varepsilon}}^{-1}\cdot\mathbf{r})\end{aligned} \tag{10.76}$$

$$\nabla \times [(\bar{\boldsymbol{\varepsilon}}^{-1}\cdot\mathbf{r})(\bar{\boldsymbol{\varepsilon}}^{-1}\cdot\mathbf{r})] = -(\bar{\boldsymbol{\varepsilon}}^{-1}\cdot\mathbf{r})\times\bar{\boldsymbol{\varepsilon}}^{-1}$$

$$\nabla \times \frac{(\mathbf{r}\times\hat{\mathbf{c}})(\mathbf{r}\times\hat{\mathbf{c}})}{(\mathbf{r}\times\hat{\mathbf{c}})^2} = -\frac{\hat{\mathbf{c}}(\mathbf{r}\times\hat{\mathbf{c}})}{(\mathbf{r}\times\hat{\mathbf{c}})^2}$$

and

$$\nabla \times \left\{\frac{1}{(\mathbf{r}\times\hat{\mathbf{c}})^2}\left[\bar{\mathbf{I}} - \hat{\mathbf{c}}\hat{\mathbf{c}} - 2\,\frac{(\mathbf{r}\times\hat{\mathbf{c}})(\mathbf{r}\times\hat{\mathbf{c}})}{(\mathbf{r}\times\hat{\mathbf{c}})^2}\right]\right\} = \bar{\mathbf{0}}$$

Hence,

$$\begin{aligned}
&\nabla \times \left[\left(1 - \frac{1}{ik_0 r_e} - \frac{1}{k_0^2 r_e^2}\right)\frac{\exp(ik_0 r_e)}{r_e}\,\bar{\boldsymbol{\varepsilon}}^{-1}\right] \\
&\quad = \frac{[3 - i3(k_0 r_e) - 2(k_0 r_e)^2 + i(k_0 r_e)^3]\exp(ik_0 r_e)}{k_0^2 r_e^4}\,\frac{\varepsilon_\perp \varepsilon_\parallel(\bar{\boldsymbol{\varepsilon}}^{-1}\cdot\mathbf{r})}{r_e}\times\bar{\boldsymbol{\varepsilon}}^{-1} \\
&\nabla \times \left[\left(1 - \frac{3}{ik_0 r_e} - \frac{3}{k_0^2 r_e^2}\right)\frac{\exp(ik_0 r_e)}{r_e^3}(\bar{\boldsymbol{\varepsilon}}^{-1}\cdot\mathbf{r})(\bar{\boldsymbol{\varepsilon}}^{-1}\cdot\mathbf{r})\right] \\
&\quad = \frac{[3 - i3(k_0 r_e) - (k_0 r_e)^2]\exp(ik_0 r_e)}{k_0^2 r_e^5}(\bar{\boldsymbol{\varepsilon}}^{-1}\cdot\mathbf{r})\times\bar{\boldsymbol{\varepsilon}}^{-1} \\
&\nabla \times \left[\frac{\exp(ik_0 r_e)}{r_e}\,\frac{(\mathbf{r}\times\hat{\mathbf{c}})(\mathbf{r}\times\hat{\mathbf{c}})}{(\mathbf{r}\times\hat{\mathbf{c}})^2}\right] \\
&\quad = \frac{\varepsilon_\perp \varepsilon_\parallel(ik_0 r_e - 1)\exp(ik_0 r_e)}{r_e^3}(\bar{\boldsymbol{\varepsilon}}^{-1}\cdot\mathbf{r})\times\frac{(\mathbf{r}\times\hat{\mathbf{c}})(\mathbf{r}\times\hat{\mathbf{c}})}{(\mathbf{r}\times\hat{\mathbf{c}})^2} - \frac{\exp(ik_0 r_e)}{r_e}\,\frac{\hat{\mathbf{c}}(\mathbf{r}\times\hat{\mathbf{c}})}{(\mathbf{r}\times\hat{\mathbf{c}})^2} \\
&\nabla \times \left[\frac{\exp(ik_0 r_0)}{r_0}\,\frac{(\mathbf{r}\times\hat{\mathbf{c}})(\mathbf{r}\times\hat{\mathbf{c}})}{(\mathbf{r}\times\hat{\mathbf{c}})^2}\right] \\
&\quad = \frac{\varepsilon_\perp(ik_0 r_0 - 1)\exp(ik_0 r_0)}{r_0^3}\,\frac{[\mathbf{r}\times(\mathbf{r}\times\hat{\mathbf{c}})](\mathbf{r}\times\hat{\mathbf{c}})}{(\mathbf{r}\times\hat{\mathbf{c}})^2} - \frac{\exp(ik_0 r_0)}{r_0}\,\frac{\hat{\mathbf{c}}(\mathbf{r}\times\hat{\mathbf{c}})}{(\mathbf{r}\times\hat{\mathbf{c}})^2} \\
&\nabla \times \left\{\frac{\exp(ik_0 r_e)}{(\mathbf{r}\times\hat{\mathbf{c}})^2}\left[\bar{\mathbf{I}} - \hat{\mathbf{c}}\hat{\mathbf{c}} - 2\,\frac{(\mathbf{r}\times\hat{\mathbf{c}})(\mathbf{r}\times\hat{\mathbf{c}})}{(\mathbf{r}\times\hat{\mathbf{c}})^2}\right]\right\} \\
&\quad = \frac{ik_0 \varepsilon_\perp \varepsilon_\parallel \exp(ik_0 r_e)}{r_e(\mathbf{r}\times\hat{\mathbf{c}})^2}(\bar{\boldsymbol{\varepsilon}}^{-1}\cdot\mathbf{r})\times\left[\bar{\mathbf{I}} - \hat{\mathbf{c}}\hat{\mathbf{c}} - 2\,\frac{(\mathbf{r}\times\hat{\mathbf{c}})(\mathbf{r}\times\hat{\mathbf{c}})}{(\mathbf{r}\times\hat{\mathbf{c}})^2}\right]
\end{aligned} \tag{10.77}$$

and

$$\nabla \times \left\{ \frac{\exp(ik_0 r_0)}{(\mathbf{r} \times \hat{\mathbf{c}})^2} \left[\bar{\mathbf{I}} - \hat{\mathbf{c}}\hat{\mathbf{c}} - 2\frac{(\mathbf{r} \times \hat{\mathbf{c}})(\mathbf{r} \times \hat{\mathbf{c}})}{(\mathbf{r} \times \hat{\mathbf{c}})^2} \right] \right\}$$
$$= \frac{ik_0 \varepsilon_\perp \exp(ik_0 r_0)}{r_0 (\mathbf{r} \times \hat{\mathbf{c}})^2} \mathbf{r} \times \left[\bar{\mathbf{I}} - \hat{\mathbf{c}}\hat{\mathbf{c}} - 2\frac{(\mathbf{r} \times \hat{\mathbf{c}})(\mathbf{r} \times \hat{\mathbf{c}})}{(\mathbf{r} \times \hat{\mathbf{c}})^2} \right]$$

Substituting Eq. (10.77) into Eq. (10.73) and noting that

$$(\varepsilon_\perp \varepsilon_\parallel \bar{\boldsymbol{\varepsilon}}^{-1} \cdot \mathbf{r}) \times \left[\bar{\boldsymbol{\varepsilon}}^{-1} - \frac{(\mathbf{r} \times \hat{\mathbf{c}})(\mathbf{r} \times \hat{\mathbf{c}})}{\varepsilon_\perp (\mathbf{r} \times \hat{\mathbf{c}})^2} \right] = -\frac{(\mathbf{r} \times \hat{\mathbf{c}})[\mathbf{r} \times (\mathbf{r} \times \hat{\mathbf{c}})]}{(\mathbf{r} \times \hat{\mathbf{c}})^2}$$

$$\varepsilon_\parallel \hat{\mathbf{c}}(\mathbf{r} \times \hat{\mathbf{c}}) - (\varepsilon_\perp \varepsilon_\parallel \bar{\boldsymbol{\varepsilon}}^{-1} \cdot \mathbf{r}) \times \left[\bar{\mathbf{I}} - \hat{\mathbf{c}}\hat{\mathbf{c}} - 2\frac{(\mathbf{r} \times \hat{\mathbf{c}})(\mathbf{r} \times \hat{\mathbf{c}})}{(\mathbf{r} \times \hat{\mathbf{c}})^2} \right]$$
$$= \frac{\varepsilon_\perp (\mathbf{r} \cdot \hat{\mathbf{c}})}{(\mathbf{r} \times \hat{\mathbf{c}})^2} \{ [\hat{\mathbf{c}} \times (\mathbf{r} \times \hat{\mathbf{c}})](\mathbf{r} \times \hat{\mathbf{c}}) + (\mathbf{r} \times \hat{\mathbf{c}})[\hat{\mathbf{c}} \times (\mathbf{r} \times \hat{\mathbf{c}})] \} \tag{10.78}$$

and

$$\hat{\mathbf{c}}(\mathbf{r} \times \hat{\mathbf{c}}) - \mathbf{r} \times \left[\bar{\mathbf{I}} - \hat{\mathbf{c}}\hat{\mathbf{c}} - 2\frac{(\mathbf{r} \times \hat{\mathbf{c}})(\mathbf{r} \times \hat{\mathbf{c}})}{(\mathbf{r} \times \hat{\mathbf{c}})^2} \right]$$
$$= \frac{\mathbf{r} \cdot \hat{\mathbf{c}}}{(\mathbf{r} \times \hat{\mathbf{c}})^2} \{ [\hat{\mathbf{c}} \times (\mathbf{r} \times \hat{\mathbf{c}})](\mathbf{r} \times \hat{\mathbf{c}}) + (\mathbf{r} \times \hat{\mathbf{c}})[\hat{\mathbf{c}} \times (\mathbf{r} \times \hat{\mathbf{c}})] \}$$

we find the magnetic field in a coordinate-free form:

$$\mathbf{H}(\mathbf{r}) = \frac{-i\omega\sqrt{\varepsilon_\perp}}{4\pi} \left\{ (1 - ik_0 r_e) \frac{\varepsilon_\parallel \exp(ik_0 r_e)}{r_e^2} \frac{(\mathbf{r} \times \hat{\mathbf{c}})[\mathbf{r} \times (\mathbf{r} \times \hat{\mathbf{c}})]}{r_e (\mathbf{r} \times \hat{\mathbf{c}})^2} \right.$$
$$- (1 - ik_0 r_0) \frac{\varepsilon_\perp \exp(ik_0 r_0)}{r_0^2} \frac{[\mathbf{r} \times (\mathbf{r} \times \hat{\mathbf{c}})](\mathbf{r} \times \hat{\mathbf{c}})}{r_0 (\mathbf{r} \times \hat{\mathbf{c}})^2} + \left[\frac{\exp(ik_0 r_e)}{r_e} - \frac{\exp(ik_0 r_0)}{r_0} \right]$$
$$\left. \cdot \frac{(\mathbf{r} \cdot \hat{\mathbf{c}})[\hat{\mathbf{c}} \times (\mathbf{r} \times \hat{\mathbf{c}})](\mathbf{r} \times \hat{\mathbf{c}}) + (\mathbf{r} \times \hat{\mathbf{c}})[\hat{\mathbf{c}} \times (\mathbf{r} \times \hat{\mathbf{c}})]}{(\mathbf{r} \times \hat{\mathbf{c}})^4} \right\} \cdot \mathbf{p}_0 \tag{10.79}$$

Sometimes, it is more convenient to express the coefficient in the last term in Eq. (10.79) as

$$\frac{1}{(\mathbf{r} \times \hat{\mathbf{c}})^2} \left[\frac{\exp(ik_0 r_e)}{r_e} - \frac{\exp(ik_0 r_0)}{r_0} \right]$$
$$= \frac{ik_0(\varepsilon_\parallel - \varepsilon_\perp) \exp[ik_0(r_e + r_0)/2]}{2r_0 r_e} \left\{ \frac{\sin[k_0(r_e - r_0)/2]}{k_0(r_e - r_0)/2} + i\frac{\cos[k_0(r_e - r_0)/2]}{k_0(r_e + r_0)/2} \right\} \tag{10.80}$$

From Eqs. (10.70) and (10.79) we conclude first that the arguments of the exponential function in $\mathbf{E}(\mathbf{r})$ and $\mathbf{H}(\mathbf{r})$ depend only on r_e and r_0, not on $\mathbf{r} \cdot \hat{\mathbf{c}}$. Hence the field vectors do not depend on the sign of $\mathbf{r} \cdot \hat{\mathbf{c}}$ as anticipated. Secondly, although the results were derived on the assumption that $\varepsilon_{\parallel} > 0$, they must also apply to the case of $\varepsilon_{\parallel} < 0$ which occurs in a uniaxial plasma when $\omega < \omega_p$. In this case we must choose r_e to be positive and purely imaginary for those field points for which r_e^2 in Eq. (10.63) is negative. Then, the factor $\exp(ik_0 r_e)$ characterizes either a propagating or a nonpropagating (or an evanescent) wave according to whether the field point lies inside or outside the conical surface whose generators pass through the origin and make the angle $\tan^{-1}\sqrt{|\varepsilon_{\perp}/\varepsilon_{\parallel}|}$ with the optic axis. The magnitudes of the field vectors become infinitely large as the surface of the cone is approached. Finally, we conclude that when $\varepsilon_{\perp} = \varepsilon_{\parallel}$, the familiar results, Eqs. (9.61) and (9.63), of a dipole in an isotropic medium are recovered (see Prob. 10.11).

10.5 SOME SPECIAL CASES

Having obtained the electric and magnetic fields for the general case of an electric dipole radiating in a uniaxial medium, we shall now consider several interesting special cases.

Case 1. Near field Very close to the dipole ($k_0 r_e \ll 1$ and $k_0 r_0 \ll 1$), the terms in Eqs. (10.70) and (10.79) that are proportional to $1/r_e^3$ predominate. They are

$$\mathbf{E}(\mathbf{r}) = \frac{\varepsilon_{\parallel}\sqrt{\varepsilon_{\perp}}}{4\pi\varepsilon_0 r_e^3}\left[3\,\frac{\varepsilon_{\perp}\varepsilon_{\parallel}(\bar{\boldsymbol{\varepsilon}}^{-1}\cdot\mathbf{r})(\bar{\boldsymbol{\varepsilon}}^{-1}\cdot\mathbf{r})}{r_e^2} - \bar{\boldsymbol{\varepsilon}}^{-1}\right]\cdot\mathbf{p}_0 \tag{10.81}$$

$$\mathbf{H}(\mathbf{r}) = \mathbf{0}$$

Here we note that the electric field is present even at zero frequency and hence is often referred to as the electrostatic solution.

Case 2. Far field On the other hand, far from the dipole ($k_0 r_e \gg 1$ and $k_0 r_0 \gg 1$), the terms that are proportional to $1/r_e$ and $1/r_0$ predominate. In this case, Eqs. (10.70) and (10.79) reduce to

$$\begin{aligned}\mathbf{E}(\mathbf{r}) = \frac{\omega^2\mu_0\sqrt{\varepsilon_{\perp}}}{4\pi}\Bigg\{&\frac{\varepsilon_{\parallel}\exp(ik_0 r_e)}{r_e}\,\frac{[\mathbf{r}\times(\mathbf{r}\times\hat{\mathbf{c}})][\mathbf{r}\times(\mathbf{r}\times\hat{\mathbf{c}})]}{r_e^2(\mathbf{r}\times\hat{\mathbf{c}})^2}\\ &+\frac{\exp(ik_0 r_0)}{r_0}\,\frac{(\mathbf{r}\times\hat{\mathbf{c}})(\mathbf{r}\times\hat{\mathbf{c}})}{(\mathbf{r}\times\hat{\mathbf{c}})^2}\\ &+\frac{\varepsilon_{\perp}-\varepsilon_{\parallel}}{\varepsilon_{\perp}}\,\frac{\exp[ik_0(r_e+r_0)/2]}{r_e+r_0}\,\frac{\sin[k_0(r_e-r_0)/2]}{k_0(r_e-r_0)/2}\\ &\cdot\frac{[\hat{\mathbf{c}}\times(\mathbf{r}\times\hat{\mathbf{c}})][\hat{\mathbf{c}}\times(\mathbf{r}\times\hat{\mathbf{c}})]-(\mathbf{r}\times\hat{\mathbf{c}})(\mathbf{r}\times\hat{\mathbf{c}})}{(\mathbf{r}\times\hat{\mathbf{c}})^2}\Bigg\}\cdot\mathbf{p}_0\end{aligned} \tag{10.82}$$

and

$$\mathbf{H}(\mathbf{r}) = \frac{\omega k_0 \sqrt{\varepsilon_\perp}}{4\pi} \left\{ \frac{\varepsilon_\perp \exp(ik_0 r_0)}{r_0} \frac{[\mathbf{r} \times (\mathbf{r} \times \hat{\mathbf{c}})](\mathbf{r} \times \hat{\mathbf{c}})}{r_0(\mathbf{r} \times \hat{\mathbf{c}})^2} \right.$$

$$- \frac{\varepsilon_\parallel \exp(ik_0 r_e)}{r_e} \frac{(\mathbf{r} \times \hat{\mathbf{c}})[\mathbf{r} \times (\mathbf{r} \times \hat{\mathbf{c}})]}{r_e(\mathbf{r} \times \hat{\mathbf{c}})^2}$$

$$+ \frac{(\varepsilon_\parallel - \varepsilon_\perp)(\mathbf{r} \cdot \hat{\mathbf{c}}) \exp[ik_0(r_e + r_0)/2]}{r_0 r_e} \frac{\sin[k_0(r_e - r_0)/2]}{k_0(r_e - r_0)}$$

$$\left. \cdot \frac{[\hat{\mathbf{c}} \times (\mathbf{r} \times \hat{\mathbf{c}})](\mathbf{r} \times \hat{\mathbf{c}}) + (\mathbf{r} \times \hat{\mathbf{c}})[\hat{\mathbf{c}} \times (\mathbf{r} \times \hat{\mathbf{c}})]}{(\mathbf{r} \times \hat{\mathbf{c}})^2} \right\} \cdot \mathbf{p}_0$$

respectively. Equation (10.82) gives the radiation field of an electric dipole immersed in a uniaxial medium.

Case 3. Dipole parallel to optic axis In this case, $\mathbf{p}_0 = p_0 \hat{\mathbf{c}}$, and Eqs. (10.70) and (10.79) become

$$\mathbf{E}(\mathbf{r}) = \frac{\omega^2 \mu_0 \sqrt{\varepsilon_\perp}\, p_0 \exp(ik_0 r_e)}{4\pi r_e} \left[\left(\frac{3}{k_0^2 r_e^2} + \frac{3}{ik_0 r_e} - 1 \right) \frac{\varepsilon_\perp \varepsilon_\parallel (\mathbf{r} \cdot \hat{\mathbf{c}})(\bar{\varepsilon}^{-1} \cdot \mathbf{r})}{r_e^2} \right.$$

$$\left. + \left(1 - \frac{1}{ik_0 r_e} - \frac{1}{k_0^2 r_e^2} \right) \hat{\mathbf{c}} \right] \tag{10.83}$$

and

$$\mathbf{H}(\mathbf{r}) = \frac{i\omega\varepsilon_\parallel \sqrt{\varepsilon_\perp}\, p_0 \exp(ik_0 r_e)}{4\pi r_e^3} (1 - ik_0 r_e)(\mathbf{r} \times \hat{\mathbf{c}})$$

respectively. Here we note that the magnetic field lies on a plane which is perpendicular to the dipole.

PROBLEMS

10.1 Derive integral representation (10.6) of the dyadic Green function.

10.2 Consider the eigenvalue problem

$$\bar{\mathbf{A}} \cdot \mathbf{e} = \lambda \bar{\varepsilon} \cdot \mathbf{e}$$

where

$$\bar{\mathbf{A}} = -(\hat{\mathbf{k}} \times \bar{\mathbf{I}}) \cdot (\hat{\mathbf{k}} \times \bar{\mathbf{I}}) = \bar{\mathbf{I}} - \hat{\mathbf{k}}\hat{\mathbf{k}}$$

$$\bar{\varepsilon} = \varepsilon_\perp \bar{\mathbf{I}} + (\varepsilon_\parallel - \varepsilon_\perp)\hat{\mathbf{c}}\hat{\mathbf{c}}$$

and $\hat{\mathbf{k}} = \mathbf{k}/k$ is the unit vector in the direction of wave propagation, $\bar{\varepsilon}$ is the dielectric tensor of a uniaxial medium, and the eigenvalue λ corresponds to the inverse of the square of the index of refraction n, that is, $\lambda = 1/n^2 = k_0^2/k^2$. Now, define the dual vector

$$\mathbf{d} = \bar{\boldsymbol{\varepsilon}} \cdot \mathbf{e}$$

and show that

(*a*) the eigenvalues are

$$\lambda_1 = \frac{1}{\varepsilon_\perp} \qquad \lambda_2 = \frac{\hat{\mathbf{k}} \cdot \bar{\boldsymbol{\varepsilon}} \cdot \hat{\mathbf{k}}}{\varepsilon_\perp \varepsilon_\parallel} \qquad \lambda_3 = 0$$

(*b*) the corresponding eigenvectors are

$$\mathbf{e}_1 = \hat{\mathbf{k}} \times \hat{\mathbf{c}} \qquad \mathbf{e}_2 = \bar{\boldsymbol{\varepsilon}}^{-1} \cdot [\hat{\mathbf{k}} \times (\hat{\mathbf{k}} \times \hat{\mathbf{c}})] \qquad \mathbf{e}_3 = \hat{\mathbf{k}}$$

(*c*) the dual vectors are

$$\mathbf{d}_1 = \varepsilon_\perp (\hat{\mathbf{k}} \times \hat{\mathbf{c}}) \qquad \mathbf{d}_2 = \hat{\mathbf{k}} \times (\hat{\mathbf{k}} \times \hat{\mathbf{c}}) \qquad \mathbf{d}_3 = \bar{\boldsymbol{\varepsilon}} \cdot \hat{\mathbf{k}}$$

(*d*) and hence the normalization factors are

$$N_1^2 = \varepsilon_\perp (\hat{\mathbf{k}} \times \hat{\mathbf{c}})^2 \qquad N_2^2 = \frac{\hat{\mathbf{k}} \cdot \bar{\boldsymbol{\varepsilon}} \cdot \hat{\mathbf{k}}}{\varepsilon_\perp \varepsilon_\parallel} (\hat{\mathbf{k}} \times \hat{\mathbf{c}})^2 \qquad N_3^2 = \hat{\mathbf{k}} \cdot \bar{\boldsymbol{\varepsilon}} \cdot \hat{\mathbf{k}}$$

such that

$$\mathbf{e}_i \cdot \mathbf{d}_j = N_i^2 \delta_{ij} \qquad \text{(no sum)}$$

10.3 Using the results of Prob. 10.2, establish Eq. (10.12) for a uniaxial medium.

10.4 Verify Eq. (10.13).

10.5 Verify Eq. (10.32).

10.6 Verify Eqs. (10.37) and (10.39).

10.7 Verify Eq. (10.58).

10.8 Establish closure relation (10.66).

10.9 Verify Eqs. (10.70) and (10.79).

10.10 Starting from Eqs. (10.70) and (10.79), obtain the electric and magnetic fields when the dipole is perpendicular to the optic axis.

10.11 Starting from Eqs. (10.70) and (10.79), show that they reduce to Eqs. (9.61) and (9.63), respectively, when the medium becomes isotropic.

10.12 Verify Eq. (10.83).

10.13 According to the method of stationary phase, the major contribution to the double integral

$$I = \iint f(x, y) e^{ik\phi(x,y)}\, dx\, dy$$

for large values of k comes from the vicinity of points where the phase ϕ is stationary, that is, at points where $\partial\phi/\partial x = \partial\phi/\partial y = 0$. Expanding the phase $\phi(x, y)$ in a Taylor series about a stationary point (x_0, y_0), show that the integral may be approximated by

$$I \sim \frac{2\pi i \sigma}{|K|^{1/2}} f(x_0, y_0) \frac{\exp[ik\phi(x_0, y_0)]}{k}$$

where

$$\sigma = \begin{cases} +1 & \text{for } \alpha\beta > \gamma^2 \quad \alpha > 0 \\ -1 & \text{for } \alpha\beta > \gamma^2 \quad \alpha < 0 \\ -i & \text{for } \alpha\beta < \gamma^2 \end{cases}$$

and $\alpha = \partial^2\phi/\partial x^2$, $\beta = \partial^2\phi/\partial y^2$, and $\gamma = \partial^2\phi/\partial x\, \partial y$, and $K = \alpha\beta - \gamma^2$ is the gaussian curvature of the surface $z = \phi(x, y)$. In the above, all α, β, γ, and K are evaluated at the stationary point.

10.14 Show that if the gaussian curvature K of a surface described by the equation $z = f(x, y)$ is

$$K = \frac{f_{xx} f_{yy} - f_{xy}^2}{(1 + f_x^2 + f_y^2)^2}$$

where $f_x = \partial f/\partial x, f_{xx} = \partial^2 f/\partial x^2, f_{xy} = \partial^2 f/\partial x\ \partial y$, etc., then

$$K = \frac{\sum_{x,y,z} [F_x^2(F_{yy} F_{zz} - F_{yz}^2) + 2F_y F_z(F_{xy} F_{zx} - F_{xx} F_{yz})]}{(F_x^2 + F_y^2 + F_z^2)^2}$$

is the gaussian curvature of the surface described by the equation $F(x\ y, z) = 0$. The summation is over cyclic permutations of the partial derivatives corresponding to the indices x, y, z.

10.15 In the far-zone, only the asymptotic form of Green's function is needed. For this purpose, let us evaluate an integral of the form

$$g(\mathbf{R}) = \frac{1}{(2\pi)^3} \int_{-\infty}^{+\infty} \frac{N(\mathbf{k})}{D(\mathbf{k})} e^{i\mathbf{k}\cdot\mathbf{R}}\, d^3k$$

for large values of $R = |\mathbf{R}|$. The function $N(\mathbf{k})$ has no singularities for finite $\mathbf{k}$. In general, $D(\mathbf{k})$ will have several zeros corresponding to the various branches of the dispersion equation $D(\mathbf{k}) = 0$ or $k_z = k_z^j(k_x, k_y, \omega)$. (*a*) Performing the contour integration with respect to k_z by the residue theorem and making use of the radiation condition, show that

$$g(\mathbf{R}) = \frac{i}{(2\pi)^2} \iint_{-\infty}^{+\infty} \sum_j \frac{N[k_x, k_y, k_z^j(k_x, k_y)]}{\partial D/\partial k_z\ |_{k_z = k_z^j(k_x, k_y)}} \exp\,[iR\phi^j(k_x, k_y)]\, dk_x\, dk_y$$

where

$$\phi^j(k_x, k_y) = \mathbf{k}^j \cdot \hat{\mathbf{R}}$$
$$= \frac{k_x(x - x') + k_y(y - y') + k_z^j(k_x, k_y)(z - z')}{R}$$

and the summation is over the various branches of the dispersion equation $k_z = k_z^j(k_x, k_y)$. (*b*) Using the method of stationary phase (see Prob. 10.13), show that

$$g(\mathbf{R}) \sim \frac{-1}{2\pi R} \sum_{j,s} \frac{N(\mathbf{k}_j^s)\,\sigma_j^s \exp\,(i\mathbf{k}_j^s \cdot \mathbf{R})}{|K_j^s|^{1/2}\ \partial D/\partial k_z|_j^s}$$

where the index s indicates the points of stationary phase on the jth branch, that is, at a point $\mathbf{k}_j^s$, $\partial\phi^j/\partial k_x = \partial\phi^j/\partial k_y = 0$. K_j^s is the gaussian curvature of surface $\phi = \phi^j(k_x, k_y)$ evaluated at the point $\mathbf{k}_j^s$. (*c*) Show that in the far-zone, $g(\mathbf{R})$ can be expressed in the following coordinate-free form:

$$g(\mathbf{R}) \sim \frac{-1}{2\pi R} \sum_{j,s} \frac{N(\mathbf{k}_j^s)\,\sigma_j^s \exp\,(i\mathbf{k}_j^s \cdot \mathbf{R})}{|\kappa_j^s|^{1/2}\ |\partial D/\partial \mathbf{k}|_j^s}$$

where κ_j^s is the gaussian curvature of the wave vector surface $k_z = k_z^j(k_x, k_y)$ evaluated at the point $\mathbf{k}_j^s$. (*d*) Finally, give a physical interpretation of the result of part (*c*).

10.16 To illustrate the technique described in Prob. 10.15, let us consider the scalar Green function (9.24):

$$g(\mathbf{R}) = \frac{1}{(2\pi)^3} \int_{-\infty}^{+\infty} \frac{e^{i\mathbf{k}\cdot\mathbf{R}}}{k^2 - k_0^2\,\varepsilon}\, d^3k$$

Show that (*a*) the point of stationary phase occurs at $\mathbf{k}^s = k_0\sqrt{\varepsilon}\hat{\mathbf{R}}$; (*b*) $\alpha < 0$ and $\alpha\beta - \gamma^2 > 0$, hence $\sigma^s = -1$; (*c*) $|\kappa^s|^{1/2} = 1/k_0\sqrt{\varepsilon}$; (*d*) $|\partial D/\partial \mathbf{k}|^s = 2k_0\sqrt{\varepsilon}$; (*e*) $\mathbf{k}^s \cdot \mathbf{R} = k_0\sqrt{\varepsilon}\, R$; (*f*) the asymptotic form is

$$g(\mathbf{R}) \sim \frac{\exp(ik_0\sqrt{\varepsilon}R)}{4\pi R}$$

which is the same as the exact expression (9.40). However, this is not to be expected in general.

10.17 The technique of asymptotic approximation to an integral developed in Prob. 10.15 can be extended to evaluate dyadic Green's function of an anisotropic medium. Consider the integral (10.6):

$$\bar{\mathbf{G}}(\mathbf{R}) = \frac{1}{(2\pi)^3}\int_{-\infty}^{+\infty} \frac{\text{adj}\ \bar{\mathbf{W}}(\mathbf{k})}{D}\, e^{i\mathbf{k}\cdot\mathbf{R}}\, d^3k$$

where $D = |\bar{\mathbf{W}}(\mathbf{k})|$ is the determinant of the wave matrix $\bar{\mathbf{W}}(\mathbf{k})$. The poles of the integrand occur on the wave vector surface $D = 0$ or $k_z = k_z^j(k_x, k_y)$. Carrying out the approximation as in Prob. 10.15, show that the far-zone dyadic Green function of an anisotropic medium can be expressed in the following coordinate-free form

$$\bar{\mathbf{G}}(\mathbf{R}) \sim \frac{-1}{2\pi R}\sum_{j,s} \frac{\text{adj}\ \bar{\mathbf{W}}(\mathbf{k}_j^s)\,\sigma_j^s \exp(i\mathbf{k}_j^s\cdot\mathbf{R})}{|\kappa_j^s|^{1/2}\,|\partial D/\partial\mathbf{k}|_j^s}$$

where κ_j^s is the gaussian curvature of the wave vector surface evaluated at a stationary phase point $\mathbf{k}_j^s$, and the summation is over all the points of $D = 0$ at which the phase is stationary.

10.18 Consider an aperture A in an infinite ground plane of vanishing thickness. Figure 10.2 shows the geometry of the problem. Assume that sources in $z < 0$ give a field distribution $\mathbf{E}_a(\mathbf{r}')$ over the aperture, and an anisotropic medium characterized by the dielectric tensor $\bar{\varepsilon}$ occupies the region $z > 0$. Since there is no source in $z > 0$, the field satisfies the homogeneous equation

$$(\nabla\nabla - \nabla^2\bar{\mathbf{I}} - k_0^2\bar{\varepsilon})\cdot\mathbf{E}(\mathbf{r}) = \mathbf{0}$$

in the region and subjects to the boundary condition that

$$\mathbf{E}(\mathbf{r}) = \begin{cases} \mathbf{E}_a(\mathbf{r}') & \text{on the aperture} \\ \mathbf{0} & \text{on the rest of the plane} \end{cases}$$

over the $z = 0$ plane and the radiation condition at infinity.

(*a*) Using the Fourier transform, show that the field at any point P in the anisotropic medium can be expressed as

$$\mathbf{E}(\mathbf{r}) = \iint_A \bar{\mathbf{G}}^{ap}(\mathbf{R})\cdot\mathbf{E}_a(\mathbf{r}')\, ds'$$

where $\mathbf{R} = \mathbf{r} - \mathbf{r}'$, and

$$\bar{\mathbf{G}}^{ap}(\mathbf{R}) = \frac{1}{(2\pi)^2}\iint_{-\infty}^{+\infty}\sum_j \frac{\hat{\mathbf{a}}_j(\mathbf{k})\hat{\mathbf{b}}_j(\mathbf{k})}{\hat{\mathbf{a}}_j(\mathbf{k})\cdot\hat{\mathbf{b}}_j(\mathbf{k})}\, e^{iR\phi^j(k_x,k_y)}\, dk_x\, dk_y$$

is the aperture dyadic Green function and

$$\phi^j(k_x, k_y) = \mathbf{k}^j\cdot\hat{\mathbf{R}} = \frac{k_x(x - x') + k_y(y - y') + k_z^j(k_x, k_y)z}{R}$$

In the integrand of $\bar{\mathbf{G}}^{ap}(\mathbf{R})$, $\hat{\mathbf{a}}_j(\mathbf{k})$ is the normalized eigenvector corresponding to the jth branch of the dispersion equation $k_z = k_z^j(k_x, k_y, \omega)$ and $\hat{\mathbf{b}}_m(\mathbf{k})$ is the unit vector orthogonal to $\hat{\mathbf{a}}_j(\mathbf{k})$ for $m \neq j$ (cf. Prob. 1.19). The summation is over the various branches of the dispersion equation.

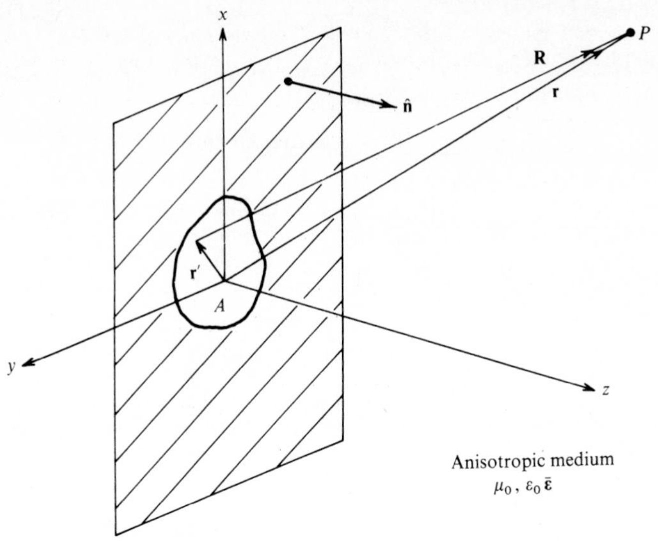

Figure 10.2 Diffraction of waves by a plane aperture.

(*b*) Show that in the far-zone, $\bar{\mathbf{G}}^{ap}(\mathbf{R})$ may be approximated asymptotically by the following coordinate-free form:

$$\bar{\mathbf{G}}^{ap}(\mathbf{R}) \sim \frac{i\hat{\mathbf{n}} \cdot \hat{\mathbf{R}}}{2\pi R} \sum_{j,s} \frac{\hat{\mathbf{a}}_j^s \, \hat{\mathbf{b}}_j^s \, \sigma_j^s \exp(i\mathbf{k}_j^s \cdot \mathbf{R})}{|\kappa_j^s|^{1/2} \, \hat{\mathbf{a}}_j^s \cdot \hat{\mathbf{b}}_j^s}$$

where $\hat{\mathbf{n}}$ is a unit vector normal to the aperture plane and κ_j^s is the gaussian curvature of the wave vector surface. All quantities are evaluated at the stationary phase points $\mathbf{k}_j^s$. The constant σ_j^s takes the values $+1$, -1, or $-i$ as stated in Prob. 10.13.

10.19 If the anisotropic medium in Prob. 10.18 is uniaxial, (*a*) show that the normalized eigenvectors corresponding to the two branches of the dispersion equation are:

Ordinary wave:

$$k^2 = k_0^2 \varepsilon_\perp \qquad \hat{\mathbf{o}}(\mathbf{k}) = \frac{\mathbf{k} \times \hat{\mathbf{c}}}{\sqrt{(\mathbf{k} \times \hat{\mathbf{c}})^2}}$$

Extraordinary wave:

$$\mathbf{k} \cdot \bar{\varepsilon} \cdot \mathbf{k} = k_0^2 \varepsilon_\perp \varepsilon_{||} \qquad \hat{\mathbf{e}}(\mathbf{k}) = \frac{k_0^2 \varepsilon_\perp \hat{\mathbf{c}} - (\mathbf{k} \cdot \hat{\mathbf{c}})\mathbf{k}}{\sqrt{[k_0^2 \varepsilon_\perp \hat{\mathbf{c}} - (\mathbf{k} \cdot \hat{\mathbf{c}})\mathbf{k}]^2}}$$

and

$$\hat{\mathbf{o}} \cdot \hat{\mathbf{e}} = 0 \qquad \hat{\mathbf{o}} \cdot \mathbf{k} = 0$$

(*b*) For the ordinary wave, show that the point of stationary phase occurs at $\mathbf{k}^{so} = k_0\sqrt{\varepsilon_\perp}\hat{\mathbf{R}}$ and $\sigma^{so} = -1$; $|\kappa^{so}|^{1/2} = 1/k_0\sqrt{\varepsilon_\perp}$; $\mathbf{k}^{so} \cdot \mathbf{R} = k_0\sqrt{\varepsilon_\perp}\, R$; $\hat{\mathbf{o}}(\mathbf{k}^{so}) = (\hat{\mathbf{R}} \times \hat{\mathbf{c}})/\sqrt{(\hat{\mathbf{R}} \times \hat{\mathbf{c}})^2}$. (*c*) For the extraordinary wave, show that the point of stationary phase occurs at

$$\mathbf{k}^{se} = \frac{k_0 \varepsilon_\perp \varepsilon_{||}}{R_e} (\bar{\varepsilon}^{-1} \cdot \mathbf{R}) = k_0 \nabla R_e$$

where
$$R_e = \sqrt{\varepsilon_\perp \varepsilon_{||} (\mathbf{R} \cdot \bar{\varepsilon}^{-1} \cdot \mathbf{R})}$$
and $\sigma^{se} = -1$; $|\kappa^{se}|^{1/2} = R_e^2/k_0 \varepsilon_{||}\sqrt{\varepsilon_\perp} R^2$; $\mathbf{k}^{se} \cdot \mathbf{R} = k_0 R_e$. (*d*) Hence, show that the far-zone aperture dyadic Green function for a uniaxial medium takes the form:

$$\bar{\mathbf{G}}^{ap}(\mathbf{R}) \sim \frac{k_0\sqrt{\varepsilon_\perp}(\hat{\mathbf{n}} \cdot \hat{\mathbf{R}})\hat{\mathbf{o}}(\mathbf{k}^{so})\hat{\mathbf{o}}(\mathbf{k}^{so}) \exp(ik_0\sqrt{\varepsilon_\perp}R)}{2\pi i R} + \frac{k_0 \varepsilon_{||}\sqrt{\varepsilon_\perp}(\hat{\mathbf{n}} \cdot \mathbf{R})\hat{\mathbf{e}}(\mathbf{k}^{se})\hat{\mathbf{e}}(\mathbf{k}^{se}) \exp(ik_0 R_e)}{2\pi i R_e^2}$$

Substitution of the above into

$$\mathbf{E}(\mathbf{r}) = \iint_A \bar{\mathbf{G}}^{ap}(\mathbf{R}) \cdot \mathbf{E}_a(\mathbf{r}')\, ds'$$

gives the electric field at an observation point P in the uniaxial medium. In other words, the problem of wave diffraction by an aperture in a uniaxial medium may be formally regarded as waves radiated by the equivalent source distribution $\mathbf{E}_a(\mathbf{r}')$ over the aperture (Huygen's principle for a uniaxial medium).

REFERENCES

Bergstein, L., and T. Zachos: "A Huygens' Principle for Uniaxially Anisotropic Media," *J. Opt. Soc. Am.*, vol. 56, no. 7, 1966, PP. 931–937.

Besieris, I. M.: "Transient Dipole Radiation in an Absorbing Uniaxial Crystal," *J. Franklin Inst.*, vol. 297, no. 4, 1974, pp. 243–258.

Bunkin, F. V.: "On Radiation in Anisotropic Media," *Sov. Phys., JETP*, vol. 5, no. 2, 1957, pp. 277–283.

Chen, H. C.: "Spectral Resolution and Solution of Field Problems in an Anisotropic and Compressible Medium," *IEEE Trans. Antennas Propagation*, vol. AP-16, no. 6, 1968, pp. 741–745.

Chen, H. C.: "Radiation Characteristic of an Electric Dipole in a Warm, Anisotropic Plasma," *J. Appl. Phys.*, vol. 40, no. 10, 1969, pp. 4068–4073.

Chen, H. C.: "Dyadic Green's Function and Radiation in a Uniaxially Anisotropic Medium," *Int. J. Electronics*, vol. 35, no. 5, 1973, pp. 633–640.

Chow, Y.: "A Note on Radiation in a Gyro-Electric-Magnetic Medium—an Extension of Bunkin's Calculation," *IRE Trans. Antennas Propagation*, vol. AP-10, no. 4, 1962, pp. 464–469.

Clemmow, P. C.: *The Plane Wave Spectrum Representation of Electromagnetic Fields*, Pergamon Press, New York, 1966, chap. 8.

Deschamps, G. A., and O. Kesler: "Radiation Field of an Arbitrary Antenna in a Magnetoplasma," *IEEE Trans. Antennas Propagation*, vol. AP-12, no. 6, 1964, pp. 783–785.

Dysthe, K. B.: "Nonlinear Interaction between Two Beams of Plane Electromagnetic Waves in an Anisotropic Medium," in J. Brown (ed.), *Electromagnetic Wave Theory*, Pergamon Press, New York, 1967, pt. 2, pp. 597–612.

Felsen, L. B., and N. Marcuvitz: *Radiation and Scattering of Waves*, Prentice-Hall, Inc., Englewood Cliffs, N.J., 1973.

Harrington, R. F., and A. T. Villeneuve: "Reciprocity Relationships for Gyrotropic Media," *IRE Trans. Microwave Theory Tech.*, vol. MTT-6, no. 3, 1958, pp. 308–310.

Hurd, R. A.: "A Note on the Field of a Dipole in a Uniaxial Medium," *Can. J. Phys.*, vol. 43, 1965, pp. 684–688.

Kogelnik, H.: "On Electromagnetic Radiation in Magneto-Ionic Media," *J. Res. NBS-D. Radio Propagation*, vol. 64D, no. 5, 1960, pp. 515–523.

Kuehl, H. H.: "Electromagnetic Radiation from an Electric Dipole in a Cold Anisotropic Plasma," *Phys. Fluids*, vol 5, no. 9, 1962, pp. 1095–1103.

Lax, M., and D. F. Nelson: "Crystal Electrodynamics," in E. Burstein (ed.), *Atomic Structure and Properties of Solids*, Academic Press, New York, 1972, pp. 48–67.

Lax, M., and D. F. Nelson: "Adventures in Green's Land: Light Scattering in Anisotropic Media," in L. Mandel and E. Wolf (eds.), *Coherence and Quantum Optics*, Plenum Press, New York, 1973, pp. 415–445.

Lighthill, M. J.: "Studies on Magneto-Hydrodynamic Waves and Other Anisotropic Wave Motions," *Phil. Trans. Roy. Soc.* (*London*) *Ser. A252*, 1960, pp. 397–430.

Meecham, W. C.: "Source and Reflection Problems in Magneto-Ionic Media," *Phys. Fluids*, vol. 4, no. 12, 1961, pp. 1517–1524.

Ogg, N. R.: "A Huygen's Principle for Anisotropic Media," *J. Phys. A: Gen. Phys.*, vol. 4, 1971, pp. 382–388.

Rao, B. R., and T. T. Wu: "On the Applicability of Image Theory in Anisotropic Media," *IEEE Trans. Antennas Propagation*, vol. AP-13, no. 5, 1965, pp. 814–815.

Shafranov, V. D.: "Electromagnetic Waves in a Plasma," in M. A. Leontovich (ed.), *Reviews of Plasma Physics*, Consultants Bureau, New York, 1967, vol. 3, pp. 1-157.

Stamnes, J. J., and G. C. Sherman: "Radiation of Electromagnetic Fields in Uniaxially Anisotropic Media," *J. Opt. Soc. Am.*, vol. 66, no. 8, 1976, pp. 780–788.

Wait, J. R.: "Theory of Radiation from Sources Immersed in Anisotropic Media," *J. Res. NBS-B. Math. and Math. Phys.*, vol. 68B, no. 3, 1964, pp. 119–136.

Yeh, K. C., and C. H. Liu: *Theory of Ionospheric Waves*, Academic Press, New York, 1972, chap. 2 and app. B.

Whole issues of the following journals are devoted to waves in anisotropic media.

"Special Issue on Electromagnetic Wave Propagation in Anisotropic Media," *Radio Science*, vol. 2, no. 8, 1967.

"URSI Symposium on Electromagnetic Waves Held at Stresa, Italy, June, 1968," *Alta Frequenza*, 1969.

Jordon, E. C. (ed.): "Electromagnetic Theory and Antennas, " *Proceedings of a Symposium Held at Copenhagen, Denmark, June 1962*, Pergamon Press, New York, 1963, pt. 1, sec. B.

CHAPTER

ELEVEN

RADIATION IN A MOVING MEDIUM

In Chap. 8, based on Minkowski's relativistic theory of electrodynamics, we presented the basic equations governing the propagation of electromagnetic waves in a moving medium. We derived the expressions for phase and group velocities. We also found the index of refraction in a moving medium as it depended on the direction of wave propagation, the velocity of motion of the medium, and other parameters. Also treated in that chapter were the phenomena of reflection and refraction of waves in the presence of a moving medium.

In this chapter we are concerned primarily with the presence of sources in a moving medium. We will begin by deriving the equation satisfied by the dyadic Green function in such a medium. The solution obtained will enable us to determine the fields of a given source in a moving medium. Finally, the radiation from an electric dipole in such a medium will also be discussed.

11.1 DYADIC GREEN'S FUNCTION OF A MOVING MEDIUM

The Maxwell equations with sources have the same mathematical form for moving media as for stationary media. After taking the Fourier transform with respect to t according to Eq. (9.3), we obtain

$$\nabla \times \mathbf{E} = i\omega\mathbf{B} \tag{11.1}$$

$$\nabla \times \mathbf{H} = -i\omega\mathbf{D} + \mathbf{J} \tag{11.2}$$

The field vectors for a moving isotropic medium are connected by Minkowski's constitutive relations which were derived in Sec. 8.1. They are

$$\mathbf{B} = -\frac{\mathbf{m} \times \mathbf{E}}{c} + \mu_0 \mu \bar{\boldsymbol{\alpha}} \cdot \mathbf{H} \tag{11.3}$$

$$\mathbf{D} = \varepsilon_0 \varepsilon \bar{\boldsymbol{\alpha}} \cdot \mathbf{E} + \frac{\mathbf{m} \times \mathbf{H}}{c} \tag{11.4}$$

where ε and μ denote, respectively, the dielectric constant and permeability of the medium at rest, which is assumed to be lossless, and c is the speed of light in free space. The symmetric tensor $\bar{\boldsymbol{\alpha}}$ is defined by

$$\bar{\boldsymbol{\alpha}} = \alpha \bar{\mathbf{I}} + (1 - \alpha)\hat{\mathbf{v}}\hat{\mathbf{v}} \tag{11.5}$$

where $\hat{\mathbf{v}}$ is a unit vector in the direction of motion of the medium and

$$\alpha = \frac{1 - \beta^2}{1 - \mu\varepsilon\beta^2} \qquad \beta = \frac{v}{c} \tag{11.6}$$

The vector $\mathbf{m}$ is given by

$$\mathbf{m} = m\hat{\mathbf{v}} \tag{11.7}$$

where

$$m = \frac{\beta(\mu\varepsilon - 1)}{1 - \mu\varepsilon\beta^2} \tag{11.8}$$

Substituting Eqs. (11.3) and (11.4) into Eqs. (11.1) and (11.2) respectively, we find the Maxwell-Minkowski equations for a moving isotropic medium:

$$(\nabla + ik_0 \mathbf{m}) \times \mathbf{E} = i\omega\mu_0 \mu \bar{\boldsymbol{\alpha}} \cdot \mathbf{H} \tag{11.9}$$

$$(\nabla + ik_0 \mathbf{m}) \times \mathbf{H} = -i\omega\varepsilon_0 \varepsilon \bar{\boldsymbol{\alpha}} \cdot \mathbf{E} + \mathbf{J} \tag{11.10}$$

Elimination of $\mathbf{H}$ from Eqs. (11.9) and (11.10) yields the complex wave equation in $\mathbf{E}$:

$$\{[(\nabla + ik_0 \mathbf{m}) \times \bar{\mathbf{I}}] \cdot \bar{\boldsymbol{\alpha}}^{-1} \cdot [(\nabla + ik_0 \mathbf{m}) \times \bar{\mathbf{I}}] - k_0^2 \mu\varepsilon \bar{\boldsymbol{\alpha}}\} \cdot \mathbf{E}(\mathbf{r}) = i\omega\mu_0 \mu \mathbf{J}(\mathbf{r}) \tag{11.11}$$

Because of the linearity of Eq. (11.11), we may again express the general solution as

$$\mathbf{E}(\mathbf{r}) = i\omega\mu_0 \mu \int_{-\infty}^{+\infty} \bar{\mathbf{G}}(\mathbf{r}, \mathbf{r}') \cdot \mathbf{J}(\mathbf{r}') \, d^3r' \tag{11.12}$$

Following the same procedure as in Sec. 9.1, we find that dyadic Green's function $\bar{\mathbf{G}}(\mathbf{r}, \mathbf{r}')$ of a moving medium satisfies the equation

$$\{[(\nabla + ik_0 \mathbf{m}) \times \bar{\mathbf{I}}) \cdot \bar{\boldsymbol{\alpha}}^{-1} \cdot [(\nabla + ik_0 \mathbf{m}) \times \bar{\mathbf{I}}] - k_0^2 \mu\varepsilon \bar{\boldsymbol{\alpha}}\} \cdot \bar{\mathbf{G}}(\mathbf{r}, \mathbf{r}') = \bar{\mathbf{I}}\, \delta(\mathbf{r} - \mathbf{r}') \tag{11.13}$$

Here we see again that the dyadic Green function satisfies the same complex wave equation (11.11), except that the source term is now replaced by the delta function.

In order to solve Eq. (11.13), we expand the dyadic Green function $\bar{\mathbf{G}}(\mathbf{r}, \mathbf{r}')$ and the delta function as Fourier integrals according to Eqs. (9.15) and (9.16), i.e.,

$$\bar{\mathbf{G}}(\mathbf{r}, \mathbf{r}') = \frac{1}{(2\pi)^3} \int_{-\infty}^{+\infty} \bar{\mathbf{G}}(\mathbf{k}, \mathbf{r}') e^{i\mathbf{k}\cdot\mathbf{r}} \, d^3k$$

and

$$\delta(\mathbf{r} - \mathbf{r}') = \frac{1}{(2\pi)^3} \int_{-\infty}^{+\infty} e^{i\mathbf{k}\cdot(\mathbf{r}-\mathbf{r}')} \, d^3k \tag{11.14}$$

where $\bar{\mathbf{G}}(\mathbf{k}, \mathbf{r}')$ is the Fourier transform of $\bar{\mathbf{G}}(\mathbf{r}, \mathbf{r}')$. Substitution of these expansions into Eq. (11.13) yields

$$\bar{\mathbf{W}}_m(\mathbf{p}) \cdot \bar{\mathbf{G}}(\mathbf{k}, \mathbf{r}') = \bar{\mathbf{I}} e^{-i\mathbf{k}\cdot\mathbf{r}'} \tag{11.15}$$

where $\bar{\mathbf{W}}_m(\mathbf{p})$ is the wave matrix of a moving medium and is given by

$$\bar{\mathbf{W}}_m(\mathbf{p}) = -(\mathbf{p} \times \bar{\mathbf{I}}) \cdot \bar{\boldsymbol{\alpha}}^{-1} \cdot (\mathbf{p} \times \bar{\mathbf{I}}) - k_0^2 \mu\varepsilon \, \bar{\boldsymbol{\alpha}} \tag{11.16}$$

and

$$\mathbf{p} = \mathbf{k} + k_0 \mathbf{m} \tag{11.17}$$

Using identity (1.120),

$$(\mathbf{p} \times \bar{\mathbf{I}}) \cdot (\text{adj } \bar{\boldsymbol{\alpha}}) \cdot (\mathbf{p} \times \bar{\mathbf{I}}) = \tilde{\bar{\boldsymbol{\alpha}}} \cdot \mathbf{pp} \cdot \tilde{\bar{\boldsymbol{\alpha}}} - (\mathbf{p} \cdot \bar{\boldsymbol{\alpha}} \cdot \mathbf{p})\tilde{\bar{\boldsymbol{\alpha}}}$$

and noting that $\bar{\boldsymbol{\alpha}}$ is symmetric, we may also express the wave matrix as

$$\bar{\mathbf{W}}_m(\mathbf{p}) = \frac{1}{|\bar{\boldsymbol{\alpha}}|} \bar{\boldsymbol{\alpha}} \cdot [(\mathbf{p} \cdot \bar{\boldsymbol{\alpha}} \cdot \mathbf{p} - k_0^2 \mu\varepsilon \, |\bar{\boldsymbol{\alpha}}|)\bar{\mathbf{I}} - \mathbf{pp} \cdot \bar{\boldsymbol{\alpha}}] \tag{11.18}$$

Premultiplying Eq. (11.15) by the inverse matrix $\bar{\mathbf{W}}_m^{-1}(\mathbf{p})$ and then substituting the result of $\bar{\mathbf{G}}(\mathbf{k}, \mathbf{r}')$ back into Eq. (11.14), we obtain

$$\bar{\mathbf{G}}(\mathbf{r}, \mathbf{r}') = \frac{1}{(2\pi)^3} \int_{-\infty}^{+\infty} \bar{\mathbf{W}}_m^{-1}(\mathbf{p}) e^{i\mathbf{k}\cdot(\mathbf{r}-\mathbf{r}')} \, d^3k \tag{11.19}$$

Now, using the results of Example 1.2 and Eq. (11.18), we find the explicit forms of the determinant and adjoint of the wave matrix:

$$|\bar{\mathbf{W}}_m(\mathbf{p})| = -\frac{k_0^2 \mu\varepsilon}{|\bar{\boldsymbol{\alpha}}|} (\mathbf{p} \cdot \bar{\boldsymbol{\alpha}} \cdot \mathbf{p} - k_0^2 \mu\varepsilon \, |\bar{\boldsymbol{\alpha}}|)^2 \tag{11.20}$$

and

$$\text{adj } \bar{\mathbf{W}}_m(\mathbf{p}) = \frac{1}{|\bar{\boldsymbol{\alpha}}|} (\mathbf{p} \cdot \bar{\boldsymbol{\alpha}} \cdot \mathbf{p} - k_0^2 \mu\varepsilon \, |\bar{\boldsymbol{\alpha}}|)(\mathbf{pp} - k_0^2 \mu\varepsilon \text{ adj } \bar{\boldsymbol{\alpha}}) \tag{11.21}$$

Hence

$$\bar{\mathbf{W}}_m^{-1}(\mathbf{p}) = \frac{\text{adj } \bar{\mathbf{W}}_m(\mathbf{p})}{|\bar{\mathbf{W}}_m(\mathbf{p})|} = \frac{k_0^2 \mu\varepsilon \text{ adj } \bar{\boldsymbol{\alpha}} - \mathbf{pp}}{k_0^2 \mu\varepsilon\alpha[(\mathbf{p} \cdot \bar{\boldsymbol{\alpha}} \cdot \mathbf{p})/\alpha - k_0^2 \mu\varepsilon\alpha]} \tag{11.22}$$

Substituting the above into Eq. (11.19) and changing the integration variable from **k** to **p** according to Eq. (11.17), we have

$$\bar{\mathbf{G}}(\mathbf{R}) = \frac{e^{-ik_0 \mathbf{m}\cdot\mathbf{R}}}{(2\pi)^3 k_0^2 \mu\varepsilon\alpha} \int_{-\infty}^{+\infty} \frac{(k_0^2 \mu\varepsilon \operatorname{adj} \bar{\boldsymbol{\alpha}} - \mathbf{pp}) e^{i\mathbf{p}\cdot\mathbf{R}}}{(\mathbf{p}\cdot\bar{\boldsymbol{\alpha}}\cdot\mathbf{p})/\alpha - k_0^2 \mu\varepsilon\alpha} \, d^3p \tag{11.23}$$

where $\mathbf{R} = \mathbf{r} - \mathbf{r}'$ and $k_0 = \omega/c$. Again because of the exponential term inside the integral sign, the vector **p** in the numerator may be replaced by $-i\nabla$. Interchanging the order of differentiation and integration, we obtain

$$\bar{\mathbf{G}}(\mathbf{R}) = e^{-ik_0 \mathbf{m}\cdot\mathbf{R}} \left(\bar{\boldsymbol{\alpha}}^{-1} + \frac{1}{k_0^2 \mu\varepsilon |\bar{\boldsymbol{\alpha}}|} \nabla\nabla \right) g_m(\mathbf{R}, \omega) \tag{11.24}$$

where

$$g_m(\mathbf{R}, \omega) = \frac{\alpha}{(2\pi)^3} \int_{-\infty}^{+\infty} \frac{e^{i\mathbf{p}\cdot\mathbf{R}}}{(\mathbf{p}\cdot\bar{\boldsymbol{\alpha}}\cdot\mathbf{p})/\alpha - (\mu\varepsilon\alpha/c^2)\omega^2} \, d^3p \tag{11.25}$$

We may also express the above integration over **k**-space. To do so, we note that

$$\beta m = \alpha - 1$$
$$\frac{m^2}{\alpha} + 1 - \mu\varepsilon\alpha = -\frac{m}{\alpha\beta} = -\gamma^2(\mu\varepsilon - 1) \tag{11.26}$$

where $\gamma^2 = 1/(1 - \beta^2)$. Thus the denominator of the integrand in Eq. (11.25) becomes

$$\frac{\mathbf{p}\cdot\bar{\boldsymbol{\alpha}}\cdot\mathbf{p}}{\alpha} - k_0^2 \mu\varepsilon\alpha = k^2 - k_0^2 - \gamma^2(\mu\varepsilon - 1)[k_0 - \beta(\mathbf{k}\cdot\hat{\mathbf{v}})]^2 \tag{11.27}$$

Substituting the above into Eq. (11.25) and using Eq. (11.17), we find

$$g_m(\mathbf{R}, \omega) = \alpha e^{ik_0 \mathbf{m}\cdot\mathbf{R}} G_m(\mathbf{R}, \omega) \tag{11.28}$$

where

$$\begin{aligned} G_m(\mathbf{R}, \omega) &= \frac{1}{(2\pi)^3} \int_{-\infty}^{+\infty} \frac{e^{i\mathbf{k}\cdot\mathbf{R}}}{(\mathbf{p}\cdot\bar{\boldsymbol{\alpha}}\cdot\mathbf{p})/\alpha - (\mu\varepsilon\alpha/c^2)\omega^2} \, d^3k \\ &= \frac{1}{(2\pi)^3} \int_{-\infty}^{+\infty} \frac{e^{i\mathbf{k}\cdot\mathbf{R}}}{k^2 - k_0^2 - \gamma^2(\mu\varepsilon - 1)[k_0 - \beta(\mathbf{k}\cdot\hat{\mathbf{v}})]^2} \, d^3k \end{aligned} \tag{11.29}$$

Once the integral (11.25) or (11.29) is evaluated, the dyadic Green function for a moving medium can then be calculated from Eq. (11.24).

11.2 EVALUATIONS OF $g_m(\mathbf{R}, t)$ AND $G_m(\mathbf{R}, t)$ IN THE TIME DOMAIN

To evaluate the scalar Green function, we take the Fourier inverse transform of Eq. (11.25) according to Eq. (9.4) and consider the following integral:

$$g_m(\mathbf{R}, t) = \frac{c^2}{(2\pi)^4 \mu\varepsilon} \int_{-\infty}^{+\infty} I_m(\mathbf{p}, t) e^{i\mathbf{p}\cdot\mathbf{R}}\, d^3p \tag{11.30}$$

where

$$I_m(\mathbf{p}, t) = \int_{-\infty}^{+\infty} \frac{e^{-i\omega t}}{c^2(\mathbf{p}\cdot\bar{\alpha}\cdot\mathbf{p})/\alpha^2\mu\varepsilon - \omega^2}\, d\omega \tag{11.31}$$

By direct substitution, we can easily verify that the scalar Green function $g_m(\mathbf{R}, t)$ in the time domain satisfies the following wave equation:

$$\left(\frac{1}{\alpha^2} \nabla\cdot\bar{\alpha}\cdot\nabla - \frac{\mu\varepsilon}{c^2}\frac{\partial^2}{\partial t^2}\right) g_m(\mathbf{R}, t) = -\delta(\mathbf{R})\,\delta(t) \tag{11.32}$$

Since Eq. (11.32) for the scalar Green function is a partial differential equation of an order higher than one, it has several linearly independent solutions. From the integral representation of $g_m(\mathbf{R}, t)$ in Eq. (11.30), it is possible to obtain all these solutions by properly avoiding poles that lie on the path of integration. The poles are located at those values of k and ω for which dispersion equation (8.40) is satisfied. Thus these poles correspond in reality to the condition for the existence of electromagnetic waves in the moving medium.

We shall first perform integration with respect to ω. The treatment of poles is dictated by the causality principle stated in Sec. 9.2, requiring that $g_m(\mathbf{R}, t)$ must vanish for $t < 0$. This implies the path of integration shown in Fig. 9.1. Hence, by the Cauchy residue theorem, we obtain

$$I_m(\mathbf{p}, t) = \frac{2\pi\, \alpha\sqrt{\mu\varepsilon}}{c\sqrt{\mathbf{p}\cdot\bar{\alpha}\cdot\mathbf{p}}} \sin\left(\frac{ct\sqrt{\mathbf{p}\cdot\bar{\alpha}\cdot\mathbf{p}}}{\alpha\sqrt{\mu\varepsilon}}\right) U(t) \tag{11.33}$$

where $U(t)$ is the unit step function. Next, we insert Eq. (11.33) into Eq. (11.30). To carry out integration over the entire $\mathbf{p}$-space, we introduce a cylindrical coordinate system whose z axis is directed along the velocity of the moving medium: $\hat{\mathbf{z}} = \hat{\mathbf{v}}$. We denote the component of the vector $\mathbf{p}$ along the z axis by p_z and its projection on a plane perpendicular to the z axis by $\mathbf{p}_\rho$. We likewise denote the corresponding projections of the vector $\mathbf{R}$ by z and $\boldsymbol{\rho}$ respectively (see Fig. 11.1). Thus

$$\begin{aligned} \mathbf{p} &= \mathbf{p}_\rho + p_z\hat{\mathbf{z}} \\ \mathbf{R} &= \boldsymbol{\rho} + z\hat{\mathbf{z}} \\ d^3p &= p_\rho\, dp_\rho\, dp_z\, d\phi \end{aligned} \tag{11.34}$$

and

$$\begin{aligned} \mathbf{p}\cdot\mathbf{R} &= \mathbf{p}_\rho\cdot\boldsymbol{\rho} + p_z z = p_\rho \rho \cos\phi + p_z z \\ \mathbf{p}\cdot\bar{\alpha}\cdot\mathbf{p} &= \alpha p_\rho^2 + p_z^2 \end{aligned} \tag{11.35}$$

After performing the integration in Eq. (11.30) over the angle ϕ according to formula (10.22), we get

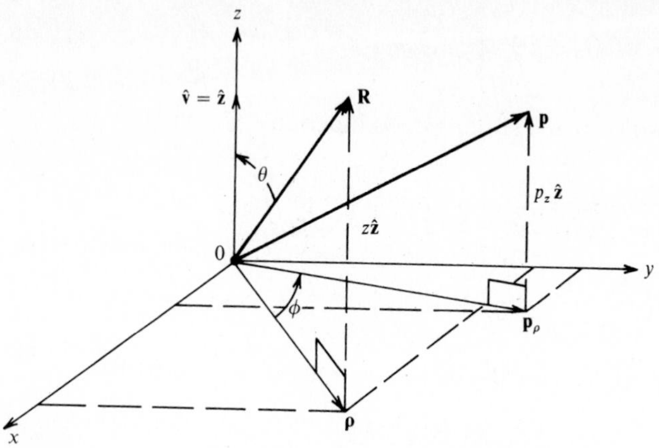

Figure 11.1 Coordinate system used in evaluating the integral.

$$g_m(\mathbf{R}, t) = \frac{\sqrt{\alpha}cU(t)}{(2\pi)^2\sqrt{\mu\varepsilon}} \int_{-\infty}^{+\infty} e^{izp_z}\, dp_z$$

$$\cdot \int_0^\infty \frac{J_0(\rho p_\rho) \sin [(ct/\sqrt{\alpha\mu\varepsilon})\sqrt{p_\rho^2 + (1/\alpha)p_z^2}]p_\rho\, dp_\rho}{\sqrt{p_\rho^2 + (1/\alpha)p_z^2}} \tag{11.36}$$

The integration over the variable p_ρ can be carried out by means of Sonine's discontinuous integral [see Appendix C],

$$\int_0^\infty \frac{J_0(cx) \sin a\sqrt{x^2 + b^2}}{\sqrt{x^2 + b^2}}\, x\, dx = \frac{\cos b\sqrt{a^2 - c^2}}{\sqrt{a^2 - c^2}}\, U(a - c) \qquad c > 0$$

where $U(a - c)$ is the unit step function. As a result, Eq. (11.36) becomes

$$g_m(\mathbf{R}, t) = \frac{\sqrt{\alpha}c}{(2\pi)^2\sqrt{\mu\varepsilon}\sqrt{c^2t^2/\alpha\mu\varepsilon - \rho^2}} \int_{-\infty}^{+\infty} e^{izp_z} \cos\left(\frac{p_z}{\sqrt{\alpha}}\sqrt{\frac{c^2t^2}{\alpha\mu\varepsilon} - \rho^2}\right) dp_z \tag{11.37}$$

where $$\frac{ct}{\sqrt{\alpha\mu\varepsilon}} > \rho > 0$$

We may also write Eq. (11.37) as

$$g_m(\mathbf{R}, t) = \frac{\sqrt{\alpha}c}{8\pi^2\sqrt{\mu\varepsilon}\,\sqrt{c^2t^2/\alpha\mu\varepsilon - \rho^2}} \left\{\int_{-\infty}^{+\infty} \exp\left[i\left(z + \frac{1}{\sqrt{\alpha}}\sqrt{\frac{c^2t^2}{\alpha\mu\varepsilon} - \rho^2}\right)p_z\right] dp_z\right.$$

$$\left. + \int_{-\infty}^{+\infty} \exp\left[i\left(z - \frac{1}{\sqrt{\alpha}}\sqrt{\frac{c^2t^2}{\alpha\mu\varepsilon} - \rho^2}\right)p_z\right] dp_z\right\} \tag{11.38}$$

Using the integral representation for the delta function,

$$\delta(x) = \frac{1}{2\pi} \int_{-\infty}^{+\infty} e^{i\lambda x} \, d\lambda$$

we obtain the final expression for the scalar Green function

$$g_m(\mathbf{R}, t) = \frac{\sqrt{\alpha}c}{4\pi\sqrt{\mu\varepsilon}\sqrt{c^2t^2/\alpha\mu\varepsilon - \rho^2}} \left[\delta\left(z + \frac{1}{\sqrt{\alpha}} \sqrt{\frac{c^2t^2}{\alpha\mu\varepsilon} - \rho^2} \right) + \delta\left(z - \frac{1}{\sqrt{\alpha}} \sqrt{\frac{c^2t^2}{\alpha\mu\varepsilon} - \rho^2} \right) \right] \tag{11.39}$$

By means of the following relation for the delta function (see Appendix B):

$$\delta[f(x)] = \sum_i \frac{\delta(x - x_i)}{|f'(x_i)|} \tag{11.40}$$

where $f'(x) = df/dx$ and where x_i is the ith root of the equation $f(x) = 0$, we obtain, in the special case when $f(x) = x^2 - a^2$,

$$\delta(x^2 - a^2) = \frac{\delta(x + a) + \delta(x - a)}{2|a|} \tag{11.41}$$

Using this result, we may express the scalar Green function (11.39) in an alternative form:

$$g_m(\mathbf{R}, t) = \frac{c}{2\pi\sqrt{\mu\varepsilon}} \delta\left(R_v^2 - \frac{c^2t^2}{\alpha^2\mu\varepsilon} \right) \tag{11.42}$$

where

$$R_v = \sqrt{\mathbf{R} \cdot \bar{\alpha}^{-1} \cdot \mathbf{R}} = \sqrt{z^2 + \frac{\rho^2}{\alpha}} \tag{11.43}$$

or, in terms of t according to Eq. (11.40),

$$g_m(\mathbf{R}, t) = \frac{\alpha}{4\pi R_v} \left[\delta\left(t - \frac{\alpha\sqrt{\mu\varepsilon}R_v}{c} \right) + \delta\left(t + \frac{\alpha\sqrt{\mu\varepsilon}R_v}{c} \right) \right] \tag{11.44}$$

To find $G_m(\mathbf{R}, t)$, we use relation (11.28) which connects $G_m(\mathbf{R}, \omega)$ and $g_m(\mathbf{R}, \omega)$ in the frequency domain. Thus

$$\begin{aligned} g_m(\mathbf{R}, t) &= \frac{1}{2\pi} \int_{-\infty}^{+\infty} g_m(\mathbf{R}, \omega) e^{-i\omega t} \, d\omega \\ &= \frac{\alpha}{2\pi} \int_{-\infty}^{+\infty} G_m(\mathbf{R}, \omega) e^{-i\omega(t - \mathbf{m}\cdot\mathbf{R}/c)} \, d\omega \\ &= \alpha G_m\left(\mathbf{R}, t - \frac{\mathbf{m} \cdot \mathbf{R}}{c} \right) \end{aligned} \tag{11.45}$$

or

$$G_m(\mathbf{R}, t) = \frac{1}{\alpha} g_m\left(\mathbf{R}, t + \frac{\mathbf{m} \cdot \mathbf{R}}{c} \right) \tag{11.46}$$

In other words, to obtain $G_m(\mathbf{R}, t)$ from $g_m(\mathbf{R}, t)$ we divide Eq. (11.44) by α and replace t in it by $t + \mathbf{m} \cdot \mathbf{R}/c$. Since the scalar Green function differs from zero only for those values of the variables $\mathbf{R}$ and t for which the argument of the delta function vanishes and also in accordance with the causality principle, only real positive values of time t need to be considered [for $t < 0$, $G_m(\mathbf{R}, t) = 0$]. We thus obtain $G_m(\mathbf{R}, t)$ in the time domain:

$$G_m(\mathbf{R}, t) = \frac{1}{4\pi R_v} \left[\tfrac{1}{2}(1 + \operatorname{sgn} t_1)\, \delta(t - t_1) + \tfrac{1}{2}(1 + \operatorname{sgn} t_2)\, \delta(t - t_2)\right] \tag{11.47}$$

where

$$t_1 = -\frac{\mathbf{m} \cdot \mathbf{R} - \alpha\sqrt{\mu\varepsilon}R_v}{c} \qquad t_2 = -\frac{\mathbf{m} \cdot \mathbf{R} + \alpha\sqrt{\mu\varepsilon}R_v}{c} \tag{11.48}$$

are the values of time t at which the argument of the delta function vanishes, and only real positive values of t_1 and t_2 have physical meaning. The sign function in Eq. (11.47) is defined by

$$\operatorname{sgn} \alpha = \frac{\alpha}{|\alpha|} = \begin{cases} +1 & \text{if } \alpha > 0 \\ -1 & \text{if } \alpha < 0 \end{cases} \tag{11.49}$$

The multipliers before the delta functions in Eq. (11.47), of the form $\frac{1}{2}(1 + \operatorname{sgn} t_{1,2})$, automatically "switch off" the δ function when the value of t_1 (or of t_2) becomes negative.

The scalar Green function $G_m(\mathbf{R}, t)$ describes the field at $\mathbf{r}$ due to an instantaneous point source located at $\mathbf{r}'$ and switched on at the instant $t = 0$. It possesses the following symmetry property, which can easily be observed from Eqs. (11.47) and (11.48). If we interchange the field and source points and at the same time reverse the direction of motion of the medium, Green's function remains unchanged:

$$G_m(\mathbf{R}, t, \mathbf{v}) = G_m(-\mathbf{R}, t, -\mathbf{v}) \tag{11.50}$$

In the case of a stationary medium ($\beta = v/c = 0$) or when $\mu\varepsilon = 1$, we have $m = 0$ and $\alpha = 1$. The Green function (11.47) for a moving medium reduces to

$$G_m(\mathbf{R}, t) = \frac{1}{4\pi R}\, \delta\left(t - \frac{\sqrt{\mu\varepsilon}R}{c}\right) \tag{11.51}$$

which is the Green function (9.42) for a stationary medium.

11.3 EXPLICIT FORMS OF SCALAR GREEN'S FUNCTION $G_m(\mathbf{R}, t)$

Depending on how the velocity v of the moving medium compares with the phase velocity $c/\sqrt{\mu\varepsilon}$ of the wave, Green's function (11.47) takes different forms. We shall consider the following three cases:

Case 1. The velocity of the moving medium is less than the phase velocity of a wave in the stationary medium (or $\beta < 1/\sqrt{\mu\varepsilon}$) In this case, both α and m are positive, R_v is real and greater than $(\mathbf{R} \cdot \hat{\mathbf{v}})$. Since $\alpha\sqrt{\mu\varepsilon} - m$ is positive, of the two roots in Eq. (11.48), root t_1 is positive and root t_2 is negative. Therefore, the scalar Green function takes the form

$$G_m(\mathbf{R}, t) = \frac{1}{4\pi R_v}\,\delta(t - t_1) \tag{11.52}$$

where

$$t_1 = -\frac{m(\mathbf{R} \cdot \hat{\mathbf{v}}) - \alpha\sqrt{\mu\varepsilon}R_v}{c} \tag{11.53}$$

Green's function (11.52) differs from zero only on a surface whose equation at any fixed instant is given by Eq. (11.53), or

$$R_v^2 = \frac{[ct + m(\mathbf{R} \cdot \hat{\mathbf{v}})]^2}{\alpha^2\mu\varepsilon} \tag{11.54}$$

This is an equation of an ellipsoid of revolution whose axis of symmetry lies along the direction of motion of the medium. To see this, we rewrite Eq. (11.54) as

$$\frac{(\mathbf{R} \times \hat{\mathbf{v}})^2}{a_0^2} + \frac{(\mathbf{R} \cdot \hat{\mathbf{v}} - z_0)^2}{b_0^2} = 1 \tag{11.55}$$

or, after letting $\hat{\mathbf{v}} = \hat{\mathbf{z}}$ and $\mathbf{R} = \boldsymbol{\rho} + z\hat{\mathbf{z}}$,

$$\frac{\rho^2}{a_0^2} + \frac{(z - z_0)^2}{b_0^2} = 1 \tag{11.56}$$

where the semiaxes of the ellipse are

$$a_0 = \frac{ct\sqrt{1 - \beta^2}}{\sqrt{\mu\varepsilon - \beta^2}} \qquad b_0 = \frac{ct\sqrt{\mu\varepsilon(1 - \beta^2)}}{\mu\varepsilon - \beta^2} \tag{11.57}$$

(when $\mu\varepsilon > 1$, $a_0 > b_0$), and where the center of the ellipse is located along the direction of motion of the medium at a distance z_0 away from the source point:

$$z_0 = \frac{(\mu\varepsilon - 1)vt}{\mu\varepsilon - \beta^2} \tag{11.58}$$

From the above equation we conclude that the surface on which Green's function is different from zero expands in space as its center shifts along the direction of motion of the medium with velocity

$$v_0 = \frac{dz_0}{dt} = \frac{\mu\varepsilon - 1}{\mu\varepsilon - \beta^2}\,v \tag{11.59}$$

This velocity is different from velocity v of the moving medium. At nonrelativistic velocities ($\beta^2 \ll 1$),

$$v_0 = \left(1 - \frac{1}{\mu\varepsilon}\right)v \tag{11.60}$$

That is, the velocity at which the center of the ellipsoid moves is equal to the velocity of the moving medium multiplied by the Fresnel drag coefficient [cf. Eq. (8.62)].

We recall that Green's function may be interpreted as a disturbance due to a source of unit strength located at the source point and switched on at $t = 0$. The surface carrying this disturbance expands in all directions. As is evident from Eqs. (11.56) to (11.58), the velocity of this expansion is different in different directions. If the medium were not moving, the surface would be a sphere (see Fig. 11.2*a*),

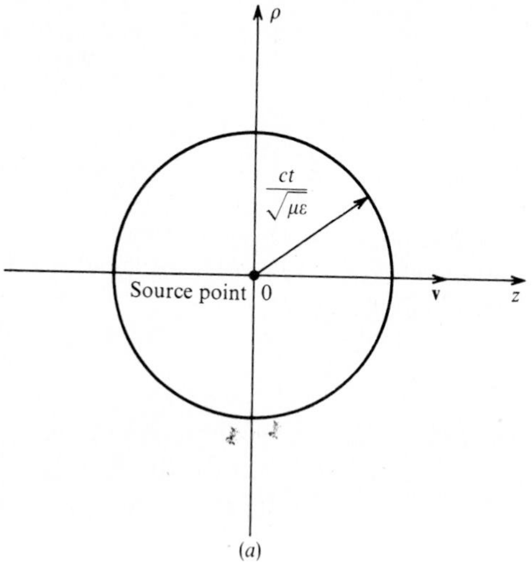

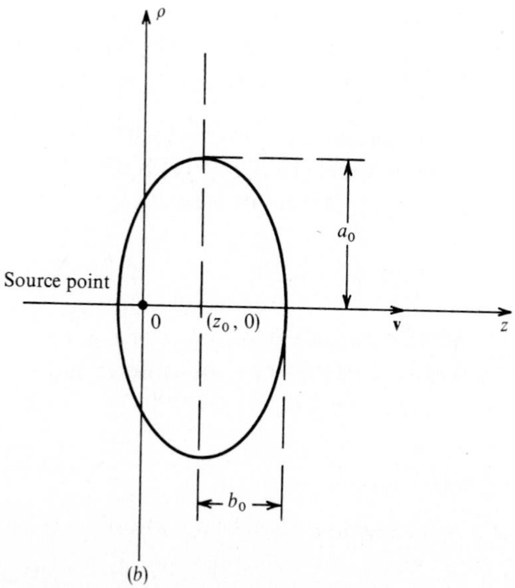

Figure 11.2 Surfaces on which Green's function is different from zero. (*a*) Propagation of a disturbance from a point source in a stationary medium, (*b*) propagation of a disturbance from a point source in a moving medium when $0 < v < c/\sqrt{\mu\varepsilon}$.

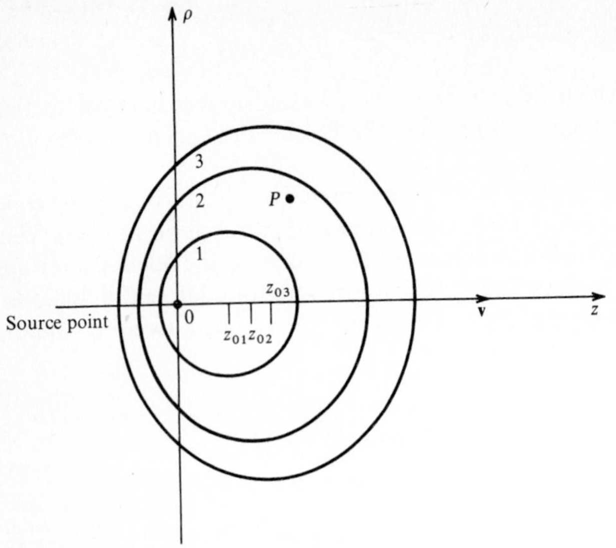

Figure 11.3 Propagation of a disturbance from a point source in a moving medium when $0 < \beta < 1/\sqrt{\mu\varepsilon}$. The numbers 1, 2, and 3 indicate the corresponding surfaces that carry the disturbance at three successive instants. P is the observation point. The source is located at the origin 0 and is switched on at the time $t = 0$.

that is, the motion of the medium contracts this sphere in the direction of motion and thus deforms it into an ellipsoid, as shown in Fig. 11-2*b*. In this case (β) $< 1/\sqrt{\mu\varepsilon}$), $b_0 > z_0$; hence the source always lies inside the expanding surface that carries the disturbance (the source is located at the origin 0 in Fig. 11.2). Eventually, the disturbance will reach an observer located at a point P in space. In other words, before the arrival of a disturbance, an observer is outside the surface, but after receiving the disturbance, one will be inside it and will remain inside thereafter (see Fig. 11.3).

Case 2. The velocity of the moving medium is equal to the phase velocity of a wave in the stationary medium (or $\beta = 1/\sqrt{\mu\varepsilon}$) As β approaches $1/\sqrt{\mu\varepsilon}$, both α and t_2 become infinities, R_v approaches z, and t_1 in Eq. (11.48) reduces to

$$t_1 = \frac{\sqrt{\mu\varepsilon}}{2c}\left[\left(1 + \frac{1}{\mu\varepsilon}\right)z + \frac{\rho^2}{z}\right] \tag{11.61}$$

In this case we see that the sign of t_1 depends on the value of z. For $z < 0$, t_1 is negative. Hence, Green's function vanishes identically. On the other hand, for $z > 0$, t_1 is positive. Green's function (11.47) becomes

$$G_m(\mathbf{R}, t) = \frac{1}{4\pi z}\,\delta(t - t_1) \tag{11.62}$$

Here $G_m(\mathbf{R}, t)$ is different from zero on a surface defined by the equation

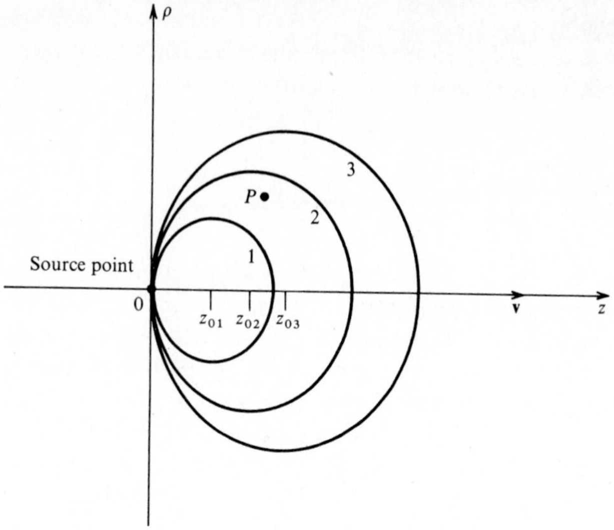

Figure 11.4 Propagation of a disturbance from a point source in a moving medium at three successive instants when $\beta = 1/\sqrt{\mu\varepsilon}$.

$$\frac{\rho^2}{a_0^2} + \frac{(z - z_0)^2}{b_0^2} = 1. \tag{11.63}$$

where

$$\begin{aligned} a_0 &= \frac{ct}{\sqrt{1 + \mu\varepsilon}} \\ b_0 = z_0 &= \frac{ct}{\sqrt{\mu\varepsilon}(1 + 1/\mu\varepsilon)} \end{aligned} \tag{11.64}$$

Again, Eq. (11.63) represents an ellipsoid of revolution about the direction of motion of the medium. Since $b_0 = z_0$, the surface of the ellipsoid expands in such a way that at any instant it is tangent to the plane $z = 0$. In other words, the surface on which Green's function is different from zero is tangent to the plane perpendicular to **v** at the source point as shown in Fig. 11.4.

Case 3. The velocity of the moving medium is greater than the phase velocity of a wave in the stationary medium (or $\beta > 1/\sqrt{\mu\varepsilon}$) In this case both α and m are negative. R_v can either be real or imaginary depending on whether $-\rho^2/\alpha$ is smaller or larger than z^2.

In the region of space when $-\rho^2/\alpha > z^2$, R_v is purely imaginary. The values of t_1 and t_2 in Eq. (11.48) become complex, hence Green's function $G_m(\mathbf{R}, t)$ defined in Eq. (11.47) vanishes identically.

On the other hand, when $-\rho^2/\alpha < z^2$, R_v is real and less than z. Here we have two possibilities depending on whether z is positive or negative. Since

$-m + \alpha\sqrt{\mu\varepsilon} > 0$, in the region $z < 0$ both roots t_1 and t_2 in Eq. (11.48) are negative, and consequently Green's function $G_m(\mathbf{R}, t)$ again vanishes identically. But in the region $z > 0$ both roots t_1 and t_2 are positive, and Green's function (11.47) takes the form

$$G_m(\mathbf{R}, t) = \frac{1}{4\pi R_v}[\delta(t - t_1) + \delta(t - t_2)] \tag{11.65}$$

where t_1 and t_2 are given by Eq. (11.48). As in case 1, Green's function is different from zero on an ellipsoidal surface whose equation at a fixed instant has the form of Eq. (11.54). But the propagation of the disturbance in this case differs significantly from that of case 1. In case 1, the source is always inside the expanding surface that carries the disturbance. In the present case ($\beta > 1/\sqrt{\mu\varepsilon}$), $b_0 < z_0$; hence the source always lies outside the surface on which Green's function is different from zero (see Fig. 11.5). This entire surface is "dragged" by the moving medium to the right of the source point. From Fig. 11.5 we see that the projections of the ellipse on the direction of motion lie in the range

$$z_0 - b_0 < z < z_0 + b_0 \tag{11.66}$$

where both lower and upper bounds of the inequality are positive and increase with time [see Eqs. (11.57) and (11.58)]. As the surface expands and moves in the

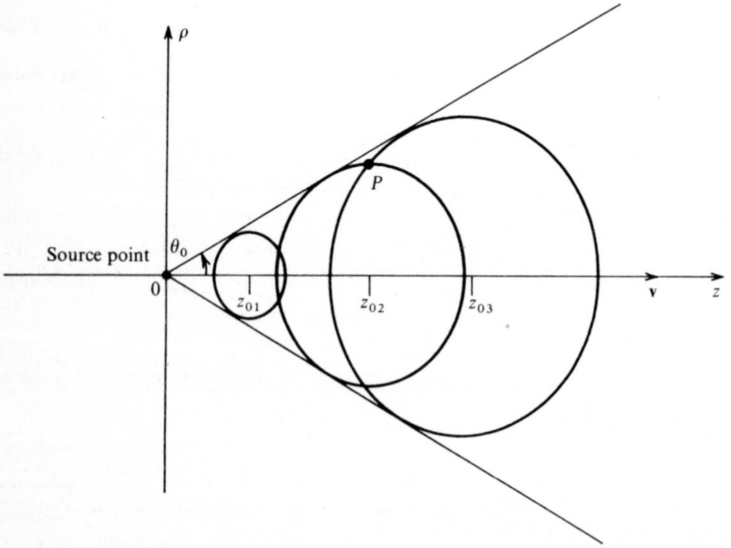

Figure 11.5 Propagation of a disturbance from a point source in a moving medium at three successive instants when $\beta > 1/\sqrt{\mu\varepsilon}$. θ_0 is the half-angle of the cone within which the disturbance is nonzero.

direction of motion of the medium, it stays at all times inside a cone of half-angle θ_0 defined by (see Fig. 11.5)

$$\tan \theta_0 = \sqrt{\frac{1 - \beta^2}{\mu\varepsilon\beta^2 - 1}} \tag{11.67}$$

The generators of this cone are tangent to all the surfaces defined by Eq. (11.55). When $\beta = 1/\sqrt{\mu\varepsilon}$, the cone is actually the entire half-space to the right of the source point. For $\beta > 1/\sqrt{\mu\varepsilon}$, the half-angle of the cone decreases as the velocity of the moving medium increases. Of course, when $\beta < 1/\sqrt{\mu\varepsilon}$, there is no cone; the disturbance propagates away from the source in all directions.

Now we let a source "flash" at the source point. An observer located outside the cone will not detect any disturbance. However, if the observer is inside the conical region defined by

$$\rho < \sqrt{\frac{1 - \beta^2}{\mu\varepsilon\beta^2 - 1}}\, z \tag{11.68}$$

he will detect the disturbance twice. First, at instant t_1 the forward part of the surface passes him and then at instant $t_2 > t_1$ the other part does (see Fig. 11.5). It is evident from Fig. 11.5 that only on the surface of the cone described by

$$\rho = \sqrt{\frac{1 - \beta^2}{\mu\varepsilon\beta^2 - 1}}\, z \tag{11.69}$$

does the disturbance pass the observer once.

11.4 GREEN'S FUNCTION IN THE FREQUENCY DOMAIN

In Sec. 11.2 we derived the scalar Green function $G_m(\mathbf{R}, t)$ in the time domain. To obtain Green's function in the frequency domain or Green's function for a monochromatic field, we take the Fourier transform of Eq. (11.47) with respect to t according to Eq. (9.3), and obtain

$$\begin{aligned} G_m(\mathbf{R}, \omega) &= \int_{-\infty}^{+\infty} G_m(\mathbf{R}, t) e^{i\omega t}\, dt \\ &= \frac{1}{4\pi R_v} \left[\tfrac{1}{2}(1 + \operatorname{sgn} t_1) e^{i\omega t_1} + \tfrac{1}{2}(1 + \operatorname{sgn} t_2) e^{i\omega t_2}\right] \end{aligned} \tag{11.70}$$

where

$$\begin{aligned} t_1 &= -\frac{\mathbf{m} \cdot \mathbf{R} - \alpha\sqrt{\mu\varepsilon} R_v}{c} \\ t_2 &= -\frac{\mathbf{m} \cdot \mathbf{R} + \alpha\sqrt{\mu\varepsilon} R_v}{c} \end{aligned} \tag{11.71}$$

are functions of $\mathbf{R}$ and $R_v = \sqrt{\mathbf{R} \cdot \bar{\boldsymbol{\alpha}}^{-1} \cdot \mathbf{R}}$. We recall that if the values of t_1 or t_2 become complex, the corresponding term in Eq. (11.70) must be set equal to zero.

Again, for various values of the velocity of the moving medium Green's function (11.70) takes different forms. As in Sec. 11.3, we shall consider the following three cases:

Case 1. $\beta < 1/\sqrt{\mu\varepsilon}$ In this case, of the two roots in Eq. (11.71), t_1 is positive and t_2 is negative. Hence Green's function (11.70) becomes

$$G_m(\mathbf{R}, \omega) = \frac{1}{4\pi R_v} \exp\,[ik_0(\alpha\sqrt{\mu\varepsilon}R_v - \mathbf{m}\cdot\mathbf{R})] \tag{11.72}$$

Also, from Eq. (11.28),

$$g_m(\mathbf{R}, \omega) = \frac{\alpha}{4\pi R_v} \exp\,(ik_0\,\alpha\sqrt{\mu\varepsilon}R_v) \tag{11.73}$$

To cast Eq. (11.72) in an alternative form, we refer to Fig. 11.1 and obtain

$$R_v = R\sqrt{1 - \xi\beta^2 \sin^2\theta} \tag{11.74}$$

where $\xi = (\mu\varepsilon - 1)\gamma^2$. Substituting Eq. (11.74) into Eq. (11.72), we get

$$G_m(\mathbf{R}, \omega) = \frac{e^{ik_0 n_{ef}(\theta)R}}{4\pi R\sqrt{1 - \xi\beta^2 \sin^2\theta}} \tag{11.75}$$

where the "effective index of refraction" $n_{ef}(\theta)$ is given by

$$n_{ef}(\theta) = \frac{-\beta\xi\cos\theta + \sqrt{\mu\varepsilon}\sqrt{1 - \xi\beta^2\sin^2\theta}}{1 - \xi\beta^2} \tag{11.76}$$

Since $n_{ef}(\theta)$ is always positive, the scalar Green function $G_m(\mathbf{R}, \omega)$ in the time domain describes a wave that propagates outward in all directions from the source. A comparison of Eq. (11.76) with Eq. (8.57) shows that $n_{ef}(\theta)$ is not equal to the index of refraction $n(\theta)$ of the moving medium in the direction of observation θ. However, in a stationary medium $n_{ef}(\theta)$ coincides with $n(\theta)$ and Eq. (11.76) becomes

$$n_{ef}(\theta) = \sqrt{\mu\varepsilon}$$

as expected.

Case 2. $\beta = 1/\sqrt{\mu\varepsilon}$ In this case, the Fourier transform of Eq. (11.62) gives

$$G_m(\mathbf{R}, \omega) = \frac{1}{4\pi z} e^{i\omega t_1} \tag{11.77}$$

where

$$t_1 = \frac{\sqrt{\mu\varepsilon}}{2c}\left[\left(1 + \frac{1}{\mu\varepsilon}\right)z + \frac{\rho^2}{z}\right] \tag{11.78}$$

Again, referring to Fig. 11.1, we may rewrite Eq. (11.77) as

$$G_m(\mathbf{R}, \omega) = \frac{1}{4\pi R\cos\theta} \exp\left[i\,\frac{k_0\sqrt{\mu\varepsilon}R}{2\cos\theta}\left(1 + \frac{\cos^2\theta}{\mu\varepsilon}\right)\right] \tag{11.79}$$

Equation (11.79) describes a wave that propagates outward from the source. However, the condition for t_1 to be positive (that is, $z > 0$) restricts the radiation in the directions $\theta < \pi/2$.

Case 3. $\beta > 1/\sqrt{\mu\varepsilon}$ In this case, the Fourier transform of Eq. (11.65) gives

$$G_m(\mathbf{R}, \omega) = \frac{1}{4\pi R_v}\,(e^{i\omega t_1} + e^{i\omega t_2}) \tag{11.80}$$

where t_1 and t_2 are positive and are determined by Eq. (11.71). Here we recall that when $\beta > 1/\sqrt{\omega\varepsilon}$, Green's function $G_m(\mathbf{R}, \omega)$ is different from zero only in the region

$$\rho\sqrt{\frac{\mu\varepsilon\beta^2 - 1}{1 - \beta^2}} < z \tag{11.81}$$

Also, from Eq. (11.28),

$$g_m(\mathbf{R}, \omega) = \frac{\alpha}{2\pi R_v}\cos\,(k_0\,\alpha\sqrt{\mu\varepsilon}R_v) \tag{11.82}$$

Now, referring to Fig. 11.1, we may rewrite Eq. (11.80) as

$$G_m(\mathbf{R}, \omega) = \frac{\exp\,[ik_0\,n_{1ef}(\theta)R] + \exp\,[ik_0\,n_{2ef}(\theta)R]}{4\pi R\sqrt{1 - \xi\beta^2\sin^2\theta}} \tag{11.83}$$

where the effective indices of refraction are

$$n_{1ef}(\theta) = \frac{\xi\beta\cos\theta - \sqrt{\mu\varepsilon}\sqrt{1 - \xi\beta^2\sin^2\theta}}{\xi\beta^2 - 1}$$

and

$$n_{2ef}(\theta) = \frac{\xi\beta\cos\theta + \sqrt{\mu\varepsilon}\sqrt{1 - \xi\beta^2\sin^2\theta}}{\xi\beta^2 - 1} \tag{11.84}$$

Substitution of Eq. (11.84) into Eq. (11.83) yields

$$G_m(\mathbf{R}, \omega) = \frac{\exp\left(i\,\dfrac{k_0\,\xi\beta\cos\theta}{\xi\beta^2 - 1}\,R\right)\cos\left(\dfrac{k_0\sqrt{\mu\varepsilon}\sqrt{1 - \xi\beta^2\sin^2\theta}}{\varepsilon\beta^2 - 1}\,R\right)}{2\pi R\sqrt{1 - \xi\beta^2\sin^2\theta}} \tag{11.85}$$

The region in which Green's function is different from zero may also be expressed in terms of θ:

$$\tan^2\theta < \frac{1 - \beta^2}{\mu\varepsilon\beta^2 - 1} \tag{11.86}$$

As is evident from Eq. (11.84), when $\beta > 1/\sqrt{\mu\varepsilon}$, both $n_{1ef}(\theta)$ and $n_{2ef}(\theta)$ are positive. Hence Green's function $G_m(\mathbf{R}, \omega)$ at any observation point represents a superposition of two waves propagating outward from the source.

11.5 DYADIC GREEN'S FUNCTION IN EXPLICIT COORDINATE-FREE FORM

In the previous three sections, we have evaluated and discussed in detail the scalar Green function in both the time and frequency domains. Now, we shall return to the dyadic Green function of a moving medium given by Eq. (11.24) and obtain the result in explicit coordinate-free form. By actually performing the differentiation, we easily establish

$$\nabla(R_v)^n = nR_v^{n-2}(\bar{\boldsymbol{\alpha}}^{-1} \cdot \mathbf{R}) \tag{11.87}$$

and

$$\nabla e^{ibR_v} = \frac{ibe^{ibR_v}}{R_v}(\bar{\boldsymbol{\alpha}}^{-1} \cdot \mathbf{R}) \tag{11.88}$$

where n is an integer and b is a constant, and $R_v = \sqrt{\mathbf{R} \cdot \bar{\boldsymbol{\alpha}}^{-1} \cdot \mathbf{R}}$. From Eqs. (11.87) and (11.88), it follows that

$$\nabla\left(\frac{e^{ibR_v}}{R_v^n}\right) = \left(\frac{ib}{R_v^{n+1}} - \frac{n}{R_v^{n+2}}\right)e^{ibR_v}(\bar{\boldsymbol{\alpha}}^{-1} \cdot \mathbf{R}) \tag{11.89}$$

and

$$\nabla\left[\left(\frac{ib}{R_v} - \frac{1}{R_v^2}\right)\frac{e^{ibR_v}}{R_v}\right] = \frac{(3 - i3bR_v - b^2R_v^2)e^{ibR_v}}{R_v^5}(\bar{\boldsymbol{\alpha}}^{-1} \cdot \mathbf{R}) \tag{11.90}$$

Hence

$$\nabla\nabla\left(\frac{e^{ibR_v}}{R_v}\right) = \frac{(3 - i3bR_v - b^2R_v^2)e^{ibR_v}}{R_v^5}(\bar{\boldsymbol{\alpha}}^{-1} \cdot \mathbf{R})(\bar{\boldsymbol{\alpha}}^{-1} \cdot \mathbf{R}) + \left(\frac{ib}{R_v} - \frac{1}{R_v^2}\right)\frac{e^{ibR_v}}{R_v}\bar{\boldsymbol{\alpha}}^{-1} \tag{11.91}$$

In the case of $\beta < 1/\sqrt{\mu\varepsilon}$, we substitute Eq. (11.73) into Eq. (11.24) and carry out the double gradient operation according to Eq. (11.91). After some simplification, we finally obtain the dyadic Green function of a moving medium in explicit coordinate-free form:

$$\begin{aligned}\bar{\mathbf{G}}(\mathbf{R}) = \alpha G_m(\mathbf{R}, \omega)&\left[\bar{\boldsymbol{\alpha}}^{-1} - \frac{(\bar{\boldsymbol{\alpha}}^{-1} \cdot \mathbf{R})(\bar{\boldsymbol{\alpha}}^{-1} \cdot \mathbf{R})}{R_v^2}\right] \\ &+ i\frac{G_m(\mathbf{R}, \omega)}{k_0\sqrt{\mu\varepsilon}R_v}\left[\bar{\boldsymbol{\alpha}}^{-1} - \frac{3(\bar{\boldsymbol{\alpha}}^{-1} \cdot \mathbf{R})(\bar{\boldsymbol{\alpha}}^{-1} \cdot \mathbf{R})}{R_v^2}\right] \\ &- \frac{G_m(\mathbf{R}, \omega)}{k_0^2\,\alpha\mu\varepsilon R_v^2}\left[\bar{\boldsymbol{\alpha}}^{-1} - \frac{3(\bar{\boldsymbol{\alpha}}^{-1} \cdot \mathbf{R})(\bar{\boldsymbol{\alpha}}^{-1} \cdot \mathbf{R})}{R_v^2}\right]\end{aligned} \tag{11.92}$$

where $G_m(\mathbf{R}, \omega)$ is given by Eq. (11.72). As we can see, Eq. (11.92) contains three terms. The first, which varies as $1/R_v$, is the radiation term since it dominates at distances far from the source. The second, which varies as $1/R_v^2$, dominates at an intermediate distance from the source. At a distance very close to the source, the last term, which varies as $1/R_v^3$, dominates.

Making use of the results in Eqs. (8.47), (11.34), and (11.43), we obtain

$$\bar{\alpha}^{-1} - \frac{(\bar{\alpha}^{-1}\cdot\mathbf{R})(\bar{\alpha}^{-1}\cdot\mathbf{R})}{R_v^2} = \frac{(\mathbf{R}\times\hat{\mathbf{v}})(\mathbf{R}\times\hat{\mathbf{v}})}{\alpha(\mathbf{R}\times\hat{\mathbf{v}})^2} + \frac{[\mathbf{R}\times(\mathbf{R}\times\hat{\mathbf{v}})][\mathbf{R}\times(\mathbf{R}\times\hat{\mathbf{v}})]}{\alpha(\mathbf{R}\times\hat{\mathbf{v}})^2 R_v^2} \tag{11.93}$$

Substitution of the above into the first term in Eq. (11.92) yields the far-zone dyadic Green's function of a moving medium:

$$\bar{\mathbf{G}}_f(\mathbf{R}) = \frac{G_m(\mathbf{R},\omega)}{(\mathbf{R}\times\mathbf{v})^2}\left\{(\mathbf{R}\times\hat{\mathbf{v}})(\mathbf{R}\times\hat{\mathbf{v}}) + \frac{[\mathbf{R}\times(\mathbf{R}\times\hat{\mathbf{v}})][\mathbf{R}\times(\mathbf{R}\times\hat{\mathbf{v}})]}{R_v^2}\right\} \tag{11.94}$$

which in turn determines the far fields of sources radiating in a moving medium. In the limiting case of a stationary medium, Eq. (11.92) reduces to Eq. (9.51) as expected.

For the complementary case where $\beta > 1/\sqrt{\mu\varepsilon}$, the explicit form of the dyadic Green function can be found in much the same way as when $\beta < 1/\sqrt{\mu\varepsilon}$.

11.6 RADIATION FROM AN OSCILLATING ELECTRIC DIPOLE

Let us now consider the field radiated by an oscillating electric dipole of frequency ω for the case $\beta < 1/\sqrt{\mu\varepsilon}$. Let the dipole be located at the origin. If $\mathbf{p}_0$ is its dipole moment, then the current distribution of the dipole is

$$\mathbf{J}(\mathbf{r}') = -i\omega\,\mathbf{p}_0\,\delta(\mathbf{r}') \tag{11.95}$$

Substituting Eq. (11.95) into Eq. (11.12), we obtain the electric field of the dipole radiating in a moving medium:

$$\mathbf{E}(\mathbf{r}) = \omega^2\mu_0\,\mu\,\bar{\mathbf{G}}(\mathbf{r},\mathbf{0})\cdot\mathbf{p}_0 \tag{11.96}$$

where the dyadic Green function is evaluated at $\mathbf{r}' = \mathbf{0}$. Accordingly, with the aid of Eq. (11.92) we may write Eq. (11.96) as

$$\begin{aligned}\mathbf{E}(\mathbf{r}) = {} & \omega^2\mu_0\,\mu\alpha\,G_m(\mathbf{r},\omega)\left[(\bar{\alpha}^{-1}\cdot\mathbf{p}_0) - \frac{\mathbf{p}_0\cdot\bar{\alpha}^{-1}\cdot\mathbf{r}}{r_v^2}(\alpha^{-1}\cdot\mathbf{r})\right] \\ & + i\sqrt{\frac{\mu_0\,\mu}{\varepsilon_0\,\varepsilon}}\,\frac{\omega G_m(\mathbf{r},\omega)}{r_v}\left[(\bar{\alpha}^{-1}\cdot\mathbf{p}_0) - \frac{3(\mathbf{p}_0\cdot\bar{\alpha}^{-1}\cdot\mathbf{r})}{r_v^2}(\bar{\alpha}^{-1}\cdot\mathbf{r})\right] \\ & - \frac{G_m(\mathbf{r},\omega)}{\alpha\varepsilon_0\,\varepsilon r_v^2}\left[(\bar{\alpha}^{-1}\cdot\mathbf{p}_0) - \frac{3(\mathbf{p}_0\cdot\bar{\alpha}^{-1}\cdot\mathbf{r})}{r_v^2}(\bar{\alpha}^{-1}\cdot\mathbf{r})\right]\end{aligned} \tag{11.97}$$

where

$$G_m(\mathbf{r},\omega) = \frac{\exp\,[ik_0(\alpha\sqrt{\mu\varepsilon}r_v - \mathbf{m}\cdot\mathbf{r})]}{4\pi r_v} \tag{11.98}$$

and
$$r_v = \sqrt{\mathbf{r}\cdot\bar{\alpha}^{-1}\cdot\mathbf{r}} \tag{11.99}$$

The corresponding magnetic field can be found from Maxwell equation (11.9).

In the far-zone, the results are considerably simplified. According to Eq. (11.94), the electric field intensity becomes

$$\mathbf{E}(\mathbf{r}) = \frac{\omega^2\mu_0\mu G_m(\mathbf{r},\omega)}{(\mathbf{r}\times\mathbf{v})^2}\left\{(\mathbf{r}\cdot\hat{\mathbf{v}}\times\mathbf{p}_0)(\mathbf{r}\times\hat{\mathbf{v}}) + \frac{(\mathbf{r}\cdot\hat{\mathbf{v}})(\mathbf{r}\cdot\mathbf{p}_0) - r^2(\mathbf{p}_0\cdot\hat{\mathbf{v}})}{r_v^2}[\mathbf{r}\times(\mathbf{r}\times\hat{\mathbf{v}})]\right\} \tag{11.100}$$

The far-zone magnetic field can be determined from Maxwell equation (11.9). Since $\mathbf{p}_0$ is a constant vector, Eq. (11.9) together with Eq. (11.96) gives

$$\mathbf{H}(\mathbf{r}) = -i\omega\bar{\alpha}^{-1}\cdot\{[\nabla\times\bar{\mathbf{G}}_f(\mathbf{r}) + ik_0\mathbf{m}\times\bar{\mathbf{G}}_f(\mathbf{r})]\cdot\mathbf{p}_0\} \tag{11.101}$$

To find the curl of the far-zone dyadic Green function, we need

$$\nabla G_m(\mathbf{r},\omega) = G_m(\mathbf{r},\omega)\left[-ik_0\mathbf{m} + \left(\frac{ik_0\alpha\sqrt{\mu\varepsilon}}{r_v} - \frac{1}{r_v^2}\right)(\bar{\alpha}^{-1}\cdot\mathbf{r})\right] \tag{11.102}$$

and

$$\nabla\left[\frac{G_m(\mathbf{r},\omega)}{r_v^2}\right] = G_m(\mathbf{r},\omega)\left[\frac{-ik_0\mathbf{m}}{r_v^2} + \left(\frac{ik_0\alpha\sqrt{\mu\varepsilon}}{r_v^3} - \frac{3}{r_v^4}\right)(\bar{\alpha}^{-1}\cdot\mathbf{r})\right] \tag{11.103}$$

which can be verified by carrying out the differentiations directly. Hence, making use of identity (10.74) and the results of Eqs. (11.102) and (11.103), we obtain

$$\begin{aligned}\nabla\times\bar{\mathbf{G}}_f(\mathbf{r}) &= \nabla\times\left\{\alpha G_m(\mathbf{r},\omega)\left[\bar{\alpha}^{-1} - \frac{(\bar{\alpha}^{-1}\cdot\mathbf{r})(\bar{\alpha}^{-1}\cdot\mathbf{r})}{r_v^2}\right]\right\}\\ &= -ik_0\mathbf{m}\times\bar{\mathbf{G}}_f(\mathbf{r}) + \frac{ik_0\alpha^2\sqrt{\mu\varepsilon}G_m(\mathbf{r},\omega)}{r_v}(\bar{\alpha}^{-1}\cdot\mathbf{r})\times\bar{\alpha}^{-1}\end{aligned} \tag{11.104}$$

Substitution of Eq. (11.104) into Eq. (11.101) yields the far-zone magnetic field:

$$\mathbf{H}(\mathbf{r}) = \frac{\alpha^2\omega^2\mu_0\mu G_m(\mathbf{r},\omega)}{\eta r_v}\bar{\alpha}^{-1}\cdot\{[(\bar{\alpha}^{-1}\cdot\mathbf{r})\times\bar{\alpha}^{-1}]\cdot\mathbf{p}_0\} \tag{11.105}$$

or, after expanding the products in Eq. (11.105) with $\bar{\alpha}^{-1}$ given by Eq. (8.47),

$$\mathbf{H}(\mathbf{r}) = \frac{\omega^2\mu_0\mu G_m(\mathbf{r},\omega)}{\eta r_v}\left\{\frac{1}{\alpha}(\mathbf{r}\times\mathbf{p}_0) + \left(1-\frac{1}{\alpha}\right)[(\mathbf{r}\cdot\hat{\mathbf{v}})(\hat{\mathbf{v}}\times\mathbf{p}_0) + (\hat{\mathbf{v}}\cdot\mathbf{r}\times\mathbf{p}_0)\hat{\mathbf{v}} + (\mathbf{p}_0\cdot\hat{\mathbf{v}})(\mathbf{r}\times\hat{\mathbf{v}})]\right\} \tag{11.106}$$

where $\eta = \sqrt{\mu_0\mu/\varepsilon_0\varepsilon}$. Equations (11.100) and (11.106) clearly show that as a result of the motion of the medium, the electromagnetic field of an oscillating dipole depends upon the orientation of the dipole with respect to the direction of motion. In the case where $\mathbf{p}_0$ is parallel to $\mathbf{v}$, that is, $\mathbf{p}_0 = p_0\,\hat{\mathbf{v}}$, Eqs. (11.100) and (11.106) reduce to

$$\mathbf{E}(\mathbf{r}) = -\frac{\omega^2 \mu_0\, \mu p_0\, G_m(\mathbf{r}, \omega)}{r_v^2}\,[\mathbf{r} \times (\mathbf{r} \times \hat{\mathbf{v}})] \tag{11.107}$$

and
$$\mathbf{H}(\mathbf{r}) = \frac{\omega^2 \mu_0\, \mu p_0\, G_m(\mathbf{r}, \omega)}{\eta r_v}\,(\mathbf{r} \times \hat{\mathbf{v}}) \tag{11.108}$$

respectively. On the other hand, when $\mathbf{p}_0$ is perpendicular to $\mathbf{v}$, that is, $\mathbf{p}_0 \cdot \hat{\mathbf{v}} = 0$, we find that the far-zone fields become

$$\mathbf{E}(\mathbf{r}) = \frac{\omega^2 \mu_0\, \mu G_m(\mathbf{r}, \omega)}{(\mathbf{r} \times \hat{\mathbf{v}})^2}\left\{(\mathbf{r} \cdot \hat{\mathbf{v}} \times \mathbf{p}_0)(\mathbf{r} \times \hat{\mathbf{v}}) + \frac{(\mathbf{r} \cdot \hat{\mathbf{v}})(\mathbf{p}_0 \cdot \mathbf{r})}{r_v^2}\,[\mathbf{r} \times (\mathbf{r} \times \hat{\mathbf{v}})]\right\} \tag{11.109}$$

and

$$\mathbf{H}(\mathbf{r}) = \frac{\omega^2 \mu_0\, \mu G_m(\mathbf{r}, \omega)}{\eta r_v}\left\{\frac{1}{\alpha}(\mathbf{r} \times \mathbf{p}_0) + \left(1 - \frac{1}{\alpha}\right)\left[(\mathbf{r} \cdot \hat{\mathbf{v}})(\hat{\mathbf{v}} \times \mathbf{p}_0) + (\mathbf{r} \cdot \mathbf{p}_0 \times \hat{\mathbf{v}})\hat{\mathbf{v}}\right]\right\} \tag{11.110}$$

It is interesting to note that the phase contained in $G_m\,(\mathbf{r}, \omega)$ of the field vectors in each case is given by

$$\phi = k_0\,(\alpha\sqrt{\mu\varepsilon}\,r_v - \mathbf{m} \cdot \mathbf{r}) - \omega t \tag{11.111}$$

in the time domain. The surfaces of constant phase defined by $\phi =$ const are given by Eq. (11.54) or Eq. (11.56). They are oblate spheroids whose axes of symmetry lie along the direction of motion of the medium. If the medium were not moving, the surfaces would be spheres. In other words, the motion of the medium contracts these spheres in the direction of motion and thus deforms them into oblate spheroids. As time increases, a spheroidal surface of constant phase expands and is dragged by the moving medium along the direction of motion.

PROBLEMS

11.1 Verify Eqs. (11.11) and (11.13).

11.2 Derive the complex wave equation for **H** of a moving isotropic medium.

11.3 Verify Eq. (11.15).

11.4 Show that the determinant and adjoint of the wave matrix (11.18) are given by Eqs. (11.20) and (11.21) respectively.

11.5 Verify Eq. (11.24).

11.6 Verify Eq. (11.27).

11.7 Verify Eq. (11.32).

11.8 Verify Eq. (11.33).

11.9 Verify Eq. (11.42).

11.10 An oscillating electric dipole is located at the origin of a cartesian coordinate system. If a medium is moving along the z direction, show that the far-zone electric and magnetic fields expressed in spherical coordinates are given by

$$\mathbf{E}(\mathbf{r}) = \omega^2 \mu_0 \mu G_m(\mathbf{r}, \omega)\left(p_\phi \hat{\boldsymbol{\phi}} + \frac{\alpha p_\theta}{\sin^2\theta + \alpha\cos^2\theta}\hat{\boldsymbol{\theta}}\right)$$

and

$$\mathbf{H}(\mathbf{r}) = \frac{\omega^2 \mu_0 \mu \sqrt{\alpha} G_m(\mathbf{r}, \omega)}{\eta\sqrt{\sin^2\theta + \alpha\cos^2\theta}}(p_\theta \hat{\boldsymbol{\phi}} - p_\phi \hat{\boldsymbol{\theta}})$$

where

$$G_m(\mathbf{r}, \omega) = \frac{\sqrt{\alpha}\exp[ik_0 r(\sqrt{\alpha\mu\varepsilon}\sqrt{\sin^2\theta + \alpha\cos^2\theta} - m\cos\theta)]}{4\pi r\sqrt{\sin^2\theta + \alpha\cos^2\theta}}$$

In the special case when the dipole is parallel to the direction of motion, the fields become

$$\mathbf{E}(\mathbf{r}) = -\frac{\omega^2 \mu_0 \mu \alpha p_0 \sin\theta}{\sin^2\theta + \alpha\cos^2\theta} G_m(\mathbf{r}, \omega)\,\hat{\boldsymbol{\theta}}$$

and

$$\mathbf{H}(\mathbf{r}) = -\frac{\omega^2 \mu_0 \mu \sqrt{\alpha} p_0 \sin\theta}{\eta\sqrt{\sin^2\theta + \alpha\cos^2\theta}} G_m(\mathbf{r}, \omega)\hat{\boldsymbol{\phi}}$$

On the other hand, when the dipole is perpendicular to the direction of motion (for example, $\mathbf{p}_0 = p_0\hat{\mathbf{x}}$), the fields reduce to

$$\mathbf{E}(\mathbf{r}) = -\omega^2 \mu_0 \mu p_0 G_m(\mathbf{r}, \omega)\left(\sin\phi\hat{\boldsymbol{\phi}} - \frac{\alpha\cos\theta\cos\phi}{\sin^2\theta + \alpha\cos^2\theta}\hat{\boldsymbol{\theta}}\right)$$

and

$$\mathbf{H}(\mathbf{r}) = \frac{\omega^2 \mu_0 \mu \sqrt{\alpha} p_0 G_m(\mathbf{r}, \omega)}{\eta\sqrt{\sin^2\theta + \alpha\cos^2\theta}}(\cos\theta\cos\phi\hat{\boldsymbol{\phi}} + \sin\phi\hat{\boldsymbol{\theta}})$$

From the above expressions we see that the time-averaged Poynting vector is purely real and purely radial.

11.11 Define the two four-dimensional antisymmetric tensors:

$$[G_{\alpha\beta}] = \begin{bmatrix} 0 & \mathscr{H}_z & -\mathscr{H}_y & -ic\mathscr{D}_x \\ -\mathscr{H}_z & 0 & \mathscr{H}_x & -ic\mathscr{D}_y \\ \mathscr{H}_y & -\mathscr{H}_x & 0 & -ic\mathscr{D}_z \\ ic\mathscr{D}_x & ic\mathscr{D}_y & ic\mathscr{D}_z & 0 \end{bmatrix} = \begin{bmatrix} -\mathscr{H}\times\bar{\mathbf{I}} & -ic\mathscr{D} \\ ic\mathscr{D} & 0 \end{bmatrix}$$

and

$$[F_{\alpha\beta}] = \begin{bmatrix} 0 & c\mathscr{B}_z & -c\mathscr{B}_y & -i\mathscr{E}_x \\ -c\mathscr{B}_z & 0 & c\mathscr{B}_x & -i\mathscr{E}_y \\ c\mathscr{B}_y & -c\mathscr{B}_x & 0 & -i\mathscr{E}_z \\ i\mathscr{E}_x & i\mathscr{E}_y & i\mathscr{E}_z & 0 \end{bmatrix} = \begin{bmatrix} -c\mathscr{B}\times\bar{\mathbf{I}} & -i\mathscr{E} \\ i\mathscr{E} & 0 \end{bmatrix}$$

(*a*) Show that the two Maxwell equations

$$\nabla \times \mathscr{H} - \frac{\partial \mathscr{D}}{\partial t} = \mathscr{J}$$

$$\nabla \cdot \mathscr{D} = \rho$$

can be expressed as

$$\frac{\partial G_{\alpha\beta}}{\partial x_\beta} = s_\alpha \qquad (\alpha = 1, 2, 3, 4)$$

where summation over the repeated Greek indices from 1 to 4 is implied. $[x_\alpha] = (\mathbf{r}, ict)$ is the 4-position vector, and $[s_\alpha] = (\mathscr{J}, ic\rho)$ is the 4-current vector (cf. Prob. 2.20).

(*b*) Show that the other two Maxwell equations

$$\nabla \times \mathscr{E} = -\frac{\partial \mathscr{B}}{\partial t}$$

$$\nabla \cdot \mathscr{B} = 0$$

can be written as

$$\frac{\partial F_{\alpha\beta}}{\partial x_\gamma} + \frac{\partial F_{\beta\gamma}}{\partial x_\alpha} + \frac{\partial F_{\gamma\alpha}}{\partial x_\beta} = 0 \qquad (\alpha,\beta,\gamma = 1, 2, 3, 4)$$

(*c*) Show that the relations

$$\mathscr{E} = -\nabla\phi - \frac{\partial \mathscr{A}}{\partial t}$$

$$\mathscr{B} = \nabla \times \mathscr{A}$$

can be written in the form

$$F_{\alpha\beta} = c\left(\frac{\partial \Phi_\beta}{\partial x_\alpha} - \frac{\partial \Phi_\alpha}{\partial x_\beta}\right)$$

where

$$[\Phi_\alpha] = \left(\mathscr{A}, \frac{i}{c}\phi\right)$$

is the 4-potential vector.

11.12 Continuing Prob. 11.11, (*a*) show that the constitutive relation

$$\mathscr{D} + \frac{1}{c^2}(\mathbf{v} \times \mathscr{H}) = \varepsilon_0 \varepsilon (\mathscr{E} + \mathbf{v} \times \mathscr{B})$$

of a moving medium can be written in four-dimensional form as

$$\frac{1}{c} G_{\alpha\beta} u_\beta = \varepsilon_0 \varepsilon F_{\alpha\beta} u_\beta \qquad (\alpha = 1, 2, 3, 4)$$

where $[u_\alpha] = (\gamma\mathbf{v}, ic\gamma)$ is the 4-velocity vector, and

$$\mathscr{B} - \frac{1}{c^2}(\mathbf{v} \times \mathscr{E}) = \mu_0 \mu(\mathscr{H} - \mathbf{v} \times \mathscr{D})$$

as

$$\frac{1}{c}(F_{\alpha\beta} u_\nu + F_{\beta\nu} u_\alpha + F_{\nu\alpha} u_\beta) = \mu_0 \mu(G_{\alpha\beta} u_\nu + G_{\beta\nu} u_\alpha + G_{\nu\alpha} u_\beta)$$

(*b*) Show that the two four-dimensional forms of the constitutive relations given in (*a*) can be combined to yield

$$G_{\alpha\beta} = \frac{1}{\mu_0 \mu c}\left[F_{\alpha\beta} + \frac{\mu\varepsilon - 1}{c^2}(u_\alpha F_{\beta\nu} u_\nu - u_\beta F_{\alpha\nu} u_\nu)\right]$$

(*c*) Show that the relations

$$\mathscr{J} = \sigma\gamma(\mathscr{E} + \mathbf{v} \times \mathscr{B})$$

and

$$\rho = \frac{\sigma\gamma}{c^2}(\mathbf{v} \cdot \mathscr{E})$$

may be expressed in four-dimensional form as

$$s_\alpha = \frac{\sigma}{c} F_{\alpha\beta} u_\beta \qquad (\alpha = 1, 2, 3, 4)$$

11.13 (*a*) Starting with the four-dimensional forms of the Maxwell equations and the constitutive relation of Probs. 11.11 and 11.12, show that the 4-potential satisfies the wave equation:

$$\left[\frac{\partial^2}{\partial x_\beta^2} - \frac{\mu\varepsilon - 1}{c^2}\left(u_\beta \frac{\partial}{\partial x_\beta}\right)^2\right]\Phi_\alpha = -\mu_0 \mu s_\alpha - \frac{\mu\varepsilon - 1}{\varepsilon_0 \varepsilon c^4} u_\nu s_\nu s_\alpha$$

provided that the following generalized Lorentz condition is valid:

$$\frac{\partial \Phi_\beta}{\partial x_\beta} - \frac{\mu\varepsilon - 1}{c^2} u_\beta u_\nu \frac{\partial \Phi_\nu}{\partial x_\beta} = 0$$

(*b*) Show that the results of part (*a*) lead to the following three-dimensional forms for the vector and scalar potentials of an isotropic medium moving at a velocity **v** with respect to the source (cf. Prob. 2.21):

$$\mathscr{L}\mathscr{A} = -\mu_0 \mu \mathscr{J} - \frac{(\mu\varepsilon - 1)\gamma^2}{\varepsilon_0 \varepsilon c^4}(\mathbf{v} \cdot \mathscr{J} - c^2\rho)\mathbf{v}$$

$$\mathscr{L}\phi = -\mu_0 \mu c^2 \rho - \frac{(\mu\varepsilon - 1)\gamma^2}{\varepsilon_0 \varepsilon c^2}(\mathbf{v} \cdot \mathscr{J} - c^2\rho)$$

where

$$\mathscr{L} \equiv \nabla^2 - \frac{1}{c^2}\frac{\partial^2}{\partial t^2} - \frac{(\mu\varepsilon - 1)\gamma^2}{c^2}\left(\frac{\partial}{\partial t} + \mathbf{v} \cdot \nabla\right)^2$$

Here the scalar and vector potentials are related by

$$\nabla \cdot \mathscr{A} + \frac{1}{c^2}\frac{\partial \phi}{\partial t} - \frac{(\mu\varepsilon - 1)\gamma^2}{c^2}\left(\frac{\partial}{\partial t} + \mathbf{v} \cdot \nabla\right)(\mathbf{v} \cdot \mathscr{A} - \phi) = 0$$

11.14 (*a*) Using the results of the previous problem, show that in a source-free region, the wave vector of a monochromatic plane wave in a moving medium satisfies the following dispersion equation [cf. Eq. (8.60)]:

$$\mathbf{k}^2 - \frac{\omega^2}{c^2} - \frac{(\mu\varepsilon - 1)\gamma^2}{c^2}(\mathbf{k} \cdot \mathbf{v} - \omega)^2 = 0$$

(*b*) Show also that the amplitude vector $\mathbf{A}_0$ of the vector potential is not perpendicular to **k** but to the vector

$$\mathbf{k} + \frac{\mu\varepsilon - 1}{c^2}(\omega - \mathbf{k} \cdot \mathbf{v})\gamma^2 \mathbf{v}$$

In the special cases of a stationary medium ($\mathbf{v} = \mathbf{0}$) and of free space ($\mu = 1, \varepsilon = 1$), it reduces to the well-known result, $\mathbf{A}_0 \perp \mathbf{k}$.

11.15 For a monochromatic source immersed in a moving simple medium, show that the vector and scalar potentials of Prob. 11.13 can be written as

$$\mathbf{A}(\mathbf{r}) = \bar{\mathbf{M}} \cdot \int_V \mathbf{J}(\mathbf{r}')G_m(\mathbf{R})\, d^3r'$$

and
$$\phi(\mathbf{r}) = \mathbf{K} \cdot \int_V \mathbf{J}(\mathbf{r}')G_m(\mathbf{R})\, d^3r'$$

respectively, where

$$\bar{\mathbf{M}} = \mu_0 \mu \bar{\mathbf{I}} + \frac{(\mu\varepsilon - 1)\gamma^2}{\varepsilon_0 \varepsilon c^4}\left(\mathbf{vv} + \frac{ic^2}{\omega}\mathbf{v}\nabla\right)$$

$$\mathbf{K} = \frac{\mu_0 \mu c^2}{i\omega}\left[1 - \frac{(\mu\varepsilon - 1)\gamma^2}{\mu\varepsilon c}\right]\nabla + \frac{(\mu\varepsilon - 1)\gamma^2}{\varepsilon_0 \varepsilon c^3}\mathbf{v}$$

and the scalar Green function $G_m(\mathbf{R})$ is given by Eq. (11.29). With a knowledge of these potentials, the electric and magnetic fields can thus be calculated according to

$$\mathbf{E}(\mathbf{r}) = i\omega\mathbf{A}(\mathbf{r}) - \nabla\phi(\mathbf{r})$$

$$\mathbf{B}(\mathbf{r}) = \nabla \times \mathbf{A}(\mathbf{r})$$

11.16 In the far-zone, the scalar Green function (11.25) can be evaluated asymptotically by the method described in Prob. 10.15. Show that

(*a*) the points of stationary phase occur at $\mathbf{p}^s = \pm k_0 \alpha\sqrt{\mu\varepsilon}(\bar{\alpha}^{-1} \cdot \mathbf{R})/R_v = \pm k_0 \alpha \sqrt{\mu\varepsilon}\, \nabla R_v$, where $R_v = (\mathbf{R} \cdot \bar{\alpha}^{-1} \cdot \mathbf{R})^{1/2}$;

(*b*) $|\kappa^s|^{1/2} = R_v^2/k_0\sqrt{\mu\varepsilon}\, R^2$;

(*c*) $|\partial D/\partial \mathbf{p}|^s = 2k_0\sqrt{\mu\varepsilon}\, R/R_v$;

(*d*) $\mathbf{p}^s \cdot \mathbf{R} = k_0 \alpha \sqrt{\mu\varepsilon}\, R_v$;

(*e*) for $\alpha > 0$, the asymptotic form is

$$g_m(\mathbf{R}, \omega) \sim \frac{\alpha \exp(ik_0 \alpha \sqrt{\mu\varepsilon}\, R_v)}{4\pi R_v}$$

which happens to be the same as the exact expression (11.73) and for $\alpha < 0$, it also coincides with Eq. (11.82).

REFERENCES

Besieris, I. M.: "Analysis of Electromagnetic Radiation in the Presence of a Uniformly Moving, Uniaxially Anisotropic Medium," *J. Math. Phys.*, vol. 10, no. 7, 1969, pp. 1156–1167.

Bolotovskii, B. M., and S. N. Stolyarov: "Current Status of the Electrodynamics of Moving Media," *Sov. Phys., USP.*, vol. 17, no. 6, 1975, pp. 875–895.

Compton, R. T., Jr.: "The Time-Dependent Green's Function for Electromagnetic Waves in Moving Simple Media," *J. Math. Phys.*, vol. 7, no. 12, 1966, pp. 2145–2152.

Compton, R. T. Jr., and C. T. Tai: "Radiation from Harmonic Sources in a Uniformly Moving Medium," *IEEE Trans. Antennas Propagation*, AP–13, no. 4, 1965, pp. 574–577.

Daly, P., K. S. H. Lee, and C. H. Papas: "Radiation Resistance of an Oscillating Dipole in a Moving Medium," *IEEE Trans. Antennas Propagation*, vol. AP-13, no. 4, 1965, pp. 583–587.

Lee, K. S. H., and C. H. Papas: "Electromagnetic Radiation in the Presence of Moving Simple Media," *J. Math. Phys.*, vol. 5, no. 12, 1964, pp. 1668–1672.

Lee, K. S. H., and C. H. Papas: "Antenna Radiation in a Moving Dispersive Medium," *IEEE Trans. Antennas Propagation*, vol. 13, no. 5, 1965, pp. 799–804.

McKenzie, J. F.: "Dipole Radiation in Moving Media," *Proc. Phys. Soc.*, vol. 91, 1967, pp. 537–551.

Sen Gupta, N. D.: "Electrodynamics of Moving Media and the Cerenkov Radiation," *J. Phys. A* (*Proc. Phys. Soc.*), ser. 2., vol. 1, 1968, pp. 340–349.

Tai, C. T.: "The Dyadic Green's Function for a Moving Isotropic Medium," *IEEE Trans. Antennas Propagation*, vol. AP-13, no. 2, 1965, pp. 322–323.

APPENDIX

A

CARTESIAN TENSORS

In vector analysis we define a vector as a quantity which possesses both a magnitude and a direction. Geometrically, it is represented by a directed line segment. With respect to a given rectangular cartesian coordinate system, a vector $\mathbf{u}$ may be represented by a column matrix with components u_1, u_2, and u_3. If a new cartesian coordinate system is introduced, the same vector $\mathbf{u}$ may be represented by a different set of components, u'_1, u'_2, and u'_3. Since both sets represent the same vector $\mathbf{u}$, they must be related in a definite manner. In this appendix, we will show how vectors and tensors may be defined in terms of the manner in which their components transform under a coordinate transformation.

Consider two rectangular cartesian coordinate systems, the unprimed and the primed, having the same origin (see Fig. A.1). Denote the mutually orthogonal unit vectors with respect to the unprimed system by $\hat{\mathbf{e}}_1$, $\hat{\mathbf{e}}_2$, $\hat{\mathbf{e}}_3$ and those with respect to the primed system by $\hat{\mathbf{e}}'_1$, $\hat{\mathbf{e}}'_2$, and $\hat{\mathbf{e}}'_3$. Then

$$\hat{\mathbf{e}}_i \cdot \hat{\mathbf{e}}_j = \delta_{ij} \tag{A.1}$$

and

$$\hat{\mathbf{e}}'_i \cdot \hat{\mathbf{e}}'_j = \delta_{ij} \tag{A.2}$$

The radius vector from the origin to a given point in space may either be expressed as

$$\mathbf{r} = x_j \hat{\mathbf{e}}_j \tag{A.3}$$

or

$$\mathbf{r}' = x'_j \hat{\mathbf{e}}'_j \tag{A.4}$$

where x_1, x_2, x_3 are the components of the radius vector referring to the unprimed coordinate system and x'_1, x'_2, x'_3 are the components of the same vector referring to the primed coordinate system. In Eqs. (A.3) and (A.4), summation over the repeated indices is implied. Since $\mathbf{r}'$ in Eq. (A.4) denotes the same radius

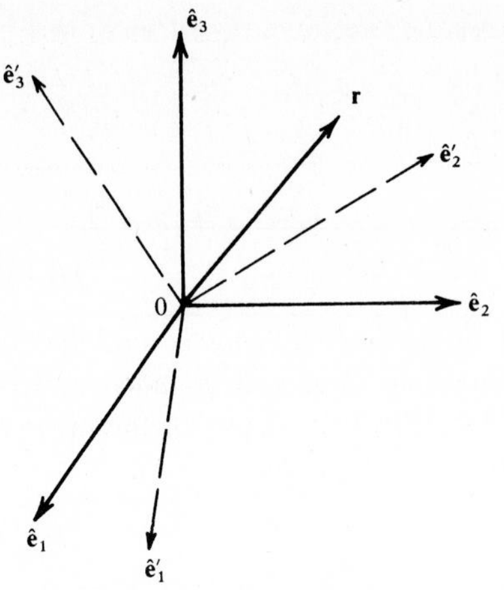

Figure A.1 The primed and unprimed rectangular cartesian coordinate systems.

vector **r**, only the components have been changed. Thus

$$x'_j \hat{\mathbf{e}}'_j = x_j \hat{\mathbf{e}}_j \tag{A.5}$$

To find how the components of the radius vector transform when the coordinate system is changed, we dot-multiply both sides of Eq. (A.5) by $\hat{\mathbf{e}}'_i$ and use Eq. (A.2) to give

$$x'_i = x_j \hat{\mathbf{e}}'_i \cdot \hat{\mathbf{e}}_j \tag{A.6}$$

Now, we define the direction cosines

$$\hat{\mathbf{e}}'_i \cdot \hat{\mathbf{e}}_j = s_{ij} \tag{A.7}$$

Transformation (A.6) becomes

$$x'_i = s_{ij} x_j \tag{A.8}$$

or, in direct form,

$$\mathbf{r}' = \bar{\mathbf{S}} \cdot \mathbf{r} \tag{A.9}$$

Equation (A.8) or Eq. (A.9) is the transformation law relating components of the same radius vector with respect to different rectangular cartesian coordinate systems.

We shall next examine some properties of the transformation matrix $\bar{\mathbf{S}}$. Since the radius vector remains unchanged no matter which coordinate system is used to represent it, the magnitude of the vector must be the same in both systems. Thus

$$\mathbf{r}'^2 = \mathbf{r}^2 \tag{A.10}$$

Substituting Eq. (A.9) into (A.10), we obtain

$$\mathbf{r}'^2 = \mathbf{r}' \cdot \mathbf{r}' = \mathbf{r} \cdot \tilde{\bar{\mathbf{S}}} \cdot \bar{\mathbf{S}} \cdot \mathbf{r} = \mathbf{r}^2 = \mathbf{r} \cdot \bar{\mathbf{I}} \cdot \mathbf{r} \tag{A.11}$$

Here, to insure that the product in $\mathbf{r}'^2$ is permissible, we have rewritten $\mathbf{r}'$ in Eq. (A.9) as

$$\mathbf{r}' = \bar{\mathbf{S}} \cdot \mathbf{r} = \mathbf{r} \cdot \tilde{\bar{\mathbf{S}}}$$

But since $\mathbf{r}$ is arbitrary, Eq. (A.11) implies that

$$\tilde{\bar{\mathbf{S}}} \cdot \bar{\mathbf{S}} = \bar{\mathbf{I}} \tag{A.12}$$

A matrix which satisfies Eq. (A.12) is called an *orthogonal matrix* and linear transformation (A.9), characterized by the orthogonal matrix $\bar{\mathbf{S}}$, is called an *orthogonal transformation.* A comparison of Eq. (A.12) with Eq. (1.69) for the inverse matrix shows that an orthogonal matrix may also be defined as

$$\tilde{\bar{\mathbf{S}}} = \bar{\mathbf{S}}^{-1} \tag{A.13}$$

That is, a matrix is orthogonal if its transpose is its inverse.

Taking the determinant on both sides of Eq. (A.12), noting that $|\tilde{\bar{\mathbf{S}}} \cdot \bar{\mathbf{S}}| = |\tilde{\bar{\mathbf{S}}}||\bar{\mathbf{S}}|$ and $|\tilde{\bar{\mathbf{S}}}| = |\bar{\mathbf{S}}|$, we obtain

$$|\bar{\mathbf{S}}|^2 = 1$$

and

$$|\bar{\mathbf{S}}| = \pm 1 \tag{A.14}$$

An orthogonal transformation represents a pure rotation when the determinant of the transformation matrix is $+1$ and a rotation with a reflection of the coordinate system when $|\bar{\mathbf{S}}| = -1$.

As an example, let us consider the following two transformation matrices:

$$\bar{\mathbf{R}} = \begin{bmatrix} \cos\theta & \sin\theta & 0 \\ -\sin\theta & \cos\theta & 0 \\ 0 & 0 & 1 \end{bmatrix} \qquad \bar{\mathbf{L}} = \begin{bmatrix} \cos\theta & \sin\theta & 0 \\ -\sin\theta & \cos\theta & 0 \\ 0 & 0 & -1 \end{bmatrix}$$

which are easily shown to be orthogonal. The determinant of $\bar{\mathbf{R}}$ is $+1$ and the determinant of $\bar{\mathbf{L}}$ is -1. Orthogonal transformation (A.9) with matrix $\bar{\mathbf{R}}$ represents a rotation of the coordinate system about vector $\hat{\mathbf{e}}_3$ through an angle θ in the positive direction. Similarly, the transformation characterized by $\bar{\mathbf{L}}$ denotes the same rotation, followed by a reflection in the plane which is perpendicular to the vector $\hat{\mathbf{e}}_3$. If $\theta = 0$ in the latter case, the transformation is a simple reflection; if $\theta = \pi$, it is an inversion.

Based on the transformation law for the radius vector, we may now proceed to define vectors, scalars and tensors as follows.

A vector (or a tensor of rank 1) is a quantity consisting of three components u_1, u_2, and u_3 which transform under a change of coordinate system according to the law

$$u_i' = s_{ij} u_j \tag{A.15}$$

or, in direct form

$$\mathbf{u}' = \bar{\mathbf{S}} \cdot \mathbf{u} \tag{A.16}$$

where $\mathbf{u}' = [u_i']$ and $\mathbf{u} = [u_i]$ denote the same vector represented in the primed and unprimed coordinate systems respectively. This definition shows that a vector transforms as does the radius vector. If we know the components of a vector in one (cartesian) coordinate system, we can use Eq. (A.15) to determine its components in any other coordinate system such that the new components always determine the same vector.

Example A.1 If $\phi = \phi(\mathbf{r})$, the gradient of ϕ with components $\partial\phi/\partial x_i$ is a vector. Since by the chain rule for partial differentiation

$$\frac{\partial \phi}{\partial x_i'} = \frac{\partial \phi}{\partial x_j}\frac{\partial x_j}{\partial x_i'} = \frac{\partial \phi}{\partial x_j}\frac{\partial}{\partial x_i'}(s_{kj}x_k') = s_{ij}\frac{\partial \phi}{\partial x_j}$$

the components of $\nabla\phi$ indeed transform according to Eq. (A.15), and in this sense form a vector.

A scalar (or a tensor of rank 0) is a quantity which remains invariant under any coordinate transformation, i.e., which does not change in value when the coordinate system is changed. Thus if ϕ is the value of a scalar in the unprimed system and ϕ' its value in the primed system, then $\phi' = \phi$.

Example A.2 The dot product of two vectors is a scalar. For

$$\mathbf{u}' \cdot \mathbf{v}' = \mathbf{u} \cdot \tilde{\bar{\mathbf{S}}} \cdot \bar{\mathbf{S}} \cdot \mathbf{v} = \mathbf{u} \cdot \bar{\mathbf{I}} \cdot \mathbf{v} = \mathbf{u} \cdot \mathbf{v}$$

as was asserted.

Finally, a tensor of rank 2 is a quantity consisting of 3^2 (nine) components a_{ij} $(i, j = 1, 2, 3)$, which transform under a change of coordinate system according to the law

$$a_{ij}' = s_{im}\, s_{jn}\, a_{mn} \tag{A.17}$$

or, in direct form,

$$\bar{\mathbf{A}}' = \bar{\mathbf{S}} \cdot \bar{\mathbf{A}} \cdot \tilde{\bar{\mathbf{S}}} \tag{A.18}$$

where $\bar{\mathbf{A}}' = [a_{ij}']$ and $\bar{\mathbf{A}} = [a_{ij}]$ represent the same tensor with components referring to the primed and unprimed coordinate systems respectively. In other words, a tensor of rank 2 is a 3×3 matrix whose nine components transform as do the products of the components of two vectors under an orthogonal transformation of coordinates.

In summary, we note the distinction between a tensor and a matrix. The concept of a matrix is essentially mathematical; matrices are arrays of numbers which obey certain rules of addition, subtraction, multiplication, and equality. On the other hand, the concept of a tensor is geometrical or physical; a tensor is independent of coordinate systems. It may be represented in a cartesian coordinate system by a matrix (cartesian tensor), but for a matrix to be a tensor, it must be transformed according to a definite law when the coordinate system is changed. In other words, all tensors can be represented by matrices; but all matrices do not necessarily represent tensors.

Example A.3 The unit matrix defined by $\bar{\mathbf{I}} = [\delta_{ij}]$ is a tensor of rank two (called a unit tensor). For

$$\bar{\mathbf{I}}' = \bar{\mathbf{I}} = \bar{\mathbf{S}} \cdot \tilde{\bar{\mathbf{S}}} = \bar{\mathbf{S}} \cdot \bar{\mathbf{I}} \cdot \tilde{\bar{\mathbf{S}}}$$

$\bar{\mathbf{I}}$ transforms according to Eq. (A.18). Hence it is a tensor of rank 2.

Example A.4 In a conducting crystal, the electric current density **J** is related to the electric field intensity **E** by Ohm's law:

$$\mathbf{J} = \bar{\boldsymbol{\sigma}} \cdot \mathbf{E}$$

Since **J** and **E** are vectors, they transform according to Eq. (A.16). Hence

$$\mathbf{J} = \bar{\mathbf{S}}^{-1} \cdot \mathbf{J}' = \tilde{\bar{\mathbf{S}}} \cdot \mathbf{J}'$$

$$\mathbf{E} = \bar{\mathbf{S}}^{-1} \cdot \mathbf{E}' = \tilde{\bar{\mathbf{S}}} \cdot \mathbf{E}'$$

Here we have used the result of Eq. (A.13). Substitution of **J** and **E** into Ohm's law yields

$$\bar{\mathbf{S}}^{-1} \cdot \mathbf{J}' = \bar{\boldsymbol{\sigma}} \cdot \tilde{\bar{\mathbf{S}}} \cdot \mathbf{E}'$$

Dot-multiplying both sides from the left by $\bar{\mathbf{S}}$, we obtain

$$\mathbf{J}' = \bar{\boldsymbol{\sigma}}' \cdot \mathbf{E}' = \bar{\mathbf{S}} \cdot \bar{\boldsymbol{\sigma}} \cdot \tilde{\bar{\mathbf{S}}} \cdot \mathbf{E}'$$

Since vector $\mathbf{E}'$ is arbitrary,

$$\bar{\boldsymbol{\sigma}}' = \bar{\mathbf{S}} \cdot \bar{\boldsymbol{\sigma}} \cdot \tilde{\bar{\mathbf{S}}}$$

That is, $\bar{\boldsymbol{\sigma}}$ transforms according to Eq. (A.18); hence it is a tensor of rank 2 and is called the conductivity tensor.

REFERENCES

Aris, R.: *Vectors, Tensors, and the Basic Equations of Fluid Mechanics*, Prentice-Hall, Inc., Englewood Cliffs, N. J., 1962.

Borisenko, A. I., and I. E. Tarapov: *Vector and Tensor Analysis with Applications*, Prentice-Hall, Inc., Englewood Cliffs, N. J., 1968.

Bourne, D. E., and P. C. Kendall: *Vector Analysis and Cartesian Tensors*, 2d ed., Academic Press, New York, 1977.

Chadwick, P.: *Continuum Mechanics*, Halsted Press, John Wiley & Sons, New York, 1976, chap. 1.

Goodbody, A. M.: *Cartesian Tensors: With Applications to Mechanics, Fluid Mechanics and Elasticity*, Halsted Press, John Wiley & Sons, New York, 1982.

Jeffreys, H.: *Cartesian Tensors*, Cambridge University Press, New York, 1961.

Nye, J. F.: *Physical Properties of Crystals*, Oxford University Press, Oxford, England, 1957.

APPENDIX

B

THE DELTA FUNCTION AND ITS PROPERTIES

The concept of a physical point in space or an instant in time is a very useful one. The mathematical representation of such a concept is conveniently accomplished by the delta function. In this appendix we shall review some of its properties.

In one dimension, the delta function denoted by $\delta(x)$ is defined by

$$\delta(x) = \begin{cases} 0 & \text{if } x \neq 0 \\ \infty & \text{if } x = 0 \end{cases} \tag{B.1}$$

so that

$$\int_{-\infty}^{+\infty} \delta(x)\, dx = 1 \tag{B.2}$$

That is, the function $\delta(x)$ has a very sharp peak at $x = 0$, but the area under the peak is 1. It follows that, given any smooth function $f(x)$, we have

$$\int_{-\infty}^{+\infty} f(x)\, \delta(x)\, dx = f(0) \int_{-\infty}^{+\infty} \delta(x)\, dx$$

since $\delta(x) \neq 0$ only when $x = 0$, and there $f(x) = f(0)$. Hence

$$\int_{-\infty}^{+\infty} \delta(x) f(x)\, dx = f(0) \tag{B.3}$$

Property (B.3) implies all the characteristics contained in Eqs. (B.1) and (B.2), and can be taken as the definition of the delta function.

A close examination of Eqs. (B.1) and (B.2) shows that $\delta(x)$ is not a function in the ordinary mathematical sense, since if a function is zero everywhere except at the origin, and the integral of this function exists, the value of the integral must necessarily be zero.† It is more appropriate to regard $\delta(x)$ as the limit of a sequence of functions $\delta_\alpha(x)$ such that

$$\lim_{\alpha\to 0}\int_{-\infty}^{+\infty} \delta_\alpha(x) f(x)\, dx = f(0) \tag{B.4}$$

Comparing Eq. (B.3) and Eq. (B.4), we write then, by definition,

$$\delta(x) = \lim_{\alpha\to 0} \delta_\alpha(x) \tag{B.5}$$

For instance, one such sequence as shown in Fig. B.1 is

$$\delta_\alpha(x) = \frac{\alpha}{\pi(x^2+\alpha^2)} \tag{B.6}$$

which possesses the desired properties, namely,

For $x \neq 0$: $$\delta(x) = \lim_{\alpha\to 0} \frac{\alpha}{\pi x^2} = 0$$

For $x = 0$: $$\delta(0) = \lim_{\alpha\to 0} \frac{1}{\pi\alpha} = \infty$$

and

$$\int_{-\infty}^{+\infty} \delta(x)\, dx = \lim_{\alpha\to 0}\int_{-\infty}^{+\infty} \frac{\alpha}{\pi(x^2+\alpha^2)}\, dx$$

$$= \lim_{\alpha\to 0}\left[\frac{1}{\pi}\tan^{-1}\frac{x}{\alpha}\right]_{-\infty}^{+\infty} = 1$$

We next list some useful identities of the δ function:

$$\begin{aligned}
&(a) && \delta(-x) = \delta(x) \\
&(b) && \delta'(x) = -\delta'(-x) \\
&(c) && x\,\delta(x) = 0 \\
&(d) && x\,\delta'(x) = -\delta(x) \\
&(e) && \delta(ax) = \frac{1}{|a|}\,\delta(x) \\
&(f) && g(x)\,\delta(x-a) = g(a)\,\delta(x-a) \\
&(g) && \delta[g(x)] = \sum_i \frac{\delta(x-x_i)}{|g'(x_i)|}
\end{aligned} \tag{B.7}$$

† For a rigorous treatment of the delta function, the reader is referred to the works listed at the end of this appendix.

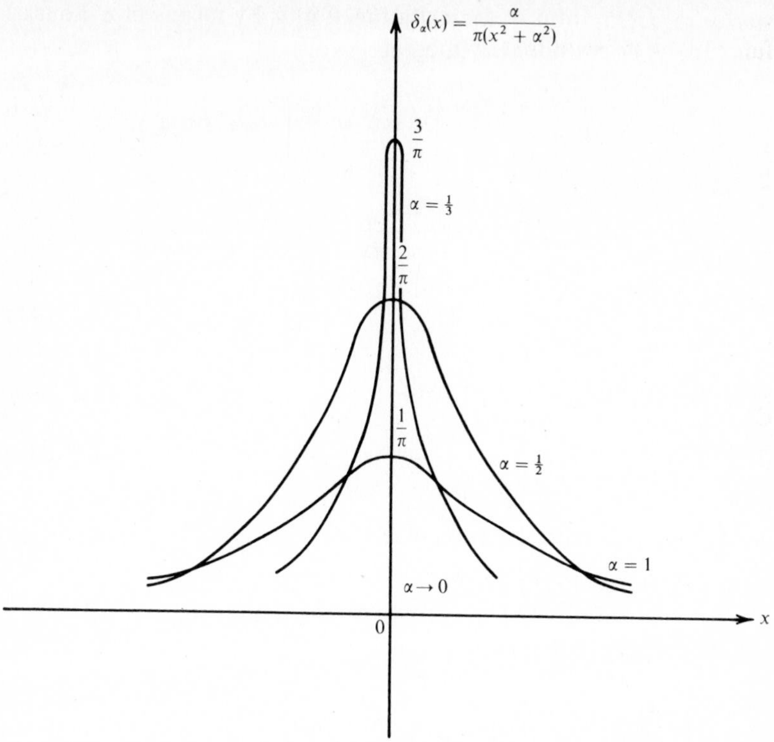

Figure B.1 The sequence of functions $\delta_\alpha(x) = \alpha/[\pi(x^2 + \alpha^2)]$ with three different values of α.

where $g'(x) = dg/dx$ and x_i is the ith root of the equation $g(x) = 0$. As a special case of Eq. (B.7g), we consider $g(x) = x^2 - a^2$; hence

$$(h) \qquad \delta(x^2 - a^2) = \frac{\delta(x + a) + \delta(x - a)}{2|a|}$$

Relations in Eq. (B.7) mean that the two sides give equivalent results when they are multiplied by a smooth function $f(x)$ and are integrated with respect to x. Thus, for example, to prove (f) we consider the integral

$$\int_{-\infty}^{+\infty} f(x)g(x)\,\delta(x - a)\,dx = f(a)g(a)$$

$$= g(a)\int_{-\infty}^{+\infty} f(x)\,\delta(x - a)\,dx = \int_{-\infty}^{+\infty} f(x)g(a)\,\delta(x - a)\,dx$$

and the result is immediate.

An alternative representation of the delta function is by means of a Fourier integral. To this end, let us consider the integral

$$\frac{1}{2\pi}\int_{-\infty}^{+\infty} e^{ikx-\alpha|k|}\,dk = \frac{1}{2\pi}\left(\int_{-\infty}^{0} e^{(ix+\alpha)k}\,dk + \int_{0}^{\infty} e^{(ix-\alpha)k}\,dk\right)$$

$$= \frac{1}{2\pi}\left(\frac{1}{ix+\alpha} - \frac{1}{ix-\alpha}\right)$$

$$= \frac{\alpha}{\pi(x^2+\alpha^2)} \tag{B.8}$$

A comparison of Eq. (B.8) with Eq. (B.6) gives

$$\delta_\alpha(x) = \frac{1}{2\pi}\int_{-\infty}^{+\infty} e^{ikx-\alpha|k|}\,dk \tag{B.9}$$

Thus we obtain the integral representation of $\delta(x)$:

$$\delta(x) = \lim_{\alpha\to 0} \delta_\alpha(x)$$

$$= \lim_{\alpha\to 0} \frac{1}{2\pi}\int_{-\infty}^{+\infty} e^{ikx-\alpha|k|}\,dk$$

$$= \frac{1}{2\pi}\int_{-\infty}^{+\infty} e^{ikx}\,dk \tag{B.10}$$

If the argument of the δ function is the time t, we may write Eq. (B.10) with the aid of Eq. (B.7a) as

$$\delta(t) = \frac{1}{2\pi}\int_{-\infty}^{+\infty} e^{-i\omega t}\,d\omega \tag{B.11}$$

That is, $\delta(t)$ may be regarded as a superposition of monochromatic waves over all frequencies.

So far we have considered a space of one dimension only, but the definition may easily be extended to a space of three dimensions. Then

$$\delta(\mathbf{r}) = \delta(x)\,\delta(y)\,\delta(z)$$

$$= \left(\frac{1}{2\pi}\int_{-\infty}^{+\infty} e^{ik_x x}\,dk_x\right)\left(\frac{1}{2\pi}\int_{-\infty}^{+\infty} e^{ik_y y}\,dk_y\right)\left(\frac{1}{2\pi}\int_{-\infty}^{+\infty} e^{ik_z z}\,dk_z\right) \tag{B.12}$$

If the radius vector $\mathbf{r} = x\hat{\mathbf{x}} + y\hat{\mathbf{y}} + z\hat{\mathbf{z}}$ and the wave vector $\mathbf{k} = k_x\hat{\mathbf{x}} + k_y\hat{\mathbf{y}} + k_z\hat{\mathbf{z}}$ are introduced, we may express Eq. (B.12) in a more compact form:

$$\delta(\mathbf{r}) = \frac{1}{(2\pi)^3}\int_{-\infty}^{+\infty} e^{i\mathbf{k}\cdot\mathbf{r}}\,d^3k \tag{B.13}$$

where $d^3k = dk_x\, dk_y\, dk_z$. Equation (B.13) may be interpreted as the plane wave expansion of the δ function.

A four-dimensional space-time delta function may similarly be expanded in terms of plane monochromatic waves:

$$\delta(\mathbf{r})\,\delta(t) = \frac{1}{(2\pi)^4}\int_{-\infty}^{+\infty} e^{i(\mathbf{k}\cdot\mathbf{r}-\omega t)}\, d^3k\, d\omega \tag{B.14}$$

which may be interpreted as the superposition of plane waves that travel in all directions in such a manner that their amplitudes add to zero at all points and at all times except at the origin, where they add to infinity at the one instant $t = 0$.

REFERENCES

Arsac, J.: *Fourier Transforms and the Theory of Distributions*, Prentice-Hall, Inc., Englewood Cliffs, N.J., 1966.

Butkov, E.: *Mathmetical Physics*, Addison-Wesley Publishing Company, Inc., Reading, Mass., 1968, chap. 6.

Gel'fand, I. M., and G. E. Shilov: *Generalized Functions*, vol. 1, *Properties and Operations*, Academic Press, New York, 1964.

Hoskins, R. F.: *Generalized Functions*, John Wiley & Sons, New York, 1979.

Jones, D. S.: *Generalized Functions*, McGraw-Hill Book Company, New York, 1966.

Lighthill, M. J.: *Introduction to Fourier Analysis and Generalized Functions*, Cambridge University Press, New York, 1960.

Schwartz, L.: *Mathematics for the Physical Sciences*, Addison-Wesley Publishing Company, Inc., Reading, Mass., 1966.

APPENDIX

C

BESSEL FUNCTIONS AND SOME USEFUL INTEGRALS

The second order Bessel differential equation

$$x^2 \frac{d^2y}{dx^2} + x \frac{dy}{dx} + (x^2 - \nu^2)y = 0 \tag{C.1}$$

admits a solution of the form

$$J_\nu(x) = \sum_{k=0}^{\infty} \frac{(-1)^k}{k!\,\Gamma(\nu + k + 1)} \left(\frac{x}{2}\right)^{\nu + 2k} \tag{C.2}$$

where Γ stands for the gamma function. This series solution is called the Bessel function of the first kind of order ν. If ν is not an integer, the second independent solution is obtained by replacing ν by $-\nu$:

$$J_{-\nu}(x) = \sum_{k=0}^{\infty} \frac{(-1)^k}{k!\,\Gamma(-\nu + k + 1} \left(\frac{x}{2}\right)^{-\nu + 2k} \tag{C.3}$$

For example, if $\nu = \frac{1}{2}$, Eq. (C.2) becomes

$$J_{1/2}(x) = \frac{(x/2)^{1/2}}{\Gamma(\frac{3}{2})} \left(1 - \frac{x^2}{3!} + \frac{x^4}{5!} - \cdots\right) \tag{C.4}$$

But $\Gamma(\frac{3}{2}) = \frac{1}{2}\Gamma(\frac{1}{2}) = \frac{1}{2}\sqrt{\pi}$; hence Eq. (C.4) may be written

$$J_{1/2}(x) = \sqrt{\frac{2}{\pi x}} \left(x - \frac{x^3}{3!} + \frac{x^5}{5!} - \frac{x^7}{7!} + \cdots\right) = \sqrt{\frac{2}{\pi x}} \sin x \tag{C.5}$$

Likewise, we may show that

$$J_{-1/2}(x) = \sqrt{\frac{2}{\pi x}} \cos x \tag{C.6}$$

However, if $\nu = n$ is a positive integer, the gamma function reduces to $\Gamma(n + k + 1) = (n + k)!$. Equation (C.2) becomes

$$J_n(x) = \sum_{k=0}^{\infty} \frac{(-1)^k}{k!\,(n+k)!} \left(\frac{x}{2}\right)^{n+2k} \tag{C.7}$$

which is related to $J_{-n}(x)$ by

$$J_{-n}(x) = (-1)^n J_n(x) \tag{C.8}$$

Hence $J_n(x)$ and $J_{-n}(x)$ are not linearly independent. In this case a second solution may be obtained by a limiting procedure. By convention we define another solution of Bessel's equation as

$$Y_\nu(x) = \frac{\cos \nu\pi \, J_\nu(x) - J_{-\nu}(x)}{\sin \nu\pi} \tag{C.9}$$

Thus

$$Y_n(x) = \lim_{\nu \to n} Y_\nu(x) \tag{C.10}$$

This limit gives a second solution to Bessel's equation when $\nu = n$ is an integer. The $Y_\nu(x)$ is called the Bessel function of the second kind of order ν.

In wave propagation problems, it is useful to define linear combinations of Bessel's functions:

$$H_\nu^{(1)}(x) = J_\nu(x) + iY_\nu(x) \tag{C.11}$$

and

$$H_\nu^{(2)}(x) = J_\nu(x) - iY_\nu(x) = H_\nu^{(1)*}(x) \tag{C.12}$$

which are known as Hankel functions of the first and second kind, respectively. For large values of x (that is, as $x \to \infty$), these functions are described by the asymptotic formulas:

$$\begin{aligned}
J_\nu(x) &\cong \sqrt{\frac{2}{\pi x}} \cos\left(x - \frac{\pi}{2}\nu - \frac{\pi}{4}\right) \\
Y_\nu(x) &\cong \sqrt{\frac{2}{\pi x}} \sin\left(x - \frac{\pi}{2}\nu - \frac{\pi}{4}\right) \\
H_\nu^{(1)}(x) &\cong \sqrt{\frac{2}{\pi x}} \exp\left[i\left(x - \frac{\pi}{2}\nu - \frac{\pi}{4}\right)\right] \\
H_\nu^{(2)}(x) &\cong \sqrt{\frac{2}{\pi x}} \exp\left[-i\left(x - \frac{\pi}{2}\nu - \frac{\pi}{4}\right)\right]
\end{aligned} \tag{C.13}$$

Hence they allow us to think conveniently of Bessel functions as trigonometric or exponential functions with a factor $x^{-1/2}$.

Bessel functions of imaginary argument are solutions of the differential equation

$$x^2 \frac{d^2y}{dx^2} + x \frac{dy}{dx} - (x^2 + \nu^2)y = 0 \tag{C.14}$$

We define in this context the modified Bessel functions of the first and second kind:

$$I_\nu(x) = e^{-\nu\pi i/2} J_\nu(ix) \tag{C.15}$$

$$K_\nu(x) = \frac{\pi i}{2} e^{\nu\pi i/2} H_\nu^{(1)}(ix) = K_{-\nu}(x) \tag{C.16}$$

In particular, when $\nu = \frac{1}{2}$, we have

$$K_{1/2} = \sqrt{\frac{\pi}{2x}}\, e^{-x} = K_{-1/2} \tag{C.17}$$

For large values of x (that is, as $x \to \infty$), these functions behave as

$$I_\nu(x) \cong \frac{e^x}{\sqrt{2\pi x}} \qquad K_\nu(x) \cong \sqrt{\frac{\pi}{2x}}\, e^{-x} \tag{C.18}$$

As an alternative approach we find that, if $e^{xt/2}$ and $e^{-x/2t}$ are expanded by the Laurent series in powers of t and the two series are multiplied together, the coefficients of the resulting series are Bessel functions. Thus

$$e^{xt/2} = \sum_{r=0}^{\infty} \frac{1}{r!}\left(\frac{xt}{2}\right)^r \qquad e^{-x/2t} = \sum_{s=0}^{\infty} \frac{1}{s!}\left(-\frac{x}{2t}\right)^s \quad t \neq 0 \tag{C.19}$$

Multiplying these two series, we obtain

$$e^{(x/2)(t-t^{-1})} = \sum_{r=0}^{\infty} \sum_{s=0}^{\infty} \frac{(-1)^s}{r!\,s!}\left(\frac{x}{2}\right)^{r+s} t^{r-s} \tag{C.20}$$

In Eq. (C.20), the coefficient of the t^n term is found to be $J_n(x)$ defined in Eq. (C.7). Hence, Eq. (C.20) may be written as

$$e^{(x/2)(t-t^{-1})} = J_0(x) + \sum_{n=1}^{\infty} J_n(x)[t^n + (-1)^n t^{-n}] \tag{C.21}$$

Substitution of $t = ie^{i\theta}$ in Eq. (C.21) yields

$$e^{ix\cos\theta} = J_0(x) + 2\sum_{n=1}^{\infty} i^n J_n(x)\cos n\theta \tag{C.22}$$

We can take advantage of the orthogonality properties of cosine and sine functions,

$$\begin{aligned} &\int_{\alpha}^{\alpha+2\pi} \sin m\theta \cos n\theta \, d\theta = 0 \\ &\int_{\alpha}^{\alpha+2\pi} \cos m\theta \cos n\theta \, d\theta = \begin{cases} \pi\delta_{mn} & m \neq 0 \\ 2\pi & m = n = 0 \end{cases} \end{aligned} \tag{C.23}$$

to derive several useful integral formulas. Multiplying both sides of Eq. (C.22) by $\cos m\theta \, d\theta$ and integrating between limits α and $2\pi + \alpha$, all terms of the right vanish except one, and we obtain

$$\int_{\alpha}^{\alpha+2\pi} \cos n\theta \, e^{ix\cos\theta} \, d\theta = i^n 2\pi J_n(x) \tag{C.24}$$

In particular,

$$\int_{\alpha}^{\alpha+2\pi} e^{ix\cos\theta} \, d\theta = 2\pi J_0(x) \tag{C.25}$$

Similarly, multiplying Eq. (C.22) by $\sin m\theta \, d\theta$ and integrating, we have

$$\int_{\alpha}^{\alpha+2\pi} \sin m\theta \, e^{ix\cos\theta} \, d\theta = 0 \tag{C.26}$$

Furthermore, a change of variable followed by application of the results of Eqs. (C.24) to (C.26) gives

$$\int_0^{2\pi} e^{ix\cos(\Phi-\phi)} \, d\Phi = 2\pi J_0(x) \tag{C.27}$$

$$\int_0^{2\pi} \cos^2\Phi \, e^{ix\cos(\Phi-\phi)} \, d\Phi = 2\pi\left[\cos^2\phi \, J_0(x) - \frac{\cos 2\phi \, J_1(x)}{x}\right] \tag{C.28}$$

$$\int_0^{2\pi} \sin\Phi\cos\Phi \, e^{ix\cos(\Phi-\phi)} \, d\Phi = \pi\sin 2\phi\left[J_0(x) - \frac{2J_1(x)}{x}\right] \tag{C.29}$$

$$\int_0^{2\pi} \sin^2\Phi \, e^{ix\cos(\Phi-\phi)} \, d\Phi = 2\pi\left[\sin^2\phi J_0(x) + \frac{\cos 2\phi \, J_1(x)}{x}\right] \tag{C.30}$$

In deriving the above results, we also used the recurrence formula:

$$J_2(x) = \frac{2}{x} J_1(x) - J_0(x) \tag{C.31}$$

We shall now list some standard integrals which have been used in the text. The integrals which contain Bessel's function can be evaluated either by using the series representation of the Bessel function and integrating term by term, or by replacing the Bessel function by an integral representation and interchanging the order of integration. For a detailed derivation, the reader is referred to the works listed at the end of this appendix.

$$\int_{-\infty}^{+\infty} \frac{e^{izt}}{t^2 + b^2}\, dt = \frac{\pi}{b}\, e^{-b|z|} \qquad \operatorname{Re} b > 0 \tag{C.32}$$

$$\int_{-\infty}^{+\infty} K_0(\rho\sqrt{t^2 - b^2})\, e^{izt}\, dt = \frac{\pi}{\sqrt{\rho^2 + z^2}} \exp\,(ib\sqrt{\rho^2 + z^2}) \tag{C.33}$$

$$\int_0^{\infty} \frac{J_0(at)}{t^2 + b^2}\, t\, dt = K_0(ab) \qquad \begin{matrix} a > 0 \\ \operatorname{Re} b > 0 \end{matrix} \tag{C.34}$$

Sonine's discontinuous integral:

$$\int_0^{\infty} \frac{\sin a\sqrt{x^2 + b^2}}{\sqrt{x^2 + b^2}}\, J_0(cx)x\, dx = \frac{\cos b\sqrt{a^2 - c^2}}{\sqrt{a^2 - c^2}}\, U(a - c) \tag{C.35}$$

where $U(x)$ is the unit step function defined by

$$U(x) = \begin{cases} 1 & \text{if } x > 0 \\ 0 & \text{if } x < 0 \end{cases} \tag{C.36}$$

Sommerfeld's integral:

$$\int_0^{\infty} \frac{\exp\,(-|z|\sqrt{x^2 - b^2})}{\sqrt{x^2 - b^2}}\, J_0(tx)x\, dx = \frac{\exp\,(ib\sqrt{t^2 + z^2})}{\sqrt{t^2 + z^2}} \tag{C.37}$$

Based on a known integral, we can derive other integrals. For example, multiplying both sides of Eq. (C.37) by $t\, dt$ and integrating between limits 0 and ρ, we have

$$\int_0^{\rho} t\, dt \int_0^{\infty} \frac{\exp\,(-|z|\sqrt{x^2 - b^2})}{\sqrt{x^2 - b^2}}\, J_0(tx)x\, dx = \int_0^{\rho} \frac{\exp\,(ib\sqrt{t^2 + z^2})}{\sqrt{t^2 + z^2}}\, t\, dt$$

Interchanging the order of integration on the left and making use of the results,

$$\int_0^{\rho} J_0(xt)t\, dt = \frac{\rho}{x}\, J_1(\rho x) \tag{C.38}$$

and

$$\int_0^{\rho} \frac{\exp\,(ib\sqrt{t^2 + z^2})}{\sqrt{t^2 + z^2}}\, t\, dt = \frac{1}{ib}\,[\exp\,(ib\sqrt{\rho^2 + z^2}) - \exp\,(ibz)] \tag{C.39}$$

we obtain

$$\int_0^\infty \frac{\exp\left(-|z|\sqrt{x^2 - b^2}\right)}{\sqrt{x^2 - b^2}} J_1(\rho x)\, dx = \frac{1}{ib\rho}\left[\exp\left(ib\sqrt{\rho^2 + z^2}\right) - \exp\left(ibz\right)\right] \tag{C.40}$$

which has been used in Chap. 10.

REFERENCES

Gradshteyn, I. S., and I. M. Ryzhik: *Table of Integrals, Series, and Products*, Academic Press, New York, 1980.

Lebedev, N. H.: *Special Functions and Their Applications*, Prentice-Hall, Inc., Englewood Cliffs, N. J., 1965, chap. 5.

Tranter, C. J.: *Bessel Functions with Some Physical Applications*, Hart Publishing Company, New York, 1969.

Watson, G. N.: *A Treatise on the Theory of Bessel Functions*, Cambridge University Press, Cambridge, England, 1944.

BIBLIOGRAPHY

References cited at the end of each chapter include articles or books on specific topics. The following books are general ones that are relevant to more than one chapter.

Introductory books

Bradshaw, M. D., and W. J. Byatt: *Introductory Engineering Field Theory*, Prentice-Hall, Inc., Englewood Cliffs, N.J., 1967.

Cook, D. M.: *The Theory of the Electromagnetic Field*, Prentice-Hall, Inc., Englewood Cliffs, N.J., 1975.

Durney, C. H., and C. C. Johnson: *Introduction to Modern Electromagnetics*, McGraw-Hill Book Company, New York, 1969.

Griffiths, D. J.: *Introduction to Electrodynamics*, Prentice-Hall, Inc., Englewood Cliffs, N.J., 1981.

Hayt, W. H.: *Engineering Electromagnetics*, 4th ed., McGraw-Hill Book Company, New York, 1981.

Holt, C. A.: *Introduction to Electromagnetic Fields and Waves*, John Wiley & Sons, New York, 1966.

Jefimenko, O. D.: *Electricity and Magnetism*, Appleton-Century-Crofts, New York, 1966.

Kraus, J. D., and K. R. Carver: *Electromagnetics*, 2d ed., McGraw-Hill Book Company, New York, 1973.

Magid, L. M.: *Electromagnetic Fields, Energy, and Waves*, John Wiley & Sons, New York, 1972.

Neff, H. P., Jr.: *Basic Electromagnetic Fields*, Harper & Row, Publishers, New York, 1981.

Owen, G. E.: *Introduction to Electromagnetic Theory*, Allyn & Bacon, Inc., Boston, 1963.

Paris, D. T., and F. K. Hurd: *Basic Electromagnetic Theory*, McGraw-Hill Book Company, New York, 1969.

Paul, C. R., and S. A. Nasar: *Introduction to Electromagnetic Fields*, McGraw-Hill Book Company, New York, 1982.

Plonus, M. A.: *Applied Electromagnetics*, McGraw-Hill Book Company, New York, 1978.

Popovic, B. D.: *Introductory Engineering Electromagnetics*, Addison-Wesley Publishing Company, Reading, Mass., 1971.

Rojansky, V.: *Electromagnetic Fields and Waves*, Dover Publications, Inc., New York, 1979.

Scott, W. T.: *The Physics of Electricity and Magnetism*, 2d ed., John Wiley & Sons, New York, 1966.

Silvester, P.: *Modern Electromagnetic Fields*, Prentice-Hall, Inc., Englewood Cliffs, N.J., 1968.

Skilling, H. H.: *Fundamentals of Electric Waves*, R. E. Krieger Publishing Co., Huntington, New York, 1974.

Skitek, G. G., and S. V. Marshall: *Electromagnetic Concepts and Applications*, Prentice-Hall, Inc., Englewood Cliffs, N.J., 1982.

Stuart, R. D.: *Electromagnetic Field Theory*, Addison-Wesley Publishing Company, Reading, Mass., 1965.

Wangsness, R. K.: *Electromagnetic Fields*, John Wiley & Sons, New York, 1979.

Whitmer, R. M.: *Electromagnetics*, 2d ed., Prentice-Hall, Inc., Englewood Cliffs, N.J., 1962.

Zahn, M.: *Electromagnetic Field Theory: A Problem Solving Approach*, John Wiley & Sons, New York, 1979.

Intermediate books

Adler, R. B., L. J. Chu, and R. M. Fano: *Electromagnetic Energy Transmission and Radiation*, John Wiley & Sons, Inc., New York, 1960.

Dearholt, D. W., and W. R. McSpadden: *Electromagnetic Wave Propagation*, McGraw-Hill Book Company, New York, 1973.

Della Torre, E., and C. V. Longo: *The Electromagnetic Field*, Allyn & Bacon, Inc., Boston, 1969.

Elliott, R. S.: *Electromagnetics*, McGraw-Hill Book Company, New York, 1966.

Fano, R. M., L. J. Chu, and R. B. Adler: *Electromagnetic Fields, Energy, and Forces*, John Wiley & Sons, Inc., New York, 1960.

Hauser, W.: *Introduction to the Principles of Electromagnetism*, Addison-Wesley Publishing Company, Reading, Mass., 1971.

Javid, M., and P. M. Brown: *Field Analysis and Electromagnetics*, McGraw-Hill Book Company, New York, 1963.

Johnk, C. T. A.: *Engineering Electromagnetic Fields and Waves*, John Wiley & Sons, Inc., New York, 1975.

Jordan, E. C., and K. G. Balmain: *Electromagnetic Waves and Radiating Systems*, 2d ed., Prentice-Hall, Inc., Englewood Cliffs, N.J., 1968.

Langmuir, R. V.: *Electromagnetic Fields and Waves*, McGraw-Hill Book Company, New York, 1961.

Lorrain, P., and D. R. Corson: *Electromagnetic Fields and Waves*, 2d ed., W. H. Freeman and Company, San Francisco, 1970.

Marion, J. B., and M. A. Heald: *Classical Electromagnetic Radiation*, 2d ed., Academic Press, New York, 1980.

Moon, P., and D. E. Spencer: *Foundations of Electrodynamics*, D. Van Nostrand Company, Inc., Princeton, N.J., 1960.

Nussbaum, A.: *Electromagnetic Theory for Engineers and Scientists*, Prentice-Hall, Inc., Englewood Cliffs, N. J., 1965.

Plonsey, R., and R. E. Collin: *Principles and Applications of Electromagnetic Fields*, McGraw-Hill Book Company, New York, 1961.

Portis, A. M.: *Electromagnetic Fields: Sources and Media*, John Wiley & Sons, New York, 1978.

Ramo, S., J. R. Whinnery, and T. Van Duzer: *Fields and Waves in Communication Electronics*, John Wiley & Sons, Inc., New York, 1965.

Rao, N. N.: *Basic Electromagnetics with Applications*, Prentice-Hall, Inc., Englewood Cliffs, N.J., 1972.

Reitz, J. R., F. J. Milford, and R. W. Christy: *Foundations of Electromagnetic Theory*, 3d ed., Addison-Wesley Publishing Company, Reading, Mass., 1979.

Seshadri, S. R.: *Fundamentals of Transmission Lines and Electromagnetic Fields*, Addison-Wesley Publishing Company, Reading, Mass., 1971.

Shadowitz, A.: *The Electromagnetic Field*, McGraw-Hill Book Company, New York, 1975.

Tamm, I. E.: *Fundamentals of the Theory of Electricity*, MIR Publishers, Moscow, 1979.

Advanced books

Becker, R., and F. Sauter: *Electromagnetic Fields and Interactions*, vol. 1, Blaisdell Publishing Company, New York, 1964.

Born, M., and E. Wolf: *Principles of Optics*, 4th ed., Pergamon Press, New York, 1970.

Eyges, L.: *The Classical Electromagnetic Field*, Addison-Wesley Publishing Company, Reading, Mass., 1972.

Felsen, L. B., and N. Marcuwitz: *Radiation and Scattering of Electromagnetic Waves*, Prentice-Hall, Inc., Englewood Cliffs, N.J., 1973.

Harrington, R. F.: *Time-Harmonic Electromagnetic Fields*, McGraw-Hill Book Company, New York, 1961.

Jackson, J. D.: *Classical Electrodynamics*, 2d ed., John Wiley & Sons, New York, 1975.

Johnson, C. C.: *Field and Wave Electrodynamics*, McGraw-Hill Book Company, New York, 1965.

Jones, D. S.: *The Theory of Electromagnetism*, Pergamon Press, New York, 1964.

Kong, J. A.: *Theory of Electromagnetic Waves*, John Wiley & Sons, New York, 1975.

Landau, L. D., and E. M. Lifshitz: *Electrodynamics of Continuous Media*, Pergamon Press, New York, 1960.

Landau, L. D., and E. M. Lifshitz: *The Classical Theory of Fields*, 3d ed., Pergamon Press, New York, 1971.

Maxwell, J. C.: *A Treatise on Electricity and Magnetism*, vols. 1 and 2, Dover Publications, Inc., New York, 1954.

Müller, C.: *Foundations of the Mathematical Theory of Electromagnetic Waves*, Springer-Verlag, New York, 1969.

Page, L., and N. I. Adams: *Electrodynamics*, Dover Publications, Inc., New York, 1965.

Panofsky, W. K. H., and M. Phillips: *Classical Electricity and Magnetism*, 2d ed., Addison-Wesley Publishing Company, Reading, Mass., 1962.

Papas, C. H.: *Theory of Electromagnetic Wave Propagation*, McGraw-Hill Book Company, New York, 1965.

Podolsky, B., and K. S. Kunz: *Fundamentals of Electrodynamics*, Marcel Dekker, New York, 1969.

Smythe, W. R.: *Static and Dynamic Electricity*, 3d ed., McGraw-Hill Book Company, New York, 1968.

Sommerfeld, A.: *Electrodynamics*, Academic Press, New York, 1952.

Stratton, J. A.: *Electromagnetic Theory*, McGraw-Hill Book Company, New York, 1941.

Tralli, N.: *Classical Electromagnetic Theory*, McGraw-Hill Book Company, New York, 1963.

Van Bladel, J.: *Electromagnetic Fields*, McGraw-Hill Book Company, New York, 1964.

Weeks, W. L.: *Electromagnetic Theory for Engineering Applications*, John Wiley & Sons, Inc., New York, 1964.

Books on special topics

Adams, M. J.: *An Introduction to Optical Waveguides*, John Wiley & Sons, New York, 1981.

Argence, E., and T. Kahan: *Theory of Waveguides and Cavity Resonators*, Blackie & Sons, Ltd., London, 1967.

Arnaud, J. A.: *Beam and Fiber Optics*, Academic Press, New York, 1976.

Auld, B. A.: *Acoustic Fields and Waves in Solids*, vols. 1 and 2, John Wiley & Sons, New York, 1973.

Baker, B. B., and E. T. Copson: *The Mathematical Theory of Huygens' Principle*, 2d ed., Clarendon Press, Oxford, 1950.

Balanis, C. A.: *Antenna Theory*, Harper & Row, New York, 1982.

Banos, A., Jr.: *Dipole Radiation in the Presence of a Conducting Half-Space*, Pergamon Press, New York, 1966.

Barlow, H. M., and J. Brown: *Radio Surface Waves*, Oxford University Press, London, 1962.

Brandstatter, J. J.: *An Introduction to Waves, Rays and Radiation in Plasma Media*, McGraw-Hill Book Company, New York, 1963.

Brekhovskikh, L. M.: *Waves in Layered Media*, 2d ed., Academic Press, New York, 1980.

Cairo, L., and T. Kahan: *Variational Techniques in Electromagnetism*, Gordon & Breach Science Publishers, New York, 1965.

Clarke, R. H., and J. Brown: *Diffraction Theory and Antennas*, John Wiley & Sons, New York, 1980.

Collin, R. E.: *Field Theory of Guided Waves*, McGraw-Hill Book Company, New York, 1960.

Collin, R. E.: *Foundations for Microwave Engineering*, McGraw-Hill Book Company, New York, 1966.

Collin, R. E., and F. J. Zucker (eds.): *Antenna Theory*, parts I and II, McGraw-Hill Book Company, New York, 1969.

Elliott, R. S.: *Antenna Theory and Design*, Prentice-Hall, Inc., Englewood Cliffs, N.J., 1981.
Graff, K. F.: *Wave Motion in Elastic Solids*, Ohio State University Press, Columbus, Ohio, 1975.
Harrington, R. F.: *Field Computation by Moment Methods*, The Macmillan Company, New York, 1968.
James, G. L.: *Geometrical Theory of Diffraction for Electromagnetic Waves*, Peter Peregrinus Ltd., Herts, 1976.
Jones, D. S.: *Methods in Electromagnetic Wave Propagation*, Oxford University Press, New York, 1979.
Kapany, S. N., and J. J. Burke: *Optical Waveguides*, Academic Press, New York, 1972.
Kline, M., and I. W. Kay: *Electromagnetic Theory and Geometrical Optics*, Interscience-Wiley Publishers, New York, 1965.
Kurokawa, K.: *An Introduction to the Theory of Microwave Circuits*, Academic Press, New York, 1969.
Lewin, L.: *Theory of Waveguides*, John Wiley & Sons, New York, 1975.
Lighthill, J.: *Waves in Fluids*, Cambridge University Press, London, 1978.
Lisitsa, M. P., L. I. Berezhinskii, and M. Ya.Valakh: *Fiber Optics*, Keter Inc., New York, 1972.
Marcuse, D.: *Light Transmission Optics*, Van Nostrand Reinhold Company, New York, 1972.
Mentzer, J. R.: *Scattering and Diffraction of Radio Waves*, Pergamon Press, New York, 1955.
Musgrave, M. J. P.: *Crystal Acoustics*, Holden-Day, San Francisco, 1970.
Okoshi, T.: *Optical Fibers*, Academic Press, New York, 1982.
Owyang, G. H.: *Foundations of Optical Waveguides*, Elsevier, New York, 1981.
Penfield, P. Jr., and H. A. Haus: *Electrodynamics of Moving Media*, The MIT Press, Cambridge, Mass., 1967.
Silver, S. (ed.): *Microwave Antenna Theory and Design*, McGraw-Hill Book Company, New York, 1949.
Skolnik, M. I.: *Introduction to Radar Systems*, 2d ed., McGraw-Hill Book Company, New York, 1980.
Sommerfeld, A.: *Optics*, Academic Press, New York, 1949.
Stutzman, W. L., and G. A. Thiele: *Antenna Theory and Design*, John Wiley & Sons, New York, 1981.
Thomassen, K. I.: *Introduction to Microwave Fields and Circuits*, Prentice-Hall, Inc., Englewood Cliffs, N.J., 1971.
Tolstoy, I.: *Wave Propagation*, McGraw-Hill Book Company, New York, 1973.
Unger, H. G.: *Planar Optical Waveguides and Fibres*, Oxford University Press, Oxford, 1977.
Wait, J. R.: *Electromagnetic Radiation from Cylindrical Structures*, Pergamon Press, New York, 1959.
Wait, J. R.: *Electromagnetic Waves in Stratified Media*, 2d ed., Pergamon Press, New York, 1970.
Watkins, D. A.: *Topics in Electromagnetic Theory*, John Wiley & Sons, New York, 1958.
Weeks, W. L.: *Antenna Engineering*, McGraw-Hill Book Company, New York, 1968.
Whitham, G. B.: *Linear and Nonlinear Waves*, John Wiley & Sons, New York, 1974.
Wolff, E. A.: *Antenna Analysis*, John Wiley & Sons, New York, 1966.

NAME INDEX

SUBJECT INDEX

Boldface numbers in parentheses refer to problems.

ADJOINT

$$\text{adj}\,\bar{\mathbf{A}} = (\text{adj}\,\bar{\mathbf{A}})_t \bar{\mathbf{I}} - \bar{\mathbf{A}}_t \bar{\mathbf{A}} + \bar{\mathbf{A}}^2$$

$$\text{adj}\,(\lambda\bar{\mathbf{A}}) = \lambda^2\,\text{adj}\,\bar{\mathbf{A}} \qquad \text{adj}\,(\lambda\bar{\mathbf{I}}) = \lambda^2\bar{\mathbf{I}}$$

$$\text{adj}\,(\text{adj}\,\bar{\mathbf{A}}) = |\bar{\mathbf{A}}|\,\bar{\mathbf{A}} \qquad \text{adj}\,(\bar{\mathbf{A}}^{-1}) = (\text{adj}\,\bar{\mathbf{A}})^{-1} = \frac{\bar{\mathbf{A}}}{|\bar{\mathbf{A}}|}$$

$$\text{adj}\,(\bar{\mathbf{A}} + \bar{\mathbf{B}}) = \text{adj}\,\bar{\mathbf{A}} + \text{adj}\,\bar{\mathbf{B}} + \bar{\mathbf{A}} \cdot \bar{\mathbf{B}} + \bar{\mathbf{B}} \cdot \bar{\mathbf{A}} - \bar{\mathbf{A}}_t \bar{\mathbf{B}} - \bar{\mathbf{B}}_t \bar{\mathbf{A}} + \bar{\mathbf{A}}_t \bar{\mathbf{B}}_t \bar{\mathbf{I}} - (\bar{\mathbf{A}} \cdot \bar{\mathbf{B}})_t \bar{\mathbf{I}}$$

$$\text{adj}\,(\bar{\mathbf{A}} \cdot \bar{\mathbf{B}}) = (\text{adj}\,\bar{\mathbf{B}}) \cdot (\text{adj}\,\bar{\mathbf{A}})$$

$$\text{adj}\,(\mathbf{a} \times \bar{\mathbf{I}}) = \mathbf{aa} \qquad \text{adj}\,(\mathbf{ab}) = 0$$

$$\text{adj}\,(\mathbf{ab} + \mathbf{cd}) = (\mathbf{b} \times \mathbf{d})(\mathbf{a} \times \mathbf{c})$$

$$\text{adj}\,(\mathbf{al} + \mathbf{bm} + \mathbf{cn}) = (\mathbf{m} \times \mathbf{n})(\mathbf{b} \times \mathbf{c}) + (\mathbf{n} \times \mathbf{l})(\mathbf{c} \times \mathbf{a}) + (\mathbf{l} \times \mathbf{m})(\mathbf{a} \times \mathbf{b})$$

$$\text{adj}\,(\lambda\bar{\mathbf{I}} + \bar{\mathbf{A}}) = \lambda^2\bar{\mathbf{I}} + \lambda(\bar{\mathbf{A}}_t \bar{\mathbf{I}} - \bar{\mathbf{A}}) + \text{adj}\,\bar{\mathbf{A}}$$

$$\text{adj}\,(\mathbf{ab} + \bar{\mathbf{A}}) = \text{adj}\,\bar{\mathbf{A}} - (\mathbf{b} \times \bar{\mathbf{I}}) \cdot \tilde{\bar{\mathbf{A}}} \cdot (\mathbf{a} \times \bar{\mathbf{I}})$$

$$\text{adj}\,(\lambda\bar{\mathbf{I}} + \mathbf{ab} + \mathbf{cd}) = \lambda[(\lambda + \mathbf{a} \cdot \mathbf{b} + \mathbf{c} \cdot \mathbf{d})\bar{\mathbf{I}} - (\mathbf{ab} + \mathbf{cd})] + (\mathbf{b} \times \mathbf{d})(\mathbf{a} \times \mathbf{c})$$

$$\text{adj}\,(\lambda\bar{\mathbf{I}} + \mathbf{ab} + \mathbf{c} \times \bar{\mathbf{I}}) = \lambda^2\bar{\mathbf{I}} - \lambda(\mathbf{ab} - \mathbf{a} \cdot \mathbf{b}\,\bar{\mathbf{I}} + \mathbf{c} \times \bar{\mathbf{I}}) + \mathbf{cc} - (\mathbf{b} \cdot \mathbf{c})(\mathbf{a} \times \bar{\mathbf{I}}) - \mathbf{c}(\mathbf{a} \times \mathbf{b})$$

DETERMINANT

$$|\bar{\mathbf{A}}| = \frac{(\bar{\mathbf{A}}_t)^3 - 3\bar{\mathbf{A}}_t(\bar{\mathbf{A}}^2)_t + 2(\bar{\mathbf{A}}^3)_t}{6}$$

$$|\tilde{\bar{\mathbf{A}}}| = |\bar{\mathbf{A}}| \qquad |\bar{\mathbf{A}}^{-1}| = |\bar{\mathbf{A}}|^{-1} \qquad |\lambda\bar{\mathbf{A}}| = \lambda^3\,|\bar{\mathbf{A}}|$$

$$|\text{adj}\,\bar{\mathbf{A}}| = |\bar{\mathbf{A}}|^2 \qquad |\text{adj}\,(\text{adj}\,\bar{\mathbf{A}})| = |\bar{\mathbf{A}}|^4$$

$$|\bar{\mathbf{A}} + \bar{\mathbf{B}}| = |\bar{\mathbf{A}}| + |\bar{\mathbf{B}}| + [(\text{adj}\,\bar{\mathbf{A}}) \cdot \bar{\mathbf{B}}]_t + [\bar{\mathbf{A}} \cdot (\text{adj}\,\bar{\mathbf{B}})]_t$$

$$|\bar{\mathbf{A}} \cdot \bar{\mathbf{B}}| = |\bar{\mathbf{A}}|\,|\bar{\mathbf{B}}|$$

$$|\mathbf{a} \times \bar{\mathbf{I}}| = 0 \qquad |\mathbf{ab}| = 0 \qquad |\mathbf{al} + \mathbf{bm}| = 0$$

$$|\mathbf{al} + \mathbf{bm} + \mathbf{cn}| = (\mathbf{a} \cdot \mathbf{b} \times \mathbf{c})(\mathbf{l} \cdot \mathbf{m} \times \mathbf{n})$$

$$|\lambda\bar{\mathbf{I}} + \bar{\mathbf{A}}| = \lambda^3 + \bar{\mathbf{A}}_t \lambda^2 + (\text{adj}\,\bar{\mathbf{A}})_t \lambda + |\bar{\mathbf{A}}|$$

$$|\mathbf{ab} + \bar{\mathbf{A}}| = |\bar{\mathbf{A}}| + \mathbf{b} \cdot (\text{adj}\,\bar{\mathbf{A}}) \cdot \mathbf{a}$$

$$|\lambda\bar{\mathbf{I}} + \mathbf{ab} + \mathbf{cd}| = \lambda[\lambda(\lambda + \mathbf{a} \cdot \mathbf{b} + \mathbf{c} \cdot \mathbf{d}) + (\mathbf{b} \times \mathbf{d}) \cdot (\mathbf{a} \times \mathbf{c})]$$

$$|\lambda\bar{\mathbf{I}} + \mathbf{ab} + \mathbf{c} \times \bar{\mathbf{I}}| = \lambda^3 + (\mathbf{a} \cdot \mathbf{b})\lambda^2 + (\mathbf{c}^2 - \mathbf{a} \times \mathbf{b} \cdot \mathbf{c})\lambda + (\mathbf{a} \cdot \mathbf{c})(\mathbf{b} \cdot \mathbf{c})$$